Analysis and Design of Curved Steel Bridges

Analysis and Design of Curved Steel Bridges

Hiroshi Nakai

Osaka City University Professor of Civil Engineering

Chai Hong Yoo

Auburn University Gottlieb Professor of Civil Engineering

McGraw-Hill Book Company

New York St. Louis San Francisco Auckland Bogotá
Caracas Colorado Springs Hamburg Lisbon
London Madrid Mexico Milan Montreal
New Delhi Oklahoma City Panama Paris
San Juan São Paulo Singapore
Sydney Tokyo Toronto

Library of Congress Cataloging-in-Publication Data

Nakai, Hiroshi, 1935–
Analysis and design of curved steel bridges / Hiroshi Nakai, Chai Hong Yoo.
p. cm.

Includes bibliographies and index.
ISBN 0-07-045866-9
1. Bridges, Iron and steel—Design and construction. 2. Steel I-beams. 3. Thin-walled structures. I. Yoo, Chai Hong. II. Title.
TG380.N35 1988
624′.25—dc19 87-21747
CIP

1234567890 DOC/DOC 89210987

ISBN 0-07-045866-9

The editors for this book were Nadine M. Post and Frank Kotowski, Jr., the designer was Naomi Auerbach, and the production supervisor was Dianne Walber. It was set in Century Schoolbook by Datapage International Limited.

Printed and bound by R. R. Donnelley & Sons Company.

Contents

Preface

The subject of horizontally curved girders has undergone remarkable development over the past two decades. The primary reasons are the increased demand for curved roadway alignment for the smooth dissemination of congested traffic along with advancement in fabrication and erection technology, and the availability of digital computers to carry out the complex mathematical computations of the structural analysis and design of such girders. Recent advancement in CAD/CAM has been instrumental for nearly automatic design alternatives and machine-aided fabrication. When curved viaducts were first introduced into the highway system, they were generally composed of a series of straight girders used as chords in forming a curved alignment. One of the major reasons why engineers were reluctant to use curved girders in forming curved roadways was the mathematical complexities associated with such girder systems. Once this mathematical difficulty is overcome, it is envisioned that more designers and planners will utilize curved components in their structures.

The design of complex structures has often required engineers' imagination and ingenuity since research results may frequently lag behind practical needs. This book intends to remedy this difficulty, at least partially, by presenting up-to-date techniques and research for such complex systems, thereby permitting a better understanding of the structural behavior.

Although this book contains much of the material that characterizes standard textbooks on the behavior of thin-walled members, its goal is to present background needed by an engineer who will be using, writing, or modifying existing computer programs for curved bridges, or doing structural testing and inspection. Therefore, this book places strong emphasis on the underlying principles and the formulation of governing equations. Despite the fact that the market has been inundated with literature on structural analysis, structural design books, especially those that present a systematic approach to the subject of horizontally curved bridges, have been very limited.

The design specifications used are those currently given in the latest Japanese Specifications for Highway Bridges adopted by the Japanese

Road Association and the Design Code, Part 2, Steel Bridge Structures by the Hanshin Expressway Public Corporation. For problems for which the codes do not specify design guides, suitable design information has been adopted. Since bridge design and specifications in Japan use the metric system of measurements, this system has been retained in this book. A conversion table to the English system is included for ready reference.

We begin with a brief review of earlier works on the subject and the organization of the book. Then we go on to review basic theory of thin-walled beams as background material. The chapters that follow present the fundamental theory of curved girders for static and dynamic behavior, for analysis of flexural and torsional stress resultants and displacements and for buckling strength of curved girders.

Later we discuss the design codes and specifications and examine the background information whenever appropriate. Then based on these code requirements, the current Japanese practice of fabrication, details, painting, and erection procedures are introduced.

The design examples that have actually been constructed in the 1980s in Japan entail the usual procedure outlined in the text and represent three typical superstructure girder systems. Included in the Appendix are a systematic evaluation of complex cross-sectional properties and vibration parameters for speedy reference. Also included in the Appendix is a list of computer programs currently being used in Japan.

In general the book is intended for practicing engineers. The material, however, should also be useful for well-motivated senior-year college and graduate students.

Although the general equations and concepts postulated by Timoshenko, Vlasov, Dabrowski, Konishi, Kamatsu, and others are developed and subsequently expanded, if necessary, resulting in numerous new formulations and design information, we are deeply indebted to them and to many others whose writings, teaching, or personal contact have shaped our thinking and approach to the subject matter.

Hiroshi Nakai
Chai Hong Yoo

Conversion Factors

(SI to English)	(English to SI)
Length:	
1 mm = 0.03937 in	1 in = 25.4 mm
1 m = 3.281 ft	1 ft = 0.3048 m
1 km = 0.6214 mi	1 mi = 1.609 km
Area:	
$1\ \text{mm}^2 = 1.55 \times 10^{-3}\ \text{in}^2$	$1\ \text{in}^2 = 0.652 \times 10^3\ \text{mm}^2$
$1\ \text{cm}^2 = 1.55 \times 10^{-1}\ \text{in}^2$	$1\ \text{in}^2 = 6.452\ \text{cm}^2$
$1\ \text{m}^2 = 10.76\ \text{ft}^2$	$1\ \text{ft}^2 = 0.0929\ \text{m}^2$
$1\ \text{m}^2 = 1.196\ \text{yd}^2$	$1\ \text{yd}^2 = 0.836\ \text{m}^2$
Volume:	
$1\ \text{mm}^3 = 6.102 \times 10^{-5}\ \text{in}^3$	$1\ \text{in}^3 = 16.387 \times 10^3\ \text{mm}^3$
$1\ \text{cm}^3 = 6.10 \times 10^{-2}\ \text{in}^3$	$1\ \text{in}^3 = 16.387\ \text{cm}^3$
$1\ \text{m}^3 = 35.3\ \text{ft}^3$	$1\ \text{ft}^3 = 0.0283\ \text{m}^3$
$1\ \text{m}^3 = 1.308\ \text{yd}^3$	$1\ \text{yd}^3 = 0.765\ \text{m}^3$
Moment of inertia:	
$1\ \text{in}^4 = 41.62\ \text{cm}^4$	$1\ \text{cm}^4 = 0.024\ \text{in}^4$
Mass:	
1 kg = 2.205 lb	1 lb = 0.454 kg
$1\ \text{kg} = 1.102 \times 10^{-3}\ \text{ton}$	1 ton (2000 lb) = 907.2 kg
1 Mg = 1.102 ton	1 tonne (metric) = 1000 kg
Force:	
1 N = 0.2248 lbf	1 lbf = 4.448 N
1 kgf = 2.205 lbf	1 kip = 4.448 kN
Stress:	
$1\ \text{kgf/cm}^2 = 14.22\ \text{psi}$	$1\ \text{psi} = 0.0703\ \text{kgf/cm}^2$
$1\ \text{kN/m}^2 = 0.145\ \text{psi}$	$1\ \text{psi} = 6.895\ \text{kPa}\ (\text{kN/m}^2)$
$1\ \text{MN/m}^2 = 0.145\ \text{ksi}$	$1\ \text{ksi} = 6.895\ \text{MN/m}^2$

Chapter

1

Introduction

1.1 Introduction

The use of horizontally curved girders has increased considerably in recent years for highway bridges and interchanges in large urban areas. The need for the smooth dissemination of congested traffic and the limitation of right-of-way along with economic and environmental considerations dictate that bridge alignments meet the overall requirements of the highway system. Also, the current emphasis on esthetic considerations has motivated increased development of designs which utilize curved configurations.

When curved viaducts were first introduced into the highway system, they were generally composed of a series of straight girders used as chords in forming a curved alignment. Recently, however, curved girders have replaced straight girders, thus forming a continuous curved alignment. The use of curved girders continuous over several spans has become more popular for several reasons. The total cost of the curved girder system has been found to be less than that of a system employing a series of straight girders for the same bridge.[16] Although the cost of the superstructures for the curved girders may be high, the cost is reduced considerably since a substantial portion of the substructuring that would be necessary for the straight beams can be eliminated. Using continuous curved girders also permits the use of shallower sections as well as a reduction in the slab overhang of outside girders. Depending on the terrain, such shallower sections may result in a significant reduction of cost and of earthwork associated with the construction of the access roads. The continuous curved girder also provides a more esthetically pleasing structure with its streamlined appearance.

In designing a structural component that is subjected to not just one but to two or more major stresses, various interaction design techniques are utilized. Due to the geometric complexities, curved girders are subjected to not only flexural stresses but also to very significant torsional stresses (torsional shear stress and warping normal stress). These are instituted by unavoidable torsion when only gravitational loads are considered. Because of this inherent rotation characteristic, diaphragms and bracings that are usually used in straight girder systems to prevent premature lateral buckling become very important load-carrying components in curved systems. The magnitude of the rotation is a function of the torsional stiffness of the girder. The limiting rotation or vertical deflection as a serviceability design criterion cannot be uniquely determined without knowing the complete interaction up to the ultimate load range.

1.2 Earlier Work on Curved Girders

With the increasing utilization of curved girder bridges, a better understanding of the true behavior of such girders is essential. At this time, the experimental test data available on curved girders is very limited. Although several analytical techniques have been published primarily on the static behavior of such girders in the elastic range, the availability of analytical methods for the elastic bifurcation buckling and ultimate load response of such girders is also quite limited. Although it is not intended to be exhaustive or complete, this book traces the development of techniques used in the analysis and design of horizontally curved girders in the earlier days.

The primary difficulty that has discouraged designers from using curved girder bridges has been the complex mathematical problems that arise in describing their behavior. A horizontally curved girder subjected to gravity loads (loads perpendicular to the plane of curvature) not only undergoes a vertical displacement, but twists with respect to its longitudinal axis. This interaction between bending and torsion is a function of girder geometry and its rigidity ratio of flexure and torsion.

The first work on the static analysis of horizontally curved beams is credited to St. Venant[17] by Love.[14] Since then, a number of European researchers have contributed to the analysis of curved beams, as pointed out by McManus et al. in their extensive review.[15] However, the first attempt to present a theoretical treatment of curved thin-walled girders is to be found in the work of Gottfeld.[9] He investigated two girder systems interconnected by cross-bracings placed on two concentric circles and loaded perpendicular to the plane of curvature. Horizontal bracings connecting top and bottom flanges were also included in the system, but were assumed not to transmit any force, a somewhat clumsy assumption. As pointed out by Dabrowski,[7] the problem treated by Gottfeld can in fact be conceived as a special case of the combined bending and torsion of curved thin-walled members. The first complete treatment of thin-walled curved girders with a doubly symmetrical I-shaped cross section was given by Umanskii.[19] He calculated the bimoments in an I beam loaded perpendicular to the plane of curvature. The bimoment is a term for a generalized force associated with warping of the cross section due to twisting, which produces normal stresses in the cross section. Timoshenko[18] discovered this phenomenon in his earlier work on the instability of an I beam. Umanskii[19] obtained solutions for several loading conditions by means of initial parameters.

In the second edition of the well-known book by Vlasov,[20] the differential equations for the coupled bending and nonuniform torsion of a curved beam of asymmetrical open section are given. Dabrowski[4] later made some corrections to these equations in order to obtain the correct displacement in the plane of curvature. Dabrowski[5] also presented the fundamental equations for the nonuniform torsion of curved box girders with nondeformable asymmetrical cross section. In his publication, the fundamental equations are obtained based on the assumptions previously used by Benscoter.[1]

Perhaps the single most important contribution made by Dabrowski[6] is his investigation of curved thin-walled girders with deformable cross sections published in IABSE (International Association for Bridge and Structural Engineering) in 1965. In the same year, Konishi and Komatsu[13] published a paper on three-dimensional analysis of curved girders with thin-walled cross sections, in which they established a refined relation to derive the modified geometrical properties of the cross section. Unlike most curved beam theories developed, Konishi and Komatsu established the relations between strain (displacement) and stress components based on a very sharp curved beam where the elastic flexural stress distribution along the beam depth could be nonlinear.

References cited above are primarily research-oriented works on the static behavior of isolated curved girders. Although Dabrowski[7] presented a vast number of formulas, tables, and influence lines, they are not easy to apply. In the mid-sixties, U.S. Steel[12] published an approximate procedure called "V-load analysis" for determining moments and shears in horizontally curved open framed highway bridges, which was utilized extensively in the industry until numerous computer programs became available in the early and mid-seventies. In the meanwhile, similar formulas and tables were published in Japan by Shimada and Kuranishi,[27] as well as by Watanabe.[28] Currently a computer program developed by BSDI[11] (Bridge Software Development International, Ltd.) is considered one of the most advanced, in that the program optimizes a horizontally curved bridge system which uses influence surfaces based on three-dimensional finite element analysis. On the other hand, the dynamic behaviors of curved girder bridges have been analyzed by Tan and Shore,[29] as well as by Komatsu and Nakai et al.[30,31] In their analyses, free vibration, forced vibration, and impact of the horizontally curved girders due to moving vehicles were investigated. However, very little research has been published on the dynamic response due to earthquake excitation.

The determination of allowable flexural stress in the bridge girder is primarily controlled by the overall lateral torsional buckling characteristics of the member. It appears that the procedure of determining allowable flange normal stress in compression in the 1980 AASHTO Guide Specifications for Horizontally Curved Highway Bridges[10] is based on the study reported by Culver and McManus.[3] (AASHTO stands for American Association of State Highway and Transportation Officials.) As compared with studies done by Yoo in a series of papers,[22–25] their research results, especially in the range of high subtended angles, yield fairly high values. In recent years, Nakai and others attempted to analyze the lateral buckling strength of a curved I girder in the elastoplastic region[33] and also the web buckling of a curved I girder on the basis of test results of a series of experimental studies.[34] Although fragmented information in the elastic buckling behavior of horizontally curved girders under a very limited loading condition and/or boundary condition is scattered, no comprehensive results have ever been documented in the public domain.

Boulton et al.[2] and Yoo et al.[21] presented plastic collapse loads for circular arc girders based on the kinematic collapse mechanism. Fukumoto et al.[8] and Yoshida et al.[26] studied the material and geometric nonlinear response of curved I beams. Mikami[32] also presented the ultimate strength of the curved frillage girder. These researchers provided a few ultimate strength curves under limited loading cases and residual stress distribution.

1.3 Organization of the Text

In classical analytical methods, the formulation of governing equations is but a starting point, which should be followed by suitable solution techniques. On the other hand, in modern structural analysis, the formulation of governing equations is the main task since much of the solution efforts of governing equations usually reduces to the solution of discretized systems; namely numerical solution by a computer program. The more computerized structural analysis techniques are used, however, the more important it becomes to the structural analyst to be well informed regarding the underlying principles inasmuch as the mathematical modeling of the structure itself is an art.

In order to present the complex problem of curved bridges in a manner that will be most helpful to engineers who have not been exposed to the relevant information during their formal education and to the students who are using the text in a course, the basic theory of thin-walled beams is examined in Chap. 2. Apart from introductory flexural behavior, the shear lag phenomenon and effective width concepts are examined. The torsional response of structural elements can be classified into two categories, namely pure torsion and warping torsion. In the case of pure torsion, most structural engineers have some familiarity with the concept, as this type of torsion is discussed in courses examining the elementary strength of materials. Warping torsion, however, is a fairly new term. The institution of this force and the resultants are described in detail for open and closed sections, and some approximate formulas for torsional warping stresses are given.

One major topic of interest currently confronting engineers is box girders, especially on curved bridges, due to their superior torsional rigidity. Discussed in detail is the phenomenon known as cross-sectional distortion. Fundamental equations for distortion and appropriate formulas are presented for rapid evaluation of the required spacing of internal diaphragms to minimize the induced stresses caused by distortion of the section.

Following the fundamental theory of thin-walled straight beams discussed in Chap. 2, similar equations are developed for curved girders for analyzing static behavior in Chap. 3. As noted earlier, the governing equations and formulas for the resultant forces for curved beams are inherently complex due to the coupling action of flexure and torsion. Also included in Chap. 3 is the dynamic response of curved girders on the basis of the fundamental equations of the above.

Chapter 4 is primarily devoted to the solution technique of the governing equations derived in Chap. 3. Detailed description of the analytical procedure is given, which will be useful in implementing the matrix method of analysis based on discretized models.

A clear understanding of the buckling behavior of structural components and assemblages is of paramount importance in the design of high-strength metallic structures. Small member sections are utilized in these structures to take advantage of the high-strength capabilities of the material. This inevitably results in buckling frequently being the controlling design criterion. Discussed in Chap. 5 are the global and local buckling strengths of curved elements up to the ultimate load range. Subsequently, design recommendations or proposals are made for the design of flange and web plates, and longitudinal and transverse stiffeners.

Curved bridges have been built and are continuing to be constructed throughout the world. Naturally, there are a number of design codes or specifications available by both governments and trade associations. The 1980 AASHTO Guide Specifications for Horizontally Curved

Highway Bridges last revised in 1986 is an example. It is, however, not intended to examine various design codes and specifications and to make any comparative assessment. Rather it is intended to introduce in Chap. 6 the current Japanese design code and the Design Code, Part 2, Steel Bridge Structures adopted in 1980 by the Hanshin Expressway Public Corporation (HEPC) to which the senior author (H.N.) made significant contribution. At present, the design of highway bridges in Japan is entirely based on the allowable stress design method. Essentially, provisions of the Japanese Specification for Highway Bridges (JSHB) adopted by the Japanese Road Association govern wherever applicable except those specifically modified by HEPC Code.

Chapter 7 presents the current Japanese practice in fabrication, details, painting and erection of curved bridges following two current Japanese codes: JSHB and HEPC.

Although none of the current Japanese codes explicitly requires CAD, almost all of the major curved steel bridges in Japan are designed by using computer programs. Chapter 8 presents an overview of current Japanese computer applications in curved bridge design. Presented as examples are three curved bridges that were actually designed by using some of the computer programs cited and constructed as a first-class bridge. The first is a multiple curved I-girder bridge, the Okunabe Bridge, with three-span continuous curved girders, which was built in 1985 by the Japanese Ministry of Construction. The second is a curved monobox girder bridge constructed in 1981. The third is a curved twin-box girder bridge constructed by the Hanshin Expressway Public Corporation in 1986.

Included in Appendix A are theoretical formulas and numerical procedures for an automatic evaluation of complex cross-sectional properties of an arbitrary bridge section. The vibration parameters listed should be of value for rapid analysis of the dynamic behavior of curved bridges. A brief overview of computer programs for curved bridge design in Japan is also included in Appendix C.

The engineer and student will be provided with the behavior of modern curved steel highway bridges. Past theories and developments are given in detail. These theories have subsequently been extended, resulting in numerous new formulations and additional design information which should provide the designer with a more complete working knowledge of this subject.

References

1. Benscoter, S. U.: "A Theory of Torsion Bending for Multicell Beams," *Journal of Applied Mechanics*, 21(1), pp. 25–34, March 1954.
2. Boulton, N. S., and B. Boonsukha: "Plastic Collapse Loads for Circular-Arc Girders," *Proceedings of the Institute of Civil Engineering*, vol. 13, 1959.

3. Culver, C. G., and P. F. McManus: "Instability of Horizontally Curved Members, Lateral Buckling of Curved Plate Girders" (project report submitted to the Pennsylvania Department of Transportation), Department of Civil Engineering, Carnegie-Mellon University, September 1971.
4. Dabrowski, R.: "The Analysis of Curved Thin-Walled Girders of Open Section," *Der Stahlbau*, 33(12), pp. 364–372, December 1964.
5. Dabrowski, R.: "Warping Torsion of Curved Box Girders of Non-Deformable Cross-Section," *Der Stahlbau*, 34(5), pp. 135–141, May 1965.
6. Dabrowski, R.: "Approximate Analysis of the Curved Box Girder of Deformable Cross-Section," *7th IABSE Congress*, preliminary publication, Zürich, Switzerland, pp. 299–306, 1965.
7. Dabrowski, R.: "Curved Thin-Walled Girders, Theory and Analysis" (transl. from German), Cement and Concrete Assoc., London, 1968.
8. Fukumoto, Y., and S. Nishida: "Ultimate Load Behavior of Curved I-Beams," *Journal of Engineering Mechanics*, ASCE, vol. 107, no. EM2, pp. 367–385, April 1982.
9. Gottfeld, H.: "The Analysis of Spatially Curved Steel Bridges," (in German) Die Bautechnik, pp. 715, 1932.
10. AASHTO: "Guide Specifications for Horizontally Curved Highway Bridges," 1980.
11. Hall, D. H.: a personal correspondence, Bridge Software Development International, Ltd., 1986.
12. *Highway Structure Design Handbook*, vol. 1, United States Steel, ADUSS, 88-2222, 1965.
13. Konishi, I., and S. Komatsu: "Three Dimensional Analysis of Curved Girder with Thin-Walled Cross-Section," IABSE, Zürich, Switzerland, vol. 25, pp. 143–203, 1965.
14. Love, A. E. H.: *A Treatise on the Mathematical Theory of Elasticity*, 4th ed., Dover, 1927.
15. McManus, P. F., G. A. Nasir, and C. G. Culver: "Horizontally Curved Girders—State of the Art," *Journal of Structural Engineering*, ASCE, vol. 95, no. ST-5, pp. 853–870, May 1969.
16. Schmitt, W.: "Interchange Utilizes Arc Welded Horizontally Curved Girder Span," unpublished paper submitted to Lincoln Arc Welding Foundation, 1966.
17. St. Venant, B.: "Mémoire sur le calcul de la resistance et la flexion des pieces solides a simple ou a double courbure, en prenant simultanement en consideration les divers efforts auxquels eiles peuvent entre soumises dans tous les sens," Compts-Rendus, l'Academie des Sciences de Paris, vol. 17, pp. 942, 1020–1031, 1843 (in French).
18. Timoshenko, S. P.: "On the Stability in Plane Bending of an I-Beam," Izvestiya St. Petersburg Politekhnicheskoqo Instituta, IV-V, 1905.
19. Umanskii, A. A.: *Spatial Structures* (in Russian), Moscow, 1948.
20. Vlasov, V. Z., Thin-Walled Elastic Beams, 2d ed., Washington, D.C., 1961.
21. Yoo, C., and C. P. Heins: "Plastic Collapse of Horizontally Curved Bridge Girders," *Journal of Structural Engineering*, ASCE, vol. 98, no. ST-4, pp. 899–914, April 1972.
22. Yoo, C.: "Stability of Curved Girders," *Proceedings of the International Conference on Finite Element Methods*, Shanghai, People's Republic of China, pp. 305–311, August 1982.
23. Yoo, C.: "Lateral Torsional Buckling of Horizontally Curved Girders," *Proceedings of Sino-American Symposium on Bridge and Structural Engineering*, part 2, Beijing, People's Republic of China, pp. 17/1–15, September 1982.
24. Yoo, C.: "Flexural Torsional Stability of Curved Beams," *Journal of the Engineering Mechanics*, ASCE, vol. 108, no. EM6, pp. 1351–1369, December 1982.
25. Yoo, C., and P. A. Pfeiffer: "Elastic Stability of Curved Members," *Journal of Structural Engineering*, ASCE, 109(12), pp. 2922–2940, December 1983.
26. Yoshida, H., and K. Maegawa: "Ultimate Strength Analysis of Curved I-Beams," *Journal of Engineering Mechanics*, ASCE, 109(1), pp. 192–214, February 1983.
27. Shimada, S., and S. Kuranishi: *Formulas for Calculation of Curved Beam*, Gihodo, Tokyo, 1966 (in Japanese).
28. Watanabe, N.: *Theory and Calculation of Curved Girder*, Gihodo, Tokyo, 1967 (in Japanese).
29. Tan, C. P., and S. Shore: "Dynamic Response of Horizontally Curved Bridges," *Journal of Structural Engineering*, ASCE, vol. 94, no. ST-4, pp. 761–781, March 1968.

30. Komatsu, S., and H. Nakai: "Study on Free Vibration of Curved Girder Bridges," *Transactions of the Japanese Society of Civil Engineers*, no. 136, pp. 27–38, December 1966.
31. Nakai, H., and H. Kotoguchi: "Dynamic Response of Horizontally Curved Girder Bridges Under Random Traffic Flow," *Proceedings of the Japanese Society of Civil Engineers*, no. 244, pp. 117–128, December 1975.
32. Yonezawa, H., and I. Mikami: "On the Limite Analysis of Curved Grillage Girder," *Transactions of the Japanese Society of Civil Engineers*, no. 132, pp. 18–26, August 1966 (in Japanese).
33. Nakai, H., and H. Kotoguchi: "A Study on Lateral Buckling Strength and Design Aid for Horizontally Curved I-Girder Bridges," *Proceedings of the Japanese Society of Civil Engineers*, no. 339, pp. 105–204, December 1983.
34. Nakai, H., T. Kitada, and R. Ohminami: "Experimental Study on Ultimate Strength of Web Panels in Horizontally Curved Girder Bridges Subjected to Bending, Shear and Their Combinations," *Proceedings of Structural Stability Research Council*, 1984, Annual Technical Session and Meeting, pp. 91–102, April 1984.

Chapter

2

Basic Theory of Thin-Walled Beams

2.1 Introduction

This chapter presents the basic theory of a thin-walled beam as a precursor to the development of the fundamental theory of a curved beam. The contents consist of four important and fundamental sections concerning thin-walled beam analysis and design: flexure, torsion, distortion, and stress distribution.

In Sec. 2.2, the flexural behavior of a thin-walled beam is described following a discussion of the thin-walled elastic beam theory. Then the procedures for estimating normal and shearing stresses in various types of girders are detailed with a few examples. Also included is an analysis of the shear lag phenomenon and the effective width of a thin-walled beam subject to various loading conditions. The fundamental equation of shear lag is derived based on a theory of elasticity incorporating Reisser's hypothesis. From this equation, stress resultants and displacements due to shear lag of simply supported girders are presented. The validity of this method is also examined by using data obtained from experimental studies. By modifying these results, the effective width of steel girder bridges is discussed through various parametric studies, and practical design formulas are recommended. Additionally, procedures for application to the box girder, other types of girders, and continuous girder bridges are also discussed.

In Sec. 2.3, the torsional behaviors, i.e., pure torsion and warping torsion of a thin-walled beam, are discussed. Then pure and warping torsional stresses and corresponding stress resultants are described. Cross-sectional properties of various types of girders such as I girders, π girders, box girders, and multiple girders are derived by considering the particular characteristics of thin-walled members. By synthesizing these analytical results, an approximate and simplified procedure to evaluate the stresses in the box girder is presented, incorporating numerical studies on the actual test data of long-span box-girder bridges.

Section 2.4 presents the distortional stress analysis of a box girder. The fundamental equations of distortion of a thin-walled box girder have been derived by the theory of elasticity and by energy methods. These equations are rearranged in a manner analogous to the torsional warping theory. Resulting equations can be categorized as equations of a beam on an elastic foundation. The analytical method is also predicated on details. Based on these analyses, an extensive parametric survey has been performed by using the actual data of long-span box-girder bridges, and distortional properties are discussed from various points of view. Also presented are the design aids for estimation of distortional warping stresses in box girders, spacing of intermediate diaphragms, proper rigidity, and strength of diaphragms.

In Sec. 2.5, formulas for cross-sectional properties and stresses for a thin-walled beam are summarized in tables, and an interaction formula is given for the combined action of these different stress components.

2.2 Flexure[1–5]

2.2.1 Flexural normal stress

Let us now consider the flexural strain ε_z in a thin-walled beam with an asymmetric cross section, as shown in Fig. 2.1, where bending moments M_x and M_y are applied with respect to the horizontal and vertical axes (x, y) through the centroid C, respectively.

The flexural strain ε_z at an arbitrary point $P(x, y)$ of cross section is given by the well-known Bernoulli-Euler hypothesis which states that

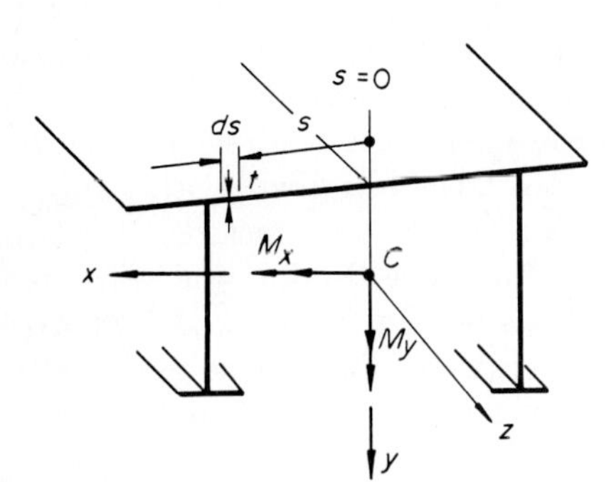

Figure 2.1 Thin-walled beam subjected to bending moments M_x and M_y.

ε_z is linearly proportional to the fiber distance from point $P(x, y)$ to the centroid C and is

$$\varepsilon_z = \alpha + \beta x + \gamma y \tag{2.2.1}$$

where α, β, and γ are constants. By denoting E as Young's modulus, the flexural normal stress σ_z can be expressed by Hooke's law:

$$\sigma_z = E(\alpha + \beta x + \gamma y) \tag{2.2.2a}$$

Introducing new constants, we find

$$\sigma_z = a + bx + cy \tag{2.2.2b}$$

Also, the relationships between bending moments M_x and M_y and the above flexural normal stress σ_z are as follows for pure bending:

$$\int_A \sigma_z \, dA = \int_A (a + bx + cy)t \, ds = N = 0 \tag{2.2.3a}$$

$$\int_A \sigma_z x \, dA = \int_A (a + bx + cy)xt \, ds = M_y \tag{2.2.3b}$$

$$\int_A \sigma_z y \, dA = \int_A (a + bx + cy)yt \, ds = M_x \tag{2.2.3c}$$

Here $A = \int_A dA = \int_A t \, ds$ is the cross-sectional area of the beam, and t and s are the thickness and curvilinear coordinates taken along the perimeter of the cross section, respectively, as illustrated in Fig. 2.1.

Since C is the centroid of the cross section, expressions for the static moments can be written as

$$\int_A xt \, ds = 0 \tag{2.2.4a}$$

$$\int_A yt \, ds = 0 \tag{2.2.4b}$$

Also, the geometric moments of inertia are given by

$$I_y = \int_A x^2 t \, ds \tag{2.2.4c}$$

$$I_x = \int_A y^2 t \, ds \tag{2.2.4d}$$

$$I_{xy} = \int_A xyt \, ds \tag{2.2.4e}$$

Substituting Eq. (2.2.4) into Eq. (2.2.3), we can generate the following set of equations:

$$a = 0 \tag{2.2.5a}$$

$$bI_y + cI_{xy} = M_y \tag{2.2.5b}$$

$$bI_{xy} + cI_x = M_x \tag{2.2.5c}$$

Solving the above set of equations for the constants a, b, and c, and substituting the values into Eq. (2.2.2b), we obtain the flexural normal stress $\sigma_b = \sigma_z$:

$$\sigma_b = \frac{M_y I_x - M_x I_{xy}}{I_x I_y - I_{xy}^2} x + \frac{M_x I_y - M_y I_{xy}}{I_x I_y - I_{xy}^2} y \tag{2.2.6}$$

This equation is convenient to use as the design formula, because we need not determine the principal axes of the cross section.

If $M_y = 0$, then Eq. (2.2.6) can be simplified to

$$\sigma_b = \frac{I_y y - I_{xy} x}{I_x I_y - I_{xy}^2} M_x \tag{2.2.7}$$

Furthermore, when the cross section of a beam is symmetric, the product of inertia I_{xy} drops out, i.e.,

$$I_{xy} = 0 \tag{2.2.8}$$

and the above equation is further simplified to

$$\sigma_b = \frac{M_x}{I_x} y \tag{2.2.9}$$

This is a familiar formula to estimate the flexural stress in a beam with a symmetric cross section.

2.2.2 Flexural shearing stress

Figure 2.2 shows a small element $dz\,ds$ in the thin-walled beam with thickness t. To know the relationships between flexural normal stress $\sigma_z = \sigma_b$ and flexural shearing stress $\tau_{sz} = \tau_b$ for this element, consider the equilibrium condition of the forces in the direction of the z axis, as shown in Fig. 2.3. Force equilibrium in the z direction results in

$$\frac{\partial \tau_b t}{\partial s} = -\frac{\partial \sigma_b t}{\partial z} \tag{2.2.10}$$

The substitution in Eq. (2.2.6) into the above equation and the

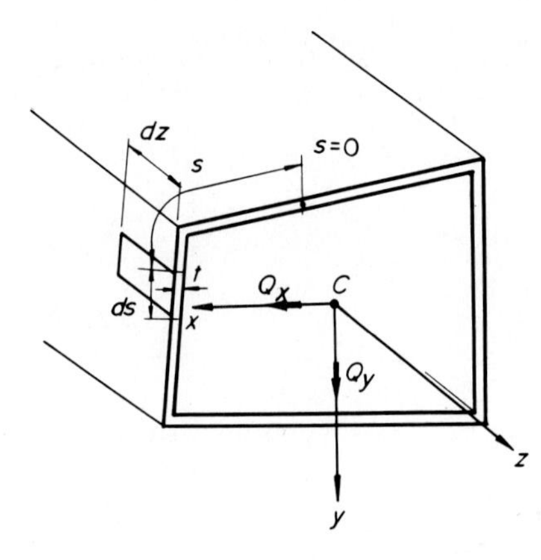

Figure 2.2 An element $dz\,ds$ cut off by the beam.

definitions of shearing forces Q_y and Q_x, expressed by the equations

$$\frac{\partial M_x}{\partial z} = Q_y \tag{2.2.11a}$$

and

$$\frac{\partial M_y}{\partial z} = Q_x \tag{2.2.11b}$$

gives

$$\frac{\partial \tau_b\, t}{\partial s} = -\frac{Q_x I_x - Q_y I_{xy}}{I_x I_y - I_{xy}^2}\, xt - \frac{Q_y I_y - Q_x I_{xy}}{I_x I_y - I_{xy}^2}\, yt \tag{2.2.12}$$

By integrating both sides of Eq. (2.2.12) and defining

$$q_b = \tau_b t \tag{2.2.13}$$

where q_b is the shear flow due to bending, the following equation is obtained:

$$q_b = \tau_b t = -\frac{Q_x I_x - Q_y I_{xy}}{I_x I_y - I_{xy}^2} \int_0^s xt\, ds - \frac{Q_y I_y - Q_x I_{xy}}{I_x I_y - I_{xy}^2} \int_0^s yt\, ds + C \tag{2.2.14}$$

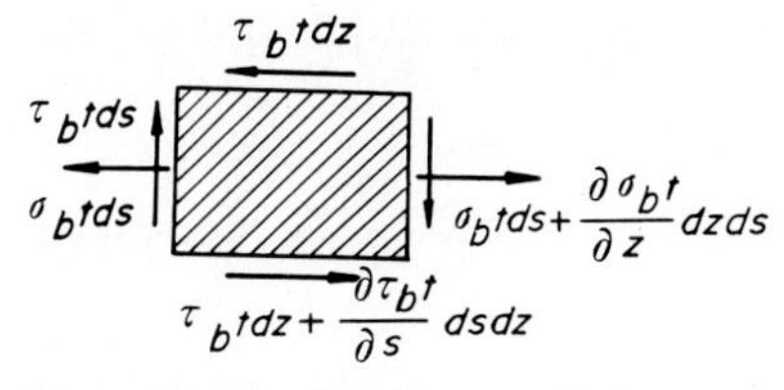

Figure 2.3 Equilibrium condition of forces on element $dz\,ds$.

The strain energy due to the flexural stresses is given by

$$\Pi = \frac{1}{2}\int_A \left(\frac{\sigma_b^2}{E} + \frac{\tau_b^2}{G}\right) t\, ds \tag{2.2.15a}$$

For stable equilibrium,

$$\frac{\partial \Pi}{\partial C} = 0 \tag{2.2.15b}$$

and this condition leads to

$$\oint \frac{q_b}{Gt}\, ds = 0 \tag{2.2.15c}$$

where $\oint ds$ is the contour integral around the closed section, and

$$C = \left[\frac{Q_x I_x - Q_y I_{xy}}{I_x I_y - I_{xy}^2} \oint \left(\int_0^s xt\, ds\right) ds + \frac{Q_y I_y - Q_x I_{xy}}{I_x I_y - I_{xy}^2} \oint \left(\frac{1}{Gt}\int_0^s yt\, ds\right) ds\right] \left(\oint \frac{ds}{Gt}\right)^{-1} \tag{2.2.16}$$

By introducing the static moments S_x and S_y given by

$$S_y = \int_0^s xt\, ds - \frac{\oint [1/(Gt)]\left(\frac{1}{Gt}\int_0^s xt\, ds\right) ds}{\oint ds/(Gt)} \tag{2.2.17a}$$

$$S_x = \int_0^s yt\, ds - \frac{\oint [1/(Gt)]\left(\int_0^s yt\, ds\right) ds}{\oint ds/(Gt)} \tag{2.2.17b}$$

substituting Eqs. (2.2.16) and (2.2.17) into Eq. (2.2.14) and simplifying, Eq. (2.2.14) reduces to

$$q_b = -\frac{Q_x I_x - Q_y I_{xy}}{I_x I_y - I_{xy}^2} S_y - \frac{Q_y I_y - Q_x I_{xy}}{I_x I_y - I_{xy}^2} S_x \tag{2.2.18}$$

For the case where the vertical shearing force Q_y acts only on the beam, Eq. (2.2.18) reduces to

$$q_b = -\frac{I_y S_x - I_{xy} S_y}{I_x I_y - I_{xy}^2} Q_y \tag{2.2.19}$$

When a beam has a vertical axis of symmetry, the above equation results in the following well-known formula:

$$q_b = -\frac{Q_y}{I_x} S_x \tag{2.2.20}$$

Let us now show some numerical examples of shear flow q_b for various cross sections. In the first example, consider the I beam shown in Fig. 2.4*a*. For this cross section, the geometric moment of inertia I_x is represented by

$$I_x \cong 2t_f b_f \left(\frac{h}{2}\right)^2 + \frac{t_w h^3}{12} = \frac{A_f h^2}{2}\left(1 + \frac{1}{6}\frac{A_w}{A_f}\right) \tag{2.2.21}$$

where $A_f = t_f b_f$ = cross-sectional area of flange plate
$A_w \cong t_w h$ = cross-sectional area of web plate

The shear flow q_b does not exist at points 1, 3, 5, and 6, since the cross section is open to the air. The shear flow q (the subscript b is omitted in foregoing discussions for the sake of simplicity) at the remaining points is calculated as follows:

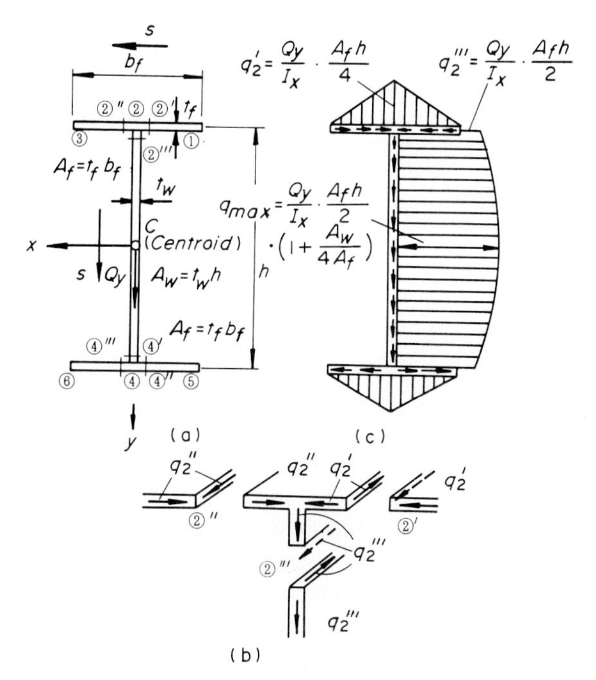

Figure 2.4 Examples of flexural shear flow in an I beam: (*a*) cross section, (*b*) shear flow in corner, (*c*) distribution.

Section 1–2′:

$$q_{1-2'} = -\frac{Q_y}{I_x}\int_{-b_f/2}^{x}\left(-\frac{h}{2}\right)t_f\,dx$$

$$= \frac{Q_y}{I_x}\frac{ht_f}{2}\left(\frac{b_f}{2}+x\right) \tag{2.2.22a}$$

at 2′, $x = 0$, so

$$q_{2'} = \frac{Q_y}{I_x}\frac{A_f h}{4} \tag{2.2.22b}$$

Points 3 and 2″:

$$q_3 = -\frac{Q_y}{I_x}\int_{0}^{b_f/2}\left(-\frac{h}{2}\right)t_f\,dx + q_{2''} \equiv 0 \tag{2.2.22c}$$

Therefore,

$$q_{2''} = -\frac{Q_y}{I_x}\frac{A_f h}{4} \tag{2.2.22d}$$

Accordingly, the shear flow $q_{2'''}$ at point 2‴ can be estimated according to the horizontal equilibrium condition of the axial forces at points 2′, 2″, and 2‴, shown in Fig. 2.4*b*, as follows:

$$q_{2'''} = q_{2'} + q_{2''} = \frac{Q_y}{I_x}\frac{A_f h}{2} \tag{2.2.22e}$$

Section 2‴–4′:

$$q_{2'''-4'} = -\frac{Q_y}{I_x}\int_{-h/2}^{y} y t_w\,dy + q_{2'''}$$

$$= -\frac{Q_y}{I_x}\left[\frac{y^2}{2}\right]_{-h/2}^{y} t_w + \frac{Q_y}{I_x}\frac{A_f h}{2}$$

$$= \frac{Q_y}{I_x}\left[\frac{t_w}{2}\left(\frac{h^2}{4} - y^2\right) + \frac{A_f h}{2}\right] \tag{2.2.22f}$$

The corresponding shear flows at the lower flange can be calculated in a manner similar to that above. These shear flows are in the direction indicated by the arrows in Fig. 2.4*c*. In general, the shear flows converge and add at the junction point of compression members and diverge at the junction point of tensile members, such as one stream branching into two streams.

The maximum shear flow q_{max} always occurs at the centroid $C(y = 0)$, and $q_b = q_{max}$ at $y = 0$ is given by

$$q_{\max} = \frac{Q_y}{I_x}\left(\frac{t_w h^2}{8} + \frac{A_f h}{2}\right)$$

$$= \frac{Q_y}{I_x}\frac{A_f h}{2}\left(1 + \frac{1}{4}\frac{A_w}{A_f}\right) \qquad (2.2.22g)$$

Therefore, the maximum flexural shearing stress $\tau_{\max}$ can be estimated by using Eqs. (2.2.13) and (2.2.21) as follows:

$$\tau_{\max} = \frac{q_{\max}}{t_w} = \frac{Q_y}{I_x t_w}\frac{A_f h}{2}\left(1 + \frac{1}{4}\frac{A_w}{A_f}\right)$$

$$= \frac{Q_y}{A_w}\frac{1 + \frac{1}{4}A_w/A_f}{1 + \frac{1}{6}A_w/A_f} \qquad (2.2.23a)$$

This equation can also be simplified by neglecting the effects of the cross-sectional area ratio A_w/A_f in the above equation through the parametric survey of actual I beams, which gives

$$\tau_{\max} = \frac{Q_y}{A_w} \qquad (2.2.23b)$$

The second example deals with the distribution of shear flow in a box beam with a bisymmetric cross section, as shown in Fig. 2.5*a*. For this box beam, clearly $q_1 = 0$ because of the symmetry of the cross section, and the shear flow throughout the cross section can be plotted easily, as illustrated in Fig. 2.5*b*. The approximate maximum shearing stress

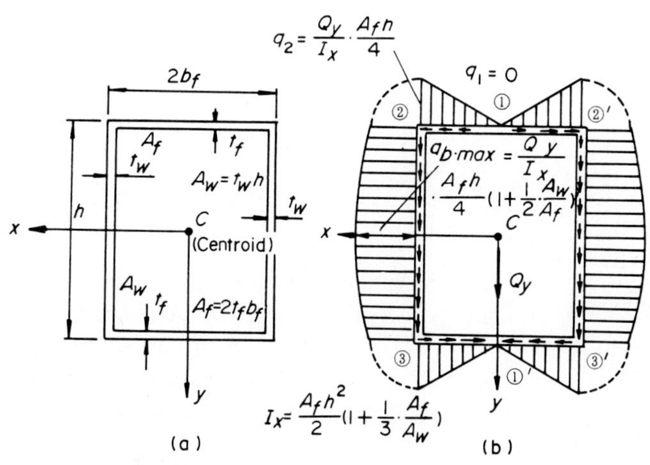

Figure 2.5 Flexural shear flow in box beam: (*a*) bisymmetric cross section, (*b*) distribution.

τ_{max} can be estimated by

$$\tau_{max} = \frac{q_{max}}{t_w} = \frac{Q_y A_f h/4}{A_f h^2 t_w/2} \frac{1 + \frac{1}{2}A_w/A_f}{1 + \frac{1}{3}A_w/A_f} \cong \frac{Q_y}{2A_w} \tag{2.2.24}$$

In the case of the box beam with an asymmetric cross section, we must, however, consider the influence due to the integration constant C, as indicated in Eq. (2.2.14). To obtain this integration constant C as simply as possible, a slit is inserted into an arbitrary point of the closed cross section, as illustrated in Fig. 2.6. Then, by denoting the shear flow for this opening with the slit q_o and introducing a statically indeterminate shear flow X, due to the closed section, which is constant across the cross section, the actual shear flows q_b can be given by superposition of these two shear flows:

$$q_b = q_o - X \tag{2.2.25}$$

The statically indeterminate shear flow X can be obtained easily by substituting this equation into Eq. (2.2.15c):

$$X = \frac{\oint (q_o/Gt)\, ds}{\oint ds/Gt} \tag{2.2.26}$$

The procedure can also be applied to a multicellular box beam, as shown in Fig. 2.7. The slits and the corresponding statically indeterminate shear flows X_1, X_2, and X_3 are inserted into each box cell, giving

$$q_b = q_o - X_1 - X_2 - X_3 \tag{2.2.27}$$

and X_1, X_2, and X_3 can be determined by solving the following

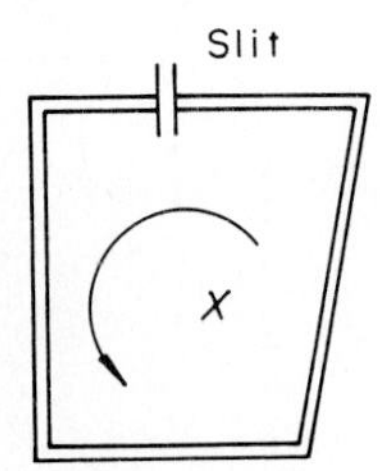

Figure 2.6 Statically indeterminate shear flow in monobox beam.

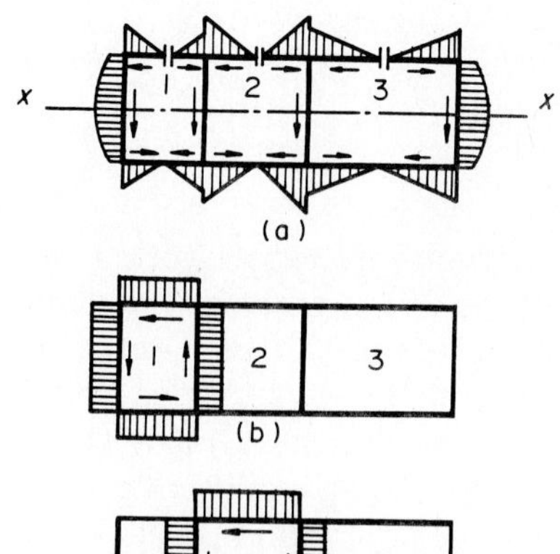

Figure 2.7 Statically indeterminate shear flows in multicellular box beam: (*a*) q_0, $X=0$; (*b*) $q_1=1$, X_1; (*c*) $q_2=1$, X_2.

simultaneous equations:

$$\delta_{11}X_1 + \delta_{12}X_2 = \delta_{10} \tag{2.2.28a}$$

$$\delta_{21}X_1 + \delta_{22}X_2 + \delta_{23}X_3 = \delta_{20} \tag{2.2.28b}$$

$$\delta_{33}X_2 + \delta_{33}X_3 = \delta_{30} \tag{2.2.28c}$$

where $\delta_{ii} = \oint_i \frac{1}{Gt}\, ds$ (contour integral for ith cell) (2.2.29*a*)

$\delta_{ij} = \int_{i,j} \frac{1}{Gt}\, ds$ (integral for boundary wall ith and jth cells) (2.2.29*b*)

$\delta_{io} = \oint_i \frac{q_o}{Gt}\, ds$ (contour integral for ith cell) (2.2.29*c*)

Finally, let us consider the shear flow in a beam made of channel section, as shown in Fig. 2.8. The resultant force of shear flow $q_{2\text{-}3}$ in the web plate must equal the shearing force Q_y; thus

$$\int_{-h/2}^{h/2} q_{2\text{-}3}\, dy = Q_y \tag{2.2.30}$$

The resultant force H of shear flow $q_{1\text{-}2}$ acting on the flange plate is given by

$$H = \int_0^{b_f} q_{1\text{-}2}\, dx = \frac{Q_y}{I_x}\frac{A_f b_f h}{4} \tag{2.2.31}$$

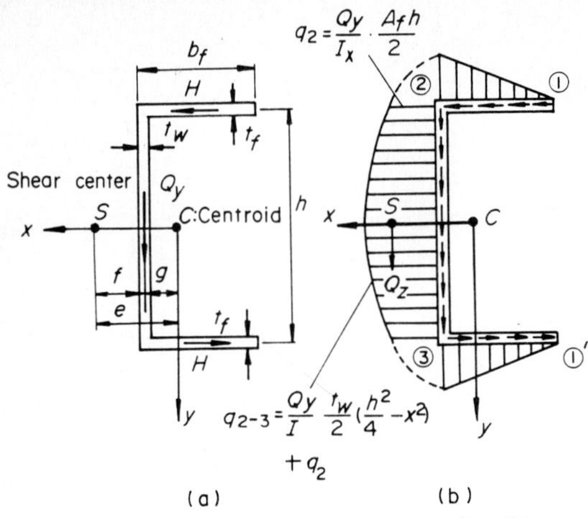

Figure 2.8 Shear flow and shear center of an [beam: (*a*) cross section, (*b*) distribution.

If the shearing force Q_y is applied to the midplane of the web plate of the channel section, the beam will always be subjected to torsion. To realize pure bending, the shearing force Q_y must be applied at point S apart from the midplane of web plate with eccentricity f. This point, S, is referred to as the *shear center* of the cross section and can be estimated by the equilibrium condition of torque $M_S = 0$ at point S due to the resultant forces, Q_z and H, as follows:

$$\sum M_S = Q_y f - \frac{Q_y}{I_x}\frac{A_f b_f h}{4} h \equiv 0 \tag{2.2.32a}$$

Therefore,

$$f = \frac{A_f b_f h^2}{4I_x} \tag{2.2.32b}$$

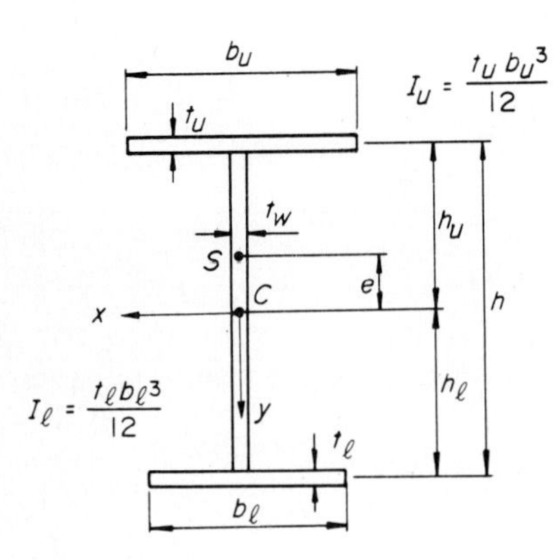

Figure 2.9 Location of shear center in an I beam.

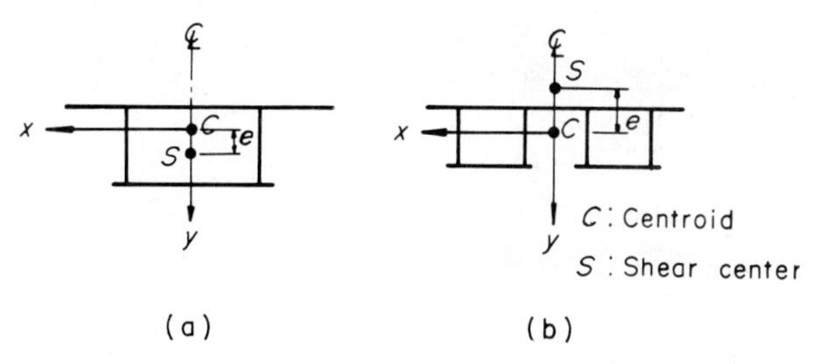

Figure 2.10 Location of shear center in (*a*) monobox girder and (*b*) twin-box girder.

Accordingly, torsion will always occur unless the shearing force Q_y is applied at a point $e = f + g$ from centroid C.

It is easily shown that the location of the shear center S for an I beam, as shown in Fig. 2.9, is a distance e from the centroid C, where e is given by

$$e = \frac{I_u h_u - I_l h_l}{I_u + I_l} \tag{2.2.33}$$

In general, the shear center S is situated inside the box for a monobox girder, whereas the shear center S is located outside the deck plate for a π girder or twin-box girder, as illustrated in Fig. 2.10.

An approximate method to determine the shear center is detailed in the following section.

2.2.3 Shear lag phenomenon and effective width[6,8–11]

(1) Shear lag phenomenon. The normal stress distribution $\sigma_z(x)$ in the flange plate of a π beam made of thin plates does not have a constant value, as is obtained by the elementary beam theory, but varies in the direction of coordinate x axis, as illustrated in Fig. 2.11. Then the maximum flexural normal stress $\sigma_{z,\max}$ occurs at the junction point of flange and web plates, and this result is significantly different from σ_z = constant, which is calculated by elementary beam theory. This phenomenon is caused by the lag of shear strain in the flange plate between the web plates and is referred to as the *shear lag phenomenon.*

The shear lag phenomenon can be analyzed on the basis of the theory of elasticity by assuming that the flange plate can be analyzed as a plane-stress problem.[8] Figure 2.12 shows the stresses σ_z, σ_x, and τ_{xy}, τ_{zx} in a small element $dx \cdot dz$, removed from the flange plate of a beam shown in Fig. 2.11, so that the equilibrium condition of stresses in the

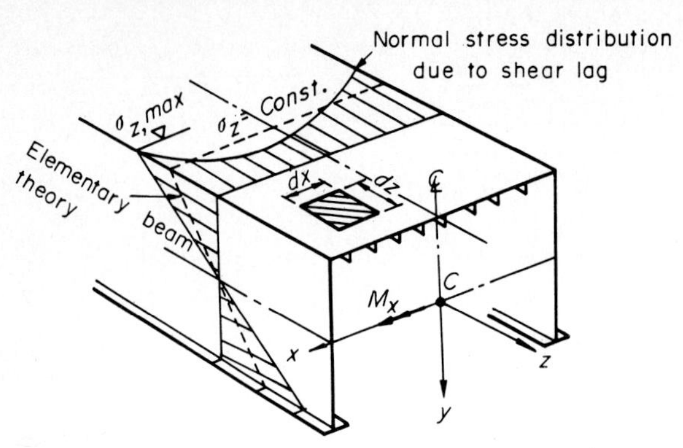

Figure 2.11 Actual bending normal stress distribution in a thin-walled beam.

direction of the z axis can be written

$$\frac{\partial \sigma_z}{\partial z} + \frac{\partial \tau_{xz}}{\partial x} = 0 \tag{2.2.34a}$$

$$\frac{\partial \sigma_y}{\partial x} + \frac{\partial \tau_{zx}}{\partial z} = 0 \tag{2.2.34b}$$

The relationships between displacements u and v in the direction of coordinate axes (z, x) and strains ε_z, ε_x, and γ_{zx} can be written as

$$\varepsilon_z = \frac{\partial u}{\partial z} \tag{2.2.35a}$$

$$\varepsilon_x = \frac{\partial v}{\partial x} \tag{2.2.35b}$$

$$\gamma_{zx} = \frac{\partial u}{\partial x} + \frac{\partial v}{\partial z} \tag{2.2.35c}$$

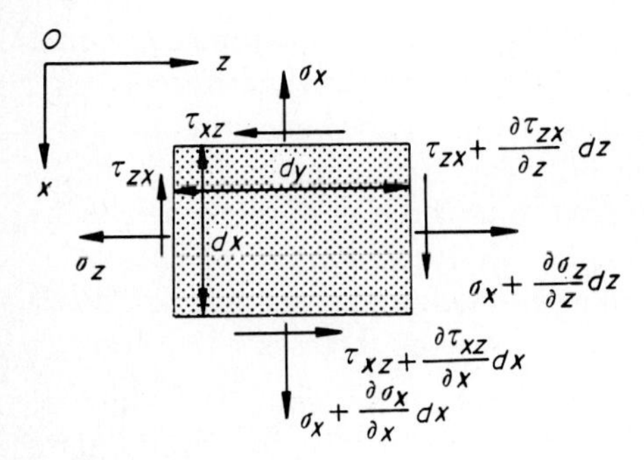

Figure 2.12 Stress situation in small element $dx\,dz$.

In addition, the compatibility equation for plane stress elasticity is given by

$$\frac{\partial^2 \varepsilon_z}{\partial x^2} + \frac{\partial^2 \varepsilon_x}{\partial z^2} = \frac{\partial^2 \gamma_{zx}}{\partial z\, \partial x} \tag{2.2.35d}$$

Stress-strain relationships for two-dimensional elasticity are

$$\varepsilon_z = \frac{1}{E}(\sigma_z - \mu\sigma_x) \tag{2.2.35e}$$

$$\varepsilon_x = \frac{1}{E}(\sigma_x - \mu\sigma_z) \tag{2.2.35f}$$

$$\gamma_{zx} = \frac{\sigma_{zx}}{G} \tag{2.2.35g}$$

where μ is Poission's ratio and $G = E/[2(1+\mu)]$. Substituting the stress-strain relationships in Eqs. (2.2.35*e*) to (2.2.35*g*) and the equilibrium condition in Eq. (2.2.34) into the compatibility equation, Eq. (2.2.35*d*), we can derive a general compatibility condition in terms of stress:

$$\left(\frac{\partial^2}{\partial z^2} + \frac{\partial^2}{\partial x^2}\right)(\sigma_z + \sigma_x) = 0 \tag{2.2.35h}$$

Thus, the shear lag problem is reduced to solving for the stresses σ_z, σ_x, and τ_{zx}, such that equilibrium and compatibility conditions are met.

These procedures can be much simplified by introducing Airy's stress function $\Phi(z, x)$:

$$\sigma_z = \frac{\partial^2 \Phi}{\partial x^2} \tag{2.2.36a}$$

$$\sigma_x = \frac{\partial^2 \Phi}{\partial z^2} \tag{2.2.36b}$$

$$\tau_{zx} = -\frac{\partial^2 \Phi}{\partial z\, \partial x} \tag{2.2.36c}$$

Then Eq. (2.2.35*h*) can readily be rewritten as

$$\frac{\partial^4 \Phi}{\partial z^4} + 2\frac{\partial^4 \Phi}{\partial z^2\, \partial x^2} + \frac{\partial^4 \Phi}{\partial x^4} = 0 \tag{2.2.37}$$

Let us now introduce some examples of shear lag analysis.[9] The first example is an analysis of shear lag phenomenon in the flange plate stiffened by many ribs with span l on which the line load $p(z)$ in the direction of the coordinate z axis:

$$p(z) = p \sin \frac{\pi z}{l} \tag{2.2.38}$$

This is applied as shown in Fig. 2.13. Taking the spacing of ribs as $2b$ and rectangular coordinate axes (z, x) on the deck plate, we see that the axial displacement u at the junction edges $x = \pm b$ of the ribs and flange plate is

$$u = -u_o \cos \frac{\pi z}{l} \tag{2.2.39}$$

Thus the axial strain ε_z gives

$$\varepsilon_z = \frac{\partial u}{\partial z} = u_o \frac{\pi}{l} \sin \frac{\pi z}{l} = \varepsilon_o \sin \frac{\pi z}{l} \tag{2.2.40a}$$

where the strain $\varepsilon_o = u_o \pi / l$ can be determined as in the proceeding discussion [cf. Eq. (2.2.51)].

Thus, the flange plate undergoes not only axial displacement but also deflection in the vertical direction due to bending. When the flange plate is composed of very thin plates, however, it is thought that

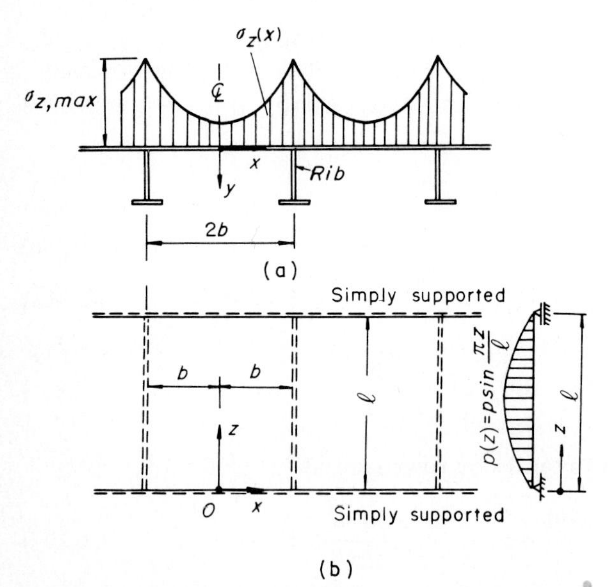

Figure 2.13 Shear lag analysis of flange plate stiffened by ribs: (*a*) cross section and normal stress distribution, (*b*) plane of deck and load condition.

the influence of bending strain in the direction of thickness of the flange plate is small enough to be ignored and that the deck plate is nearly equal to the in-plane stress situation. Then the stress distribution can be obtained by applying Eq. (2.2.37).

For this problem, the stress function $\Phi(z, x)$ can be assumed to be

$$\Phi(z, x) = \phi(x) \sin \frac{\pi z}{l} \tag{2.2.40b}$$

The substitution of the above equation into Eq. (2.2.37) gives the following differential equation concerning the unknown function $\phi(x)$:

$$\frac{d^4\phi}{dx^4} - 2\alpha^2 \frac{d^2\phi}{dx^2} + \alpha^4\phi = 0 \tag{2.2.40c}$$

where

$$\alpha = \frac{\pi}{l} \tag{2.2.40d}$$

Clearly the solution of this differential equation must include double roots $\pm\alpha$ and four integration constants A, B, C, and D, which gives

$$\phi = (A + B\alpha x) \sinh \alpha x + (C + D\alpha x) \cosh \alpha x \tag{2.2.40e}$$

Since $\phi(x)$ is a symmetric function with respect to the x axis, the final expression for the stress function $\Phi(z, x)$ must be reduced to

$$\Phi(z, x) = (B\alpha x \sinh \alpha x + C \cosh \alpha x) \sin \alpha z \tag{2.2.41}$$

Therefore, the stress components σ_z, σ_x, and τ_{zx} can be evaluated from Eq. (2.2.36) as follows:

$$\sigma_z = \alpha^2[B\alpha x \sinh \alpha x + (2B + C) \cosh \alpha x] \sin \alpha z \tag{2.2.42a}$$

$$\sigma_x = -\alpha^2(B\alpha x \sinh \alpha x + C \cosh \alpha x) \sin \alpha z \tag{2.2.42b}$$

$$\tau_{zx} = -\alpha^2[(B + C) \sinh \alpha x + B\alpha x \cosh \alpha x] \cos \alpha z \tag{2.2.42c}$$

But the strain components ε_z and γ_{zx} can also be estimated from Eqs. (2.2.35a) and (2.2.35c) as follows:

$$\varepsilon_z = \frac{\alpha^2}{E} \{(1 + \mu)B\alpha x \sinh \alpha x + [2B + (1 + \mu)C] \cosh \alpha x\} \sin \alpha z \tag{2.2.43a}$$

$$\gamma_{zx} = \frac{\alpha^2}{G} [(B + C) \sinh \alpha x + B\alpha x \cosh \alpha x] \cos \alpha z \tag{2.2.43b}$$

Now, the displacement v must have a constant value in the direction of the z axis at the junction edges $x = \pm b$ of the ribs and flange plate;

i.e., the conditions $\partial v/\partial z = 0$ and $\gamma_{zx} = \partial u/\partial x + \partial v/\partial z = \partial u/\partial x$ should lead to the following compatibility conditions for the strains γ_{zx} and ε_z at these edges ($x = \pm b$):

$$\left(\frac{\partial \gamma_{zx}}{\partial z}\right)_{x=\pm b} = \left(\frac{\partial \varepsilon_z}{\partial x}\right)_{x=\pm b} \tag{2.2.44}$$

The substitution of Eq. (2.2.41) into the above equation determines the integration constants B and C as follows:

$$B = F \sinh \frac{\pi b}{l} \tag{2.2.45a}$$

$$C = F\left(\frac{1-\mu}{1+\mu} \sinh \frac{\pi b}{l} - \frac{\pi b}{l} \cosh \frac{\pi b}{l}\right) \tag{2.2.45b}$$

in which case

$$F = \frac{2E\varepsilon_o(l/\pi)^2}{(3-\mu)\sinh(2\pi b/l) - 2(1+\mu)2\pi b/l} \tag{2.2.45c}$$

Once all the unknown integration constants have been determined, the normal stress distribution can be plotted as shown in Fig. 2.13*a*, and the shear lag phenomenon can be discussed in detail.

Girkmann[9] has analyzed the shear lag behavior of a beam subjected to arbitrary loads, where the bending moment M_x is represented by

$$M_x = \sum_{n=1}^{\infty} m_n \sin \alpha_n z \tag{2.2.46}$$

$$\alpha_n = \frac{n\pi}{l} \qquad n = 1, 2, 3, \ldots \tag{2.2.47}$$

and the corresponding stress function was given by

$$\Phi = \sum_{n=1}^{\infty} \frac{1}{\alpha_n^2}(A_n \cosh \alpha_n x + B_n \alpha_n x \sinh \alpha_n x + C_n \sinh \alpha_n x + D_n \alpha_n x \cosh \alpha_n x) \sin \alpha_n z \tag{2.2.48}$$

The corresponding results are discussed in the next section.

(2) Definition of effective width. If the normal stress distribution $\sigma_z(x)$ in the flange plate is determined by the preceding procedure, it is convenient to define the effective width of the flange plate for practical design use. Figure 2.14 illustrates a normal stress distribution in the deck plate of a π beam. In this figure, the most important stress in our

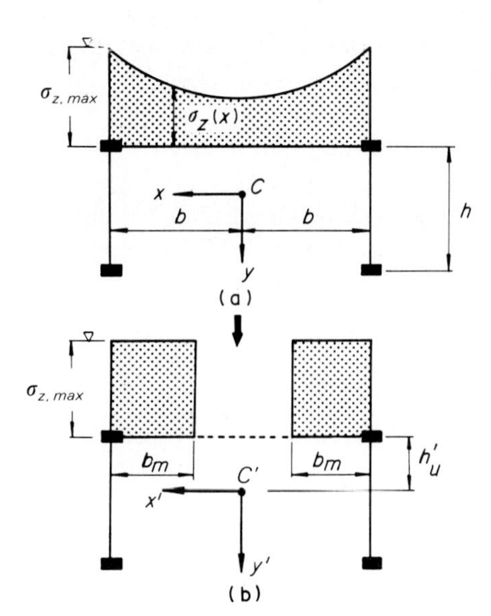

Figure 2.14 Definition of effective width: (*a*) original stress distributions, (*b*) idealized stress distribution and effective width b_m.

design calculations is the maximum stress $\sigma_{z,\max}$ at the junction point of the web and flange plates, so we try to obtain the same stress on the basis of elementary beam theory. For this purpose, it is assumed that a middle part of the flange plate does not cooperate with the cross section of π beam but that the flange plates in a region b_m are only effective as shown in Fig. 2.14*b*.

However, since the normal forces acting on the flange plate of Fig. 2.14*a* and 2.14*b* must have an identical value, the following equilibrium conditions should be satisfied:

$$\int_0^b \sigma_z(x)\,dx = b_m \sigma_{z,\max} \tag{2.2.49a}$$

Accordingly, the effective width b_m can be estimated by[7]

$$b_m = \frac{\int_0^b \sigma_x(x)\,dx}{\sigma_{z,\max}} \tag{2.2.49b}$$

When a new centroidal point C' and the corresponding geometric moment of inertia I'_x and I'_y and product of inertia I'_{xy} are calculated by taking into account the effective width b_m, as illustrated in Fig. 2.14*b*, the flexural normal stress $\sigma_{z,\max}$ including the shear lag phenomenon can be estimated on the basis of the elementary beam theory, as formulated in Eq. (2.2.6). From this, we see that a conservative and rational stress analysis can be conducted by introducing the concept of the effective width of the flange plate.

For example, the variations of the effective-width ratio b_m/b in the direction of span for a T beam with the flange subjected to various loads have been summarized, as shown in Fig. 2.15 by Girkmann.[9] Moreover, the value of b_m/b for a beam, shown in Fig. 2.13, can be obtained from Eqs. (2.2.42*a*) and (2.2.49) together with Eqs. (2.2.45) and (2.2.46) as follows:

$$\frac{b_m}{b} = \frac{4l \sinh^2(\pi b/l)}{\pi(1+\mu)[(3-\mu)\sinh(2\pi b/l) - 2(1+\mu)\pi b/l]} \tag{2.2.50a}$$

For the larger value of b/l, this equation can also be expressed by

$$\frac{b_m}{b} = \frac{2l}{\pi(1+\mu)(3-\mu)} \tag{2.2.50b}$$

and the numerical result of this equation, $b_m/b = 0.182$, gotten by applying Poisson's ratio, $\mu = 0.3$, coincides well with the results obtained by Girkmann,[9] as shown in Fig. 2.15*d*.

By the way, the unknown strain ε_o in Eqs. (2.2.40*a*) and (2.2.45*c*) can be determined by

$$\varepsilon_o = -\frac{M_o}{EI'_x} h'_u \tag{2.2.51}$$

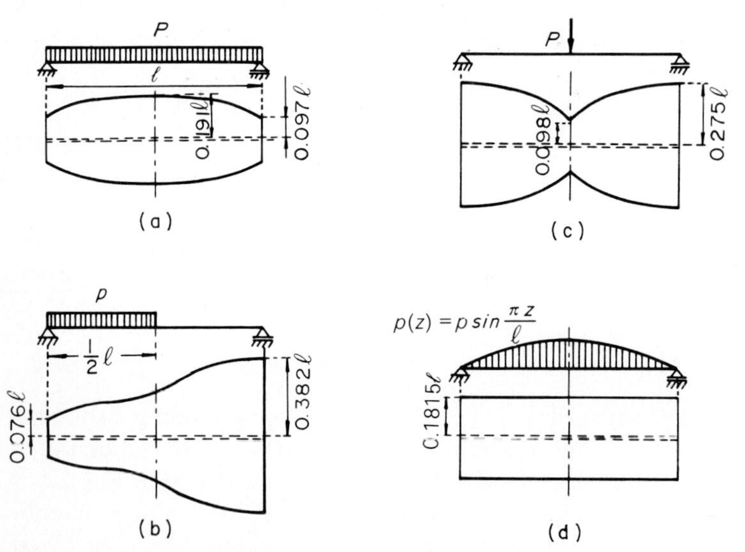

Figure 2.15 Effective width of T beam with a wide flange plate under various loading conditions: (*a*) uniformly distributed load (full load), (*b*) uniformly distributed load (half load), (*c*) concentrated load, (*d*) sinusoidal load.

where h'_u and I'_x are, respectively, the fiber distance from centroid C' to the flange plate and the geometric moment of inertia with respect to the centroidal x' axis by taking into consideration the effective width b_m, and M_o is the maximum bending moment on the beam due to the load given by Eq. (2.2.38).

(3) Effective-width formulas for various plate girders. It is not practical and it is too complex to apply the above method for the analysis of shear lag phenomena of various plate girders such as π and box girders. Therefore, Reissner's hypothesis,[6] which states that the distribution of the flexural normal stress in the flange plate of girders can be approximated by a second-order parabolic curve, was compared to experimental studies. Then the shear lag phenomenon and the corresponding effective width of various types of the plate girders with steel decks were analyzed on the basis of Reissner's hypothesis. Figure 2.16 illustrates an example of experimental and calculated results for the model box girder in which the analytical method is detailed as follows.[12]

a. **Cross section of a girder.** For the sake of convenience, let us now

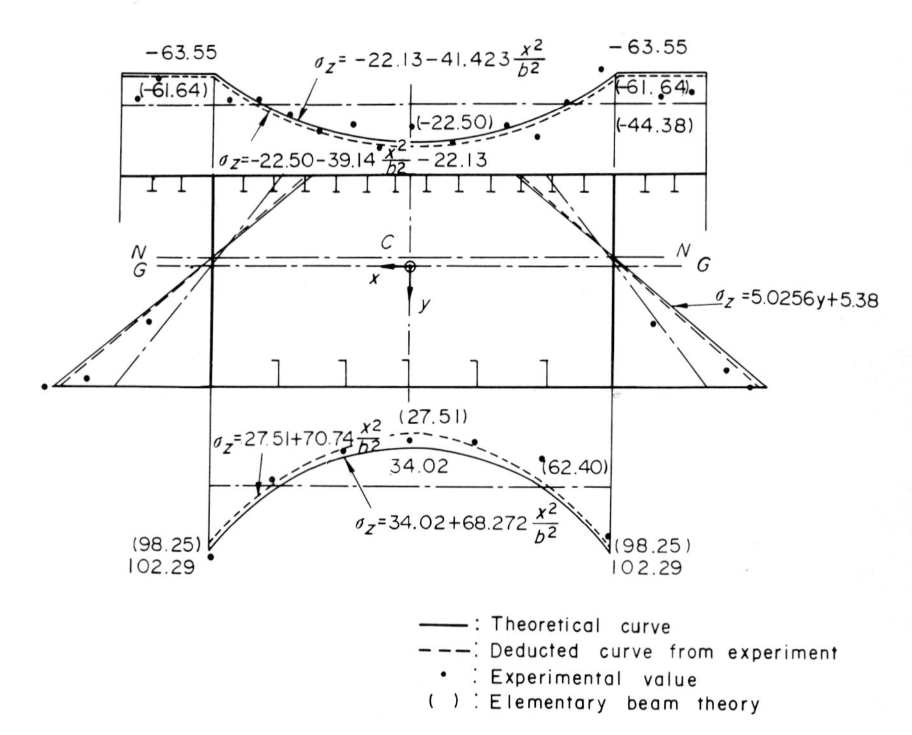

Figure 2.16 Example of shear lag phenomenon in box girder.

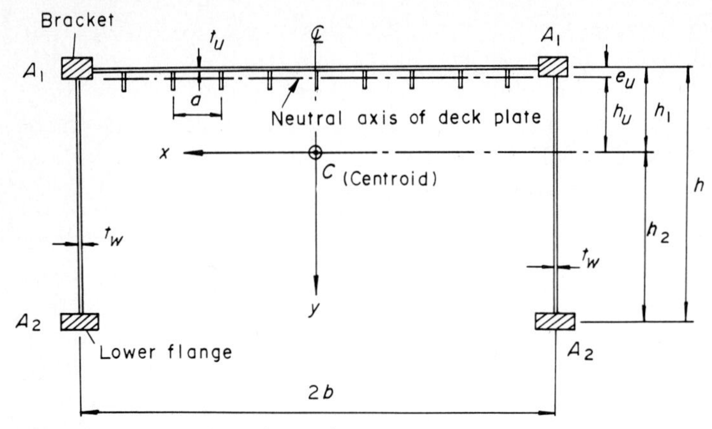

Figure 2.17 Dimensions of π section.

consider the π-shaped section shown in Fig. 2.17 with the following notation:

h = girder height

$2b$ = distance between web plates

t_u = thickness of deck plate

t_w = thickness of web plate

A_1 = cross-sectional area of bracket

A_2 = cross-sectional area of lower flange

The cross-sectional properties of the girder can be computed as follows. Half of the total cross-sectional area of girder is

$$A = A_u + A_l + A_w + A_2 \tag{2.2.52a}$$

where A_u and A_w equal one-half the cross-sectional area of the deck and web plates, which can be estimated by considering Poisson's ratio μ and the theory of elasticity[8] as follows:

$$A_u = b\left(\frac{t_u}{1-\mu^2} + \frac{A_R}{a}\right) = b\bar{t}_u \tag{2.2.52b}$$

$$A_w = \frac{t_w h}{1-\mu^2} \tag{2.2.52c}$$

where a is the rib spacing and A_R is the cross-sectional area of one rib,

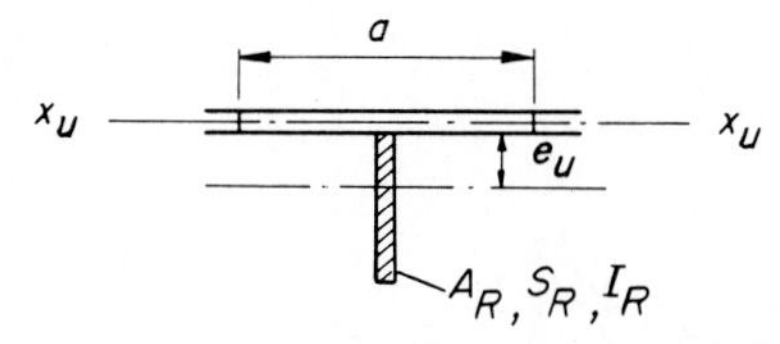

Figure 2.18 Cross-sectional value of a rib.

as shown in Fig. 2.18, and the thickness $\bar{t}_u$ can be defined as the equivalent thickness for the deck plate.

The geometric moment of inertia is

$$I_x = 2\left[A_u h_u^2 + (\tfrac{1}{3}h^2 - h_1 h_2)A_w + \frac{b}{a} I_R + A_1 h_1^2 + A_2 h_2^2\right] \quad (2.2.53)$$

in which the location of the centroid C_n is found to be

$$h_1 = \frac{1}{2A}\left[(2A_2 + A_w)h + \frac{2b}{a} S_R\right] \quad (2.2.54a)$$

$$h_2 = h - h_1 \quad (2.2.54b)$$

where S_R and I_R, respectively, designate the static and geometric moments of inertia with respect to the x_u axis, as shown in Fig. 2.18. The location of the neutral h_u axis of the deck plate relative to the centroid C can be determined from

$$h_u = h_1 - e_u = h_1 - \frac{S_R}{a\bar{t}_u} \quad (2.2.54c)$$

***b.* Derivation of fundamental equation for shear lag**

(i) Equilibrium condition for girder elements. If we assume that the displacements in the transverse direction of the deck plate can be effectively prevented by the transverse ribs and diaphragms, then the normal stress σ_z and the shearing stress τ_{zx} and τ_{zy} acting on a small element $dz\,dx$ or $dz\,dy$, as illustrated in Fig. 2.19, are as follows:
For the deck plate:

$$\sigma_z = \frac{E}{1-u^2}\left(\frac{\partial u_u}{\partial z} + y_u \frac{d\varphi_x}{dz}\right) \quad (2.2.55a)$$

$$\tau_{zx} = G\frac{\partial u_u}{\partial x} \quad (2.2.55b)$$

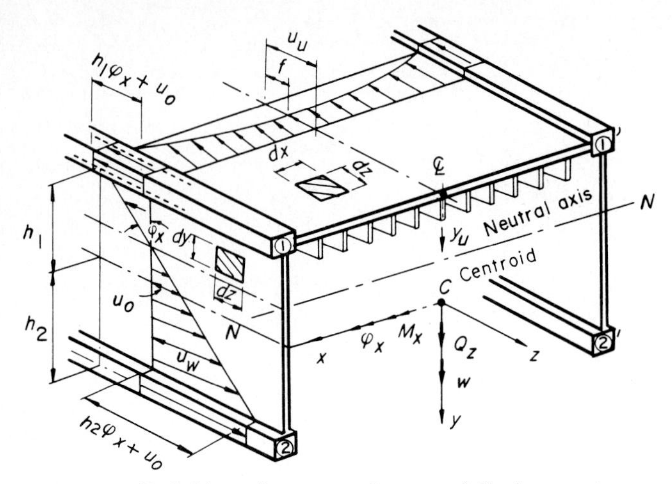

Figure 2.19 Definition of stress resultants and displacements.

For the web plate:

$$\sigma_z = \frac{E}{1-\mu^2}\frac{\partial u_w}{\partial z} \tag{2.2.55c}$$

$$\tau_{zy} = G\frac{\partial u_w}{\partial y} \tag{2.2.55d}$$

where u_u = axial displacement of deck plate
u_w = axial displacement of web plate

$$\varphi_x(z) = -\frac{dw(z)}{dz} \tag{2.2.56}$$

$\varphi_x(z)$ = deflection angle with respect to x axis
$w(z)$ = deflection in the direction of y axis

and y_u is the subcoordinate axis taken in the vertical downward direction from the centerline of the deck plate, as shown in Fig. 2.19.

Now consider the difference between the normal force N_z per unit length and the shearing force T_{zx} per unit length, as induced in a small element of the deck plate, shown in Fig. 2.20. The equilibrium condition for these forces, acting in the direction of the z axis, gives the following differential equation:

$$\frac{\partial N_z}{\partial z} + \frac{\partial T_{zx}}{\partial x} = 0 \tag{2.2.57a}$$

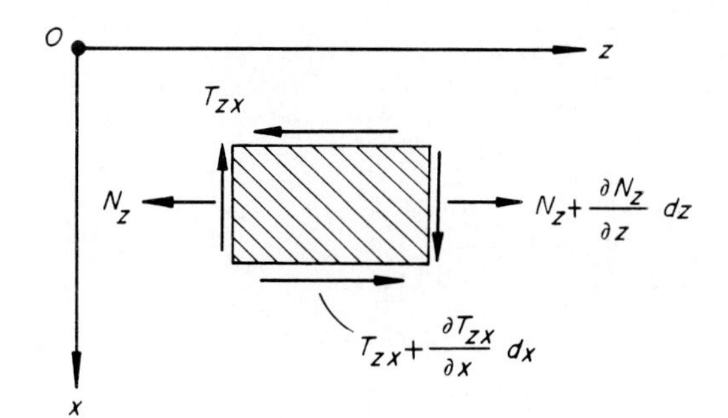

Figure 2.20 Equilibrium condition for small element.

Substitution of

$$N_z = \int_0^1 \sigma_z t_u\,dx = E\left(\bar{t}_u \frac{\partial u_u}{\partial z} + \frac{S_R}{a}\frac{d\varphi_x}{dz}\right) \tag{2.2.57b}$$

and

$$T_{zx} = \int_0^1 \tau_{zx} t_u\,dz = Gt_u \frac{\partial u_u}{\partial x} \tag{2.2.57c}$$

into Eqs. (2.2.55*a*) and (2.2.55*b*) yields

$$E\frac{\bar{t}_u}{t_u}\frac{\partial^2 u_u}{\partial z^2} + G\frac{\partial^2 u_u}{\partial x^2} + E\frac{S_R}{at_u}\frac{d^2\varphi_x}{dz^2} = 0 \tag{2.2.57d}$$

In a similar manner, the differential equation for a small element $dz \cdot dy$ of web plate can be expressed by

$$\frac{E}{1-\mu^2}\frac{\partial^2 u_w}{\partial z^2} + G\frac{\partial^2 u_w}{\partial y^2} = 0 \tag{2.2.58}$$

The equilibrium conditions for the corner part 1 can be found by examining Fig. 2.21, which gives

$$E\left[\frac{\partial^2 u_u}{\partial z^2}\right]_{x=b} - \frac{Gt_u}{A_1}\left[\frac{\partial u_u}{\partial x}\right]_{x=b} + \frac{Gt_w}{A_1}\left[\frac{\partial u_w}{\partial y}\right]_{y=-h_1} = 0 \tag{2.2.59}$$

Similarly for corner part 2,

$$E\left[\frac{\partial_2^2 u_w}{\partial z^2}\right]_{y=h_2} - \frac{Gt_2}{A_2}\left[\frac{\partial u_w}{\partial y}\right]_{y=h_2} = 0 \tag{2.2.60}$$

(ii) Assumption of displacement function. By combining beam theory and Reissner's hypothesis,[6] the displacements on each girder

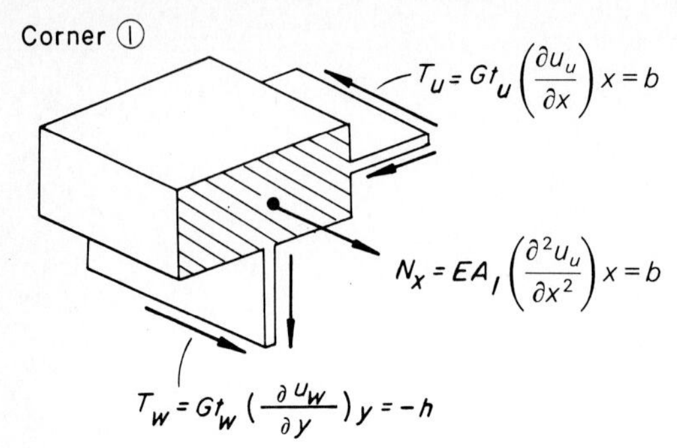

Figure 2.21 Equilibrium condition at corner 1.

element can be rationally written by the following approximate formulas as seen in Fig. 2.19:
For the desk plate:

$$u_u(z, x) = -\varphi_x(z)h_1 + \left[1 - \left(\frac{x}{b}\right)^2\right]f(z) + u_0(z) \qquad (2.2.61a)$$

For the web plate:

$$u_w(z, y) = \varphi_x(z)y + u_0(z) \qquad (2.2.61b)$$

These displacements will, of course, satisfy the boundary condition for the shearing stress $\tau_{zx} = 0$ at the origin ($x = 0$)—that is,

$$\left[\frac{\partial u_u}{\partial x}\right]_{x=0} = 0 \qquad (2.2.62a)$$

and the continuity condition for displacements u_u and u_w at the corner 1 and 1′—thus

$$[u_u]_{x=\pm b} = [u_w]_{y=-h_1} \qquad (2.2.62b)$$

(iii) Derivation of fundamental differential equation for displacement. The normal stress σ_z can be rewritten from Eqs. (2.2.55) and (2.2.61) as follows:
For the deck plate:

$$\sigma_z = \frac{E}{1-\mu^2}\left\{-(h_1 - y_u)\frac{d\varphi_x(z)}{dz} + \left[1 - \left(\frac{x}{b}\right)^2\right]\frac{df(z)}{dz} + \frac{du_o(z)}{dz}\right\} \qquad (2.2.63a)$$

For the web plate:

$$\sigma_z = \frac{E}{1-\mu^2}\left[x\frac{d\varphi_x(z)}{dz} + \frac{du_0(z)}{dz}\right] \qquad (2.2.63b)$$

For the corner part 1:

$$\sigma_z = E\left[-h_1 \frac{d\varphi_x(z)}{dx} + \frac{du_o(z)}{dz}\right] \tag{2.2.63c}$$

For the corner part 2:

$$\sigma_z = E\left[h_2 \frac{d\varphi_x(z)}{dz} + \frac{du_o(z)}{dz}\right] \tag{2.2.63d}$$

Now from the equilibrium condition,

$$\int_A \sigma_z \, dA = 0 \tag{2.2.64}$$

which must be satisfied, since there is no axial force present, that is, $N_z = 0$. Through substitution of Eq. (2.2.63) into the above equation, we have

$$\frac{2}{3} A_u \frac{df(z)}{dz} + A \frac{du_o(z)}{dz} = 0 \tag{2.2.65}$$

By integrating this equation with respect to z (and neglecting the integration constant, which represents the rigid-body displacement), the following important relationship between $u_o(z)$ and $f(z)$ can be obtained:

$$u_o(z) = -\frac{2}{3}\frac{A_u}{A} f(z) \tag{2.2.66}$$

Also, the following equilibrium condition

$$\int_A \sigma_z y \, dA = M_x(z) \tag{2.2.67}$$

must be fulfilled for the given bending moment $M_x(z)$. This condition can also be rewritten by substituting Eq. (2.2.63) into Eq. (2.2.67) and using the relationship given by Eq. (2.2.66), to get

$$I_x \frac{d\varphi_x(z)}{dz} - \frac{4}{3} A_u h_u \frac{df(z)}{dz} = \frac{M_x(z)}{E} \tag{2.2.68}$$

The evaluation of shear displacement is more difficult, so the following energy method based on the generalized Galerkin method is used. This method makes the error due to the assumed function $f(z)$ as small as possible and is

$$\sum_i \int_0^{s_i} \varepsilon_i(z, s_i)\lambda(s_i)t_i \, ds_i = 0 \tag{2.2.69}$$

where the error function $\varepsilon_i(z, s)$ and corresponding coordinate function $\lambda_i(s_i)$ can be written as follows[12] for different sections of the girder.

For the deck plate, substituting Eq. (2.2.61*a*) into the equilibrium equation (2.2.57*d*) and considering Eq. (2.2.66), we see that the error function $\varepsilon_u(z, x)$ is given by

$$\varepsilon_u(z, x) = E\frac{\bar{t}_u}{t_u}\left\{-h_1\frac{d^2\varphi_x(z)}{dz^2} + \left[1 - \left(\frac{x}{b}\right)^2 - \frac{2}{3}\frac{A_u}{A}\right]\frac{d^2f(z)}{dz}\right\} - \frac{2G}{b^2}f(z) + E\frac{S_R}{at_u}\frac{d^2\varphi_x(z)}{dx^2} \quad (2.2.70a)$$

where the coordinate function is

$$\lambda_u(x) = 1 - \left(\frac{x}{b}\right)^2 - \frac{2}{3}\frac{A_u}{A} \quad (2.2.70b)$$

In a similar manner, the following functions are obtained for the web plate:

$$\varepsilon_w(z, y) = \frac{E}{1-\mu^2}\left[y\frac{d^2\varphi_x(z)}{dz^2} - \frac{2}{3}\frac{A_u}{A}\frac{d^2f(z)}{dz^2}\right] \quad (2.2.70c)$$

$$\lambda_w(y) = -\frac{2}{3}\frac{A_u}{A} \quad (2.2.70d)$$

Also, for corner part 1,

$$\varepsilon_1(z) = E\left[-h_1\frac{d^2\varphi_x(z)}{dz^2} - \frac{2}{3}\frac{A_u}{A}\frac{d^2f(z)}{dz^2}\right] + \frac{2Gt_u}{A_1 b}f(z) + \frac{Gt_w}{A_1}\varphi_x(z) \quad (2.2.70e)$$

and

$$\lambda_1 = -\frac{2}{3}\frac{A_u}{A} \quad (2.2.70f)$$

For corner part 2,

$$\varepsilon_2(z) = E\left[h_2\frac{d^2\varphi_x(z)}{dz^2} - \frac{2}{3}\frac{Au}{A}\frac{d^2f(z)}{dz^2}\right] - \frac{Gt_w}{A_2}\varphi_x(z) \quad (2.2.70g)$$

and

$$\lambda_2 = -\frac{2}{3}\frac{A_u}{A} \quad (2.2.70h)$$

Equation (2.2.69) can now be rewritten as

$$\sum_i \int_0^{s_i} \varepsilon(z, s_i)\lambda_i(s_i)t_i \, ds_i = \int_{-b}^{b} \varepsilon_u(z, x)\lambda_u(x)t_u \, dx + 2\left[\varepsilon_1(z)\lambda_1 A_1 + \int_{-h_1}^{h_2} \varepsilon_w(z, y)\lambda_w(y)t_w \, dy + \varepsilon_2(z)\lambda_2 A_2\right] = 0 \tag{2.2.71}$$

which can be reduced to

$$A_u h_u \frac{d^2\varphi_x(z)}{dz^2} - A_u\left(\frac{4}{5} - \frac{2}{3}\frac{A_u}{A}\right)\frac{d^2 f(z)}{dz^2} + \frac{2Gt_u}{Eb} f(z) = 0 \tag{2.2.72}$$

c. Solution for typical loading conditions

(i) Stress resultants and displacements. The three fundamental equations (2.2.66), (2.2.68), and (2.2.72) representing the three unknown displacements $u_o(z)$, $\varphi_x(z)$, and $f(z)$ have been derived, so the shear lag problem for π-shaped girders can be determined by these equations, when the bending moment $M_x(z)$ or shearing force $Q_y(z)$ is given.

Figure 2.22 shows the loading conditions for a simply supported beam. The differential equations representing the shearing force Q_y and bending moment M_x are given by the following well-known formulas:

$$\frac{dQ_z(z)}{dz} = \begin{cases} -q & \text{uniformly distributed load } q \quad (2.2.73a) \\ -P\delta(z-c) & \text{concentrated load } P \text{ where } \delta(z-c) \text{ is Dirac's delta function} \quad (2.2.73b) \end{cases}$$

$$\frac{dM_x(z)}{dz} = Q_y(z) \tag{2.2.74}$$

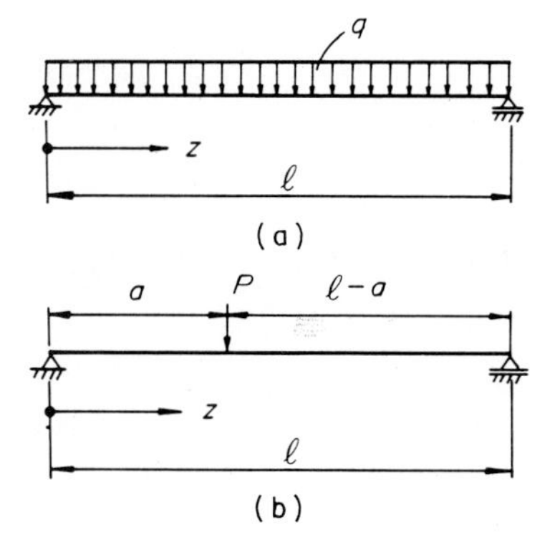

Figure 2.22 Load conditions: (*a*) uniformly distributed load q, (*b*) concentrated load P.

TABLE 2.1 $Q_y(z)$ and $M_x(z)$

Load	Shearing force $Q_y(z)$		Bending moment $M_x(z)$	
Uniform load q	$\frac{ql}{2}\left(1-2\frac{z}{l}\right)$		$\frac{ql^2}{2}\left(\frac{z}{l}\right)\left(1-\frac{z}{l}\right)$	
Concentrated load P	$P\left(1-\frac{a}{l}\right)$	$0 \leqq z \leqq a$	$Pl\left(1-\frac{a}{l}\right)\frac{z}{l}$	$0 \leqq z \leqq a$
	$-P\frac{a}{l}$	$a \leqq z \leqq l$	$Pl\left(1-\frac{z}{l}\right)\frac{a}{l}$	$a \leqq z \leqq l$

By solving Eqs. (2.2.73) and (2.2.74), the values of $Q_y(z)$ and $M_x(z)$ can be determined; they are summarized in Table 2.1.

The differential equation for $f(z)$ can now be determined by differentiating Eq. (2.2.68) with respect to z and substituting into Eq. (2.2.74). By eliminating the term $d^2\varphi_x(z)/dz^2$ from Eq. (2.2.72), the following differential equation is obtained:

$$\frac{d^2f(z)}{dz^2} - \alpha^2 f(z) = \frac{\beta h_u}{EI_x} Q_y(z) \tag{2.2.75}$$

where the parameter α can be estimated by

$$\alpha = \frac{1}{b}\sqrt{\frac{\beta}{\omega}} \tag{2.2.76}$$

provided that

$$\beta = \frac{1.5}{1.2 - \kappa} \tag{2.2.77}$$

$$\omega = \frac{\bar{t}_u}{t_u}(1+\mu) = \frac{1}{1-\mu} + \frac{A_R}{at_u}(1+\mu) \tag{2.2.78}$$

$$\kappa = \frac{A_u}{A} + \frac{I_u}{I_x} \tag{2.2.79}$$

and

$$I_u = 2A_u h_u^2 \tag{2.2.80}$$

By solving Eq. (2.2.75) for loads q and P under the boundary conditions

$$\left[\frac{df(z)}{dz}\right]_{z=0} = 0 \tag{2.2.81a}$$

and

$$\left[\frac{df(z)}{dz}\right]_{z=l} = 0 \tag{2.2.81b}$$

TABLE 2.2 Solution of $f(z)$

Load	Displacement due to shear lag $f(z)$	
Uniform load q	$\frac{qh_u\beta}{EI_y\alpha^3}\left[\alpha\left(z-\frac{l}{2}\right)-\frac{\sinh\alpha(z-l/2)}{\cosh(\alpha l/2)}\right]$	
Concentrated load P	$\frac{Ph_u\beta}{EI_x\alpha^2}\left[\frac{\sinh\alpha(l-a)}{\sinh\alpha l}\cosh\alpha z-\left(1-\frac{a}{l}\right)\right]$	$0 \leqq z \leqq a$
	$\frac{Ph_u\beta}{EI_x\alpha^2}\left[-\frac{\cosh\alpha(l-z)}{\sinh\alpha l}\sinh\alpha a+\frac{a}{l}\right]$	$a \leqq z \leqq l$

which make the stress at both ends of the simply supported girder equal to zero, as is observed from Eqs. (2.2.63) and (2.2.66), the solutions of $f(z)$ obtained are shown in Table 2.2.

Finally, the differential equation for deflection $w(z)$ can be rewritten by using Eqs. (2.2.56) and (2.2.68) as follows:

$$\frac{d^2w(z)}{dz^2} = -\frac{M_x}{EI_x} - \frac{\gamma}{h_u}\frac{df(z)}{dz} \tag{2.2.82}$$

where the parameter γ is defined by

$$\gamma = \frac{4}{3}\frac{A_u h_u^2}{I_x} \tag{2.2.83}$$

Then, by integrating Eq. (2.2.82) twice with respect to z, the deflection $w(z)$ is

$$w(z) = -\int_0^z \left(\int \frac{M_x(z)}{EI_x}\,dz\right)dz - \frac{\gamma}{h_u}\int_0^z f(z)\,dz + C_1 z + C_2 \tag{2.2.84}$$

where the integration constants C_1 and C_2 can be determined from the boundary conditions

$$[w(z)]_{z=0} = 0 \tag{2.2.85a}$$

and

$$[w(z)]_{z=l} = 0 \tag{2.2.85b}$$

The final solutions of $w(z)$ for loads q and P are shown in Table 2.3.

(ii) Stress distribution due to shear lag. The stress distribution in a girder bridge is given by Eq. (2.2.63), which can be rewritten by using the information in Tables 2.1 and 2.2. In this case, let

$$m(z) = \frac{EI_x}{h_u}\frac{df(z)}{dz} \tag{2.2.86}$$

TABLE 2.3 Solutions of $w(z)$

Load	Deflection $w(z)$	
Uniform load q	$\frac{ql^4}{24EI_x}\frac{z}{l}\left[1-2\left(\frac{z}{l}\right)^2+\left(\frac{z}{l}\right)^3\right]$ $+\frac{q\beta\gamma}{EI_x\alpha^4}\left\{\frac{(\alpha l)^2}{2}\frac{z}{l}\left(1-\frac{z}{l}\right)-\left[1-\frac{\cosh\alpha(z-l/2)}{\cosh(\alpha l/2)}\right]\right\}$	
Concentrated load P	$\frac{Pl^3}{6EI_y}\frac{z}{l}\left(1-\frac{a}{l}\right)\left[2\frac{a}{l}-\left(\frac{a}{l}\right)^2-\left(\frac{z}{l}\right)^2\right]$ $+\frac{P\beta\gamma}{EI_x\alpha^3}\left[\left(1-\frac{a}{l}\right)\alpha z-\frac{\sinh\alpha(l-a)}{\sinh\alpha l}\sinh\alpha z\right]$	$0\leqq z\leqq a$
	$\frac{Pl^3}{6EI_y}\frac{a}{l}\left(1-\frac{z}{l}\right)\left[2\frac{z}{l}-\left(\frac{z}{l}\right)^2-\left(\frac{a}{l}\right)^2\right]$ $+\frac{P\beta\gamma}{EI_x\alpha^3}\left[\left(1-\frac{z}{l}\right)\alpha z-\frac{\sinh\alpha(l-z)}{\sinh\alpha l}\sinh\alpha a\right]$	$a\leqq z\leqq l$

which is an additional moment due to shear lag. The curvature of the girder, given by Eq. (2.2.68), can be written as

$$\frac{d\varphi_x(z)}{dz}=\frac{1}{EI_x}\left[M_x(z)+\gamma m(z)\right] \tag{2.2.87}$$

Then the final stress can be obtained by using Eqs. (2.2.63) and (2.2.66):
For the deck plate:

$$\sigma_z=-\frac{1}{W_u}\left\{\frac{h_1-y_u}{h_u}M_x(z)-\left[1-\left(\frac{x}{b}\right)^2-\frac{2}{3}\frac{A_u}{A}-\frac{h_1-z_u}{h_u}\gamma\right]m(z)\right\} \tag{2.2.88a}$$

For the web plate:

$$\sigma_z=\frac{1}{W_y}\left[M_x(z)+\gamma m(z)\right]-\frac{2}{3}\frac{A_u}{A}\frac{m(z)}{W_u} \tag{2.2.88b}$$

where W is the section modulus, defined as

$$W_u=\frac{I_x}{h_u}(1-\mu^2) \tag{2.2.89a}$$

$$W_y=\frac{I_x}{y}(1-\mu^2) \tag{2.2.89b}$$

The stress, Eq. (2.2.88a), can be simplified when $y_u = e_u$, that is, the location at the neutral axis $h_u = h_1 - e_u$ of the deck plate, as follows: For the midpoint of the deck plate,

$$\sigma_m = -\frac{1}{W_u}[M_x(z) - (1 - \tfrac{2}{3}\kappa)m(z)] \tag{2.2.90a}$$

For the junction point of the deck and web plates,

$$\sigma_e = -\frac{1}{W_u}[M_x(z) + \tfrac{2}{3}\kappa m(z)] \tag{2.2.90b}$$

Accordingly, if we let

$$\sigma_s = \sigma_e - \sigma_m \tag{2.2.91}$$

then

$$\sigma_s = -m(z)/W_u \tag{2.2.92}$$

The stress σ_z at an arbitrary point with eccentricity x from the midpoint of the deck plate can be expressed as

$$\sigma_z(x) = \sigma_m + \sigma_s\left(\frac{x}{b}\right)^2 \tag{2.2.93}$$

These equations may be utilized for design in the estimation of stresses at the top of the deck plate if the section modulus W_f at the top of the deck plate is used instead of W_u.

Table 2.4 gives the results of the additional bending moment $m(x)$ due to shear lag for typical loading conditions.

***d.* Effective width for π girders**[12]

(i) Derivations of effective-width formulas. The substitution of Eq. (2.2.93) into the definition of Eq. (2.2.49) gives

$$b(\sigma_m + \tfrac{1}{3}\sigma_s) = b_m\sigma_e \tag{2.2.94}$$

so that

$$\frac{b_m}{b} = 1 - \frac{\tfrac{2}{3}m(z)}{M_x(z) + \tfrac{2}{3}\kappa m(z)} \tag{2.2.95}$$

From this general equation, the effective width of girder bridges subjected to various loads can be determined, as discussed in the next section.

In addition to these developments, Eq. (2.2.94) can be solved for the stress σ_s. Recall that

$$\sigma_s = \sigma_e - \sigma_m \tag{2.2.91}$$

TABLE 2.4 Solution of $m(z)$

Load	Additional moment $m(z)$	
Uniform load q	$qb^2\omega\left[1 - \dfrac{\cosh \alpha(z - l/2)}{\cosh (\alpha l/2)}\right]$	
Concentrated load P	$Pb\sqrt{\omega\beta}\,\dfrac{\sinh \alpha(l - a)}{\sinh \alpha l} \sinh \alpha z$	$0 \leqq z \leqq a$
	$Pb\sqrt{\omega\beta}\,\dfrac{\sinh \alpha(l - z)}{\sinh \alpha l} \sinh \alpha a$	$a \leqq z \leqq l$

From Eq. (2.2.94) the following simple formula can be obtained:

$$\sigma_s = \frac{3}{2}\left(1 - \frac{b_m}{b}\right)\sigma_e \tag{2.2.96}$$

Then substitution of this equation into Eq. (2.2.93) yields the following formula, which can be used to estimate the stress distributions $\sigma_z(x)$:

$$\sigma_z(x) = \sigma_e\left\{1 - \frac{3}{2}\left(1 - \frac{b_m}{b}\right)\left[1 - \left(\frac{x}{b}\right)^2\right]\right\} \tag{2.2.97}$$

This equation is similar in form to the design formula given in BS5400, part 3.[20] The physical meaning of σ_z in Eq. (2.2.97) is as indicated in Fig. 2.23.

Figure 2.24 shows the loading conditions imposed on the simply supported girder bridge, where the following parameters are introduced:

Section location,

$$z = \psi l \tag{2.2.98}$$

Loading point of the concentrated load P,

$$a = \varphi l \tag{2.2.99}$$

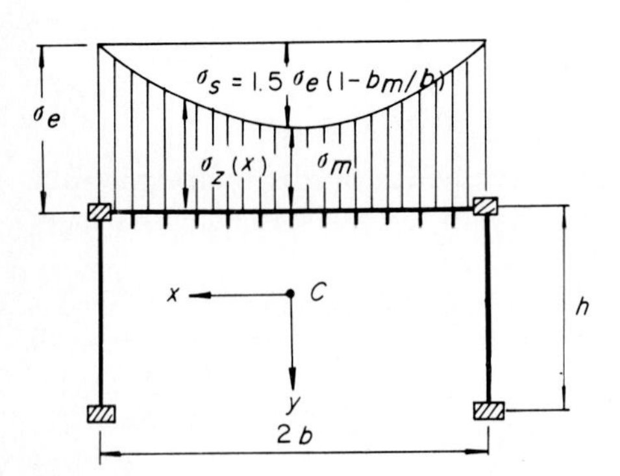

Figure 2.23 Stress distribution in deck plate.

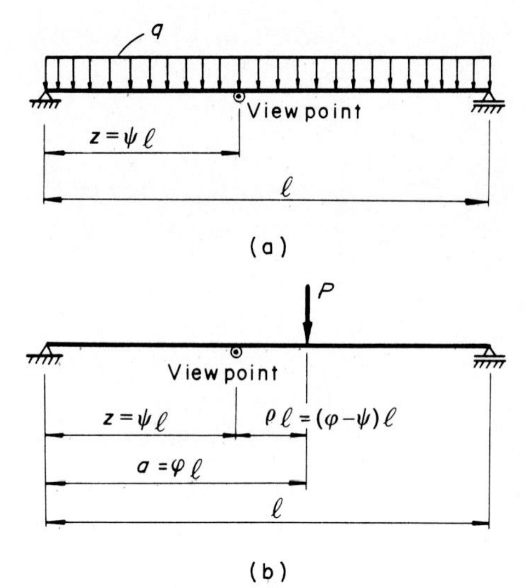

Figure 2.24 Fraction parameters ψ, φ, and ρ: (*a*) uniformly distributed load q and (*b*) concentrated load P.

and the distance between the location and loading point of the following fraction parameter is introduced:

$$\rho = \varphi - \psi \qquad 0 \leqq \psi \leqq \varphi \tag{2.2.100}$$

Substituting the bending moment $M(z)$, given in Table 2.1, and the additional moment $m(z)$, given in Table 2.4, into Eq. (2.2.95) and utilizing Eqs. (2.2.98) to (2.2.100), we can define the effective-width ratio b_m/b at an arbitrary position by the following formulas:
For uniformly distributed load q,

$$\frac{b_m}{b} = 1 - \frac{2\omega\chi}{1.5\psi(1-\psi)(l/b)^2 + 2\kappa\omega\chi} \tag{2.2.101a}$$

For a concentrated load P,

$$\frac{b_m}{b} = 1 - \frac{\sqrt{1.5\omega/(1.2-\kappa)}\,\lambda}{3(1-\varphi)\psi l/b + \kappa\sqrt{1.5\omega/(1.2-\kappa)}\,\lambda} \tag{2.2.101b}$$

For combined loads of q and P,

$$\frac{b_m}{b} = 1 - \frac{\sqrt{\dfrac{1.5\omega}{1.2-\kappa}}\lambda + \dfrac{2ql}{p}\dfrac{1}{l/b}\omega\chi}{\left[3(1-\varphi)\psi + 1.5\psi(1-\psi)\dfrac{ql}{P}\right]\dfrac{l}{b} + \kappa\left(\sqrt{\dfrac{1.5\omega}{1.2-\kappa}}\lambda + \dfrac{2q}{P}\dfrac{1}{l/b}\omega\chi\right)} \tag{2.2.101c}$$

where the parameters χ and λ are the variables given by the following hyperbolic functions:

$$\chi = 1 - \frac{\cosh\left[\left(\psi - \frac{1}{2}\right)\frac{lb}{\omega}\sqrt{\frac{1.5\omega}{1.2-\kappa}}\right]}{\cosh\left(\frac{lb}{2\omega}\sqrt{\frac{1.5\omega}{1.2-\kappa}}\right)} \tag{2.2.102a}$$

$$\lambda = 2\frac{\sinh\left[(1-\varphi)\frac{lb}{\omega}\sqrt{\frac{1.5\omega}{1.2-\kappa}}\right]}{\sinh\left(\frac{lb}{\omega}\sqrt{\frac{1.5\omega}{1.2-\kappa}}\right)}\sinh\left(\psi\frac{lb}{\omega}\sqrt{\frac{1.5\omega}{1.2-\kappa}}\right) \tag{2.2.102b}$$

When $l/b > 5$, the following approximate formulas can be used:

$$\chi = 1 - \exp\left(-\psi\frac{lb}{\omega}\sqrt{\frac{1.5\omega}{1.2-\kappa}}\right) \tag{2.2.103a}$$

$$\lambda = \exp\left(-\rho\frac{lb}{\omega}\sqrt{\frac{1.5\omega}{1.2-\kappa}}\right) \tag{2.2.103b}$$

(ii) Values of parameters ω and κ. As illustrated by Eqs. (2.2.101*a*) and (2.2.101*b*), the effective width is affected by the values of ω and κ with the exception of l/b. Therefore, these parameters should be investigated by using the actual dimension of bridges. A study of several bridges is shown in Fig. 2.25.

From these numerical calculations, we conclude that the range of these parameters is given by[18]

$$\omega = 1.5\text{–}2.5 \tag{2.2.104a}$$

and

$$\kappa = 0.25\text{–}1.0 \tag{2.2.104b}$$

(iii) Variations of effective width due to fraction parameters κ and φ. Assuming $\omega = 2.0$ and $\kappa = 0.75$ which are typical values for actual bridges, the variations of effective width b_m in the direction of span can be plotted, as shown in Fig. 2.26. In this figure, the effective width has a minimum value at the loading point and rapidly increases toward the end supports, as in the case of the concentrated load. But the effective width has maximum values at midspan and gradually decreases toward the end supports, as in the case of a uniformly distributed load.

Next the variations of the effective width at section ψl for uniformly distributed loads and at section ρl for concentrated loads with $\varphi = 0.2$ have been plotted as shown in Fig. 2.27. This figure shows that the effective width for a uniformly distributed load will be affected by the

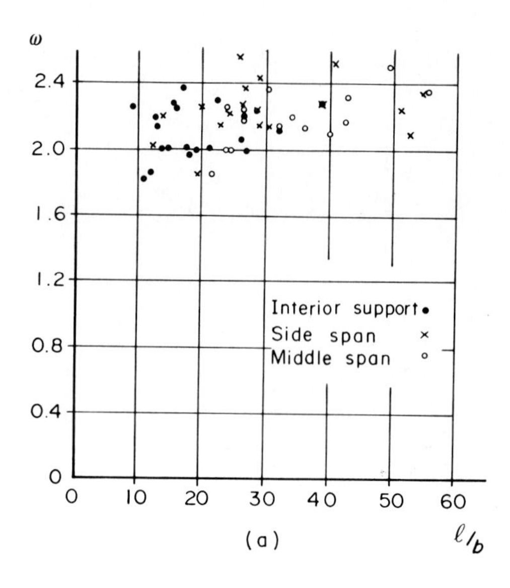

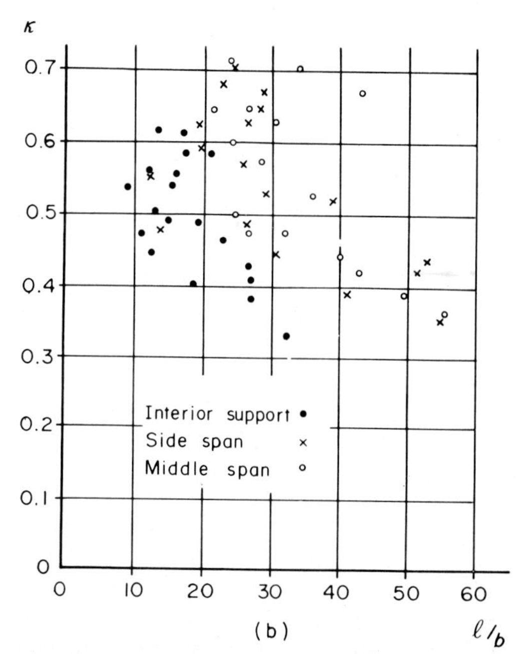

Figure 2.25 Variations of (a) ω and (b) κ in actual plate girder bridges.

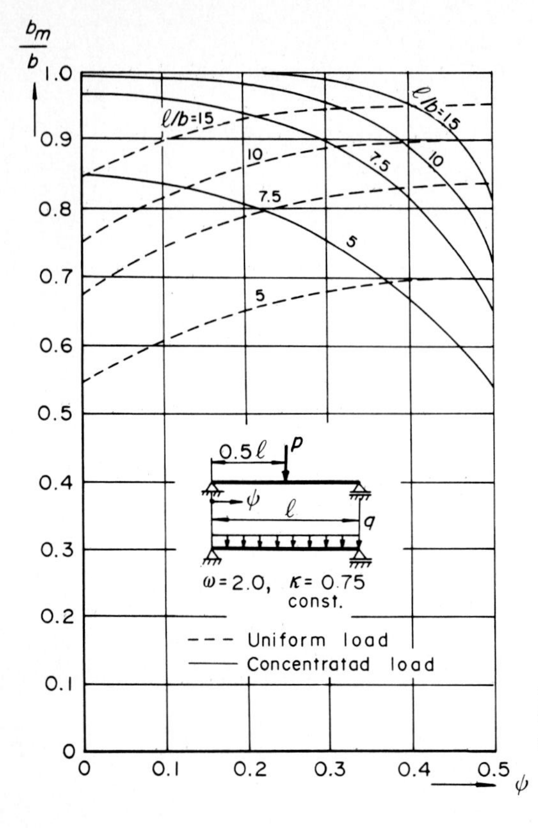

Figure 2.26 Variation of effective width in the direction of the span.

cross section, but that the difference of effective widths can be ignored at the small values near the midspan of the girder.

The effective widths near the end supports are not important in design. Therefore, we direct our attention to the effective width near the midspan of the girder. By letting $z = a$, that is, $\varphi = \psi$ for the concentrated load, the fractional variables of bending moment in Fig. 2.22 and Table 2.1 can be assumed to be

$$\psi(1-\varphi) = \psi(1-\psi) \tag{2.2.105}$$

for both loads.

For example, the value of $\psi(1-\psi)$ reduces to 0.25 at the midspan, where $\psi = 0.5$. Accordingly, the effective width at other than $\psi = 0.5$ can be estimated by replacing the span-width ratio l/b as follows[12]:

$$\frac{l}{b} = 4\psi(1-\psi)\frac{l}{b} \tag{2.2.106}$$

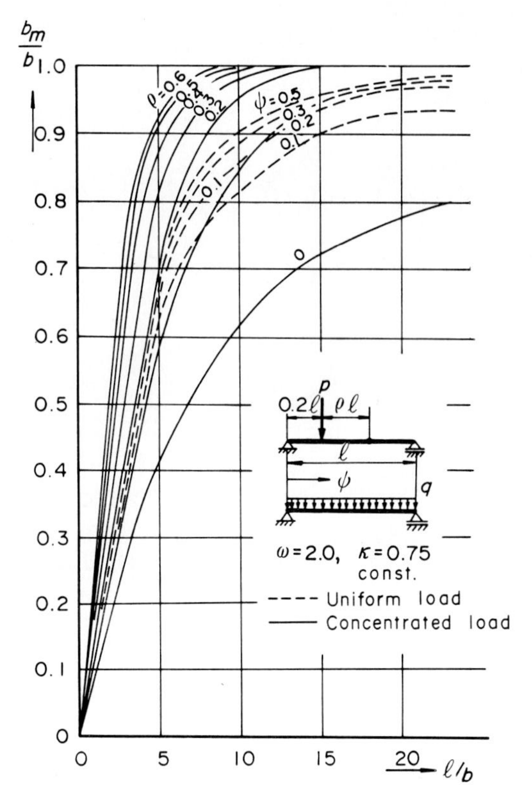

Figure 2.27 Variation of effective width due to the cross section.

The following formulas can be used to determine the effective width: For a uniformly distributed load q,

$$\frac{b_m}{b} = 1 - \frac{2\omega\chi}{(0.375l/b)(l/b) + 2\kappa\omega\chi} \tag{2.2.107a}$$

For a concentrated load P,

$$\frac{b_m}{b} = 1 - \frac{\sqrt{1.5\omega/(1.2-\kappa)\,\lambda}}{0.75l/b + \kappa\sqrt{1.5\omega/(1.2-\kappa)\,\lambda}} \tag{2.2.107b}$$

For combined loads of q and P,

$$\frac{b_m}{b} = 1 - \frac{\sqrt{\dfrac{1.5\omega}{1.2-\kappa}\lambda} + \dfrac{2ql}{p}\dfrac{1}{l/b}\omega\chi}{(0.75 + 0.375ql/p)(l/b) + \kappa\left(\sqrt{\dfrac{1.5\omega}{1.2-\kappa}\lambda} + \dfrac{2ql}{p}\dfrac{1}{lb}\omega\chi\right)} \tag{2.2.107c}$$

Parameters χ and λ can also be simplified by

$$\chi = \begin{cases} 1 & \dfrac{l}{b} \geqq 5 \quad (2.2.108a) \\ \tanh \dfrac{l/b}{2\omega}\sqrt{\dfrac{1.5\omega}{1.2-\kappa}} & \dfrac{l}{b} < 5 \quad (2.2.108b) \end{cases}$$

$$\lambda = \begin{cases} 1 & \dfrac{l}{b} \geqq 10 \quad (2.2.108c) \\ 1 - \operatorname{sech} \dfrac{l/b}{2\omega}\sqrt{\dfrac{1.5\omega}{1.2-\kappa}} & \dfrac{l}{b} < 10 \quad (2.2.108d) \end{cases}$$

(iv) Variations of effective width due to parameters ω, κ, and l/b. The values of parameters ω and κ have been given by Eq. (2.2.104). The variation of effective width due to these parameters is now examined.

Figure 2.28 shows the variations of the effective width due to

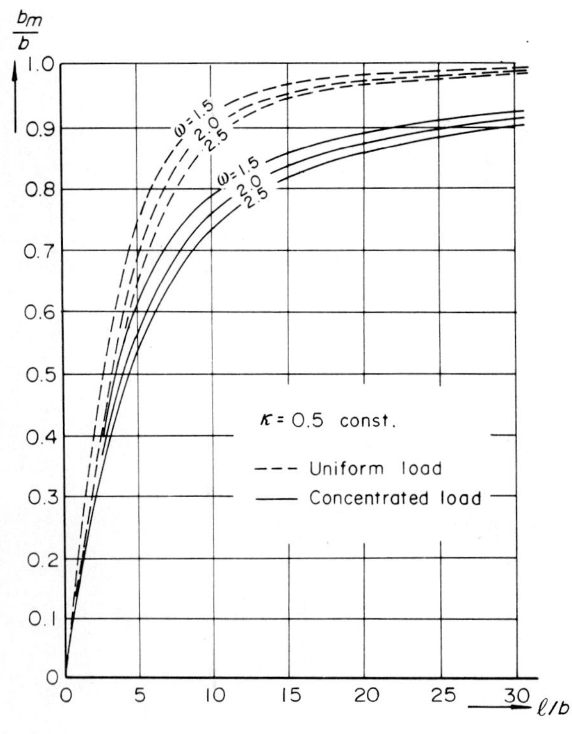

Figure 2.28 Variation of effective width due to parameter ω.

parameters ω and l/b for a constant value of $\kappa = 0.5$. The influence of parameter ω indicates that the effective width decreases for an increasing value of ω. Since the parameter ω, that is, the orthotropical parameter of the flange plate, depends on Poisson's ratio and the cross-sectional properties of ribs, given by

$$\omega = \frac{1}{1-\mu} + \frac{A_R}{at_u}(1+\mu) \tag{2.2.78}$$

the effective width of the steel deck plate ($A_R \neq 0$) will become smaller than that of the flat plate without ribs ($A_R = 0$).

In addition, the variations of the effective width due to parameters κ and l/b are plotted, as shown in Fig. 2.29, for a constant value of $\omega = 2.0$. Figure 2.29 also shows that parameter κ causes a greater variation in the case of the concentrated load and that the effective width decreases for the larger values of κ. Recall that κ was given as

$$\kappa = \frac{A_u}{A} + \frac{I_u}{I_x} \tag{2.2.79}$$

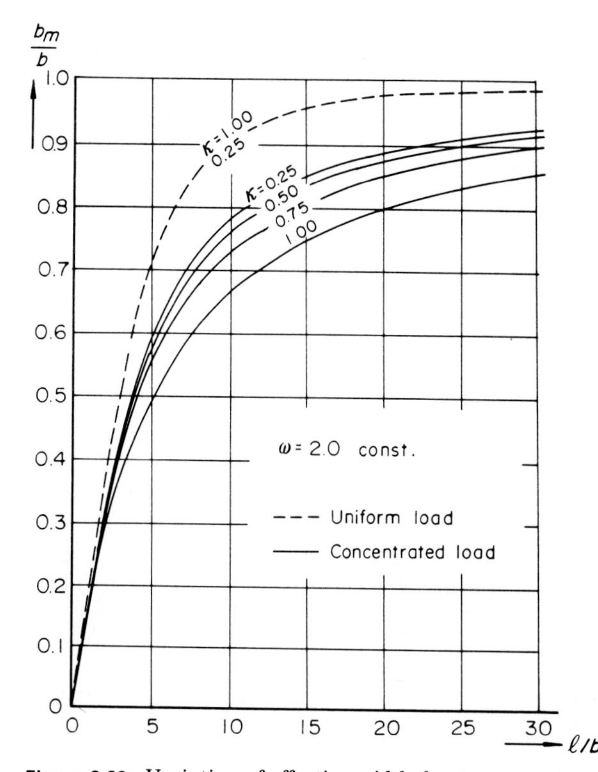

Figure 2.29 Variation of effective width due to parameter κ.

and this parameter represents a cross-sectional shape parameter. Therefore, when the cross-sectional area A_u and geometric moment of inertia I_u of the deck plate have values such as in the case of a shallow girder with a wide deck and low web plates, the value of κ nearly equals 1.0. Thus the effective width becomes small.

For the uniform load, however, the difference of effective width due to the parameter κ is negligible. Parameter κ can be set equal to 0.5 to 0.75 for use in design.

Finally, Fig. 2.30 shows the variations of the effective width due to the parameter ql/P for constant values of $\omega = 2.0$ and $\kappa = 0.75$. In this figure, the curve for $ql/P = 0$ corresponds to the case of the concentrated load, and that for $ql/P = \infty$ represents the case of a uniform load. Generally, uniform loads govern the design of long-span bridges, which alters the effective width.

(v) Proposition of effective width in design calculation and comparisons with some specifications. The validity of this theory has been examined by means of experimental studies (as an example, see Fig. 2.16). Moreover, Table 2.5 shows the comparisons of this method with the results of Moffatt and Dowling by using the finite-element

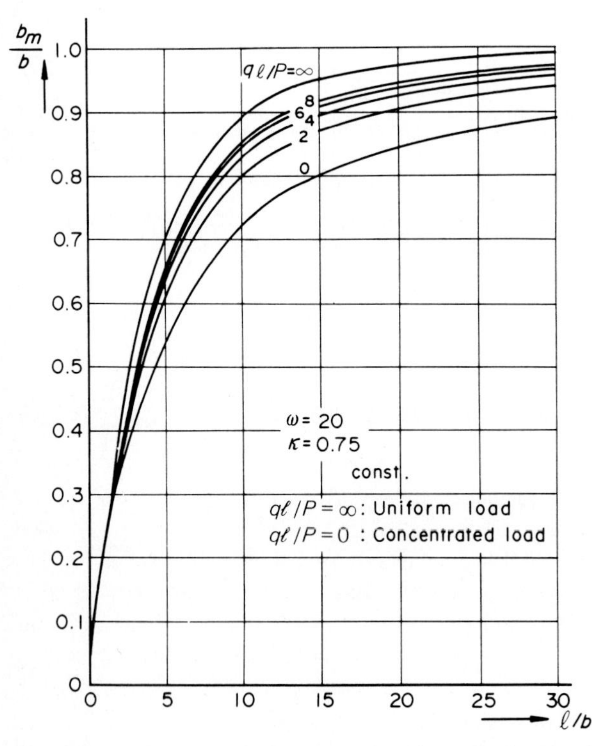

Figure 2.30 Variation of effective width due to load ratio ql/P.

TABLE 2.5 Comparisons of Effective-Width Ratio $\frac{b_m}{b}$ by This Method and Moffatt's Results

Load	$\frac{l}{b}$	$\frac{nA_R}{2bt_f}$	Effective-width ratio $\frac{b_m}{b}$: Present method	Effective-width ratio $\frac{b_m}{b}$: Moffatt's results	Remarks
Uniform load	20	0	0.98	0.98	Cross section
	10		0.93	0.95	
	5		0.77	0.81	
	2.5		0.48	0.50	
Concentrated load	20		0.83	0.83	
	10		0.71	0.71	
	5		0.55	0.53	$b = h = 6$ ft
	2.5		0.37	0.33	$t_f = t_w = 0.5$ in. without rib
Uniform load	20	1	0.96	0.97	Cross section
	10		0.87	0.89	
	5		0.64	0.67	
	2.5		0.36	0.35	
Concentrated load	20		0.78	0.75	
	10		0.63	0.59	
	5		0.46	0.40	The same as the above except eight ribs with flat plate 4.5 in. × 1 in.
	2.5		0.29	0.23	

method.[15,16] These results must now be modified for use in design and then compared to the results with other foreign specifications.[13,19,20]

The Japanese Specification for the Design of Highway Bridges[34] gives the following intensities of live loads (L loading):

Concentrated load: $$P = 5 \text{ tf/m} \tag{2.2.109a}$$

(tf/m = tons of force per meter)

Uniform load: $$q = 350 \text{ kgf/m}^2 \tag{2.2.109b}$$

Thus,

$$q/P = 0.07/\text{m} \tag{2.2.110}$$

Applying these values to actual bridges (see Fig. 2.25) and using Eq. (2.2.107c), we see that the effective widths at the midspan of the girders, where $\psi = \varphi = 0.5$, are reduced to the values summarized in Table 2.6. Comparison of the values of b_m/b for each l/b indicates that the error is less than 3 percent. Then, by taking the average of the theoretical values, a dashed curve can be plotted, as shown in Fig. 2.31.

The design curves of the effective width provided by other foreign

TABLE 2.6 Numerical Examples of the $\frac{b_m}{b}$ Ratio for L Loading Provided by the Japanese Specification

	$\frac{l}{b}$								
Bridge	3	4	5	7	10	15	20	25	30
Save bridge	0.400	0.543	0.581	0.704	0.812	0.898	0.936	0.957	0.969
Morinomiya bridge	0.421	0.513	0.585	0.689	0.781	0.862	0.904	0.930	0.946
Model girder	0.388	0.477	0.561	0.679	0.787	0.877	0.921	0.945	0.959
Model girder	0.399	0.492	0.578	0.695	0.799	0.885	0.925	0.948	0.961
Jogashima bridge	0.409	0.505	0.584	0.697	0.796	0.883	0.923	0.946	0.960

specifications have also been plotted in Fig. 2.31.[20,34] Comparing the theoretical values obtained by this method with the specified values given in DIN 1078 for composite girders, we find that the effective width of the steel girders having steel deck plates seems somewhat smaller than that of composite girders with concrete slabs.

These differences are caused by the variation in the materials and cross-sectional properties of the top plates, such as

For steel deck:

$$\mu = 0.3 \qquad A_R \neq 0 \tag{2.2.111a}$$

For concrete slab:

$$\mu = 1/6 \qquad A_R = 0 \tag{2.2.111b}$$

Then the effective width of a steel deck becomes smaller than that of the composite girders, which have smaller values of ω.

However, it seems that the value specified by BS 5400[20] is entirely based on the effective-width ratio of the uniformly distributed load, where the orthotropical parameter of the flange plate ω is taken as $nA_R/(2bt_f)$.

In either case, the effective-width ratio becomes $b_m/b = 1$ for large values of l/b. For instance, this limit of l/b is taken as $l/b = 20$ in the case of DIN 1078. Then the limit of l/b can be set equal to 30 for steel deck plates, since the values of b_m/b are almost equal to 1.0, as seen in Table 2.7. Thus, by assuming that the theoretical curve is horizontal at $b_m/b = 1.0$ for $l/b = 30$, a practical design curve is proposed by the solid line, shown in Fig. 2.31.

The effective width is nearly a straight line for the range where the values of l/b are less than 3. Therefore the slope of the curve will result

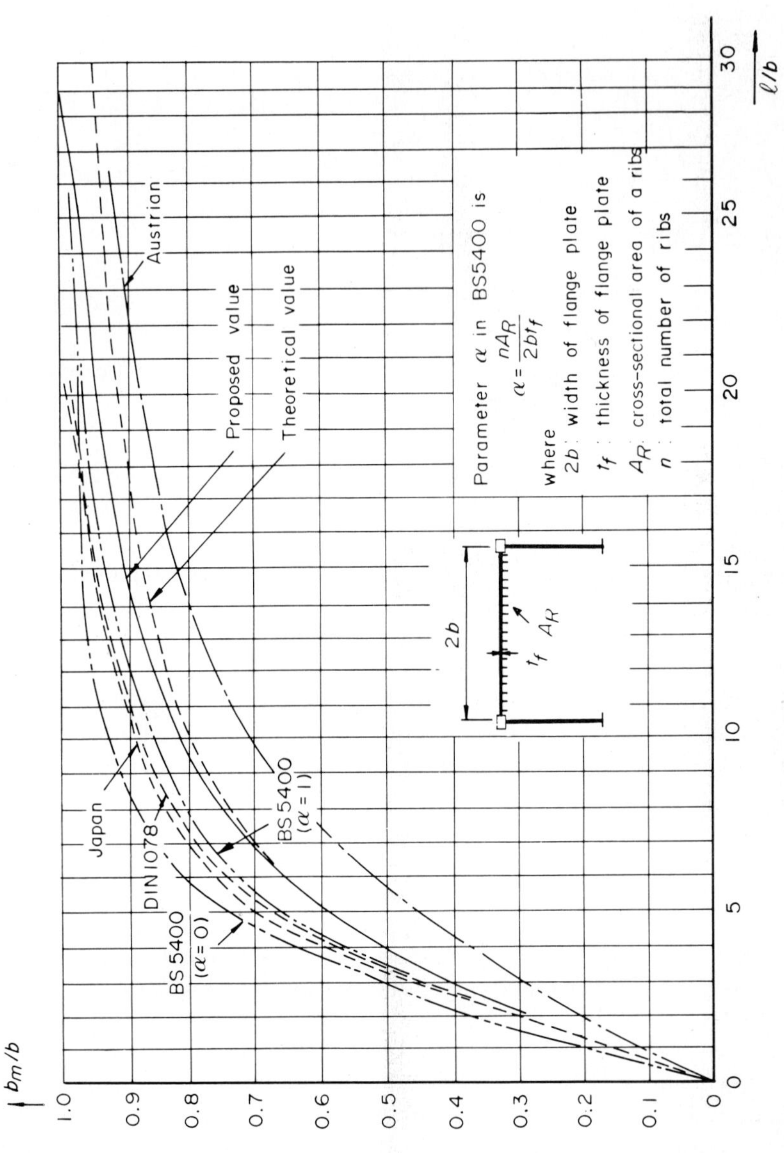

Figure 2.31 Comparisons of effective-width design curves among various other foreign specifications (simply supported plate girder).

TABLE 2.7 Proposed Values for the Effective Width of Steel Girder Bridges

$\frac{l}{b}$	3	4	5	7	10	15	20	25	30
$\frac{b}{l}$	0.333	0.250	0.200	0.143	0.100	0.067	0.050	0.040	0.033
$\frac{b_m}{b}$	0.41	0.51	0.59	0.70	0.81	0.90	0.95	0.98	1.00

in the following practical formula:

$$b_m = 0.137l \qquad \frac{l}{b} < 3 \tag{2.2.112a}$$

Table 2.7 shows the proposed values of b_m/b for $3 < l/b < 30$. For values of $l/b > 30$, the following formula can be used:

$$b_m = b \qquad \frac{l}{b} \geqq 30 \tag{2.2.112b}$$

***e.* Application to various box girders.**[12] Although the theoretical studies concerning the shear lag of box-girder bridges have been developed by K. Kondo et al.,[12] the analytical techniques from these studies are complicated. Therefore, an approximate method applying the formula given in this section is developed here.

Figure 2.32*a* shows a cross section of a box-girder bridge.[12] To determine the effective width of the top plate of this bridge, the bottom plate is cut with respect to the centerline of the cross section. The cross-sectional areas $2A_l$ of the bottom plate are then divided into the cross-sectional area A_l of each lower flange, the π-shaped section shown in Fig. 2.32*b*. Thus, the box girder can be replaced by a π section, and then the effective width $b_{m,u}$ of the top plate can be estimated by the design formula in Eq. (2.2.107).

The effective width of the bottom plate $b_{m,l}$ can be evaluated in the same manner for a reversed π section, as shown in Fig. 2.32*c*.

The effective width of the floor deck between twin boxes will be replaced with π sections, as shown in Fig. 2.33*b*, by taking the effective widths $b_{m,u}$ and $b_{m,l}$ for each box section on the basis of the above method. In a similar manner, the effective width of the cantilever deck of the monobox, shown in Fig. 2.33*c*, can now be replaced by a T section, as shown in Fig. 2.33*d*. However, this section is one-half of the π section, shown in Fig. 2.33*b*, connected with both sides of the free edges of the cantilever deck.

Thus, the effective width of the π section can be determined by using Eq. (2.2.107). The design value of the effective width, using BS5400,

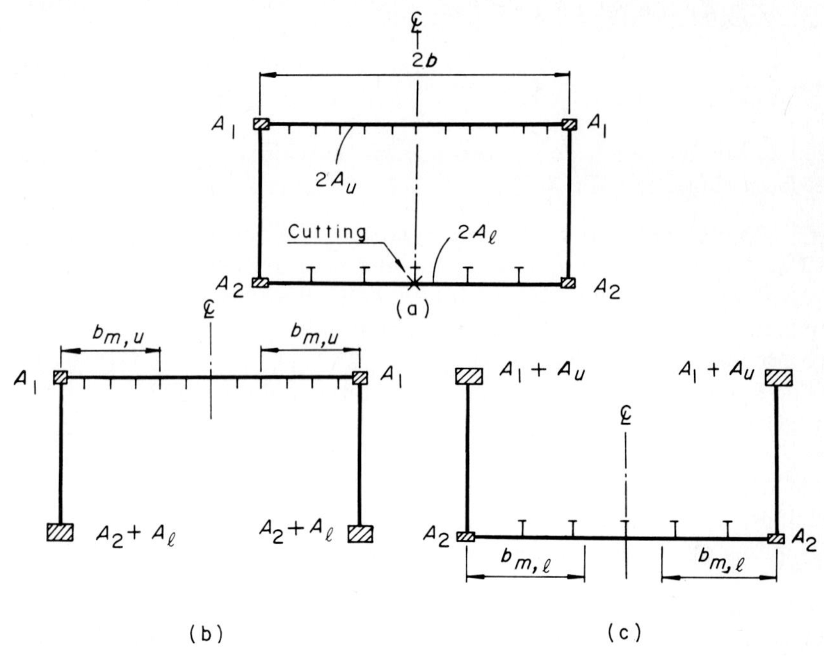

Figure 2.32 Approximate method for determining the effective width of a box girder: (*a*) box section, (*b*) π section, (*c*) reversed π section.

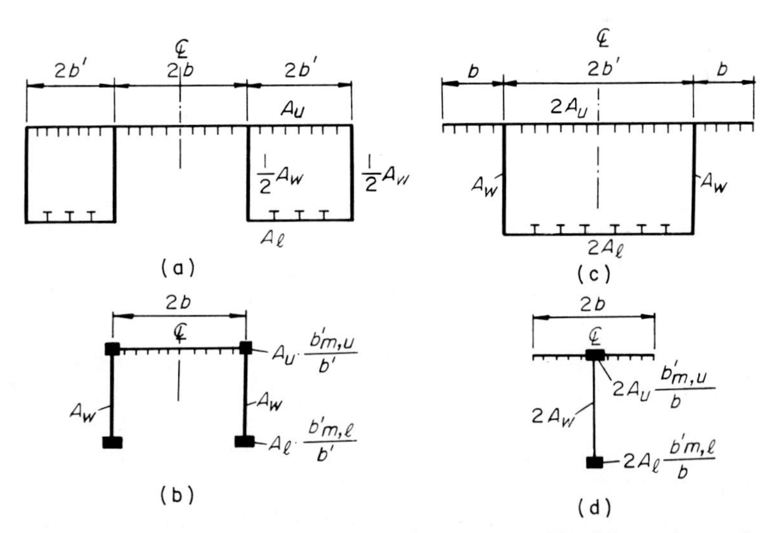

Figure 2.33 (*a*) Floor deck between twin boxes replaced by (*b*) π section, and (*c*) cantilevered deck for monobox girder replaced by (*d*) T section.

Part 3,[20] uses the reduction criterion of

$$\left[1 - 0.15\left(\frac{b}{l}\right)\right]\frac{b_m}{b} \qquad (2.2.113)$$

because the boundary conditions of the T section are somewhat different from those of the π section.

***f.* Effective width for continuous girders.**[17] The design formula, given by Eq. (2.2.107), is applicable to continuous girder bridges provided an equivalent simple span and loading condition are used that will make the bending moments of the equivalent simple span equal to the bending moments of the continuous span.

For example, Fig. 2.34*a* shows an arbitrary distributed load q imposed on a continuous girder. Then the bending moment diagram can be plotted as shown in Fig. 2.34*b*. The equivalent span for the negative-moment region is equal to the span between points A and B, where the moments are equal to zero. Thus, the simple beam of equivalent span AB subjected to the uniformly distributed load q and concentrated load $P = R_r$, where R_r is the internal reaction at support r, can be determined as shown in Fig. 2.34*c* and 2.34*d*. The resulting bending moment of this simple beam has the same values as that of the continuous girder bridge, so that the effective width can be estimated by the design formulas given in Eq. (2.2.107).

For the positive-moment regions, the equivalent span BC can be determined, as shown in Fig. 2.34*b*. Then the effective width of the

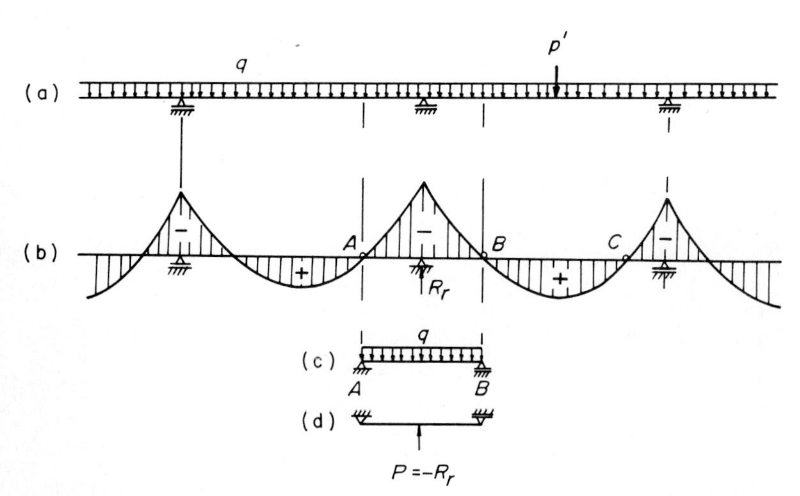

Figure 2.34 Equivalent span for continuous girder bridge: (*a*) continuous girder with arbitrary load, (*b*) bending moment diagram, (*c*) simple girder with uniform load, (*d*) simple girder with concentrated load.

simple beam BC subjected to a uniform load q
P can then be determined in the same manne

This approximate method, in addition to be
ous girder bridge with variable cross sectior
types of girder bridges.

Typical continuous girder bridges of tv
spans (span ratio of 1:1.25:1) and their e...
analyzed by a computer program based on the transfe...
The loads (L loading) are applied to these bridges such that the be...
ing moments at the side span ($0.4l$), internal support, or center span ($0.5l$) are as large as possible. The cross-sectional quantities used are given in Fig. 2.35, because parameters ω and κ do not greatly affect the effective width of these bridges.

Figures 2.36 and 2.37 show the variations of the effective width due to parameter l/b for the two- and three-span continuous girder bridges. In these figures, the span l is fixed as follows:

$$\text{For side span:} \qquad l = l_1 \qquad (2.2.114a)$$

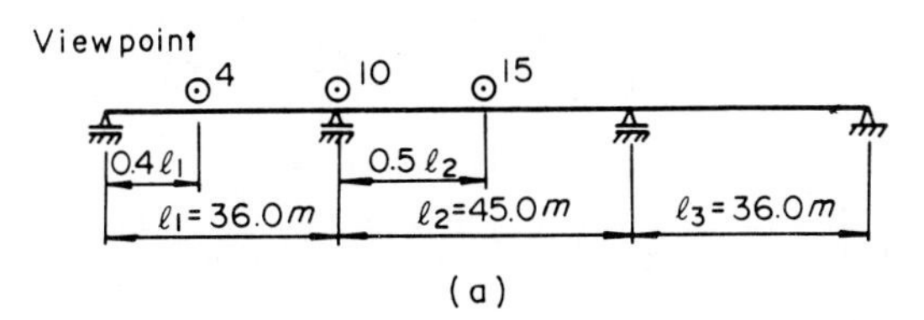

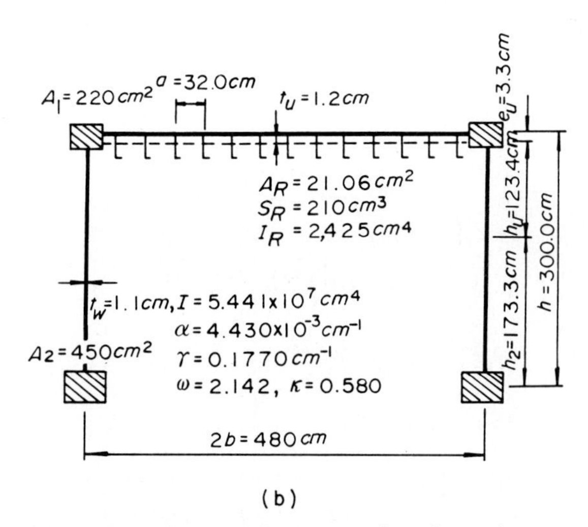

Figure 2.35 (*a*) Span and (*b*) cross section of a continuous girder bridge.

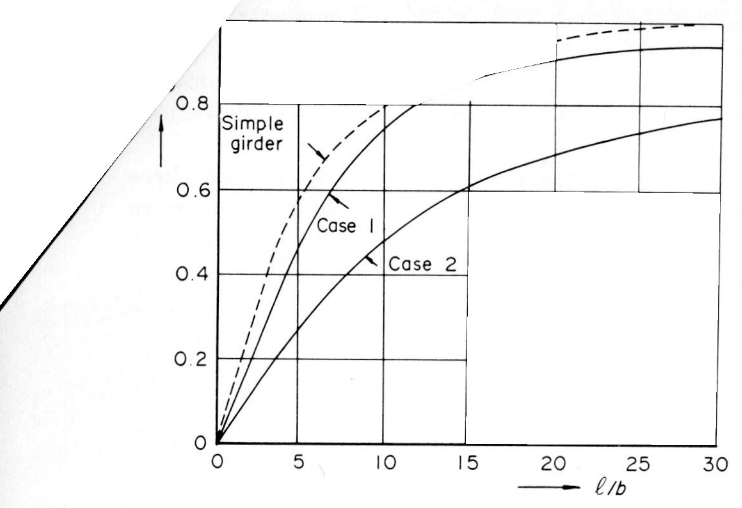

Two

Figure 2.36 Effective width of two-span continuous girder bridge.

For internal support: $l = {}^1/_2(l_1 + l_2)$ (2.2.114*b*)

For center span: $l = l_2$ (2.2.114*c*)

Comparing these values with those of simple girder bridges, we see that the effective width for the side span (case 1 and case 3) is nearly equal to that of the simple girder bridge with a span of $0.85l$. The effective width at the internal supports (case 2 and case 4) coincides

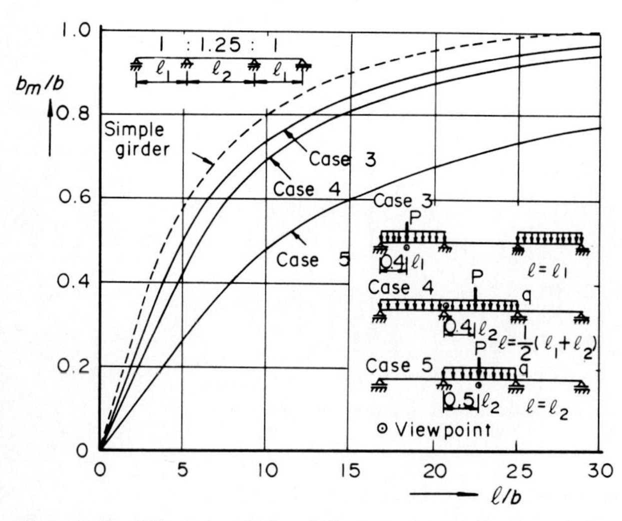

Figure 2.37 Effective width of three-span continuous girder bridge.

TABLE 2.8 Proposed Value to Determine the Equivalent Simple Span

Item	Viewpoint		
	$0.4l_1$ (side span)	$1.0l_1$ (internal support)	$0.5l_2$ (center span)
l_i	l_1	$\frac{1}{2}(l_1 + l_2)$	l_2
β	0.85	0.40	0.70

with that of a simple girder bridge with a span of $0.4l$. Moreover, the effective width of the center span (case 5) is the same as the case of a simple girder bridge with a span of $0.70l$.

Accordingly, the equivalent simple span,[34] which approximates a continuous girder bridge by a simple girder bridge, is proposed by the simple formula

$$l = \beta_e l_i \tag{2.2.115}$$

where the span l_i and a coefficient β_e can be taken as shown in Table 2.8.

The variations of the effective width in the direction of two equal spans and three-span continuous bridges have been investigated and are shown in Figs. 2.38 and 2.39. Examination of these figures shows

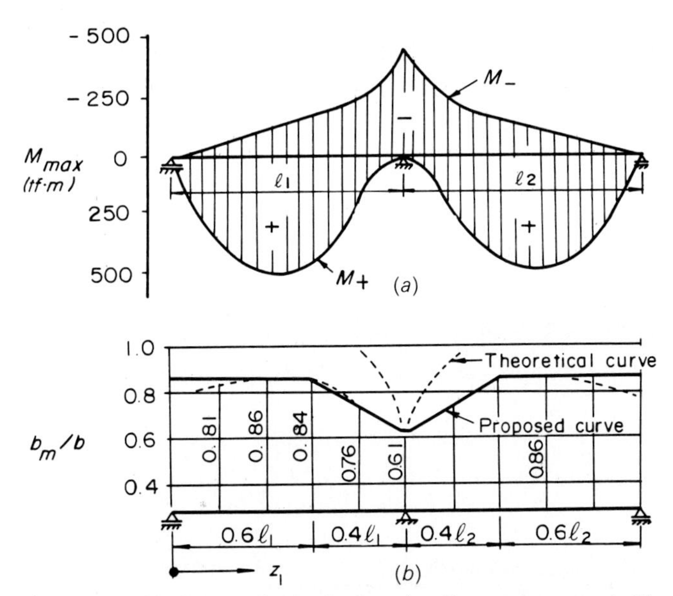

Figure 2.38 Variation of (*a*) absolute bending moment and (*b*) effective width for two equal-span continuous girder bridges. M_- = absolute maximum negative moment; M_+ = absolute maximum positive moment.

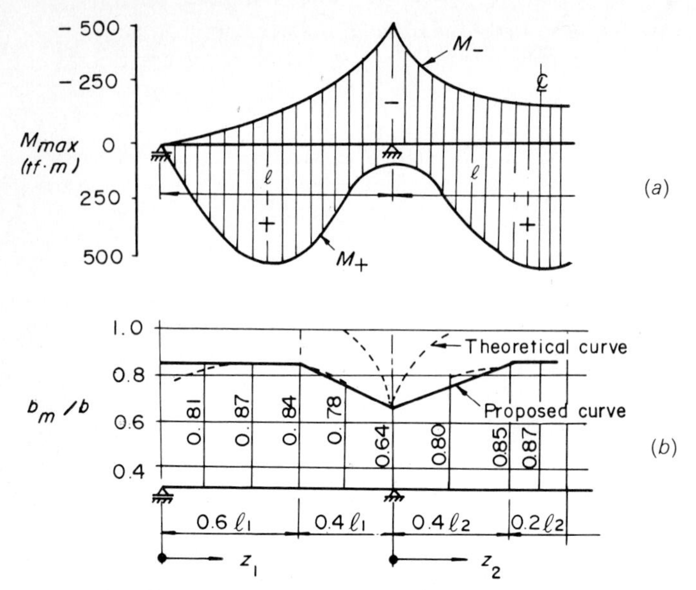

Figure 2.39 Variation of (*a*) absolute bending moment and (*b*) effective width for three-span continuous girder bridge. M_- = absolute maximum negative moment; M_+ = absolute maximum positive moment.

that the effective width for a positive moment is a maximum at section of $0.4l_1$ or $0.5l_2$ and varies gradually toward the interior and end supports. This tendency is similar to that given for the uniform load, shown in Fig. 2.26. The effective width near the internal supports, however, decreases and has minimum values at the interior supports. This tendency is similar to the effective width caused by a concentrated load, shown in Fig. 2.27, because the greatest reactions occur at the internal supports.

To establish an appropriate curve for these variations, an envelope can be established. Thus, the linear approximation of the variation of effective width in the direction of span is proposed, as shown by the solid line in Figs. 2.38 and 2.39. This curve can also be written in the following expressions by taking coordinates z_1 and z_2, as shown in Figs. 2.38 and 2.39.

For a two-equal-span continuous girder bridge,

$$\frac{b_m}{b} = \begin{cases} \left(\dfrac{b_m}{b}\right)_{0.4l_1} & \text{for} \quad 0 \leqq z_1 \leqq 0.6l_1 \qquad (2.2.116a) \\[2ex] \left(\dfrac{b_m}{b}\right)_{0.4l_1} - \dfrac{(b_m/b)_{0.4l_1} - (b_m/b)_{1.0l_1}}{0.4l_1}(z_1 - 0.6l_1) & \\ & \text{for} \quad 0.6l_1 \leqq z_1 \leqq 1.0l_1 \qquad (2.2.116b) \end{cases}$$

For three-span continuous girder bridges,

$$\frac{b_m}{b} = \begin{cases} \left(\frac{b_m}{b}\right)_{0.4l_1} & \text{for} \quad 0 \leqq z_1 \leqq 0.6l_1 \qquad (2.2.117a) \\ \left(\frac{b_m}{b}\right)_{0.4l_1} - \frac{(b_m/b)_{0.4l_1} - (b_m/b)_{1.0l_1}}{0.4l_1}(z_1 - 0.6l_1) & \\ & \text{for} \quad 0.6l_1 \leqq z_1 \leqq 1.0l_1 \qquad (2.2.117b) \\ \left(\frac{b_m}{b}\right)_{1.0l_1} + \frac{(b_m/b)_{0.5l_2} - (b_m/b_1)_{1.0l_1}}{0.4l_2} z_2 & \\ & \text{for} \quad 0 \leqq z_2 \leqq 0.4l_2 \qquad (2.2.117c) \\ \left(\frac{b_m}{b}\right)_{0.5l_2} & \text{for} \quad 0.4l_2 \leqq z_2 \leqq 0.6z_2 \qquad (2.2.117d) \end{cases}$$

where $(b_m/b)_{0.4l_1}$, $(b_m/b)_{1.0l_1}$, and $(b_m/b)_{0.5l_2}$ can be calculated as simple beams by adopting the equivalent span length proposed in Table 2.7.

2.3 Torsion[1–5,8,21–24]

2.3.1 Pure torsion

(1) Phenomenon of pure torsion. When a beam consisting of a round or prismatic bar, pipe, or rectangular box section with a constant thickness is subjected to twisting forces, the phenomenon of pure torsion results. In addition, a prismatic beam with a general cross section, which is allowed to warp freely during twisting, is also in a state of pure torsion. These phenomena are referred to as *St. Venant's torsion.*

Let us now consider the pure torsional behavior of a pipe with diameter $2r$. The torsional moment T (henceforth denoted T_s to indicate St. Venant's torsion) is applied to both ends of the pipe. This pipe will rotate through an angle θ, known as the *twisting angle* (unit: radian), around the shear center S due to the pure torsional moment T_s, as shown in Fig. 2.40. The change of twisting angle ϑ in the direction of the z coordinate axis is denoted

$$\vartheta = \frac{d\theta}{dz} \qquad (2.3.1)$$

Then, a point 1 in the cross section located a distance z from an end of the pipe will be transformed to a new point, 1′, after deformation, as seen in Fig. 2.41. The corresponding displacements u, v, and w in the

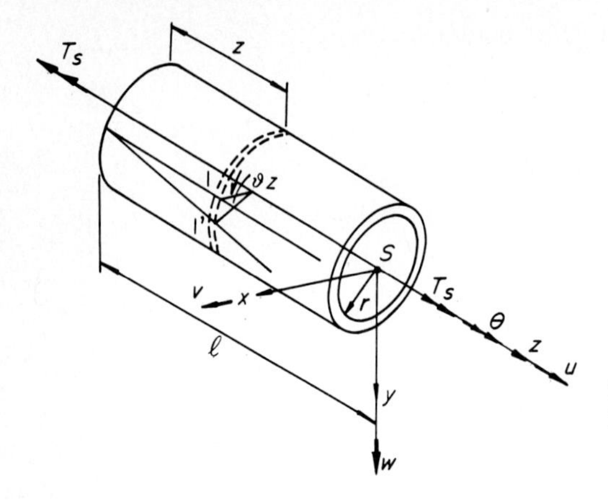

Figure 2.40 Pure torsion of pipe.

direction of the z, x, and y coordinate axes, respectively, with respect to the shear center S are as follows:

$$u = 0 \tag{2.3.2a}$$

$$v = -\vartheta zy \tag{2.3.2b}$$

$$w = \vartheta zx \tag{2.3.2c}$$

Therefore, the relationships between these displacements and stresses will be expressed by means of Hooke's law, which gives[8]

$$\sigma_z = E\varepsilon_z = E\frac{\partial u}{\partial z} = 0 \tag{2.3.3a}$$

$$\sigma_x = E\varepsilon_x = E\frac{\partial v}{\partial x} = 0 \tag{2.3.3b}$$

$$\sigma_y = E\varepsilon_y = E\frac{\partial w}{\partial y} = 0 \tag{2.3.3c}$$

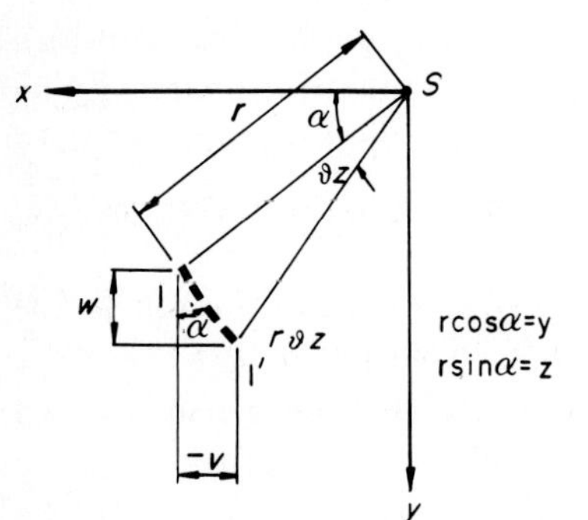

Figure 2.41 Displacements u, v, and w.

$$\gamma_{xy} = G\left(\frac{\partial v}{\partial y} + \frac{\partial w}{\partial x}\right) = 0 \tag{2.3.3d}$$

and

$$\tau_{xz} = G\gamma_{xz} = G\left(\frac{\partial u}{\partial x} + \frac{\partial v}{\partial z}\right) = -G\vartheta y \tag{2.3.3e}$$

$$\tau_{yz} = G\gamma_{yz} = G\left(\frac{\partial u}{\partial y} + \frac{\partial w}{\partial z}\right) = G\vartheta x \tag{2.3.3f}$$

It is obvious from the above equation that the phenomenon of pure torsion does not induce the normal stress.

However, the equilibrium condition of these stresses can also be obtained from the above equations:

$$\frac{\partial \tau_{xz}}{\partial z} = 0 \tag{2.3.4a}$$

$$\frac{\partial \tau_{yz}}{\partial z} = 0 \tag{2.3.4b}$$

and it follows from $\partial \tau_{xz}/\partial x = 0$ and $\partial \tau_{yz}/\partial y = 0$ that

$$\frac{\partial \tau_{xz}}{\partial x} + \frac{\partial \tau_{yz}}{\partial y} = 0 \tag{2.3.5}$$

i.e., stresses τ_{xz} and τ_{yz} are not affected by the z coordinate axis.

Thus, the stress function Φ for pure torsion can be set similar to Eq. (2.2.36*c*):

$$\tau_{xz} = \frac{\partial \Phi}{\partial y} \tag{2.3.6a}$$

$$\tau_{yz} = -\frac{\partial \Phi}{\partial x} \tag{2.3.6b}$$

The derivatives of Eqs. (2.3.3*e*) and (2.3.3*f*) with respect to the y and x coordinate axes, respectively, and the summation of these equations give

$$\frac{\partial^2 \Phi}{\partial x^2} + \frac{\partial^2 \Phi}{\partial y^2} = -2G\vartheta \tag{2.3.7}$$

This fundamental equation describes the pure torsional behaviors of a thin-walled beam.

To investigate the boundary conditions of the stress function Φ, let us now focus on the stresses shown in Fig. 2.42. The shearing stress τ_{zn} in the direction of a normal line should be represented by

$$\tau_{zn} = \tau_{zx}\frac{dy}{ds} + \tau_{zy}\left(-\frac{dx}{ds}\right) \equiv 0 \tag{2.3.8}$$

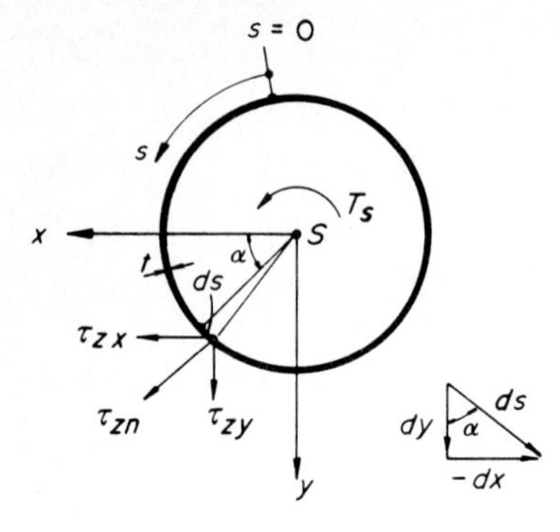

Figure 2.42 Shearing stresses τ_{zx}, τ_{zy}, and τ_{zn}.

However, since $\tau_{zx} = \tau_{xz}$ and $\tau_{zy} = \tau_{yz}$, substituting Eq. (2.3.6) into the above equation and noting that the stress function Φ is a function only of the x and y coordinate axes yield

$$\frac{\partial \Phi}{\partial y}\frac{dy}{ds} + \frac{\partial \Phi}{\partial x}\frac{dx}{ds} = \frac{\partial \Phi}{\partial s} = 0 \tag{2.3.9}$$

Therefore, the stress function $\Phi(z, x)$ must satisfy the following boundary conditions:

$$\Phi = \text{const} \tag{2.3.10a}$$

$$\Phi = 0 \tag{2.3.10b}$$

The physical phenomenon analogous to pure torsion, which is equivalent in form to the fundamental equation (2.3.7) and also satisfies the boundary conditions of Eq. (2.3.10), is well known as the *soap-film analogy*, which can be explained as follows.[8] The soap film is spread across a slit in the shape of an arbitrary cross section of a thin-walled beam. The base and top plates are located as illustrated in Fig. 2.43. When the air pressure p is applied to the soap film, the altitude $h(x, y)$ at an arbitrary point of this soap film is given by

$$\frac{\partial^2 h}{\partial x^2} + \frac{\partial^2 h}{\partial y^2} = -\frac{p}{H} \tag{2.3.11}$$

where H is the tensile force in the soap film. Hence the boundary conditions are similar to Eq. (2.3.7); that is, $h_o = 0$ and $h_i = \text{constant}$ at the inner and outer sides of the slit, respectively, so that the stress function $\Phi(x, y)$ can be easily deduced from the altitude $h(x, y)$ of the soap film. Thus, the pure torsional analysis of a beam with various cross sections will be performed on the basis of the soap-film analogy.

In this case, the torsional moment is given by the integrations of the

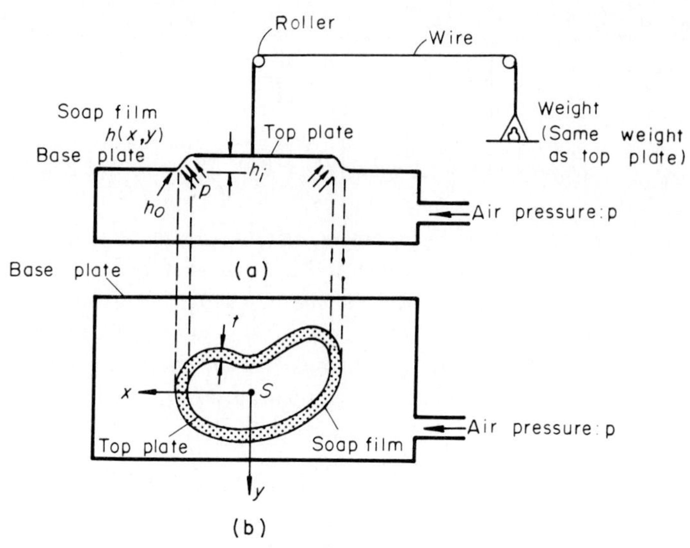

Figure 2.43 Soap-film analogy showing (*a*) side elevation and (*b*) plane.

products of τ_{zx} and τ_{xz} and the corresponding arms, y and x, respectively, as in Fig. 2.42. Thus

$$T_s = \int_A \int (-\tau_{xz}\, y + \tau_{yz} x)\, dx\, dy \tag{2.3.12}$$

By using the conditions $\tau_{zx} = \tau_{xz}$ and $\tau_{yz} = \tau_{zy}$ and Eq. (2.3.6) as well as integration by parts, the above equation leads to

$$T_s = -\int_A \int \left(\frac{\partial \Phi}{\partial y}\, y + \frac{\partial \Phi}{\partial x}\, x \right) dx\, dy \tag{2.3.13a}$$

$$= -\int [\Phi y]_{y_{\min}}^{y_{\max}}\, dx + \int_A \int \Phi\, dx\, dy$$

$$- \int [\Phi x]_{x_{\min}}^{x_{\max}}\, dy + \int_A \int \Phi\, dx\, dy \tag{2.3.13b}$$

However, the first and third terms in the above equation must vanish, since $\Phi = 0$ along the boundaries of the cross section. This results in the following important formula:

$$T_s = 2 \int_A \int \Phi(x, y)\, dx\, dy \tag{2.3.14}$$

By the way, it is easy to verify the following equations:

$$Q_x = \int_A \left[\int (-\tau_{xz})\, dx \right] dy = 0 \tag{2.3.15a}$$

$$Q_y = \int_A \left[\left(\int \tau_{xz} \right) dx \right] dy = 0 \tag{2.3.15b}$$

Recall from Sec. 2.2.2 that shearing forces Q_x and Q_y in the direction of the x and y axes, respectively, are generated by transverse loading of the beam and that these shearing forces are related to the bending moment at any given point along the axis of the beam. Equations (2.3.15) tells us that no additional shearing forces in the direction of the x and y axes are introduced when the torsional moment T_s is applied at the shear center S of the cross section. Thus, the bending and torsional behaviors can be treated independently if we take the origin of the x and y coordinate axes as the shear central axis S of the cross section.

(2) Pure torsion without warping

***a.* Rectangular cross section.** Let us first consider the pure torsion of a rectangular beam, shown in Fig. 2.44*a*, with the cross section having thickness t and width b, where the condition of a thin-walled beam, $t/b < 1/10$, is fulfilled. Obviously the stress function can be represented directly by the parabolic curve at an arbitrary section m–m, except the corner parts, as shown in Fig. 2.44*b*; thus[4]

$$\Phi = \Phi_o \left(1 - \frac{4x^2}{t^2} \right) \tag{2.3.16}$$

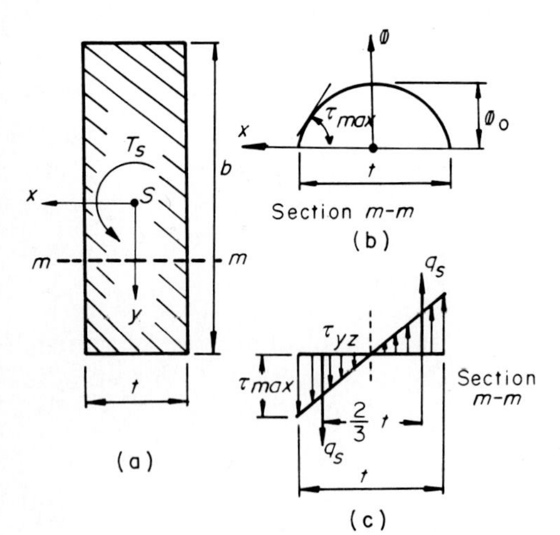

Figure 2.44 Pure torsion of rectangular beam: (*a*) cross section, (*b*) stress function Φ of section m–m, and (*c*) distribution of τ.

where Φ_o is a certain constant and can be decided by substituting the above equation into Eq. (2.3.7) as follows:

$$\Phi_o = G\vartheta \frac{t^2}{4} \tag{2.3.17}$$

Accordingly, the shearing stress τ_{yz} can be obtained from Eq. (2.3.6):

$$\tau_{yz} = 2G\vartheta x \tag{2.3.18}$$

The torsional moment T_s can also be estimated by Eq. (2.3.14) as

$$T_s = 2\int_A\int \frac{t^2}{4} G\vartheta\left(1 - \frac{4x^2}{t^2}\right) dx\, dy$$

$$= \frac{t^2}{2} G\vartheta b \int_{-t/2}^{t/2} \left(1 - \frac{4x^2}{t^2}\right) dx = \tfrac{1}{3} bt^3 G\vartheta \tag{2.3.19}$$

By introducing

$$K = \tfrac{1}{3} bt^3 \tag{2.3.20}$$

Eq. (2.3.19) can be rewritten as

$$T_s = GK\vartheta = GK\frac{\partial\theta}{\partial z} \tag{2.3.21}$$

This equation says that the torsional moment equals the torsional rigidity times the change of torsional angle, much as the bending moment equals the flexural rigidity times the change of deflection angle. Therefore, K is referred to as the *pure torsional constant.*

The stress $\tau_s = \tau_{yz}$ induced by the pure torsion, then, is given by

$$\tau_s = 2\frac{T_s}{K} x \tag{2.3.22}$$

The substitution of $x = t/2$ gives the following maximum shearing stress $\tau_{s,\max}$ due to the pure torsion:

$$\tau_{s,\max} = \frac{T_s}{K} t \tag{2.3.23}$$

The engineering implications of this fact can be detailed as shown in Fig. 2.45. The resultant force q_s due to shearing stress τ_s is evidently given by

$$q_s = \tau_{s,\max} \times \frac{t}{2} \times \frac{1}{2} = \frac{\tau_{s,\max}}{4} t = G\vartheta \frac{t^2}{4} \tag{2.3.24}$$

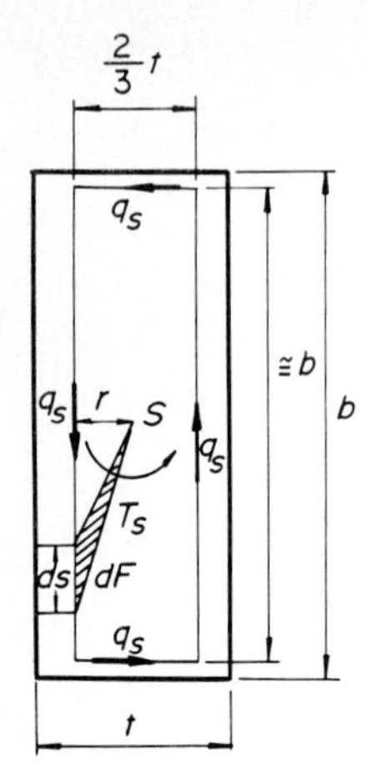

Figure 2.45 Shear flow in rectangular cross section.

And this force q_s is distributed along the cross section as illustrated in Fig. 2.45, so that the integration of the product of q_s and the perpendicular distance r from the origin S, that is, $q_s r$, around the cross section for equilibrium must be equivalent to the torsional moment T_s. Thus

$$T_s = \oint q_s r \, ds = q_s \oint r \, ds \tag{2.3.25}$$

Hence $r\,ds$ is no other than twice the area of dF, the shaded part in Fig. 2.45, and the integration is represented by $\oint r\,ds = 2F$. Then

$$\oint r \, ds = 2F \cong \tfrac{4}{3}bt \tag{2.3.26}$$

Consequently,

$$T_s = \tfrac{1}{3}bt^3 G\vartheta \tag{2.3.27}$$

and this equation coincides with Eq. (2.2.21).

When a rectangular beam consists of thick plates such as $t/b < \tfrac{1}{10}$, the width b can be approximated by $b - 0.63t$ in the above equations.[8]

***b*. Thin-walled closed section.** Let us now consider the pure torsion of a beam with the thin-walled closed section, as shown in Fig. 2.46*a*. The stress function Φ can also be deduced from the soap-film analogy as a stress hill, illustrated in Fig. 2.46*b*, where the toe of a stress hill is shown in Fig. 2.46*c*.

Then the stress function Φ can be represented by taking a new coordinate axis e from the midplane of a thin-walled member as[4]

$$\Phi = \Phi_o\left(1 - \frac{4e^2}{t^2}\right) + \frac{H}{2}\left(1 - \frac{2e}{t}\right) \tag{2.3.28}$$

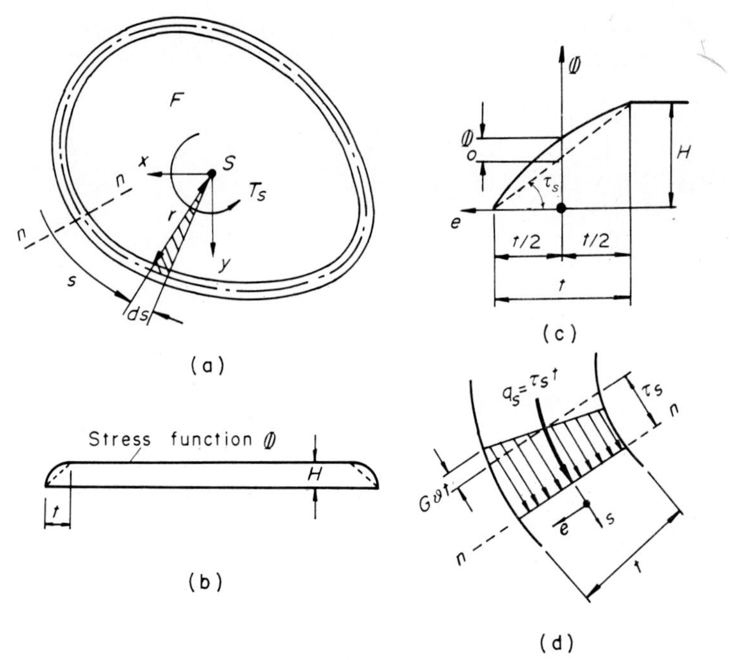

Figure 2.46 Pure torsion of thin-walled closed section: (*a*) cross section, (*b*) stress function Φ, (*c*) slope of Φ, and (*d*) distribution of τ.

where Φ_o is a constant for the solid section, as given by Eq. (2.3.17), and H is the altitude of the stress hill. To determine this height H, let us consider the average shearing stress τ_s acting on the midplane of the member, $e = 0$, at the cross section n-n, as indicated by the broken line in Fig. 2.46*d*, which gives

$$\tau_s = -\left[\frac{\partial \Phi}{\partial e}\right]_{e=0} = \frac{H}{t} \qquad (2.3.29)$$

The integration of this stress around the cross section can be expressed by

$$\oint \tau_s \, ds = 2FG\vartheta \qquad (2.3.30)$$

as detailed in the next section, where F is the area surrounded by the closed section. The substitution of Eq. (2.3.29) into the above equation reduces to

$$H = \frac{2FG\vartheta}{\oint ds/t} \qquad (2.3.31)$$

Therefore, the substitutions of Eqs. (2.3.17), (2.3.28), and (2.3.31) into (2.3.14) reduce to the following expression for the torsional moment T_s:

$$T_s = 2\int_A\int \Phi\, dx\, dy = 2\int_A\int \Phi_o\, dx\, dy + 2FH$$
$$= G\left(\frac{1}{3}\oint t^3\, ds + \frac{4F^2}{\oint ds/t}\right)\vartheta \tag{2.3.32}$$

From the definition of the pure torsional constant K,

$$K = \frac{1}{3}\oint t^3\, ds + \frac{4F^2}{\oint ds/t} \tag{2.3.33}$$

and we conclude that Eq. (2.3.21) is also valid.

However, the shearing stress $\tau_s(e)$ at an arbitrary point e apart from the midplane of the thin-walled member can be derived from Eqs. (2.3.17), (2.3.28), and (2.3.29):

$$\tau_s(e) = -\frac{\partial\Phi}{\partial e} = 2G\vartheta e + \frac{H}{t}$$
$$= G\vartheta\left(2e + \frac{2F}{t\oint ds/t}\right) \tag{2.3.34}$$

However, since $T_s = GK\vartheta$, the above equation can be rewritten as

$$\tau_s(e) = \frac{T_s}{K}\left(2e + \frac{2F}{t\oint ds/t}\right) \tag{2.3.35}$$

In general, the influence of the first term in the right-hand side of Eq. (2.3.32) or (2.3.35) is small enough to be ignored for design purposes. Let us show a much more simple and practical method to determine the shear flow q_s due to pure torsion by considering the example illustrated in Fig. 2.47. It is clear that the shear flow q_s due to the average shearing stress τ_s at the midplane of a thin-walled member can be given by

$$q_s = \tau_s t = H = \text{constant} \tag{2.2.36}$$

This shear flow is referred to as the *shear flow due to pure torsion*, and it behaves in a manner analogous to stream flow where flex (q_s) = velocity (τ_s) × cross-sectional area (t). The shear flow due to pure torsion has a constant value even in the case where the cross section

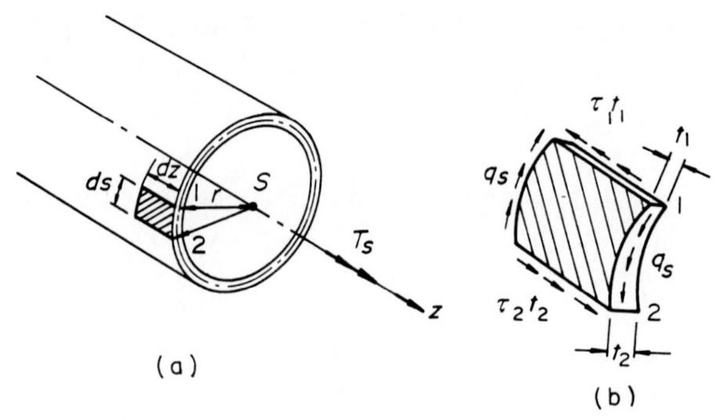

Figure 2.47 Shear flow in thin-walled closed section: (*a*) element *dz dc*, (*b*) equilibrium of shear flow.

has a varying thickness. To ensure this property, let us examine the equilibrium condition of shear flow acting on a small element *dz ds*, as shown in Fig. 2.47*a*. The shear flow q_s can be plotted as shown in Fig. 2.47*b*. Then the equilibrium condition $\Sigma H = 0$ in the direction of the z coordinate axis results in

$$\tau_1 t_1 \, dz - \tau_2 t_2 \, dz = 0 \tag{2.3.37a}$$

$$\therefore \tau_1 t_1 = \tau_2 t_2 = q_s \tag{2.3.37b}$$

Besides, the torsional moment T_s must be given by the contour integration of the products of shear flow q_s and perpendicular distance r from the shear center S. Thus

$$T_s = \oint q_s r \, ds = 2Fq_s \tag{2.3.38a}$$

$$\therefore q_s = \frac{T_s}{2F} \tag{2.3.38b}$$

This equation corresponds to Eq. (2.3.35) if we put $e = 0$, that is, $\tau_s(e) = \tau_s$, and neglect the influence of first term of the right-hand side of Eq. (2.3.32).

Finally, the change of twisting angle ϑ for a thin-walled closed section is given by

$$\vartheta = \frac{d\theta}{dz} = \frac{T_s}{GK} = \frac{T_s}{4GF^2} \oint \frac{ds}{t} \tag{2.3.39}$$

Equations (2.3.38*b*) and (2.3.39) are called *Bredt's* and *Batho's formula*, respectively.

For example, the maximum torsional shearing stress $\tau_{\max}$ in a pipe with diameter $2r$ and thickness t can be calculated by

$$\tau_{\max} = \frac{T_s}{K}\left(r + \frac{t}{2}\right) \tag{2.3.40}$$

where the torsional constant K is the same as the polar moment of inertia with respect to the centroid.

(3) Pure torsion with warping. In the case of pure torsion having a general cross section, the axial displacement u is not always zero. This displacement u is referred to as the *warping due to torsion* and is the function of the x and y coordinate axes without being affected by the z coordinate axis.

If this warping is not restrained, the following equation always holds:

$$\sigma_z = E\varepsilon_z = E\frac{\partial u}{\partial z} = 0 \tag{2.3.41}$$

and the generality of the fundamental equations of pure torsion in the previous section is not violated. Now, let us discuss the pure torsion of a thin-walled closed section with an arbitrary shape, as sketched in Fig. 2.48*a*. A small element 1, 2, 3, 4 cut off by the segment $dz\,ds$, shown in this figure, will be moved to the new positions 1′, 2′, 3′, 4′ after deformation by the action of pure torsional moment T_s, as illustrated in Fig. 2.48*b*.

Therefore, devoting our attention to this shear deformation, we see that

$$\tan^{-1}\alpha = r\vartheta \tag{2.3.42a}$$

$$\tan^{-1}\beta = \frac{\partial u}{\partial s} \tag{2.3.42b}$$

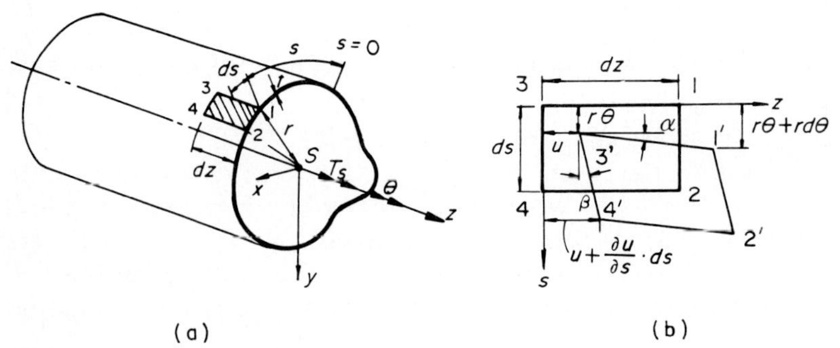

Figure 2.48 Displacements in pure torsion with warping: (*a*) coordinate axis and cross section, (*b*) displacement of element.

The shearing stresses τ_s can be written readily from their definitions:

$$\tau_s = G\gamma = G\left(r\vartheta + \frac{\partial u}{\partial s}\right) \tag{2.3.43}$$

From the above equation, the derivative $\partial u/\partial s$ can be expressed as

$$\frac{\partial u}{\partial s} = -\vartheta r + \frac{\tau_s}{G} \tag{2.3.44}$$

Then the integration of both sides of the above equation gives

$$u(s) = u_o - \vartheta \int_0^s r\,ds + \int_0^s \frac{\tau_s}{G}\,ds \tag{2.3.45}$$

where u_o is the warping at the origin $s = 0$.

For a cross section with a single cell, the contour integration can be given by

$$\oint \frac{\partial u}{\partial s}\,ds = -\vartheta \oint r\,ds + \oint \frac{\tau_s}{G}\,ds = u_1 - u_2 \tag{2.3.46}$$

Then if the warping u_1 and u_2 at the origin and terminus, respectively, do not coincide with each other, some displacements in the direction of z coordinate axis at the origin and terminus may occur. This contradiction must lead to

$$\oint \frac{\partial u}{\partial s}\,ds = 0 \tag{2.3.47}$$

Accordingly, this condition can be rewritten as Eq. (2.3.30) after some rearrangements.

(4) Calculations of pure torsional stress in various girders

***a.* Design formula for an open section.** The maximum shearing stress $\tau_{s,\max}$ in an open section can be estimated easily by Eq. (2.3.23), so that only the method for determining the torsional constant K is presented in the preceding discussion.

Figure 2.49*a* shows an I girder as an example of an open section. For this cross section, the shear flow q_s is distributed within the I girder, as in Fig. 2.49*b*. Moreover, this shear flow can be approximately calculated by separating the section into individual flange and web plates, as in Fig. 2.49*c*. Thus, the torsional constant K is nothing but the sum of the torsional constants for the individual members

$$K = \tfrac{1}{3} \sum_i b_i t_i^3 \tag{2.3.48a}$$

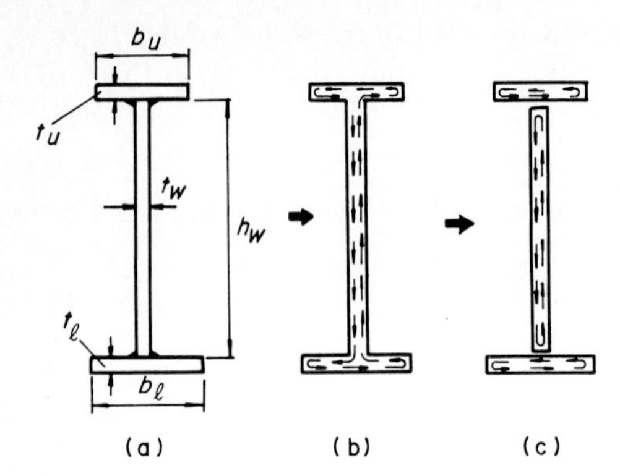

Figure 2.49 Pure torsion of I girder. (*a*) I girder as open section. (*b*) Distribution of shear flow q_s. (*c*) Separation of flange and web plates.

or more concisely

$$K = \tfrac{1}{3}(b_u t_u^3 + h_w t_w^3 + b_l t_l^3) \tag{2.3.48b}$$

where Σ_i means the summation of all the members.

This equation can directly be applied to the π girder shown in Fig. 2.50 and gives

$$K = \tfrac{1}{3}(2Bt_u^3 + nh_R t_R^3 + 2ht_w^3 + 2b_l t_l^3) \tag{2.3.49}$$

where n is the number of ribs.

***b.* Design formula for a closed section.** The shearing stress τ_s for a box girder with a single cell can be estimated by Eqs. (2.3.36) and (2.3.38*b*):

$$\tau_s = \frac{q_s}{t} = \frac{T_s}{2Ft} \tag{2.3.50}$$

The corresponding torsional constant K can also be evaluated from Eq. (2.3.39) by neglecting the contribution of Eq. (2.3.48) due to the open section. Thus

$$K = \frac{4F^2}{\oint ds/t} \tag{2.3.51}$$

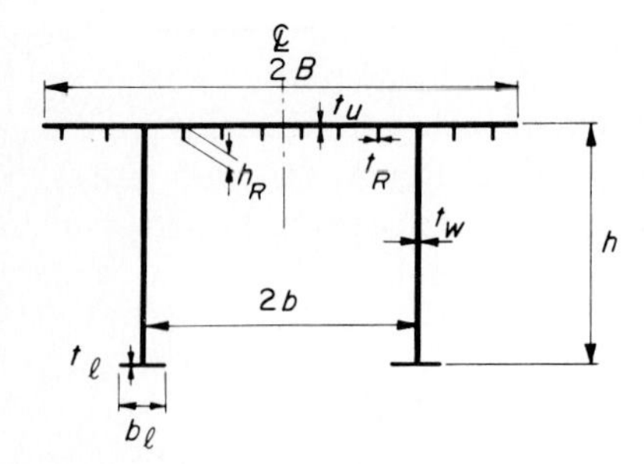

Figure 2.50 Dimension of π girder.

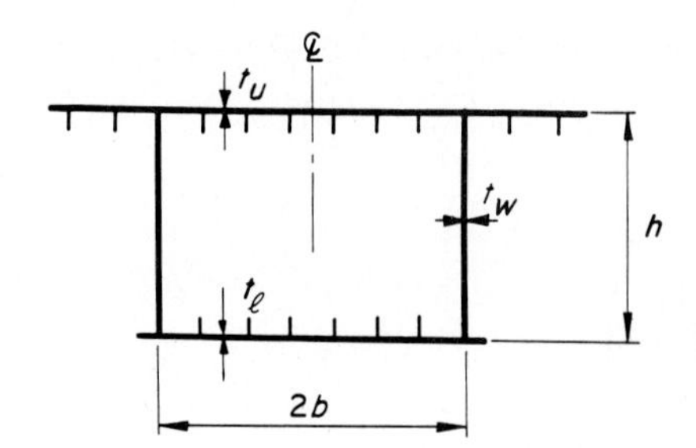

Figure 2.51 Dimension of monobox girder.

For instance, Fig. 2.51 shows a monobox girder, and τ_s and K are

$$\tau_s = \frac{T_s}{4bht} \tag{2.3.52a}$$

$$K = \frac{8b^2h^2}{b/t_u + h/t_w + b/t_l} \tag{2.3.52b}$$

In a case of a multicellular box girder, as in Fig. 2.52, additional calculations must be performed to evaluate the shear flow q_s and torsional constant K.

Let the shear flows for each box cell be denoted q_1, q_2, and q_3, and let us omit the subcript s. Obviously the shear flows are $q_{1,2} = q_1 - q_2$ and $q_{2,1} = q_2 - q_1$ for the intermediate wall $\overline{12}$ and $q_{2,3} = q_2 - q_3$ and $q_{3,2} = q_3 - q_2$ for the intermediate wall $\overline{23}$. (The overbar represents the interior wall between cells 1 and 2 and between cells 2 and 3, respectively.)

Accordingly, the generalized equation for a multicellular box girder having cells $i = 1, 2, \ldots, n$ can be represented from Eq. (2.3.30) as

$$\oint_i \frac{\tau_s}{G} ds = \oint_i \frac{q_i}{Gt} ds = 2F_i\vartheta \qquad i = 1, 2, \ldots, n \tag{2.3.53}$$

To simplify this equation, let us define $\tilde{q}_i$ as

$$q_i = G\vartheta\tilde{q}_i \tag{2.3.54}$$

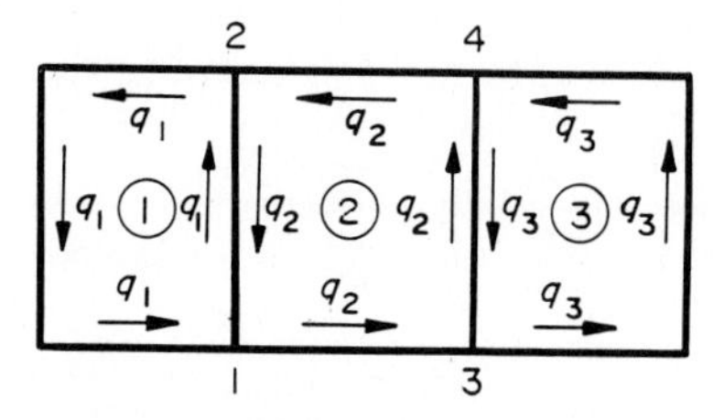

Figure 2.52 Multicellular box girder.

and denote it as the pure torsional function (with units of square centimeters). Thus, the following simultaneous equations can be obtained from the above two equations:

$$\tilde{q}_i\delta_{ii} - \tilde{q}_j\delta_{ij} = 2F_i \qquad i = 1, 2, \ldots, n \tag{2.3.55}$$

where the coefficients δ_{ii} and δ_{ij} are given by

$$\delta_{ii} = \oint_i \frac{ds}{t} \qquad \text{contour integration around } i\text{th cell} \tag{2.3.56a}$$

$$\delta_{ij} = \int_{i,j} \frac{ds}{t} \qquad \text{integration for common wall of } i\text{th and } j\text{th cells} \tag{2.3.56b}$$

For example, Eq. (2.3.55) can be readily written for the box girder, shown in Fig. 2.52, as

$$\tilde{q}_1\delta_{11} - \tilde{q}_2\delta_{12} = 2F_1 \tag{2.3.57a}$$

$$-\tilde{q}_1\delta_{21} + \tilde{q}_2\delta_{22} - \tilde{q}_3\delta_{23} = 2F_2 \tag{2.3.57b}$$

$$-\tilde{q}_2\delta_{32} + \tilde{q}_3\delta_{33} = 2F_3 \tag{2.3.57c}$$

By solving these simultaneous equations for $\tilde{q}_1$, $\tilde{q}_2$ and $\tilde{q}_3$ and multiplying $G\vartheta$ by Eq. (2.3.54), the shear flow q_i for ith cell can easily be estimated, so that the pure torsional shearing stress can be found from $\tau_s = q_i/t$ by referring to Fig. 2.52.

The corresponding pure torsional moment T_s is no other than the sum of the stress resultants for each cell. Note that the following equation can be obtained from Eq. (2.3.54):

$$T_s = \sum_{i=1}^{n} \oint_i q_i r\, ds = \sum_{i=1}^{n} 2q_iF_i = 2G\vartheta \sum_{i=1}^{n} \tilde{q}_iF_i \tag{2.3.58}$$

and the torsional constant K will be

$$K = 2\sum_{i=1}^{n} \tilde{q}_iF_i \tag{2.3.59}$$

c. Multiple-I, box, and quasi box girders. The torsional constant for multiple girders, as illustrated in Fig. 2.53, is obtained by summing the torsional constants for each girder, $j = 1, 2, \ldots, m$:

$$K = \sum_{j=1}^{m} K_j \tag{2.3.60}$$

In the case of multiple-I girders with sway and lateral bracings, the

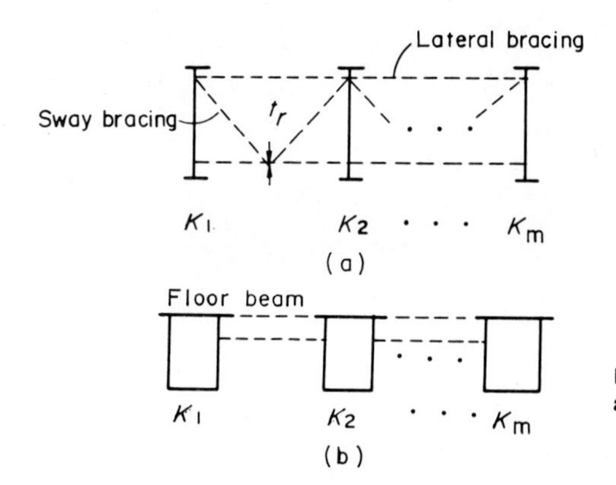

Figure 2.53 (*a*) Multiple-I girder and (*b*) multiple-box girder.

upper and lower lateral bracings can contribute to the pure torsion in a manner similar to the top and bottom flange plates of a box girder.

A box girder with lateral bracing is referred to as a *quasi box girder*. The equivalent thickness t_r of the top and bottom flange plates of a quasi box girder, which is estimated by equating the shear deformation of a truss to that of a thin plate, can be evaluated by Table 2.9 according to the type of lateral bracings.[24]

TABLE 2.9 Equivalent Thickness for Quasi Box Girder

Type of lateral bracing	Equivalent thickness t_r
λ, $2b$, d, A_d, A_f, $S_d = qd$	$\dfrac{E}{G}\dfrac{2\lambda b}{d^3/A_d + 2\lambda^3/(3A_f)}$
λ, $2b$, d, A_d, A_f, A_v, $S_d = qd$	$\dfrac{E}{G}\dfrac{2\lambda b}{2d^3/A_d + 4b^3/A_v + \lambda^3/(6A_f)}$
λ, $2b$, d, A_d, A_v, A_f, $S_d = \dfrac{qd}{2}$	$\dfrac{E}{G}\dfrac{2\lambda b}{d^3/(2A_d) + \lambda^3/(6A_f)}$
λ, $2b$, d, A_d, A_v, A_f, $S_d = qd$	$\dfrac{E}{G}\dfrac{2\lambda b}{d^3/A_d + 8d^3/A_v + \lambda^3/(6A_f)}$

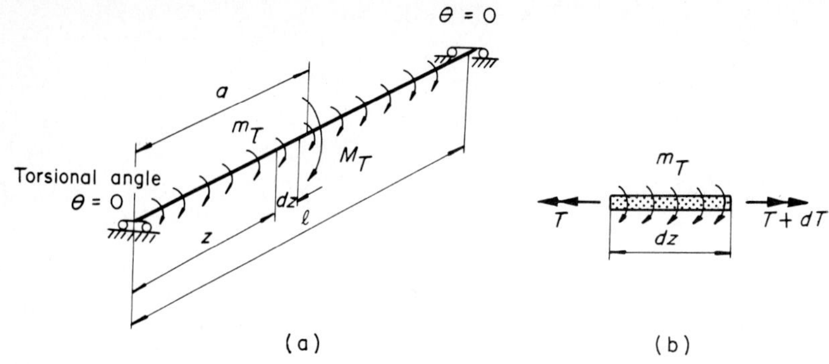

Figure 2.54 Simple beam subjected to torque m_T and M_T: (*a*) applied torque and boundary condition, (*b*) equilibrium of torque.

(5) Stress resultant and displacement for pure torsion.[3] Figure 2.54*a* illustrates a beam in which both ends are supported so as to satisfy the torsional angle $\theta = 0$ at the ends. When a uniformly distributed torque m_T or concentrated torque M_T is applied to this beam, the equilibrium condition of these torques and torsional moment T for a small element dz removed from the beam, as seen in Fig. 2.54*b*, give the following equations:

$$\frac{dT}{dz} = \begin{cases} -m_T & \text{uniform torque } m_T \quad (2.3.61a) \\ -M_T\delta(z-a) & \text{concentrated torque } M_T \quad (2.3.61b) \end{cases}$$

From the above equations, the torsional moment $T(z)$ at an arbitrary section z can be summarized as in Table 2.10. Examination of this table indicates that the distribution of torsional moment is the same as that of the shearing force $Q_y(z)$ in a simple beam, provided that the load terms q and p are changed by m_T and M_T in Table 2.1. Also the

TABLE 2.10 Solution of $T(z)$ and $\theta(z)$ for a Simple Beam

Torque	Torsional moment $T(z)$		Torsional angle $\theta(z)$	
Uniform torque m_T	$\frac{m_T l}{2}\left(1-2\frac{z}{l}\right)$		$\frac{m_T l^2}{2GK}\left(\frac{z}{l}\right)\left(1-\frac{z}{l}\right)$	
Concentrated torque M_T	$M_T\left(1-\frac{a}{l}\right)$	$0 \leqq z \leqq a$	$\frac{M_T}{GK}l\left(1-\frac{a}{l}\right)\frac{z}{l}$	$0 \leqq z \leqq a$
	$-M_T\frac{a}{l}$	$a \leqq z \leqq l$	$\frac{M_T}{GK}l\left(1-\frac{z}{l}\right)\frac{a}{l}$	$a \leqq z \leqq l$

torsional angle $\theta(z)$ can be found by

$$\frac{d\theta}{dz} = \frac{T}{GK} \tag{2.3.39}$$

This equation is analogous to Eq. (2.2.74) for determining the bending moment $M_x(z)$ by dividing by the factor GK, so that these results can also be summarized as in Table 2.10.

2.3.2 Torsional warping

(1) Phenomenon of torsional warping. Let us now consider the warping stresses in a thin-walled beam with an arbitrary cross section subject to twisting. For this case, if the axial displacement u is restrained, the following normal stress σ_ω will occur in the direction of the z coordinate axis:

$$\sigma_\omega = E\varepsilon_z = E\frac{\partial u}{\partial z} \tag{2.3.62}$$

Corresponding to the changes of this normal stress σ_ω in the z coordinate axis, the secondary shearing stress τ_ω is induced, as shown in Fig. 2.55a and 2.55b which leads to

$$\frac{\partial \sigma_\omega t}{\partial z} + \frac{\partial \tau_\omega t}{\partial s} = 0 \tag{2.3.63}$$

based on the equilibrium condition of these stresses in the direction of the z coordinate axis.

To solve for σ_ω let us adopt the warping u as derived in Eq. (2.3.45) and introduce the following torsional warping function ω (in units of

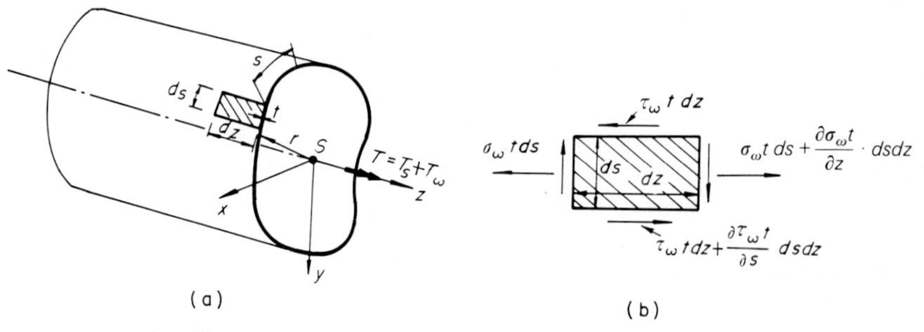

Figure 2.55 (a) Thin-walled beam subjected to torsional warping, resulting in (b) equilibrium of stress.

square centimeters) for the sake of simplicity:

$$u = \omega(s)\vartheta = \omega(s)\frac{d\theta}{dz} \tag{2.3.64}$$

Then, using Eq. (2.3.54), we can write the torsional warping function ω as

$$\omega = \omega_o - \int_0^s r\,ds + \int_0^s \frac{\tilde{q}}{t}\,ds \tag{2.3.65}$$

where ω_o is an unknown torsional warping function at the origin $s = 0$.

The secondary torsional shear flow q_ω can be defined similarly to the shear flow q_b due to bending as follows:

$$q_\omega = \tau_\omega t \tag{2.3.66}$$

The substitution of this equation into Eq. (2.3.63) and integration with respect to the s coordinate axis give

$$q_\omega = -\int_0^s \frac{\partial \sigma_\omega t}{\partial z}\,ds = -E\frac{d^3\theta}{dz^3}\left(\int_0^s \omega t\,ds + C\right) \tag{2.3.67}$$

where C is an integration constant and can be chosen by the following equation in the same manner as bending:

$$\oint \frac{q_\omega}{Gt}\,ds = 0 \tag{2.3.68}$$

For instance, C is given by

$$C = -\frac{\oint [1/(Gt)]\left(\int_0^s \omega t\,ds\right)ds}{\oint ds/(Gt)} \tag{2.3.69}$$

for a box girder with a single cell. This procedure can also apply to a multicellular box girder by considering $q_o = -E(d^3\theta/dx^3)\int_0^s \omega t\,ds$ for an open section and adopting Eqs. (2.2.27) through (2.2.29).

But the products of the secondary torsional shear flow q_ω and perpendicular distance r from the shear center S, as seen in Fig. 2.55, represent a sort of torsional moment T_ω, and this is referred to as the *secondary torsional moment*, or *Wagner's torsional moment* (in units of Newton-centimeters). The final expression for T_ω is

$$T_\omega = \int_A q_\omega r\,ds = -E\frac{d^3\theta}{dz^3}\int_A \omega^2 t\,ds \tag{2.3.70}$$

where

$$I_\omega = \int_A \omega^2 t\,ds \tag{2.3.71}$$

And I_ω is called the *torsional warping constant* (in units of cm^6) and represents the geometric moment of inertia with respect to the warping function ω. Therefore, Eq. (2.2.70) can be simplified to

$$T_\omega = -EI_\omega \frac{d^3\theta}{dz^3} \tag{2.3.72}$$

The expression in parentheses on the right-hand side of Eq. (2.3.67) can also be represented through an analogy to the definition of the static moment as

$$S_\omega = \int_0^s \omega t\, ds - \frac{\oint [1/(Gt)]\left(\int_0^s \omega t\, ds\right) ds}{\oint ds/(Gt)} \tag{2.3.73}$$

Here S_ω is referred to as the *static moment* with respect to the warping function (in units of cm^4). Thus, the shearing stress τ_ω due to T_ω is given by the same form as the flexural shear flow q_b, so that

$$\tau_\omega = \frac{q_\omega}{t} = \frac{T_\omega}{I_\omega t} S_\omega \tag{2.3.74}$$

Moreover, the substitution of Eq. (2.3.64) into Eq. (2.3.62) gives the following equation, which is similar to the flexural normal stress σ_b:

$$\sigma_\omega = E\omega \frac{d^2\theta}{dz^2} = \frac{M_\omega}{I_\omega}\omega \tag{2.3.75}$$

where

$$M_\omega = EI_\omega \frac{d^2\theta}{dz^2} \tag{2.3.76}$$

And M_ω is referred to as the *torsional warping moment* or bimoment (in units of N·cm^2). The integration of Eq. (2.3.75) throughout the cross-sectional area A of a thin-walled beam must equal zero, since there is no axial force. Thus

$$\int_A \sigma_\omega t\, ds = \frac{M_\omega}{I_\omega} \int_A \omega t\, ds \equiv 0 \tag{2.3.77}$$

and

$$\int_A \omega t\, ds = 0 \tag{2.3.78}$$

offers a condition where the volume of the warping function must always equal zero. By utilizing this equation, an unknown warping function ω_o at the origin $s = 0$ in Eq. (2.3.65) can easily be determined to be

$$\omega_o = \frac{1}{A}\left[\int_A t\left(\int_0^s r\, ds\right) ds - \int_A t\left(\int_0^s \frac{\tilde{q}}{t}\, ds\right)\right] ds \tag{2.3.79}$$

In general, note that $\omega_o = 0$ if we take the origin $s = 0$ on the shear central axis.

Finally, the torsional moment T should consist of the sum of St. Venant's torsional moment T_s and Wagner's torsional moment T_ω in the case where the warping is restrained. Then,

$$T = T_s + T_\omega \tag{2.3.80}$$

Consequently, the fundamental equation for the torsional warping phenomenon can be set by substituting Eqs. (2.3.21) and (2.3.72) into the above equation, which yields

$$EI_\omega \frac{d^3\theta}{dz^3} - GK\frac{d\theta}{dz} = -T \tag{2.3.81}$$

The solution of this equation is discussed in detail in the following section.

(2) Physical meaning of torsional warping theory. Let us consider the I girder, shown in Fig. 2.56*a*, to understand the physical meaning of torsional behavior in detail. If the torsional moment T is applied to an end of this beam and the other end is supported under the condition $\theta = 0$ and free to warp, then the torsional angle $\theta(z)$ must have a constant value throughout the I girder, as illustrated in Fig. 2.56*b*, and this phenomenon is nothing but pure torsion. But if the end of the I girder is fixed and warping is restrained, then the torsional angle $\theta(z)$,

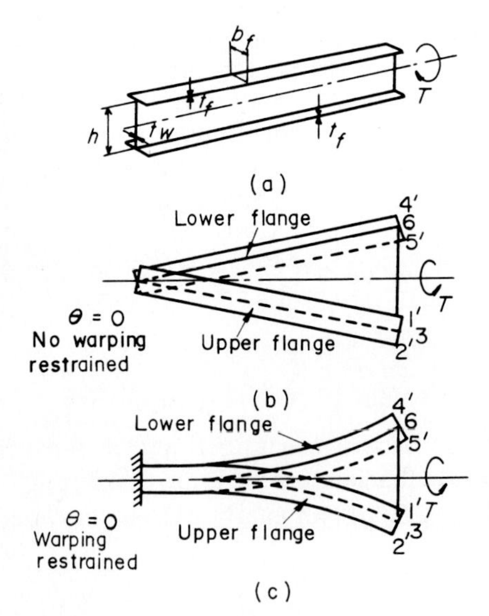

Figure 2.56 Differences between pure torsion and torsional warping: (*a*) I girder (showing dimensions) subjected to (*b*) pure torsion and (*c*) torsional warping.

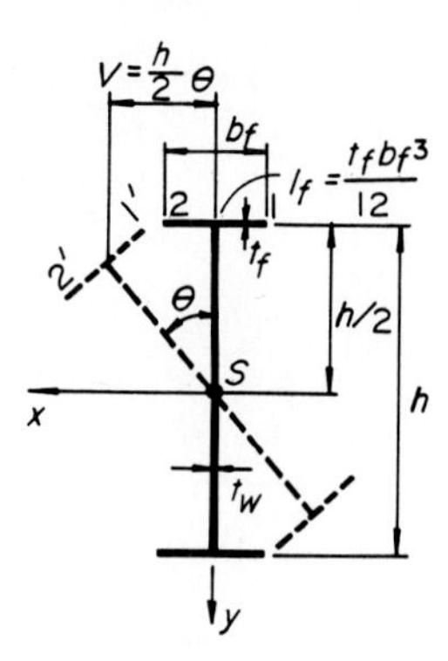

Figure 2.57 Displacement of I girder subjected to torsional warping.

will vary in the direction of the girder axis, as shown in Fig. 2.56c, so that this behavior is referred to as *nonuniform torsion* or torsional warping.[23,24]

Figure 2.57 shows the deformation of the cross section of the I girder subjected to twisting, where the displacement v of the upper flange in the direction of the x coordinate axis is given by

$$v = \frac{h\theta}{2} \tag{2.3.82}$$

The warping u due to the slope of the displacement $dv/dz = (h/2)(d\theta/dz)$, as illustrated in Fig. 2.58, can now be represented as

$$u = \omega \frac{d\theta}{dz} \tag{2.3.83}$$

Thus the meaning of Eq. (2.3.64) can be clearly understood.

It is obvious from Eq. (2.3.65) that $q = 0$ for the open section and $\omega_o = 0$ on the y axis. Thus the torsional warping function ω of the I girder can be estimated from Eq. (2.3.65) as

$$\omega = -\int_0^s r\,ds \tag{2.3.84}$$

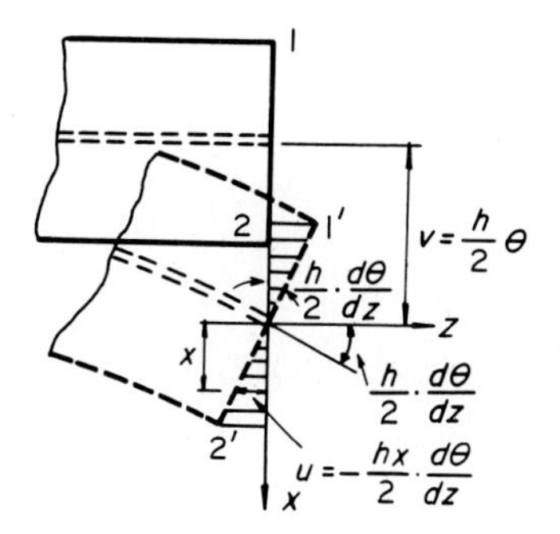

Figure 2.58 Displacement of upper flange of I girder subjected to torsional warping.

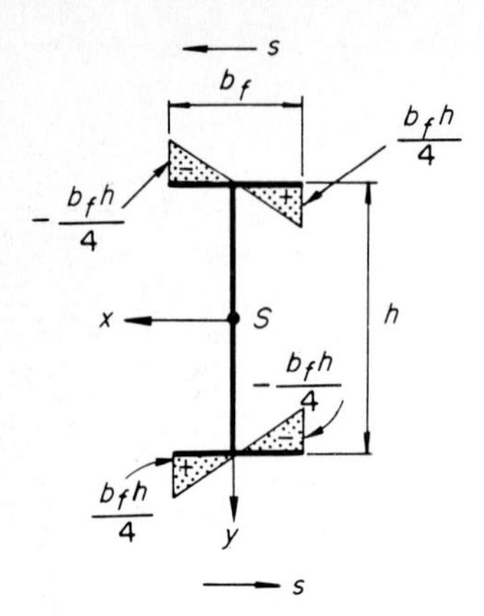

Figure 2.59 Torsional warping function of I girder.

For example, $\omega = -\int_0^{b_f/2} (h/2)\, dx = -b_f h/4$ for the left end of the upper flange, where b_f and h are the width of the flange plate and the depth of the I girder, respectively. Thus, the torsional warping function can be plotted as shown in Fig. 2.59.

The stress resultants, which cause the displacement in Fig. 2.56*c*, can be replaced by the equivalent in-plane bending moment M_f and shearing force Q_f of the flange plates, as shown in Fig. 2.60; i.e.,

$$M_f = \mp EI_f \frac{d^2 v}{dz^2} = \mp EI_f \frac{h}{2} \frac{d^2\theta}{dz^2} \qquad (2.3.85a)$$

$$Q_f = \frac{dM_f}{dz} = \mp EI_f \frac{h}{2} \frac{d^3\theta}{dz^3} \qquad (2.3.85b)$$

Note that double signs in the above equations correspond to the upper and lower flanges, respectively, and I_f is the geometric moment of

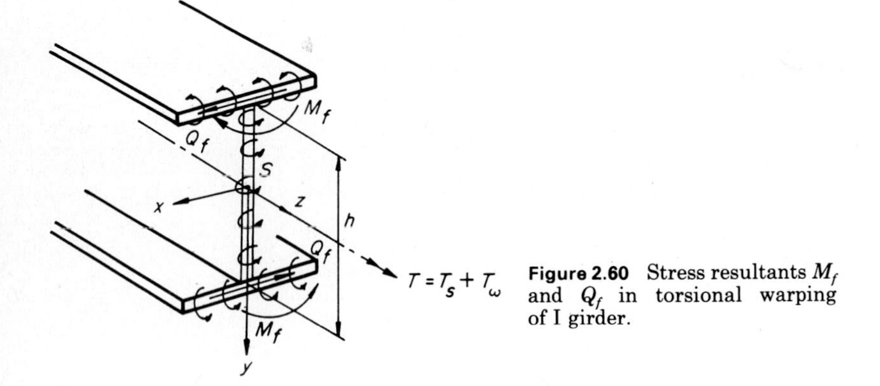

Figure 2.60 Stress resultants M_f and Q_f in torsional warping of I girder.

inertia of the upper and lower flange plates with respect to the y coordinate axis.

Therefore, the equilibrium condition of the applied torque T and these stress resultants gives

$$T = T_s + T_w = GK\frac{d\theta}{dz} + Q_f h$$

$$= GK\frac{d\theta}{dz} - EI_f\frac{h^2}{2}\frac{d^3\theta}{dz^3} \tag{2.3.86}$$

where
$$I_\omega = I_f\frac{h^2}{2} \tag{2.3.87}$$

for a bisymmetric cross section, as demonstrated previously, so that the above equation coincides with Eq. (2.3.81). Furthermore, if Eq. (2.3.80) is rewritten by using Eq. (2.3.85*a*), the torsional warping moment M_ω can be rewritten as

$$M_\omega = M_f h \tag{2.3.88}$$

The term M_ω is referred to as *bimoment* (in units of N·cm^2) and is calculated from the product of the in-plane bending moments M_f of flange plates and the girder depth h.

The normal stress σ_ω at the left end of the upper flange due to this torsional warping moment M_ω can be obtained from $\omega = b_f h/4$ and Eq. (2.3.88) as well as Eq. (2.3.87), which gives

$$\sigma_\omega = \frac{M_\omega}{I_\omega}\omega = -\frac{M_f h}{I_f h^2/2}\frac{b_f h}{4} = -\frac{M_f}{I_f}\frac{b_f}{2} \tag{2.3.89}$$

This equation also coincides with the flexural normal stress produced by M_f with respect to the y axis.

S_ω can be estimated from Eq. (2.3.73) by neglecting the second term, because this girder is an open section. Then the diagram of S_ω can be plotted as shown in Fig. 2.61. In this case, since the secondary torsional moment T_ω is

$$T_\omega = Q_f h \tag{2.3.90}$$

the maximum shearing stress due to this torsional moment occurs at the midpoint of the flange plate and reduces to

$$\tau_\omega = \frac{T_\omega}{I_\omega t_f}S_\omega = \frac{Q_f h}{(t_f b_f^3/12)(h^2/2)t_f}\frac{h t_f b_f^2}{16}$$

$$= \frac{3}{2}\frac{Q_f}{A_f} \tag{2.3.91}$$

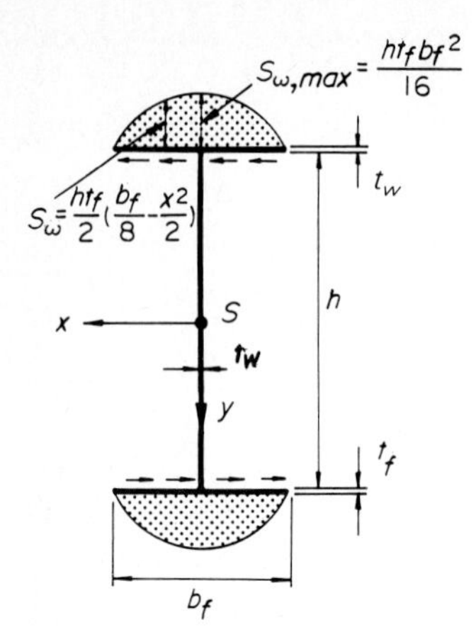

Figure 2.61 Static moment S_ω.

Since $A_f = t_f b$ is the cross-sectional area of the flange plate, this stress τ_ω also coincides with the flexural shearing stress τ_b caused by the shearing force Q_f in the direction of the x axis.

(3) Calculations of cross-sectional quantities for various girders

***a. π* girder.** Consider a thin-walled beam with a π section as illustrated in Fig. 2.62. These are frequently utilized as the main girder of steel bridges. It is necessary to determine the location of the shear center S, since all the cross-sectional quantities with the torsional warping are derived with respect to the shear center. For this purpose, the flexural shear flow q_b is calculated by applying the shearing force Q_x in the direction of the x coordinate axis at the centroid C. Based on the distributions of q_b, the location of the shear center can easily be found by means of the method presented in Sec. 2.2.2. In general, the shear center of the π girder is located at an upward point e_y away from the midplane of the deck plate.

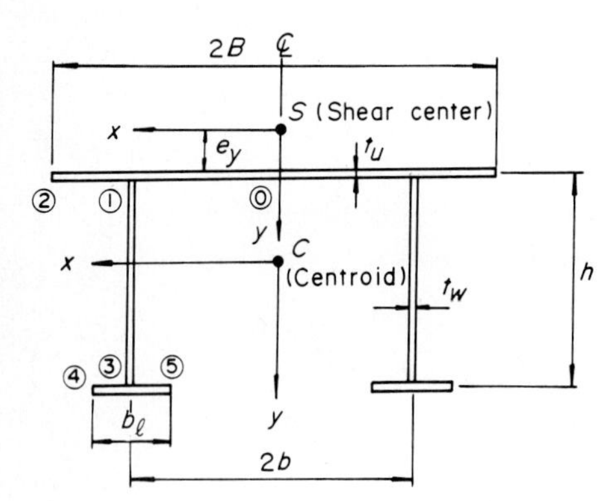

Figure 2.62 Cross section of π girder.

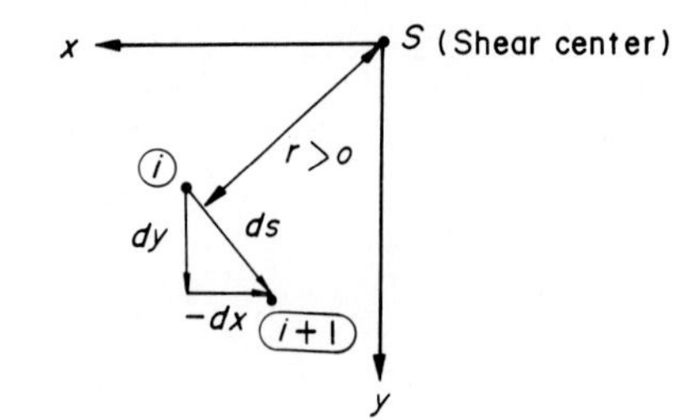

Figure 2.63 Definitions of r and s.

The warping function ω can be calculated by noting that the perpendicular distance r is greater than zero when the perimeter coordinate s proceeds in a counterclockwise direction around the shear center S, as illustrated in Fig. 2.63. The integration $\int r\,ds$ is given by

$$\int r\,ds = \int x\,dy - \int y\,dx \tag{2.3.92}$$

thus

$$\omega_{i+1} = \begin{cases} \omega_i - x\,\Delta_y & \text{proceeds in positive direction of } x \text{ axis} \quad (2.3.93a) \\ \omega_i + y\,\Delta_x & \text{proceeds in positive direction of } y \text{ axis} \quad (2.3.93b) \end{cases}$$

where Δ_x and Δ_y are the length of segments i and $i+1$ in the direction of the x and y coordinate axes, respectively. Their signs correspond to the positive directions of the x or y coordinate axis.

It is convenient to estimate the warping function ω by the numerical processes shown in Table 2.11, and the ω diagram can be plotted, as shown in Fig. 2.64.

TABLE 2.11 Calculation of ω

	$-\int r\,ds$			
Segment	$-x\,\Delta y$	$y\,\Delta x$	i	ω_i
0–1	—	$e_y b$	1	$e_y b$
1–2	—	$e_y(B-b)$	2	$e_y B$
1–3	$-bh$	—	3	$e_y b - bh$
3–4	—	$\dfrac{(e_y+h)b_l}{2}$	4	$e_y b - bh + \dfrac{(e_y+h)b_l}{2}$
3–5	—	$-\dfrac{(e_y+h)b_l}{2}$	5	$e_y b - bh - \dfrac{(e_y+h)b_l}{2}$

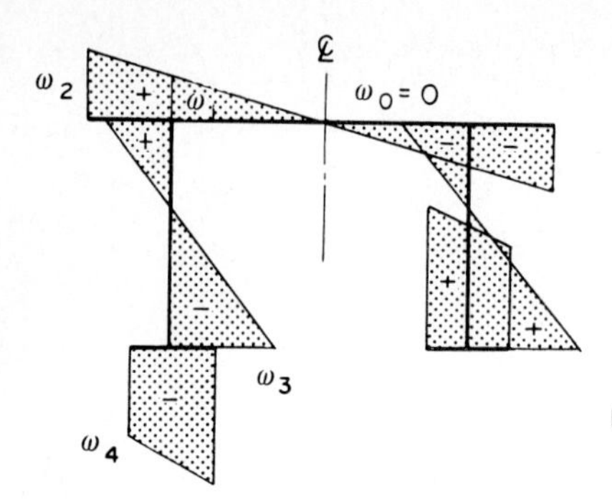

Figure 2.64 ω diagram.

S_ω can be obtained by integrating according to the formula $S_\omega = \int_0^s \omega t \, ds$. In this calculation, the warping function is always a linear function of the perimeter coordinate s, so that the numerical integration can be easily performed by multiplying Eq. (2.3.93) by the thickness t and taking Δ_x and Δ_y as explained above and on the basis of the definition of S_ω in Eq. (2.3.73). This results in

$$S_{\omega,i+1} = S_{\omega,i} + t \, \Delta_y \, (\omega_i - \tfrac{1}{2} x \, \Delta_y) \qquad (2.3.94a)$$

$$S_{\omega,i+1} = S_{\omega,i} + t \, \Delta_x \, (\omega_i + \tfrac{1}{2} y \, \Delta_x) \qquad (2.3.94b)$$

where $S_{\omega,i}$ is the static moment at point i. However, $S_{\omega,i} = 0$ at the end of the open section such as at points 2, 4, and 5. The numerical procedure can be summarized as shown in Table 2.12. Figure 2.65 shows the S_ω diagram, where the direction of S_ω is indicated by the arrows, since the shear flow q_ω is proportional to S_ω as seen from Eq. (2.3.67).

TABLE 2.12 Calculation of S_ω

	0	1	2	3	0 + 3
Segment	$S_{\omega,i}$	ω_i	$-\frac{1}{2} \times x \, \Delta y$ or $\frac{1}{2} \times y \, \Delta x$	$(1+2) \times t_i \, \Delta z$ or $\times \, t_i \, \Delta y$	$S_{\omega,i+1}$
$2\text{–}1_l$	0	ω_2	$\frac{1}{2} e_y[-(B-b)]$	$(1+2) \times t_u(B-b)$	$S_{1,l}$
$4\text{–}3_l$	0	ω_4	$\frac{1}{2}\frac{(e_y+h)-b_l}{2}$	$(1+2) \times \frac{t_l b_l}{2}$	$S_{3,l}$
$5\text{–}3_r$	0	ω_5	$\frac{1}{2}\frac{(e_y+h)b_l}{2}$	$(1+2) \times \frac{t_l b_l}{2}$	$S_{3,r}$
$3_o\text{–}1_u$	$S_{3,l} + S_{3,r}$	ω_3	$-\frac{1}{2} b(-h)$	$(1+2) \times t_w h$	$S_{1,u}$
$1_r\text{–}0$	$S_{1,l} + S_{1,u}$	ω_1	$\frac{1}{2} e_y(-b)$	$(1+2) \times t_u b$	S_o

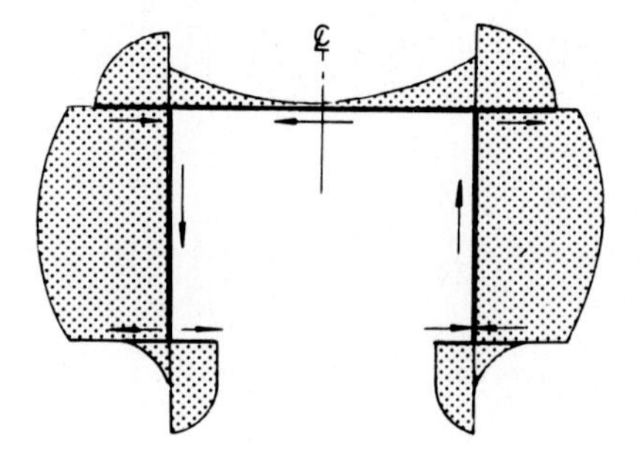

Figure 2.65 S_ω diagram.

Finally, the torsional warping constant I_ω can be determined by the integration of ω_i at the segments i through $(i+1)$, as follows:

$$I_\omega = \tfrac{1}{3} \sum_A (\omega_i^2 + \omega_i \omega_{i+1} + \omega_{i+1}^2) t_i \, \Delta_i \tag{2.3.95}$$

where Σ_A means the summation over all the segments and Δ_i is the absolute length of segment i. Table 2.13 shows these numerical procedures for the π girder.

***b.* Box girder.** For the sake of simplicity, let us now evaluate the cross-sectional quantities of a box girder with a single cell and one axis of symmetry, as shown in Fig. 2.66. The location of the shear center S can be determined by a procedure similar to that used for the π girder including the statically indeterminate shear flow, which is illustrated in Fig. 2.6 and described in Eqs. (2.2.25) and (2.2.26).

TABLE 2.13 Calculation of I_ω

Segment			1	2	1 × 2
$i \sim i+1$	ω_i	ω_{i+1}	$\omega_i^2 + \omega_i\omega_{i+1} + \omega_{i+1}^2$	$t_i \, \Delta_i$	
0–1	0	ω_1	ω_1^2	$t_u b$	W_1
1–2	ω_1	ω_2	$\omega_1^2 + \omega_1\omega_2 + \omega_2^2$	$t_u(B-b)$	W_2
1–3	ω_2	ω_3	$\omega_2^2 + \omega_2\omega_3 + \omega_3^2$	$t_\omega h$	W_3
3–4	ω_3	ω_4	$\omega_3^2 + \omega_3\omega_4 + \omega_4^2$	$t_l b$	W_4
3–5	ω_3	ω_5	$\omega_3^2 + \omega_3\omega_5 + \omega_5^2$	$t_l b$	W_5
	$I_\omega = \tfrac{2}{3} W$			Total	W

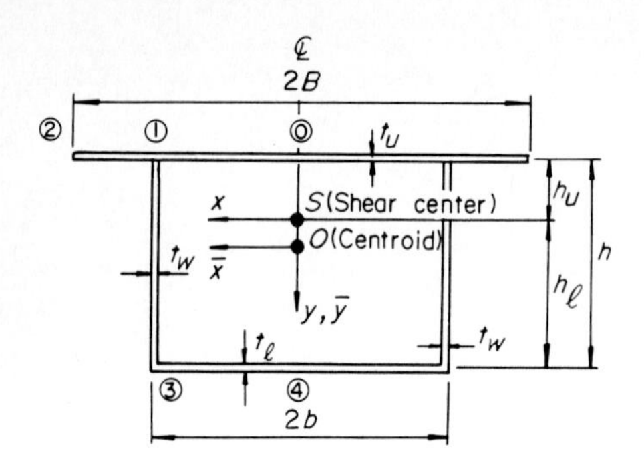

Figure 2.66 Cross section of box girder.

The shear center S of a box girder is generally located inside the box cell. Furthermore, the influence of the torsional function $\tilde{q}$ must be taken into account in evaluating the torsional warping function ω, through Eq. (2.3.65) in addition to Eq. (2.3.93). This results in

$$\int_i^{i+1} \frac{\tilde{q}}{t}\, ds = \frac{\tilde{q}}{t_i}\, \Delta_i \tag{2.3.96}$$

where

$$\tilde{q} = \frac{2F}{\oint ds/t} = \frac{2bh}{b/t_u + h/t_\omega + b/t_l} \tag{2.3.97}$$

Therefore, ω can be calculated in a manner similar to the procedure in Table 2.11 by taking into consideration the differences of the location of the shear center S and the additional terms due to the torsional function $\tilde{q}$. The calculations of S_ω and I_ω can be performed in almost the same way, as listed in Tables 2.12 and 2.13. However, note that the static moment $S_{\omega o} = 0$ for this box girder at a point 0. To make $S_{\omega o} = 0$, the slit is inserted at the point 0, and the statically indeterminate shear flow X is calculated by using Eq. (2.3.68). Note that the static moment is modified by Eq. (2.3.73). Thus, the diagrams for ω and S_ω can be plotted as in Fig. 2.67. Table 2.14 summarizes the cross-sectional quantities of a box girder with a single cell.

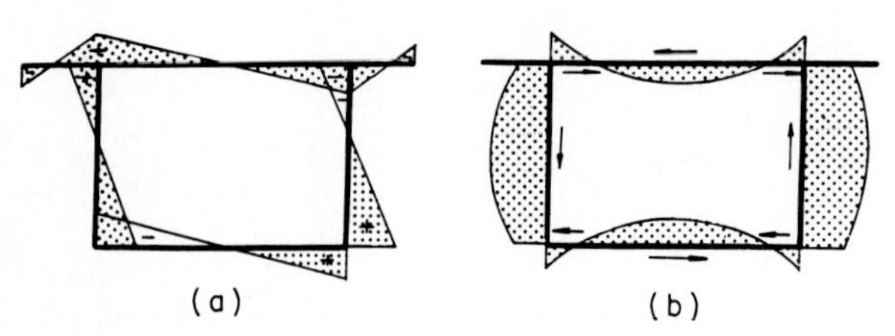

Figure 2.67 Diagram for (*a*) ω and (*b*) S_ω.

TABLE 2.14 Cross-Sectional Quantities of Box Girders with a Single Cell

Shear center
$$h_s = \frac{1}{\gamma_2^3 \bar{\beta}_u \bar{\beta}_l + 6}\left\{h(\bar{\beta}_l + 9) - \gamma_1\left[\frac{1}{2}(1 - 3\gamma_2^2)\varepsilon_u + \varepsilon_l + \frac{6}{\beta_l} + \frac{6}{\gamma_1^2}\right]\frac{\tilde{q}}{t_w}\right\}$$

Warping function
$$\omega_1 = \frac{b}{2}\left(\frac{\tilde{q}}{t_u} - h_s\right),\ \omega_2 = \frac{b}{2}\left(\frac{\tilde{q}}{t_u} - \gamma_s h_s\right),\ \omega_3 = \frac{b}{2}\left(-\frac{\tilde{q}}{t_l} + h - h_3\right)$$

Others are symmetrical with respect to the centerline of the cross section.

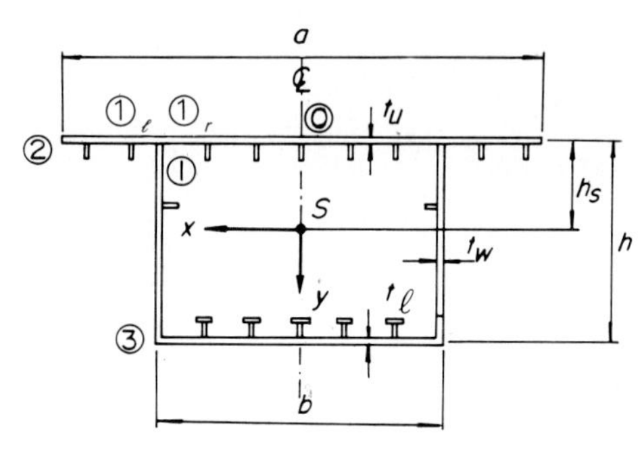

Static moment with respect to torsional warping function

$$S_{\omega,1} = (\omega_1 + \omega_2)\frac{a-b}{4}\bar{t}_u + \bar{S}_\omega \qquad S_{\omega,1l} = -(\omega_1 + \omega_2)\frac{a-b}{4}\bar{t}_u$$

$$S_{\omega,1r} = \bar{S}_\omega \quad S_{\omega,3} = -S_{\omega,1l} + (\omega_2 + \omega_3)\frac{h}{2}\bar{t}_w + \bar{S}_\omega \quad S_{\omega,2} = 0$$

where
$$\bar{S}_\omega = -\frac{1}{12}\left[3\bar{t}_u(\gamma_2 - 1)(\omega_1 + \omega_2)\left(\frac{1}{t_w} + \frac{\gamma_1}{2t_l}\right) + 3(\omega_1 + \omega_3)\frac{\bar{t}_w}{t_l} + 2(\omega_3 + 2\omega_1)\frac{\bar{t}_w}{t_w}\frac{1}{\gamma_1} + \gamma_1\left(\frac{\bar{t}_l}{t_l}\omega_3 - \frac{\bar{t}_u}{t_u}\omega_1\right)\right]\Psi$$

Others are asymmetric with respect to the centerline of the cross section.

Torsional warping function
$$I_\omega = \tfrac{1}{3}h\bar{t}_w[\omega_3^2(2 + \bar{\beta}_l) + 2\omega_1\omega_3 + \omega_1^2(2 + \bar{\beta}_u) + (\omega_1^2 + \omega_1\omega_2 + \omega_2^3)(\gamma_2 - 1)\bar{\beta}_u]$$

Equivalent thickness: $\bar{t}_u = t_u + \dfrac{iA_{ru}}{a} \qquad \bar{t}_l = t_l + \dfrac{jA_{rl}}{b} \qquad \bar{t}_w = t_w + \dfrac{kA_{rw}}{h}$

A_{ru}, A_{rl}, A_{rw} = Cross-sectional area of stiffening ribs for deck, bottom, and web plate, respectively

i, j, k = number of ribs for deck, bottom, and web plate, respectively

Parameter: $\beta_u = \dfrac{bt_u}{ht_w},\ \bar{\beta}_u = \dfrac{b\bar{t}_u}{h\bar{t}_w},\ \beta_l = \dfrac{bt_l}{ht_w},\ \bar{\beta}_l = \dfrac{b\bar{t}_l}{h\bar{t}_w},\ \gamma_1 = \dfrac{b}{h},\ \gamma_2 = \dfrac{a}{b},\ \varepsilon_u = \dfrac{\bar{t}_u t_w}{t_u \bar{t}_w},\ \varepsilon_l = \dfrac{\bar{t}_l t_w}{t_l \bar{t}_w}$

Pure torsional function: $\tilde{q} = \dfrac{2bh}{b/t_u + b/t_l + 2h/t_w}$

TABLE 2.15 Cross-Sectional Quantities for Various Cross Sections

Cross section	Cross-sectional quantities
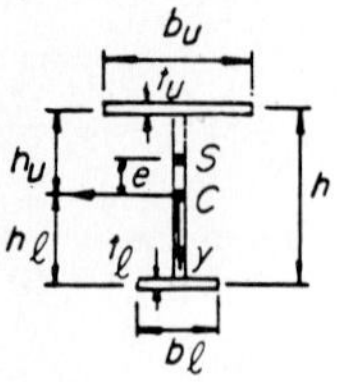	$e = \dfrac{I_u h_u - I_l h_l}{I_u + I_l}$ $I_\omega = I_u(h_u - e)^2 + I_l(h_l + e)^2$ where $I_u = \dfrac{t_u b_u^3}{12} \quad I_l = \dfrac{t_l h_l^3}{12}$
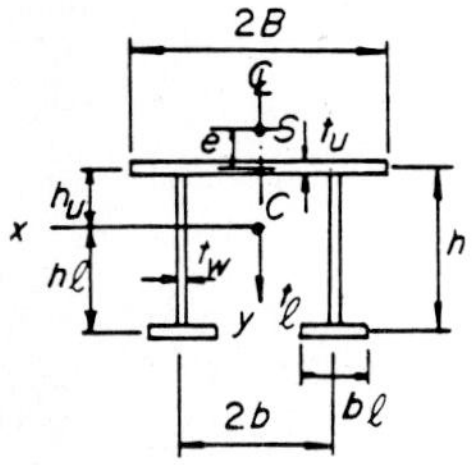	$e = 2h\dfrac{I_f}{I_y} - h_u\left(1 + b^2\dfrac{A}{I_y}\right)$ $I_\omega = b^2\left[I_x + h_u^2 A\left(1 - b^2\dfrac{A}{I_y}\right)\right] + 2h^2 I_f + 4b^2 h h_u A\dfrac{I_f}{I_y} - 4h^2\dfrac{I_f^2}{I_y}$ where $I_f = \dfrac{t_l b_l^3}{12}$ A = total cross-sectional area I_x, I_y = geometric moment of inertia with respect to x and y centroidal axes, respectively
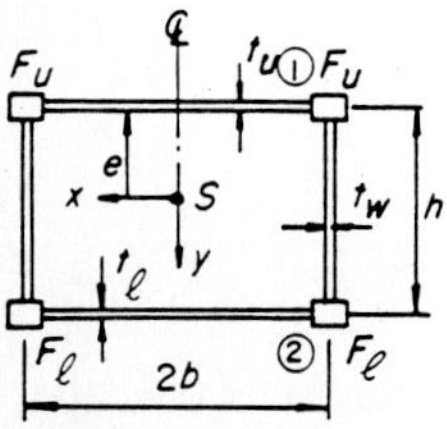	$e = \dfrac{\omega_1}{b} - \dfrac{\tilde{q}}{t_u}$ $I_\omega = \dfrac{h t_w}{3}[\omega_1^2(2 + \beta_u) + 2\omega_1\omega_2 + \omega_2^2(2 + \beta_l)]$ $K = \dfrac{4b^2h^2}{b/t_u + h/t_w + b/t_l}$ where $\beta_u = \dfrac{2bt_u + 6F_u}{t_w h} \quad \beta_l = \dfrac{2bt_l + F_l}{t_w h}$ Torsional function: $\tilde{q} = 2bt_w\dfrac{\varepsilon}{\beta_u + \beta_l + \varepsilon} \quad \varepsilon = 2\dfrac{t_u t_l}{t_w^2}$ Torsional warping function: $\omega_1 = -\Delta\omega\dfrac{\beta_l + 3}{\beta_u + \beta_l + 6}$ $\omega_2 = \Delta\omega\dfrac{\beta_u + 3}{\beta_u + \beta_l + 6}$ where $\Delta\omega = bh + \tilde{q}\dfrac{h}{t_w}$

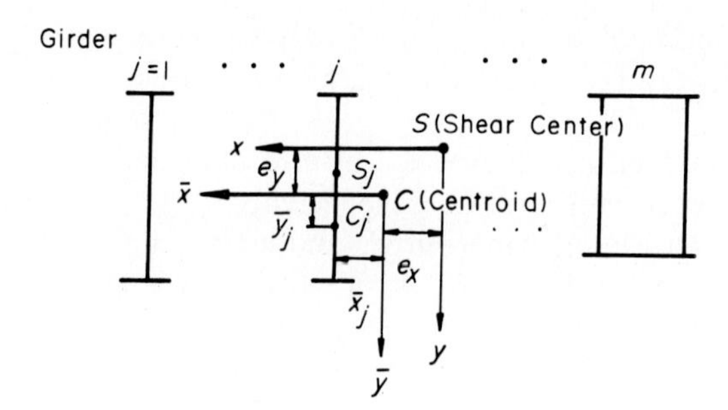

Figure 2.68 Multiple-plate girder.

c. **Approximate formulas.** Location of the center S and the torsional warping constant I_ω are too complex to evaluate exactly for a girder with a variable cross section, but an approximate formula, derived by Resinger,[4] can be applied for design purposes. These formulas are summarized in Table 2.15.

By expanding the formula for the I girder, the location of the shear center S and the warping constant I_ω of a multiple-I or box girder with girders, $j = 1, 2, \ldots, m$, connected by rigid sway and lateral bracings, as illustrated in Fig. 2.68, can be estimated as follows:
Location of the shear center:

$$e_x = \frac{\sum_{j=1}^{m} I_{\bar{x},j}\bar{x}_j}{\sum_{j=1}^{m} I_{\bar{x},j}} \tag{2.3.98a}$$

$$e_y = \frac{\sum_{j=1}^{m} I_{\bar{y},j}\bar{y}_i}{\sum_{j=1}^{m} I_{\bar{y},j}} \tag{2.3.98b}$$

Torsional warping function:

$$\omega_S = \omega_{S,j} \begin{cases} -x_j y & \text{in direction of } y \text{ axis} \quad (2.3.99a) \\ +y_j x & \text{in direction of } x \text{ axis} \quad (2.3.99b) \end{cases}$$

Pure torsional constant:

$$K = \sum_{j=1}^{m} K_j \tag{2.3.100}$$

Torsional warping constant:

$$I_\omega \cong \sum_{j=1}^{m} (I_{\omega,j} + x_j^2 I_{\bar{x},j} + y_j^2 I_{\bar{y},j}) \tag{2.3.101}$$

where $I_{\bar{x},j}$, $I_{\bar{y},j}$ = geometric moment of inertia with respect to horizontal and vertical axis of centroid C_j, jth girder, respectively

$\bar{x}_j$, $\bar{y}_j$ = horizontal and vertical distance from global centroid C_n to individual centroid C_j, respectively

e_x, e_y = eccentricity of global centroid C_n and shear center S

ω_S = torsional warping function with respect to global shear center S

$\omega_{S,j}$ = torsional warping function with respect to shear center S_j for jth girder

y_j, z_j = horizontal and vertical distances from global shear center to individual shear center S_j, respectively

K_j = pure torsional constant for jth girder

$I_{\omega,j}$ = torsional warping constant for jth girder

(4) Stress resultants and displacement for torsional warping. In the case of a uniformly distributed torque m_T, the fundamental equation for determining the torsional angle $\theta(z)$ can be taken from Eqs. (2.3.61) and (2.3.81) as

$$EI_\omega \frac{d^4\theta}{dz^4} - GK\frac{d^2\theta}{dz^2} = m_T \tag{2.3.102}$$

or

$$\frac{d^4\theta}{dz^4} - \alpha^2 \frac{d^2\theta}{dz^2} = \frac{m_T}{EI_\omega} \tag{2.3.103}$$

where the parameter α is given by

$$\alpha = \sqrt{\frac{GK}{EI_\omega}} \tag{2.3.104}$$

The solution of Eq. (2.3.102) is

$$\theta = A \sinh \alpha z + B \cosh \alpha z + Cz + D + \frac{m_T}{2GK} z(l - z) \tag{2.3.105}$$

The four integration constants A, B, C, and D can be determined by the boundary conditions imposed on the beam. For instance, these boundary conditions can be written for a beam, shown in Fig. 2.54, as follows:

Geometric boundary conditions (that is, torsional angle θ)

$$\theta = 0 \qquad \text{for } z = 0 \text{ and } z = l \tag{2.3.106a}$$

Static boundary conditions (that is, $\sigma_\omega = EI_\omega\, d^2\theta/dz^2$)

$$\frac{d^2\theta}{dz^2} = 0 \qquad \text{for } z = 0 \text{ and } z = l \tag{2.3.106b}$$

Once the unknown integration constants have been determined, the following stress resultants can be obtained:
Pure torsional moment,

$$T_s = GK\frac{d\theta}{dz} \qquad \text{kN·cm} \qquad (2.3.107a)$$

Torsional warping moment,

$$M_\omega = EI_\omega \frac{d^2\theta}{dz^2} \qquad \text{kN·cm}^2 \qquad (2.3.107b)$$

Secondary torsional moment,

$$T_\omega = -EI_\omega \frac{d^3\theta}{dz^3} \qquad \text{kN·cm} \qquad (2.3.107c)$$

a. **Simple beam.**[3] The solution for the torsional angle $\theta(z)$ and corresponding stress resultants T_s, M_ω, and T_ω for a simple beam subjected to uniformly distributed torque m_T and concentrated torque M_T can be summarized as shown in Table 2.16.

b. **Continuous beam.**[3] Let us now estimate the torsional angle and stress resultants of a continuous beam having constant cross-sectional quantities for each span, as illustrated in Fig. 2.69*a*. This beam is subdivided into simple beams by introducing hinges at each support, as in Fig. 2.69*b*. The applications of the statically indeterminate torsional warping moments $\ldots, X_{r-1}, X_r, \ldots$ for these intermediate supports $\ldots, r-1, r, \ldots$ give the following equation of the torsional angle θ_r for rth span of a continuous beam:

$$\theta_r = \theta_{o,r} + \frac{1}{GK_r}\left\{X_{r-1}\left[\frac{\sinh \alpha_r(l_r - z_r)}{\sinh \alpha_r l_r} - \frac{l_r - z_r}{l_r}\right] + X_r\left(\frac{\sinh \alpha_r z_r}{\sinh \alpha_r l_r} - \frac{z_r}{l_r}\right)\right\} \qquad (2.3.108)$$

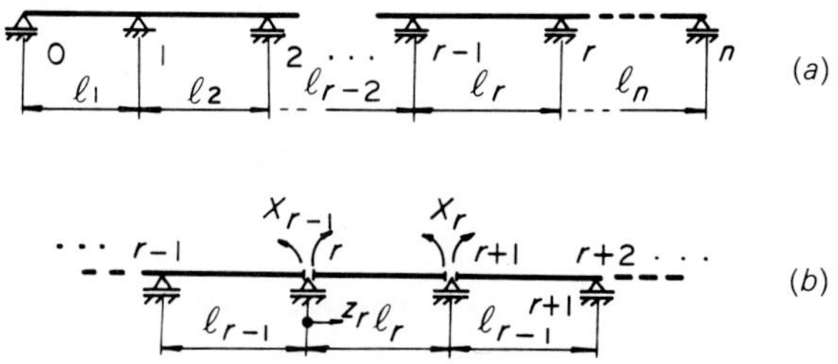

Figure 2.69 Torsional warping of (*a*) continuous beam and (*b*) statically indeterminate moments.

TABLE 2.16 Torsional Angle $\theta(z)$ and Corresponding Stress Resultants of a Simple Beam

Loading condition	Item	Torsional angle and stress resultants
m_T, z, ℓ	θ	$\theta(z) = \dfrac{m_T}{GK\alpha^2}\left[\dfrac{\sinh \alpha z + \sinh \alpha(l-z)}{\sinh \alpha l} - 1 + \dfrac{\alpha^2}{2} z(l-z)\right]$
	T_s	$T_s(z) = \dfrac{m_T}{\alpha}\left[\dfrac{\cosh \alpha z - \cosh \alpha(l-z)}{\sinh \alpha l} + \alpha\left(\dfrac{l}{2} - z\right)\right]$
	M_ω	$M_\omega(z) = \dfrac{m_T}{\alpha^2}\left[\dfrac{\sinh \alpha z + \sinh \alpha(l-z)}{\sinh \alpha l} - 1\right]$
	T_ω	$T_\omega(z) = -\dfrac{m_T}{\alpha}\left[\dfrac{\cosh \alpha z - \cosh \alpha(l-z)}{\sinh \alpha l}\right]$
M_T, a, z, ℓ	θ	$\theta(z) = -\dfrac{M_T}{GK\alpha}\left[\dfrac{\sinh \alpha(l-a)}{\sinh \alpha l}\sinh \alpha z - \dfrac{l-a}{l}\alpha z\right] \quad 0 \leqq z \leqq a$
		$\theta(z) = -\dfrac{M_T}{GK\alpha}\left[\dfrac{\sinh \alpha a}{\sinh \alpha l}\sinh \alpha(l-z) - \dfrac{l-z}{l}\alpha a\right] \quad a \leqq z \leqq l$
	T_s	$T_s(z) = -M_T\left[\dfrac{\sinh \alpha(l-a)}{\sinh \alpha l}\cosh \alpha z - \dfrac{l-a}{l}\right] \quad 0 \leqq z \leqq a$
		$T_s(z) = M_T\left[\dfrac{\sinh \alpha a}{\sinh \alpha l}\cosh \alpha(l-z) - \dfrac{a}{l}\right] \quad a \leqq z \leqq l$
	M_ω	$M_\omega(z) = -\dfrac{M_T}{\alpha}\dfrac{\sinh \alpha(l-a)}{\sinh \alpha l}\sinh \alpha z \quad 0 \leqq z \leqq a$
		$M_\omega(z) = -\dfrac{M_T}{\alpha}\dfrac{\sinh \alpha a}{\sinh \alpha l}\sinh \alpha(l-z) \quad a \leqq z \leqq l$
	T_ω	$T_\omega(z) = M_T\dfrac{\sinh \alpha(l-a)}{\sinh \alpha l}\cosh \alpha z \quad 0 \leqq z \leqq a$
		$T_\omega(z) = -M_T\dfrac{\sinh \alpha a}{\sinh \alpha l}\cosh \alpha(l-z) \quad a \leqq z \leqq l$

where $\theta_{o,r}$ is the torsional angle for rth span if we regard the continuous beam as a simple beam.

The derivative of the above equation with respect to z_r leads to the change of torsional angle $d\theta_r/dz_r$ as

$$\frac{d\theta_r}{dz_r} = \frac{d\theta_{or}}{dz_r} + \frac{1}{GK_r}\left\{X_{r-1}\left[-\frac{\alpha_r \cosh \alpha_r(l_r - z_r)}{\sinh \alpha_r l_r} + \frac{1}{l_r}\right] + X_r\left(\frac{\alpha_r \cosh \alpha_r z_r}{\sinh \alpha_r l_r} - \frac{1}{l_r}\right)\right\} \quad (2.3.109)$$

This change of torsional angle must satisfy the continuity conditions at the right end of the rth span and at the left end of the $(r+1)$th span. Thus

$$\left[\frac{d\theta_r}{dz_r}\right]_{z_r = l_r} - \left[\frac{d\theta_{r+1}}{dz_{r+1}}\right]_{z_r+1} = 0 \quad (2.3.110)$$

By using the above condition, the statically indeterminate torsional warping moments X_r $(r = 1, 2, \ldots, n-1)$ can be determined by the following set of three moment equations for torsional warping:

$$\begin{aligned} & X_{r-1}\left(1 - \frac{\alpha_r l_r}{\sinh \alpha_r l_r}\right) \\ & + X_r\left[\alpha_r l_r \coth \alpha_r l_r - 1 + \frac{K_r}{K_{r+1}}\frac{l_r}{l_{r+1}}(\alpha_{r+1} l_r + \coth \alpha_{r+1} l_{r+1} - 1)\right] \\ & + X_{r+1}\frac{K_r}{K_{r+1}}\frac{l_r}{l_{r+1}}\left(1 - \frac{\alpha_{r+1} l_{r+1}}{\sinh \alpha_{r+1} l_{r+1}}\right) \\ & = l_r\left(\frac{K_r}{K_{r+1}}[T_{so,r+1}]_{z_r+1=0} - [T_{so,r}]_{z_r = l_r}\right) \end{aligned} \quad (2.3.111)$$

where $[T_{so,r+1}]_{r+1=0}$ and $[T_{so,r}]_{z_r = l_r}$ are the pure torsional moments at the left and right ends of $(r+1)$th and rth spans, respectively.

Final expressions for the stress resultants in the rth span of a continuous beam can be written as

$$T_{s,r} = T_{so,r} + \frac{1}{l_r}X_{r-1}\left[1 - \frac{\alpha_r l_r \cosh \alpha_r(l_r - z_r)}{\sinh \alpha_r l_r}\right] + X_r\left(\alpha_r l_r \frac{\cosh \alpha_r z_r}{\sinh \alpha_r l_r} - 1\right) \quad (2.3.112a)$$

$$M_{\omega,r} = M_{\omega o,r} + X_{r-1}\frac{\sinh \alpha_r l_r(l_r - z_r)}{\sinh \alpha_r l_r} + X_r\frac{\sinh \alpha_r z_r}{\sinh \alpha_r l_r} \quad (2.3.112b)$$

$$T_{\omega,r} = T_{\omega o,r} + X_{r-1}\frac{\alpha_r \cosh \alpha_r(l_r - z_r)}{\sinh \alpha_r l_r} - X_r\frac{\alpha_r \cosh \alpha_r z_r}{\sinh \alpha_r l_r} \quad (2.3.112c)$$

where $T_{so,r}$, $M_{\omega o,r}$, and $T_{\omega o,r}$ are the stress resultants for the span given the continuous beam as a simple beam. These stress resultants can be determined easily from Table 2.16.

(5) Approximate method for determining the torsional warping stress in a box girder[25,27]

***a.* Variations of stress resultants due to the torsional warping parameter.** The stress resultants due to torsional warping will be affected by the parameter α in Eq. (2.3.104) or the following torsional warping constant κ:

$$\kappa = \alpha l = l\sqrt{\frac{GK}{EI_\omega}} \tag{2.3.113}$$

Now, let us look into the variations in stress resultants of a simply supported beam and a continuous beam, from Table 2.16 and the method described in Sec. 2.3.2, when we include torsional warping.

Figures 2.70 and 2.71, respectively, show the variations of stress

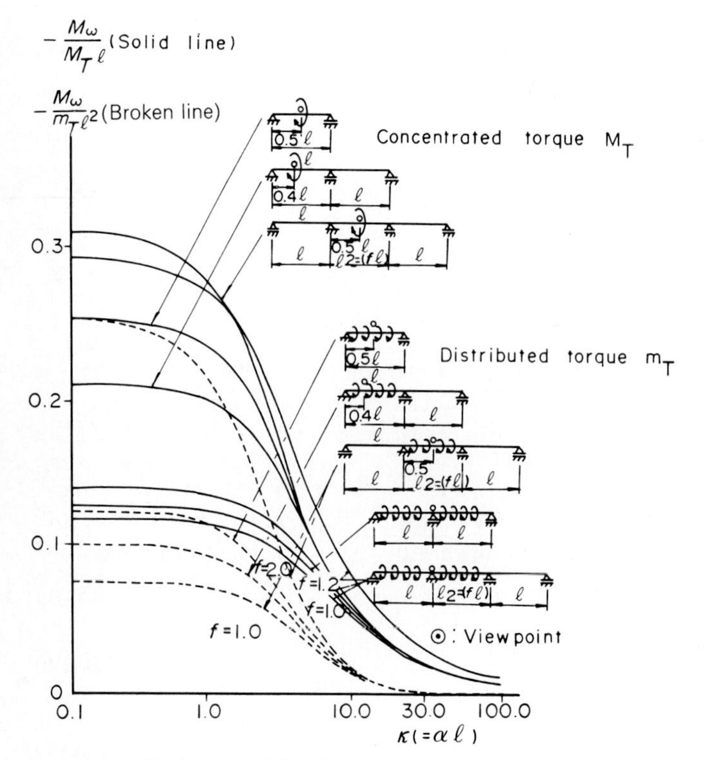

Figure 2.70 Variations of M_ω due to κ.

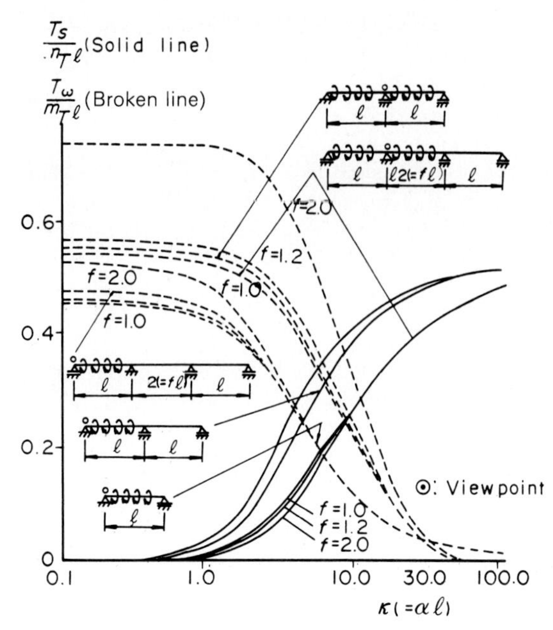

Figure 2.71 Variations of T_s and T_ω due to κ.

resultants M_ω, T_s, and T_ω due to the parameter.[26] Examinations of these figures show that

$$T_\omega \cong 0 \qquad \text{for } \kappa \leqq 0.4 \tag{2.3.114}$$

and

$$T_s \cong T \qquad \text{for } \kappa \geqq 30 \tag{2.3.115}$$

For $\kappa < 0.4$, $GK = 0$, and so Eq. (2.3.102) can be written as

$$\frac{d^4\theta}{dz^4} = \frac{m_T}{EI_\omega} \tag{2.3.116}$$

This equation is the same form as the fundamental equation of bending, provided that the torsional angle θ, distributed torque m_T, and torsional warping rigidity EI_ω are substituted for deflection w, distributed load q, and flexural rigidity EI_x, respectively. Thus

$$M_{\omega,\kappa=0} = -EI_x \frac{d^2w}{dz^2} = M_x \tag{2.3.117a}$$

and

$$T_{\kappa=0} = T_s + T_\omega = -EI_x \frac{d^3w}{dz^3} = Q_y \tag{2.3.117b}$$

But Eq. (2.3.115) suggests the applicability of the following equation:

$$\frac{d^2\theta}{dz^2} = -\frac{m_T}{GK} \tag{2.3.118}$$

However, there still remain the stress resultants M_ω and T_ω, as is seen from Figs. 2.70 and 2.71 even in a case where $\kappa > 30$. Therefore, an approximate method based on Eq. (2.3.118) has been developed for a condition where

$$\kappa \geqq 30 \tag{2.3.119}$$

This condition has, in fact, been investigated through a parametric survey[18] of plate girder bridges with the monobox girder, as in Fig. 2.72.

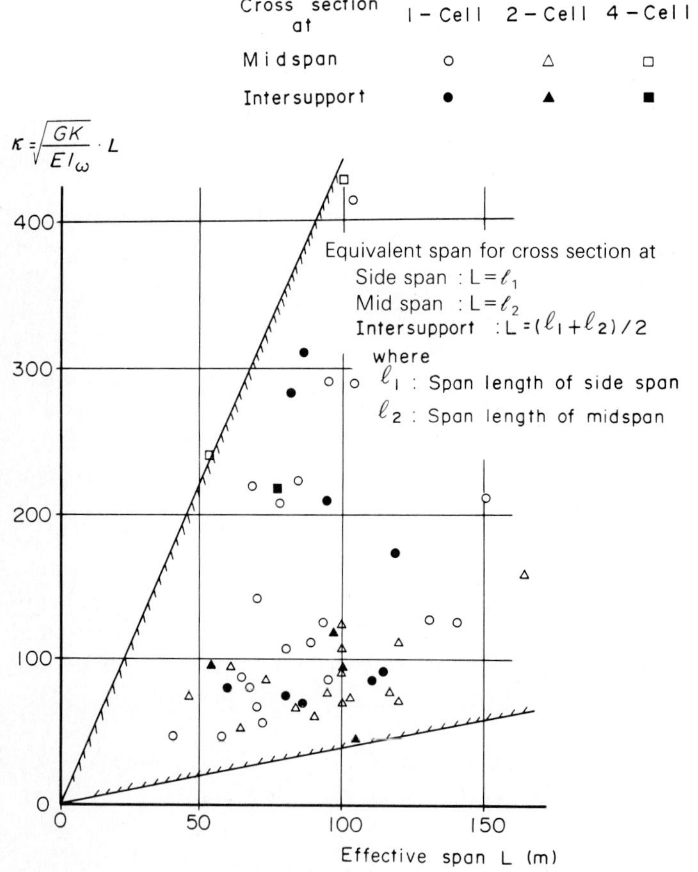

Figure 2.72 Relationships between κ and effective span L.

***b.* Stress resultants due to torsional warping of the box girder.**[27] To obtain an approximate formula for stress resultants in the case where $\kappa = 30$, Fig. 2.73 shows the distributions of stress resultants in a box girder subjected to concentrated torque M_T. It is obvious that the torsional moment $T = T_s + T_\omega$ will be distributed as shown in Fig. 2.73*a*, which is already explained in Table 2.10.

The bimoment has its peak $M_{\omega,\max}$ at the loading point of M_T, and this value decreases rapidly away from the loading point, as shown in Fig. 2.73*b*. To obtain the distribution of the torsional moment ΔT,

$$\Delta T = M_T \tag{2.3.120}$$

is produced at a point $a = l/2$. The corresponding bimoment $M_{\omega,\Delta T}$ at point z' apart from this point can be evaluated for $\kappa \geqq 30$ as follows:

$$\begin{aligned} M_{\omega,\Delta T} &= -\frac{\Delta T}{\alpha}\frac{\sinh(\alpha l/2)}{\sinh \alpha l}\sinh \alpha\left(\frac{l}{2} - z'\right) \\ &\cong -\frac{\Delta T}{\alpha}\frac{e^{\alpha l/2}}{e^{\alpha l}}\frac{1}{2}e^{\alpha(l/2 - z')} \\ &= -\frac{\Delta T}{2\alpha}e^{-\alpha z'} \end{aligned} \tag{2.3.121}$$

In the general cases where $a \neq l/2$ and $a \neq 0$, the bimoment can be plotted as shown in Fig. 2.73*c*, and the maximum value $M_{\omega,\max,\Delta T}$ can approximately be written as

$$M_{\omega,\max,\Delta T} = -\frac{\Delta T}{2\alpha} \tag{2.3.122}$$

Thus, the generalized formula for the bimoment $M_{\omega,\Delta T}$ can be obtained from

$$M_{\omega,\Delta T} = M_{\omega,\max,\Delta T}\, e^{-\alpha z'} \tag{2.3.123}$$

Also, the secondary torsional moment T_ω can be determined from Table 2.16 as follows:

$$T_{\omega,\Delta T} = T_{\omega,\max(\text{or min}),\Delta T}\, e^{-\alpha z'} \tag{2.3.124}$$

where

$$T_{\omega,\max(\text{or min}),\Delta T} = \pm\frac{\Delta T}{2} \tag{2.3.125}$$

and the distribution of $T_{\omega,\Delta T}$ can be illustrated, as in Fig. 2.73*d*.

Furthermore, the pure torsional moment T_s is given by

$$T_s = T - T_{\omega,\Delta T} = T \mp \frac{\Delta T}{2}e^{-\alpha z'} \tag{2.3.126}$$

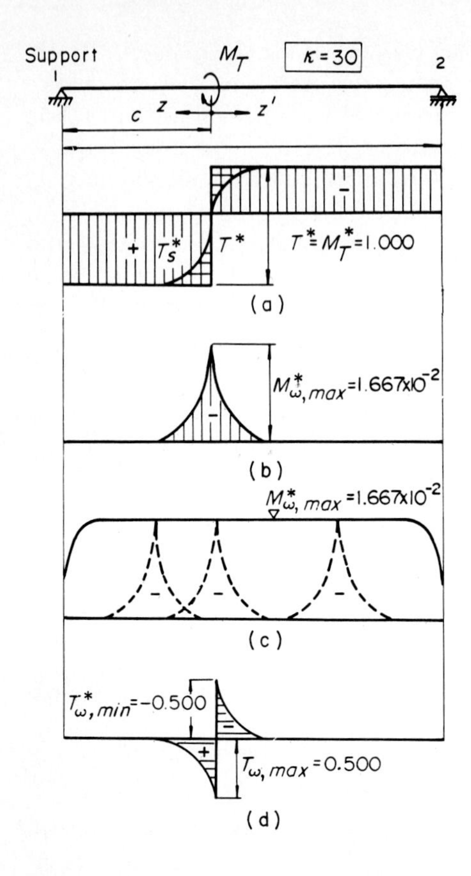

Figure 2.73 Distribution of stress resultants due to torsional warping (concentrated torque M_T): (*a*) Torsional moment diagram $T(=T/M_T)$. (*b*) Bimoment diagram with $M^*_{\omega,\Delta T}$ $(=M_\omega/M_T l)$ and $M^*_{\omega,\max} = 1.667 \times 10^{-2}$. (*c*) Maximum bimoment $M_{\omega,\max,\Delta T}$. (*d*) Secondary torsional moment diagram $T^*_{\omega,\Delta T}\,(=T_\omega/M_T)$.

where the sign of ΔT is positive for $T > 0$ and negative for $T < 0$. The distribution of pure torsional moment T_s can also be plotted, as shown in Fig. 2.73*a*.

In a similar manner, the distributions of the stress resultants for the uniformly distributed torque m_T can be investigated from Table 2.16 and are plotted in Fig. 2.74. From this figure, the pure torsional moment T_s is almost the same as the torsional moment T. Thus

$$T_{s,m_T} \cong T_{m_T} \tag{2.3.127a}$$

$$T_{\omega,m_T} \cong 0 \tag{2.3.127b}$$

and the bimoment can be approximated from Table 2.16 for $z = l/2$ as

$$M_{\omega,m_T} = -\frac{m_T}{\alpha^2}\left(1 - \frac{1}{\cosh(\alpha l/2)}\right) \cong -\frac{m_T}{\alpha^2} \tag{2.3.128}$$

except at the edges of the beam.

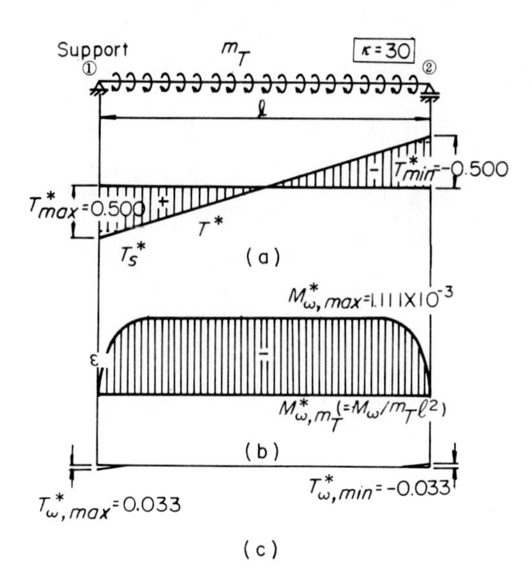

Figure 2.74 Distribution of stress resultants due to torsional warping (uniformly distributed torque m_T): (*a*) torsional moment diagram T^*_{m1} $(=T/m_T l)$, (*b*) bimoment diagram, (*c*) secondary torsional moment diagram T^*_{ω,m_T} $(=T_\omega/m_T l)$.

For example, the above formulas can be applied directly to the analysis of stress resultants for the torsional warping of a continuous beam subjected to a distributed torque m_T along the first and second span, as illustrated in Fig. 2.75. The pure torsional moment T_s can be plotted easily on the basis of pure torsional theory, as in Fig. 2.75*a*. Then the bimoment M_ω can be estimated by Eq. (2.3.128) for the middle parts of the first and second spans. At intermediate supports 1 and 2, there occurs the step of pure torsional moment ΔT_1 and ΔT_2, respectively, so that the bimoments $M_{\omega,\Delta T}$ due to these concentrated torques ΔT_1 and ΔT_2 can be obtained directly from Eqs. (2.2.122) and (2.2.123). Thus, the bimoment diagram can be plotted as shown in Fig. 2.75*b*. Finally, the secondary torsional moment diagram T_ω can be plotted from Eqs. (2.3.125) and (2.3.127*b*), as illustrated in Fig. 2.75*c*.

The errors between this approximate method and the exact solution were checked by various numerical calculations and were within a few percentage points. Therefore, the above approximate method can be utilized for the torsional warping stress analysis of a box girder, as detailed in the next section.

c. Approximate formula for torsional warping stress in a box girder[27]

(i) Normal stress due to torsional warping. Let us now derive an approximate formula for normal stress due to the steps of pure torsional moment ΔT, such as a concentrated torque M_T in a simple box girder or those ΔT_1, ΔT_2, . . . in intermediate supports of a continuous box girder.

In general, the normal stress due to torsional warping has the

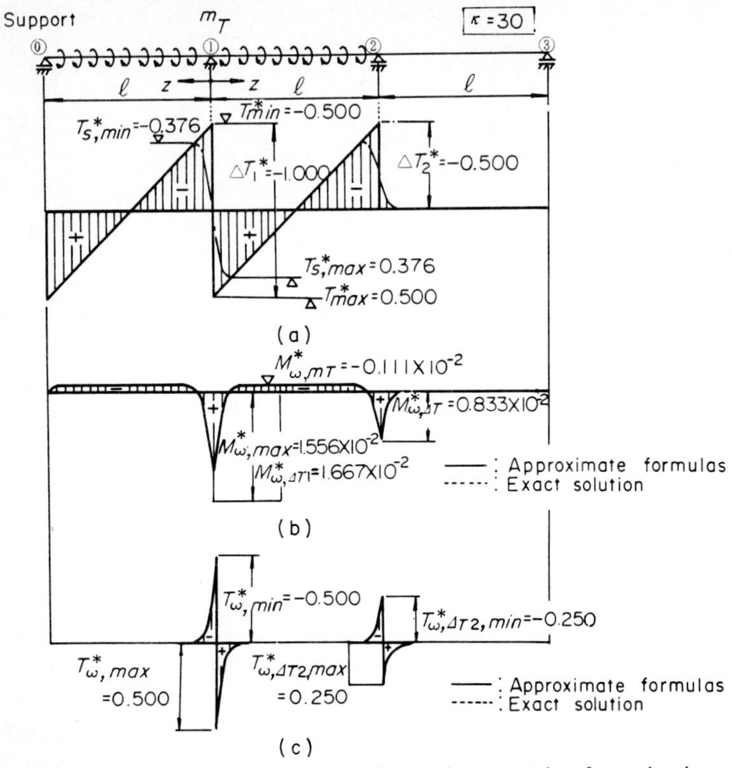

Figure 2.75 Distribution of stress resultants due to torsional warping in a continuous beam: (*a*) torsional moment diagram T^* $(=T/m_T l)$, (*b*) bimoment diagram M_ω^* $(=M_\omega/m_T l^2)$, and (c) secondary torsional moment diagram T_ω^* $(=T_\omega/m_T l)$.

maximum value at a junction point 3 of the bottom plate and the web plate, as shown in Fig. 2.76. So let us direct our attention to this stress $\sigma_{\omega,3}$, and we denote the corresponding warping function ω_3 as

$$\omega_3 = \lambda \frac{bh}{4} \tag{2.3.129a}$$

or

$$\lambda = \frac{4\omega_3}{bh} \tag{2.3.129b}$$

where λ is a nondimensional parameter that depends on the cross-sectional shape and dimensions of a box girder, such as a/b, b/h, t_u/t_ω, and t_l/t_ω etc.

The stress $\sigma_{\omega,3,\Delta T}$ can be expressed from Eqs. (2.3.75), (2.3.122),

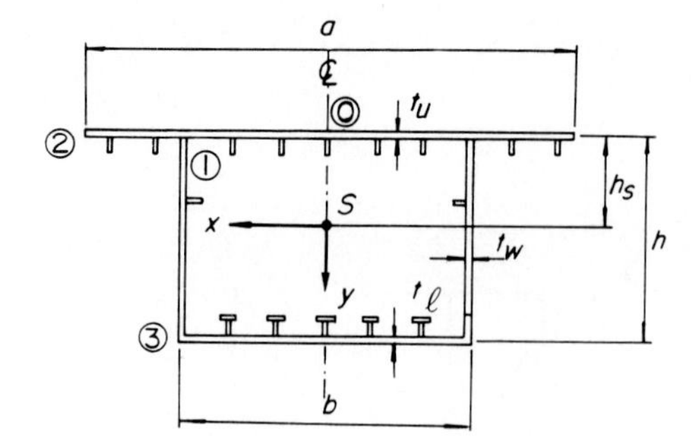

Figure 2.76 Cross section of box girder.

(2.3.123), (2.3.104), and (2.3.129) as

$$\sigma_{\omega,3,\Delta T} = \frac{M_{\omega,\Delta T}}{I_\omega}\,\omega_3 = -\frac{\Delta T}{2\alpha}\,e^{-\alpha z'}\,\frac{1}{I_\omega}\,\frac{\lambda bh}{4}$$

$$= -\frac{\Delta T}{8I_\omega}\,\frac{\lambda bh}{\sqrt{GK/(EI_\omega)}}\,e^{-\sqrt{[GK/(EI_\omega)]}z'} \tag{2.3.130}$$

By introducing two dimensionless parameters

$$\eta = \lambda b\sqrt{\frac{K}{I_\omega}} \tag{2.3.131}$$

and

$$\xi = \sqrt{\frac{G}{E}\,\frac{\eta}{\lambda}} \tag{2.3.132}$$

and the material properties for steel

$$\sqrt{\frac{G}{E}} = \sqrt{\frac{8.1\times 10^5}{2.1\times 10^6}} \cong 0.62 \tag{2.3.133}$$

we can simplify Eq. (2.3.130) as follows:

$$\sigma_{\omega,3,\Delta T} \cong -\frac{\eta h}{5K}\,\Delta T\,e^{-\xi z'/b} \tag{2.3.134}$$

where the values of parameters η and ξ can be plotted as shown in Figs. 2.77 and 2.78 from a parametric survey of box-girder bridges.[18] Examinations of these figures show that the values of η can be categorized into pattern $A(\omega_3 > 0)$ and pattern $B(\omega_3 < 0)$, and they reduce to

$$\eta \cong \begin{cases} 3.5 & \text{for } \dfrac{b}{h} < 2.3 \quad (2.3.135a) \\[2ex] -5.0 & \text{for } \dfrac{b}{h} > 1.8 \quad (2.3.135b) \end{cases}$$

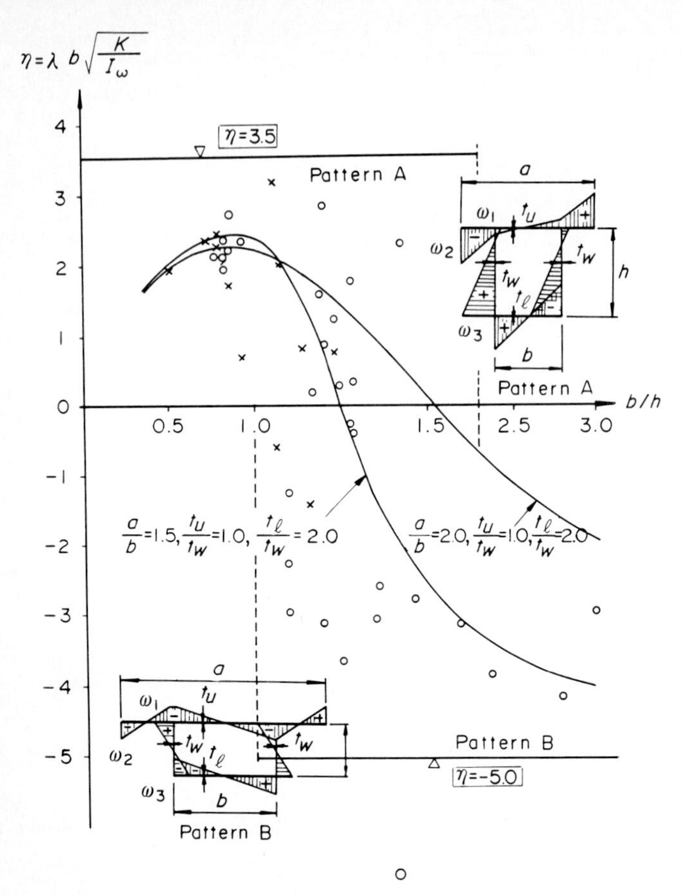

Figure 2.77 Distribution of parameter η.

Corresponding values of ξ give

$$\xi \cong \begin{cases} 2.0 & \text{for } \dfrac{b}{h} < 2.3 \qquad (2.3.136a) \\ 4.0 & \text{for } \dfrac{b}{h} > 1.8 \qquad (2.3.136b) \end{cases}$$

In a similar manner, the torsional warping normal stress due to a uniformly distributed torque m_T can be estimated as

$$\sigma_{\omega,3,m_T} = \frac{M_{\omega.m_T}}{I_\omega}\omega_3 = -\frac{E}{4G}\lambda\frac{bh}{K}m_T$$

$$\cong -\frac{3}{5}\frac{\lambda h}{K}bm_T \tag{2.3.137}$$

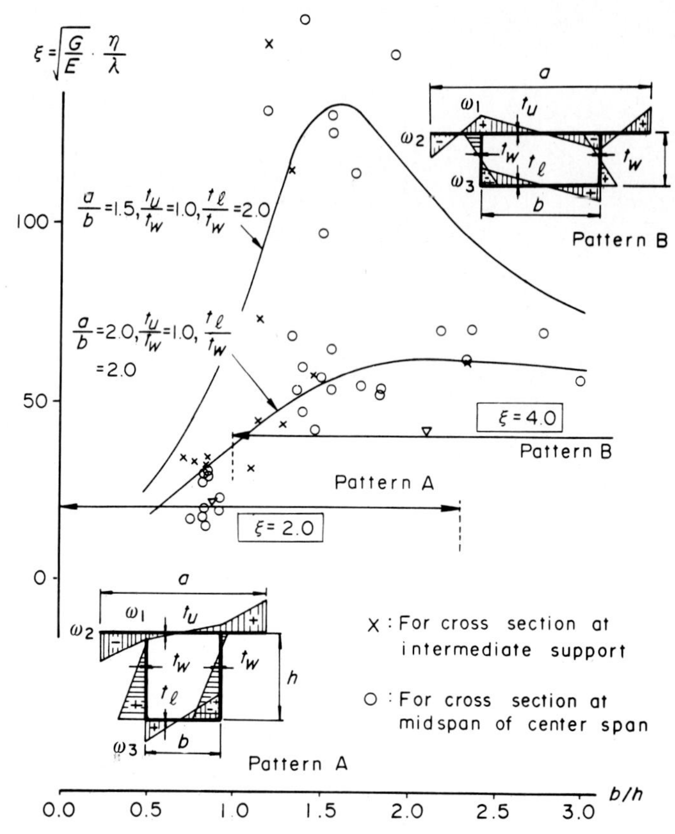

Figure 2.78 Distribution of parameter ξ.

The parameter λ, given by Eq. (2.3.129*b*), can also be plotted as shown in Fig. 2.79 through a parametric survey of box-girder bridges[18] and can be defined as

$$\lambda = \begin{cases} 0.8 & \text{for } \dfrac{b}{h} < 2.3 \qquad (2.3.138a) \\ -0.6 & \text{for } \dfrac{b}{h} > 1.8 \qquad (2.3.138b) \end{cases}$$

(ii) Shearing stress due to torsional warping. The shearing stress due to pure torsion τ_s can be expressed through Eqs. (2.3.50) and (2.3.126) as

$$\tau_s = \frac{T_s}{2Ft} = \frac{1}{2Ft}\left(T \mp \frac{\Delta T}{2} e^{-\xi z'/b}\right) \qquad (2.3.139a)$$

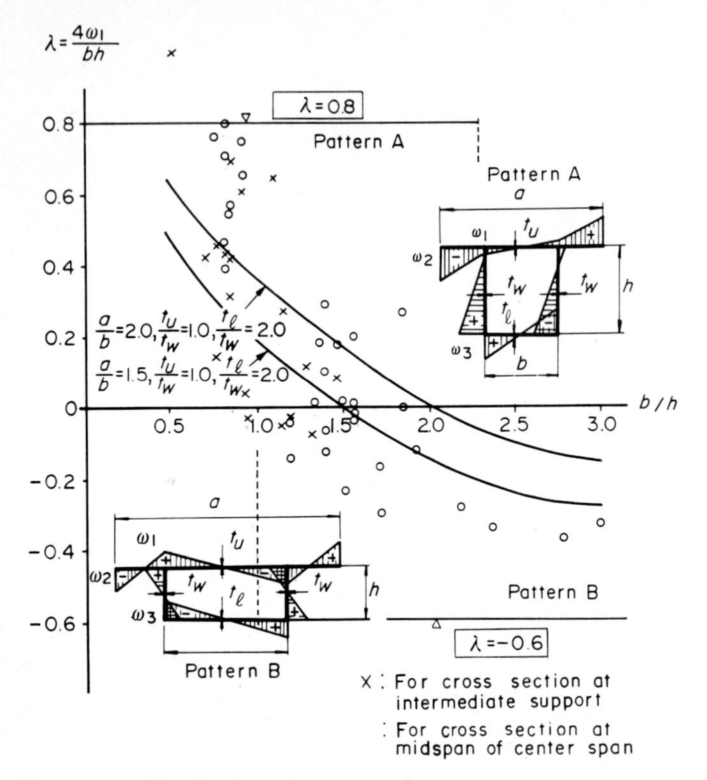

Figure 2.79 Distribution of parameter λ.

However, the shearing stress τ_ω due to torsional warping can also be evaluated by Eqs. (2.3.74), (2.3.124) and (2.3.125) as

$$\tau_\omega = \frac{S_\omega}{I_\omega t} T_\omega = \frac{\chi}{2Ft} T_\omega = \pm \frac{\chi}{2Ft} \frac{\Delta T}{2} e^{-\xi z'/b} \qquad (2.3.139b)$$

where a new dimensionless parameter χ is

$$\chi = \frac{2FS_\omega}{I_\omega} \qquad (2.3.140)$$

Accordingly, the sum of Eqs. (2.2.139*a*) and (2.2.139*b*) gives

$$\tau_{s+\omega} = \tau_s + \tau_\omega = \frac{1}{2Ft}\left[T \pm (\chi - 1)\frac{\Delta T}{2} e^{-\xi z'/b}\right] \qquad (2.3.141)$$

Thus, the induced total shearing stress $\tau_{s+\omega}$ can easily be estimated on the basis of pure torsional theory, provided that the value of parameter χ is known. To investigate this value, the distribution

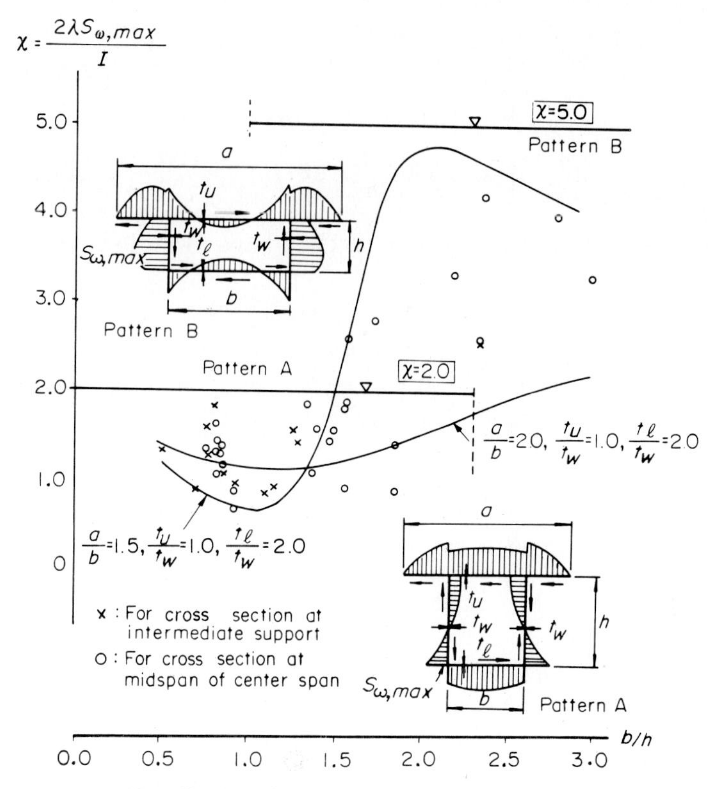

Figure 2.80 Distribution of parameter χ.

patterns of S_ω in Eq. (2.3.140), which is directly proportional to τ_ω, are plotted and shown in Fig. 2.80.[18] By examining the directions of flow of shearing stresses indicated by the arrows in Figs. 2.80 and 2.81, which are, respectively, shear stresses caused by bending and pure torsion, we see that the maximum shearing stress occurs at the junction point of the bottom and web plates for pattern A and web plate for pattern

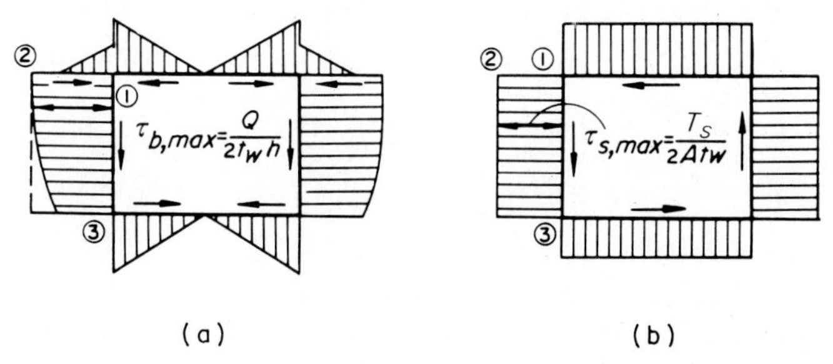

Figure 2.81 Shear stress in box girder due to (*a*) bending and (*b*) torsion.

B. For these solutions,

$$\chi = \begin{cases} 2.0 & \text{for } \dfrac{b}{h} < 2.3 \qquad (2.3.142a) \\ 5.0 & \text{for } \dfrac{b}{h} > 1.8 \qquad (2.3.142b) \end{cases}$$

(iii) Design formulas for box girders. The above formulas can be summarized as follows:

$$\sigma_{\omega,3} = -\frac{h}{5K}(\eta\ \Delta T\, e^{-\xi z'/b} + 3\lambda b m_T) \qquad (2.3.143)$$

with
$$\tau_{s+\omega} = \tau_s + \tau_\omega = \frac{T}{2Ft_w}\left[1 + \frac{|\Delta T|}{2|T|}(\chi - 1)e^{-\xi z'/b}\right] \qquad (2.3.144)$$

where

T = pure torsional moment

ΔT = step of pure torsional moment

m_T = intensity of uniformly distributed torque

$K = \dfrac{4b^2h^2}{b/t_u + b/t_l + 2h/t_w}$ = pure torsional constant

b = spacing of web plate

h = depth of box girder

t_u, t_l = thickness of top and bottom flange plates, respectively

t_w = thickness of web plate

$F = bh$ = area surrounded by thin-walled plates

z' = distance from ΔT to point of interest in direction of girder axis

And the parameters are

$\eta = 3.5$ pattern A, $b/h < 2.5$ (2.3.145*a*)

$\eta = -5.0$ pattern B, $b/h > 1.5$ (2.3.145*b*)

$\xi = 2.0$ pattern A, $b/h < 2.5$ (2.3.145*c*)

$\xi = 4.0$ pattern B, $b/h > 1.5$ (2.3.145*d*)

$\lambda = 0.8$ pattern A, $b/h < 2.5$ (2.3.145*e*)

$\lambda = -0.6$ pattern B, $b/h > 1.5$ (2.3.145*f*)

$\chi = 2.0$ pattern A, $b/h < 2.5$ (2.3.145*g*)

$\chi = 5.0$ pattern B, $b/h > 1.5$ (2.3.145*h*)

Note that patterns A and B overlap in the case where $b/h = 1.5$ to 2.5.

Furthermore, the torsional warping normal stress at points 1 and 2 in Fig. 2.76 can be obtained by modifying $\sigma_{\omega,3}$ as follows:

$$\sigma_{\omega,1} = \frac{1}{3\gamma_2^2 + 6/\gamma_1 - 1}\left[\frac{(\gamma_2 - 1)^2(2\gamma_2 + 1)}{\lambda} - 2\left(\frac{3}{\gamma_1} + 1\right)\right]\sigma_{\omega,3} \tag{2.3.146a}$$

$$\sigma_{\omega,2} = -\frac{1}{3\gamma_2^2 + 6/\gamma_1 - 1}\left[\frac{(\gamma_2 - 1)(\gamma_2^2 + \gamma_2 + 6/\gamma_1)}{\lambda} + 2\left(\frac{3}{\gamma_1} + 1\right)\right]\sigma_{\omega,3} \tag{2.3.146b}$$

where the additional parameters γ_1 and γ_2 are given by

$$\gamma_1 = \frac{b}{h} \tag{2.3.147a}$$

$$\gamma_2 = \frac{a}{b} \tag{2.3.147b}$$

2.4 Distortion[28–33]

2.4.1 Phenomenon of distortion

When a box girder having a symmetric cross section is subjected to a vertical distributed load p with an eccentricity e from the centerline of the cross section, the applied force can be resolved easily into a flexural force p and torsional force $m = pe$. The torsional force will be divided into pure torsion and distortion, as shown in Fig. 2.82c and d, respectively. This induced stress can obviously be estimated by

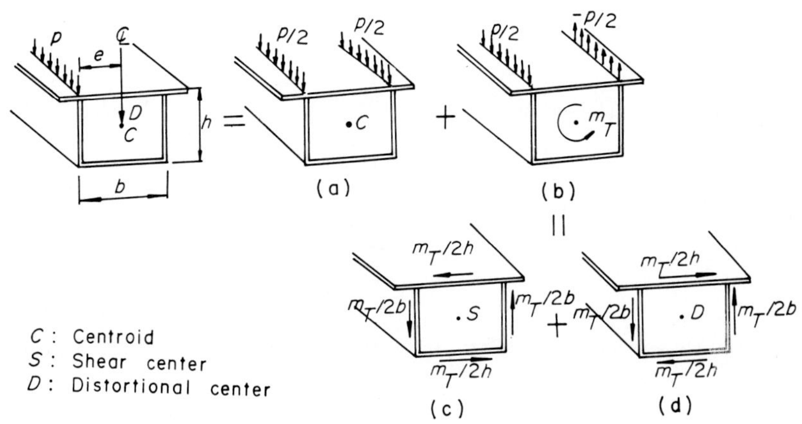

Figure 2.82 Box girder subjected to eccentric loading, where (a) flexure, plus (b) torsion equals (c) pure torsion plus (d) distortional warping.

analyzing flexure (Fig. 2.82*b*), pure torsion (Fig. 2.82*c*), and distortion (Fig. 2.82*d*).[30]

Since the analytical methods for flexure and pure torsion have already been discussed, the method for analyzing distortional warping is briefly detailed here.

Figure 2.83 shows the cross section of a box girder under consideration

where a = breadth of bracket
b = spacing of web plate
h = girder depth
h_u, h_l = distances from distortional center D to top and bottom flange plate, respectively
t_w = thickness of web plate
t_u, t_l = thickness of top and bottom flange plate, respectively

$$A_u = t_u(2a + b) = \text{cross-sectional area of top flange plate} \tag{2.4.1a}$$

$$A_w = t_w h = \text{cross-sectional area of web plate} \tag{2.4.1b}$$

$$A_l = t_l b = \text{cross-sectional area of bottom flange plate} \tag{2.4.1c}$$

Furthermore, the rectangular coordinate axes (x, y, z) are taken at the distortional center D, and the curvilinear coordinate axis s is also adopted along the perimeter of cross section, as shown in Fig. 2.83.

The cross section of the box girder will deform and produce the angular distortion Θ at each corner of the box section, as shown in

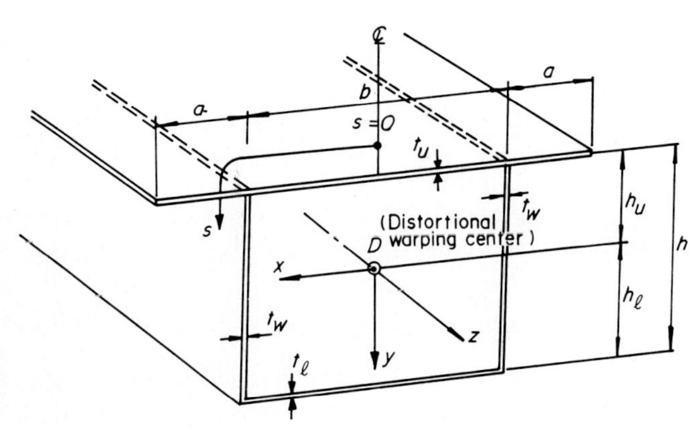

Figure 2.83 Cross section of box girder.

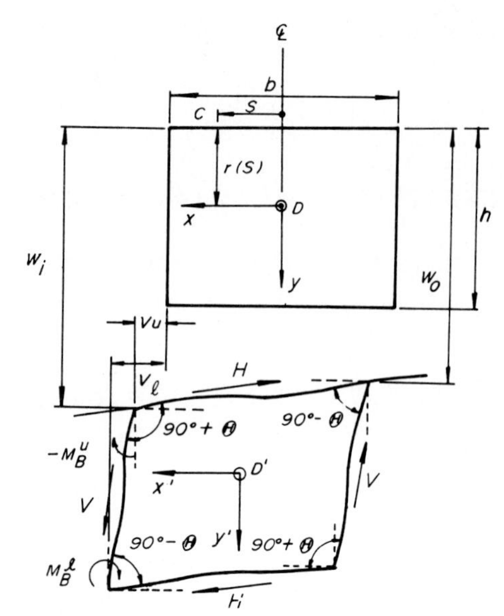

Figure 2.84 Definition of distortional angle Θ.

Fig. 2.84, due to the distortional forces (Fig. 2.82*d*). This angular distortion Θ is defined by the following geometric relationships:

$$\Theta = \frac{v_l - v_u}{h} + \frac{w_i - w_o}{b} \tag{2.4.2}$$

provided that the displacements v and w in the direction of the x and y coordinate axes, having their origin at the center of distortion D, are known. Furthermore, the derivative of the angular distortion $\Theta' = d\Theta/dz$ will yield the displacements in the direction of the z coordinate axis, as in Fig. 2.84, analogous to the definition of torsional warping theory. Therefore,

$$u = \omega_D \Theta' \tag{2.4.3}$$

where ω_D is defined as the warping function due to the distortion. For the case where the displacement u is restrained, normal and shearing stresses will occur in the box section. Then the corresponding stress resultants can be defined as shown in Table 2.17 and in Fig. 2.85. Note that these are similar in definition to the torsional warping theory.

2.4.2 Derivation of the fundamental equation for distortion

(1) Displacement due to distortion. Let the displacement components v and w in the directions of the x and y coordinate axes with reference to the distortional center D of a box section be denoted by a

TABLE 2.17 Stress Resultants Due to Distortion

Notation	Definition
$M_{D\omega}$	Bimoment due to distortion
$T_{D\omega}$	Torsional moment due to distortion

representative symbol v. From this simplification, the normal strain ε and the shearing strain γ can be expressed easily, according to the fundamental theory of elasticity,[28] by

Normal strain:
$$\varepsilon \cong \frac{\partial u}{\partial z} \tag{2.4.4a}$$

Shearing strain:
$$\gamma \cong \frac{\partial u}{\partial s} + \frac{\partial v}{\partial z} \tag{2.2.4b}$$

Now, the following two approximate assumptions are introduced to determine the displacement functions u and v. The first assumption is that the shearing strain γ is considered to be infinitesimal

$$\gamma \cong 0 \tag{2.4.5}$$

The second assumption is that the displacement v can be separated into two variables and written as the product of the angular distortion $\Theta(z)$ and the perpendicular distance $r(s)$ from the distortional center D of the cross section to a point C shown in Fig. 2.84:

$$v(z, s) = \Theta(z)r(s) \tag{2.4.6}$$

Accordingly, the substitution of Eq. (2.4.6) into Eq. (2.4.4b) and Eq. (2.4.5) gives

$$\frac{\partial u}{\partial s} = -\frac{\partial}{\partial z}[\Theta(z)r(s)] = -r(s)\frac{\partial \Theta(z)}{\partial z} \tag{2.4.7}$$

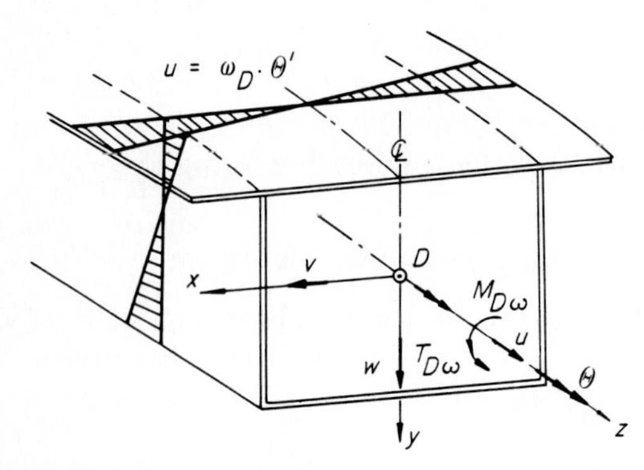

Figure 2.85 Definition of stress resultants and displacement due to distortional warping.

The integration of Eq. (2.4.7) by the perimeter coordinate s reduces to

$$u(z, s) = \frac{\partial \Theta}{\partial z}\left[-\int_0^s r(s)\,ds + C_l\right] \tag{2.4.8}$$

where C_1 is an integration constant. The distortional warping function $\omega_D(s)$ can thus be denoted

$$\omega_D(s) = -\int_0^s r(s)\,ds + C_1 \tag{2.4.9}$$

Consequently, Eq. (2.4.8) becomes

$$u(z, s) = \frac{\partial \Theta(z)}{\partial z}\,\omega_D(s) \tag{2.4.10}$$

(2) Normal stress and the corresponding strain energy to distortion. Substituting Eq. (2.4.10) into Eq. (2.4.4), we see that the distortional warping stress $\sigma_{D\omega}$ can be given by Hooke's law:

$$\sigma_{D\omega} = E\varepsilon = E\frac{\partial u}{\partial z} = E\frac{\partial^2 \Theta}{\partial z^2}\,\omega_D \tag{2.4.11}$$

Here E is Young's modulus. Since the distortion does not produce any additional axial force N_x or bending moments M_y and M_z, the following equations must be satisfied:

$$N_z = \int_A \sigma_{D\omega}\,dA = 0 \tag{2.4.12a}$$

$$M_x = \int_A \sigma_{D\omega}\,y\,dA = 0 \tag{2.4.12b}$$

$$M_y = \int_A \sigma_{D\omega}x\,dA = 0 \tag{2.4.12c}$$

Substitution of Eqs. (2.4.9) and (2.4.11) into Eq. (2.4.12) determines the distribution of the distortional warping function ω_D. This is shown in Fig. 2.86. The values of ω_D at corner points 1, 2, and 3 can easily be determined by

$$\omega_{D,1} = \frac{bh}{4}\frac{1}{1+\beta} \tag{2.4.13a}$$

$$\omega_{D,2} = \frac{bh}{4}\left(1+\frac{2a}{b}\right)\frac{1}{1+\beta} = \left(1+\frac{2a}{b}\right)\omega_{D,1} \tag{2.4.13b}$$

$$\omega_{D,3} = -\frac{bh}{4}\frac{\beta h}{1+\beta} = \beta\omega_{D,1} \tag{2.4.13c}$$

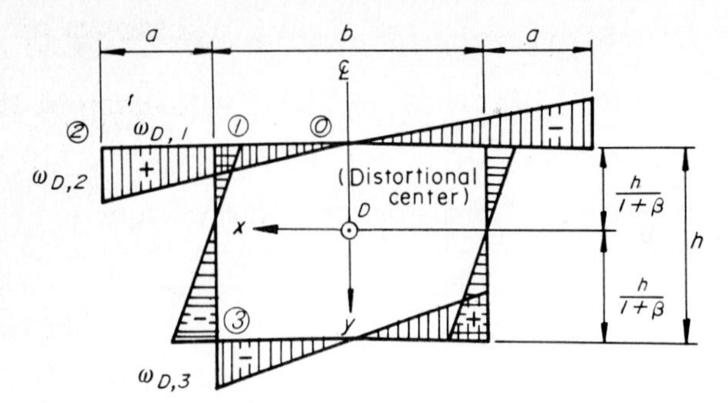

Figure 2.86 Warping function due to distortion.

where parameter β is

$$\beta = \frac{A_u(1 + 2a/b)^2 + 3A_w}{A_l + A_w} \tag{2.4.14}$$

Next, the strain energy U_σ due to the distortional warping stress $\sigma_{D\omega}$ can be found directly from Eq. (2.4.11):

$$\begin{aligned} U_\sigma &= \frac{1}{2E}\int_0^l \int_A \sigma_{Dz}^2 \, dA \, dz \\ &= \frac{1}{2}\int_0^l \int_A \left(\frac{\partial u}{\partial z}\right)^2 dA \, dz \\ &= \frac{E}{2}\int_A \omega_D^2 \, dA \int_0^l \left(\frac{d^2\Theta}{\partial z^2}\right)^2 dz \end{aligned} \tag{2.4.15}$$

where l is the span length of the girder.

By using the diagram of the warping function ω_D shown in Fig. 2.84, the distortional warping constant $I_{D\omega}$ can be rationally defined as

$$I_{D\omega} = \int_A \omega_D^2 \, dA = \frac{b^2h^2}{48(1+\beta)}[A_l\beta + A_w(2\beta - 1)] \tag{2.4.16}$$

Consequently, Eq. (2.4.15) reduces to

$$U_\sigma = \frac{EI_{D\omega}}{2}\int_0^l \left(\frac{\partial^2\Theta}{\partial z^2}\right)^2 dz \tag{2.4.17}$$

Now, the bimoment $M_{D\omega}$ due to the distortion will, moreover, be denoted by the following equation, analogous to the definitions developed in the torsional warping theory:

$$M_{D\omega} = EI_{D\omega}\frac{d^2\Theta}{dz^2} \tag{2.4.18}$$

From the above equation, Eq. (2.4.11) can be rewritten in the same form as the torsional warping theory. Thus

$$\sigma_{D\omega} = \frac{M_{D\omega}}{I_{D\omega}} \omega_D \tag{2.4.19}$$

(3) Shearing stress and the corresponding strain energy. The shearing stress is reduced by the difference in the normal stress in the axial direction of the box girder. An equilibrium condition for the axial direction of the small element $dz \cdot dc$ can be expressed by

$$\frac{\partial \sigma_{D\omega} t}{\partial z} + \frac{\partial \tau_{D\omega} t}{\partial s} = 0 \tag{2.4.20}$$

Substituting Eq. (2.4.19) into Eq. (2.4.20), we define the shear flow $q_{D\omega}$ due to the distortion as

$$q_{D\omega} = \tau_{D\omega} t = \frac{S_{D\omega}}{I_{D\omega}} T_{D\omega} \tag{2.4.21}$$

in which $T_{D\omega}$ is the torsional moment due to distortion, defined by

$$T_{D\omega} = -\frac{\partial M_{D\omega}}{\partial z} = -EI_{D\omega} \frac{\partial^3 \Theta}{\partial z^3} \tag{2.4.22}$$

and $S_{D\omega}$ is the static moment of the distortional warping function ω_D, given by

$$S_{D\omega} = \int_0^s \omega_D t \, ds + C_2 = \bar{S}_{D\omega} + C_2 \tag{2.4.23}$$

The integration constant C_2 can be directly determined so as to fulfill the condition that the torsional moment T_z due to distortion be zero; thus

$$T_z = \oint q_{D\omega} r \, ds = 0 \tag{2.4.24}$$

Substitution of Eqs. (2.4.22) and (2.4.23) into (2.4.24) gives

$$\begin{aligned} C_2 &= -\frac{1}{2bh} \oint \bar{S}_{D\omega} r \, ds \\ &= -\frac{\omega_{D,1}}{4} \left[A_u \left(1 + \frac{2a}{b} \right) - \frac{A_l}{3} \beta + \frac{A_w}{3} (5 - 4\beta) \right] \end{aligned} \tag{2.4.25}$$

The final diagram of $S_{D\omega}$ is shown in Fig. 2.87.

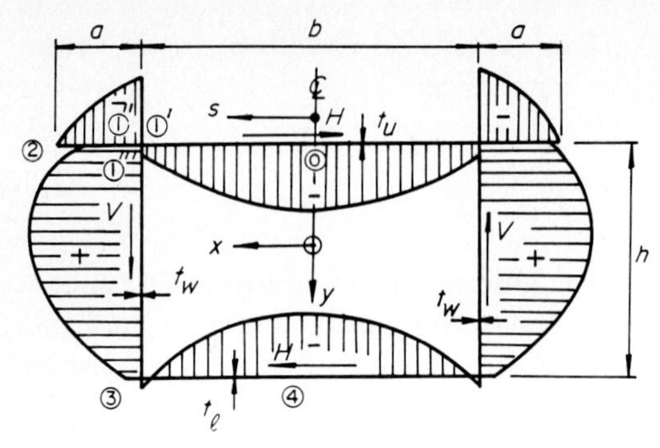

$$S_{D\omega,0} = C_2$$

$$S_{D\omega,1'} = \frac{b}{4} t_u \omega_{D,1} + C_2$$

$$S_{D\omega,1''} = -at_u\left(1 + \frac{a}{b}\right)\omega_{D,1}$$

$$S_{D\omega,1'''} = \frac{A_u}{4}\left(1 + \frac{2a}{b}\right)\omega_{D,1} + C_2$$

$$S_{D\omega,3} = -\left[\frac{\beta - 1}{2} A_w - \frac{A_u}{4}\left(1 + \frac{2a}{b}\right)\right]\omega_{D,1} + C_2$$

$$S_{D\omega,4} = C_2$$

where $C_2 = -\dfrac{\omega_{D,1}}{4} - A_u\left(1 + \dfrac{2a}{b}\right)\dfrac{A_l}{3}\beta + \dfrac{A_w}{3}(5 - 4\beta)$

Figure 2.87 Diagram of static moment with respect to distortional warping function. $A_u = t_u(2a + b)$; $A_w = t_w h$; $A_l = t_l b$.

Because of these shear flows $q_{D\omega}$, the web and flange plates will undergo a vertical force V and horizontal force H, as follows:
For the left web plate:

$$V = -\int_{b/2}^{(b/2)+h} q_{D\omega}\, ds = -\frac{T_{D\omega}}{I_{D\omega}} \int_{b/2}^{(b/2)+h} S_{D\omega}\, ds$$

$$= -\frac{T_{D\omega}}{I_{D\omega}} \int_{b/2}^{(b/2)+h} \left\{\left[\left(\frac{1+\beta}{2h} s^2 - s\right)t_w - \frac{Au}{4}\left(1 + \frac{2a}{b}\right)\right]\omega_{D,1} + C_2\right\} ds$$

$$= \frac{T_{D\omega}}{b} \tag{2.4.26}$$

For the bottom plates:

$$H = -\int_{(b/2)+h}^{\frac{3}{2}b+h} q_{D\omega}\, ds = -\frac{T_{D\omega}}{I_{D\omega}} \int_{(b/2)+h}^{\frac{3}{2}b+h} S_{D\omega}\, ds$$

$$= -\frac{T_{D\omega}}{I_{D\omega}} \int_{b/2}^{\frac{3}{2}+h} \left\{ \left[\left(\frac{s^2}{b} - s \right) \beta t_1 - \frac{\beta - 1}{2} A_w + \frac{A_u}{4} \left(1 + \frac{2a}{b} \right) \right] \omega_{D,1} + C_2 \right\} ds$$

$$= \frac{T_{D\omega}}{h} \tag{2.4.27}$$

These forces V and H, shown in Fig. 2.88*a*, will make the distortion of the box section as illustrated in Fig. 2.88*c*. Thus, the angular distortion Θ will be produced at each corner of the box section. To determine this angular distortion Θ, hinges are inserted at each corner of the box section and unit moments $M = \pm 1$ are applied to these points, as shown in Fig. 2.88*b*.

The angular distortion Θ can thereby be evaluated by applying the principle of virtual work:

$$\Theta = \frac{1}{4} \oint M\bar{M}\, ds = \frac{T_{D\omega}}{K_{D\omega}} \tag{2.4.28}$$

or

$$T_{D\omega} = K_{D\omega} \Theta \tag{2.4.29}$$

where

$$K_{D\omega} = \frac{24EI_w}{\alpha_o h} \tag{2.4.30}$$

and $K_{D\omega}$ is the stiffness of the box section against the distortion. Also α_o, given by

$$\alpha_o = 1 + \frac{2b/h + 3(I_u + I_l)/I_w}{(I_u + I_l)/I_w + (6h/b)(I_u I_l / I_w^2)} \tag{2.4.31}$$

is a cross-sectional parameter, and I_u, I_w, and I_l are the geometric moments of inertia for the top, web, and bottom plate elements, respectively, and are given by

$$I_u = \frac{t_u^3}{12(1 - \mu^2)} \tag{2.4.32a}$$

$$I_w = \frac{t_w^3}{12(1 - \mu^2)} \tag{2.4.32b}$$

$$I_l = \frac{t_l^3}{12(1 - \mu^2)} \tag{2.4.32c}$$

$$X_1 = \frac{3\dfrac{h}{I_W} + \dfrac{b}{I_1}}{b^2\left(\dfrac{1}{I_1} + \dfrac{1}{I_U}\right) + 6\,\dfrac{bh}{I_W}}\,T_{D\omega}$$

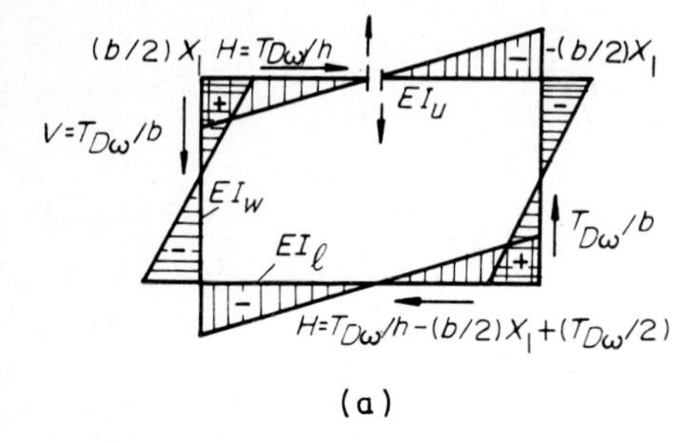

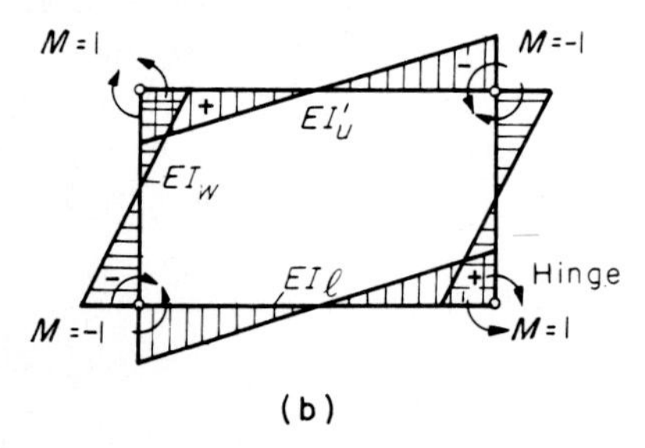

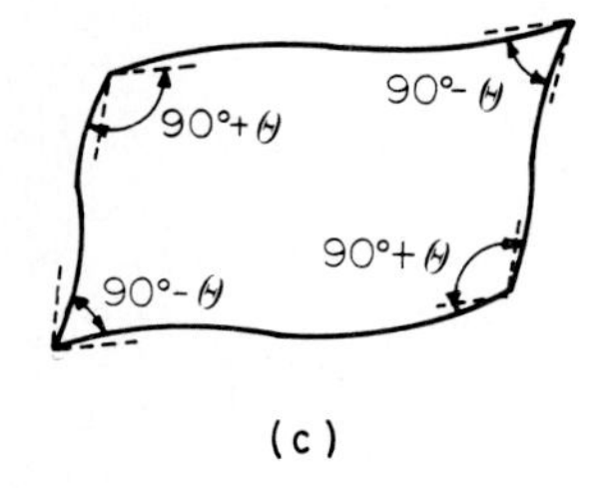

Figure 2.88 Derivation of angular distortion: (*a*) distortional forces H and V and corresponding bending moment diagram, (*b*) bending moment diagram due to virtual moments $M \pm 1$, and (*c*) angular distortion Θ.

where μ is Poisson's ratio. The strain energy U_τ due to the shearing stress $\tau_{D\omega}$ can therefore be obtained as follows.

$$U_\tau = \frac{1}{2}\int_0^l T_{D\omega}\Theta\,dz = \frac{K_{D\omega}}{2}\int_0^l \Theta^2\,dz \tag{2.4.33}$$

(4) Work due to an external force. The distortional forces q_V and q_H along the unit length of the box-girder section are shown in Fig. 2.89.

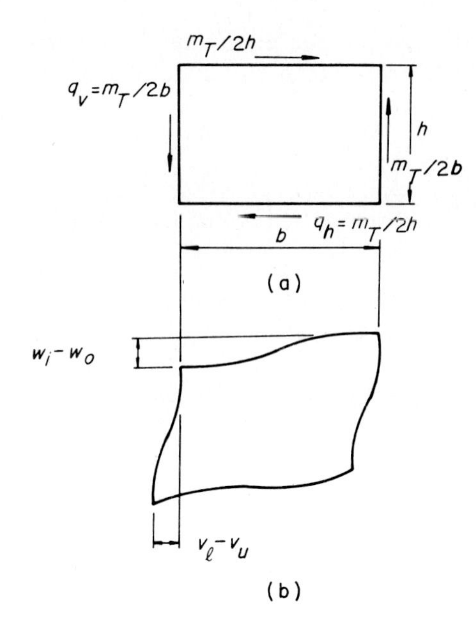

Figure 2.89 (*a*) Applied force and (*b*) distortional displacement.

The magnitude of these forces can be estimated easily by using Bredt's formula:

$$q_V = qh = \frac{m_T}{2b} \tag{2.4.34a}$$

$$q_H = qb = \frac{m_T}{2h} \tag{2.4.34b}$$

Since the distortional displacements corresponding to these forces are illustrated in Fig. 2.89*b*, the work done by distortional forces can obviously be evaluated as follows:

$$\begin{aligned} V_m &= -\int_0^l [q_V(w_i - w_o) + q_H(v_l - v_u)]\,dz \\ &= -\int_0^l \frac{m_T}{2}\left(\frac{w_i - w_o}{b} + \frac{v_l - v_u}{h}\right) dz \end{aligned} \tag{2.4.35}$$

However, the term in parentheses of the above equation is nothing but angular distortion Θ, as previously defined by Eq. (2.4.2). Accordingly, Eq. (2.4.35) can be simplified to

$$V_m = -\int_0^l \frac{m_T}{2}\Theta\,dz \tag{2.4.36}$$

(5) The fundamental differential equation for distortion. By summing Eqs. (2.4.15), (2.4.33), and (2.4.36), the total potential energy Π can be given as

$$\Pi = U_\sigma + U_\tau + V_m$$
$$= \frac{EI_{D\omega}}{2}\int_0^l \left(\frac{d^2\Theta}{dz^2}\right)^2 dz + \frac{K_{D\omega}}{2}\int_0^l \Theta^2\,dz - \int_0^l \frac{m_T}{2}\Theta\,dz \tag{2.4.37}$$

Now, a condition, which makes the total potential energy Π a minimum, can be written by means of the variational calculus as

$$\delta\Pi = EI_{D\omega}\int_0^l \frac{d^2\Theta}{dz^2}\,\delta\left(\frac{d^2\Theta}{dz^2}\right) dz + K_{D\omega}\int_0^l \Theta\delta\Theta\,dz - \int_0^l \frac{m_T}{2}\,\delta\Theta\,dz$$
$$= \left[M_{D\omega}\delta\left(\frac{d\Theta}{dz}\right)\right]_0^l - \int_0^l EI_{D\omega}\frac{d^3\Theta}{dz^3}\,\delta\left(\frac{d\Theta}{dz}\right) dz$$
$$+ \int_0^l \left(K_{D\omega}\Theta - \frac{m_T}{2}\right)\delta\Theta\,dz$$
$$= [M_{D\omega}\delta\Theta]_0^l - [T_{D\omega}\delta\Theta]_0^l + \int_0^l \left(EI_{D\omega}\frac{d^4\Theta}{dz^4} + K_{D\omega}\Theta - \frac{m_T}{2}\right)\delta\Theta\,dz \equiv 0$$
$$\tag{2.4.38}$$

The boundary conditions at both ends of a box girder can be found, as shown in Table 2.18.

Since Eq. (2.4.38) is valid for any set of boundary conditions, the following equation must be satisfied:

$$EI_{D\omega}\frac{d^4\Theta}{dz^4} + K_{D\omega}\Theta = \frac{m_T}{2} \tag{2.4.39}$$

2.4.3 Beam-on-elastic-foundation analogy

Equation (2.4.39) is similar to the equation describing the deflection w of a beam on an elastic foundation with flexural rigidity EI, spring

TABLE 2.18 Boundary Conditions

Supports	Boundary conditions
Simple	$\Theta = 0,\ M_{D\omega} = 0$
Rigid	$\Theta = 0,\quad \Theta' = 0$
Free	$M_{D\omega} = 0,\ T_{D\omega} = 0$

constant k, and uniformly distributed load q by substituting Θ, $EI_{D\omega}$, $K_{D\omega}$, and $m_T/2$, respectively, as follows:

$$EI\frac{d^4w}{dz^4} + kw = q \tag{2.4.40}$$

Therefore, this theory is referred to as the *beam-on-elastic-foundation* (BEF) analogy.[28]

Namely, the girder is subjected to the distortional force $m_T/2$ of Fig. 2.90*a* and is elastically supported not only by the elastic foundation with stiffness $K_{D\omega}$, as defined in Eq. (2.4.30), but also by the diaphragms with stiffness K_D, as we discuss later.

Thus, the fundamental differential equation for distortion is

$$\frac{d^4\Theta}{dz^4} + 4\lambda^4\Theta = \frac{1}{EI_{D\omega}}\frac{m_T}{2} \tag{2.4.41}$$

where

$$\lambda = \sqrt[4]{\frac{K_{D\omega}}{4EI_{D\omega}}} \tag{2.4.42}$$

As an example, a solution for the distortional angle Θ in the case of uniformly distributed load $m_T/2$ can be written as

$$\Theta = A \sin \lambda z \sinh \lambda z + B \sin \lambda z \cosh \lambda z$$

$$+ C \cos \lambda z \sinh \lambda z + D \cos \lambda z \cosh \lambda z + \frac{1}{K_{D\omega}}\frac{m_T}{2} \tag{2.4.43}$$

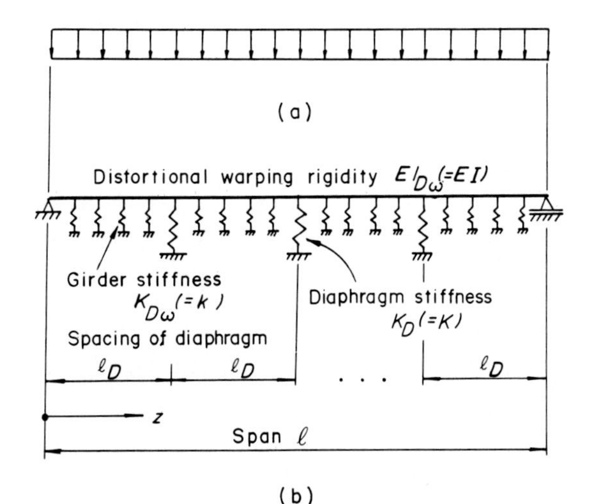

Figure 2.90 BEF analogy: (*a*) distortional load ($q = m_T/2$), (*b*) equivalent beam.

where A, B, C, and D are the integration constants and can be determined by the boundary conditions at both ends, that is, $z = 0$ and $z = l$,

$$\Theta = 0 \tag{2.4.44a}$$

and

$$M_{D\omega} = 0 \tag{2.4.44b}$$

for a simply supported girder.

The distortional warping moment $T_{D\omega}$ must change abruptly when the intermediate diaphragms are attached to the girder, as in Fig. 2.90, and these values can be obtained by

$$T^s_{D\omega} = K_D \Theta \tag{2.4.45}$$

Here K_D is the stiffness of the intermediate diaphragms against the distortional warping and can be estimated according to the type of diaphragm. Figure 2.91 illustrates various types of diaphragms, and their related stiffnesses are as follows:
For a plate-type diaphragm,

$$K_D = Gt_D bh \tag{2.4.46}$$

For a truss-type diaphragm,

$$K_D = \begin{cases} \dfrac{2EA_b b^2 h^2}{l_b^3} & X \text{ type} \quad (2.4.47a) \\[2ex] \dfrac{EA_b b^2 h^2}{2l_b^3} & V \text{ type} \quad (2.4.47b) \end{cases}$$

where A_b = cross-sectional area of truss member
l_b = length of truss member

For frame-type diaphragm,

$$K_D = \frac{24EI_w}{\alpha_o h} \tag{2.4.48}$$

In Eq. (2.4.48), α_o can be evaluated from Eq. (2.4.31), provided that the geometric moments of inertia I_u, I_w and I_l are determined in the same manner as those of the frame, by taking into account the effective width b_m according to the pitch of the intermediate diaphragm l_D, as follows [see Eq. (2.2.50)]:

$$b_m = \begin{cases} \dfrac{d}{3} & l_D \geqq \dfrac{d}{3} \quad (2.4.49a) \\[2ex] l_D & l_D < \dfrac{d}{3} \quad (2.4.49b) \end{cases}$$

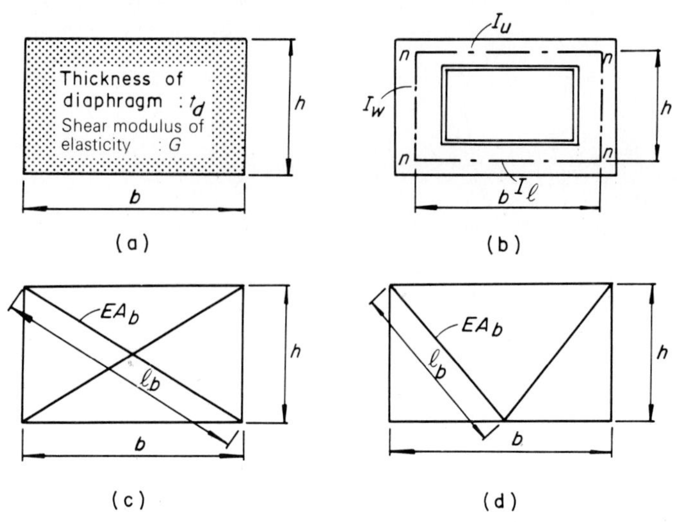

Figure 2.91 Types of intermediate diaphragm: (*a*) solid-plate type, (*b*) frame type, (*c*) truss *X* type, (*d*) truss *V* type.

Here d is the smaller value of the spacing of the web plate b or the girder height h.

For instance, the variations of normal stress $\sigma_{D\omega}$ due to the distortional warping of a box girder can be plotted by solving Eqs. (2.4.43) through (2.4.46) as shown in Fig. 2.92, where there are nine diaphragms so that the ratio of the diaphragm spacing l_D to span l is taken as $l_D/l = 1/10$.

The line load q along the span and a concentrated load P are applied to one of the web plates (see Fig. 2.82). Since the distortional warping stress varies in the direction of span, $\sigma_{D\omega}$ is made a dimensionless value by the absolute maximum distortional warping stress $|(\sigma_{D\omega})_{KD=\infty}|_{\max}$ with the stiffness $K_D = \infty$ of the intermediate diaphragm, where $\gamma = K_D/(K_{D\omega}l_D)$ is the stiffness parameter of the intermediate diaphragm.

Examination of Fig. 2.92*a* shows that the variations of $\sigma_{D\omega}$ for the line load p can be approximated as being in the same form as the flexural stress distribution of a continuous beam supported by the intermediate supports. Then, in cases where γ has comparatively large values, the distortional stress $\sigma_{D\omega}$ alternates from a positive to a negative value at the diaphragm and middiaphragm, respectively, and the maximum distortional stress $\sigma_{D\omega}$ occurs at the end panel. The same kind of behavior can be observed in the case of a concentrated load P, as illustrated in Fig. 2.92*b*.

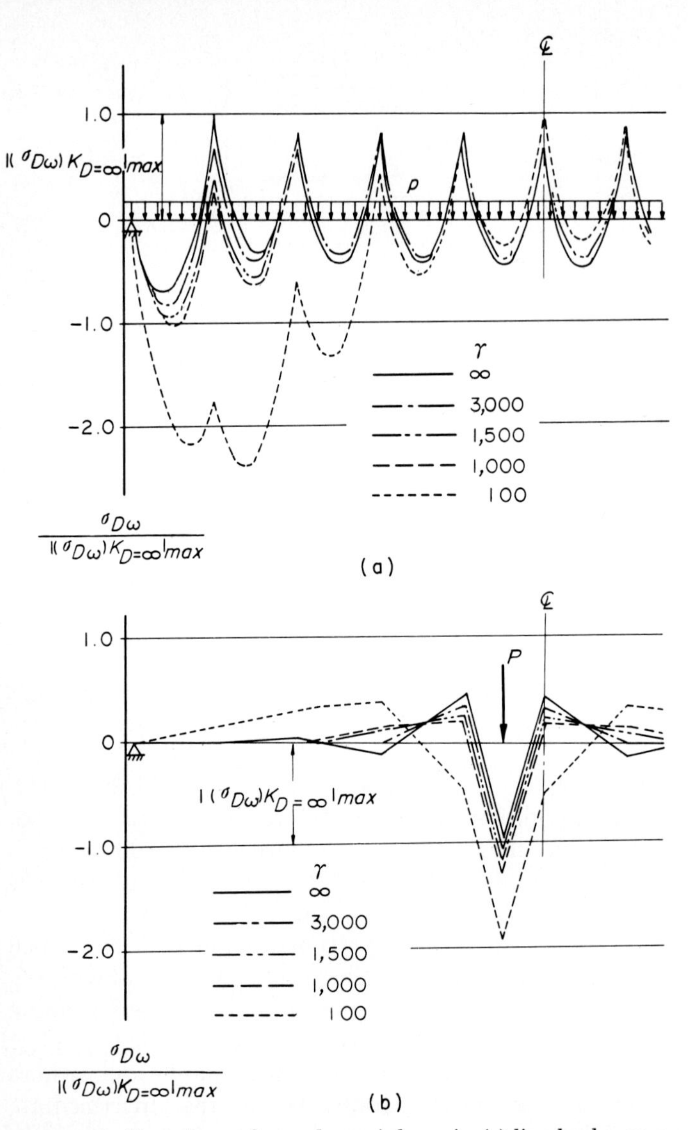

Figure 2.92 Variations of $\sigma_{D\omega}$ along girder axis: (*a*) line load *p* on a web plate of box girder, (*b*) concentrated load *P* on a web plate of box girder at $x = 9l/20$.

The diaphragm undergoes the distortional moments $T^s_{D\omega}$ shown in Eq. (2.4.45). The induced normal stress σ and shearing stress τ can be evaluated as follows:
Plate-type diaphragm,

$$\tau = \frac{T^s_{D\omega}}{bht_D} \tag{2.4.50}$$

Truss-type diaphragm,

$$\sigma = \frac{N}{A_b} \tag{2.4.51}$$

in which the following hold:
X-type truss,

$$N = \frac{T^s_{D\omega}}{4bh} \tag{2.4.52a}$$

V-type truss,

$$N = \frac{T^s_{D\omega}}{2bh} \tag{2.4.52b}$$

Frame-type diaphragm,

$$\sigma = \frac{M_D}{W} \tag{2.4.53}$$

$$\tau_1 = \frac{2M^u_D}{bF_w} \tag{2.4.54a}$$

$$\tau_2 = \frac{M^u_D + M^l_D}{hF_w} \tag{2.4.54b}$$

where $M_D = (M^u_D, M^l_D)$ are bending moments in the transverse direction, as shown in Fig. 2.93, and are given by

$$M^u_D = \frac{T^s_{D\omega}}{4}\left[1 + \frac{I_u - I_l}{I_u + I_l + 6(h/b)(I_u I_l / I^2_w)}\right] \tag{2.4.55a}$$

$$M^l_D = \frac{T^s_{D\omega}}{4}\left[1 + \frac{I_l - I_u}{I_u + I_l + 6(h/b)(I_u I_l / I^2_w)}\right] \tag{2.4.55b}$$

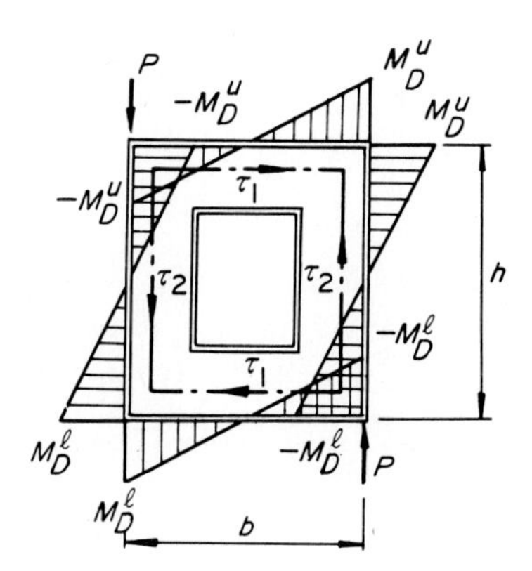

Figure 2.93 Transverse bending of frame member and induced stresses.

in which W = section modulus of frame members by taking into account the effective width of flange plates, given in Eq. (2.4.49)

F_w = cross-sectional area of web plate of frame member

2.4.4 Parametric analysis of distortion

(1) Distortional warping parameter.[18] To investigate the characteristic of distortional behavior, a parametric survey was performed on the basis of actual data of long-span box-girder bridges. First, the relationships between the distortional warping constant $I_{D\omega}$ [see Eq. (2.4.16), with units of m^6] and the geometric moment of inertia I_x of the cross section [see Eq. (2.4.57), with units of m^4] with respect to the horizontal axis can be plotted as shown in Fig. 2.94 and are approximated by

$$I_{D\omega} = \left(\frac{bh}{150} + \frac{b^2}{30}\right)\frac{I_x}{100} \tag{2.4.56}$$

where

$$I_x = A_u h_u^2 + A_l h_l^2 + 2A_w\left[\frac{h^2}{12} + \left(\frac{h}{2} - h_u\right)^2\right] \tag{2.4.57}$$

Next parameters β [see Eq. (2.4.14)] and λ [see Eq. (2.4.42)] have wide values, because they depend on the cross-sectional dimension a/b (see Figs. 2.95 and 2.96) and are distributed within the following ranges:

$$\beta = 1.5\text{–}7.0 \qquad \text{(nondimensional value)} \tag{2.4.58}$$

$$\lambda = (1.5\text{–}3.5) \times 10^{-2}\ \text{m}^{-1} \tag{2.4.59}$$

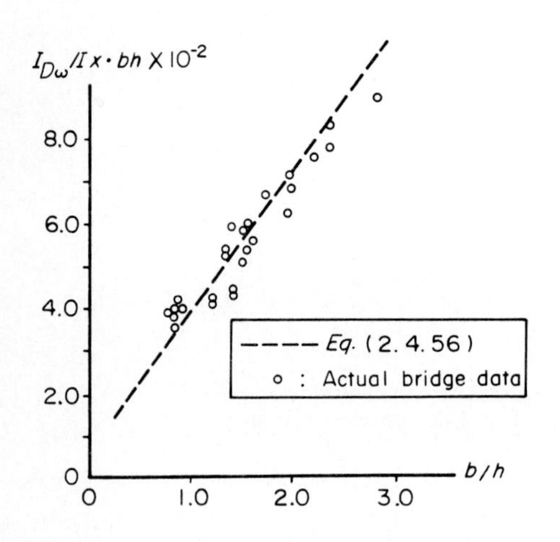

Figure 2.94 Relationships between $I_{D\omega}$ and I_x.

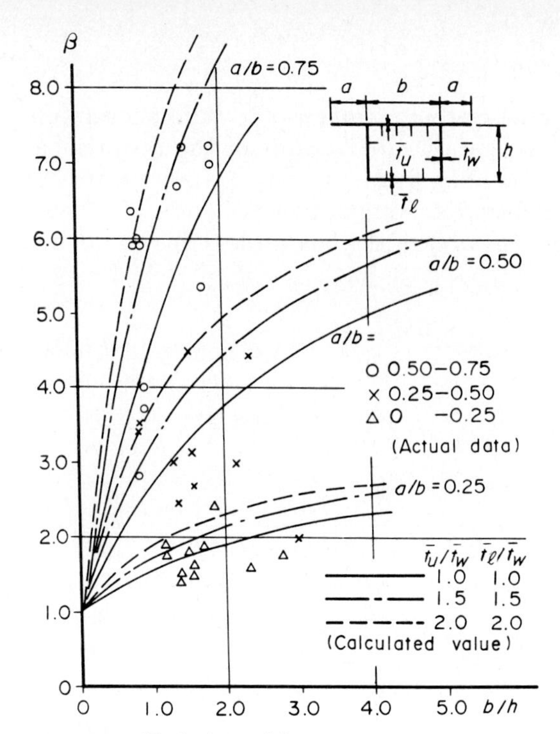

Figure 2.95 Variations of β.

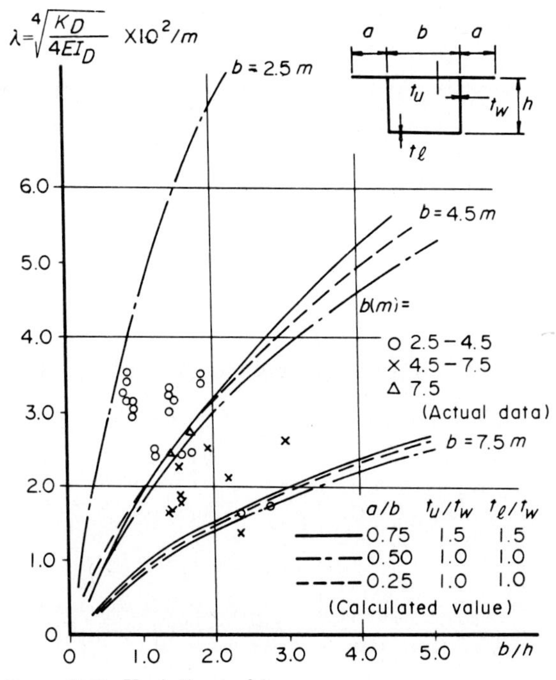

Figure 2.96 Variations of λ.

(2) Variations of distortional warping stress $\sigma_{D\omega}$.[33] Parametric studies were performed to investigate the variations of the distortional warping stress $\sigma_{D\omega}$ due to the diaphragm spacing l_D, cross-sectional quantities L/b, and the rigidity of the diaphragm γ.

The transverse bending stress σ_{Db} in the box girders due to the distortion is

$$\sigma_{Db} = \frac{6M_B}{(1-\mu^2)t} \tag{2.4.60}$$

where M_B is distributed as shown in Fig. 2.97, and M_B^u and M_B^l are given by

$$M_B^u = \frac{K_{D\omega}\Theta}{4}\left[1 + \frac{I_u - I_l}{I_u + I_l + (6h/b)(I_u I_l / I_w^2)}\right] \tag{2.4.61a}$$

$$M_B^l = \frac{K_{D\omega}\Theta}{4}\left[1 + \frac{I_l - I_u}{I_u + I_l + (6h/b)(I_u I_l / I_w^2)}\right] \tag{2.4.61b}$$

and t is taken as t_u, t_l, or t_w (see Fig. 2.83). In these analyses, σ_{Db}, M_B^u, and M_B^l are ignored, since they are very small values in comparison with the distortional warping stresses $\sigma_{D\omega}$.

Moreover, the loading conditions are (1) a line load p in the direction of the girder axis and (2) a concentrated load P. These loads, p and P, were applied on the left web of the box girder.

Finally, the distortional warping stresses $\sigma_{D\omega}$ are fixed at the junction point 3, shown in Fig. 2.86, and are taken as the extreme values in the direction of the girder axis. These values are also nondimensionalized by the following flexural stresses σ_b due to the bending moment M_x:

$$\sigma_b = \frac{M_x}{W_l} \tag{2.4.62}$$

where W_l = section modulus at the junction point of web and bottom

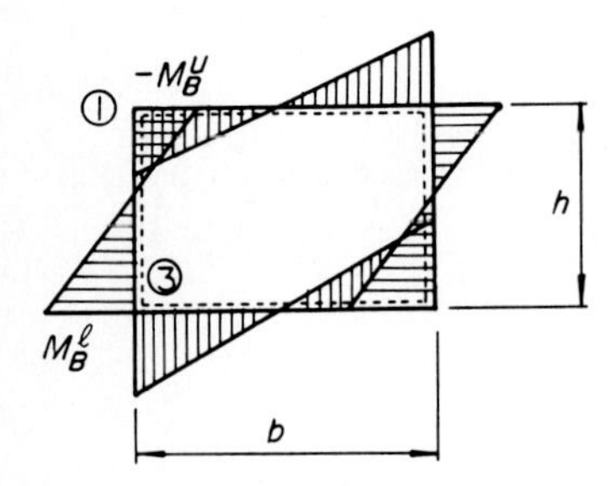

Figure 2.97 Transverse bending of box section.

plates, and the following formulas are introduced for the bending moment:

$$M_x = \begin{cases} \dfrac{pl^2}{8} & \text{for line load } p \text{ along the span} \quad (2.4.63a) \\ \dfrac{P_l}{4} & \text{for concentrated load } P \quad (2.4.63b) \end{cases}$$

The results of numerical calculations show that the normal stress $\sigma_{D\omega}$ is significantly affected by the spacing of the intermediate diaphragm l_D; moreover, it is obvious from the parametric studies that the stress ratio $\sigma_{D\omega}/\sigma_b$ is proportional to the spacing ratios $(l_D/l)^2$ and l_D/l for the line load p and the concentrated load P, respectively. Therefore, the approximate formulas for determining the stress ratio $\sigma_{D\omega}/\sigma_b$ can be found by the least-squares method on the basis of the numerical results of the actual bridge data, summarized in Table 2.19.

(3) Influence of the stiffness parameter γ.[33] In the previous discussions, the rigidities of the diaphragm were assumed to be infinitely large. To determine the effects of the stiffness of the diaphragm on the distortional warping stress, the stiffness K_D of Eqs. (2.4.46) through (2.4.48) is expressed by a dimensionless parameter as

$$\gamma = \frac{K_D}{K_{D\omega} l_D} \quad (2.4.64)$$

where $K_{D\omega}$ is given by Eq. (2.4.30) and l_D is the spacing of diaphragms.

The effect of parameter γ on the distributions of stresses $\sigma_{D\omega}$ in the direction of the span is plotted and shown in Fig. 2.92 under the conditions $l_D/l = 1/10$, where the ordinate is also nondimensionalized by the absolute maximum distortional warping stress with $K_D = \infty$, or $|(\sigma_{D\omega})_{K_D=\infty}|_{\max}$.

TABLE 2.19 Approximation of $\dfrac{\sigma_{D\omega}}{\sigma_b}$

Loading conditions	Viewpoint	
	Midpoint of intermediate diaphragm	At intermediate diaphragm
Line load p along span	$-1.2\left(\dfrac{l_D}{l}\right)^2$	$1.8\left(\dfrac{l_D}{l}\right)^2$
Concentrated load P	$-1.85\left(\dfrac{l_D}{l}\right)$	$1.0\left(\dfrac{l_D}{l}\right)$

These figures show that the concentrations of stresses can clearly be observed to occur at the section near the support and the midspan for the line load p and concentrated load P, respectively, in accordance with the decrease of the rigidities of the diaphragms γ.

For the reasons listed above, the stress ratio $\sigma_{D\omega}/|(\sigma_{D\omega})_{K_D=\infty}|_{\max}$ was examined under the conditions $l_D/l = 10$. These results are plotted in Fig. 2.98. Observing the relationships between the stress ratio $\sigma_{D\omega}/|(\sigma_{D\omega})_{K_D=\infty}|_{\max}$ and the rigidity parameter γ of Fig. 2.98a, we see that the absolute value of the stress $\sigma_{D\omega}$ at the section at middiaphragm is almost equal to that at middiaphragm for $z = l/10$, when the stiffness parameter has the value of

$$\gamma \geqq 1500 \tag{2.4.65}$$

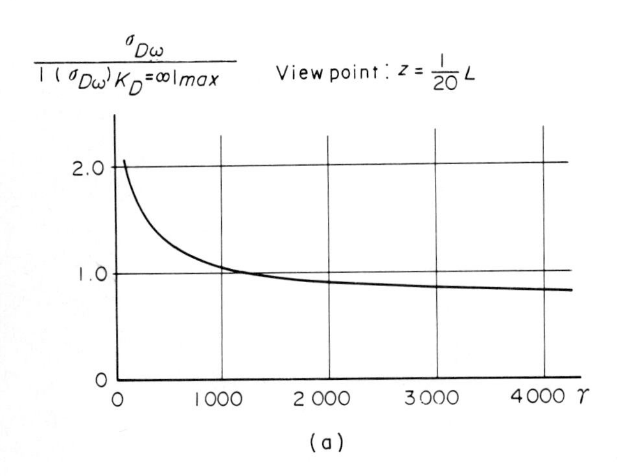

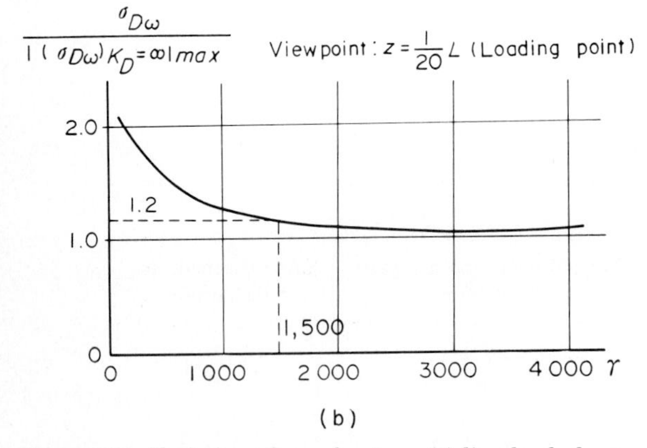

Figure 2.98 Variation of $\sigma_{D\omega}$ due to γ: (a) line load along span p, (b) concentrated load P.

for the line load along the bridge axis. Within these conditions, estimations of the distortional warping stress $\sigma_{D\omega}$ can be found from the approximate formula for the section at the diaphragm (right-hand side of Table 2.19) instead of the formula for the section at middiaphragm. Otherwise, the condition of Eq. (2.4.65) is not satisfied, the distortional warping stress for the section at middiaphragm will be greater than that indicated in Table 2.19, and additional exact analysis will be required for the distortional stress.

For the concentrated load of Fig. 2.98*b*, however, the stress ratio is reduced to

$$\frac{\sigma_{D\omega}}{\left|(\sigma_{D\omega})_{K_D=\infty}\right|_{\max}} \cong 1.2 \qquad (2.4.66)$$

2.4.5 Design formula for distortion[33]

On the basis of the above parametric analysis, the design formulas for the distortional warping stress and the corresponding information relative to both spacing and strength of intermediate diaphragms in a box girder can be summarized as follows:

(1) Distortional warping stress $\sigma_{D\omega}$. For a distributed live load P_l (kN/cm^2),

$$\sigma_{D\omega} = 1.8\frac{p_l B l_D^2}{16 W_l} \qquad (2.4.67)$$

For a knife-edge live load P_l (kN/cm),

$$\sigma_{D\omega} = 2.2\frac{P_l B l_D}{8 W_l} \qquad (2.4.68)$$

where B = clear width of roadway (cm)
l_D = spacing of intermediate diaphragm (cm)
W_l = section modulus at bottom plate of box girder (cm^3)

And loads p_l and P_l are applied to the half-width of B, as shown in Fig. 2.99, and the stress at locations other than point 3 can be estimated according to the distribution of the distortional warping function, as shown in Fig. 2.86.

(2) Appropriate intermediate diaphragm spacing. Let us now consider the combinations of live loads p_l and P_l shown in Fig. 2.99. The

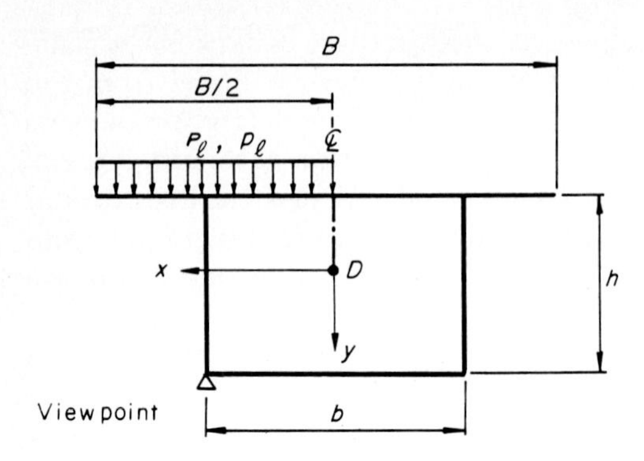

Figure 2.99 Load condition of live loads.

corresponding stresses are obtained by adding Eqs. (2.4.67) and (2.4.68):

$$\frac{\sigma_{D\omega}}{\sigma_b} = \frac{1.8p_l l_D^2 + 4.4P_l l_D}{p_l l^2 + 2P_l l} \tag{2.4.69}$$

From this equation, a condition which makes the distortional warping stress $\sigma_{D\omega}$ less than 5 percent of the flexural stress σ_b can be derived numerically:

$$\frac{\sigma_{D\omega}}{\sigma_b} \leqq 0.05 \tag{2.4.70}$$

For example, if we take the live-load intensities as

$$p_l = 0.35 \text{ tf/m}^2 \tag{2.4.71a}$$

$$P_l = 5.0 \text{ tf/m} \tag{2.4.71b}$$

according to the Japanese Specification of Highway Bridges, the relationship between span l and diaphragm spacing l_D can be plotted as shown in Fig. 2.100.

This indicates that the diaphragm spacing l_D (m) can be decided by the following equations:

$$l_D < 6 \qquad (l < 60 \text{ m}) \tag{2.4.72a}$$

$$l_D \leqq 0.14l - 2.4 \qquad (60 \text{ m} \leqq l \leqq 160 \text{ m}) \tag{2.4.72b}$$

$$l_D = 20 \qquad (l > 160 \text{ m}) \tag{2.4.72c}$$

for a straight box girder as indicated by Sakai and Nagai,[31] where l (m) is the span length of the box-girder bridges.

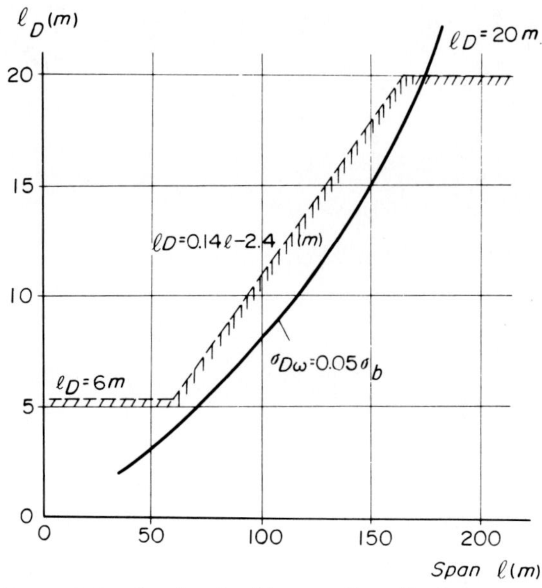

Figure 2.100 Spacing l_D of intermediate diaphragm.

(3) Required stiffness of diaphragms. The approximate formulas (2.4.67) and (2.4.68) should be adopted under the condition that

$$\gamma = \frac{K_D}{K_{D\omega} l_D} \geqq 1500 \tag{2.4.73}$$

where K_D = diaphragm rigidity [Eqs. (2.4.46) to (2.4.48), in kN·m]
$K_{D\omega}$ = distortional constant [Eqs. (2.4.30) and (2.4.31) in kilonewtons]
l_D = diaphragm spacing (m)

According to the results of the parametric survey based on actual bridges, the stiffness parameter γ can be plotted as shown in Fig. 2.101.[18] Since a few bridges are in the situation where $\gamma < 1500$, it is necessary to take care of these points in designing the box girders.

(4) Stress checks for diaphragms. The diaphragm should be designed with not only enough stiffness but also sufficient strength against the distortional stresses. Conservatively, the distortional warping moment $T^s_{D\omega}$ (in kN·m) can be approximately expressed by

$$T^s_{D\omega} = \frac{P_l b}{4} + \frac{p_l b l_D}{4} \tag{2.4.74}$$

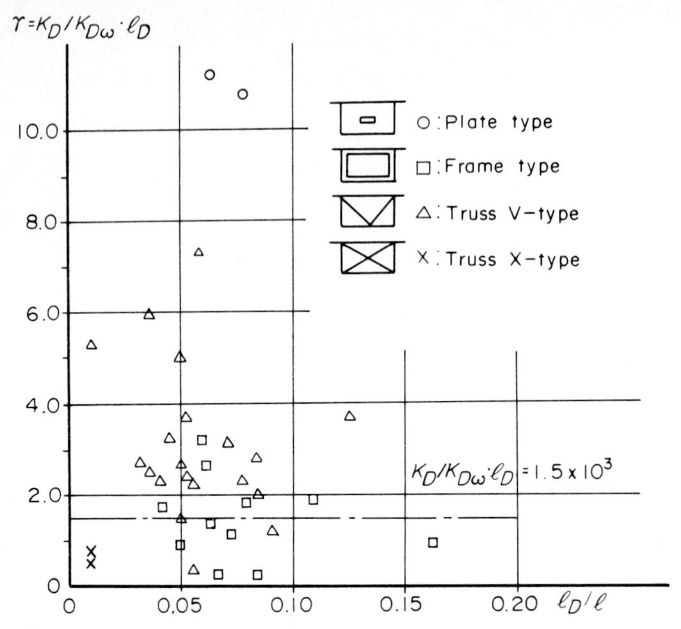

Figure 2.101 Variations of rigidity parameter γ.

By considering the loading condition of Fig. 2.99, the loads are

$$P = \frac{P_l B}{2} \tag{2.4.75}$$

and

$$p = \frac{p_l B}{2} \tag{2.4.76}$$

Therefore, Eq. (2.4.74) reduces to

$$T^s_{D\omega} = \frac{Bb}{8}(P + pl_D) \tag{2.4.77}$$

Thus, the stress check can easily be performed with Eqs. (2.4.50) to (2.4.55).

2.5 Stresses in Thin-Walled Beams

2.5.1 Cross-sectional quantities

The cross-sectional quantities of a thin-walled beam can be rearranged as shown in Table 2.20. By examining the analogy between these cross-

TABLE 2.20 Cross-Sectional Quantities

Item	Origin of coordinate	Static behavior	Cross-sectional quantities	Unit	Notation or formula	Analogy
Flexure	Centroid C	Bending and shear	Coordinate	cm	x, y	1
			Cross-sectional area	cm^2	$A = \int_A t\,ds$	2
			Static moment	cm^3	$S_x = \int_0^s yt\,ds$	3
			Moment of inertia	cm^4	$I_x = \int_A y^2 t\,ds$	4
Torsion	Shear center S	Pure torsion	Pure torsional function	cm^2	$\bar{q}_i$	
			Pure torsional constant	cm^4	$K = 2 \sum_{i=1}^{n} \bar{q}_i F_i$	2
		Torsional warping	Warping function	cm^2	ω	1
			Static moment	cm^4	$S_\omega = \int_0^s \omega t\,ds$	3
			Torsional warping constant	cm^6	$I_\omega = \int_A \omega^2 t\,ds$	4
Distortion	Distortional center D	Distortional warping	Warping function	cm^2	ω_D	1
			Static moment	cm^4	$S_{D\omega} = \int_0^s \omega_D t\,ds$	3
			Distortional warping constant	cm^6	$I_\omega = \int_A \omega_D^2 t\,ds$	4

sectional quantities, the warping functions ω and ω_D are entirely dependent on coordinates x and y, so that ω and ω_D are with respect to the same x and y coordinate axes as indicated by the mark 1 in Table 2.20. The warping functions ω and ω_D are also referred to as the *sector area coordinates*. In category 2 of Table 2.20, the formula for the cross-sectional area $A = \Sigma_{j=1}^{m} A_j$ is analogous to that of the torsional constant $K = \Sigma_{j=1}^{m} K_j$. In the remaining cross-sectional quantities 3 and 4, the y coordinate axis is replaced in the warping function by ω or ω_D, as shown in Table 2.20. Note that the following must be taken into account in calculating the cross-sectional quantities of a girder having the cross section illustrated in Fig. 2.102: For a girder with the steel deck, shown in Fig. 2.102*a*, it is difficult to evaluate the cross-sectional quantities exactly, because there are so many stiffening ribs on the top and bottom plates. In this case, the steel deck is replaced by a flat plate with the equivalent thickness $\bar{t}_u$, as determined by[3]

$$\bar{t}_u = t_u + \frac{mA_R}{2B} \tag{2.5.1}$$

where t_u = thickness of deck plate
A_R = cross-sectional area of a rib
m = total number of ribs
B = width of deck plate

The bottom plate can also be treated in the same manner as the deck plate. Thus the cross-sectional quantities for flexure can be estimated by these equivalent thicknesses. However, the original thickness t_u must be used in evaluating the cross-sectional quantities for pure torsion and statically indeterminate shear flow due to flexure and torsional warping.

In the case of the composite girder, shown in Fig. 2.102*b*, the equivalent thickness $\bar{t}_u$ should be categorized as follows:

1: For cross-sectional quantities for flexure and torsional warping,

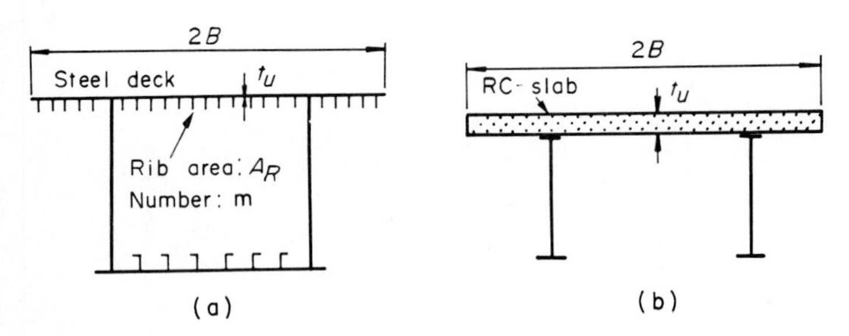

Figure 2.102 Cross section of (*a*) steel box girder and (*b*) composite girder.

except the calculation of statically indeterminate shear flow,

$$\bar{t}_u = \frac{t_u}{n} \tag{2.5.2}$$

2: For cross-sectional quantities for pure torsion and statically indeterminate shear flows due to flexure and torsional warping,

$$\bar{t}_u = \frac{t_u}{n_g} \tag{2.5.3}$$

where E = Young's modulus, G = shear modulus of elasticity, subscript s is for steel, and subscript c is for concrete, the Young's and shear modulus ratios are

$$n = \frac{E_s}{E_c} \tag{2.5.4a}$$

$$n_g = \frac{G_s}{G_c} \tag{2.5.4b}$$

That is, all the cross-sectional quantities are converted to the steel material.

2.5.2 Stress formula

The stress in a thin-walled beam can be summarized for the simple case where $I_{xy} = 0$, as shown in Table 2.21. The stress formulas for torsional and distortional warping are almost the same form as those for flexure.

In this table, since all the cross-sectional quantities are converted to

TABLE 2.21 Stress Formula

Item	Normal stress	Shearing stress	Analogy
Flexural stress	$\sigma_b = \frac{M_x}{nI_x} y$	$\gamma_b = \frac{Q_y}{nI_x t} S_x$	←
Pure torsional stress		$\gamma_{s,\max} = \frac{T_s}{K} t$ open section $\gamma_s = \frac{T_\omega}{2Ft}$ monobox	
Torsional warping stress	$\sigma_\omega = \frac{M_\omega}{nI_\omega} \omega$	$\gamma_\omega = \frac{T_\omega}{nI_\omega t} S_\omega$	←
Distortional warping stress	$\sigma_{D\omega} = \frac{M_\omega}{nI_{D\omega}} \omega_D$		←

the steel material, the value of Young's modulus ratio n should be taken as

$$n = \begin{cases} 1 & \text{steel material} \quad (2.5.5a) \\ \dfrac{E_s}{E_c} & \text{concrete material} \quad (2.5.5b) \end{cases}$$

2.5.3 Design criteria for stress combinations

In general, the elements of a thin-walled beam are subjected to not only biaxial normal stresses σ_z and σ_x in the longitudinal and transverse directions, respectively, but also shearing stress τ_{zx}, as shown in Fig. 2.103. For these stress situations, the combined stress σ_V for the steel material can be expressed by the von Mises' hypothesis as

$$\sigma_V = \sqrt{\sigma_z^2 + \sigma_x^2 - \sigma_z\sigma_x + 3\tau_{zx}^2} \tag{2.5.6}$$

where σ_z, σ_x, and τ_{zx}, are given from Table 2.21 as

$$\sigma_z = \sigma_b + \sigma_\omega + \sigma_{D\omega} \tag{2.5.7a}$$

$$\sigma_x = \sigma_{bt} + \sigma_{Db} \tag{2.5.7b}$$

$$\tau_{zx} = \tau_b + \tau_s + \tau_\omega \tag{2.5.7c}$$

in which σ_{bt} is the transverse bending stress of a beam, taking into account the effective width shown in Fig. 2.15 and Eq. (2.2.50), and σ_{Db} is the transverse distortional warping stress defined in Eq. (2.4.60).

By utilizing Eq. (2.5.6), the design formula can be gotten for a thin-walled beam. For example, the yield criteria $\sigma_V < \sigma_f$, including a certain factor of safety to be applied to Eq. (2.5.6), can be rewritten on the basis of the allowable stress method as

$$\left(\frac{\sigma_z}{\sigma_a}\right)^2 - \left(\frac{\sigma_z}{\sigma_a}\right)\left(\frac{\sigma_x}{\sigma_a}\right) + \left(\frac{\sigma_x}{\sigma_a}\right)^2 + \left(\frac{\tau}{\tau_a}\right)^2 \leqq 1.0 \tag{2.5.8}$$

for

$$\sigma_a = \frac{\sigma_f}{\nu} \tag{2.5.9}$$

$$\tau_a = \frac{\sigma_a}{\sqrt{3}} \tag{2.5.10}$$

where σ_f = yield stress
ν = factor of safety (=1.5–1.7)
σ_a = allowable normal stress
τ_a = allowable shearing stress

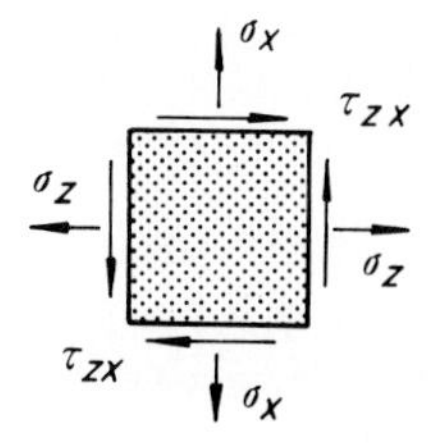

Figure 2.103 Stress in thin-walled beam.

In the case of $\sigma_x = 0$, the design criteria of Eq. (2.5.8) reduce to

$$\left(\frac{\sigma}{\sigma_a}\right)^2 + \left(\frac{\tau}{\tau_a}\right)^2 \leqq 1.0 \qquad (2.5.11)$$

where

$$\sigma = \sigma_b + \sigma_\omega + \sigma_{D\omega} \qquad (2.5.12a)$$

$$\tau = \tau_b + \tau_s + \tau_\omega \qquad (2.5.12b)$$

References

1. Vlasov, V. Z.: *Thin-Walled Elastic Beam*, National Science Foundation, Washington, D.C., 1967.
2. Galombos, T. V.: *Structural Members and Frames*, Prentice-Hall, Englewood Cliffs, N.J., 1968.
3. Komatsu, S.: *Theory and Calculation of Thin-Walled Structures*, Sankaido, 1969 (in Japanese).
4. Kollbrunner, Hajdin: *Dünnwandige Stäbe, Bd.1, Stabe mit undeformierbaren Querschnitten*, Springer-Verlag, Berlin, 1972.
5. Heins, C. P.: *Bending and Torsional Design in Structural Members*, Lexington Books, Lexington, Mass., 1975.
6. Reissner, E.: "Analysis of Shear Lag in Box Beams by Principle of Minimum Potential Energy," *Quarterly of Applied Mathematics*, 4(3):268–278, 1946.
7. Chwalla, E.: *Uber das Problem der voll mittragender Breite von Gurt und Rippen*, Alfons-Leon-Gederschrift, Wein, 1952.
8. Timoshenko, S. P., and J. N. Goodier: *Theory of Elasticity*, 2d ed., McGraw-Hill, New York, 1952.
9. Girkmann, K.: *Flachentragwerke, Dritte Auflage*, Springer-Verlag, Wein, 1954.
10. Kuhn, P.: *Stress in Aircraft and Shell Structure*, McGraw-Hill, New York, 1956.
11. Hawranek, Steinhardt: *Theorie und Berechnung der Stahlbrücken*, Springer-Verlag, Berlin, 1958.
12. Kondo, K., S. Komatsu, and H. Nakai: "Theoretical and Experimental Researches on Effective Width of Girder Bridges with Steel Deck Plates," *Transactions of the Japanese Society of Civil Engineers*, no. 86, pp. 1–17, Oct. 1962 (in Japanese).
13. Deutchen Stahlbau Verbund: *Ein Handbuch für Studium und Praxis*, Bd. 3, Stahlbau-Verlag, Köln (in German), 1962.
14. Abdel-Samad, S. R.: "Effective Width of Steel Deck Plates," *Proceedings of the American Society of Civil Engineers*, vol. 93, no. ST7, pp. 1459–1473, July 1967.
15. Moffatt, K. R.: "Finite Element Analysis of Box Girder Bridges," Ph.D. thesis, Imperial College of Science and Technology, London, 1974.
16. Moffatt, K. R., and P. J. Dowling: "Shear Lag in Steel Box Girder Bridges," *Structural Engineer*, 53(10):439, 448, Oct. 1975.
17. Nakai, H., and H. Kotoguchi: "Studies on Shear Lag and Effective Width of Continuous Girder Bridge by Transfer Matrix Method," *Proceedings of the Japanese Society of Civil Engineers*, no. 252, pp. 29–44, Aug. 1976 (in Japanese).

18. Japanese Construction Consultants Association, Kinki Branch: *Survey on Long Span Box Girder Bridges*, Report no. 515-33, March 1979 (in Japanese).
19. Federal Highway Administration Office of Research and Development: *Proposed Specification for Steel Box Girder Bridges*, FHWA-80-205, Jan. 1980.
20. British Standard Institution: *Code of Practice for Design of Steel Bridges*, BS5400, part 3, April 1982.
21. Bleich, F.: *Buckling Strength of Metal Structure*, McGraw-Hill, New York, 1952.
22. Stüssi, F.: *Entwurf und Berechnung von Stahlbauten, Erster Bd.*, Springer-Verlag, Berlin, 1958.
23. Timoshenko, S. P., and J. M. Gere: *Theory of Elastic Stability*, 2d ed., McGraw-Hill, New York, 1961.
24. Kollbrunner, C. F., and K. Basler: *Torsion*, Springer-Verlag, Berlin, 1966 (in German).
25. Flint, A. R., and M. R. Horne: "Conclusion of Research Programme and Summary of Parametric Studies, Steel Box Girder Bridges," *Proceedings of the International Conference Organized by ICE*, London, 1973, pp. 173–191.
26. Nakai, H., and H. Kotoguchi: "Applications of Transfer Matrix Method for Analyzing the Stress-Resultants and Displacements of Thin-Walled Girder Bridges Subjected to Non-uniform Torsion," *Proceedings of the Japanese Society of Civil Engineers*, no. 223, pp. 1–15, Jan. 1975 (in Japanese).
27. Nakai, H., and T. Tani: "An Approximate Method for the Evaluation of Torsional and Warping Stress in Box Girder Bridges," *Proceedings of the Japanese Society of Civil Engineers*, no. 277, pp. 41–55, Sept, 1978 (in Japanese).
28. Wright, R. N., S. R. Abdel-Samad, and A. R. Robinson: "BEF Analogy for Analysis of Box Girder," *Proceedings of American Society of Civil Engineers*, 97(7):1719–1743, July 1968.
29 Abdel-Samad, S. R., R. N. Wright, and A. R. Robinson: "Analysis of Box Girder with Diaphragms," *Proceedings of American Society of Civil Engineers*, vol. 94, no. ST-10, pp. 2231–2256, Oct. 1968.
30. Dabrowski, R.: *Gekrümmte dünnwandige Träger*, Springer-Verlag, Berlin, 1968.
31. Sakai, F., and M. Nagai: "A Recommendation on the Design of Intermediate Diaphragms in Steel Box Girder Bridges," *Proceedings of Japanese Society of Civil Engineers*, no. 261, pp. 21–34, Aug. 1977 (in Japanese).
32. Nakai, H., and T. Miki: "Theoretical and Experimental Research on Distortion of Thin-Walled Horizontally Curved Box Girder," *Journal of Civil Engineering Design*, 2(1):63–101, 1980.
33. Nakai, H., and T. Murayama: "Distortional Stress Analysis and Design Aid for Horizontally Curved Girder Bridges," *Proceedings of the Japanese Society of Civil Engineers*, no. 277, pp. 25–39, Sept. 1978 (in Japanese).
34. Japanese Road Association: *The Japanese Specification for Highway Bridges*, Feb. 1973 (in Japanese).

Chapter

3

Fundamental Theory of Curved Girders for Analyzing Static and Dynamic Behavior

3.1 Introduction

This chapter presents the fundamental theory and various behavior, including static and dynamic responses of a thin-walled curved beam as described by the fundamental equations presented in Chap. 2.

First, the analytical methods of curved beams based on the coupled behavior of flexure and pure torsion or flexure and warping torsion are introduced according to the thin-walled elastic beam theory. Next, the fundamental equation for the stresses due to flexure, pure torsion, and torsional warping as well as the corresponding cross-sectional quantities are rigorously derived by taking into consideration the characteristics of a curved beam. In addition to these topics, the basic equation for the shear lag phenomenon in a curved beam is derived by means of Reissner's hypothesis and the generalized Galerkin method. Furthermore, the fundamental equation for distortion in a curved box beam is derived by virtue of the beam-on-elastic-foundation (BEF) analogy. Then this method is compared to experimental studies. Finally, the theory of sector plates is introduced to discuss the static behavior of deck plates and the stability problems of plate members in the curved girders.

Section 3.3 deals with the combined interactions of flexural, torsional, and distortional behavior in a curved beam subjected to static loads. Applying an approximate method to evaluate torsional warping and distortional warping stresses as well as deflections, we select a few important parameters through various numerical calculations, using data from actual curved girder bridges. It is suggested that in the regions where the estimations of warping stress can be approximated by the simplified method adopting the pure torsional theory, the torsional warping stresses are related to the values of torsional parameters and the central angle of curved girders. From the analyses of distortional warping stresses, valuable information concerning the estimation of distortional warping stress, spacing of the intermediate diaphragms in connection with the central angle, and appropriate rigidity or strength of the diaphragms for a curved girder is proposed in the form of simplified design criteria. Analyses of deflections show that the torsional-to-flexural rigidities ratio is an important parameter in determining the relationships between span length and radius of curvature of curved bridges with cross sections such as monobox, twin-box, or I girders.

The dynamic response of a curved beam is distinguished from that of straight girders because the flexural vibrations are always coupled with the torsional ones, much like the static behavior explained in Secs. 3.3 and 3.4. In Sec. 3.4, the fundamental equations for the vibrations of a curved beam are derived on the basis of Lagrange's equation by considering the coupled vibrations of bending and torsion. The basic equations, including the derivation of the effective mass, stiffness, and damping coefficients of a curved girder, are determined and then incorporated into a modal analysis of a curved girder system. Furthermore, the fundamental differential equations for free vibrations with and without damping, various forced vibrations, and coupled vibrations under the moving vehicles are developed in detail. The responses obtained from the field experimental study of a continuous curved girder bridge are simulated by an analog computer and are compared with the analytical results.

To establish a rational impact coefficient for curved girder bridges, the derivation of the dynamic amplification factor (DAF) due to a single vehicle and the effects of various parameters such as cross-sectional quantities, the plane configuration of curved girder bridges, the frequency ratio between curved girders and vehicles as well as the mass ratio, loading conditions, speed, and pulsating forces of vehicles are presented in Sec. 3.5. Based on these parametric analyses and experimental data, a realistic model of traffic flows is composed and applied to the actual curved girder bridges built in Japan having different cross-sectional shapes, span lengths, and radii of curvature to determine the values of the DAF of these bridges through a deterministic simulation method on a digital computer.

3.2 Fundamental Theory of Curved Beams under Static Forces[1–11]

3.2.1 Theory of curved beams based on flexure and pure torsion

This theory can be applied to a curved beam with the prismatic or box section where only the flexural rigidity EI_x and the pure torsional rigidity GK are taken into consideration in the analysis. In this case, the centroid C is coincident with the shear center S, and the beam axis lies along a circular arc with central angle Φ and constant radius of curvature R.

(1) Equilibrium of forces. Let us now consider the equilibrium condition of forces acting on a curved beam element

$$ds = R\,d\phi \tag{3.2.1}$$

as shown in Fig. 3.1 where

p = distributed vertical load

m_T = distributed torque around x axis

Q_y = shearing force in y axis (vertical axis)

M_x = bending moment around x axis (horizontal axis)

T_z = torsional moment around z axis (tangential axis)

w = deflection in y axis

β = rotation angle around z axis

Φ = central angle of curved beam

R = radius of curvature on centroidal beam axis

(x, y, z) = right-hand rectangular coordinate system taken at centroidal axis

ϕ = angular coordinate axis

The equilibrium of forces $\Sigma V = 0$ in the direction of the y axis, shown in Fig. 3.2, gives[1–10]

$$Q_y + dQ_y - Q_y + p\,ds = 0 \tag{3.2.2a}$$

or

$$\frac{dQ_y}{ds} = -p \tag{3.2.2b}$$

and this equation is similar to that of a straight beam.

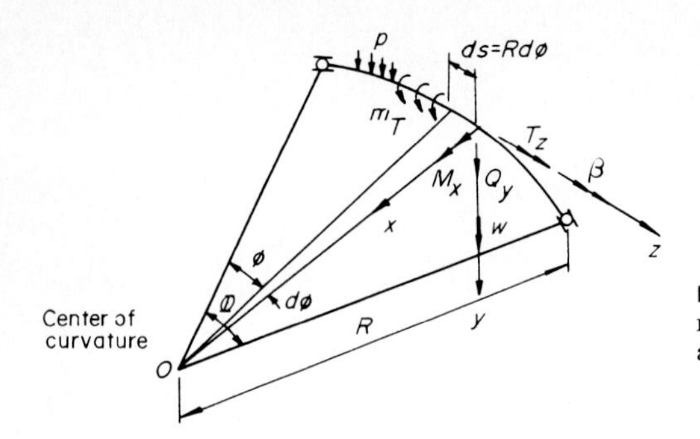

Figure 3.1 Applied load, stress resultants, and displacements of a curved beam.

The moment equilibrium $\Sigma M = 0$ leads to

$$M_x + dM_x - M_x - Q_y R\, d\phi + T_z\, d\phi + \frac{p(R\, d\phi)^2}{2} = 0 \qquad (3.2.3a)$$

or

$$\frac{dM_x}{ds} + \frac{T_z}{R} = Q_y \qquad (3.2.3b)$$

Similarly, the equilibrium condition for the torque $\Sigma T = 0$ reduces to

$$T_z + dT_z - T_z - M_x\, d\phi + m_T\, ds = 0 \qquad (3.2.4a)$$

or

$$\frac{dT_z}{ds} - \frac{M_x}{R} = -m_T \qquad (3.2.4b)$$

By differentiating Eq. (3.2.3*b*) with respect to the curvilinear coordinate s and substituting Eq. (3.2.4*b*) into this equation, the relationships among the bending moment M_x, load p, and torque m_T can be

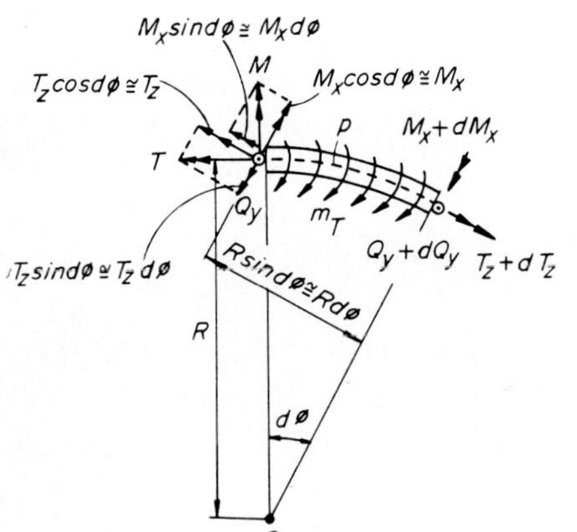

Figure 3.2 Equilibrium of forces.

expressed as

$$\frac{d^2M_x}{ds^2}+\frac{M_x}{R^2}=-p+\frac{m_T}{R} \tag{3.2.5}$$

The solution of the above equation reduces to

$$M_x = A\sin\phi + B\cos\phi - pR^2 + m_T R \tag{3.2.6}$$

and the corresponding boundary conditions to determine the two integration constants A and B are

$$[M_x]_{\phi=0} = 0 \tag{3.2.7a}$$

$$[M_x]_{\phi=\Phi} = 0 \tag{3.2.7b}$$

for a simply supported curved beam at both ends, that is, $\phi = 0$ and $\phi = \Phi$.

In general, the solution of the pure torsional moment T_z can be written

$$T_z = \int_0^s \left(\frac{M_x}{R} - m_T\right) ds + C \tag{3.2.8}$$

However, the integration constant C in the above equation cannot be determined by the equilibrium condition of forces without considering the compatibility condition of displacements.

In addition, the shearing force Q_y, bending moment M_x, and torsional moment T_z are always coupled with one another owing to the curvature of the beam axis, as indicated in Eqs. (3.2.3) to (3.2.5).

(2) Relationship between stress resultants and displacements. It is obvious from Fig. 3.3 that the deflection angle about the x axis, dw/ds,

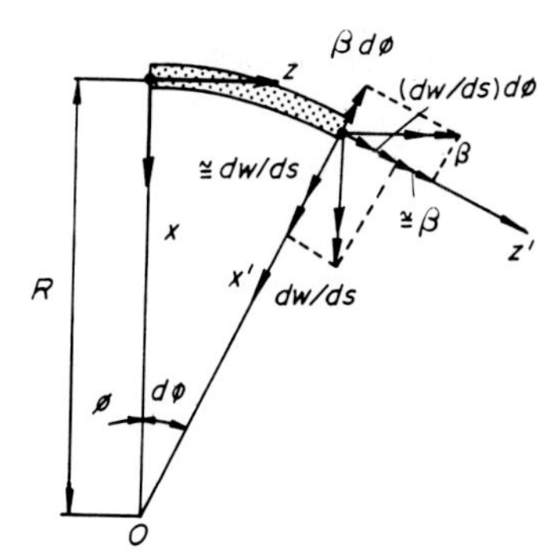

Figure 3.3 Changes in deflection angle and rotation angle in a curved beam.

and the rotation angle around the z axis, β, at section ϕ will change at section $\phi + d\phi$ as follows:

$$\text{Deflection angle around } x \text{ axis} = \frac{dw}{ds} - \beta \, d\phi$$

$$\text{Rotation angle around } z \text{ axis} = \beta + \frac{dw}{ds} \, d\phi$$

Since the derivatives of the above relationships with respect to the curvilinear coordinate s are, respectively, defined as the change of curvature and the torsional angle, the relationships between the bending moment M_x and the torsional moment T_z can be obtained:

$$\frac{d^2w}{ds^2} - \frac{\beta}{R} = -\frac{M_x}{EI_x} \tag{3.2.9a}$$

$$\frac{d\theta}{ds} = \frac{T_z}{GK} \tag{3.2.9b}$$

where EI_x = flexural rigidity
GK = pure torsional rigidity

and

$$\theta = \beta + \frac{w}{R} \tag{3.2.10}$$

is referred to as the torsional angle of the curved beam, as shown in Fig. 3.4.[4] Equation (3.2.9) can be written as

$$\frac{d^3w}{ds^3} + \frac{1}{R^2}\frac{dw}{ds} = \frac{1}{R}\frac{T_z}{GK} - \frac{1}{EI_x}\frac{dM_x}{ds} \tag{3.2.11a}$$

$$\beta = \left(\frac{d^2w}{ds^2} + \frac{M_x}{EI_x}\right) R \tag{3.2.11b}$$

Accordingly, these equations can be easily solved for the deflection w and rotation angle β, provided that the bending moment M_x and the torsional moment T_z are given as detailed in Sec. 4.2.

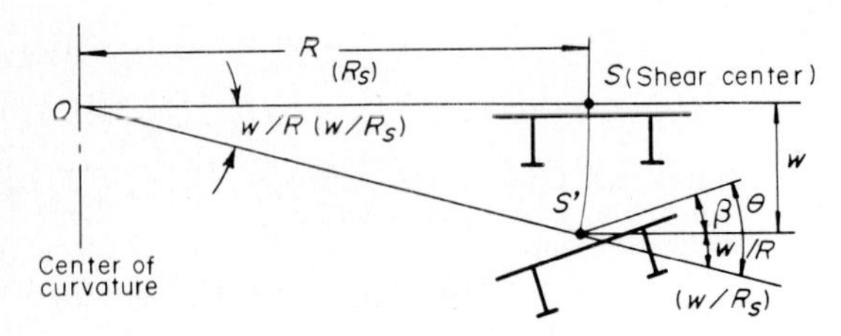

Figure 3.4 Definition of torsional angle.

3.2.2 Theory of the thin-walled curved beam based on torsional warping[4,8]

(1) Definition of notation. For a curved beam consisting of thin-walled plates, the torsional warping behavior can be determined as explained in Sec. 2.3.2. Therefore, the fundamental equation governing the behavior of a thin-walled curved beam is explained in this section.

Figure 3.5 shows a cross section of a thin-walled curved beam. For this asymmetric cross section, the centroid C does not coincide with the shear center S, so that the right-hand coordinate axis and additional subcoordinates are defined as follows:

(X, Y, Z) = horizontal, vertical, and tangential coordinate axes, respectively, taken at centroid C

(x, y, z) = horizontal, vertical, and tangential coordinate axes, respectively, taken at shear center S

$$s = R_s\phi \tag{3.2.12a}$$

= curvilinear coordinate axis along curved beam

ϕ = angular coordinate axis

R_s = radius of curvature at shear center S

R_o = radius of curvature at centroid C

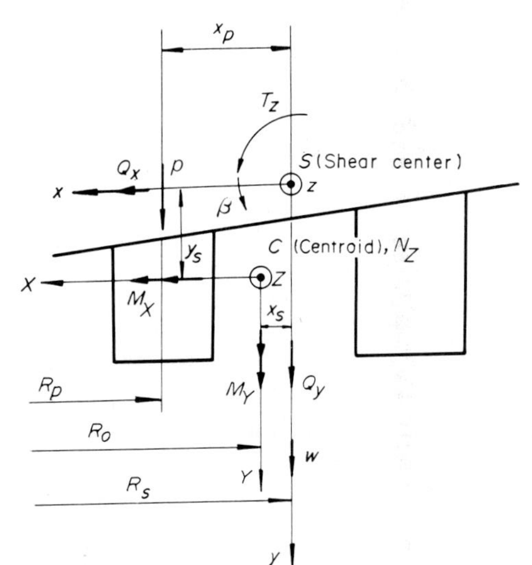

Figure 3.5 Definitions of coordinate axes, stress resultants, and displacements.

R_p = radius of curvature at loading point of concentrated load P or distributed load p

$$x_p = R_s - R_p \tag{3.2.13}$$

= eccentricity of applied load p

x_s, y_s = eccentricity of centroid C in reference to the shear center S in the direction of x and y coordinate axes

Various stress resultants occur as shown in Fig. 3.5. Among them, the bending moment and axial force should be applied to the centroid C, whereas the shearing forces Q_x and Q_y and torsional moment T_z should be applied to the shear center S to avoid the complicated coupled behaviors of these stress resultants. Thus the stress resultants can be designated by subscripts corresponding to their coordinate axes as follows:

p = distributed vertical load

M_X, M_Y = bending moment acting on the centroidal X and Y coordinate axes, respectively

N_Z = axial force in the Z centroidal axis

Q_x, Q_y = shearing force acting on the shear center x and y coordinate axes, respectively

T_z = torsional moment with respect to the shear center z axis

However, the displacements can also be defined as follows:

w = deflection in the direction of y coordinate axis

β = rotation angle with respect to z axis

$$\theta = \beta + \frac{w}{R_s} \tag{3.2.14}$$

θ = torsional angle with respect to the z axis (see Fig. 3.4)

(2) Fundamental equations for stress resultants and displacements.[51] The equilibrium of forces relative to the shearing force Q_x, bending moment M_X, and the torsional moment T_z acting on a differential segment on the shear center axis, given by

$$ds = R_s\, d\phi \tag{3.2.12b}$$

yields the following simultaneous equations:

$$\frac{dQ_y}{ds} + \frac{R_p}{R_s} p = 0 \tag{3.2.15a}$$

$$\frac{dM_X}{ds} - Q_y + \frac{T_z}{R_s} + y_s \frac{dN_Z}{ds} = 0 \tag{3.2.15b}$$

$$\frac{dT_z}{ds} - \frac{M_X}{R_s} - \frac{y_s}{R_s} N_Z + \frac{R_p}{R_s} x_p p = 0 \tag{3.2.15c}$$

When there are no loads in the direction of the Z and X coordinate axes, the following equations can be obtained for the shearing force Q_x and normal force N_Z:

$$\frac{dN_Z}{ds} - \frac{Q_x}{R_s} = 0 \tag{3.2.16a}$$

$$\frac{dQ_x}{ds} + \frac{N_Z}{R_s} = 0 \tag{3.2.16b}$$

A set of the above equations can also be written as

$$\frac{d^2N_Z}{ds^2} + \frac{N_Z}{R_s^2} = 0 \tag{3.2.17a}$$

or

$$N_Z = A \sin \phi + B \cos \phi \tag{3.2.17b}$$

In an ordinary curved girder, however, the bearing shoes at the supports allow free movement in the direction of the girder axis, so that

$$N_Z = 0 \tag{3.2.18}$$

at an arbitrary section, and the shearing force Q_x is reduced to

$$Q_x = 0 \tag{3.2.19}$$

in the case where only a vertical load acts on the curved beam. Thus, Eq. (3.2.16) can be neglected in analyzing the static behavior of the curved beam.

To solve for the stress resultants, the basic equations for the shearing force Q_y, bending moment M_x, and torsional moment T_z can also be determined from Eqs. (3.2.15*a*) to (3.2.15*c*) and (3.2.18) as follows:

$$\frac{dQ_y}{ds} = -\frac{R_p}{R_s} p \tag{3.2.20}$$

$$\frac{d^2M_X}{ds^2} + \frac{M_X}{R_s^2} = -\left(\frac{R_p}{R_s}\right)^2 p \tag{3.2.21}$$

$$\frac{dT_z}{ds} = \frac{M_X}{R_s} - \frac{R_p}{R_s} x_p p \tag{3.2.22}$$

In addition, the basic equation for torsional warping moment (bimoment) M_ω can be expressed in the same form as Eq. (2.3.76) for a curved beam:

$$M_\omega = E_s I_\omega \frac{d^2\theta}{ds^2} \tag{3.2.23}$$

where I_ω = torsional warping constant (converted to steel material) and E_s = Young's modulus for steel. Therefore, the basic equation for the torsional angle θ can be adopted from Eq. (2.3.81), so that

$$E_s I_\omega \frac{d^4\theta}{ds^4} - G_x K \frac{d^2\theta}{ds^2} = -\frac{dT_z}{ds} \tag{3.2.24a}$$

Or the substitution of Eqs. (3.2.22) and (3.2.23) into the above equation yields

$$\frac{d^2M_\omega}{ds^2} - \alpha^2 M_\omega = \frac{R_P}{R_s} x_p p - \frac{M_X}{R_s} \tag{3.2.25}$$

where the torsional parameter α is

$$\alpha = \sqrt{\frac{G_s K}{E_s I_\omega}} \tag{3.2.26a}$$

and K = pure torsional constant (converted to steel material) and G_s = shear modulus of elasticity for steel.

Thereafter, the torsional angle θ can be obtained by integrating the bimoment M_ω with respect to the curvilinear coordinate s twice as follows:

$$\theta = \int_0^s \left(\int \frac{M_\omega}{E_s I_\omega} ds \right) ds + C_1 s + C_2 \tag{3.2.27}$$

Here C_1 and C_2 are integration constants, which can be determined from the boundary conditions of a thin-walled curved beam.

The remaining stress resultants, i.e., the pure torsional moment T_s and the secondary torsional moment T_ω can be obtained from

$$T_s = G_s K \frac{d\theta}{ds} \tag{3.2.28}$$

$$T_\omega = -E_s I_\omega \frac{d^3\theta}{ds^3} \tag{3.2.29}$$

Finally, the deflection w is described by

$$\frac{d^2w}{ds^2} + \frac{w}{R_s^2} = -\frac{M_X}{E_sI_X} + \frac{\theta}{R_s} \tag{3.2.30}$$

The solutions for these stress resultants and displacements are explained in Sec. 4.3.

3.2.3 Cross-sectional quantities and stress formulas for flexure and torsion[4,9,10,12–14]

(1) Flexure. Figure 3.6 shows a general cross section of a curved girder, where the centroid is defined by C and the vertical and horizontal axes are, respectively, denoted X and Y. The coordinate axis Z is taken along the radial axis of the curved beam. The radius of curvature at the centroid C and the corresponding central angle are designated R_o and ϕ, respectively.

Since the torsional moment T_z, which is discussed in detail in the following section, is equal to zero in the case of pure bending, the

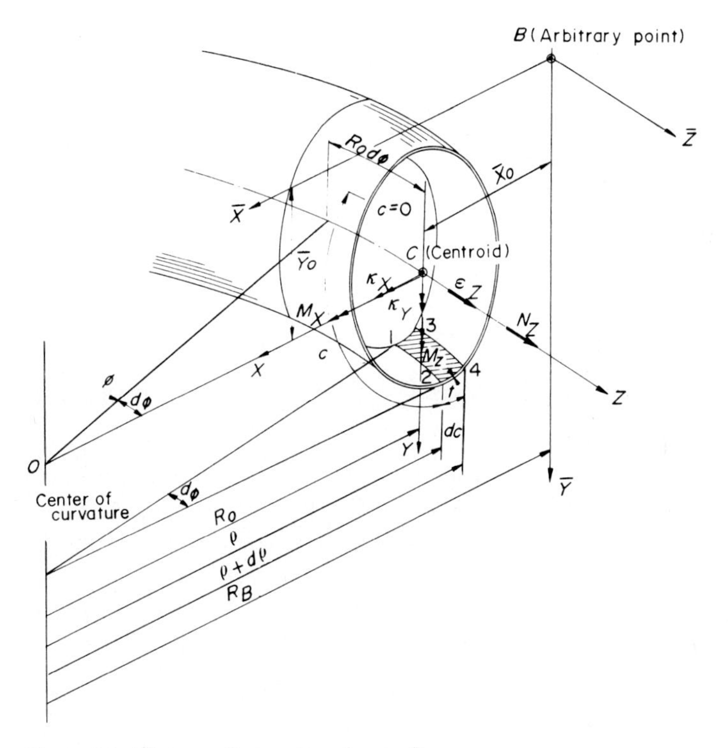

Figure 3.6 Change of curvature due to flexure.

stresses due to the bending moment M_X or M_Y and the normal force N_Z will be functions of the changes in curvature κ_X and κ_Y and the axial strain ε_Z on the centroid C, as shown in Fig. 3.6.

For a small element 1,2,3,4 cut from the radius of curvature, $\rho + d\rho$ and the central angle $d\phi$, as illustrated in Fig. 3.6, the flexural normal stress σ_b is given by the product of strain $du/(\rho\, d\phi)$ and the modulus of elasticity E_s, or

$$\sigma_b = E_s \frac{du}{\rho\, d\phi} = \frac{E_s R_o}{n\rho}(\varepsilon_Z + \kappa_X Y - \kappa_Y X) \tag{3.2.31}$$

where E_s = Young's modulus for steel and E_c = Young's modulus for concrete, or

$$n = \frac{E_s}{E_c} \tag{3.2.32}$$

Now, by introducing the new curvilinear coordinate axis c along the perimeter of the cross section and the thickness t of the member, as shown in Fig. 3.6, the relationships among the flexural stress σ_b and the stress resultants M_X, M_Y, and N_Z can be satisfied by the following equilibrium conditions:

$$\int_A \sigma_b t\, dc = N_Z \tag{3.2.33a}$$

$$\int_A \sigma_b Yt\, dc = M_X \tag{3.2.33b}$$

$$\int_A \sigma_b Xt\, dc = M_Y \tag{3.2.33c}$$

Here A represents the integration across the cross-sectional area $t\, dc$.

Substitution of Eq. (3.2.31) into Eq. (3.2.33) gives

$$N_Z = E_s A_Z \varepsilon_Z \tag{3.2.34a}$$

$$M_X = E_s(I_X \kappa_X - I_{XY} \kappa_Y) \tag{3.2.34b}$$

$$M_Y = E_s(I_{XY} \kappa_X - I_Y \kappa_Y) \tag{3.2.34c}$$

where

$$A_Z = \int_A \frac{R_o}{\rho} \frac{t}{n}\, dc \tag{3.2.35a}$$

$$S_X = \int_A \frac{R_o}{\rho} \frac{Yt}{n}\, dc = 0 \tag{3.2.35b}$$

$$S_Y = \int_A \frac{R_o}{\rho} \frac{Xt}{n} dc = 0 \qquad (3.2.35c)$$

$$I_X = \int_A \frac{R_o}{\rho} \frac{Y^2 t}{n} dc \qquad (3.2.35d)$$

$$I_Y = \int_A \frac{R_o}{\rho} \frac{X^2 t}{n} dc \qquad (3.2.35e)$$

$$I_{XY} = \int_A \frac{R_o}{\rho} \frac{XYt}{n} dc \qquad (3.2.35f)$$

However, the locations of the centroid C has not been selected. So, taking an arbitrary point B with a radius of curvature R_B and denoting the $\bar{X}$, $\bar{Y}$, and $\bar{Z}$ coordinate axes at that point B, we see that the X and Y coordinate axes are given by

$$X = \bar{X} - \bar{X}_o \qquad (3.2.36a)$$

$$Y = \bar{Y} - \bar{Y}_o \qquad (3.2.36b)$$

Substitution of this equation into Eqs. (3.2.35*b*) and (3.2.35*c*) gives

$$S_X = \frac{R_c}{R_B}(S_{\bar{X}} - \bar{Y}_o A_{\bar{Z}}) = 0 \qquad (3.2.37a)$$

$$S_Y = \frac{R_c}{R_B}(S_{\bar{Y}} - \bar{X}_o A_{\bar{Z}}) = 0 \qquad (3.2.37b)$$

The centroid $(\bar{X}_o, \bar{Y}_o)$ can now be established as

$$\bar{X}_o = \frac{S_{\bar{Y}}}{A_{\bar{Z}}} \qquad (3.2.38a)$$

$$\bar{Y}_o = \frac{S_{\bar{X}}}{A_{\bar{Z}}} \qquad (3.2.38b)$$

where the radius of curvature R_o at the centroid C is given by

$$R_o = R_B - \bar{X}_o \qquad (3.2.39)$$

Accordingly, the geometric moments of inertia I_X and I_Y and the product of inertia I_{XY} can be estimated by Eqs. (3.2.35*d*) to (3.2.35*f*), which gives

$$I_X = \frac{R_c}{R_B}(I_{\bar{X}} - \bar{Y}_o^2 A_{\bar{Z}}) \tag{3.2.40a}$$

$$I_Y = \frac{R_c}{R_B}(I_{\bar{Y}} - \bar{X}_o^2 A_{\bar{Z}}) \tag{3.2.40b}$$

$$I_{XY} = \frac{R_c}{R_B}(I_{XY} - \bar{X}_o \bar{Z}_o A_{\bar{Z}}) \tag{3.2.40c}$$

where

$$A_{\bar{Z}} = \int_A \frac{R_B}{\rho}\frac{t}{n}\,dc \tag{3.2.41a}$$

$$S_{\bar{X}} = \int_A \frac{R_B}{\rho}\frac{\bar{Y}t}{n}\,dc \tag{3.2.41b}$$

$$S_{\bar{Y}} = \int_A \frac{R_B}{\rho}\frac{\bar{X}t}{n}\,dc \tag{3.2.41c}$$

$$I_{\bar{X}} = \int_A \frac{R_B}{\rho}\frac{\bar{Y}^2 t}{n}\,dc \tag{3.2.41d}$$

$$I_{\bar{Y}} = \int_A \frac{R_B}{\rho}\frac{\bar{X}^2 t}{n}\,dc \tag{3.2.41e}$$

$$I_{\bar{X}\bar{Y}} = \int_A \frac{R_B}{\rho}\frac{\bar{X}\bar{Y}t}{n}\,dc \tag{3.2.41f}$$

Solving Eqs. (3.2.34*b*) and (3.2.34*c*) for the changes in curvature κ_X and κ_Y and then substituting Eqs. (3.2.34*a*) to (3.2.34*c*) into Eq. (3.2.31) give the resultant normal stress σ_b due to the bending moments M_X and M_Y and axial force N_Z

$$\begin{aligned}\sigma_b &= \frac{R_o}{n\rho}\left[\frac{N_A}{A_Z} - E_s(\kappa_Z X - \kappa_Y Y)\right] \\ &= \frac{R_o}{n\rho}\left(\frac{N_Z}{A_Z} + \frac{M_Y I_X - M_X I_{XY}}{I_X I_Y - I_{XY}^2}X + \frac{M_X I_Y - M_Y I_{XY}}{I_X I_Y - I_{XY}^2}Y\right)\end{aligned} \tag{3.2.42}$$

For the case when $N_Z = 0$ and $M_Y = 0$, Eq. (3.2.42) can be written as

$$\sigma_b = \frac{R_o}{n\rho}\frac{I_Y Y - I_{XY} X}{I_X I_Y - I_{XY}^2} M_X \tag{3.2.43}$$

To examine the shear flow due to bending, which is defined as

$$q_b = \tau_b t \tag{3.2.44}$$

consider the equilibrium condition of the small element 1,2,3,4, as

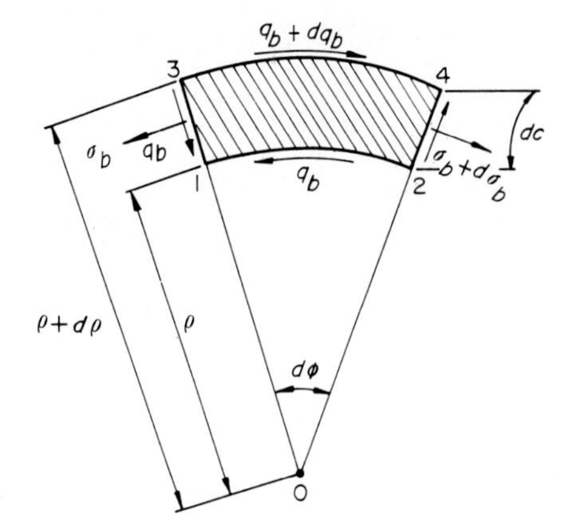

Figure 3.7 Equilibrium of forces for a small element.

shown in Fig. 3.7. Considering the equilibrium of all the forces in the axial direction, we can derive the following differential equation:

$$\frac{d\sigma_b t}{\rho\, d\phi} + 2\frac{q_b}{\rho}\frac{d\rho}{dc} + \frac{dq_b}{dc} = 0 \tag{3.2.45a}$$

and this can be rewritten as

$$\frac{d}{dc}(q_b\rho^2) = -R_s\rho\frac{d\sigma_b t}{ds} \tag{3.2.45b}$$

where

$$ds = R_s\, d\phi \tag{3.2.12b}$$

This is with respect to the curvilinear coordinates located at the shear central axis. Substitution of Eq. (3.2.43) into the right-hand side of Eq. (3.2.45*b*) gives

$$\rho\frac{d\sigma_b t}{ds} = \rho\frac{R_o}{n\rho}\frac{I_Y Y - I_{XY}X}{I_X I_Y - I_{XY}^2}\frac{dM_X}{ds}t \tag{3.2.46a}$$

$$= \frac{R_o}{n}\frac{I_Y Y - I_{XY}X}{I_X I_Y - I_{XY}^2}\bar{Q}t \tag{3.2.46b}$$

where

$$\bar{Q} = \frac{dM_X}{ds} = Q_Y - \frac{T_z}{R_s} \tag{3.2.47}$$

and this is the shearing force due to bending described in the preceding section. The shear flow q_b due to bending can now be obtained by integration of Eq. (3.2.45b) with respect to c, which gives

$$q_b = \tau_b t = -\frac{R_o R_s}{\rho^2}\left(\frac{I_Y S_{X,b} - I_{XY} S_{Y,b}}{I_X I_Y - I_{XY}^2}\,\bar{Q} - C\right) \tag{3.2.48a}$$

$$= \frac{R_o R_s}{\rho^2}(\bar{q}_b + C) \tag{3.2.48b}$$

where
$$\bar{q}_b = -\frac{I_Y S_{X,b} - I_{XY} S_{Y,b}}{I_X I_Y - I_{XY}^2}\,\bar{Q} \tag{3.2.49}$$

$$S_{X,b} = \int_0^c \frac{Yt}{n}\,dc \tag{3.2.50a}$$

$$S_{Y,b} = \int_0^c \frac{Xt}{n}\,dc \tag{3.2.50b}$$

and C is an integration constant.

As given in Konishi and Komatsu,[4] the following condition must be satisfied for the determination of C:

$$\oint \frac{q_b n_g}{\rho t}\,dc = 0 \tag{3.2.51}$$

where
$$n_g = \frac{G_s}{G_c} \tag{3.2.52}$$

and G_s = shear modulus of elasticity for steel and G_c = shear modulus of elasticity for concrete. For example, this condition will reduce to

$$C = -\frac{\oint (R_o/\rho)^3(\bar{q}_b n_g/t)\,dc}{\oint (R_o/\rho)^3(n_g/t)\,dc} \tag{3.2.53}$$

for a box girder with a single cell, and $C = 0$ for the free edge of the open cross section. The generalized formulas for girders with various cross-sectional shapes are discussed in App. A.

(2) Torsion. Figure 3.8 shows the coordinate system for the analysis of torsional behavior, where the applied torsional moment T_z acts about the tangential axis z which is taken at the shear center S.

Let us now examine the deformation of a small element 1,2,3,4 cut by $\rho\,d\phi$, $(\rho + d\rho)\,d\phi$, and dc, as illustrated in Fig. 3.8. The displacement due to the torsional moment T_z consists of two parts, i.e., the twisting angle θ with respect to the z axis and the longitudinal displacement u

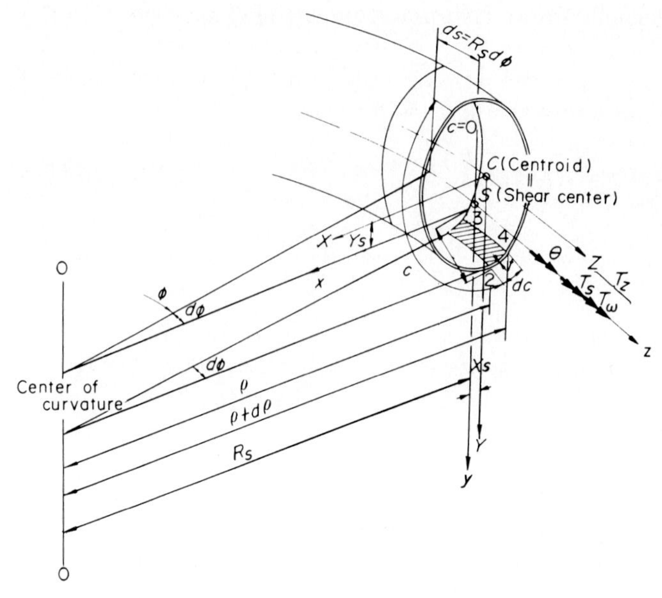

Figure 3.8 Coordinate axes for torsion.

(warping). Therefore, the small element 1,2,3,4 will be deformed to position 1′,2′,3′,4′, as shown by the shaded element in Fig. 3.9.

Accordingly, the shear strain γ can be written as

$$\gamma = \angle\alpha + \angle\beta \tag{3.2.54a}$$

$$= r_s\left(\frac{d\theta}{\rho\, d\phi}\right) + \rho\left[\frac{d(u/\rho)}{dc}\right] \tag{3.2.54b}$$

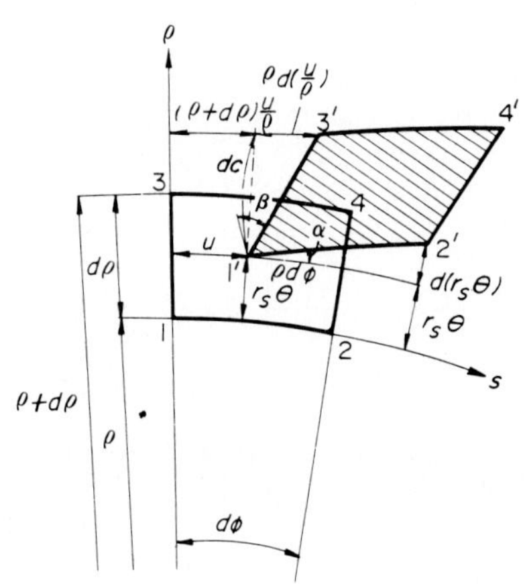

Figure 3.9 Deformation due to torsion.

where r_s is the perpendicular distance from the shear center S to the small element.

The shear flow q due to torsion can now be defined as a function of the twisting angle θ and warping u as follows:

$$q = \tau t = G_s t\gamma \tag{3.2.55a}$$

$$= G_s \frac{t}{n_g}\left[r_s \frac{d\theta}{\rho\, d\phi} + \rho \frac{d}{dc}\left(\frac{u}{\rho}\right)\right] \tag{3.2.55b}$$

a. Pure torsion. Pure torsion is a phenomenon in which normal stresses do not occur, since the cross section is allowed to warp freely. Therefore, for the case of pure torsion, only the shear flow q_s [$=q$ in Eq. (3.2.55*a*)] needs to be considered.

Figure 3.10 shows the induced shear flow q_s on the small element $d\rho\, d\phi$. Taking the equilibrium condition of the forces in the direction of the longitudinal axis gives

$$2\frac{d\rho}{\rho} = -\frac{dq_s}{q_s} \tag{3.2.56}$$

The solution of this differential equation is

$$q_s \rho^2 = \text{constant} \tag{3.2.57}$$

Therefore the shear flow $q_{s,s}$, which is defined as the shear flow with respect to the shear center S, can be written as

$$q_s = \left(\frac{R_s}{\rho}\right)^2 q_{s,s} \tag{3.2.58}$$

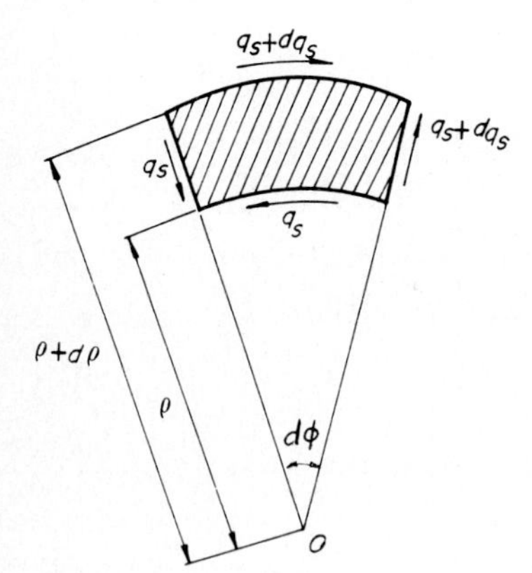

Figure 3.10 Equilibrium condition for q_s.

For the sake of simplicity, let

$$q_{s,s} = G_s \tilde{q}_s \frac{d\theta}{ds} \tag{3.2.59a}$$

or

$$q_s = G_s \left(\frac{R_s}{\rho}\right)^2 \tilde{q}_s \frac{d\theta}{ds} \tag{3.2.59b}$$

and

$$u = \frac{\rho}{R_s} \omega \frac{d\theta}{ds} \tag{3.2.60}$$

where $\tilde{q}_s$ represents the torsional function and ω is the warping function. These functions can be evaluated by substituting Eqs. (3.2.58) through (3.2.60) into Eq. (3.2.55*b*):

$$\tilde{q}_s = \frac{t}{n_g}\left[\frac{\rho}{R_s} r_s + \left(\frac{\rho}{R_s}\right)^3 \frac{d\omega}{dc}\right] \tag{3.2.61}$$

or

$$\frac{d\omega}{dc} = \left(\frac{R_s}{\rho}\right)^3 \tilde{q}_s \frac{n_g}{t} - \left(\frac{R_s}{\rho}\right)^2 r_s \tag{3.2.62}$$

For example, by taking into consideration the condition

$$\oint \frac{d\omega}{dc} dc = 0 \tag{3.2.63}$$

for a box girder with a single cell, the torsional function q_s can be estimated as

$$\tilde{q}_s = \frac{\oint (R_s/\rho)^2 r_s \, dc}{\oint (R_s/\rho)^3 (n_g/t) \, dc} \tag{3.2.64}$$

The warping function ω can also be evaluated by integrating both sides of Eq. (3.2.62) with respect to c and setting the integration constant $C = \bar{\omega}$ at the origin $c = 0$ of curvilinear coordinates, which gives

$$\omega = \tilde{\omega} + \bar{\omega} \tag{3.2.65}$$

where

$$\tilde{\omega} = \int_0^c \left(\frac{R_s}{\rho}\right)^3 \tilde{q}_s \frac{n_g}{t} dc - \int_0^c \left(\frac{R_s}{\rho}\right)^2 r_s \, dc \tag{3.2.66}$$

However, the perpendicular distance r_s is

$$r_s = \frac{R_s}{\rho} \tilde{q}_s \frac{n_g}{t} - \left(\frac{\rho}{R_s}\right)^2 \frac{d\omega}{dc} \tag{3.2.67}$$

as obtained from Eq. (3.2.61). Therefore, the St. Venant's torsional moment T_s can be calculated from Eqs. (3.2.59*b*) and (3.2.67), which gives

$$T_s = \oint q_s r_s \, dc \tag{3.2.68a}$$

$$= G_s \left[\oint \left(\frac{R_s}{\rho} \right)^2 \tilde{q}_s r_s \, dc \right] \frac{d\theta}{ds} \tag{3.2.68b}$$

$$= G_s \left\{ \oint \left(\frac{R_s}{\rho} \right)^2 \tilde{q}_s \left[\frac{R_s}{\rho} \tilde{q}_s \frac{n_g}{t} - \left(\frac{\rho}{R_s} \right)^2 \frac{d\omega}{dc} \right] dc \right\} \frac{d\theta}{ds} \tag{3.2.68c}$$

$$= G_s \left[\oint \left(\frac{R_s}{\rho} \right)^3 \tilde{q}_s^2 \frac{n_g}{t} \, dc \right] \frac{d\theta}{ds} \tag{3.2.68d}$$

or
$$T_s = G_s K \frac{d\theta}{ds} \tag{3.2.28}$$

Here K is the torsional constant defined by

$$K = \oint \left(\frac{R_s}{\rho} \right)^3 \tilde{q}_s^2 \frac{n_g}{t} \, dc \tag{3.2.69}$$

Consequently, the shearing stress τ_s due to pure torsion can be written from Eqs. (3.2.59*b*) and (3.2.68) as

$$\tau_s = \frac{q_s}{t} \tag{3.2.70a}$$

$$= \left(\frac{R_s}{\rho} \right)^2 \frac{\tilde{q}_s T_s}{Kt} \tag{3.2.70b}$$

The generalized method for determining the torsional function q_s and torsional constant K is detailed in App. A.

***b.* Torsional warping.** When the axial displacement u is restrained, the following axial stress σ_ω will occur and can be determined by the axial strain ε_ω:

$$\sigma_\omega = E_s \varepsilon_\omega = \frac{E_s}{n} \frac{R_s}{\rho} \frac{du}{ds} \tag{3.2.71}$$

Now by using the definitions of the warping function ω, in Eq. (3.2.60), Eq. (3.2.71) becomes

$$\sigma_\omega = \frac{E_s}{n} \omega \frac{d^2\theta}{ds^2} \tag{3.2.72}$$

Corresponding to this normal stress σ_ω, a shearing stress τ_ω is presented and must be accounted for. The relationship between the shear flow q_ω and the warping u of the cross section can be obtained by examining Fig. 3.9:

$$q_\omega = \tau_\omega t \tag{3.2.73a}$$

$$= G_s t \tan^{-1}\beta \tag{3.2.73b}$$

$$= \frac{G_s t}{n_g}\frac{\rho}{R_s}\left[\frac{d}{dc}\left(\frac{R_s}{\rho}u\right)\right] \tag{3.2.73c}$$

This equation is applicable in estimating the shear flow q_ω in various types of girder bridges (see App. A).

In addition to the previous conditions, the following equilibrium state must exist, similar to that in the derivation of Eq. (3.2.45*b*):

$$\frac{d}{dc}(q_\omega \rho^2) = -R_s \rho \frac{d\sigma_\omega t}{ds} \tag{3.2.74}$$

The substitution of Eq. (3.2.72) into Eq. (3.2.74) gives

$$\frac{d}{dc}\left[q_\omega\left(\frac{\rho}{R_s}\right)^2\right] = -\frac{\rho}{R_s}\frac{d}{ds}\left(\frac{E_s}{n}\omega t\frac{d^2\theta}{ds^2}\right) \tag{3.2.75a}$$

$$= -\frac{\rho}{R_s}\frac{E_s}{n}\omega t\frac{d^3\theta}{ds^3} \tag{3.2.75b}$$

The warping torsional moment (secondary or Wagner's torsional moment) T_ω can be expressed as

$$T_\omega = \int_A q_\omega r_s \, dc \tag{3.2.76a}$$

or from Eqs. (3.2.67) and (3.2.75*b*),

$$T_\omega = \int_A q_\omega \frac{R_s}{\rho}\tilde{q}_s \frac{n_g}{t}\,dc - \int_A q_\omega \left(\frac{\rho}{R_s}\right)^2\left(\frac{d\omega}{dc}\right)dc \tag{3.2.76b}$$

$$= \tilde{q}_s \int_A q_\omega \frac{R_s}{\rho}\frac{n_g}{t}\,dc - \sum_A \left[q_\omega\left(\frac{\rho}{R_s}\right)^2\omega\right]_{c_{i-1}}^{c_i} + \int_A \frac{d}{dc}\left[q_\omega\left(\frac{\rho}{R_s}\right)^2\right]\omega\,dc \tag{3.2.76c}$$

$$= \int_A \frac{d}{dc}\left[q_\omega\left(\frac{\rho}{R_s}\right)^2\right]\omega\,dc \tag{3.2.76d}$$

$$= -E_s \int_A \frac{\rho}{R_s}\frac{\omega^2 t}{n}\,dc\,\frac{d^3\theta}{ds^3} \tag{3.2.76e}$$

that is,

$$T_\omega = -E_s I_\omega \frac{d^3\theta}{ds^3} \tag{3.2.29}$$

where the warping constant is defined by

$$I_\omega = \int_A \frac{\rho}{R_s} \frac{\omega^2 t}{n} \, dc \tag{3.2.77}$$

Accordingly, Eq. (3.2.75*b*) can be rewritten as

$$q_\omega = \tau_\omega t = \left(\frac{R_s}{\rho}\right)^2 \frac{T_\omega}{I_\omega} S_\omega \tag{3.2.78}$$

where

$$S_\omega = \bar{S}_\omega + C = \int_0^c \frac{\rho}{R_s} \frac{\omega t}{n} \, dc + C \tag{3.2.79}$$

and C is the integration constant. For the free edge of an open cross section $C = 0$ and for a single cell, it is expressed by

$$C = -\frac{\oint [(R_s/\rho)^3(\bar{S}_\omega n_g/t)] \, dc}{\oint [(R_s/\rho)^3(n_g/t)] \, dc} \tag{3.2.80}$$

an expression similar to Eq. (3.2.53).

The torsional moment T_s is given by Eq. (3.2.28), and T_ω is given by Eq. (3.2.29). Then

$$G_s K \frac{d\theta}{ds} - E_s I_\omega \frac{d^3\theta}{ds^3} = T_z \tag{3.2.24b}$$

The solution of this equation will give the stress resultants and deformations of curved girder bridges.

Let us now define the *warping moment* (bimoment) as follows:

$$\frac{dM_\omega}{ds} = -T_\omega \tag{3.2.81}$$

This warping moment can now be defined from Eq. (3.2.72) as

$$M_\omega = E_s I_\omega \frac{d^2\theta}{ds^2} \tag{3.2.82a}$$

$$= E_s \int_A \frac{\rho}{R_s} \frac{\omega^2 t}{n} \, dc \, \frac{d^2\theta}{ds^2} \tag{3.2.82b}$$

$$= \int_A \frac{\rho}{R_s} \left(E_s \frac{\omega}{n} \frac{d^2\theta}{ds^2}\right) \omega t \, dc \tag{3.2.82c}$$

$$= \int_A \frac{\rho}{R_s} \sigma_\omega \omega t \, dc \tag{3.2.82d}$$

and the stress formula for σ_ω reduces to

$$\sigma_\omega = \frac{M_\omega}{nI_\omega}\,\omega \tag{3.2.83a}$$

However, the relationships between this normal stress σ_ω and the other stress resultants N_Z, M_X, and M_Y obviously should be satisfied by the following conditions:

$$N_Z = \int_A \sigma_\omega t\, dc = B\frac{d^2\theta}{ds^2} = 0 \tag{3.2.84a}$$

$$M_X = \int_A \sigma_\omega Yt\, dc = C_X\frac{d^2\theta}{ds^2} = 0 \tag{3.2.84b}$$

$$M_Y = \int_A \sigma_\omega Xt\, dc = C_Y\frac{d^2\theta}{ds^2} = 0 \tag{3.2.84c}$$

Since $d^2\theta/ds^2 \neq 0$, the coefficients B, C_X, and C_Y must equal

$$B = \int_A \frac{\omega t}{n}\, dc = 0 \tag{3.2.85a}$$

$$C_X = \int_A \frac{\omega Y}{n}\, t\, dc = 0 \tag{3.2.85b}$$

$$C_Y = \int_A \frac{\omega X}{n}\, t\, dc = 0 \tag{3.2.85c}$$

Examination of Eq. (3.2.85a) shows that the volume of warping equals zero. So the warping function $\bar{\omega}$ at the origin ($c = 0$) can be evaluated by substituting Eq. (3.2.65) into Eq. (3.2.85a)

$$\bar{\omega} = -\frac{\int_A (\tilde{\omega}t/n)\, dc}{\int_A (t/n)\, dc} \tag{3.2.86}$$

Although all the equations derived above are evaluated with respect to the shear center S, the radius of curvature R_s has not yet been located. The x and y coordinates axes with respect to the shear center S can, however, be expressed by the centroidal X and Y coordinate axes, as shown in Fig. 3.11. Thus

$$x = X - X_s \tag{3.2.87a}$$

$$y = Y - Y_s \tag{3.2.87b}$$

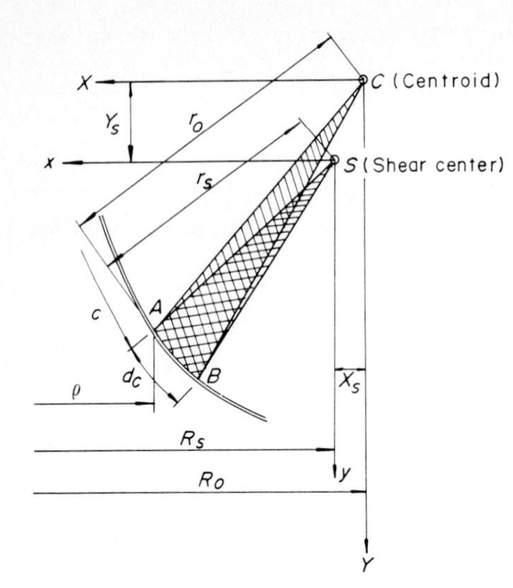

Figure 3.11 Definition of sectorial area with respect to centroid C and shear center S.

Integrating the second term of the right-hand side of Eq. (3.2.66) results in[4]

$$\int_0^c \frac{r_s}{\rho^2}\,dc = \frac{R_s}{R_o}\int_0^c \frac{r_o}{\rho^2}\,dc - \frac{X_s}{R_o}\frac{Y}{\rho} + \frac{Y_s}{R_o}\frac{X}{\rho} \tag{3.2.88a}$$

$$\oint \frac{r_s}{\rho^2}\,dc = \frac{R_s}{R_o}\oint \frac{r_o}{\rho^2}\,dc \tag{3.2.88b}$$

The sectorial area with respect to the shear center S can now be transferred to the centroid C:

$$\int_0^c \left(\frac{R_s}{\rho}\right)^2 r_s\,dc = \left(\frac{R_s}{R_o}\right)^2 \int_0^c \left(\frac{R_o}{\rho}\right)^2 r_s\,dc \tag{3.2.89a}$$

$$= \left(\frac{R_s}{R_o}\right)^2 \left[\frac{R_s}{R_o}\int_0^c \left(\frac{R_o}{\rho}\right)^2 r_s\,dc + \frac{R_o}{\rho}(Y_sX - X_sY)\right] \tag{3.2.89b}$$

where

$$\int_0^c \left(\frac{R_o}{\rho}\right)^2 r_o\,dc = \int_0^c \left(\frac{R_o}{\rho}\right)^2 X\,dY - \int_0^c \left(\frac{R_o}{\rho}\right)^2 Y\,dX \tag{3.2.90}$$

Therefore, Eq. (3.2.66) can be written as

$$\tilde{\omega} = \left(\frac{R_s}{R_o}\right)^2 \left\{\frac{R_s}{R_o}\left[\int_0^c \left(\frac{R_o}{\rho}\right)^3 \tilde{q}_o \frac{n_g}{t}\,dc - \int_0^c \left(\frac{R_o}{\rho}\right)^2 r_o\,dc\right] + \frac{R_o}{\rho}(Y_sX - X_sY)\right\}$$

$$= \left(\frac{R_s}{R_o}\right)^2 \left[\frac{R_s}{R_o}\tilde{\omega}_o + \frac{R_o}{\rho}(Y_sX - X_sY)\right] \tag{3.2.91}$$

Thus the cross-sectional quantities for torsion can be estimated with respect to the centroid C relative to a single-cell section:
Torsional constants:

$$\tilde{q}_o = \frac{\oint [(R_o/\rho)^2 r_o]\, dc}{\oint [(R_o/\rho)^3 n_g/t]\, dc} = \tilde{q}_s \tag{3.2.92}$$

$$K_o = \oint \left(\frac{R_o}{\rho}\right)^3 \tilde{q}_o^2 \frac{n_g}{t}\, dc \tag{3.2.93}$$

Warping constants:

$$\tilde{\omega}_o = \int_0^c \left(\frac{R_o}{\rho}\right)^3 \tilde{q}_o \frac{n_g}{t}\, dc - \int_0^c \left(\frac{R_o}{\rho}\right)^2 r_o\, dc \tag{3.2.94}$$

$$I_{\omega,o} = \int_A \frac{\rho}{R_o} \frac{\omega_o^2 t}{n}\, dc \tag{3.2.95}$$

Now, if we let

$$\omega_o = \tilde{\omega}_o + \bar{\omega}_o \tag{3.2.96}$$

then the value of $\bar{\omega}_o$ can be estimated by

$$\bar{\omega}_o = -\frac{\int_A (\omega_o t/n)\, dc}{\int_A (t/n)\, dc} \tag{3.2.97}$$

Accordingly, the warping function ω with respect to the shear center S can be obtained by using the warping function ω_o with respect to the centroid C:

$$\omega = \left(\frac{R_s}{R_o}\right)^2 \left[\frac{R_s}{R_o}\omega_o + \frac{R_o}{\rho}(Y_s X - X_s Y)\right] \tag{3.2.98}$$

By substituting Eq. (3.2.98) into Eqs. (3.2.85*b*) and (3.2.85*c*) and using the definitions given by Eqs. (3.2.35*b*) to (3.2.35*f*), the shear center (X_s, Y_s) can be determined from the following simultaneous equations:

$$\left(I_{XY} + \frac{C_{Y_o}}{R_o}\right)X_s - I_Y Y_s = C_{Y_o} \tag{3.2.99a}$$

$$\left(I_X + \frac{C_{X_o}}{R_o}\right)X_s - I_{XY} Y_s = C_{X_o} \tag{3.2.99b}$$

where
$$C_{X_o} = \int_A \frac{\omega_o Y t}{n}\, dc \tag{3.2.100a}$$

$$C_{Y_o} = \int_A \frac{\omega_o X t}{n}\, dc \tag{3.2.100b}$$

and
$$R_s = R_o - X_s \tag{3.2.101}$$

The solution of Eq. (3.2.99) gives

$$X_s = \frac{I_Y C_{X_o} - I_{XY} C_{Y_o}}{I_Y (I_X + C_{X_o}/R_o) - I_{XY}(I_{XY} + C_{Y_o}/R_o)} \tag{3.2.102a}$$

$$Y_s = \frac{(I_{XY} + C_{Y_o}/R_o) C_{X_o} - (I_X + C_{X_o}/R_o) C_{Y_o}}{I_Y (I_X + C_{X_o}/R_o) - I_{XY}(I_{XY} + C_{Y_o}/R_o)} \tag{3.2.102b}$$

Finally, the torsional constant K with respect to the shear center S is

$$K = \left(\frac{R_s}{R_o}\right)^3 K_o \tag{3.2.103}$$

From Eqs. (3.2.69) and (3.2.92), and by using Eqs. (3.2.77), (3.2.95), and (3.2.100) through (3.2.102), the warping constant I_ω is

$$I_\omega = \left(\frac{R_s}{R_o}\right)^4 \left(\frac{R_s}{R_o} I_{\omega o} + Y_s C_{Y_o} - X_s C_{X_o}\right) \tag{3.2.104}$$

3.2.4 Shear lag in curved girders[15–21]

In Sec. 3.2.3(1), the shear lag phenomenon is omitted in evaluating the flexural normal stress σ_b, but it is treated now.

(1) Notation.[19] By expanding the theory described in Sec. 2.2.3, the shear lag phenomenon and the corresponding effective width of a curved steel girder with a nonsymmetric cross section are analyzed for an idealized cross section, shown in Fig. 3.12, where the following notation is used:

A_1, A_3 = cross-sectional areas of brackets

A_2, A_4 = cross-sectional areas of low flanges

A_R, S_R, I_R = cross-sectional area, static moment, and geometric moment of inertia for a rib, respectively

t_u, t_w = thickness of deck and web plates, respectively

a = spacing of rib

h = girder depth

$2b$ = distance between web plates

E = Young's modulus for steel

G = shear modulus for elasticity

μ = Poisson's ratio for steel

Figure 3.13 shows the plan view of a curved girder bridge, where the radii of curvature and central angles are defined as

R_s = radius of curvature at shear center S

R_o = radius of curvature at centroid C

R_c = radius of curvature at center line of cross section

R_1, R_2 = radius of curvature at inner and outer sides of web plate, respectively

ρ = radius of curvature at viewpoint

Φ = central angle of curved girder

ϕ = angular coordinate of viewpoint

Finally, the displacements of the curved girder that need to be determined are

w = vertical deflection at shear center

β = angle of rotation

θ = twisting angle

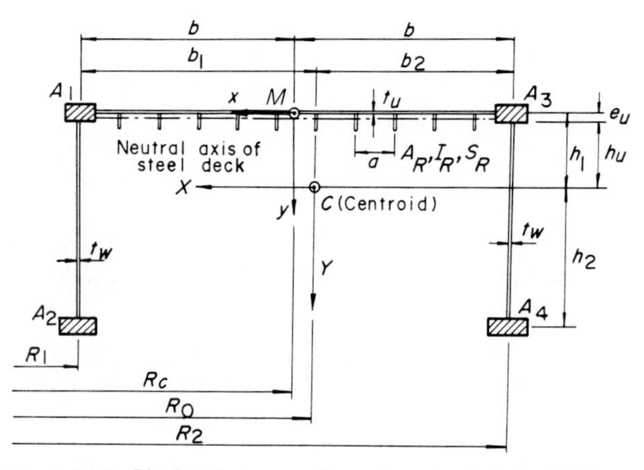

Figure 3.12 Idealized cross section of a curved girder bridge.

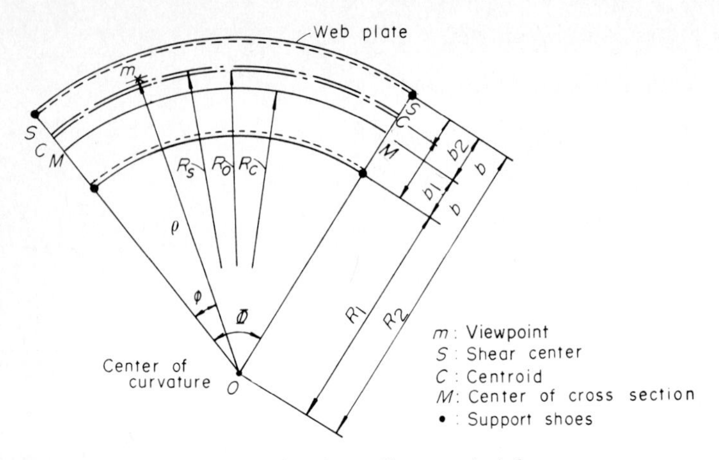

Figure 3.13 Plane shape of horizontally curved girder.

(2) Definitions of cross-sectional quantities.[19] In deriving the fundamental differential equations of the above displacements, the following cross-sectional quantities are required (see Fig. 3.12).

Relative to the centroid C:

$$h_1 = \frac{\left(\dfrac{R_o}{R_1} A_2 + \dfrac{R_o}{R_2} A_4\right)h + \dfrac{1}{2}\left(\dfrac{R_o}{R_1} + \dfrac{R_o}{R_2}\right)A_w h + 2A_u e_u}{\dfrac{R_o}{R_1}(A_1 + A_w + A_2) + 2A_u + \dfrac{R_o}{R_2}(A_3 + A_w + A_4)} \tag{3.2.105a}$$

$$h_2 = h - h_1 \tag{3.2.105b}$$

and

$$b_1 = \frac{2b[(R_o/R_2)(A_3 + A_w + A_4) + (R_o/R_c)A_u]}{(R_o/R_1)(A_1 + A_w + A_2) + 2A_u + (R_o/R_2)(A_3 + A_w + A_4)} \tag{3.2.106a}$$

$$b_2 = 2b - b_1 \tag{3.2.106b}$$

where

$$2A_u = \int_{R_1}^{R_2} \frac{R_o}{\rho} \bar{t}_u \, d\rho \tag{3.2.107a}$$

$$= R_o \bar{t}_u \ln \frac{R_2}{R_1} \tag{3.2.107b}$$

$$\cong 2b\bar{t}_u = 2b\left(\frac{t_u}{1-\mu^2} + \frac{A_R}{a}\right) \tag{3.2.107c}$$

$$A_w = \frac{t_w h}{1 - \mu^2} \tag{3.2.108}$$

$$e_u = \frac{S_R}{a\bar{t}_u} \tag{3.2.109}$$

provided that the radius of curvature R_o at the centroid is given by Eqs. (3.2.36) through (3.2.39), or

$$R_o = R_1 + b_1 \tag{3.2.110}$$

The cross-sectional area F of the curved girder is

$$F = 2A = \frac{R_o}{R_1}(A_1 + A_w + A_2) + 2A_u + \frac{R_o}{R_2}(A_3 + A_w + A_4) \tag{3.2.111}$$

where the geometric moment of inertia with respect to the centroidal X and Y axes, shown in Fig. 3.12, is

$$I_X = \frac{R_o}{R_1}(A_1 h_1^2 + I_w + A_2 h_2^2) + 2A_u\left(h_u^2 + \frac{I_R}{a\bar{t}_u}\right) + \frac{R_o}{R_2}(A_3 h_1^2 + I_w + A_4 h_2^2) \tag{3.2.112}$$

$$I_Y = \frac{R_o}{R_1} b_1^2(A_1 + A_w + A_2) + \tfrac{1}{3}\bar{t}_u(b_1^3 + b_2^3) + \frac{R_o}{R_2} b_2^2(A_3 + A_w + A_4) \tag{3.2.113}$$

$$I_{XY} = \frac{R_o}{R_1} b_1[-A_1 h_1 - \tfrac{1}{2}A_w(h_1 - h_2) + A_2 h_2]$$

$$+ \frac{R_o}{R_2} b_2\left[A_3 h_1 + \tfrac{1}{2}A_w(h_1 - h_2) - A_4 h_2\right] - A_u(b_1 - b_2)h_u \tag{3.2.114}$$

in which case

$$I_w = \frac{1}{3}\frac{t_w}{1 - \mu^2}(h_1^3 + h_2^3) \tag{3.2.115}$$

(3) Fundamental equation for shear lag[19]

***a.* Assumption of axial displacements.** Although the cross section of a curved girder is asymmetric, the displacements u_i at an arbitrary point on the girder elements $i = 1$, 2, and 3 are given as functions of the X and Y coordinate axes with respect to the centroid C. As in the case of

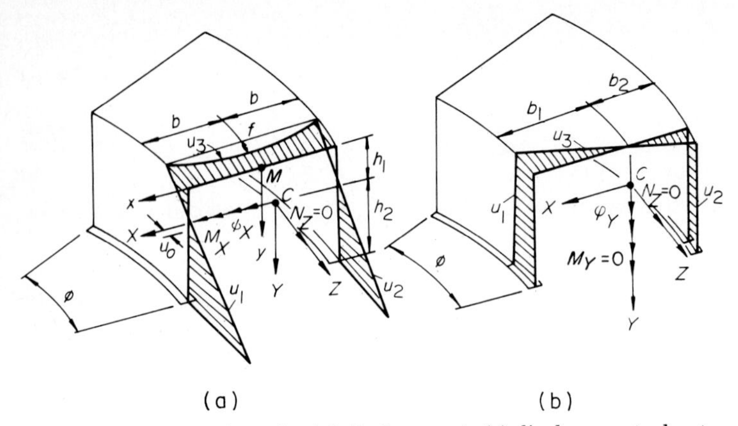

Figure 3.14 Assumption of axial displacement: (*a*) displacements due to φ_X and f, (*b*) displacement due to φ_Y.

a straight girder, the following equations can be used relating the displacement functions, as illustrated in Fig. 3.14:
For the deck plate,

$$u_3 = u_o(\phi) - (h_1 - y)\varphi_X(\phi) + \left[1 - \left(\frac{x}{b}\right)^2\right] f(\phi) - X\varphi_Y(\phi) \tag{3.2.116}$$

For the web plates and corner parts,

$$u_i = u_o(\phi) + Y\varphi_X(\phi) + b_i\varphi_Y(\phi) \qquad i = 1, 2 \tag{3.2.117}$$

Here φ_X and φ_Y, respectively, designate the deflection angle with respect to the X and Y centroidal axes.

***b.* Fundamental differential equations for flexural stress.** Analogous to the definitions of straight girders, the equilibrium conditions for the stress resultants of the curved girder are

$$\int_F \sigma_Z\, dF = N_Z = 0 \tag{3.2.118a}$$

$$\int_F \sigma_Z Y\, dF = M_X \tag{3.2.118b}$$

$$\int_F \sigma_Z X\, dF = M_Y = 0 \tag{3.2.118c}$$

Hence the stress σ_Z can be estimated from Eqs. (3.2.116) and (3.2.117) as follows:
For the deck plate,

$$\sigma_{Z,u} = -\frac{E}{1-\mu^2}\frac{R_o}{\rho}\left\{(h_1 - y)\frac{d\varphi_X(\phi)}{R_o\, d\phi} - \left[1-\left(\frac{x}{b}\right)^2\right]\frac{df(\phi)}{R_o\, d\phi} + X\frac{d\varphi_Y(\phi)}{R_o\, d\phi} - \frac{du_o(\phi)}{R_o\, d\phi}\right\} \tag{3.2.119}$$

For the web plate,

$$\sigma_{Z,w,i} = \frac{E}{1-\mu^2}\frac{R_o}{R_i}\left[Y\frac{d\varphi_X(\phi)}{R_o\, d\phi} \mp I_{b_i}\frac{d\varphi_Y(\phi)}{R_o\, d\phi} + \frac{du_o(\phi)}{R_o\, d\phi}\right] \qquad i = 1, 2 \tag{3.2.120}$$

For the corner parts,

$$\sigma_{Z,j} = -E\frac{R_o}{R_i}\left[h_1\frac{d\varphi_X(\phi)}{R_o\, d\phi} \pm b_i\frac{d\varphi_Y(\phi)}{R_o\, d\phi} - \frac{du_o(\phi)}{R_o\, d\phi}\right] \qquad i = 1, 2 \text{ for } j = 1, 3 \tag{3.2.121a}$$

$$\sigma_{Z,j} = E\frac{R_o}{R_i}\left[h_2\frac{d\varphi_X(\phi)}{R_o\, d\phi} \mp b_i\frac{d\varphi_Y(\phi)}{R_o\, d\phi} + \frac{du_o(\phi)}{R_o\, d\phi}\right] \qquad i = 1, 2 \text{ for } j = 2, 4 \tag{3.2.121b}$$

The fundamental differential equations for the unknown displacements $u_o(\phi)$, $\varphi_X(\phi)$, and $\varphi_Y(\phi)$ under the applied bending moment M_X can be derived as follows: Utilizing the condition of Eq. (3.2.118*a*), we have

$$F\frac{du_o(\phi)}{R_o\, d\phi} + F_u\frac{df(\phi)}{R_o\, d\phi} = 0 \tag{3.2.122}$$

where

$$F_u = \int_{R_1}^{R_2}\frac{R_o}{\rho}\bar{t}_u\, d\rho - \frac{1}{b^2}\int_{R_1}^{R_2}\frac{R_o}{\rho}x^2\bar{t}_u\, d\rho \cong {}^4\!/_3 A_u \tag{3.2.123}$$

Accordingly, this can also be written as

$$u_o(\phi) = -\frac{2}{3}\frac{A_u}{A}f(\phi) \tag{3.2.124}$$

from Eq. (3.2.111). The above equation agrees with the straight girder, as seen from Eq. (2.2.66).

Next the deflection angles φ_X and φ_Y, from Eqs. (3.2.118*b*) and (3.2.118*c*), are

$$I_x \frac{d\varphi_X(\phi)}{R_o\, d\phi} - I_{XY} \frac{d\varphi_Y(\phi)}{R_o\, d\phi} - \frac{4}{3} A_u h_u \frac{df(\phi)}{R_o\, d\phi} = \frac{M_X}{E} \tag{3.2.125a}$$

$$I_{XY} \frac{d\varphi_X(\phi)}{R_o\, d\phi} - I_Y \frac{d\varphi_Y(\phi)}{R_o\, d\phi} - \frac{4}{3} A_u b_u \frac{df(\phi)}{R_o\, d\phi} = 0 \tag{3.2.125b}$$

where
$$b_u = \frac{A_u(b_1 - b_2) - [\bar{t}_u/(3b^2)](b - b_1)(b_1^3 + b_2^3)}{\frac{4}{3}A_u} \tag{3.2.126}$$

If we neglect the shear lag terms and let

$$\kappa_X = \frac{d\varphi_X}{R_o\, d\phi} \tag{3.2.127a}$$

and
$$\kappa_Y = \frac{d\varphi_Y}{R_o\, d\phi} \tag{3.2.127b}$$

which represent the changes of curvature, then Eq. (3.2.125) becomes

$$I_X \kappa_X - I_{XY} \kappa_Y = \frac{M_X}{E} \tag{3.2.128a}$$

$$I_{XY} K_X - I_Y \kappa_Y = 0 \tag{3.2.128b}$$

These equations are the relationships between curvature κ_X and κ_Y and the bending moment M_X, as defined by elementary curved beam theory [see Eqs. (3.2.34*b*) and (3.2.34*c*)].

The final condition, which minimizes the error due to the assumed displacement function $f(\phi)$, can be expressed in a manner similar to Eq. (2.2.69):

$$\sum_i \int_0^{c_i} \varepsilon_i(\phi, c_i) \lambda_i(c_i) t_i \, dc_i = 0 \tag{3.2.129}$$

where $\varepsilon_i(\phi, c_i)$ and $\lambda_i(c_i)$ are, respectively, the error and coordinate functions. The corresponding coordinate function can then be obtained by substituting Eqs. (3.2.116) and (3.2.117) into the equilibrium conditions for the curved girder elements. Thus, the above condition can be simplified to

$$A_u h_u \frac{d^2\varphi_X(\phi)}{R_o^2\, d\phi^2} - A_u \left(\frac{4}{5} - \frac{2}{3} \frac{A_u}{A} \right) \frac{d^2 f(\phi)}{R_o^2\, d\phi^2} + \frac{R_c}{R_o} \frac{2Gt_u}{Eb} f(\phi) = 0 \tag{3.2.130}$$

This equation is similar to Eq. (2.2.72), which represents the straight girder condition.

c. Fundamental equations for displacements. The solutions of the simultaneous equations for $d\varphi_X(\phi)/d\phi$ and $d\varphi_Y(\phi)/d\phi$ are given by

$$\frac{d\varphi_X(\phi)}{d\phi} = \frac{R_o I_Y}{E(I_X I_Y - I_{XY}^2)} M_X(\phi) + \gamma_X \frac{df(\phi)}{d\phi} \tag{3.2.131a}$$

$$\frac{d\varphi_Y(\phi)}{d\phi} = -\frac{R_o I_{XY}}{E(I_X I_Y - I_{XY}^2)} M_X(\phi) + \gamma_X \frac{df(\phi)}{d\phi} \tag{3.2.131b}$$

where

$$\gamma_X = \frac{4}{3}\frac{A_u(h_u I_Y - b_u I_{XY})}{I_X I_Y - I_{XY}} \tag{3.2.132a}$$

and

$$\gamma_Y = \frac{4}{3}\frac{A_u(h_u I_{XY} - b_u I_X)}{I_X I_Y - I_{XY}^2} \tag{3.2.132b}$$

The substitution of Eq. (3.2.131a) into Eq. (3.2.130) gives the following basic equations for $f(\phi)$:

$$\frac{d^2 f(\phi)}{d\phi^2} - \lambda^2 f(\phi) = \frac{R_o I_Y h_u \beta}{E(I_X I_Y - I_{XY}^2)} \tag{3.2.133}$$

where

$$\lambda = \frac{1}{b}\sqrt{\frac{\beta}{\omega} R_o R_c} \tag{3.2.134}$$

$$\beta = \frac{1.5}{1.2 - \kappa} \tag{3.2.135}$$

$$\omega = \frac{\bar{t}_u}{t_u}(1 + \mu) = \frac{1}{1 - \mu} + \frac{A_R}{a t_u} \tag{3.2.136}$$

$$\kappa = \frac{A_u}{A} + \frac{2A_u h_u (I_Y h_u - I_{XY} b_u)}{I_X I_Y - I_{XY}^2} \tag{3.2.137}$$

Next, examining Fig. 3.4 reveals that the relationship among the deflection w, angle of rotation β, and twisting angle θ can be expressed as

$$\theta(\phi) = \beta(\phi) + \frac{w(\phi)}{R_s} \tag{3.2.14b}$$

The basic equation for this angle, $\theta(\phi)$, is given by the well-known formula

$$\frac{d^3\theta(\phi)}{d\phi^3} - \alpha^2 \frac{d\theta(\phi)}{d\phi} = -\frac{R_s^3}{EI_\omega} T_z(\phi) \tag{3.2.24c}$$

where $\alpha = R_s \sqrt{\dfrac{GK}{EI_\omega}}$ (3.2.26b)

GK = torsional rigidity
EI_ω = warping rigidity
$T_z(\phi)$ = torsional moment

The normal stress σ_ω due to warping can be estimated from

$$\sigma_\omega = \frac{M_\omega}{I_\omega} \tag{3.2.83b}$$

where

$$M_\omega = EI_\omega \frac{d^2\theta}{R_s^2\, d\phi^2} \tag{3.2.82b}$$

is the warping moment and ω is the warping function.

The relationship among the deflection $w(\phi)$, deflection angle $\varphi_X(\phi)$, and angle of rotation $\beta(\phi)$ is given by the well-known formula

$$\frac{d\varphi_X(\phi)}{d\phi} = -\left[\frac{1}{R_s}\frac{d^2w(\phi)}{d\phi^2} - \beta(\phi)\right] \tag{3.2.138}$$

Next, eliminating $w(\phi)$ from Eqs. (3.2.138) and (3.2.14b), we obtain the basic equation for the angle of rotation $\beta(\phi)$:

$$\frac{d^2\beta(\phi)}{d\phi^2} + \beta(x) = \frac{d^2\theta}{d\phi^2} + \frac{d\varphi_X(\phi)}{d\phi} \tag{3.2.139}$$

This equation can be solved easily when the load terms $d^2\theta\phi/d\phi^2$ and $d\varphi_Y(\phi)/d\phi$ are given by the solution of Eqs. (3.2.24c) and (3.2.131a), respectively.

Consequently, the deflection can be determined from Eq. (3.2.14b):

$$w(\phi) = R_s[\theta(\phi) - \beta(\phi)] \tag{3.2.140}$$

***d.* Solutions of $M_X(\phi)$ and $f(\phi)/d\phi$.** The formula estimating the stresses can be written from Eqs. (3.2.119), (3.2.120), and (3.2.124) as follows: For the deck plate,

$$\sigma_u = -\frac{1}{1-\mu^2}\frac{1}{\rho}\Bigg\{[(h_1 - y)\varphi_X + X\varphi_Y]M_X(\phi) - \left[1 - \frac{2}{3}\frac{A_u}{A} - \left(\frac{x}{b}\right)^2 - (h_1 - z)\gamma_X - X\gamma_Y\right]\frac{df(\phi)}{d\phi}\Bigg\} \tag{3.2.141}$$

For the web plate,

$$\sigma_{\omega,i} = \frac{1}{1-\mu^2}\frac{1}{R_i}\left[(Y\varphi_X \mp b_i\varphi_Y)M_X(\phi) - \left(\frac{2}{3}\frac{A_u}{A} - Z\gamma_X \pm b_i\gamma_Y\right)\frac{df(\phi)}{d\phi}\right] \qquad i = 1, 2 \quad (3.2.142)$$

The bending moment M_X and derivative $df(\phi)/d\phi$ are summarized in Tables 3.1 and 3.2 for the loading conditions shown in Fig. 3.15. Note that the terms L_1 and η in Tables 3.1 and 3.2 are defined by

$$L_1 = \tfrac{1}{3}(R^3_{\text{out}} - R^3_{\text{in}}) \quad (3.2.143)$$

and

$$\eta = \frac{R_o I_Y h_u \beta}{E(I_X I_Y - I^2_{XY})} \quad (3.2.144)$$

(4) Effective width of curved girders.[19] The normal stress distribution about the neutral axis of the deck plate can be computed by

$$\sigma_u = -\frac{1}{1-\mu^2}\frac{R_o}{\rho}\left\{\frac{I_Y h_u + I_{XY}X}{I_X I_Y - I^2_{XY}} M_X - \left[1 - \frac{2}{3}\frac{A_u}{A} - \left(\frac{x}{b}\right)^2 - h_u\gamma_X - X\gamma_Y\right]\right\} \quad (3.2.145)$$

obtained from Eq. (3.2.141) when $y = e_u$ and $h_u = h_1 - e_u$. Then the stress at both edges of the deck plate can be estimated as

$$\sigma_1 = [\sigma_u]_{x=b,\ X=b_1} \quad (3.2.146a)$$

$$\sigma_3 = [\sigma_u]_{x=-b,\ X=-b_2} \quad (3.2.146b)$$

These stresses are plotted in Fig. 3.16.

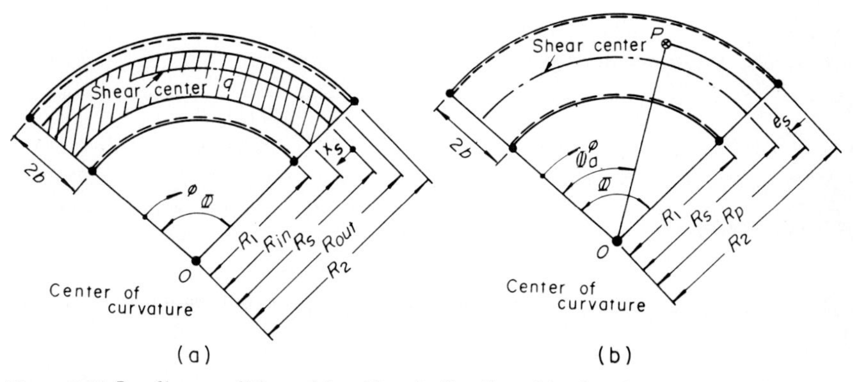

Figure 3.15 Loading conditions: (*a*) uniformly distributed load q, (*b*) concentrated load P.

TABLE 3.1 Solutions of $M(\phi)$

Load	Bending moment $M(\phi)$
Uniform load q	$qL_1\left[\dfrac{\sin\phi + \sin(\Phi - \phi)}{\sin\phi} - 1\right]$
Concentrated load P	$PR_P\dfrac{\sin(\Phi - \phi a)}{\sin\Phi}\sin\phi \qquad 0 \leqq \phi \leqq \Phi_a$

The normal stresses derived from curved beam theory are defined by $\bar{\sigma}_u$, as shown in Fig. 3.16, and can be approximated by

$$\bar{\sigma}_u = \frac{\sigma_3 - \sigma_1}{2b}x + \frac{\sigma_1 + \sigma_3}{2} \tag{3.2.147}$$

The effective widths $b_{m,1}$ and $b_{m,2}$, respectively, of the deck plate of the curved girder bridge can be defined by

$$\int_0^b \sigma_u\,dx = \int_{b - b_{m,1}}^b \bar{\sigma}_u\,dx \tag{3.2.148a}$$

$$\int_{-b}^0 \sigma_u\,dx = \int_{-b}^{-(b - b_{m,2})} \bar{\sigma}_u\,dx \tag{3.2.148b}$$

as shown in Fig. 3.16. However, it is difficult to obtain a closed-form solution, as in the case of a straight girder, since the stress formulas are expressed as not only a function of the radius of curvature ρ, but also the x and X coordinate axes, as seen in Eq. (3.2.145). Therefore, the effective widths $b_{m,1}$ and $b_{m,2}$ are usually determined by numerical integration.

TABLE 3.2 Solutions of $\dfrac{df(\phi)}{d\phi}$

Load	Derivative $\dfrac{df(\phi)}{d\phi}$
Uniform load q	$qL_1\dfrac{n}{\lambda^2+1}\left[\dfrac{\sin\phi + \sin(\Phi - \phi)}{\sin\Phi} - \dfrac{\sinh\lambda\phi + \sinh\lambda(\Phi - \phi)}{\sinh\lambda\phi}\right]$
Concentrated load P	$PR_P\dfrac{n}{\lambda^2+1}\left[\dfrac{\sin(\Phi - \phi_a)}{\sin\phi}\sin\phi + \dfrac{\lambda\sinh\lambda(\Phi - \phi_a)}{\sinh\lambda\Phi}\sinh\lambda\phi\right]$ $(0 \leqq \phi \leqq \Phi_a)$

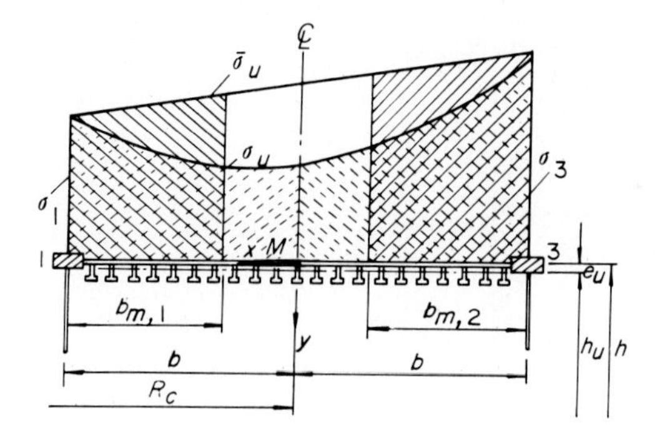

Figure 3.16 Definition of effective width $b_{m,1}$ and $b_{m,2}$.

Performing numerical integration, we can estimate the effective widths $b_{m,1}$ and $b_{m,2}$. Then the geometric moments of inertia about the axes passing through centroid C are

$$I_X = \int_A \frac{R_o}{\rho} Y^2 \, dA \tag{3.2.35g}$$

$$I_Y = \int_A \frac{R_o}{\rho} X^2 \, dA \tag{3.2.35h}$$

$$I_{XY} = \int_A \frac{R_o}{\rho} XY \, dA \tag{3.2.35i}$$

The stresses including shear lag phenomena can be estimated by

$$\sigma_b = \frac{R_o}{\rho} \frac{I_Y Y - I_{XY} X}{I_X I_Y - I_{XY}^2} M_X \tag{3.2.149}$$

Figure 3.17 shows the variation of the effective widths $b_{m,1}$ and $b_{m,2}$ in the direction of span ϕ/Φ for the curved girder in Fig. 3.18. The loading conditions are given in Fig. 3.19. These curves exhibit tendencies similar to those for a straight girder, as shown in Fig. 2.26. Also, the differences among $b_{m,1}$, $b_{m,2}$, and b_m for the straight girder estimated by Eqs. (2.2.107a) and (2.2.107b) are small and thus can be ignored.

Next, to clarify the variations of $b_{m,1}$ and $b_{m,2}$ due to various parameters L/b, a few numerical calculations have been made in Table 3.3. Examination of these data shows that the difference between $b_{m,1}$ for the inner deck and $b_{m,2}$ for the outer deck can be neglected, so the following criterion can be used:

$$b_{m,1} = b_{m,2} \tag{3.2.150}$$

The effective width for curved[19] and straight girders[18] is a function of L/b, and can be determined by the formula b_m/b from Eqs. (2.2.107a)

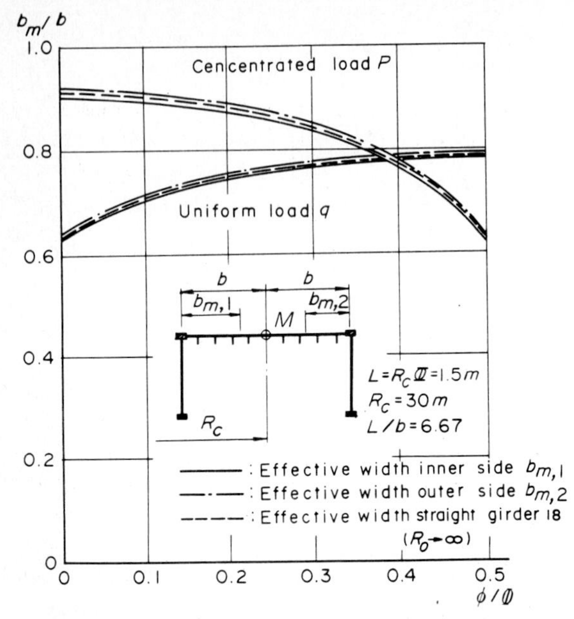

Figure 3.17 Variation of effective width in the direction of span.

and (2.2.107*b*), and results in the data shown in Table 3.3, where the span length L of the curved girder is

$$L = R_c\Phi \tag{3.2.151}$$

3.2.5 Distortion in curved box girders[22–26]

The distortional warping normal stress $\sigma_{D\omega}$ in a curved box girder is not as large as the flexural normal stress σ_b for the cases when the appropriate number of intermediate diaphragms is provided. In general, although the cross section of a curved box girder is composed of

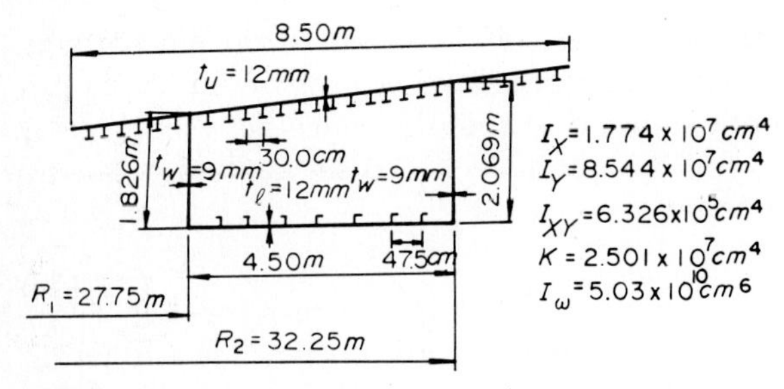

Figure 3.18 Cross section of a curved monobox-girder bridge.

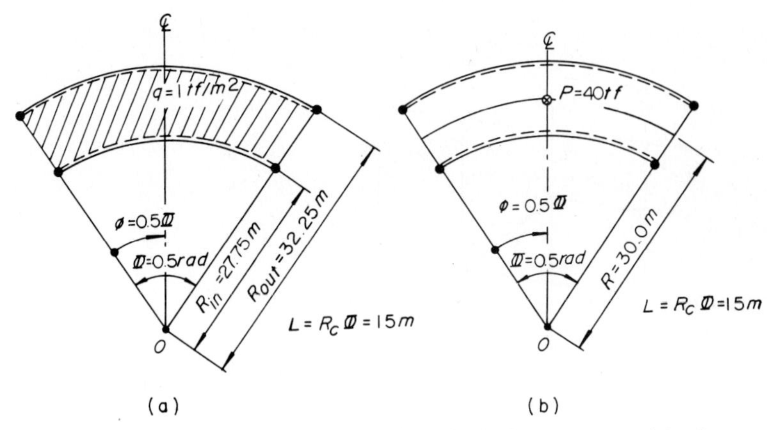

Figure 3.19 Loading conditions: (*a*) uniform load, (*b*) concentrated load.

asymmetric thin-walled members, an approximate method for evaluating the distortion of a curved box girder having a symmetric cross section is now introduced for the sake of simplicity.

For these analyses, the method developed in Sec. 2.4 can also be applied to a curved box girder. However, it is somewhat different from the distortion of a straight box girder, because the flexural and

TABLE 3.3 Variations of the Effective Width for a Curved Girder Bridge due to L/b

		Uniform Load									
		L/b									
		2	4	6	8	10	12	14	16	18	20
Curved girder	$b_{m,1}/b$	0.325	0.593	0.751	0.840	0.890	0.920	0.938	0.951	0.960	0.966
	$b_{m,2}/b$	0.355	0.601	0.760	0.846	0.895	0.923	0.941	0.933	0.962	0.968
Straight girder	b_m/b	0.356	0.600	0.760	0.848	0.898	0.927	0.946	0.958	0.967	0.972
		Concentrated Load									
		L/b									
		2	4	6	8	10	12	14	16	18	20
Curved girder	$b_{m,1}/b$	0.317	0.494	0.615	0.692	0.745	0.784	0.813	0.837	0.857	0.874
	$b_{m,2}/b$	0.318	0.501	0.623	0.701	0.753	0.791	0.821	0.844	0.863	0.880
Straight girder	b_m/b	0.320	0.500	0.621	0.677	0.748	0.784	0.811	0.832	0.849	0.863

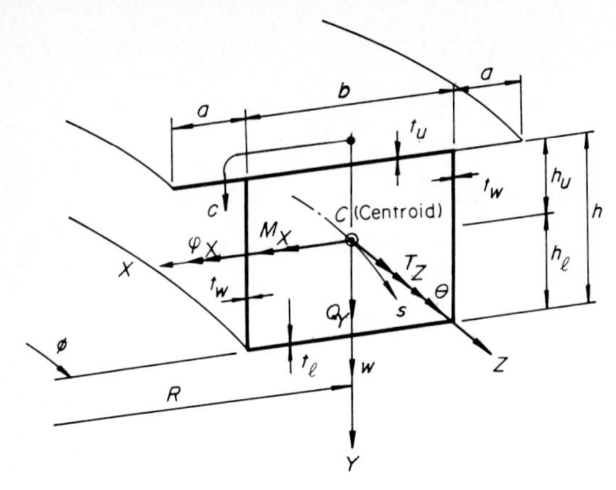

Figure 3.20 Cross section of a curved box girder.

torsional contributions to the distortion of a curved box girder cannot be treated independently.

(1) Notation.[26] Figure 3.20 shows the cross section of the curved box girder under consideration; the notation, similar to that in Sec. 2.4.1, is as follows:

a = breadth of bracket

b = web plate spacing

h = girder depth

R = radius of curvature at centroid (= shear and distortional center)

ϕ = angular coordinate

$$s = R\phi \tag{3.2.152}$$

s = curvilinear coordinate

$L = R\Phi$ = span length of curved box girder

t_u, t_l = thickness of top and bottom flange plates, respectively

t_w = thickness of web plate

$$A_u = t_u(2a + b) \tag{3.2.153a}$$

$$A_l = t_l b \tag{3.2.153b}$$

A_u, A_l = cross-sectional area of top and bottom flange plates, respectively

$$A_w = t_w h \tag{3.2.154}$$

A_w = cross-sectional area of web plate

$$A = A_u + A_l \tag{3.2.155}$$

A = total cross-sectional area of the girder

The applied forces, stress resultants, and displacements for flexure and pure torsion with respect to the X, Y, and Z coordinate axes, defined in Fig. 3.20, can be summarized as shown in Table 3.4.

In addition, the distortional angle Θ and the axial displacement (warping) $u = \omega_D \, d\Theta/ds = \omega_D \Theta'$ with respect to the distortional center D, having coordinate axes x, y, and z, can also be defined as in Sec. 2.4, as illustrated in Fig. 3.21 and Table 3.5.

(2) Fundamental equations[26]

a. **Flexure and torsion.** The coupled behavior of flexure and torsion (pure torsion) can be obtained easily from the following fundamental equations on the basis of the method described in Sec. 3.2.1:
Equilibrium equations for stress resultants,

$$\frac{dQ_Y}{ds} = -p \tag{3.2.156a}$$

$$\frac{dT_Z}{ds} - \frac{M_X}{R} = m_T \tag{3.2.156b}$$

$$-\frac{dM_X}{ds} + \frac{M_X}{R} - Q_Y = 0 \tag{3.2.156c}$$

TABLE 3.4 Applied Loads, Stress Resultants, and Displacements for Bending and Pure Torsion

Notation		Definition
Applied force	p	Distributed load in the direction of Y axis
	m_T	Distributed torque with respect to Z axis
Stress resultants	M_X	Bending moment with respect to X axis
	T_Z	Torsional moment with respect to Z axis
	Q_Y	Shearing force in the direction of Y axis
Displacement	φ	Deflection angle with respect to X axis
	θ	Torsional angle with respect to Z axis
	w	Deflection in the direction of Y axis

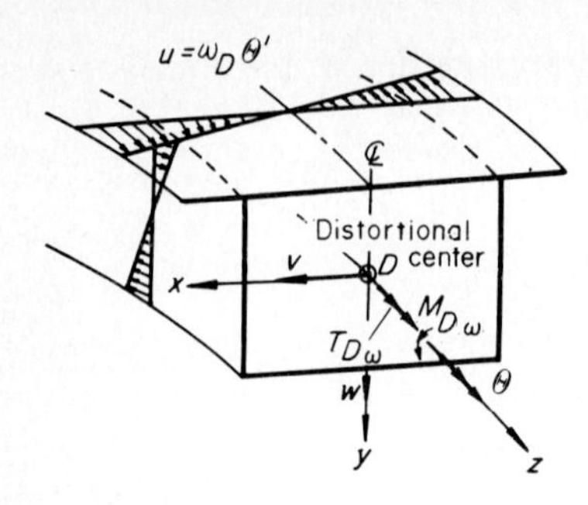

Figure 3.21 Stress resultants and displacements for distortional analysis.

Relationships between stress resultants and displacements,

$$\frac{d\theta}{ds} = \frac{T_Z}{GK} \tag{3.2.157a}$$

$$\frac{d^2w}{ds^2} + \frac{w}{R^2} = -\frac{M_X}{EI_X} + \frac{\theta}{R} \tag{3.2.157b}$$

$$\varphi_X = -\frac{dw}{ds} \tag{3.2.157c}$$

where
$$I_X = A_u h_u^2 + A_l h_l^2 + 2A_w\left[\frac{h^2}{12} + \left(\frac{h}{2} - h_u\right)^2\right] \tag{3.2.158}$$

h_u, h_l = distance from centroid C to top and bottom flange plates, respectively

I_X = geometric moment of inertia of cross section with respect to X axis

$$K = \frac{4b^2h^2}{b/t_u + 2h/t_w + b/t_l} \tag{3.2.159}$$

and K is the pure torsional constant.

The analytical procedure of developing simultaneous differential equations (3.2.156) and (3.2.157) is explained in detail in Secs. 4.2 and 4.3.

b. **Distortion.** The fundamental equation for distortion can now be derived by nearly the same procedures as those employed in Sec. 2.4. In

TABLE 3.5 Stress Resultants due to Distortion

Notation	Definition
$M_{D\omega}$	Bimoment due to distortion
$T_{D\omega}$	Torsional moment due to distortion

this case, the potential energy Π stored in a curved box girder is estimated by the following equation instead of Eq. (2.4.37) with respect to a straight box girder:

$$\Pi = U_\sigma + U_\tau + V_m + V_M \tag{3.2.160a}$$

Here U_σ, U_τ, and V_m can be approximated, respectively, by Eqs. (2.4.17), (2.4.33), and (2.4.36), by omitting the influence of the curvature of the curved girder elements. The remaining term V_M is caused by the torsional force M_X/R due to the influence of curvature, as is seen in Eq. (3.2.156*b*), and must be taken into consideration. The work done by this force can be represented by the following, which is similar to Eq. (2.4.36):

$$V_M = -\int_0^L \frac{\psi}{R} M\Theta \, ds \tag{3.2.161}$$

where ψ is a dimensionless parameter given by

$$\psi = \alpha_1 - \frac{\alpha_2}{\alpha_0} \tag{3.2.162}$$

in which

$$\alpha_0 = 1 + \frac{2b/h + 3(I_u + I_l)/I_w}{(I_u + I_l)/I_w + (6h/b)(I_u I_l/I_w^2)} \tag{3.2.163}$$

$$\alpha_1 = \frac{7h_u - 3h_l}{10I_X} h^2 t_w + \frac{A_u h_u h}{I_X} - \frac{1}{2} \tag{3.2.164a}$$

$$\alpha_2 = \frac{ht_w}{15I_X} \frac{(3h_u - 2h_l)(b + 3hI_l/I_u) + (3h_l - 2h_u)(b + 3hI_u/I_l)}{(I_u + I_l)/I_w + (6h/b)(I_u I_l/I_w^2)} \tag{3.2.164b}$$

The geometric moments of inertia for the top, web, and bottom plate elements, I_u, I_w, and I_l, respectively, are

$$I_u = \frac{t_u^3}{12(1-\mu^2)} \tag{3.2.165a}$$

$$I_w = \frac{t_w^3}{12(1-\mu^2)} \tag{3.2.165b}$$

$$I_l = \frac{t_l^3}{12(1-\mu^2)} \tag{3.2.165c}$$

Thus, the total potential energy is

$$\Pi = \frac{EI_{D\omega}}{2}\int_0^L \left(\frac{d^2\Theta}{ds^2}\right)^2 ds + \frac{K_{D\omega}}{2}\int_0^L \Theta^2 \, ds - \int_0^L \left(\frac{m_T}{2} + \psi\frac{M_X}{R}\right)\Theta \, ds \tag{3.2.160b}$$

The potential energy function Π can be minimized through variational calculus techniques, resulting in the following differential equation governing the distortional behavior of the beam:

$$EI_{D\omega}\frac{d^4\Theta}{ds^4} + K_{D\omega}\Theta = \frac{m_T}{2} + \psi\frac{M_X}{R} \tag{3.2.166}$$

Here the distortional warping constant $I_{D\omega}$ is given by

$$I_{D\omega} = \int_A \omega_D^2\, dA = \frac{b^2h^2}{48(1+\beta)}[A_l\beta + A_w(2\beta - 1)] \tag{3.2.167}$$

in which

$$\beta = \frac{A_u(1+2a/b)^2 + 3A_w}{A_l + A_w} \tag{3.2.168}$$

and $K_{D\omega}$ is the stiffness of the box section against the distortion, given by

$$K_{D\omega} = \frac{24EI_w}{\alpha_o h} \tag{3.2.169}$$

The expression for α_o is given in Eq. (3.2.163).

Since the fundamental differential equation (3.2.166) for the distortion is analogous to that of a beam on an elastic foundation, the curved box girder can be modified as shown in Fig. 3.22. Namely, the girder is

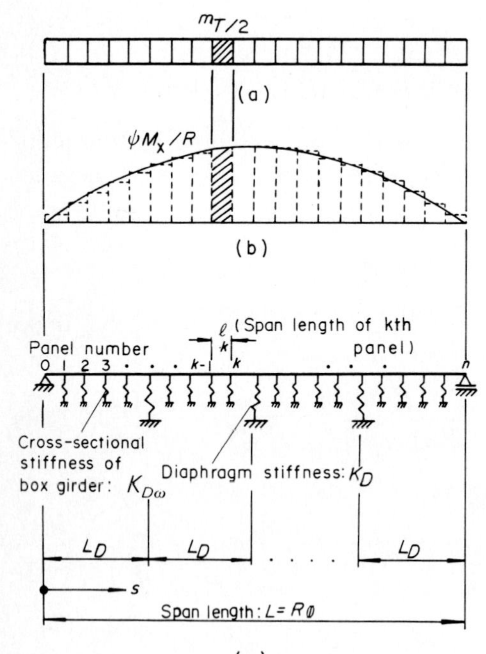

Figure 3.22 Beam-on-an-elastic-foundation analogy for a curved box girder: (*a*) distortional force $m_T/2$, (*b*) distortional force due to the bending moment, (*c*) BEF with elastic support.

subjected to the distortional force $m_T/2$ of Fig. 3.22*a* and the additional distortional force $\psi M/R$ of Fig. 3.22*b* due to the bending moment. And the girder is elastically supported by not only the elastic foundation with stiffness $K_{D\omega}$ of Eq. (3.2.169) but also the diaphragms with stiffness K_D, which can be estimated according to the type of diaphragm, as indicated in Fig. 3.23:

For plate-type diaphragms,

$$K_D = Gt_D bh \tag{3.2.170}$$

For truss-type diaphragms,

$$K_D = \begin{cases} \dfrac{2EA_b b^2 h^2}{l_b^3} & X \text{ type} \quad (3.2.171a) \\[2ex] \dfrac{EA_b b^2 h^2}{2l_b^3} & V \text{ type} \quad (3.2.171b) \end{cases}$$

where A_b is the cross-sectional area of truss member and l_b is the length of truss member.

For frame-type diaphragms,

$$K_D = \frac{24EI_w}{\alpha_o h} \tag{3.2.172}$$

Here α_o can be evaluated from Eq. (3.2.163), provided that the geometric moments of inertia I_u, I_w, and I_l are determined in the same manner as those of the frame—by taking into consideration the effective width b_m according to the pitch of the intermediate diaphragm L_D, as follows

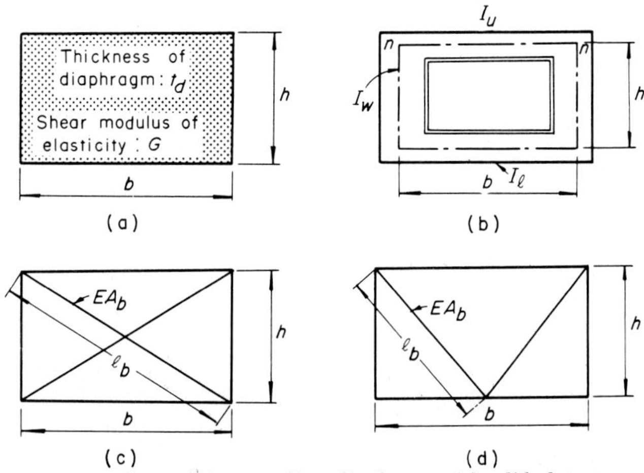

Figure 3.23 Types of intermediate diaphragm: (*a*) solid-plate type, (*b*) frame type, (*c*) truss *X* type, and (*d*) truss *V* type.

[see Eq. (2.2.49)]:

$$b_m = \frac{d}{3} \qquad L_D \geqq \frac{d}{3} \tag{3.2.173a}$$

$$b_m = L_D \qquad L_D < \frac{d}{3} \tag{3.2.173b}$$

where d is the spacing b of the web plate or the girder depth h, whichever is smaller.

The boundary conditions at both ends of a curved box girder are described in Table 3.6.

The distortional warping moment $T_{D\omega}$ must be present in the case where the intermediate diaphragms are attached to the girder (see Fig. 3.22), and these values can be obtained by

$$T^s_{D\omega} = K_D \Theta \tag{3.2.174}$$

Once Eq. (3.2.166) has been solved by applying the boundary conditions in Table 3.6 and Eq. (3.2.174), the following stress resultants can be found:

$$M_{D\omega} = EI_{D\omega} \frac{d^2\Theta}{ds^2} \tag{3.2.175}$$

$$T_{D\omega} = -\frac{dM_{D\omega}}{ds} = -EI_{D\omega} \frac{d^3\Theta}{ds^3} \tag{3.2.176}$$

Therefore, the distortional normal stress $\sigma_{D\omega}$ and the shearing stress $\tau_{D\omega}$ now can be estimated in a form similar to that of the torsional warping theory:

$$\sigma_{D\omega} = \frac{M_{D\omega}}{I_{D\omega}} \omega_D \tag{3.2.177}$$

$$q_{D\omega} = \tau_{D\omega} t = \frac{S_{D\omega}}{I_{D\omega}} T_{D\omega} \tag{3.2.178}$$

The distortional warping functions ω_D can be obtained from

TABLE 3.6 Boundary Conditions

Support	Boundary conditions	
Simple support	$\Theta = 0$,	$M_{D\omega} = 0$
Rigid support	$\Theta = 0$,	$d\Theta/dS = 0$
Free support	$M_{D\omega} = 0$,	$T_{D\omega} = 0$

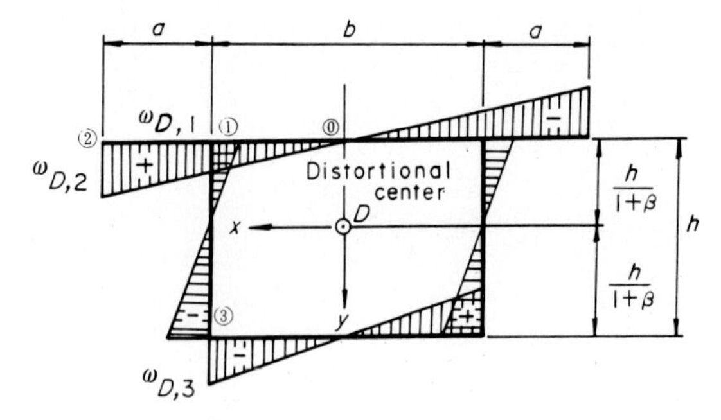

Figure 3.24 Warping function due to distortion.

Fig. 3.24, and their values are

$$\omega_{D,1} = \frac{bh}{4}\frac{1}{1+\beta} \tag{3.2.179a}$$

$$\omega_{D,2} = \frac{bh}{4}\left(1+\frac{2a}{b}\right)\frac{1}{1+\beta} = \left(1+\frac{2a}{b}\right)\omega_{D,1} \tag{3.2.179b}$$

$$\omega_{D,3} = \frac{bh}{4}\frac{\beta h}{1+\beta} = -\beta\omega_{D,1} \tag{3.2.179c}$$

and the static moment $S_{D\omega}$ with respect to ω_D can be plotted as shown in Fig. 3.25.

The diaphragm undergoes the distortional moment $T^s_{D\omega}$, as given in Eq. (3.2.174); therefore the induced normal stress σ and the shearing stress τ can be evaluated as follows:
For a plate-type diaphragm,

$$\tau = \frac{T^s_{D\omega}}{bht_D} \tag{3.2.180}$$

For a truss-type diaphragm,

$$\sigma = \frac{N}{A_b} \tag{3.2.181}$$

in which

$$N = \frac{T^s_{D\omega}}{4bh} \quad X\text{-type truss} \tag{3.2.182a}$$

$$N = \frac{T^s_{D\omega}}{2bh} \quad V\text{-type truss} \tag{3.2.182b}$$

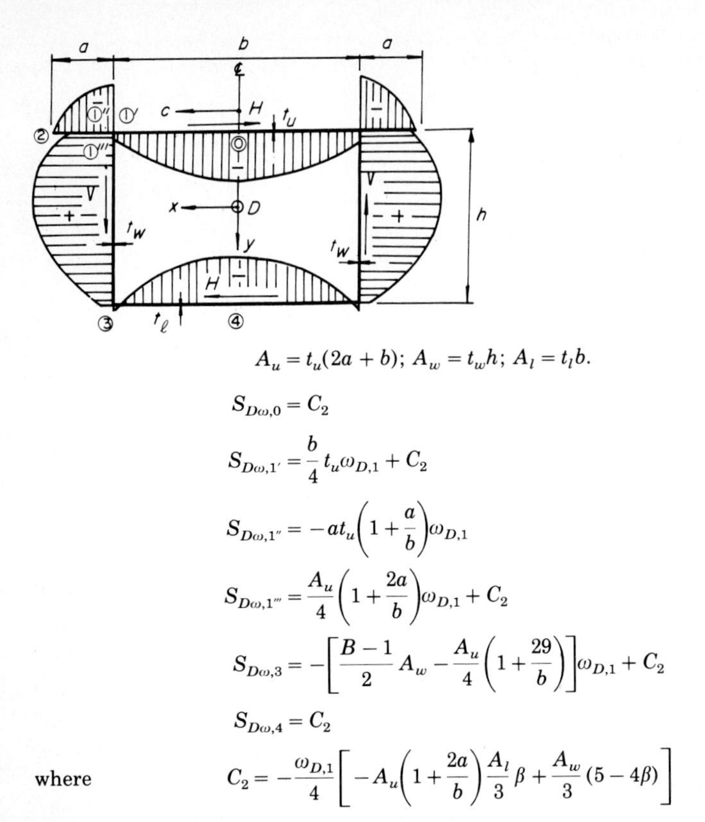

$$A_u = t_u(2a + b);\ A_w = t_w h;\ A_l = t_l b.$$

$$S_{D\omega,0} = C_2$$

$$S_{D\omega,1'} = \frac{b}{4} t_u \omega_{D,1} + C_2$$

$$S_{D\omega,1''} = -at_u\left(1 + \frac{a}{b}\right)\omega_{D,1}$$

$$S_{D\omega,1'''} = \frac{A_u}{4}\left(1 + \frac{2a}{b}\right)\omega_{D,1} + C_2$$

$$S_{D\omega,3} = -\left[\frac{B-1}{2} A_w - \frac{A_u}{4}\left(1 + \frac{29}{b}\right)\right]\omega_{D,1} + C_2$$

$$S_{D\omega,4} = C_2$$

where
$$C_2 = -\frac{\omega_{D,1}}{4}\left[-A_u\left(1 + \frac{2a}{b}\right)\frac{A_l}{3}\beta + \frac{A_w}{3}(5 - 4\beta)\right]$$

Figure 3.25 Diagram of static moment for the distortional warping function.

For a frame-type diaphragm,

$$\sigma = \frac{M_D}{W} \tag{3.2.183}$$

$$\tau_1 = \frac{2M_D^u}{bF_w} \tag{3.2.184a}$$

$$\tau_2 = \frac{M_D^u + M_D^l}{hF_w} \tag{3.2.184b}$$

Here $M_D = (M_D^u, M_D^l)$ are the bending moments in the transverse direction, shown in Fig. 3.26, and are given by

$$M_D^u = \frac{T_{D\omega}^s}{4}\left[1 + \frac{I_u - I_l}{I_u + I_l + 6(h/b)(I_u I_l / I_w^2)}\right] \tag{3.2.185a}$$

$$M_D^l = \frac{T_{D\omega}^s}{4}\left[1 + \frac{I_l - I_w}{I_u + I_l + 6(h/b)(I_u I_l / I_w^2)}\right] \tag{3.2.185b}$$

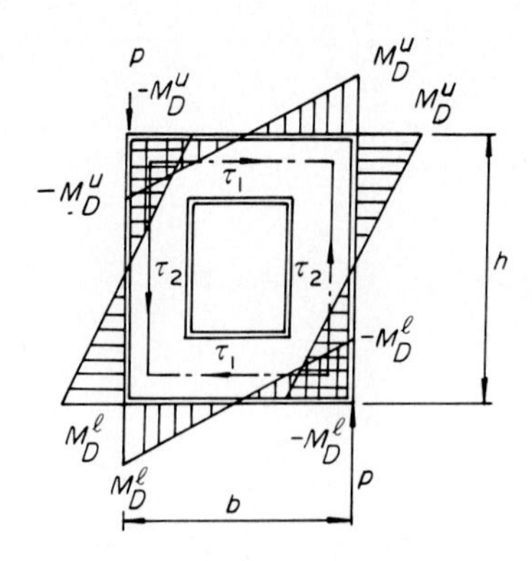

Figure 3.26 Transverse bending of frame member and induced stresses.

in which W = section modulus of the frame members [taking into account the effective width of the flange plates as prescribed by Eq. (3.2.173)] and F_w = area of the web plate of the frame members.

(3) Distortional behavior of curved box girders by model tests.[26] To verify the theoretical study and to check the accuracy of the analytical method, experimental studies were carried out on four small curved girder models with various numbers of rigid diaphragms. All the model girders were built with the same span length L = 1000 mm, radius of curvature R = 2000 mm, and central angle Φ = 0.5 rad, by making use of Plexiglas plate 2 mm thick. The numbers of diaphragms with a thickness of 1 mm varied from 1, 3, 5, and 7 for model girders CG-1 through CG-4, respectively, as shown in Fig. 3.27.

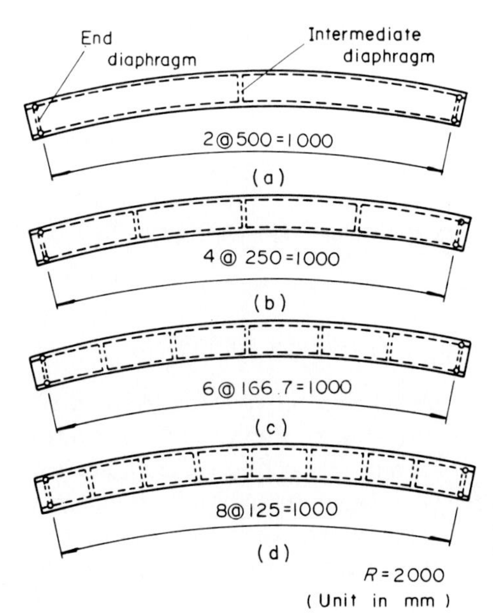

Figure 3.27 Model girder: (*a*) CG-1 model, (*b*) CG-2 model, (*c*) CG-3 model, and (*d*) CG-4 model.

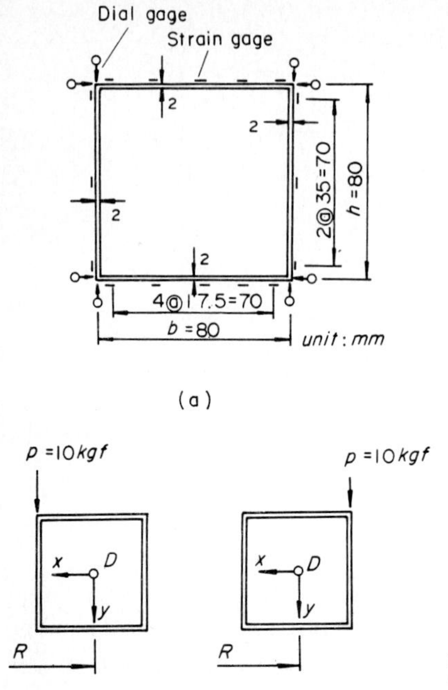

Figure 3.28 (*a*) Cross section of model girder and (*b*) load conditions.

The cross section of the model girders consisted of a square section with flange and web plates of equal thickness, as shown in Fig. 3.28*a*, to avoid the influence of torsional warping stresses.

The normal strains in the box section were measured by 16 strain gages per section, and the distortional angle was measured by eight dial gages, as shown in Fig. 3.28*a*.

The model girders were tested under simply supported end conditions. The vertical concentrated loads $P = 10$ kgf were applied as shown in Fig. 3.28*b*.

The elastic properties of the Plexiglas were determined to be

$$\text{Young's modulus} = E = 2.8 \times 10^4 \text{ kgf/cm}^2$$

$$\text{Shear modulus of elasticity} = G = 1.04 \times 10^4 \text{ kgf/cm}^2$$

Figure 3.29 shows the variation of the normal stresses in the direction of the girder axis. When the loads were applied at the section midway between diaphragms, the normal stress due to the distortion gradually decreased in accordance with the increase in the numbers of intermediate diaphragms, and the total normal stresses approached those predicted by elementary bending theory. However, the normal

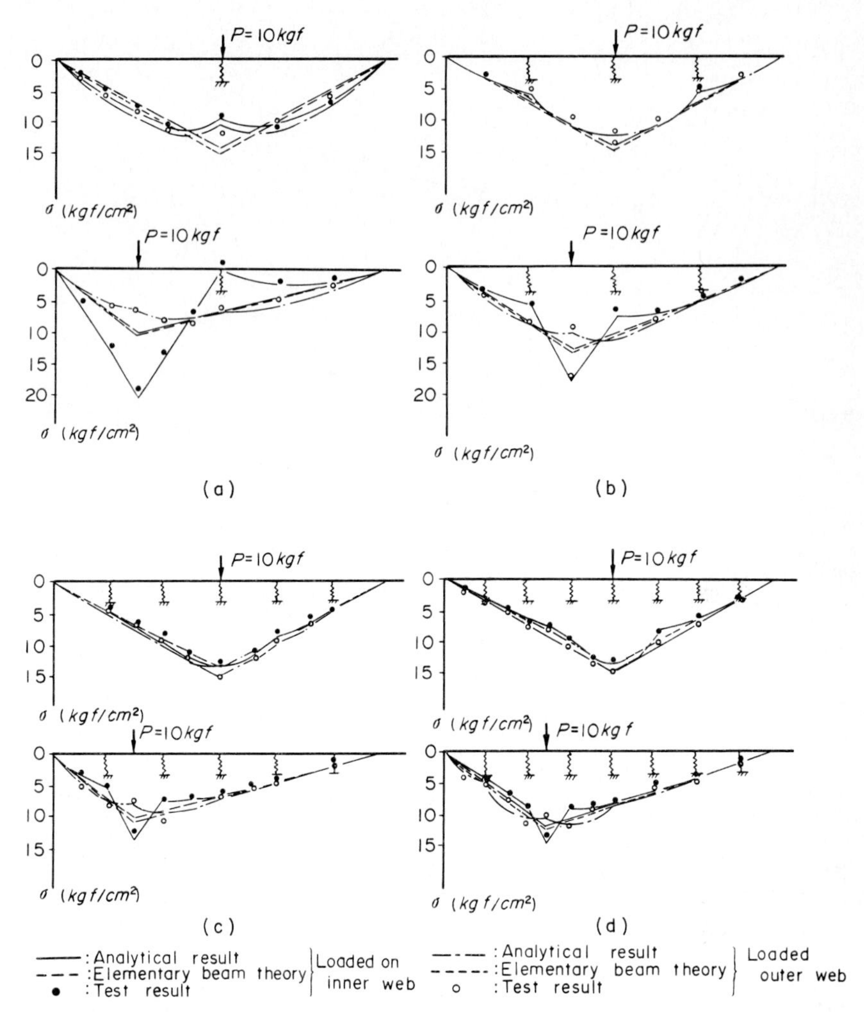

Figure 3.29 Variation of normal stress in the direction of the girder axis: (*a*) model girder with one diaphragm (CG-1), (*b*) model girder with three diaphragms (CG-2), (*c*) model girder with five diaphragms (CG-3), and (*d*) model girder with seven diaphragms (CG-4).

stress due to distortion is very small when the load acts at the section over the diaphragm. Also the normal stress when the load acts on the inner web is always greater than the normal stresses produced when the load acts on the outer web, as can be seen from Eq. (3.2.166).

The deflection curves for CG-1 and CG-4 are shown in Fig. 3.30. From these figures, the influence of the numbers of diaphragms on the deflection curves is not obvious.

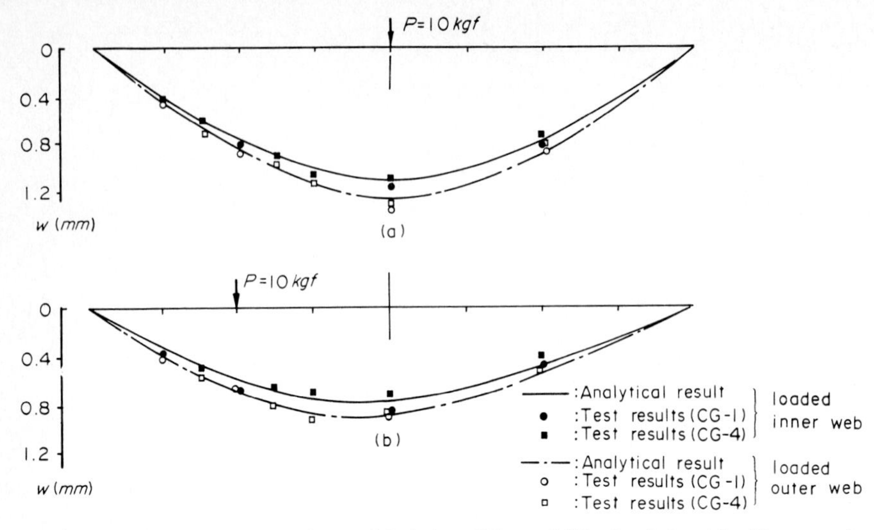

Figure 3.30 Deflection curves for model girders CG-1 and CG-4 loaded at (*a*) midspan and (*b*) quarterspan.

3.2.6 Theory of sector plates[27–32]

In evaluating the stresses in the deck plate or the buckling strength of the flange plate of a curved girder, the theory of sector plates is considered instead of the theory of flat plates due to the curvature of the girders.

(1) Isotropic sector plate. We now consider the isotropic sector plate shown in Fig. 3.31. For this sector plate, it is convenient to use the polar coordinate axes (r, ϕ) shown in Fig. 3.31*a*. The fundamental

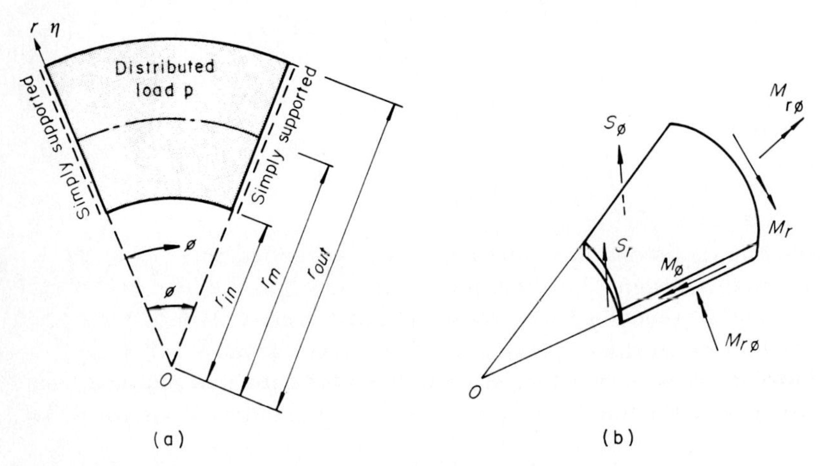

Figure 3.31 Sector plate in (*a*) coordinate axes and (*b*) stress resultants.

equation for the deflection $w(r, \phi)$ of this sector plate subjected to the distributed vertical load p is

$$\nabla^2 w = \frac{p}{D} \tag{3.2.186}$$

where $$\nabla = \frac{\partial^2}{\partial r^2} + \frac{1}{r}\frac{\partial}{\partial r} + \frac{1}{r^2}\frac{\partial^2}{\partial \phi^2} \tag{3.2.187}$$

and $$D = \frac{Eh^3}{12(1-\mu^2)} \tag{3.2.188}$$

= flexural rigidity of isotropic sector plate

E = Young's modulus
μ = Poisson's ratio
h = thickness of sector plate

The stress resultants in the sector plate can now be defined as in Fig. 3.31*b*, where

M_r, M_ϕ = bending moment in radial and tangential directions, respectively
$M_{r\phi}$ = torsional moment
S_r, S_ϕ = shearing force in radial and tangential directions, respectively

Equations for the stress resultants can be expressed as

$$M_r = -D\left[\frac{\partial^2 w}{\partial r^2} + \mu\left(\frac{1}{r}\frac{\partial w}{\partial r} + \frac{1}{r^2}\frac{\partial^2 w}{\partial \phi^2}\right)\right] \tag{3.2.189a}$$

$$M_\phi = -D\left(\frac{1}{r}\frac{\partial w}{\partial r} + \frac{1}{r^2}\frac{\partial^2 w}{\partial \phi^2} + \mu\frac{\partial^2 w}{\partial r^2}\right) \tag{3.2.189b}$$

$$M_{r\phi} = D(1-\mu)\left(\frac{1}{r}\frac{\partial^2 w}{\partial r\,\partial \phi} - \frac{1}{r^2}\frac{\partial w}{\partial \phi}\right) \tag{3.2.189c}$$

and

$$S_r = -D\left(\frac{\partial^3 w}{\partial r^3} + \frac{1}{r}\frac{\partial^2 w}{\partial r^2} - \frac{1}{r^2}\frac{\partial w}{\partial r} + \frac{1}{r^2}\frac{\partial^3 w}{\partial r\,\partial \phi^2} - \frac{2}{r^3}\frac{\partial^2 w}{\partial \phi^2}\right) \tag{3.2.190a}$$

$$S_\phi = -D\left(\frac{1}{r}\frac{\partial^3 w}{\partial r^2\,\partial \phi} + \frac{1}{r^2}\frac{\partial^2 w}{\partial r\,\partial \phi} + \frac{1}{r^2}\frac{\partial^3 w}{\partial \phi^3}\right) \tag{3.2.190b}$$

To solve Eq. (3.2.186) as exactly as possible, the dimensionless coordinate η is introduced:

$$\eta = \frac{r}{r_m} \tag{3.2.191}$$

Here r_m is the radius of curvature at the midline of the sector plate, as shown in Fig. 3.31*a*. Then the solution of deflection $w(r, \phi)$ can be expanded by the Fourier series as

$$w = \sum_{n=1,2,\ldots} R_n(\eta) \sin \frac{n\pi\phi}{\Phi} \tag{3.2.192}$$

The above equation should, of course, satisfy the following boundary conditions:

$$[w]_{\phi=0} = [w]_{\phi=\Phi} = 0 \tag{3.2.193a}$$

$$[M_\phi]_{\phi=0} = [M_\phi]_{\phi=\Phi} = 0 \tag{3.2.193b}$$

The substitution of Eq. (3.2.192) into Eq. (3.2.186) gives the characteristic equation for $R_n(\eta)$, which leads to

$$\frac{1}{r_m^2}\left[\frac{d^2}{d\eta^2} + \frac{1}{\eta}\frac{d}{d\eta} - \frac{1}{\eta^2}\left(\frac{n\pi}{\Phi}\right)^2\right]^2 R_n(\eta) = 0 \tag{3.2.194}$$

The solution can be written in the form

$$R_n(\eta) = A_n\eta^m + B_n\eta^{-m} + C_n\eta^{2+m} + D_n\eta^{2-m} \tag{3.2.195}$$

where
$$m = \frac{n\pi}{\Phi} \tag{3.2.196}$$

and A_n, B_n, C_n, and D_n are the integration constants that can be determined by the boundary conditions at both edges of the sector plate.

In addition, a concentrated vertical load P, analogous to the wheel loads of vehicle, acting on a point $\eta_p(r_p/r_m, \phi_p)$ can also be represented by a Fourier series as

$$p = \sum_{n=1,2,\ldots} \frac{2P}{r_m\eta_p\Phi} \sin m\phi_p \sin m\phi \tag{3.2.197}$$

Accordingly, the complete solution can be written by

$$w = \sum_{n=1,2,\ldots} \{A_n\eta^m + B_n\eta^{-m} + C_n\eta^{2+m} + D_n\eta^{2-m} + [A_n'\eta'^m + B_n'(\eta')^{-m} + C_n'(\eta')^{2+m} + D_n'(\eta')^{2-m}]\} \sin m\phi \tag{3.2.198}$$

provided that

$$\eta < \eta_p \qquad \text{for } (\eta')_p^m = 0 \tag{3.2.199a}$$

$$\eta > \eta_p \qquad \text{for } (\eta')_p^m = \eta^m \tag{3.2.199b}$$

To determine the four integration constants A_n', B_n', C_n', and D_n', the continuity conditions for displacements and stress resultants required that

$$[w]_{\eta_p - 0} = [w]_{\eta_p + 0} \tag{3.2.200a}$$

$$\left[\frac{\partial w}{\partial r}\right]_{\eta_p - 0} = \left[\frac{\partial w}{\partial r}\right]_{\eta_p + 0} \tag{3.2.200b}$$

$$\left[\frac{\partial^2 w}{\partial r^2}\right]_{\eta_p - 0} = \left(\frac{\partial^2 w}{\partial r^2}\right)_{\eta_p + 0} \tag{3.2.200c}$$

$$\left[\frac{\partial^3 w}{\partial r^3}\right]_{\eta_p - 0} = \left[\frac{\partial^3 w}{\partial r^3}\right]_{\eta_p + 0} + \frac{2P}{r_m \eta_p D} \sin m\phi_p \tag{3.2.200d}$$

Thus, Eq. (3.2.198) can now be rearranged in the following manner:

$$w = \sum_{n=1,2,\ldots} \left\{ A_n \eta^m + B_n \eta^{-m} + C_n \eta^{2+m} + D_n \eta^{2-m} - \frac{P \sin m\phi_p}{4 r_m \eta_p D \Phi_m (m^2 - 1)} [(m+1)\eta_p^{3-m}(\eta')_p^m + (m-1)\eta_p^{3+m}(\eta')_p^{-m} - (m-1)\eta_p^{1-m}(\eta')_p^{2+m} - (m+1)\eta_p^{1+m}(\eta')_p^{2-m}] \right\} \sin m\phi \tag{3.2.201}$$

Finally, the remaining integration constants A, B, C, and D can be determined by

$$[w]_{\phi=0} = [w]_{\phi=\Phi} \tag{3.2.202a}$$

$$[M_\phi]_{\phi=0} = [M_\phi]_{\phi=\Phi} \tag{3.2.202b}$$

where

$$M_\phi = S_r - \frac{1}{r}\frac{\partial M_r \phi}{\partial \phi} \tag{3.2.203}$$

or

$$\frac{\partial^3 w}{\partial r^3} + \frac{1}{r}\frac{\partial^2 w}{\partial r^2} - \frac{1}{r^2}\frac{\partial w}{\partial r} + \frac{2-\mu}{r^2}\frac{\partial^3 w}{\partial r\,\partial \phi^2} - \frac{3-\mu}{r^3}\frac{\partial^2 w}{\partial \phi^2} = 0 \tag{3.2.204}$$

(2) Orthotropic sector plate. In the case of an orthotropic sector plate such as the steel deck of a curved girder bridge, the differential equation for the deflection $w(r, \phi)$ can be approximated as

$$D_r\left(\frac{\partial^4 w}{\partial r^4} + \frac{2}{r}\frac{\partial^3 w}{\partial r^3}\right) + (D_{r\phi} + D_{\phi r})\left(\frac{1}{r^2}\frac{\partial^4 w}{\partial r^2\,\partial \phi^2} - \frac{1}{r^3}\frac{\partial^3 w}{\partial r\,\partial \phi^2} + \frac{1}{r^4}\frac{\partial^2 w}{\partial \phi^2}\right) + D_\phi\left(\frac{1}{r^4}\frac{\partial^4 w}{\partial^4 \phi} + \frac{2}{r^4}\frac{\partial^2 w}{\partial^2 \phi} + \frac{1}{r^3}\frac{\partial w}{\partial r} - \frac{1}{r^2}\frac{\partial^2 w}{\partial r^2}\right) = p(r, \phi) \tag{3.2.205}$$

where D_r and D_ϕ are the flexural rigidities of the orthotropic sector plate in the radial and tangential directions, respectively, and $D_{r\phi}$ and $D_{\phi r}$ are the torsional rigidities of the orthotropic sector plates in the $r\phi$ and ϕr planes, respectively.

The corresponding stress resultants can be obtained by using the solution for the deflection $w(r, \phi)$ as follows:

$$M_r = -D_r\left[\frac{\partial^2 w}{\partial r^2} + \mu_\phi\left(\frac{1}{r}\frac{\partial w}{\partial r} + \frac{1}{r^2}\frac{\partial^2 w}{\partial \phi^2}\right)\right] \tag{3.2.206a}$$

$$M_\phi = -D_\phi\left(\frac{1}{r}\frac{\partial w}{\partial r} + \frac{1}{r^2}\frac{\partial^2 w}{\partial \phi^2} + \mu_\phi\frac{\partial^2 w}{\partial r^2}\right) \tag{3.2.206b}$$

$$M_{r\phi} = D_{r\phi}\left(\frac{1}{r}\frac{\partial^2 w}{\partial r\,\partial \phi} - \frac{1}{r^2}\frac{\partial w}{\partial \phi}\right) \tag{3.2.206c}$$

$$M_{\phi r} = -D_{\phi r}\left(\frac{1}{r}\frac{\partial^2 w}{\partial r\,\partial \phi} - \frac{1}{r^2}\frac{\partial w}{\partial \phi}\right) \tag{3.2.206d}$$

where M_r, M_ϕ = bending moment in radial and tangential directions, respectively

$M_{r\phi}, M_{\phi r}$ = torsional moment in $r\phi$ and ϕr planes, respectively

μ_r, μ_ϕ = Poisson's ratio in radial and tangential directions, respectively

3.3 Static Behavior of Curved Girders[33–44]

3.3.1 Torsional warping stress

In the design of curved girder bridges, the engineer is faced with a complex stress situation, since these types of bridges are subjected to both bending and torsional forces. In general, the torsional forces consist of two parts, i.e., St. Venant's and warping. Thus the procedure for determining the induced stresses in a curved girder is difficult.

To clarify the magnitude of the torsional warping stress, the following preliminary analysis is conducted.

(1) Definition of torsional parameter κ.[41,44] The governing differential equation for the twisting angle θ of a curved beam subjected to torque m_T with $R_P = R_s$ is

$$EI_\omega\frac{d^4\theta}{ds^4} - GK\frac{d^2\theta}{ds^2} = m_T - \frac{M_X}{R_s} \tag{3.3.1}$$

The bimoment M_ω is given by the well-known formula

$$M_\omega = EI_\omega\frac{d^2\theta}{ds^2} \tag{3.3.2}$$

The differential equation (3.3.1) can be rewritten with respect to the bimoment as

$$\frac{d^2M_\omega}{ds^2} - \alpha^2\frac{M_\omega}{R_s^2} = m_T - \frac{M_X}{R_s} \tag{3.3.3}$$

where the parameter α is given by

$$\alpha = R_s\sqrt{\frac{GK}{EI_\omega}} \tag{3.3.4}$$

This parameter can be nondimensionalized by multiplying by the central angle of curved girder Φ, which yields

$$\kappa = \alpha\Phi = R\sqrt{\frac{GK}{EI_\omega}} = L\sqrt{\frac{GK}{EI_\omega}} \tag{3.3.5}$$

(2) Values of parameter κ for actual bridges.[41,44] The torsional parameters κ for various curved girder bridges with cross sections as illustrated in Fig. 3.32 were investigated by using the actual dimensions of the bridges.

In evaluating the torsional constant K and warping constant I_ω of bridges modeled as a single curved girder, exact solutions may be applied. In addition to these techniques, approximate and simple formulas can be applied for multiple-I-girder and twin-box-girder bridges. First, an arbitrary point B is chosen as the origin, as shown

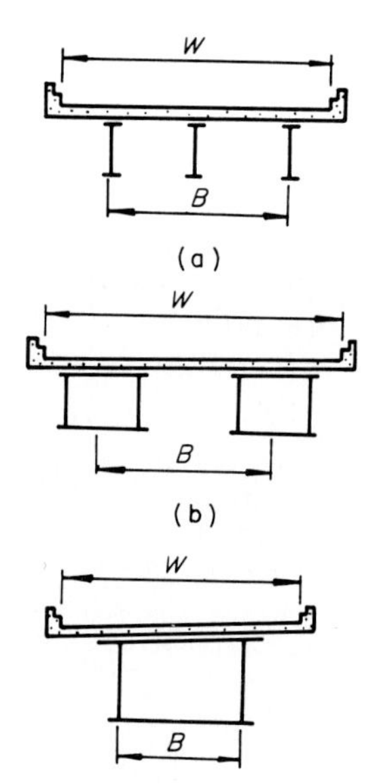

Figure 3.32 Typical cross-sectional shape of curved girder bridges: (*a*) multiple-I girder, (*b*) twin-box girder, and (*c*) monobox girder.

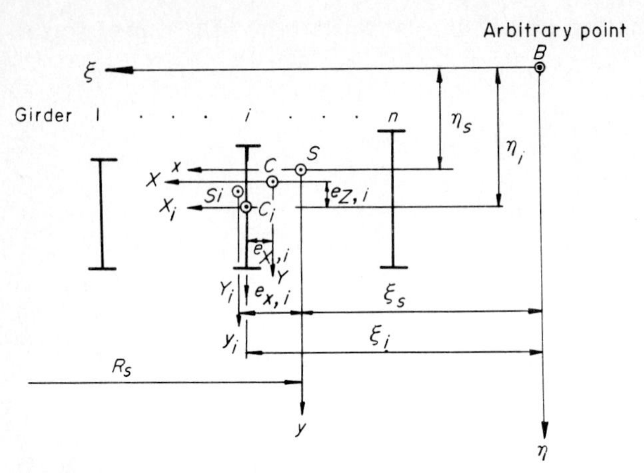

Figure 3.33 Estimation of shear center S for curved multiple-I-girder bridge.

in Fig. 3.33. If we assume horizontal and vertical axes ξ and η, respectively, the location of the shear center S for the multiple girder bridge, idealized as a single unit, can be determined from the equations[33]

$$\xi_s \cong \frac{\sum_{i=1}^{n} I_{X,i}\xi_i}{\sum_{i=1}^{n} I_{X,i}} \tag{3.3.6a}$$

$$\eta_s \cong \frac{\sum_{i=1}^{n} I_{Y,i}\eta_i}{\sum_{i=1}^{n} I_{Y,i}} \tag{3.3.6b}$$

where ξ_i, η_i = horizontal and vertical distances, respectively, between centroid C_i of ith girder and point B

$I_{X,i}, I_{Y,i}$ = moments of inertia of ith girder with respect to the centroidal X_i and Y_i axes, respectively

The torsional and warping constants can be approximated as[33]

$$K \cong \sum_{i=1}^{n} K_i \tag{3.3.7}$$

$$I_\omega \cong \sum_{i=1}^{n} (I_{\omega,i} + I_{X,i}e_{X,i}^2 + I_{Y,i}e_{Y,i}^2) \tag{3.3.8}$$

where K_i = torsional constant of ith girder
$I_{\omega,i}$ = warping constant with respect to S_i of ith girder
$e_{X,i}, e_{Y,i}$ = horizontal center S_i of ith girder and shear center S of the system

Also the centroid C for the system of curved beams can be determined[33] from

$$\xi_0 = \frac{\sum_{i=1}^{n} A_i \xi_i}{\sum_{i=1}^{n} A_i} \qquad (3.3.9a)$$

$$\eta_0 = \frac{\sum_{i=1}^{n} A_i \eta_i}{\sum A_i} \qquad (3.3.9b)$$

where A_i is the cross-sectional area of the ith girder. The corresponding moment and product of inertia, which will be required in the stress analysis, can be approximated[33] by

$$I_X = \sum_{i=1}^{n} (I_{X,i} + A_i e_{y,i}^2) \qquad (3.3.10a)$$

$$I_Y = \sum_{i=1}^{n} (I_{Y,i} + A_i e_{X,i}^2) \qquad (3.3.10b)$$

$$I_{XY} = \sum_{i=1}^{n} (I_{XY,i} + A_i e_{X,i} e_{Y,i}) \qquad (3.3.10c)$$

where $I_{X,i}, I_{Y,i}$ = geometric moments of inertia with respect to X_i and Y_i axes, respectively, of ith girder
$I_{XY,i}$ = product of inertia with respect to C_i of ith girder
$e_{X,i}, e_{Y,i}$ = horizontal and vertical distances respectively, between C_i and C

By applying these approximate formulas (more exact ones are shown in Chap. 6) to actual bridges, the interrelationship between κ and Φ has been determined; see the results in Fig. 3.34.

Examination of the trends in Fig. 3.34 indicates that Φ is not important and that the parameter κ will have the following ranges:

$\kappa = 0.5$–3 multiple-I girder (3.3.11a)

$\kappa = 3$–10 twin box (3.3.11b)

$\kappa \geqq 30$ monobox (3.3.11c)

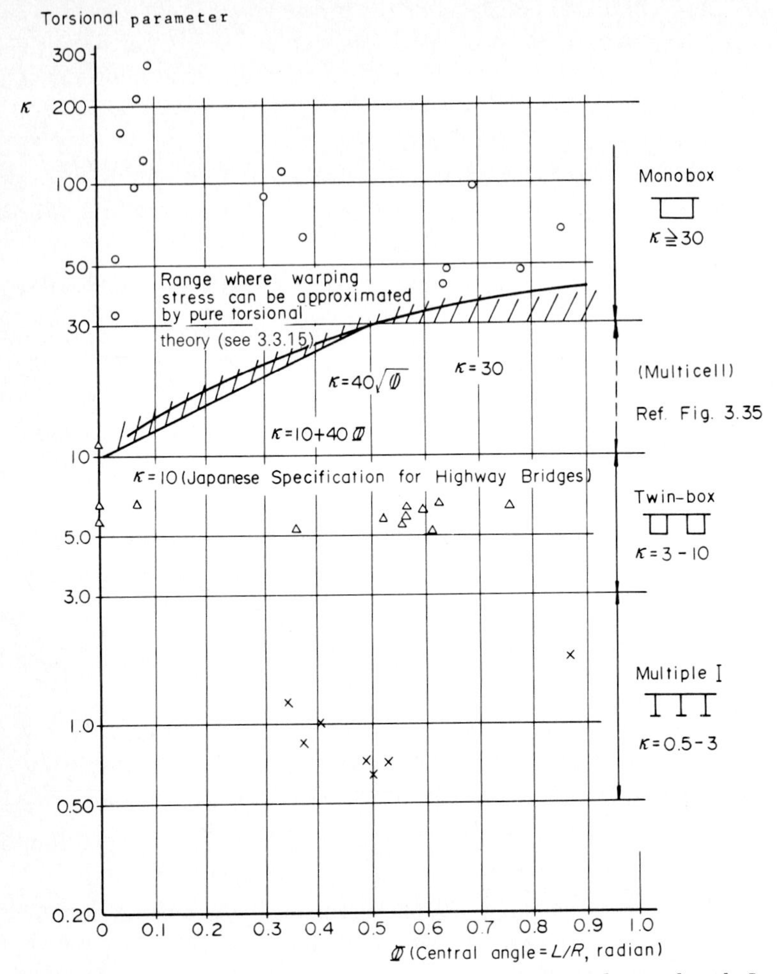

Figure 3.34 Relationships between torsional parameter κ and central angle Φ.

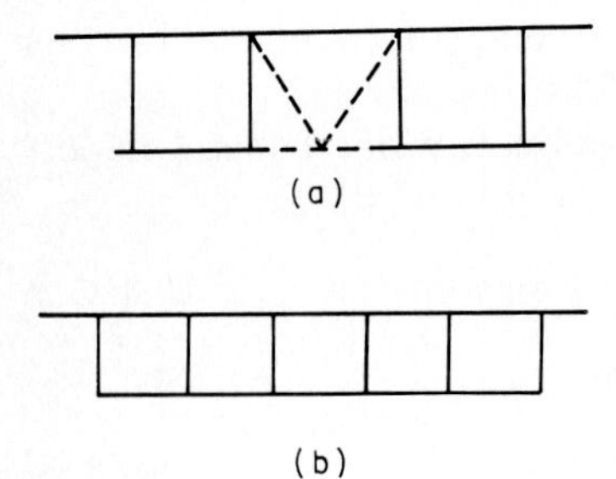

Figure 3.35 Box section with multicell: (*a*) twin box with lateral and sway bracings, (*b*) shallow multicell box.

The curved bridge geometries shown in Fig. 3.35, i.e., twin boxes and multicells, have κ values of 10 to 30.

(3) Relationships between the stress ratio σ_ω/σ_b and κ.[41,44] The design of curved girder bridges is obviously related to the various loading conditions such as dead and live loads. The purpose of this section, therefore, is to determine the most severe loading conditions that will induce the largest bending stress σ_b and warping stress σ_ω. These loading conditions can be idealized by a concentrated load P or the uniformly distributed load q, as shown in Fig. 3.36.

For a concentrated load P and a uniform load q, the induced midspan bending moments M_X can be obtained by solving Eq. (3.2.6), which gives[34]

$$M_X = \frac{P}{2} R_s \tan \frac{\Phi}{2} \qquad \text{or} \qquad M_X = qR_s^2\left(\sec \frac{\Phi}{2} - 1\right) \tag{3.3.12}$$

The corresponding bimoments M_ω can be obtained by solving Eqs. (3.3.1) and (3.3.2),[34] which results in

$$M_\omega = \frac{PR_s}{2\alpha^2} R_s \tan \frac{\Phi}{2} \qquad \text{or} \qquad M_\omega = \frac{qR_s^3}{\alpha^2}\left(\sec \frac{\Phi}{2} - 1\right) \tag{3.3.13}$$

in which the parameter $\kappa = \alpha\Phi \geqq 9$.

Next, the ratio of warping stress σ_ω to bending stress σ_b, or σ_ω/σ_b, can be estimated by applying Eq. (3.2.43) with $R_o/np = 1$ and $I_{XY} = 0$ and Eq. (3.2.83a) with $n = 1$:

$$\frac{\sigma_\omega}{\sigma_b} = \frac{(M_\omega/I_\omega)\omega}{(M_X/I_X)Y} = \frac{I_X\omega}{I_\omega Y}\frac{R_s}{\alpha^2} = \frac{I_X\omega}{I_\omega Y}\frac{L}{\alpha^2\Phi} \tag{3.3.14}$$

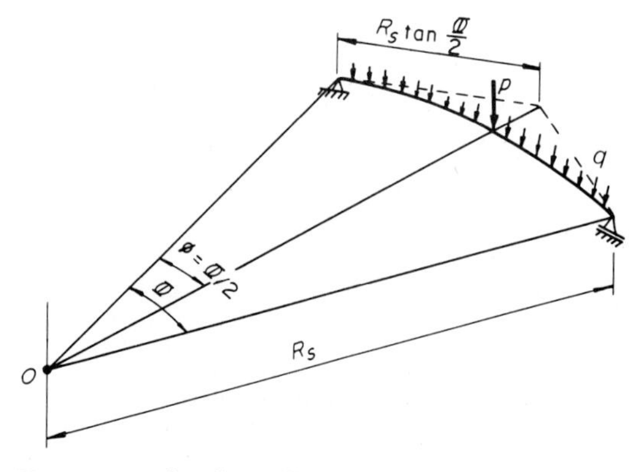

Figure 3.36 Load conditions to estimate bending stress σ_b and warping stress σ_ω.

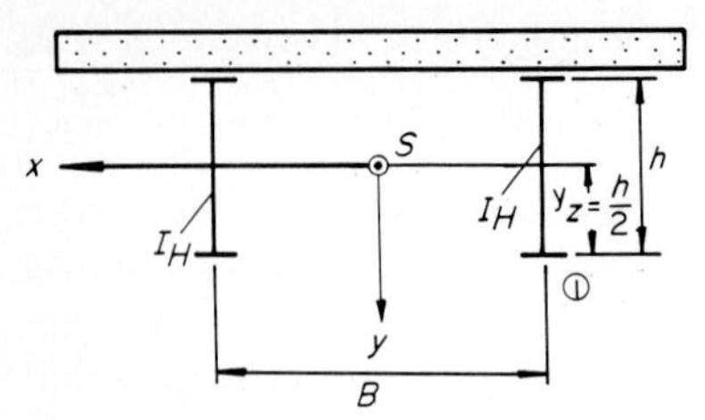

Figure 3.37 Idealized cross section for curved I-girder bridges.

where Y is the fiber distance and ω is the warping function of the cross section.

Consider now the simple, two-I-girder curved bridge idealized in Fig. 3.37. This bridge consists of two W-shaped girders and a noncomposite slab. From Eq. (3.3.10a), the geometric moment of inertia I_X of a single curved girder is twice the value of the individual girder inertia I_H; thus

$$I_X = 2I_H \tag{3.3.15}$$

The maximum fiber distance Y_1 to point 1 located on the lower flange of the I girder, is

$$Y_1 = \frac{h}{2} \tag{3.3.16}$$

where h is the girder depth.

The warping constant I_ω of a single curved girder bridge can be calculated by utilizing Eq. (3.3.8), or

$$I_\omega \cong 2I_H\left(\frac{B}{2}\right)^2 = \frac{I_H B^2}{2} = \frac{I_X B^2}{4} \tag{3.3.17}$$

where B is the spacing of the web plates. The warping function ω_1, also at point 1, can be evaluated easily from the well-known formula

$$\omega_1 = -\int_0^{h/2} r\,dc = -\left(-\frac{B}{2}\right)\left(\frac{h}{2}\right) = \frac{Bh}{4} \tag{3.3.18}$$

Therefore, the ratio of $I_X\omega/(I_\omega Y_1)$ in Eq. (3.3.14) is equal to

$$\frac{I_X\omega_1}{I_\omega Y_1} = \frac{I_X}{I_X B^2/4}\,\frac{BH/4}{h/2} = \frac{2}{B} \tag{3.3.19}$$

Now, denoting a new parameter

$$\psi = \frac{I_X\omega}{I_\omega Y}\,\frac{B}{2} \tag{3.3.20}$$

and assuming that this parameter ψ can be applied to twin-box and monobox curved girder bridges, we obtain a generalized form of Eq. (3.3.14):

$$\frac{\sigma_\omega}{\sigma_b} = 2\psi \frac{1}{\alpha^2 \Phi} \frac{L}{B} \tag{3.3.21}$$

Numerical values for the parameter ψ, given by Eq. (3.3.20), have been determined for actual bridges. This parameter can be related to the cross-sectional shape of curved girder bridges and is categorized as follows:

$$\psi = 0.5\text{–}1.5 \qquad \text{multiple-I girder} \tag{3.3.22a}$$

$$\psi = 1.5\text{–}2.5 \qquad \text{twin-box girder} \tag{3.3.22b}$$

$$\psi \geqq 1.5 \qquad \text{monobox girder} \tag{3.3.22c}$$

Also, the maximum ratio of the span L to the girder width B, as indicated in Fig. 3.32, has the limitation[46]

$$\frac{L}{B} \leqq 10 \tag{3.3.23}$$

This limitation is required because negative reactions will occur at the inner side of the supports, and thus the bearing shoes must be designed for uplift.

(4) The critical torsional parameter κ_{cr}.[41,44] It is assumed that there is a critical value of the torsional parameter κ_{cr} at which the warping stress σ_ω cannot be determined exactly. This value will occur between the twin-box and monobox curved girder bridge configuration. Therefore, to estimate the stress ratio, let $\varepsilon\,(\%) = 100\sigma_\omega/\sigma_b$. Now assume that $\psi = 2.5$, which is the upper value for a twin-box section, as given in Eq. (3.3.22), and let $L/B = 10$, as shown in Eq. (3.3.23). Then by applying Eq. (3.3.21),

$$\varepsilon\,(\%) = \frac{5000}{\alpha^2 \Phi} \tag{3.3.24}$$

Setting $\alpha\sqrt{\Phi}$ equal to various values gives the following ε values:

$$\varepsilon = \begin{cases} 2.0\% & \text{for } \alpha\sqrt{\Phi} = 50 \quad (3.3.25a) \\ 3.1\% & \text{for } \alpha\sqrt{\Phi} = 40 \quad (3.3.25b) \\ 5.6\% & \text{for } \alpha\sqrt{\Phi} = 30 \quad (3.3.25c) \end{cases}$$

If the analysis of the warping stress σ_ω is not important in comparison with the bending stress σ_b when $\varepsilon < 4$ percent, then $\alpha\sqrt{\Phi} \geqq 40$, as shown in Eq. (3.3.25*b*). Therefore, the critical torsional parameter κ_{cr} can be rewritten by using Eq. (3.3.5).

$$\kappa_{cr} \geqq 40\sqrt{\Phi} \tag{3.3.26}$$

This equation has been plotted, as shown in Fig. 3.34. Examination of this figure shows that the value of κ_{cr} increases as the value of the central angle Φ increases. For $\Phi \geqq 0.5$, however, a constant value of $\kappa_{cr} = 30$ may be assumed.

Under these considerations, a more convenient formula, for practical design purposes, can be proposed:

$$\kappa_{cr} = \begin{cases} 10 + 40\Phi & \text{for } 0 \leqq \Phi \leqq 0.5 \quad (3.3.27a) \\ 30 & \text{for } \Phi \geqq 0.5 \quad (3.3.27b) \end{cases}$$

Note that $\kappa_{cr} = 10$ when $\Phi = 0$, which represents the critical value for straight plate girders, as presented in the Japanese Specification for Highway Bridges.[46]

The proposed equations (3.3.27) are plotted in Fig. 3.34. Thus, we conclude that the evaluation of the warping stress σ_ω is not important for monobox girder bridges, whereas the evaluation of the warping stress σ_ω is required for twin-box or multicell curved girder bridges.

(5) Approximation of σ_ω and $\tau_{s+\omega}$ in curved box girders.[43,45] The warping stress σ_ω in a curved box girder is small enough that the following approximate method can be applied as described in Sec. 2.3.2(4). The warping and shear stresses are as follows:
Warping stress,

$$\sigma_{\omega,3} = \frac{h}{5K}(\eta \, \Delta T \, e^{-\xi z'/b} - 3\lambda b m_T) \tag{3.3.28}$$

Shearing stress,

$$\tau_{s+\omega} = \tau_s + \tau_\omega = \frac{T}{2F_w}\left[1 + \frac{|\Delta T|}{2|T|}(\chi - 1)e^{-\xi z'/b}\right] \tag{3.3.29}$$

where T = pure torsional moment
ΔT = step of pure torsional moment
m_T = intensity of uniformly distributed torque

$$K = \frac{4b^2h^2}{b/t_u + b/t_l + 2h/t_w} = \text{pure torsional constant}$$

b = web plate spacing

h = depth of box girder
t_u, t_l = thicknesses of top and bottom flange plates, respectively
t_w = thickness of web plate
$F = bh$ = area surrounded by thin-walled plates
z' = distance from ΔT to viewpoint in direction of girder axis

And the parameters are

$$\eta = \begin{cases} 3.5 & \text{pattern } A, \dfrac{b}{h} < 2.5 \quad (3.3.30a) \\ -5.0 & \text{pattern } B, \dfrac{b}{h} > 1.5 \quad (3.3.30b) \end{cases}$$

$$\xi = \begin{cases} 2.0 & \text{pattern } A, \dfrac{b}{h} < 2.5 \quad (3.3.30c) \\ 4.0 & \text{pattern } B, \dfrac{b}{h} > 1.5 \quad (3.3.30d) \end{cases}$$

$$\lambda = \begin{cases} 0.8 & \text{pattern } A, \dfrac{b}{h} < 2.5 \quad (3.3.30e) \\ -0.6 & \text{pattern } B, \dfrac{b}{h} > 1.5 \quad (3.3.30f) \end{cases}$$

$$\chi = \begin{cases} 2.0 & \text{pattern } A, \dfrac{b}{h} < 2.5 \quad (3.3.30g) \\ 5.0 & \text{pattern } B, \dfrac{b}{h} > 1.5 \quad (3.3.30h) \end{cases}$$

Note that both patterns A and B coexist where $b/h = 1.5$ to 2.5.

Furthermore, the torsional warping normal stress at points 1 and 2 in Fig. 3.38 can be found by modifying σ_ω as follows:

$$\sigma_{\omega,1} = \frac{1}{3\gamma_2^2 + 6/\gamma_1 - 1}\left[\frac{(\gamma_2 - 1)^2(2\gamma_2 + 1)}{\lambda} - 2\left(\frac{3}{\gamma_1} + 1\right)\right]\sigma_{\omega,3} \quad (3.3.31a)$$

$$\sigma_{\omega,2} = -\frac{1}{3\gamma_2^2 + 6/\gamma_1 - 1}\left[\frac{(\gamma_2 - 1)(\gamma_2^2 + \gamma_2 + 6/\gamma_1)}{\lambda} + 2\left(\frac{3}{\gamma_1} + 1\right)\right]\sigma_{\omega,3} \quad (3.3.31b)$$

where the additional parameters γ_1 and γ_2 are given by

$$\gamma_1 = \frac{b}{h} \quad (3.3.32a)$$

$$\gamma_2 = \frac{a}{b} \quad (3.3.32b)$$

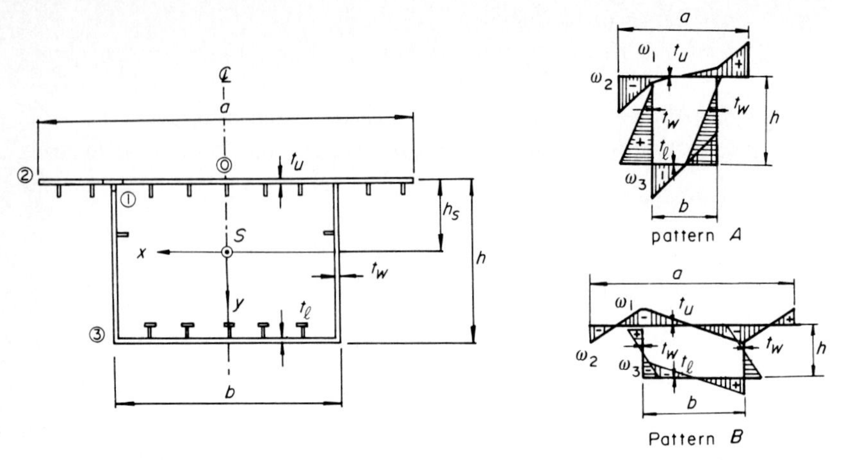

Figure 3.38 Cross section and distribution pattern of torsional warping function.

3.3.2 Distortional warping stress[22–26]

The fundamental differential equation for the distortion of curved box girders has already been given [Eq. (3.2.166)], and it can be rewritten as

$$\frac{d^4\Theta}{ds^4} 4\lambda^4\Theta = \frac{1}{EI_{D\omega}}\left(\frac{m_T}{2} + \rho\frac{M_X}{R}\right) \tag{3.3.33}$$

where

$$\lambda = \sqrt[4]{\frac{K_{D\omega}}{4EI_{D\omega}}} \tag{3.3.34}$$

(1) Parameters of distortion.[26] The distortional warping parameter λ, the parameter β [see Eq. (3.2.168)], and the distortional warping constant $I_{D\omega}$ [see Eq. (3.2.167)] occur within the following ranges (the data are the result of a parametric survey of actual bridges, as mentioned in Sec. 2.4.4):

$$\lambda = (1.5\text{–}3.5) \times 10^{-2} \qquad \text{unit: m}^{-1} \tag{3.3.35a}$$

$$\beta = 1.5\text{–}7.0 \qquad \text{dimensionless value} \tag{3.3.35b}$$

$$I_{D\omega} = \left(\frac{bh}{150} + \frac{b^2}{30}\right)\frac{I_X}{100} \qquad \text{unit: m}^6 \tag{3.3.36}$$

Furthermore, the maximum value of ψ [see Eq. (3.2.162)] is

$$\psi = 0.525 \qquad \text{dimensionless value} \tag{3.3.37}$$

as shown in Fig. 3.39.

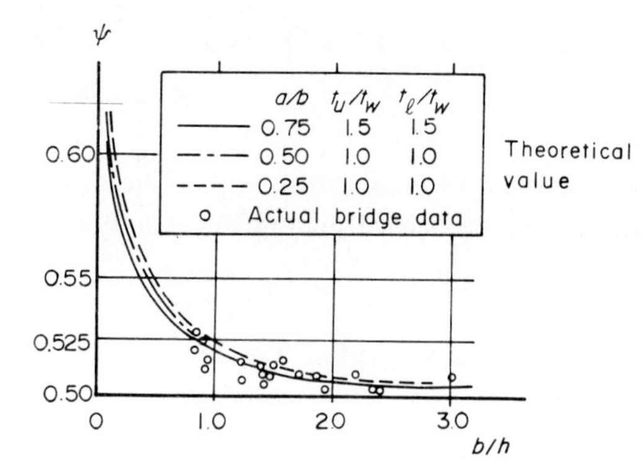

Figure 3.39 Variations of ψ due to b/h.

Although there may be many different combinations of these distortional parameters, the parametric analyses were performed with the actual data limiting the following four box-girder bridges, as indicated in Table 3.7.

Parametric studies were performed to determine the variations in the distortional warping stresses $\sigma_{D\omega}$ due to the diaphragm spacing L_D, central angle Φ, cross-sectional quantities L/b, and the rigidity parameter of the diaphragm, γ. In these analyses, the transverse bending stresses σ_{Db} in the curved box girders due to the distortion, shown in Eqs. (2.4.60) and (2.4.61) and Fig. 2.97, are ignored as being very small in comparison with the distortional warping stresses $\sigma_{D\omega}$.

Moreover, the loading conditions are as follows: a uniformly distributed load w, a line load p in the direction of the girder axis, and a concentrated load P. The p and P loads were applied on the inner web of the curved box girder bridges to make the distortional warping stresses as large as possible.

Finally, the distortional warping stresses $\sigma_{D\omega}$ are calculated at the junction point 3, shown in Fig. 3.24, and are taken as the extreme values in the direction of the bridge axis. These values are also nondimensionalized by the following flexural stresses σ_b due to the bending

TABLE 3.7 Cross-Sectional Values and Parameters of Typical Box-Girder Bridges

Bridge	h, cm	b, cm	a, cm	t_u, cm	t_w, cm	t_l, cm	L, m	I_X, m^4	$I_{D\omega}$, m^6	λ, $\times 10^{-2}$/m	β	ψ
B-1	200	410	80	1.98	1.0	1.29	60.0	0.162	0.065	3.49	2.46	0.508
B-2	250	480	110	1.82	1.0	1.39	90.0	0.306	0.306	2.57	2.44	0.504
B-3	199	594	199	1.89	1.3	2.97	120.0	0.377	0.377	2.73	2.36	0.504
B-4	400	550	225	1.86	1.1	2.77	150.0	1.471	0.665	1.71	2.56	0.512

moment M_X:

$$\sigma_b = \frac{M_X}{W_l} \tag{3.3.38}$$

where W_l is the section modulus at the junction point of the web and bottom plates. And the approximate fomulas for the bending moment of the curved bridge are

$$M_X = \frac{wL^2}{8}(1 + 0.11\Phi^2) \quad \text{for distributed load } w \tag{3.3.39a}$$

$$M_X = \frac{pL^2}{8}(1 + 0.11\Phi^2) \quad \text{for line load } p \tag{3.3.39b}$$

$$M_X = \frac{PL}{4}(1 + 0.13\Phi^2) \quad \text{for concentrated } P \tag{3.3.39c}$$

(2) Variation in the distortional warping stresses due to various parameters[26]

***a.* Effects of diaphragm spacing.** The variations of $\sigma_{D\omega}/\sigma_b$ due to the diaphragm spacing—$L_D/L = \frac{1}{20}, \frac{1}{10}, \frac{1}{8}, \frac{1}{5}$, and $\frac{1}{4}$—can be plotted as in Fig. 3.40 by assuming that $\Phi = \frac{1}{3}$ and $K_D = \infty$. From these figures the variations in $\sigma_{D\omega}/\sigma_b$ are parabolic forms for the distributed load w and line load p, in accordance with the increase in L_D/L. Also the values of $\sigma_{D\omega}/\sigma_b$ vary linearly with L_D/L for the concentrated load P.

***b.* Effects of the central angle.** The influences of the central angle Φ were examined by varying Φ from 0 to $\frac{1}{5}, \frac{1}{3}$, and $\frac{2}{3}$ radian under the conditions of $L_D/L = \frac{1}{10}$ and $K_D = \infty$. Figure 3.41 shows the results.

The variations in $\sigma_{D\omega}/\sigma_b$ due to Φ are linear for a distributed load w and a line load q but nearly constant for a concentrated load P.

***c.* Effects of the cross-sectional quantities.** The preliminary analyses revealed that the effects of the thicknesses t_u, t_w, and t_l and of the dimensions a and h on $\sigma_{D\omega}/\sigma_b$ are negligible, but the effect of the spacing of the web plate b is significant.

Therefore, the variations in $\sigma_{D\omega}/\sigma_b$ were examined by altering the spacing b and by setting L/b equal to 10, 30, and 40. Figure 3.42 shows the results where $L_D/L = 10$, $\Phi = \frac{2}{3}$, and $K_D = \infty$. For distributed and line loads p, the influences of L/b on the distortional warping stress are positive. This tendency is, however, reversed for a concentrated load.

From these analyses, the approximate formulas to evaluate $\sigma_{D\omega}/\sigma_b$ can be summarized as in Table 3.8 for $K_D = \infty$.

(3) Rigidity of intermediate diaphragms.[26] In the above discussion, the rigidities of the diaphragm were assumed to be infinitely large. To

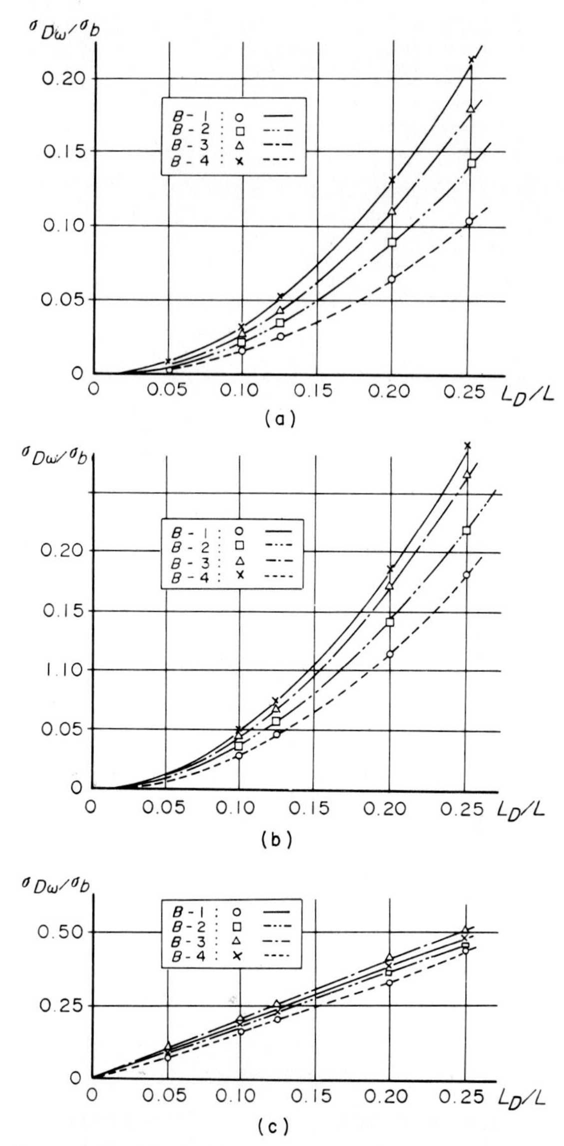

Figure 3.40 Variations of $\sigma_{D\omega}/\sigma_b$ with L_D/L: (*a*) uniformly distributed load (value at section on diaphragm), (*b*) line load along bridge axis (value at section on diaphragm), and (*c*) concentrated load (loaded and calculated at section on middiaphragm).

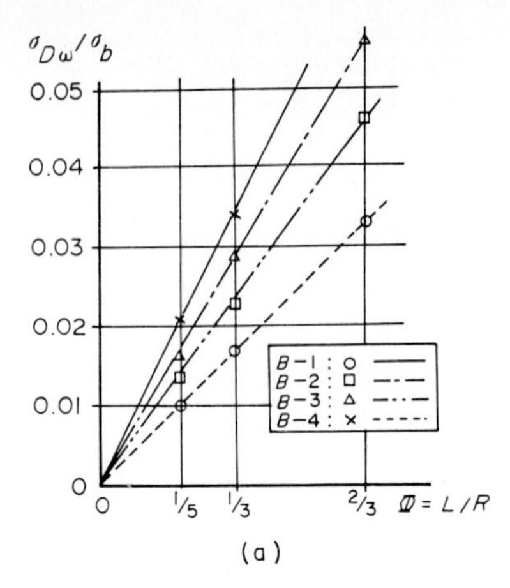

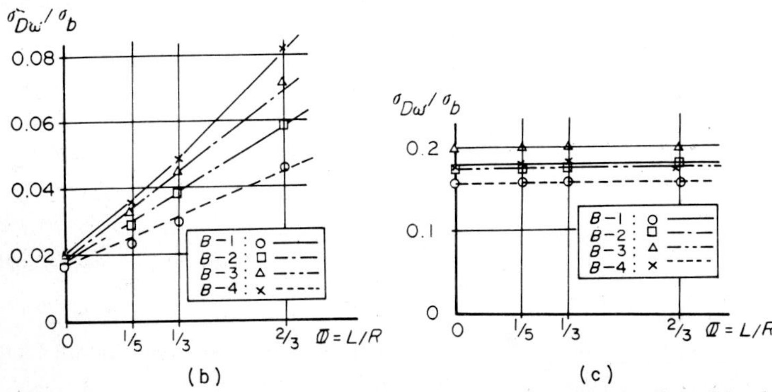

Figure 3.41 Variations of $\sigma_{D\omega}/\sigma_b$ with L/R: (*a*) uniformly distributed load (value at section on diaphragm), (*b*) line load along bridge axis (value at section on diaphragm), and (*c*) concentrated load (loaded and calculated at section on middiaphragm).

TABLE 3.8 Approximation of $\dfrac{\sigma_{D\omega}}{\sigma_b}$ $(K_D = \infty)$

Load	Viewpoint: At section on middle of diaphragm	Viewpoint: At section on diaphragm
Uniformly distributed load w	$-\left(\dfrac{L_D}{L}\right)^2 \Phi\left(0.4 + 0.16\dfrac{L}{b}\right)$	$\left(\dfrac{L_D}{L}\right)^2 \Phi\left(0.8 + 0.32\dfrac{L}{b}\right)$
Line load along bridge axis p	$-\left(\dfrac{L_D}{L}\right)^2 \left(1.2 + 0.15\Phi\dfrac{L}{b}\right)$	$\left(\dfrac{L_D}{L}\right)^2 \left(1.8 + 0.32\Phi\dfrac{L}{b}\right)$
Concentrated load P	$-1.85\dfrac{L_D}{L}$	$1.0\dfrac{L_D}{L}$

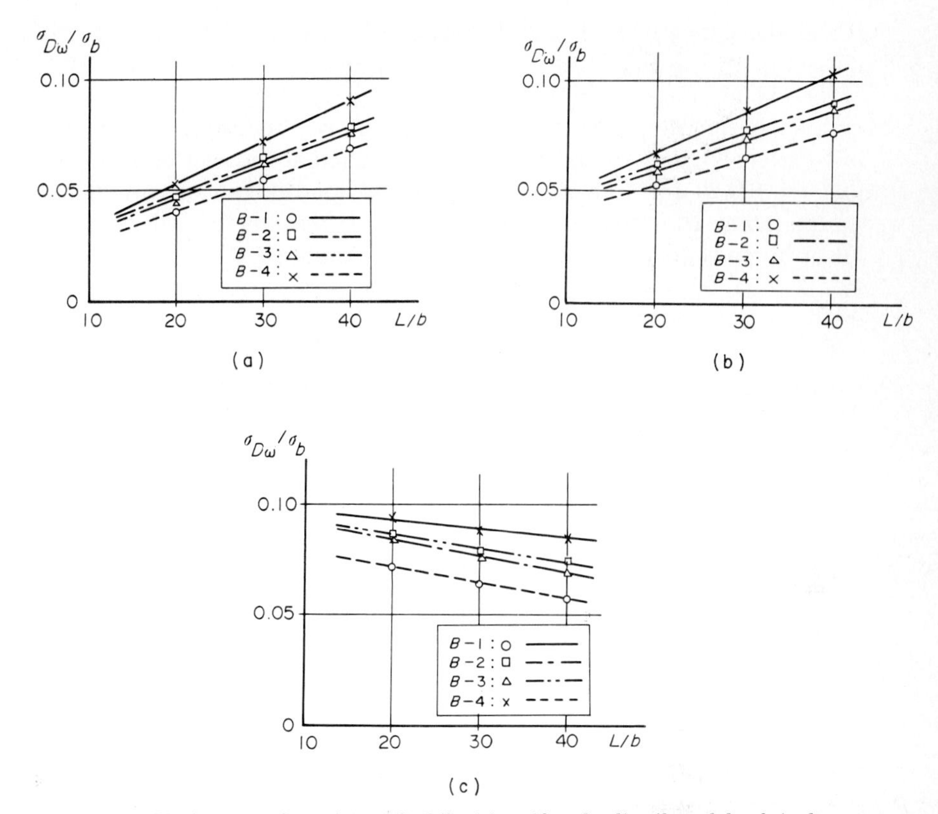

Figure 3.42 Variations of $\sigma_{D\omega}/\sigma_b$ with L/b: (*a*) uniformly distributed load (value at section on diaphragm), (*b*) line load along bridge axis (value at section on diaphragm), and (*c*) concentrated load (loaded and calculated at section on middiaphragm).

determine the effects of the rigidity of the diaphragm on the distortional warping stress, the rigidity K_D of Eqs. (3.2.170) to (3.2.172) is expressed by a dimensionless parameter as

$$\gamma = \frac{K_D}{K_{D\omega} L_D} \tag{3.3.40}$$

where $K_{D\omega}$ is given by Eq. (3.2.169) and L_D is the spacing of the diaphragms.

The effects of γ on the distributions of stresses $\sigma_{D\omega}$ in the direction of the span can be plotted, as shown in Fig. 2.92, under the conditions $L_D/L = 1/10$ and $\Phi = 0$, where the ordinate is nondimensionalized by the absolute maximum distortional warping stress with $K_D = \infty$, that is, $|(\sigma_{D\omega})_{K_D=\infty}|_{\max}$.

These figures show that the concentration of stresses can clearly be observed at the section near the support and midspan for the line load

p and the concentrated load P, respectively, in accordance with the decreases in the rigidities of the diaphragms γ.

Next, the effects of the central angle Φ corresponding to various rigidity parameters γ were examined. These results are plotted in Fig. 3.43, where $L_D/L = 10$ and viewpoints are fixed at the sections $s = L/20$ and $s = 9L/20$ for line and concentrated loads, respectively. These figures show that the effects of the rigidity parameter γ decrease in accordance with increases in the central angle Φ, so that the distortional warping stress $\sigma_{D\omega}$ of the curved box-girder bridges can safely be evaluated by setting the central angle $\Phi = 0$.

For the above reasons, the stress ratio $\sigma_{D\omega}/|(\sigma_{D\omega})_{K_D=\infty}|_{\max}$ was examined for $\Phi = 0$ and $L_D/L = 10$. These results are plotted in Fig. 3.44. Observing the relationships between the stress ratio

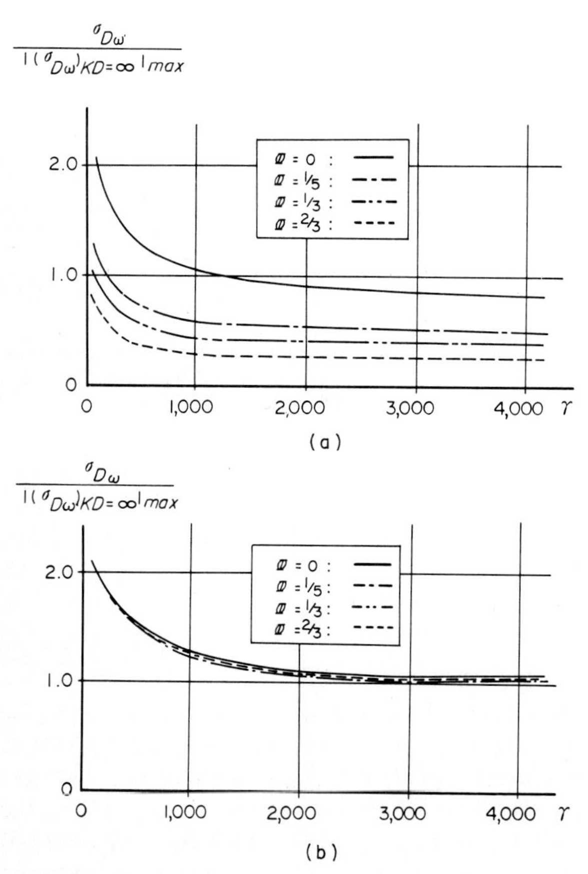

Figure 3.43 Variation of $\sigma_{D\omega}$ with Φ and γ: (*a*) Line load along bridge axis. Viewpoint: $S = \frac{1}{2}L$. (*b*) Concentrated load. Viewpoint: $S = \frac{9}{20}L$.

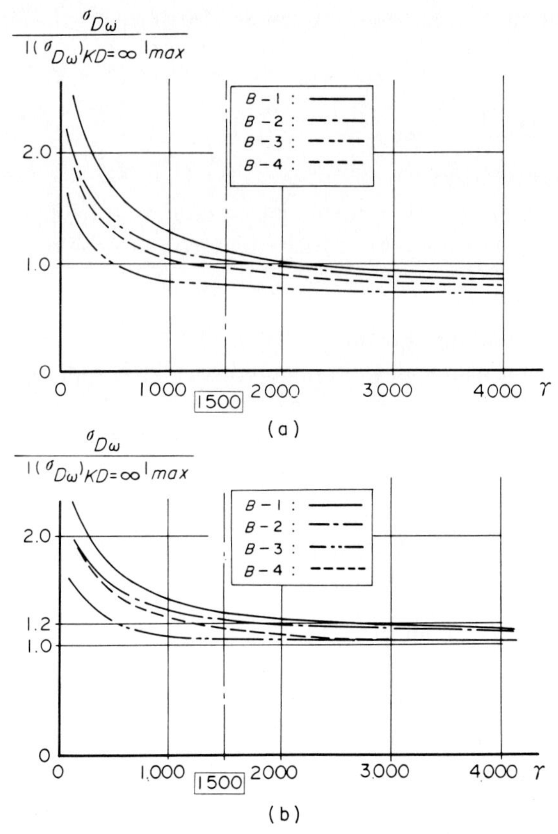

Figure 3.44 Variation of $\sigma_{D\omega}$ with γ: (*a*) line load along bridge axis and (*b*) concentrated load.

$\sigma_{D\omega}/|(\sigma_{D\omega})_{K_D=\infty}|_{\max}$ and the rigidity parameter γ of Fig. 3.44*a*, we see that the absolute value of the stress $\sigma_{D\omega}$ at the middiaphragm section is almost equal to the stress $\sigma_{D\omega}$ at the diaphragm ($s = L/10$) when the rigidity parameter of the diaphragm is

$$\gamma \geqq 1500 \tag{3.3.41}$$

for the line load along the bridge axis. Within these conditions, the distortional warping stress $\sigma_{D\omega}$ is given by the approximate formula at the section at the diaphragm (indicated on the right-hand side of Table 3.8) instead of the formula for the section at middiaphragm. Unless the condition of Eq. (3.3.41) is satisfied, the distortional warping stress at the section at middiaphragm will be greater than that of the value in Table 3.8, and an additional exact analysis will be required to determine the distortional stress.

For the concentrated load of Fig. 3.44*b*, however, the stress ratio is reduced to

$$\frac{\sigma_{D\omega}}{|(\sigma_{D\omega})_{K_D=\infty}|_{\max}} \cong 1.2 \tag{3.3.42}$$

even when Eq. (3.3.41) is fulfilled.

Finally, the results of the parametric survey based on actual bridges concerned with the rigidity parameter γ can be plotted, as shown in Fig. 2.101.

(4) Design formula for curved box girders[26]

a. Approximate formulas for distortional warping stress. The approximate formulas used to evaluate the distortional stress $\sigma_{D\omega}$, shown in Table 3.8, can be rearranged by taking into account the effects of the rigidity of the diaphragm as follows:

For a dead load w_d,

$$\sigma_{D\omega} = \frac{w_d L_D^2}{8w_1}\Phi\left(0.8 + 0.32\frac{L}{b}\right) \tag{3.3.43}$$

For a uniformly distributed live load p_l,

$$\sigma_{D\omega} = \frac{p_l B L_D}{16W_1}\left(1.8 + 0.32\Phi\frac{L}{b}\right) \tag{3.3.44}$$

For a concentrated live load P_l,

$$\sigma_{D\omega} = 2.2\frac{P_l B L_D}{8W_1} \tag{3.3.45}$$

where w_d = dead load intensity per unit length
p_l = live load intensity per unit area
P_l = live load intensity per unit width
$L = R\Phi$ = span of curved girder
R = radius of curvature
Φ = central angle (rad)
b = spacing of web plate
L_D = spacing of diaphragm
B = clear width of roadway
W_1 = sectional modulus at the bottom plate (m^3)

The corresponding loading conditions for live loads are assumed to be one-half the width of the deck, as illustrated in Fig. 3.45. Moreover, the influences of curvature in Eqs. (3.3.39*a*) to (3.3.39*c*) are small enough to be ignored, and the coefficient 2.2 in Eq. (3.3.45) is derived from Eq. (3.3.42).

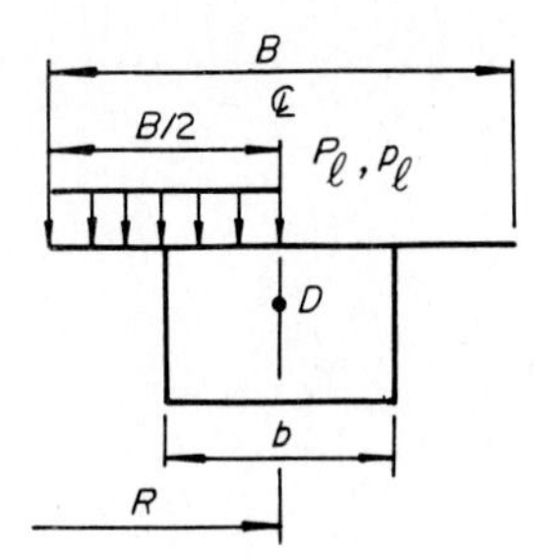

Figure 3.45 Load conditions for live load.

Note that the magnitude of $\sigma_{D\omega}$ is specified at point 3 in Fig. 3.24, and this value should be modified corresponding to the ordinate of the warping function of Fig. 3.24.

***b.* Determination of diaphragm spacing.** Let us now consider the combination of live loads p_l and P_l shown in Fig 3.45. The corresponding stresses are found by adding Eqs. (3.3.44) and (3.3.45):

$$\frac{\sigma_{D\omega}}{\sigma_b} = \frac{p_l L_D^2[1.8 + 0.32\Phi(L/b)] + 4.4 P_l L_D}{p_l L^2 + P_l L} \tag{3.3.46}$$

From this equation, a condition which makes the distortional warping stress $\sigma_{D\omega}$ less than 5 percent of the bending stress σ_b can be derived numerically, i.e.,

$$\frac{\sigma_{D\omega}}{\sigma_b} \leqq 0.05 \tag{3.3.47}$$

For example, taking the live load intensities as

$$p_l = 0.35 \text{ tf/m}^2 \tag{3.3.48a}$$

and

$$P_l = 5.0 \text{ tf/m}^2 \tag{3.3.48b}$$

by the Japanese Specification for Highway Bridges[46] and setting $b = 2.0$ m (the minimum width of the actual bridge data), we can plot the relationships among the span L, central angle Φ, and diaphragm spacing L_D (in meters) as in Fig. 3.46.

Thus the diaphragm spacing l_D (in meters) can be found from the following equations:

$$l_D < 6 \qquad l < 60 \text{ m} \tag{3.3.49a}$$

$$l_D \leqq 0.14l - 2.4 \qquad 60 \text{ m} \leqq l \leqq 160 \text{ m} \tag{3.3.49b}$$

$$l_D = 20 \qquad l > 160 \text{ m} \tag{3.3.49c}$$

for the straight box-girder bridges, as pointed out by Sakai and

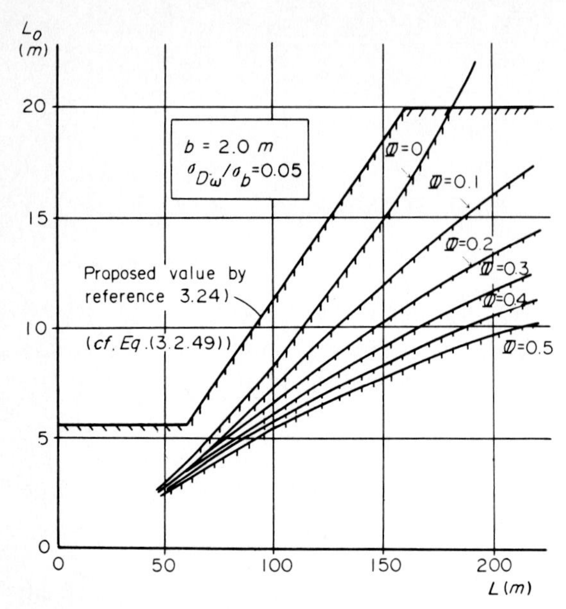

Figure 3.46 Relationships among L_D, L, and Φ.

Nagai,[24] where l (m) is the span length of the straight box-girder bridges. For the curved box-girder bridges, however, the spacing of the diaphragm should be smaller than that of Eq. (3.3.49) in accordance with the increases in the central angle Φ.

The approximate formula for the spacing of the diaphragm L_D (m) in the curved box girder with the span length L (m) is governed by the following equation:

$$L_D = l_D \zeta(\Phi, L) \tag{3.3.50}$$

where the reduction factor $\zeta(\Phi, L)$ is

$$\zeta(\Phi, L) = \begin{cases} 1.0 & L < 60 \text{ m} \quad (3.3.51a) \\ 1 - \dfrac{\sqrt{\Phi}(L-60)}{100\sqrt{2}} & 60 \text{ m} \leqq L \leqq 200 \text{ m} \quad (3.3.51b) \end{cases}$$

c. **Required rigidity of diaphragms.** The approximate formulas, Eqs. (3.3.43) to (3.3.45), should be adopted under the condition that

$$\gamma = \frac{K_D}{K_{D\omega} L_D} \geqq 1500 \tag{3.3.52}$$

where K_D = diaphragm rigidity [Eqs. (3.2.170) to (3.2.172)]
$K_{D\omega}$ = distortional constant [Eqs. (3.2.163) and (3.2.169)]
L_D = diaphragm spacing

***d.* Stress checks for diaphragms.** The diaphragm should be designed to provide not only enough rigidity but also sufficient strength against the distortional stresses. Conservatively, the distortional warping moment $T^s_{D\omega}$ can be expressed approximately by

$$T^s_{D\omega} = \frac{Pb}{4} + \frac{pbL_D}{4} + \frac{\psi M_X L_D}{R} \tag{3.3.53}$$

which is similar to Eq. (2.4.74).

Considering the loading condition of Fig. 3.45 and setting $\psi = 0.525$ [see Eq. (3.3.37)], we get

$$P = \frac{P_l B}{2} \tag{3.3.54a}$$

$$p = \frac{p_l B}{2} \tag{3.3.54b}$$

and

$$M_X = \frac{w_d L_D^2}{8} + \frac{BL_D(2P_l + p_l L_D)}{16} \tag{3.3.55}$$

Therefore, Eq. (3.3.53) reduces to

$$T^s_{D\omega} = \frac{Bb}{8}(P_l + p_l L_D) + \frac{0.525 L_D^2}{16R}(2wdL_D + 2P_l B + p_l BL_D) \tag{3.3.56}$$

Thus, the stress check can be performed by Eqs. (3.2.180) to (3.2.185).

3.3.3 Deflection of curved girder bridges[41,44]

(1) Approximate solution for deflection. In addition to designing a bridge for strength or stress, the structure must have sufficient stiffness. Stiffness is needed to ensure against dynamic loads, overall and local buckling, and any intolerable human response.

The displacements w and β can be directly determined by solving the simultaneous differential equations (3.2.14*a*) and (3.2.9*a*) with $R = R_s$ and Eq. (3.3.1) under the given bending moment M_X and external torque m_T. However, an alternate method can be used to obtain a simple and approximate formula based on the energy method.

Figure 3.47 shows the loading conditions under which a concentrated load P and a torque Γ_z acts on a curved girder bridge at an arbitrary position $s = a$. The vertical deflection w and angle of rotation β can be expressed by utilizing a Fourier series and considering a

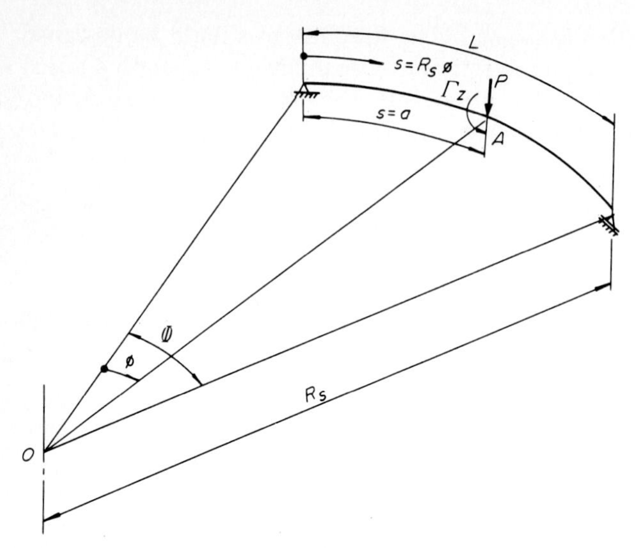

Figure 3.47 Load condition to evaluate displacements w and β.

displacement function of the form

$$w = \sum_{i=1,2,3} w_i \sin \frac{i\pi s}{L} \tag{3.3.57a}$$

$$\beta = \sum_{i=1,2,3} b_i \sin \frac{i\pi s}{L} \tag{3.3.57b}$$

which satisfies the boundary conditions for a simply supported curved girder.

The potential energy Π can be estimated by

$$\Pi = U - P[w]_{s=a} - \Gamma_z[\beta]_{s=a} \tag{3.3.58}$$

where U denotes the strain energy stored in the curved girder. This strain energy can be evaluated by applying Eq. (3.3.9a) with $R = R_s$ and Eqs. (3.2.23) and (3.2.28), which leads to

$$U = \int_0^L \left(\frac{M_X^2}{2EI_X} + \frac{M_\omega^2}{2EI_\omega} + \frac{T_s^2}{2GK} \right) ds \tag{3.3.59a}$$

$$= \frac{1}{2} \int_0^L \left[EI_X \left(\frac{d^2 w}{ds^2} - \frac{\beta}{R_s} \right)^2 + EI_\omega \left(\frac{d^2 \beta}{ds^2} + \frac{1}{R} \frac{d^2 w}{ds^2} \right)^2 + GK \left(\frac{d\beta}{ds} + \frac{1}{R} \frac{dw}{ds} \right)^2 \right] ds \tag{3.3.59b}$$

$$= \frac{L}{4} \sum_{i=1,2,3} \left\{ EI_X \left[\left(\frac{i\pi}{L} \right)^2 w_i + \frac{b_i}{R_s} \right]^2 + EI_\omega \left(\frac{i\pi}{L} \right)^4 \left(b_i + \frac{w_i}{R_s} \right)^2 \right.$$

$$\left. + GK \left(\frac{i\pi}{L} \right)^2 \left(b_i + \frac{w_i}{R_s} \right)^2 \right\} \quad (3.3.59c)$$

The unknown coefficients w_i and b_i, as given in Eq. (3.3.57), can be evaluated by applying the principle of least work such that $\partial \Pi / \partial w_i = 0$ and $\partial \Pi / \partial b_i = 0$. This gives

$$K_{ww,i} w_i + K_{w\beta,i} b_i = P \sin \frac{i\pi a}{L} \quad (3.3.60a)$$

$$K_{\beta w,i} w_i + K_{\beta\beta,i} b_i = \Gamma_z \sin \frac{i\pi a}{L} \quad (3.3.60b)$$

where K represents the stiffness coefficient and is given by

$$K_{ww,i} = \frac{EI_X}{2L^3} (i\pi)^2 [(i\pi)^2 + \gamma_i \Phi^2] \quad (3.3.61a)$$

$$K_{w\beta,i} = K_{\beta,wi} = \frac{EI_X}{2L^3} (i\pi)^2 \Phi (1 + \gamma_i) \quad (3.3.61b)$$

$$K_{\beta\beta,i} = \frac{EI_X}{2L} [\Phi^2 + \gamma_i (i\pi)^2] \quad (3.3.61c)$$

and the parameter γ_i is

$$\gamma_i = \frac{GK + EI_\omega (i\pi/L)^2}{EI_X} \quad (3.3.62)$$

Substituting w_i and b_i from Eq. (3.3.60) into Eq. (3.3.57), we find that

$$w = \sum_{i=1,2,3,\ldots} \frac{1}{K_{ww,i}} \frac{P - \Gamma_z k_{\beta,i}}{1 - k_{w,i} k_{\beta,i}} \sin \frac{i\pi a}{L} \sin \frac{i\pi s}{L} \quad (3.3.63a)$$

$$\beta = \sum_{i=1,2,3,\ldots} \frac{1}{K_{\beta\beta,i}} \frac{\Gamma_z - P_{k_{w,i}}}{1 - k_{w,i} k_{\beta,i}} \sin \frac{i\pi a}{L} \sin \frac{i\pi s}{L} \quad (3.3.63b)$$

where

$$k_{w,i} = L \frac{(1 + \gamma_i)\Phi}{(i\pi)^2 + \gamma_i \Phi^2} \quad (3.3.64a)$$

$$k_{\beta,i} = \frac{1}{L} \frac{(1 + \gamma_i)(i\pi^2)}{\Phi^2 + \gamma_i (i\pi)^2} \quad (3.3.64b)$$

(2) Definition of the deflection increment factor ν. When a concentrated load P acts at the midspan of a curved girder bridge, the vertical deflection w at $s = a = L/2$ can be written as

$$w = \frac{2PL^3}{EI_X} \sum_{i=1,2,3,\ldots} \frac{1}{(i\pi)^2[(i\pi)^2 + \gamma_i\Phi](1 - k_{w,i}k_{\beta,i})} \tag{3.3.65}$$

For a straight girder bridge, the parameters $\Phi = 0$ and $k_{w,i} = k_{\beta,i} = 0$ and the deflection of a straight bridge $w_{\Phi=0}$ reduce to

$$w_{\Phi=0} = \frac{2PL^3}{EI_X\pi^4} \sum_{i=1,3,5,\ldots} \frac{1}{i^4} \tag{3.3.66}$$

In general, this equation is accurate for $i < 3$. Therefore, by setting $i = 1$, the deflection increment factor ν for curved girders in comparison to straight girders of identical span and cross-sectional geometry can be expressed by

$$\nu = \frac{\pi^2}{(\pi^2 + \gamma\Phi^2)\{1 - [(1+\gamma)\pi\Phi]^2/[(\pi^2 + \gamma\Phi^2)(\Phi^2 + \gamma\pi^2)]\}} \tag{3.3.67}$$

where γ is defined as the ratio of the effective torsional rigidity

$$\overline{GK} = GK + EI_\omega\left(\frac{\pi}{L}\right)^2 \tag{3.3.68}$$

to the flexural rigidity EI_X, or

$$\gamma = \frac{GK + EI_\omega(\pi/L)^2}{EI_X} \tag{3.3.69}$$

The deflection incremental factor ν has also been determined for the uniformly distributed load q. However, this factor seems to depend on the torsional-flexural rigidities ratio γ and the central angle Φ of curved girder bridges.

Therefore, let us now investigate the value of γ for various types of curved girder bridges.

(3) Values of γ for actual bridges. The values of torsional-flexural rigidity ratio γ were determined from actual bridges. These results are plotted in Fig. 3.48. From this figure, the relationships between γ and Φ are given by the following inequalities:

$$\gamma > 0.5 \qquad \text{monobox girders} \tag{3.3.70a}$$

$$0.2 < \gamma \leqq 0.5 \qquad \text{twin-box girders} \tag{3.3.70b}$$

$$\gamma \leqq 0.2 \qquad \text{multiple-I girders} \tag{3.3.70c}$$

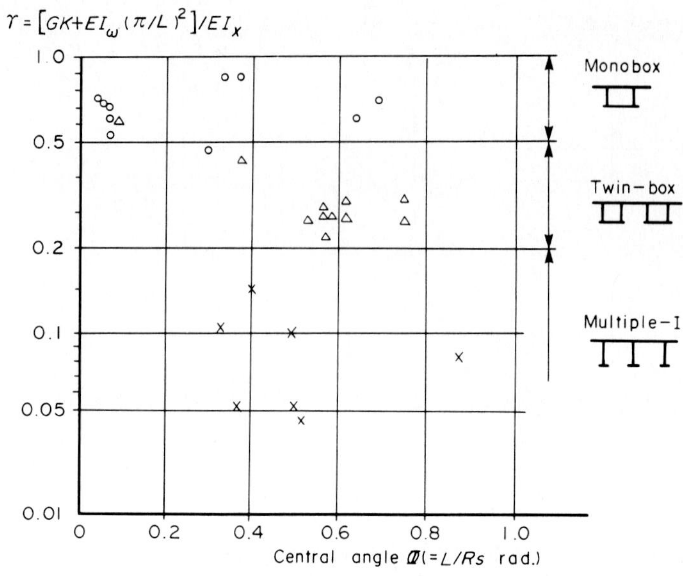

Figure 3.48 Relationships between γ and Φ.

(4) Variation in deflection due to γ and Φ. The variation in the displacement of a curved girder relative to a straight girder is given by ν. The rate of increase of the relative displacements can be expressed as

$$\varepsilon\,(\%) = 100(\nu - 1) \tag{3.3.71}$$

When ε varies in the ranges of 5 to 25 percent, relationships between γ and Φ can be obtained in Fig. 3.49.

A limiting value of the central angle Φ of the bridge can now be determined if we assume that the greatest variation between the curved and straight girder deformations is 5 percent. The angle Φ will be a function of γ and of the type of system. The results are given by the following inequalities:
For multiple-I girders,

$$\Phi \leqq 0.09 + 1.0(\gamma - 0.05) \qquad \text{where } 0.05 \leqq \gamma < 0.2 \tag{3.3.72a}$$

For twin-box girders,

$$\Phi \leqq 0.25 + 0.4(\gamma - 0.2) \qquad \text{where } 0.2 \leqq \gamma < 0.5 \tag{3.3.72b}$$

For monobox girders,

$$\Phi \leqq 0.36 + 0.12(\gamma - 0.5) \qquad \text{where } 0.5 \leqq \gamma < 1.0 \tag{3.3.72c}$$

These relationships give the minimum angle Φ for which the bridge

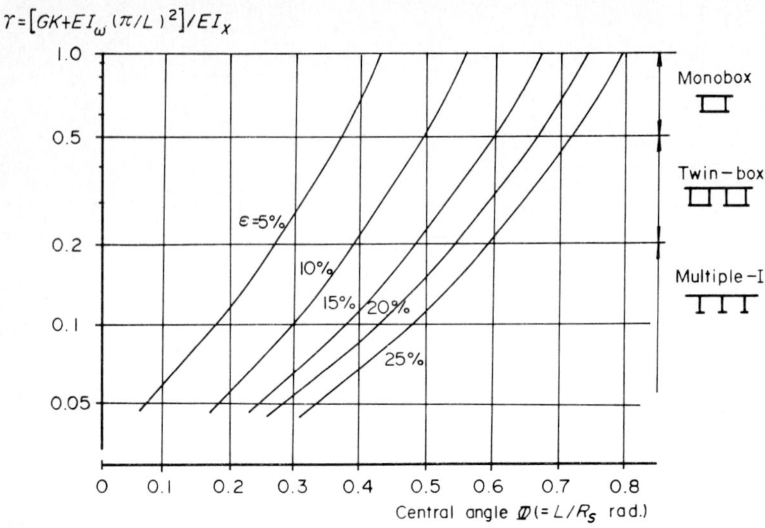

Figure 3.49 Rate of increase for deflection ε with parameters γ and Φ.

can be designed as a straight girder. However, when $\varepsilon \geqq 5$ percent for central angles Φ greater than those predicted in Eq. (3.3.72), the design should be governed by curved beam theory. However, many remaining problems need to be solved in connection with the buckling strength of curved girders when Φ takes large values. So, it is better to set a limitation on the central angle Φ, to avoid the considerable reduction of stiffness in the curved girder bridges. For instance, setting $\varepsilon = 25$ percent results in the following inequalities:
For multiple-I girders,

$$\Phi \leqq 0.33 + 0.18(\gamma - 0.05) \qquad \text{where } 0.05 \leqq \gamma \leqq 0.2 \qquad (3.3.73a)$$

For twin-box girders,

$$\Phi \leqq 0.6 + 0.4(\gamma - 0.2) \qquad \text{where } 0.2 < \gamma < 0.5 \qquad (3.3.73b)$$

For monobox girders,

$$\Phi \geqq 0.72 \qquad \text{where } \gamma \geqq 0.5 \qquad (3.3.73c)$$

3.4 Fundamental Theory of Curved Girders under Dynamic Forces[47–56]

3.4.1 Fundamental equation for forced vibrations[55]

(1) Stress resultants and displacements. As mentioned by Kamatsu and Nakai,[55] the vibration of a curved girder should be treated as the

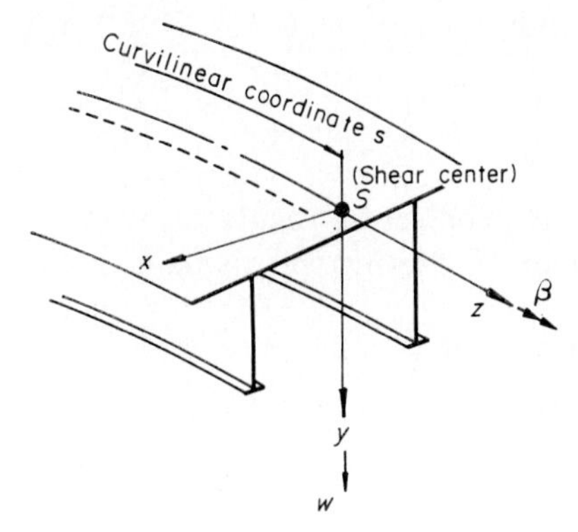

Figure 3.50 Coordinate system (x, y, z) and s.

coupled vibration of flexural vibrations v and w in the two perpendicular directions of the principal x and y axes and torsional vibrations β around the shear center z axis, shown in Fig. 3.50. In an ordinary curved girder bridge, however, vertical and horizontal axes are nearly coincident with the direction of the principal axes, and the flexural rigidity around the vertical axes, $E_s I_y$, is so large in comparison with the rigidity around the horizontal axis, $E_s I_x$, that the displacement v in the horizontal direction is considered small enough to be ignored, that is, $v \cong 0$. For these reasons, in this section we treat the dynamic behavior of a curved girder as the combined vibrations of vertical bending vibration w and torsional vibration β, for the sake of convenience.

The cross section is now situated by the curvilinear coordinate s, as shown in Fig. 3.50. When a vertical force $P(s, t)$, which is represented by a function of coordinate s and time t, is applied to a curved girder, the displacements w and β will occur, as shown in Fig. 3.51, and the

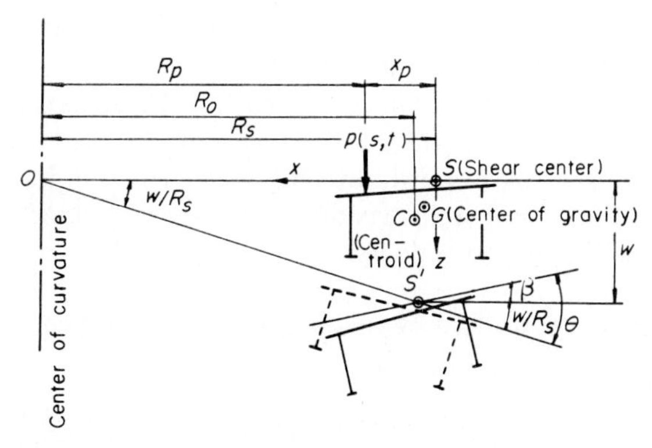

Figure 3.51 Displacement of cross section.

torsional angle θ can be given directly by

$$\theta = \beta + \frac{w}{R_s} \tag{3.4.1}$$

Corresponding to these displacements, the stress resultants can be denoted as follows: The bending moment M_x around the x axis is given by

$$M_x = -E_s I_{x'} \left(\frac{\partial^2 w}{\partial s^2} - \frac{\beta}{R_s} \right) \tag{3.4.2}$$

where $$I_{x'} = \frac{R_0}{R_s} I_x \tag{3.4.3}$$

E_x = Young's modulus for steel
I_x = geometric moment of inertia with respect to x axis
R_0, R_s = radii of curvature at centroid C and shear center S, respectively

The warping moment M_ω and torsional moment T_s are

$$M_\omega = E_s I_\omega \frac{\partial^2 \theta}{\partial s^2} \tag{3.4.4}$$

and $$T_s = G_s K \frac{\partial \theta}{\partial s} \tag{3.4.5}$$

where G_s = shear modulus of elasticity for steel
I_ω = warping constant
K = St. Venant's torsional constant

Although other stress resultants such as the shearing force Q_y and secondary torsional moment T_ω are induced, the details of these stress resultants are omitted.

(2) Derivation of the fundamental equation. Let us now separate the variables of the coordinate function s and the time function t from the displacements w and β in Eqs. (3.4.2), (3.4.4), and (3.4.5). In general, since the normal functions $\omega_i(s)$ and $b_i(s)$ for free vibration are used as the shape functions, the generalized coordinate functions for the displacements $w(s, t)$ and $\beta(s, t)$ are written as products of the corresponding time functions, $W_i(t)$ and $B_i(t)$, as follows:

$$w(s, t) = \sum_{i=1,2,3,\ldots} w_i(s, t) = \sum_{i=1,2,3,\ldots} \omega_i(s) W_i(t) \tag{3.4.6a}$$

$$\beta(s, t) = \sum_{i=1,2,3,\ldots} \beta_i(s, t) = \sum_{i=1,2,3,\ldots} b_i(s) B_i(t) \tag{3.4.6b}$$

The coordinate functions $\omega_i(s)$ and $b_i(s)$ should, of course, not only satisfy the boundary conditions of a curved girder but also fulfill the following orthogonality conditions, i.e., when $i \neq j$,

$$\int_0^L \Omega_i \Omega_j \, ds = 0 \tag{3.4.7a}$$

$$\int_0^L \Omega_i' \Omega_j' \, ds = 0 \tag{3.4.7b}$$

$$\int_0^L \Omega_i'' \Omega_j \, ds = 0 \tag{3.4.7c}$$

$$\int_0^L \Omega_i'' \Omega_j'' \, ds = 0 \tag{3.4.7d}$$

where $\Omega_i(s)$ is a representative of coordinate functions $\omega_i(s)$ and $b_i(s)$, L is the total span length of the curved girder bridge, and the symbols Q' and Q'' represent the derivatives with respect to coordinate s.

To derive the differential equations for the time functions $W_i(t)$ and $B_i(t)$ by use of Lagrange's equation, let us now evaluate the kinetic energy T due to the displacements w and β, shown in Fig. 3.51 and described in Eqs. (3.4.6) and (3.4.7).

The kinetic energy T is

$$T = \frac{1}{2} \int_0^L \frac{R_G}{R_S} \frac{\gamma_s}{g} \left[A_s \left(\frac{\partial w}{\partial t} + x_G \frac{\partial \beta}{\partial t} \right)^2 + I_G \left(\frac{\partial \beta}{\partial t} \right)^2 \right] ds \tag{3.4.8a}$$

$$= \frac{1}{2} m \sum_{i=1,2,\ldots} \left(A_s \dot{W}_i^2 \int_0^L \omega_i^2 \, ds + 2 S_y \dot{W}_i \dot{B}_i \int_0^L \omega_i b_i \, ds + I_S \dot{B}_i^2 \int_0^L b_i^2 \, ds \right) \tag{3.4.8b}$$

where the overdot means the derivative with respect to time t and

γ_s = density of steel

g = acceleration of gravity

A_s = cross-sectional area

$x_G = R_S - R_G$ = horizontal distance between shear center S and center of gravity G

$S_y = A_s x_G$ = static moment with respect to y axis

$I_S = I_G + A_s x_G^2$ = polar moment of inertia with respect to shear center S

$I_G =$ polar moment of inertia with respect to center of gravity G

$m = \dfrac{R_G}{R_s}\dfrac{\gamma_s}{g} =$ mass per unit volume with respect to shear center axis

However, the potential energy U, which is stored as the strain energy in the curved girder, will mainly consist of the stress resultants M_x, M_ω, and T_s. The contribution of the remaining stress resultants, Q_y and T_ω, is considered negligible. Thus

$$U = \frac{1}{2}\int_0^L \left[\left(\frac{R_0}{R_S}\right)^2 \frac{M_x^2}{E_s I_x'} + \frac{M_\omega^2}{E_s I_\omega} + \frac{T_S^2}{G_s K}\right] ds \tag{3.4.9a}$$

$$= \frac{1}{2}\int_0^L \left[E_s I_x'\left(\frac{\partial^2 w}{\partial s^2} - \frac{\beta}{R_S}\right)^2 + E_s I_\omega\left(\frac{\partial^2 \beta}{\partial s^2} + \frac{1}{R_S}\frac{\partial^2 w}{\partial s^2}\right)^2 + G_s K\left(\frac{\partial \beta}{\partial s} + \frac{1}{R_S}\frac{\partial w}{\partial s}\right)^2\right] ds \tag{3.4.9b}$$

The above equations can be rewritten by using Eq. (3.4.6) and the orthogonality conditions of Eq. (3.4.7) as follows:

$$U = \frac{1}{2}\sum_{i=1,2,3} E_s I_x' \left[W_i^2 \int_0^L (\omega_i'')^2\, ds - 2W_i B_i \int_0^L \frac{\omega_i b_i}{R_S}\, ds + B_i^2 \int_0^L \frac{b_i}{R_S}\, ds\right]$$
$$+ E_s I_\omega \left[W_i^2 \int_0^L \left(\frac{\omega_i''}{R_S}\right)^2 ds + 2W_i B_i \int_0^L \frac{\omega_i b_i}{R_S}\, ds + B_i^2 \int_0^L (b_i'')^2\, ds\right]$$
$$+ G_s K\left[W_i^2 \int_0^L \left(\frac{\omega_i'}{R_S}\right)^2 ds + 2W_i B_i \int_0^L \frac{\omega_i' b_i'}{R_S}\, ds + B_i^2 \int_0^L (b_i')^2\, ds\right] \tag{3.4.9c}$$

Finally, the dissipative energy D can be estimated by assuming viscous damping, and the corresponding damping coefficients are denoted d_w and d_β for bending the torsional vibrations, respectively:

$$d_w = \sum_{i=1,2,3,\ldots} d_{w,i} \tag{3.4.10a}$$

$$d_\beta = \sum_{i=1,2,3,\ldots} d_{\beta,i} \tag{3.4.10b}$$

Provided the relationships between $d_{w,i}$ and $d_{\beta,i}$ are also orthogonal, the dissipative energy D will be given, similar to Eq. (3.4.8b), as

$$D = \frac{1}{2}\int_0^L \frac{R_G}{R_S}\frac{\gamma_s}{g}\left\{A_s\left[d_w\left(\frac{\partial w}{\partial t}\right)^2 + d_\beta\left(x_G\frac{\partial \beta}{\partial t}\right)^2\right] + I_G d_\beta\left(\frac{\partial \beta}{\partial t}\right)^2\right\} ds \qquad (3.4.11a)$$

$$= \frac{1}{2}\sum_{i=1,2,3,\ldots}\left(A_s d_{\omega,i}\dot{W}_i^2\int_0^L \omega_i^2\, ds + I_s d_{\beta,i}\dot{B}_i^2\int_0^L b_i^2\, ds\right) \qquad (3.4.11b)$$

Therefore, the solutions of $W_i(t)$ and $B_i(t)$ can be obtained from Lagrange's equations, which results in a set of simultaneous differential equations for $W_i(t)$ and $B_i(t)$:

$$\frac{d}{dt}\left(\frac{\partial T}{\partial \dot{W}_i}\right) + \frac{\partial D}{\partial W_i} + \frac{\partial U}{\partial W_i} = F_i \qquad (3.4.12a)$$

$$\frac{d}{dt}\left(\frac{\partial T}{\partial \dot{B}_i}\right) + \frac{\partial D}{\partial B_i} + \frac{\partial U}{\partial B_i} = \Gamma_i \qquad (3.4.12b)$$

where F_i and Γ_i are the generalized vertical force and torque, respectively, acting at the shear center. For the loading conditions shown in Fig. 3.52, these generalized forces are

$$F_i = \int_0^L \int_{R_{P,IN}}^{R_{P,OT}} \left[\frac{R_P}{R_S}\, q(s, R_P, t)\omega_i(s)\right] dR_P\, ds \qquad (3.4.13a)$$

and

$$\Gamma_i = \int_0^L \int_{R_{P,IN}}^{R_{P,OT}} \left[\frac{R_P}{R_S}(R_P - R_S)q(s, R_P, t)b_i(s)\right] dR_P\, ds \qquad (3.4.13b)$$

where $R_{P,IN}$ and $R_{P,OT}$ designate the radii of curvature at the inner and outer edges of the applied distributed load $q(s, R_P, t)$, respectively. Consequently, a set of simultaneous differential equations for $W_i(t)$ and $B_i(t)$ can be given as follows (the overdots represent derivatives with

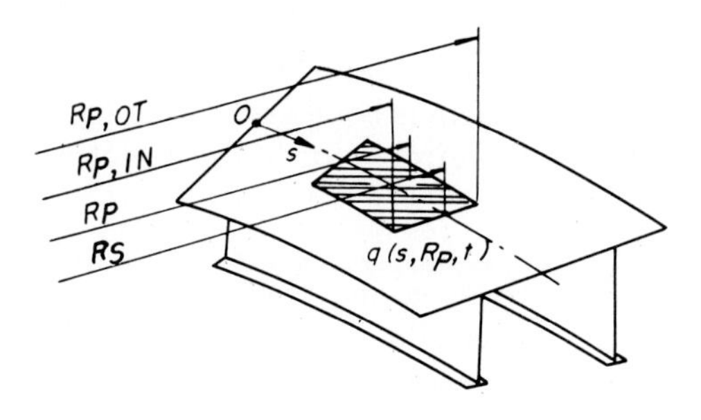

Figure 3.52 Load $q(s, R_P, t)$.

respect to time):

$$mA_s\ddot{W}_i\int_0^L \omega_i^2\,ds + mA_s d_{\omega,i}\dot{W}_i\int_0^L \omega_i^2\,ds + \Bigg[E_sI'_x\int_0^L (\omega_i'')^2\,ds$$

$$+ E_sI_\omega\int_0^L\left(\frac{\omega_i''}{R_S}\right)^2 ds + G_sK\int_0^L\left(\frac{\omega_i'}{R_S}\right)^2 ds\Bigg]W_i + mS_y\ddot{B}_i\int_0^L \omega_i b_i\,ds$$

$$+\left(-E_sI'_x\int_0^L\frac{\omega_i'' b_i}{R_S}\,ds + E_sI_\omega\int_0^L\frac{\omega_i'' b_i''}{R_S}\,ds + G_sK\int_0^L\frac{\omega_i' b_i'}{R_S}\,ds\right)B_i = F_i$$

$$(3.4.14a)$$

$$mS_y\ddot{W}_i\int_0^L \omega_i b_i\,ds + \left(-E_sI'_x\int_0^L\frac{\omega_i'' b_i}{R_S}\,ds + E_sI_\omega\int_0^L\frac{\omega_i'' b_i''}{R_S}\,ds\right.$$

$$\left.+ G_sK\int_0^L\frac{\omega_i' b_i}{R_S}\,ds\right)W_i + mI_s\ddot{B}_i\int_0^L b_i^2\,ds + mI_s d_{\beta,i}\dot{B}_i\int_0^L b_i^2\,ds$$

$$+\left[E_sI'_x\int_0^L\left(\frac{b_i}{R_S}\right)^2 ds + E_sI_\omega\int_0^L (b_i'')^2\,ds + G_sK\int_0^L (b_i')^2\,ds\right]B_i = \Gamma_i$$

$$(3.4.14b)$$

3.4.2 Approximate method using model analysis

(1) Configuration of curved girder bridges.[51] The fundamental equation has been derived on the assumption that a curved girder has a constant radius of curvature and a uniform cross section. If these quantities vary continuously along the axial direction, it becomes much more difficult to solve the fundamental equation of motion.

However, the radius of curvature of the usual continuous curved girder bridge may be assumed approximately constant in each span, as shown in Fig. 3.53, where the radii of curvature in the spans are denoted by $R_{S,1}$, $R_{S,2}$, . . . , $R_{S,r}$, . . . , $R_{S,n}$; the central angles by Φ_1, Φ_2, . . . , Φ_r, . . . , Φ_n; and the span lengths by L_1, L_2, . . . , L_r, . . . , L_n. The subcoordinates s_r having their origin $s_r = 0$ at support r will be properly used as the curvilinear coordinates in the rth span. The signs of $R_{S,r}$ and Φ_r are positive if the girder in the rth span is bent toward the right side as we proceed from the origin along the girder axis in the direction of coordinate s.

(2) Cross-sectional quantities.[51] Strictly speaking, the cross section of an ordinary curved girder bridge varies somewhat; however, the mean values of all the cross-sectional quantities may be adopted for

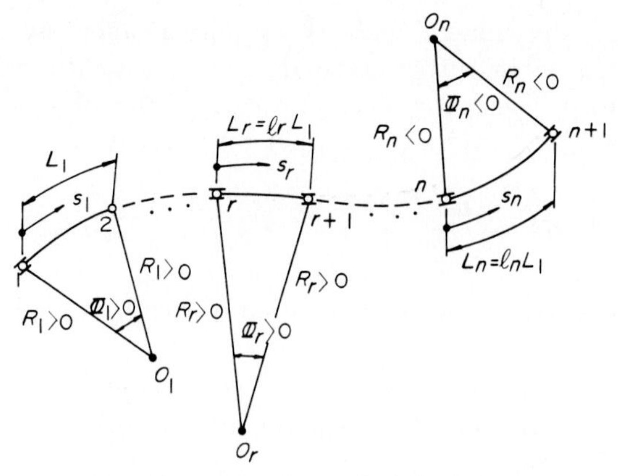

Figure 3.53 Geometry of curved girder bridge.

convenience. The problem of forced vibration is not a microscopic one of investigating the local deformation or the stress distribution, but a macroscopic one of clarifying the dynamic behavior throughout the bridge. Given this fact, we expect that the errors included in the approximate solution for the response will not be serious, even in the extreme case of a girder with a varying height.

Accordingly, if $Q_{W,r}(s_r)$ represents all the varying cross-sectional quantities in the rth span, then the mean values $\bar{Q}_W$ of $Q_{W,r}(s_r)$ are

$$\bar{Q}_{W,r} = \frac{1}{L_r} \int_0^{L_r} Q_{W,r}(s_r)\, ds \tag{3.4.15}$$

So the mean value Q^*_W across the entire length of the bridge can be found by

$$Q^*_W = \frac{\sum\limits_{r=1}^{n} l_r \bar{Q}_{W,r}}{\sum\limits_{r=1}^{n} l_r} \tag{3.4.16}$$

Hence the weight l_r is the ratio between the rth span and the reference span, or

$$l_1 = 1 \qquad l_2 = \frac{L_2}{L_1} \qquad \ldots \qquad l_r = \frac{L_r}{L_1} \qquad l_n = \frac{L_n}{L_1} \tag{3.4.17}$$

Thus, the mean values I^*_G, K^*, and I^*_ω (where the asterisk indicates the mean value for the entire length of the bridge) can be found directly from Eqs. (3.4.15) and (3.4.16).

In determining the cross-sectional area A_s of a bridge, one must take into account not only the area of the main girder but also the area of the slab and of the secondary members, such as the floor beams, diaphragms, etc., since the area of the latter is usually not smaller than that of the former. Therefore, the cross-sectional area of the bridge may be replaced by the dead load intensity $w_{d,r}$ per unit length, or

$$A_{s,r,}(s_r) = \frac{w_{d,r}(s_r)}{\gamma_s} \tag{3.4.18}$$

Now, the mean value of the rth span can be found easily by using Eq. (3.4.15).

Also the radius of curvature (R_S, R_0, or R_G) should be included in the formulas for the mean values m^* and $I_x'^*$, as given in the following equations instead of Eq. (3.4.16):

$$m^* = \frac{\gamma_s \sum_{r=1}^{n} \frac{R_{G,r}}{R_{S,r}} l_r}{g \sum_{r=1}^{n} l_r} \tag{3.4.19a}$$

$$I_x'^* = \frac{\sum_{r=1}^{n} \frac{R_{S,r}}{R_{0,r}} I_x l_r}{\sum_{r=1}^{n} l_r} \tag{3.4.19b}$$

Furthermore, the mean polar moment of inertia I_S^* with respect to the shear center can be written as

$$I_S^* = \frac{\sum_{r=1}^{n} \frac{l_r}{L_r} \int_0^{L_r} [I_{G,r}(s_r) + A_{s,r}(s_r) x_{G,r}^2(s_r)]\, ds_r}{\sum_{r=1}^{n} l_r} \tag{3.4.20}$$

If the mean values of each span are used, I_S^* may be approximated by

$$I_S^* = I_G^* + \frac{\sum_{r=1}^{n} \bar{A}_{s,r} \bar{x}_{G,r}^2 l_r}{\sum_{r=1}^{n}} \tag{3.4.21}$$

In a similar manner, the mean values A_y^* and S_y^* are

$$A^* = \frac{\sum_{r=1}^{n} \bar{A}_{s,r} l_r}{\sum_{r=1}^{n} l_r} \tag{3.4.22}$$

$$S_y^* = \frac{\sum_{r=1}^{n} \pm (\bar{A}_{s,r}\bar{x}_{G,r})l_r}{\sum_{r=1}^{n} l_r} \tag{3.4.23a}$$

$$S_y'^* = \frac{\sum_{r=1}^{n} \pm \left(\bar{A}_{s,r}\bar{x}_{G,r} - \frac{\bar{I}_{s,r}}{R_{s,r}}\right)l_r}{\sum_{r=1}^{n} l_r} \tag{3.4.23b}$$

The sign of $\bar{x}_{G,r}$ is illustrated in Fig. 3.54, in which two cross sections a and b are exactly the same shape and size, but are bent in opposite directions to each other. Therefore, if the positive direction of the rectangular coordinate s_r in Eq. (3.4.15) coincides with the positive direction of coordinate s, then the coordinate $x_{G,r}$ of G_r must have a minus sign in Fig. 3.54b. Consequently, the $\pm$ sign in Eqs. (3.4.23a) and (3.4.23b) must be positive (or negative) if $R_{S,r}$ is positive (or negative).

(3) Shape functions.[51] Let us consider the shape function for the rth span of the curved girder, shown in Fig. 3.53. The approximate shape function may be used as the solution to our problem, but all the boundary conditions and the conditions of orthogonality should be satisfied by these shape functions. Therefore, the common function $\Omega_{i,r}$ is adopted:

$$\omega_{i,r} = b_{i,r} = \Omega_{i,r} \tag{3.4.24a}$$

$$= K_{i,r} \sin k_i s_r + \Lambda_{i,r} \cos k_i s_r + M_{i,r} \sinh k_i s_r + N_{i,r} \cosh k_i s_r \tag{3.4.24b}$$

The symbol k_i is a parameter describing the span ratio l_r, and $K_{i,r}$, $\Lambda_{i,r}$, $M_{i,r}$, and $N_{i,r}$ are coefficients determined by the boundary conditions.

These boundary conditions are rewritten in terms of parameters

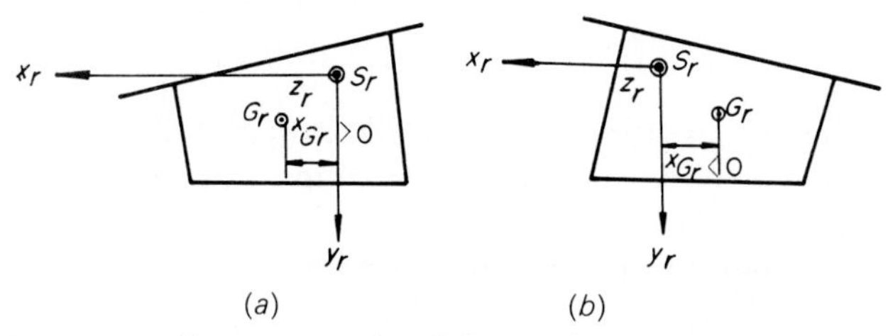

Figure 3.54 Two cross-sectional shapes of curved girder bridges (a) and (b).

illustrated in Fig. 3.53:

$$[\Omega_{i,r}]_{s_r=0} = [\Omega_{i,r}]_{s_r=L_r} = 0 \tag{3.4.25}$$

at all the supports including the end supports, $r = 1$ and $r = n + 1$. The conditions of continuity at the $(r + 1)$th support can be written as

$$\left[\frac{d\Omega_{i,r}}{ds_r}\right]_{s_r=L_r} = \left[\frac{d\Omega_{i,r+1}}{ds_{r+1}}\right]_{s_{r+0}=0} \tag{3.4.26a}$$

$$\left[\frac{d^2\Omega_{i,r}}{ds_r^2}\right]_{s_r=L_r} = \left[\frac{d^2\Omega_{i,r+1}}{ds_{r+1^2}}\right]_{s_{r+1}=0} \tag{3.4.26b}$$

and the conditions at both ends, $r = 1$ and $r = n + 1$, are

$$\left[\frac{d^2\Omega_{i,1}}{ds_1^2}\right]_{s_l=0} = 0 \tag{3.4.27a}$$

$$\left[\frac{d^2\Omega_{i,n}}{ds_n^2}\right]_{s_n=L_n} = 0 \tag{3.4.27b}$$

From Eqs. (3.4.25) through (3.4.27), we find the approximate shape functions for the following two cases.

a. **Simply supported girder bridges.** In a simply supported girder bridge, $n = 1$ and $\Lambda_{i,1} = M_{i,1} = N_{i,1} = 0$, so that

$$\Omega_{i,1} = K_{i,1} \sin k_i s_1 \tag{3.4.28}$$

where

$$k_i L_1 = i\pi \qquad i = 1, 2, 3, \ldots \tag{3.4.29a}$$

$$K_{i,1} = 1 \tag{3.4.29b}$$

b. **Continuous girder bridges.** The shape function of the rth span is given by

$$\Omega_{i,r} = \Lambda_{i,r}\left[\frac{\sin k_i(L_r - s_r)}{\sin k_i L_r} - \frac{\sinh k_i(L_r - s_r)}{\sinh k_i L_r}\right] + \Lambda_{i,r+1}\left(\frac{\sin k_i s_r}{\sin k_i L_r} - \frac{\sinh k_i s_r}{\sinh k_i L_r}\right) \tag{3.4.30}$$

Clearly all the boundary conditions, Eqs. (3.4.25) through (3.4.27), are satisfied by the above function, where the shape parameters k_i ($i = 1, 2, 3, \ldots$) can be determined by solving the following $n - 1$ simultaneous equations (dynamic three-moment equation):

$$\Lambda_{i,r} l_r \psi_{i,r} + \Lambda_{i,r+1}(l_r \chi_{i,r} + l_{r+1}\chi_{i,r+1}) + \Lambda_{i,r+2} l_{r+1}\psi_{i,r+1} = 0$$

$$n = 1, 2, \ldots, n - 1 \tag{3.4.31}$$

From the end conditions (3.4.27), it is obvious that

$$\Lambda_{i,1} = 0 \tag{3.4.32a}$$

$$\Lambda_{i,n+1} = 0 \tag{3.4.32b}$$

and the variables $\psi_{i,r}$ and $\chi_{i,r}$ are a function of $l_r k_i L_1$ (see App. B);

$$\psi_{i,r} = \frac{1}{l_r k_i L_1} (\csc l_r k_i L_1 - \operatorname{csch} l_r k_i L_1) \tag{3.4.33a}$$

$$\chi_{i,r} = \frac{1}{l_r k_i L_1} (\coth l_r k_i L_1 - \cot l_r k_i L_1) \tag{3.4.33b}$$

Since Eq. (3.4.31) is a set of linear homogeneous equations, a transcendental equation $f(k_i L_1) = 0$ can be obtained by setting the determinant built by the coefficients of $\Lambda_{i,2}, \Lambda_{i,3}, \ldots, \Lambda_{i,r}, \ldots$, and $\Lambda_{i,n}$ equal to zero. After the roots $k_i L_1$ have been determined, the ratios among $\Lambda_{i,2}$, $\Lambda_{i,3}, \ldots, \Lambda_{i,3}, \ldots$, and $\Lambda_{i,n}$ can be determined by substituting them again into Eq. (3.3.31). Thus, this ratio can be denoted by $\lambda_{i,r}$, as follows:

$$\lambda_{i,1} = 0 \qquad \lambda_{i,2} = 1 \qquad \ldots \qquad \lambda_{i,r} = \frac{\Lambda_{i,r}}{\Lambda_{i,2}} \qquad \ldots \qquad \lambda_{i,n+1} = 0 \tag{3.4.34}$$

To find the roots $k_i L_i$, the trial-and-error method is suitable, but it requires lots of numerical calculations. However, when the variables $\psi_{i,r}$ and $\chi_{i,r}$ are given in the form of a table, it is comparatively easy to find the first approximate solution $[k_i L_1]_1$. When much more exact values, that is, $[k_i L_1]_2, [k_i L_1]_3, \ldots, [k_i L_1]_h, \ldots$ are required, the following formula works:

$$[k_i L_1]_{h+1} = [k_i L_1]_h + [\Delta k_i \, L_1]_h \tag{3.4.35}$$

Hence the notation $[\Delta k_i \, L_1]_h$ is a compensating value for making the value $[k_i L_1]_h$ more accurate, and it can be estimated by the Newton-Raphson method as

$$[\Delta k_i \, L_1]_h = -[f(k_i L_1)]_h \bigg/ \left[\frac{df(k_i L_1)}{d(k_i L_1)}\right]_h \tag{3.4.36a}$$

where the derivatives

$$\frac{d\psi_{i,r}}{d(k_i L_1)} = \psi'_{i,r} \qquad \text{and} \qquad \frac{d\chi_{i,r}}{d(k_i L_1)} = \chi'_{i,r} \tag{3.4.36b}$$

involved in the function

$$\frac{df(k_i L_1)}{d(k_i L_1)} \tag{3.4.36c}$$

can be found easily by applying the following formulas:

$$\psi'_{i,r} = \frac{\zeta_{i,r}}{l_r} - \frac{\psi_{i,r}}{k_i L_1} \tag{3.4.37a}$$

$$\chi'_{i,r} = k_i L_1 \psi_{i,r} \varphi_{i,r} - \frac{\chi_{i,r}}{k_i L_1} \tag{3.4.37b}$$

The new variables $\zeta_{i,r}$ and $\varphi_{i,r}$ are both a function of $l_r k_i L_1$ (see App. B), which gives

$$\zeta_{i,r} = (\text{csch } l_r k_i L_1)(\coth l_r k_i L_1) - (\csc l_r k_i L_1)(\cot l_r k_i L_1) \tag{3.4.38a}$$

$$\varphi_{i,r} = \csc l_r k_i L_1 + \text{csch } l_r k_i L_1 \tag{3.4.38b}$$

Specifically, we can show that the values $k_i L_1$ for a two-equal-span continuous girder bridge are

$$k_i L_1 = (2i - 1)\pi \qquad i = 1, 3, \ldots \tag{3.4.39a}$$

$$= 3.927, 7.069, \ldots, \left(\frac{i}{2} + 0.25\right)\pi \qquad i = 2, 4, \ldots \tag{3.4.39b}$$

Accordingly, the odd modes are the same as in a simply supported girder and are antisymmetric, whereas the even modes are symmetric.

Before we estimate the terms for mass, stiffness, etc., the approximate shape functions obtained above should be checked to see whether they satisfy the conditions of orthogonality.

It is clear that all the conditions in Eqs. (3.4.7*a*) to (3.4.7*d*) are satisfied for a simply supported girder. Therefore, the shape function (3.4.28) gives a set of exact solutions.

In the case of a continuous girder, taking derivatives of $\Omega_{i,r}$ in Eq. (3.4.30) twice and 4 times with respect to s_r yields

$$\frac{d^2\Omega_{i,r}}{ds_r^2} = -k_i^2 \left\{ \lambda_{i,r} \left[\frac{\sin k_i(L_r - s_r)}{\sin k_i L_r} + \frac{\sinh k_i(L_r - s_r)}{\sinh k_i L_r} \right] + \lambda_{i,r+1} \left(\frac{\sin k_i s_r}{\sin k_i L_r} + \frac{\sinh k_i s_r}{\sinh k_i L_r} \right) \right\} \tag{3.4.40a}$$

$$\frac{d^4\Omega_{i,r}}{ds_r^4} = k_i^4 \Omega_{i,r} \tag{3.4.40b}$$

Integrating by parts, we have

$$\int_0^L \frac{d^4\Omega_i}{ds^4}\Omega_j\,ds = \sum_{r=1}^{n}\left(\frac{d^3\Omega_{i,r}}{ds_r^3}\Omega_{j,r} - \frac{d^3\Omega_{j,r}}{ds_r^3}\Omega_{i,r}\right)_0^{L_r} - \sum_{r=1}^{n-1}\left\{\left[\left(\frac{d^2\Omega_{i,r}}{ds_r^2}\frac{d\Omega_{j,r}}{ds_r}\right)\right]_{s_r=L_r}\right.$$
$$-\left(\frac{d^2\Omega_{i,r+1}}{ds_{r+1}^2}\frac{d\Omega_{j,r+1}}{ds_{r+1}}\right)_{s_{r+1}=0}\Bigg] - \left[\left(\frac{d^2\Omega_{j,r}}{ds_r^2}\frac{d\Omega_{i,r}}{ds_r}\right)_{s_r=L_r}\right.$$
$$\left.\left.-\left(\frac{d^2\Omega_{j,r+1}}{ds_{r+1}^2}\frac{d\Omega_{i,r+1}}{ds_{r+1}}\right)_{s_{r+1}=0}\right]\right\} - \left\{\left[\left(\frac{d^2\Omega_{i,1}}{ds_1^2}\frac{d\Omega_{j,1}}{ds_1}\right)_{s_1=0}\right.\right.$$
$$-\left(\frac{d^2\Omega_{i,n}}{ds_n^2}\frac{d\Omega_{j,n}}{ds_n}\right)_{s_n=L_n}\Bigg] - \left[\left(\frac{d^2\Omega_{j,1}}{ds_1^2}\frac{d\Omega_{i,1}}{ds_1}\right)_{s_1=0}\right.$$
$$\left.\left.-\left(\frac{d^2\Omega_{j,n}}{ds_n^2}\frac{d\Omega_{i,n}}{ds_n}\right)_{s_n=L_n}\right]\right\} + \int_0^L \frac{d^4\Omega_j}{ds^4}\Omega_i\,ds \qquad (3.4.41)$$

The first term on the right-hand side of Eq. (3.4.41) equals zero from the boundary conditions (3.4.25), the second term vanishes from the condition of continuity, Eq. (3.4.26), and the third one also disappears from the end conditions of Eqs. (3.4.27). Therefore, when Eq. (3.3.40*b*) is substituted into Eq. (3.4.41),

$$k_i^4\int_0^L \Omega_i\Omega_j\,ds = k_j^4\int_0^L \Omega_j\Omega_i\,ds \qquad (3.4.42a)$$

is obtained. For $i \neq j$ and $k_i \neq k_j$, the following orthogonality relationships can be obtained:

$$\int_0^L \Omega_i\Omega_j\,ds = 0 \qquad (3.4.7a)$$

$$\int_0^L \frac{d^4\Omega_i}{ds^4}\Omega_j\,ds = 0 \qquad (3.4.7e)$$

Similarly,

$$\int_0^L \frac{d^2\Omega_i}{ds^2}\Omega_j\,ds = \sum_{r=1}^{n}\left(\frac{d\Omega_{i,r}}{ds_r}\Omega_{j,r} - \frac{d\Omega_{j,r}}{ds_r}\Omega_{i,r}\right)_0^{L_r} + \int_0^L \frac{d^2\Omega_j}{ds^2}\Omega_i\,ds \qquad (3.4.42b)$$

is given, but the first term on the right-hand side must equal zero to satisfy the boundary conditions of Eqs. (3.4.26) and (3.4.27). Then,

$$\int_0^L \frac{d^2\Omega_i}{ds^2}\Omega_j\,ds = \int_0^L \frac{d^2\Omega_j}{ds^2}\Omega_i\,ds \qquad (3.4.7c)$$

However, clearly another condition (3.4.7*b*) is not satisfied, if Eq. (3.4.30) is substituted into Eq. (3.4.7*b*).

To fulfill this condition perfectly, the shape function should take a different form from Eq. (3.4.24). Although the above approximate function Ω_i is quite proper for the conditions of orthogonality in Eqs. (3.4.7*a*), (3.4.7*c*), and (3.4.7*d*), errors will be created in the remaining condition of Eq. (3.4.7*b*). However, it is important to examine the response due to several modes of vibration for our purposes. Accordingly, we deal with only the lower frequencies for the sake of simplicity.

(4) Basic equation of motion.[55] By substituting Eq. (3.4.28) or (3.4.30) into Eqs. (3.4.14*a*) and (3.4.14*b*) and considering Eqs. (3.4.6) and (3.4.13), the equations of motion for the deflection $w_i(t)$ and the angle of rotation $\beta_i(t)$ for the ith mode of vibration reduce to

$$M_{B,i}\ddot{w}_i + D_{B,i,w}\dot{w}_i + K_{B,i,ww}w_i + N_{B,i}\ddot{\beta}_i + K_{B,i,\beta w}\beta_i = \frac{F_i}{\Omega_i(s)} \tag{3.4.43a}$$

$$N_{B,i}\ddot{w}_i + K_{B,i,w\beta}w_i + I_{B,i}\ddot{\beta}_i + D_{B,i,\beta}\dot{\beta}_i + K_{B,i,\beta\beta}\beta_i = \frac{\Gamma_i}{\Omega_i(s)} \tag{3.4.43b}$$

where $M_{B,i}$, $I_{B,i}$, and $N_{B,i}$ can be designated by

$$M_{B,i} = m^*A_s^*L_1 \frac{\sum_{r=1}^{n} \nabla_{i,r}}{\Omega_i^2(s)} \tag{3.4.44a}$$

$$I_{B,i} = m^*I_s^*L_1 \frac{\sum_{r=1}^{n} \nabla_{i,r}}{\Omega_i^2(s)} \tag{3.4.44b}$$

$$N_{B,i} = m^*S_y^*L_1 \frac{\sum_{r=1}^{n} \nabla_{i,r}}{\Omega_i^2(s)} \tag{3.4.44c}$$

that is, $M_{B,i}$ and $I_{B,i}$ represent the effective mass and rotational inertia for the ith mode of vibration, respectively, $N_{B,i}$ being the effective mass or the rotational inertia term due to the eccentricity x_G of mass, $M_{B,i}$.

Now $K_{B,i,ww}$, $K_{B,i,\beta w}$, $K_{B,i,w\beta}$, and $K_{B,i,\beta\beta}$ are the equivalent spring constants given by

$$K_{B,i,ww} = \frac{E_sI_x'^*}{L_1^3}\left[(k_iL_1)^4 \sum_{r=1}^{n} \nabla_{i,r} + \frac{I_\omega^*}{I_x'^*L_1^2}(k_iL_1)^4 \sum_{r=1}^{n}\left(\frac{\Phi_r}{l_r}\right)^2 \nabla_{i,r} + \frac{G_sK^*}{E_sI_x'^*}(k_iL_1)^2 \sum_{r=1}^{n}\left(\frac{\Phi_r}{l_r}\right)^2 \Delta_{i,r}\right][\Omega_i^2(s)]^{-1} \tag{3.4.45a}$$

$$K_{B,i,w\beta} = K_{B,i,\beta w} = \frac{E_s I_x'^* + G_s K^*}{L_1^2}\left[\frac{E_s I_\omega^*}{(E_s I_x'^* + G_s K^*)L_1^2}(k_i L_1)^4 \sum_{r=1}^{n}\left(\pm\frac{\Phi_r}{l_r}\right)\nabla_{i,r} + (k_i L_1)^2 \sum_{r=1}^{n}\left(\pm\frac{\Phi_r}{l_r}\right)\Delta_{i,r}\right][\Omega_i^2(s)]^{-1} \quad (3.4.45b)$$

$$K_{B,i,\beta\beta} = \frac{E_s I_\omega^*}{L_1^3}\left[(k_i L_1)^4 \sum_{r=1}^{n} \nabla_{i,r} + \frac{G_s K^* L_1^2}{E_s I_\omega^*}(k_i L_1)^2 \sum_{r=1}^{n} \Delta_{i,r} + \frac{I_x'^* L_1^2}{I_\omega^*}\sum_{r=1}^{n}\left(\frac{\Phi_r}{l_r}\right)^2 \nabla_{i,r}\right][\Omega_i^2(s)]^{-1} \quad (3.4.45c)$$

Here $K_{B,i,ww}$ means the required force to give a unit downward displacement at the shear center S, and $K_{B,i,\beta\beta}$ designates the necessary torque to induce a unit rotational angle around the shear center. Note that $K_{B,i,w\beta} = K_{B,i,\beta w}$, and these are the corresponding required torques and forces at the shear center, respectively. In Eq. (3.4.45), Φ_r is the central angle of the rth span with length L_r and radius of curvature $R_{s,r}$ of a continuous curved girder, which gives

$$\Phi_r = \frac{L_r}{R_{s,r}} \quad (3.4.46a)$$

$$L_r = l_r L_1 \quad (3.4.46b)$$

where the sign of Φ_r is taken as indicated in Fig. 3.53.

Finally, $D_{B,i,w}$ and $D_{B,i,\beta}$ are the damping coefficients for bending and rotational vibrations, given by

$$D_{B,i,w} = M_{B,i} d_{i,w} \quad (3.4.47a)$$

$$D_{B,i,\beta} = I_{B,i} d_{i,\beta} \quad (3.4.47b)$$

In Eqs. (3.4.44) and (3.4.45) the functions $\nabla_{i,r}$ and $\Delta_{i,r}$ are given by

$$\nabla_{i,r} = \frac{1}{L_r}\int_0^{L_r} [\Omega_{i,r}(s_r)]^2\, ds_r \quad (3.4.48a)$$

$$\Delta_{i,r} = \frac{1}{k_i^2 L_r}\int_0^{L_r} [\Omega_{i,r}''(s_r)]^2\, ds \quad (3.4.48b)$$

For a simply supported girder,

$$\nabla_{i,1} = \Delta_{i,1} = \tfrac{1}{2} \quad (3.4.49a,b)$$

For a continuous girder,

$$\nabla_{i,r} = l_r[(\lambda_{i,r}^2 + \lambda_{i,r+1}^2)\mu_{i,r} + \lambda_{i,r}\lambda_{i,r+1}\kappa_{i,r}] \tag{3.4.50a}$$

$$\Delta_{i,r} = l_r[(\lambda_{i,r}^2 + \lambda_{i,r+1}^2)\nu_{i,r} + \lambda_{i,r}\lambda_{i,r+1}\Theta_{i,r}] \tag{3.4.50b}$$

Four new parameters are introduced here. They are a function of $l_r k_i L_1$ and are given by

$$\mu_{i,r} = \tfrac{1}{2}(l_r k_i L_1 \psi_{i,r}\varphi_{i,r} - \chi_{i,r}) \tag{3.4.51a}$$

$$\kappa_{i,r} = \zeta_{i,r} - \psi_{i,r} \tag{3.4.51b}$$

$$\nu_{i,r} = \tfrac{1}{2}\left[\csc^2 l_r k_i L_1 + \operatorname{csch}^2 l_r k_i L_1 - \frac{1}{l_r k_i L_1}(\coth l_r k_i L_1 + \cot l_r k_i L_1)\right] \tag{3.4.51c}$$

$$\Phi_{i,r} = \frac{\varphi_{i,r}}{l_r k_i L_1} - (\operatorname{csch} l_r k_i L_1 \coth l_r k_i L_1 + \csc l_r k_i L_1 \cot l_r k_i L_1) \tag{3.4.51d}$$

The equivalent mechanical vibration system of Eq. (3.4.43) can be illustrated in Fig. 3.55*a*, where the effective mass $M_{B,i}$ is concentrated at the center of gravity G of the cross section and the effective rotational mass is given by

$$I_{B,G,i} = m^* I_G^* L_1 \frac{\sum_{r=1}^{n} \nabla_{i,r}}{\Omega_i^2(s)} \tag{3.4.52}$$

This mass is concentrated at the shear center S. These masses are connected by a massless rigid bar with a horizontal eccentricity x_G and are supported by vertical springs $K_{B,i,ww}$ and $K_{B,i,\beta w}$ and rotational springs $K_{B,i,\beta\beta}$ and $K_{B,i,\beta w}$, as in Fig. 3.55*a*.

When the external forces F_i/Ω_i and Γ_i/Ω_i are applied to this model, the dashpots with damping coefficient $D_{B,i,w}$ for vertical vibration and $D_{B,i,\beta}$ for rotational vibration are attached to this vibration system, to obtain damping forces proportional to the velocity of the displacements.

The validity of this mechanical model can be confirmed by the equilibrium conditions for vertical forces and torques by virtue of d'Alembert's principle, as illustrated in Fig. 3.55*b*. So the forced vibrations of a curved girder bridge can be determined easily by the model shown in Fig. 3.55*a*.

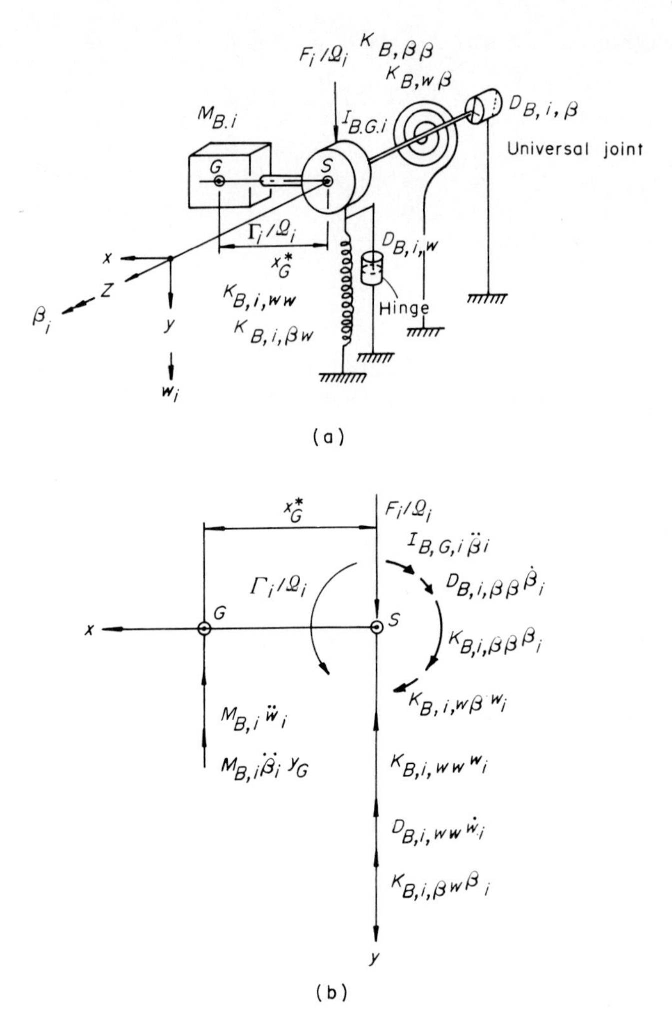

Figure 3.55 (*a*) Equivalent mechanical vibration system and (*b*) corresponding equilibrium condition.

3.4.3 Analysis for various dynamic forces

(1) Free vibrations with damping.[55] In a freely vibrating curved girder, the forcing terms equal zero:

$$F_i = \Gamma_i = 0 \tag{3.4.53}$$

Thus the frequency equation can be found as indicated below.

For the sake of simplicity, let us consider first free vibration without damping effects. The division of Eq. (3.4.14*b*) by R_s and the subtraction from Eq. (3.4.14*a*) give

$$A_y^*(\ddot{w}_i + p_{i,ww}^2 w_i) + S_y'^*(\ddot{\beta}_i + p_{i,\beta w}^2 \beta_i) = 0 \tag{3.4.54a}$$

$$S_y^*(\ddot{w}_i + p_{i,w\beta}^2 w_i) + I_s^*(\ddot{\beta}_i + p_{i\beta\beta}^2 \beta_i) = 0 \tag{3.4.54b}$$

where

$$A_y = A_s - \frac{S_y}{R_s} \tag{3.4.55a}$$

and

$$S_y' = S_y - \frac{I_s}{R_s} \tag{3.4.55b}$$

The mean values A_y^* and $S_y'^*$ are estimated as previously indicated. Moreover, the following notations are introduced:

$$p_{i,ww}^2 = \frac{E_s I_x'^*}{m^* A_y^* L_1^4}\left[(k_i L_1)^4 - (k_i L_1)^2 \frac{\sum_{r=1}^{n} (\Phi_r/l_r)^2 \Delta_{i,r}}{\sum_{r=1}^{n} \nabla_{i,r}}\right] \tag{3.4.56a}$$

$$p_{i,\beta w}^2 = \frac{E_s I_x'^*}{m^* S_y'^* L_1^3}\left[(k_i L_1)^2 \frac{\sum_{r=1}^{n} (\pm\Phi_r/l_r)\Delta_{i,r}}{\sum_{r=1}^{n} \nabla_{i,r}} - \frac{\sum_{r=1}^{n} (\pm\Phi_r/l_r)^3 \nabla_{i,r}}{\sum_{r=1}^{n} \nabla_{i,r}}\right] \tag{3.4.56b}$$

$$p_{i,w\beta}^2 = \frac{E_s I_\omega^*}{m^* S_y^* L_1^5}\left[(k_i L_1)^4 \frac{\sum_{r=1}^{n} (\pm\Phi_r/l_r)\nabla_{i,r}}{\sum_{r=1}^{n} \nabla_{i,r}} + \frac{L_1^2}{I_\omega^*}\left(\frac{G_s K^*}{E_s} + I_x'^*\right)(k_i L_1)^2 \frac{\sum_{r=1}^{n} (\pm\Phi_r/l_r)\Delta_{i,r}}{\sum_{r=1}^{n} \nabla_{i,r}}\right] \tag{3.4.56c}$$

$$p_{i,\beta\beta}^2 = \frac{E_s I_\omega^*}{m^* I_s^* L_1^4}\left[(k_i L_1)^4 + \frac{G_s K^* L_1^2}{E_s I_\omega^*}(k_i L_1)^2 \frac{\sum_{r=1}^{n} \Delta_{i,r}}{\sum_{r=1}^{n} \nabla_{i,r}} + \frac{I_x'^* L_1^2}{I_\omega^*} \frac{\sum_{r=1}^{n} (\Phi_r/l_r)^2 \nabla_{i,r}}{\sum_{r=1}^{n} \nabla_{i,r}}\right] \tag{3.4.56d}$$

In these equations, $p_{i,ww}$ and $p_{i,\beta\beta}$ represent the uncoupled circular frequency for bending and torsional vibrations, respectively.

The solution of Eq. (3.4.54) can be found by setting $w_i = \bar{w}_i e^{\zeta t}$ and $\beta_i = \bar{\beta}_i e^{\zeta t}$ and setting the determinant of the $\bar{w}_i$ and $\bar{\beta}_i$ coefficient matrix equal to zero:

$$\begin{vmatrix} A_y^*(p_{i,ww}^2 - p_i^2) & S_y'^*(p_{i,w\beta}^2 - p_i^2) \\ S_y^*(p_{i,\beta w}^2 - p_i^2) & I_s^*(p_{i,\beta\beta}^2 - p_i^2) \end{vmatrix} = 0 \tag{3.4.57}$$

From this equation, two coupled frequencies $p_{i,\mathrm{I}}$ and $p_{i,\mathrm{II}}$ can be obtained:

$$p_{i,\mathrm{I\ or\ II}} = \sqrt{\frac{b_i \pm \sqrt{b_i^2 - 4ac_i}}{2a}} \tag{3.4.58}$$

where $a = A_y^* I_S^*$ (3.4.59a)

$$b_i = A_y^* I_S^*(p_{i,ww}^2 + p_{i,\beta\beta}^2) - S_y^* S_y'^*(p_{i,\beta w}^2 + p_{i,w\beta}^2) \tag{3.4.59b}$$

$$c_i = A_y^* I_S^* p_{i,ww}^2 p_{i,\beta\beta}^2 - S_y^* S_y'^* p_{i,\beta w}^2 p_{i,w\beta}^2 \tag{3.4.59c}$$

Strictly speaking, the horizontal bending vibration will be coupled with vibrations due to the unsymmetric shape of the cross section of the curved girder bridge, so that the free vibrations become much more complicated. The details of these phenomena have, however, been studied by Komatsu and Nakai,[51] and the results completely coincide with the expressions presented above, provided that the vibration coupled with the horizontal bending displacement is ignored.

Consider next the free vibration with damping effects. For this phenomenon, the following equations can be derived from Eqs. (3.4.43) through (3.4.47) and (3.4.56):

$$\ddot{w}_i + \tilde{d}_{i,ww}\dot{w}_i + \tilde{p}_{i,ww}^2 w_i - \tilde{d}_{i,\beta w}\dot{\beta}_i - \tilde{p}_{i,\beta w}^2 \beta_i = 0 \tag{3.4.60a}$$

$$-\tilde{d}_{i,w\beta}\dot{w}_i + \tilde{p}_{i,w\beta}^2 w_i + \ddot{\beta}_i + \tilde{d}_{i,\beta\beta}\dot{\beta}_i + \tilde{p}_{i,\beta\beta}^2 \beta_i = 0 \tag{3.4.60b}$$

where

$$\tilde{p}_{i,ww}^2 = \frac{I_{B,i}K_{B,i,ww} - N_{B,i}K_{B,i,w\beta}}{M_{B,i}I_{B,i} - N_{B,i}^2} = \frac{A_y^* I_S^* p_{i,ww}^2 - S_y^* S_y'^* p_{i,w\beta}^2}{A_s^* I_G^*} \tag{3.4.61a}$$

$$\tilde{p}_{i,\beta w}^2 = \frac{N_{B,i}K_{B,i,\beta,\beta} - I_{B,i}K_{B,i,\beta w}}{M_{B,i}I_{B,i} - N_{B,i}^2} = \frac{S_y'^* I_S^*(p_{i,\beta\beta}^2 - p_{i,\beta w}^2)}{A_s^* I_G^*} \tag{3.4.61b}$$

$$\tilde{p}^2_{i,w\beta} = \frac{M_{B,i}K_{B,i,w\beta} - N_{B,i}K_{B,i,ww}}{M_{B,i}I_{B,i} - N^2_{B,i}}$$

$$= \frac{A^*_y S^*_y (p^2_{i,w\beta} - p^2_{i,ww})}{A^*_s I^*_G} \tag{3.4.61c}$$

$$\tilde{p}^2_{i,\beta\beta} = \frac{M_{B,i}K_{B,i,\beta\beta} - N_{B,i}K_{B,i,\beta w}}{M_{B,i}I_{B,i} - N^2_{B,i}}$$

$$= \frac{A^*_y I^*_S p^2_{i,\beta\beta} - S^*_y S'^*_y p^2_{i,\beta w}}{A^*_s I^*_G} \tag{3.4.61d}$$

and

$$\tilde{d}_{i,ww} = \frac{I_{B,i}D_{B,i,w}}{M_{B,i}I_{B,i} - N^2_{B,i}} = \frac{I^*_S}{I^*_G} d_{i,w} \tag{3.4.62a}$$

$$\tilde{d}_{i,\beta w} = \frac{N_{B,i}D_{B,i,\beta}}{M_{B,i}I_{B,i} - N^2_{B,i}} = \frac{S^*_y I^*_S}{A^*_s I^*_G} d_{i,\beta} \tag{3.4.62b}$$

$$\tilde{d}_{i,w\beta} = \frac{N_{B,i}D_{B,i,w}}{M_{B,i}I_{B,i} - N^2_{B,i}} = \frac{S^*_y}{I^*_G} d_{i,w} \tag{3.4.62c}$$

$$\tilde{d}_{i,\beta\beta} = \frac{M_{B,i}D_{B,i,\beta}}{M_{B,i}I_{B,i} - N^2_{B,i}} = \frac{I^*_S}{I^*_G} d_{i,\beta} \tag{3.4.62d}$$

The substitution of $w_i = \bar{w}_i e^{\zeta t}$ and $\beta_i = \bar{b}_i e^{\zeta t}$ into Eq. (3.4.60) gives a set of simultaneous equations relative to the unknown amplitudes $\bar{w}_i$ and $\bar{b}_i$:

$$(\zeta^2 + \tilde{d}_{i,ww}\zeta + \tilde{p}^2_{i,ww})\bar{w}_i - (\tilde{d}_{i,\beta w}\zeta + \tilde{p}^2_{i,\beta w})\bar{b}_i = 0 \tag{3.4.63a}$$

$$-(\tilde{d}_{i,w\beta}\zeta - \tilde{p}^2_{i,w\beta})\bar{w}_i + (\zeta^2 + d_{i,\beta\beta}\zeta + \tilde{p}^2_{i,\beta\beta})\bar{b}_i = 0 \tag{3.4.63b}$$

When both amplitudes do not vanish and $\bar{w}_i \neq 0$ and $\bar{b}_i \neq 0$, the following fourth-order characteristic equation with respect to the variable applies:

$$(\zeta^2 + \tilde{d}_{i,ww}\zeta + \tilde{p}^2_{i,ww})(\zeta^2 + \tilde{d}_{i,\beta\beta}\zeta + \tilde{p}^2_{i,\beta\beta}) - (\tilde{d}_{i,w\beta}\zeta + \tilde{p}^2_{i,w\beta})(\tilde{d}_{i,\beta w}\zeta - \tilde{p}^2_{i,\beta w}) = 0 \tag{3.4.63c}$$

The roots of this equation are

$$\zeta_{1,2} = -d_{i,\mathrm{I}} \pm jp_{i,\mathrm{I}'} \tag{3.4.64a}$$

$$\zeta_{3,4} = -d_{i,\mathrm{II}} \pm jp_{i,\mathrm{II}'} \tag{3.4.64b}$$

where $j = \sqrt{-1}$ and the damping coefficients are assumed to be small.

Therefore, the solutions of Eq. (3.4.63) yield

$$w_i = e^{-d_{i,\mathrm{I}}t}(A_i \sin p'_{i,\mathrm{I}}t + B_i \cos p'_{i,\mathrm{I}}t) + e^{-d_{i,\mathrm{II}}t}(C_i \sin p'_{i,\mathrm{II}}t + D_i \cos p'_{i,\mathrm{II}}t) \quad (3.4.65a)$$

$$\beta_i = e^{-d_{i,\mathrm{I}}t}(E_i \sin p'_{i,\mathrm{I}}t + F_i \cos p'_{i,\mathrm{I}}t) + e^{-d_{i,\mathrm{II}}t}(G_i \sin p'_{i,\mathrm{II}}t + H_i \cos p'_{i,\mathrm{II}}t) \quad (3.4.65b)$$

The eight integration constants A_i through H_i can be determined by four relationships between the amplitude ratio $\bar{w}_i$ and $\bar{b}_i$, when Eq. (3.4.65) is substituted into Eq. (3.4.60) with the following four initial conditions

$$w_i(0) \qquad \dot{w}_i(0) \qquad \beta_i(0) \qquad \dot{\beta}_i(0) \quad (3.4.65c)$$

Thus, the natural frequency $f'_{i,\mathrm{I}}$ or $f'_{i,\mathrm{II}}$ (in hertz) can be obtained by

$$f'_{i,\mathrm{I(or\ II)}} = \frac{p'_{i,\mathrm{I(or\ II)}}}{2\pi} \quad (3.4.66a)$$

However, the effect of the damping coefficients on the natural frequency is generally small, and an approximate solution is

$$f_{i,\mathrm{I(or\ II)}} = \frac{p_{i,\mathrm{I(or\ II)}}}{2\pi} \quad (3.4.66b)$$

Accordingly, we conclude that two frequencies exist for the ith mode of coupled vibrations of bending and torsion, and the relationships between uncoupled frequencies $p_{i,ww}$ and $p_{i,\beta\beta}$ and coupled frequencies $p_{i,\mathrm{I}}$ and $p_{i,\mathrm{II}}$ can be represented by

$$p_{i,\mathrm{I}} < p_{i,ww} < p_{i,\beta\beta} < p_{i,\mathrm{II}} \quad (3.4.67)$$

The damping coefficients $d_{i,\mathrm{I}}$ and $d_{i,\mathrm{II}}$ can be represented approximately by

$$d_{i,\mathrm{I(or\ II)}} = \frac{D_{i,\mathrm{I(or\ II)}}\, p'_{i,\mathrm{I(or\ II)}}}{\pi} \quad (3.4.68a)$$

when the logarithmic damping coefficients $D_{i,\mathrm{I}}$ and $D_{i,\mathrm{II}}$ corresponding to frequencies $p_{i,\mathrm{I}}$ and $p_{i,\mathrm{II}}$, respectively, are known from experimental studies or other data. Substituting these values into Eqs. (3.4.64) and (3.4.60), we get a set of simultaneous equations relative to four unknown damping coefficients $d_{i,ww}$, $d_{i,\beta\beta}$, $d_{i,w\beta}$, and $d_{i,\beta w}$. The damping coefficients $d_{i,w}$ and $d_{i,\beta}$ for bending and torsional vibrations can then be determined from the above solutions and Eq. (3.3.62).

However, from the approximations $p'_{i,\mathrm{I}} = p_{i,\mathrm{I}}$ and $p'_{i,\mathrm{II}} = p_{i,\mathrm{II}}$ as well as $d_{i,\mathrm{I}} \cong d_{i,w}$ and $d_{i,\mathrm{II}} \cong d_{i,\beta}$, as previously explained, the following equations can be generated:

$$d_{i,w} = \frac{D_{i,\mathrm{I}} p_{i,\mathrm{I}}}{\pi} = 2D_{i,\mathrm{I}} f_{i,\mathrm{I}} \tag{3.4.68b}$$

$$d_{i,\beta} = \frac{D_{i,\mathrm{II}} p_{i,\mathrm{II}}}{\pi} = 2D_{i,\mathrm{II}} f_{i,\mathrm{II}} \tag{3.4.68c}$$

Therefore, the approximate values for $d_{i,ww}$, $d_{i,\beta w}$, $d_{i,w\beta}$, and $d_{i,\beta\beta}$ can be directly determined by Eq. (3.4.62).

(2) Various forced vibrations.[55] The differential equation for various forced vibrations of a curved girder bridge can easily be given by adding the forcing terms:

$$\ddot{w}_i + \tilde{d}_{i,ww}\dot{w}_i + \tilde{p}^2_{i,ww} w_i - \tilde{d}_{i,\beta w}\dot{\beta}_i - \tilde{p}^2_{i,\beta w}\beta_i = \delta_i \tag{3.4.69a}$$

$$-\tilde{d}_{i,w\beta}\dot{w}_i + \tilde{p}^2_{i,w\beta} w_i + \ddot{\beta}_i + \tilde{d}_{i,\beta\beta}\dot{\beta}_i + \tilde{p}^2_{i,\beta\beta}\beta_i = \gamma_i \tag{3.4.69b}$$

Here the forcing terms are represented by

$$\delta_i = \frac{I_{B,i} F_i/\Omega_i(s) - N_{B,i}\Gamma_i/\Omega_i(s)}{M_{B,i} I_{B,i} - N^2_{B,i}}$$

$$= \frac{\Omega_i^2(s)}{m^* L_1 \sum_{r=1}^{n} \nabla_{i,r}} \; \frac{I^*_S F_i/\Omega_i(s) - S^*_y \Gamma_i/\Omega_i(s)}{A^*_s I^*_G} \tag{3.4.70a}$$

and

$$\gamma_i = \frac{M_{B,i}\Gamma_i/\Omega_i(s) - N_{B,i} F_i/\Omega_i(s)}{M_{B,i} I_{B,i} - N^2_{B,i}}$$

$$= \frac{\Omega_i^2(s)}{m^* L_1 \sum_{r=1}^{n} \nabla_{i,r}} \; \frac{A^*_s \Gamma_i/\Omega_i(s) - S^*_y F_i/\Omega_i(s)}{A^*_s I^*_G} \tag{3.4.70b}$$

The generalized forces F_i and Γ_i can be categorized as follows:

***a.* Indicial forces.** When a concentrated vertical force P_o is applied to the section at $s = a$ with a radius of curvature R_P at time $t = t_o$, the generalized forces F_i and Γ_i from Eq. (3.4.13) are

$$F_i = \frac{R_P}{R_S} P_o [\Omega_i(s)]_{s=a} U(t - t_o) \tag{3.4.71a}$$

$$\Gamma_i = \frac{R_P}{R_S} P_o x_P [\Omega_i(s)]_{s=a} U(t - t_o) \tag{3.4.71b}$$

where $U(t - t_o)$ is the unit step function and x_P is the eccentricity of the force P_o, given by

$$x_P = R_P - R_S \tag{3.4.72}$$

b. **Sinusoidal forces.** For a sinusoidal force with circular frequency ω_o, that is, $P_o \sin \omega_o t$, the generalized forces F_i and Γ_i can be found:

$$F_i = \frac{R_P}{R_S} P_o[\Omega_o(s)]_{s=a} \sin \omega_o t \tag{3.4.73a}$$

$$\Gamma_i = \frac{R_P}{R_S} P_o x_P[\Omega_i(s)]_{s=a} \sin \omega_o t \tag{3.4.73b}$$

c. **Moving sinusoidal forces.** When a sinusoidal force moves with constant speed V_o from the origin $s = 0$ to section $s = a$, we have

$$a = \frac{R_S}{R_P} V_o t \tag{3.4.74}$$

d. **Moving constant force.** In this case, we set

$$\sin \omega_o t = 1 \tag{3.4.75}$$

The simultaneous differential equations (3.4.69) relative to w_i and β_i can now be solved analytically with a Laplace transformation by letting

$$\hat{f}(\zeta) = \mathscr{L}f(t) = \int_0^\infty e^{-\zeta t} f(t)\, dt \tag{3.4.76}$$

and setting the initial conditions as

$$w_i(0) = \dot{w}_i(0) = 0 \qquad \beta_i(0) = \dot{\beta}_i(0) = 0 \tag{3.4.77a–d}$$

which yields

$$\hat{w}_i(\zeta) = \frac{\hat{\delta}_i(\zeta)(\zeta^2 + \tilde{d}_{i,ww}\zeta + \tilde{p}^2_{i,ww}) + \gamma_i(\zeta)(\tilde{d}_{i,w\beta}\zeta - \tilde{p}^2_{i,w\beta})}{(\zeta^2 + \tilde{d}_{i,ww}\zeta + \tilde{p}^2_{i,ww})(\zeta^2 + \tilde{d}_{i,\beta\beta}\zeta + \tilde{p}^2_{i,\beta\beta}) + \eta(\zeta) - (\tilde{d}_{i,\beta w}\zeta + \tilde{p}^2_{i,\beta w})(\tilde{d}_{i,w\beta}\zeta - \tilde{p}^2_{i,\beta\beta})} \tag{3.4.78a}$$

$$\hat{\beta}_i(\zeta) = \frac{\hat{\gamma}_i(\zeta)(\zeta^2 + \tilde{d}_{i,ww}\zeta + \tilde{p}^2_{i,ww}) - \hat{\delta}_i(\zeta)(\tilde{d}_{i,\beta w}\zeta - \tilde{p}^2_{i,\beta w})}{(\zeta^2 + \tilde{d}_{i,ww}\zeta + \tilde{p}^2_{i,ww})(\zeta^2 + \tilde{d}_{i,\beta\beta}\zeta + \tilde{p}^2_{i,\beta\beta}) + \eta(\zeta) - (\tilde{d}_{i,\beta w}\zeta + \tilde{p}^2_{i,\beta w})(\tilde{d}_{i,w\beta}\zeta - \tilde{p}^2_{i,\beta\beta})} \tag{3.4.78b}$$

The solutions of $w_i(t)$ and $\beta_i(t)$ are obtained by using the inverse

Laplace transformation:

$$w_i(t) = \mathscr{L}^{-1}\{\hat{w}_i(\zeta)\} = \frac{1}{2\pi j}\int_{\gamma - j\infty}^{\gamma + j\infty} e^{\zeta t}\hat{w}_i(\zeta)\, d\zeta \qquad (3.4.79a)$$

$$\beta_i(t) = \mathscr{L}^{-1}\{\hat{\beta}_i(\zeta)\} = \frac{1}{2\pi j}\int_{\gamma - j\infty}^{\gamma + j\infty} e^{\zeta t}\hat{\beta}_i(\zeta)\, d\zeta \qquad (3.4.79b)$$

The final results of the deflection $w(s, t)$ and the rotation angle $\beta(s, t)$ due to the sinusoidal force $P_o \sin \omega_o t$ can be written as

$$w(s, t) = \sum_{i=1,2,3,\ldots} \frac{\sin \omega_o(t - \varphi_{i,w})}{\sqrt{a_{i,w}^2 + b_{i,w}^2}} \qquad (3.4.80a)$$

$$\beta(s, t) = \sum_{i=1,2,3,\ldots} \frac{\sin \omega_o(t - \varphi_{i,\beta})}{\sqrt{a_{i,\beta}^2 + b_{i,\beta}^2}} \qquad (3.4.80b)$$

Thus, the deflection w_x at an arbitrary point with eccentricity x from the shear center S can be estimated by

$$w_x = w + x\beta \qquad (3.4.81)$$

where the a_i and b_i coefficients are

$$a_{i,w} = \frac{A_i C_{i,w} + B_i D_{i,w}}{C_{i,w}^2 + D_{i,w}^2} \qquad (3.4.82a)$$

$$b_{i,w} = \frac{B_i C_{i,w} - A_i D_{i,w}}{C_{i,w}^2 + D_{i,w}^2} \qquad (3.4.82b)$$

$$a_{i,\beta} = \frac{A_i C_{i,\beta} + B_i D_{i,\beta}}{C_{i,\beta}^2 + D_{i,\beta}^2} \qquad (3.4.82c)$$

$$b_{i,\beta} = \frac{B_i C_{i,\beta} - A_i D_{i,\beta}}{C_{i,\beta}^2 + D_{i,\beta}^2} \qquad (3.4.82d)$$

The phase angle φ is

$$\varphi_{i,w} = \tan^{-1} \frac{b_{i,w}}{a_{i,w}} \qquad (3.4.83a)$$

$$\varphi_{i,\beta} = \tan^{-1} \frac{b_{i,\beta}}{a_{i,\beta}} \qquad (3.4.83b)$$

in which

$$A_i = (\tilde{p}_{i,ww} - \omega_o^2)(\tilde{p}_{i,\beta\beta}^2 - \omega_o^2) - \omega_o^2(\tilde{d}_{i,ww}\tilde{d}_{i,\beta\beta} - \tilde{d}_{i,\beta w}\tilde{d}_{i,w\beta}) + \tilde{p}_{i,\beta w}^2\tilde{p}_{i,w\beta}^2 \qquad (3.4.84a)$$

$$B_i = \omega_o[\tilde{d}_{i,ww}(\tilde{p}_{i,\beta\beta}^2 - \omega_o^2) - \tilde{d}_{i,\beta\beta}(\tilde{p}_{i,ww}^2 - \omega_o^2) + \tilde{d}_{i,\beta w}\tilde{p}_{i,w\beta}^2 - \tilde{d}_{i,w\beta}\tilde{p}_{i,\beta w}^2] \qquad (3.4.84b)$$

$$C_{i,w} = q_{Po,i}(\tilde{p}^2_{i,\beta\beta} - \omega_o^2) + \tau_{Po,i}\tilde{p}^2_{i,\beta w} \tag{3.4.84c}$$

$$D_{i,w} = \omega_o(q_{Po,i}\tilde{d}_{i,\beta\beta} + \tau_{Po,i}\tilde{d}_{i,\beta w}) \tag{3.4.84d}$$

$$C_{i,\beta} = \tau_{Po,i}(\tilde{p}^2_{i,ww} - \omega_o^2) - q_{Po,i}\tilde{p}^2_{i,w\beta} \tag{3.4.84e}$$

$$D_{i,\beta} = \omega_o(\tau_{Po,i}\tilde{d}_{i,ww} - q_{Po,i}\tilde{d}_{i,w\beta}) \tag{3.4.84f}$$

The two terms representing the forced terms $q_{Po,i}$ and $\tau_{Po,i}$ can be evaluated as follows:

$$q_{Po,i} = \frac{R_P}{R_S}\frac{I_{B,i} - N_{B,i}x_P}{M_{B,i}I_{B,i} - N^2_{B,i}} P_o \frac{\Omega_i(a)}{\Omega_i(s)} \tag{3.4.85a}$$

$$= \frac{R_P}{R_S}\frac{\Omega_i^2(s)}{m^*L_1 \sum_{r=1}^{n} \nabla_{ir}} \frac{I^*_G - S^*_y x_P}{A^*_s I^*_G} P_o \frac{\Omega_i(a)}{\Omega_i(s)} \tag{3.4.85b}$$

$$\tau_{Po,i} = \frac{R_P}{R_S}\frac{M_{B,i}x_P - N_{B,i}}{M_{B,i}I_{B,i} - N^2_{B,i}} P_o \frac{\Omega_i(a)}{\Omega_i(s)} \tag{3.4.85c}$$

$$= \frac{R_P}{R_S}\frac{\Omega_i^2(s)}{m^*L_1 \sum_{r=1}^{n} \nabla_{i,r}} \frac{A^*_s x_P - S^*_y}{A^*_s I^*_G} P_o \frac{\Omega_i(a)}{\Omega_i(s)} \tag{3.4.85d}$$

For the special case of a static force, a set of simultaneous equations can be obtained by setting $w_i(s, t) = w_{st,i}(s)$ and $\beta_i(s, t) = \beta_{st,i}(s)$ and neglecting the time-dependent function in Eq. (3.4.43), which gives

$$K_{B,i,ww}w_{st,i} + K_{B,i,\beta w}\beta_{st,i} = \frac{R_P}{R_S} P_o\Omega_i(a) \tag{3.4.86a}$$

$$K_{B,i,w\beta}w_{st,i} + K_{B,i,\beta\beta}\beta_{st,i} = \frac{R_P}{R_S} P_o x_P\Omega_i(a) \tag{3.4.86b}$$

Therefore, we have

$$w_{st} = \sum_{i=1,2,3,\ldots} w_{st,i}$$

$$= \frac{R_P}{R_S} P_o \sum_{i=1,2,3,\ldots} \frac{K_{B,i,\beta\beta} - x_P K_{B,i,w\beta}}{K_{B,i,ww}K_{B,i,\beta\beta} - K^2_{B,i,w\beta}} \frac{\Omega_i(a)}{\Omega_i(s)} \tag{3.4.87a}$$

and $\beta_{st} = \sum_{i=1,2,3,\ldots} \beta_{st,i}$

$$= \frac{R_P}{R_S} P_o \sum_{i=1,2,3,\ldots} \frac{x_P K_{B,i,ww} - K_{B,i,\beta w}}{K_{B,i,ww}K_{B,i,\beta\beta} - K^2_{B,i,w\beta}} \frac{\Omega_i(a)}{\Omega_i(s)} \tag{3.4.87b}$$

The deflection $w_{st,x}$ with eccentricity x from the shear center S is

$$w_{st,x} = w_{st} + x\beta_{st} \tag{3.4.81b}$$

The corresponding stress resultants M_x, M_w, and T_s can now be determined from Eqs. (3.4.2), (3.4.4), and (3.4.5).

Now Eqs. (3.4.87*a*) and (3.4.87*b*) can be rewritten from Eqs. (3.4.61), (3.4.85*a*), and (3.4.85*b*):

$$w_{st} = \sum_{i=1,2,3,\ldots} \frac{q_{Po,i}\tilde{p}^2_{i,\beta\beta} + \tau_{Po,i}\tilde{p}^2_{i,\beta w}}{\tilde{p}^2_{i,ww}\,\tilde{p}^2_{i,\beta\beta} + \tilde{p}^2_{i,\beta w}\,\tilde{p}^2_{i,w\beta}} \tag{3.4.87c}$$

$$\beta_{st} = \sum_{i=1,2,3,\ldots} \frac{\tau_{Po,i}\tilde{p}^2_{i,ww} - q_{Po,i}\tilde{p}^2_{i,w\beta}}{\tilde{p}^2_{i,ww}\,p^2_{i,\beta\beta} + \tilde{p}^2_{i,\beta w}\,\tilde{p}^2_{i,w\beta}} \tag{3.4.87d}$$

This equation reduces to

$$q_{Po,i} = \tilde{p}^2_{i,ww}w_{st,i} - \tilde{p}^2_{i,\beta w}\beta_{st,i} \tag{3.4.88a}$$

$$\tau_{Po,i} = \tilde{p}_{i,w\beta}w_{st,i} + \tilde{p}^2_{i,\beta\beta}\beta_{st,i} \tag{3.4.88b}$$

if the time-dependent functions, given in Eq. (3.4.69), are neglected. Accordingly, the values of $q_{Po,i}$ and $\tau_{Po,i}$ can be estimated by utilizing the solutions of $w_{st,i}$ and $\beta_{st,i}$ and Eqs. (3.4.87*a*) and (3.4.87*b*):

$$q_{Po,i} = \frac{R_P}{R_S}P_o\left(\frac{K_{B,i,\beta\beta} - x_P K_{B,i,w\beta}}{K_{B,i,ww}K_{B,i,\beta\beta} - K^2_{B,i,w\beta}}\tilde{p}^2_{i,ww} - \frac{x_P K_{B,i,ww} - K_{B,i,\beta w}}{K_{B,i,ww}K_{B,i,\beta\beta} - K^2_{B,i,w\beta}}\tilde{p}^2_{i,ww}\right)\frac{\Omega_i(c)}{\Omega_i(s)} \tag{3.4.88c}$$

$$\tau_{Po,i} = \frac{R_P}{R_S}P_o\left(\frac{K_{B,i,\beta\beta} - x_P K_{B,i,w\beta}}{K_{B,i,ww}K_{B,i,\beta\beta} - K^2_{B,i,w\beta}}\tilde{p}^2_{i,w\beta} + \frac{x_P K_{B,i,ww} - K_{B,i,\beta w}}{K_{B,i,ww}K_{B,i,\beta\beta} - K^2_{B,i,w\beta}}\tilde{p}^2_{i,\beta\beta}\right)\frac{\Omega_i(c)}{\Omega_i(s)} \tag{3.4.88d}$$

3.4.4 Response due to moving vehicles[55–62]

(1) Modeling of the vehicle and the curved girder.[55] The dynamic response of a curved girder bridge under a moving vehicle is a complicated problem, because both responses are coupled to each other. To analyze this behavior, the following assumptions are introduced:

For the vehicle:

1. Vehicle can be represented by a simple vibration system with one degree of freedom having a mass M_V and a spring K_V.

2. The mass M_V of the vehicle consists of a sprung mass $M_{V,u}$ and an unsprung mass $M_{V,l}$: that is, only the sprung mass $M_{V,u}$ is considered to be supported by the spring K_V of the vehicle.
3. The damping effect D_V is, of course, taken into consideration.
4. The surface of the bridge slab is completely flat, and the unsprung mass $M_{V,1}$ can smoothly move along the dynamic deflection curve of the curved girder.

For the curved girder bridge:

1. The curved girder bridge can be treated as a vibration system with two degrees of freedom that is capable of bending and torsional vibrations for the ith mode of vibration.
2. The damping effects are taken into account.
3. The effective mass of the curved girder bridge is affected by the unsprung mass of the vehicle, but this effect is considered negligible.
4. The forcing terms due to the vehicle are thus composed of the total mass of the vehicle M_V and an additional dynamic force due to the displacements of the spring δ as well as the dashpot of the vehicle.

(2) Response of the vehicle.[55] The free vibration of a vehicle, shown in Fig. 3.56, is given by the well-known formula

$$M_{V,u}\ddot{y}_V + D_V\dot{y}_V + K_V y_V = 0 \tag{3.4.89}$$

where $M_V = M_{V,1} + M_{V,u}$ = total mass of vehicle
K_V = stiffness of vehicle
D_V = damping coefficient of vehicle

The solution of Eq. (3.4.89) yields

$$y_V = e^{-n_v t}(A \sin \sqrt{p_v^2 - n_v^2}\,t + B \cos \sqrt{p_v^2 - n_v^2}\,t) \tag{3.4.90}$$

where

$$p_v^2 = \frac{K_v}{M_{V,u}} \tag{3.4.91a}$$

$$n_v = \frac{D_V}{2M_{V,u}} \tag{3.4.91b}$$

The integration constants A and B can be determined from the following two initial conditions:

$$y_V(0) \tag{3.4.92a}$$

$$\dot{y}_V(0) \tag{3.4.92b}$$

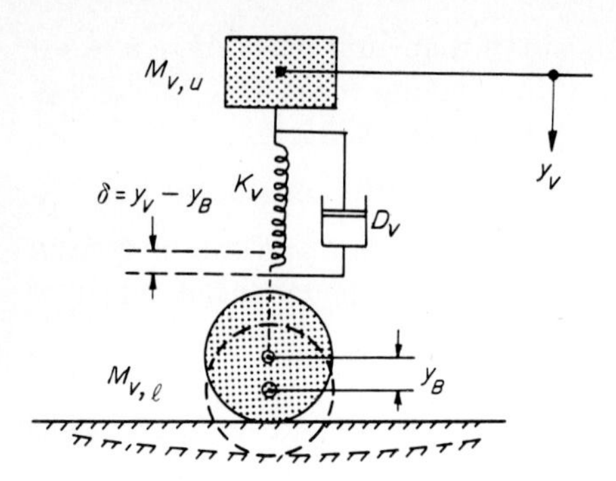

Figure 3.56 Vibration system for a vehicle.

In general, since $p_v \gg n_v$, the natural frequency of the vehicle (in hertz) f_v is given by

$$f_v = \frac{p_v}{2\pi} \tag{3.4.93}$$

The damping coefficient d_v is

$$d_v = \frac{D_V}{M_{V,u}} = \frac{D_V p_v}{\pi} = 2D_V f_v \tag{3.4.94a–c}$$

provided that the logarithmic damping coefficient D_v is known.

Consider now that this vehicle is moving on a curved girder bridge. If the deflection of the curved girder bridge is denoted y_B, the displacement δ of the spring and dashpot of the vehicle will be

$$\delta = y_V - y_B \tag{3.4.95}$$

Thus, Eq. (3.4.89) can be rewritten as

$$\ddot{y}_V + d_v \dot{\delta} + p_v^2 \delta = 0 \tag{3.4.96}$$

From this equation, it is clear that the dynamic response of the vehicle is coupled with that of the curved girder.

(3) Coupled response of the vehicle and the curved girder.[55] Let us now consider the dynamic response of the vehicle on the rth span of a continuous curved girder, as illustrated in Fig. 3.57, in which the vehicle is loaded at the section $s_r = a_r$ from the left support r. The deflection y_{B,a_r} of the curved girder due to the moving vehicle with eccentricity $x_{P,r}$ from the shear center can be represented, from Eqs.

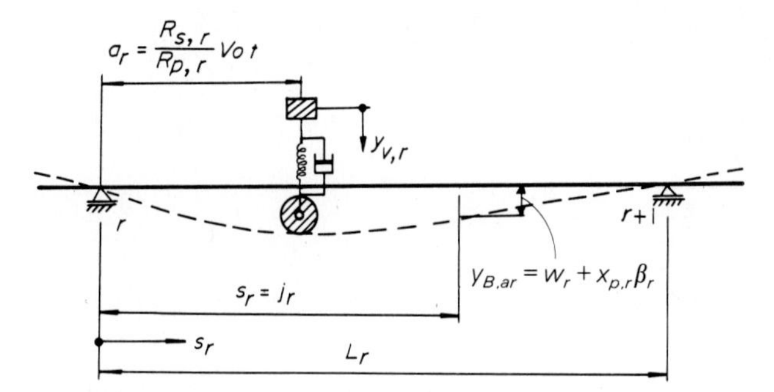

Figure 3.57 Deflection of continuous curved girder in rth span.

(3.4.6) and (3.4.81a), as

$$y_{B,a_r} = \sum_{i=1,2,\ldots} w_{i,a_r}(t) + x_{P,r} \sum_{i=1,2,\ldots} \beta_{i,a_r}(t)$$

$$= \sum_{i=1,2,\ldots} W_i(t)[\Omega_{i,r}]_{s_r=a_r} + x_{P,r} \sum_{i=1,2,\ldots} B_i(t)[\Omega_{i,r}]_{s_r=a_r} \qquad (3.4.97a)$$

At the same time, the deflection $w_{i,j_r}(t)$ and rotation angle $\beta_{i,j_r}(t)$ of the curved girder for a viewpoint $s = j_r$ can be expressed by using Eqs. (3.4.6) and (3.4.24a) as

$$w_{i,j_r}(t) = W_i(t)[\Omega_{i,r}]_{s_r=j_r} \qquad (3.4.97b)$$

$$\beta_{i,j_r}(t) = B_i(t)[\Omega_{i,r}]_{s_r=j_r} \qquad (3.4.97c)$$

Elimination of the time functions $W_i(t)$ and $B_i(t)$ from the above two equations yields

$$y_{B,a_r} = \sum_{i=1,2,\ldots} w_{i,j_r}\Upsilon_{i,j_r} + x_{P,r} \sum_{i=1,2,\ldots} \beta_{i,j_r}\Upsilon_{i,j_r} \qquad (3.4.97d)$$

where the function Υ_{i,j_r} is

$$\Upsilon_{i,j_r} = \frac{[\Omega_{i,r}]_{s_r=a_r}}{[\Omega_{i,r}]_{s_r=j_r}} \qquad (3.4.98)$$

When the vehicle is moving with a constant speed V_o on a line with radius of curvature $R_{P,r}$, we have

$$a_r = \frac{R_{S,r}}{R_{P,r}} V_o t \qquad (3.4.99)$$

Thus the relative displacement δ_r of the vehicle, i.e., Eq. (3.4.95), can be rewritten as

$$\delta_r = y_{V,r} - \left[\sum_{i=1,2,\ldots} w_{i,j_r} \Upsilon_{i,j_r}(v_{o,i,r}t) + x_{P,r} \sum_{i=1,2,\ldots} \beta_{i,j_r} \Upsilon_{i,j_r}(v_{o,i,r}t) \right] \tag{3.4.100}$$

where

$$v_{o,i,r} = \frac{R_{S,r}}{R_{P,r}} k_i V_o \tag{3.4.101}$$

designates the velocity parameter and k_i is the eigenvalue given by Eqs. (3.4.29*a*) and (3.4.31).

Finally, the force terms, which act on the curved girder bridge, can be given by

$$P_o = M_V g + K_V \delta_r + d_V \dot{\delta}_r \tag{3.4.102}$$

as mentioned above. Substitution of this equation into Eq. (3.4.70) gives the following forcing terms:

$$\delta_{i,r} = q_{V,i,j_r}\left(1 + \frac{K_V}{M_V g}\delta_r + \frac{d_V \dot{\delta}_r}{M_V g}\right)\Upsilon_{i,j_r}(v_{o,i,r}t) \tag{3.4.103a}$$

$$\gamma_{i,r} = \tau_{V,i,j_r}\left(1 + \frac{K_V}{M_V g}\delta_r + \frac{d_V \dot{\delta}_r}{M_V g}\right)\Upsilon_{i,j_r}(v_{o,i,r}t) \tag{3.4.103b}$$

In Eq. (3.4.103), q_{v,i,j_r} and τ_{r,i,j_r} are defined as

$$\begin{aligned} q_{V,i,j_r} &= \frac{R_{P,r}}{R_{S,r}} \frac{I_{B,i} - N_{B,i} x_{P,r}}{M_{B,i} I_{B,i} - N_{B,i}^2} M_V g \\ &= \frac{R_{P,r}}{R_{S,r}} \left(\frac{K_{B,i,\beta\beta} - x_{P,r} K_{B,i,w\beta}}{K_{B,i,ww} K_{B,i,\beta\beta} - K_{B,i,w\beta}^2} \tilde{p}_{i,ww}^2 \right. \\ &\quad \left. - \frac{x_{P,r} K_{B,i,ww} - K_{B,i,\beta w}}{K_{B,i,ww} K_{B,i,\beta\beta} - K_{B,i,w\beta}^2} \tilde{p}_{i,\beta w}^2 \right) M_V g \\ &= \frac{R_{P,r}}{R_{S,r}} \frac{\Omega_{i,r}^2(j_r)}{m^* L_1 \sum_{r=1}^{n} \nabla_{i,r}} \frac{I_S^* - S_y^* x_{P,r}}{A_s^* I_G^*} M_V g \end{aligned} \tag{3.4.104a}$$

and

$$\begin{aligned} \tau_{v,i,j_r} &= \frac{R_{P,r}}{R_{S,r}} \frac{M_{B,i} x_{P,r} - N_{B,i}}{M_{B,i} I_{B,i} - N_{B,i}^2} M_V g \\ &= \frac{R_{P,r}}{R_{S,r}} \left(\frac{K_{B,i,\beta\beta} - x_{P,r} K_{B,iw\beta}}{K_{B,i,ww} K_{B,i,\beta\beta} - K_{B,i,w\beta}^2} \tilde{p}_{i,w\beta}^2 \right. \end{aligned}$$

$$+\frac{x_{P,r}K_{B,i,ww}-K_{B,i,\beta w}}{K_{B,i,ww}K_{B,i,\beta\beta}-K^2_{B,i,w\beta}}\tilde{p}^2_{i,\beta\beta}\Bigg)M_Vg$$

$$=\frac{R_{P,r}}{R_{S,r}}\frac{\Omega^2_{i,r}(j_r)}{m^*L_1\sum_{r=1}^{n}\nabla_{i,r}}\frac{A^*_s x_{P,r}-S^*_y}{A^*_s I^*_G}M_Vg \tag{3.4.104b}$$

Consequently, the coupled vibrations of the curved girder and the vehicle can be summarized by this set of simultaneous differential equations with respect to the displacements w, β, and y_V:

$$\ddot{w}_{i,j_r}+\tilde{d}_{i,ww}\dot{w}_{i,j_r}+\tilde{p}^2_{i,ww}w_{i,j_r}-\tilde{d}_{i,\beta w}\dot{\beta}_{i,j_r}-\tilde{p}^2_{i,\beta w}\beta_{i,j_r}$$
$$=q_{V,i,j_r}\left(1+\frac{K_V}{M_Vg}\delta_r+d_V\dot{\delta}_r\right)\Upsilon_{i,j_r}(v_{o,i,r}t) \tag{3.4.105a}$$

$$\ddot{\beta}_{i,j_r}+\tilde{d}_{i,\beta\beta}\dot{\beta}_{i,j_r}+\tilde{p}^2_{i,\beta\beta}\beta_{i,j_r}-\tilde{d}_{i,w\beta}\dot{w}_{i,j_r}+\tilde{p}^2_{i,w\beta}w_{i,j_r}$$
$$=\tau_{V,i,j_r}\left(1+\frac{K_V}{M_Vg}\delta_r+d_V\dot{\delta}_r\right)\Upsilon_{i,j_r}(v_{o,i,r}t) \qquad i=1,2,\ldots \tag{3.4.105b}$$

$$\ddot{y}_{V,r}+d_v\dot{\delta}_r+p^2_v\delta_r=0 \tag{3.4.105c}$$

$$\delta_r=y_{V,r}-\left[\sum_{i=1,2,\ldots}w_{i,j_r}\Upsilon_{i,j_r}(v_{o,i,r}t)+x_{P,r}\sum_{i=1,2,\ldots}\beta_{i,j_r}\Upsilon_{i,j_r}(v_{o,i,r}t)\right] \tag{3.4.105d}$$

The solutions of these equations are difficult to obtain analytically; however, numerical integration methods can be employed with initial conditions

$$w_{j_r}(0) \tag{3.4.106a}$$

$$\dot{w}_{j_r}(0) \tag{3.4.106b}$$

$$\beta_{j_r}(0) \tag{3.4.106c}$$

$$\dot{\beta}_{j_r}(0) \tag{3.4.106d}$$

$$y_{V,r}(0) \tag{3.4.106e}$$

$$\dot{y}_{V,r}(0) \tag{3.4.106f}$$

The deflection $w_{j_r}(x_{j_r},t)$ at a point with eccentricity x_{j_r} from the shear center is estimated by

$$w_{j_r}(x_{j_r},t)=\sum_{i=1,2,\ldots}w_{i,j_r}+x_{j_r}\sum_{i=1,2,\ldots}\beta_{i,j_r} \tag{3.4.107}$$

The procedure of solving these equations by using a computer is explained in detail in the next section.

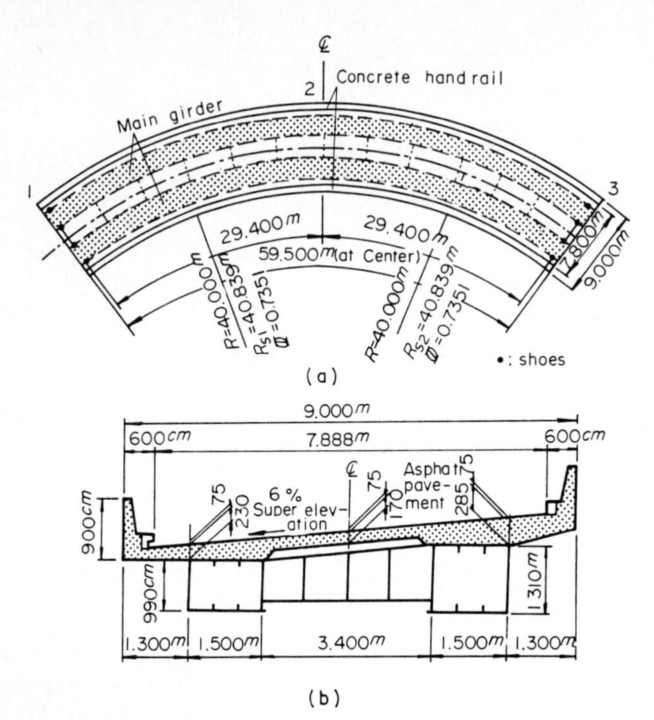

Figure 3.58 Details of Nishimomiya Bridge: (*a*) general plane and (*b*) typical cross section.

(4) Analytical and experimental examples.[59] To compare the analytical procedure with actual behavior, field tests have been conducted on a two-equal-span continuous curved box-girder bridge built at the Nishino-Miya interchange of Highway Kobe-Nagoya. Figure 3.58 shows the span and cross section of this bridge.

***a*. Cross-sectional quantities.** The mean cross-sectional quantities of this bridge can be summarized as in Table 3.9 by taking into consideration the composite action of the girder and the concrete slab with the ratio of Young's modulus n equal to $E_S/E_C = 7$. The stiffness estimations at the measured point, j_1 in Fig. 3.58, i.e., the cross-section values I'_x, K, and I_ω and radii of curvature R_S and R_o, are also indicated in this table.

***b*. Values of $\Omega_{i,r}$, $\nabla_{i,r}$, and $\Delta_{i,r}$.** Since this bridge is a two-span continuous curved girder bridge with equal span length, the modes of free vibration can easily be approximated as follows:
Antisymmetric modes,

$$\Omega_{i,r} = \frac{\sin i\pi s_r}{L_r} \tag{3.4.108a}$$

$$\nabla_{i,r} = \Delta_{i,r} = \tfrac{1}{2} \qquad i = 1, 3, 5, \ldots; \; r = 1, 2 \tag{3.4.108b}$$

TABLE 3.9 Cross-Sectional Quantities

Cross-sectional values	Mean value	Measured point
A_S (cm^4)	1.2713×10^4	
S_y (cm^3)	5.3531×10^5	
S'_y (cm^3)	4.3197×10^5	
I'_x (cm^4)	1.3834×10^7	1.1770×10^7
I_S (cm^4)	4.2203×10^8	
K (cm^4)	9.9898×10^6	8.1128×10^6
I_ω (cm^6)	8.2144×10^{11}	6.9465×10^{11}
R_S (cm)	4.0839×10^3	4.0786×10^3
R_0 (cm)	4.0265×10^3	4.0241×10^3
m [kgf/(cm^2·s^2)]	7.8976×10^{-6}	
L (cm)	3.0021×10^3	

Symmetric modes,

$$\Omega_{i,1} = \frac{\sin k_i(L_1 - s_1)}{\sin k_i L_1} - \frac{\sinh k_i(L_1 - s_1)}{\sinh k_i L_1} \qquad (3.4.109a)$$

$$k_2 L_1 = 3.927 \qquad k_4 L_1 = 7.069 \qquad \ldots \qquad i = 2, 4, 6, \ldots \qquad (3.4.109b)$$

In this case, the values of $\nabla_{i,r}$ and $\Delta_{i,r}$ can be found from Eq. (3.4.50):

$$\nabla_{i,r} = \mu_{i,r} \qquad (3.4.110a)$$

$$\Delta_{i,r} = \nu_{i,r} \qquad i = 2, 4, 6, \ldots; r = 1, 2 \qquad (3.4.110b)$$

Table 3.10 lists these values.

c. **Static behavior.** The stiffnesses $K_{B,i,ww}$, $K_{B,i,w\beta}$, and $K_{B,i,\beta\beta}$ can be estimated from Eqs. (3.4.45), which give the values shown in Table 3.11. To ensure the validity of these values, static load tests have been carried out. Figure 3.59 shows the loading points A, B, and C where deflections at viewpoints a, b, and c, respectively, were measured.

TABLE 3.10 Vibration Modes Ω_{i,j_1}

i	Ω_{i,j_1}	$\sum_{r=1}^{2} \nabla_{i,r}$	$\sum_{r=1}^{2} \Delta_{i,r}$
1	1.0000	1.0000	1.0000
2	−1.4420	1.9984	1.4922
3*	0	—	—
4	−0.5705	1.9938	1.7119
5	−1.0000	1.0000	1.0000

*Third mode is meaningless for measured point.

TABLE 3.11 Stiffness K_B

i	$K_{B,i,ww}$ (kgf/cm)	$K_{B,i,w\beta}$ (kgf·cm/cm)	$K_{B,i,\beta\beta}$ (kgf/cm)
1	9.0629×10^4	2.6487×10^7	3.0933×10^{10}
2	2.2571×10^5	3.1162×10^7	4.0334×10^{10}
4	1.4058×10^7	8.7267×10^7	1.4250×10^{12}
5	7.2486×10^6	3.3096×10^8	6.2419×10^{11}

For these loading conditions, the theoretical deflection $w_{st,i}$ and rotation angle $\beta_{st,i}$ due to a concentrated load $P_o = 1$ kgf are shown in Table 3.12. Note that the values of $w_{st,i}$ and $\beta_{st,i}$ converge rapidly with an increase in the mode number i. For $i = 1$ to 5, the final results of $w_{st} = \Sigma_{i=1}^{5} \omega_{st,i}$ and $\beta_{st} = \Sigma_{i=1}^{5} \beta_{st,i}$ are summarized in the last row of Table 3.12. Thus the theoretical influence surface for the deflection can be plotted by the solid lines in Fig. 3.60.

In addition, static load tests have been undertaken by applying a truck weighing 19.9 tf to load points A, B, and C. The measured deflections at viewpoints a, b, and c are shown by the dotted lines in Fig. 3.60.

These results indicate good agreement between theoretical values and measured ones. Therefore, we conclude that the method for evaluating the stiffnesses $K_{B,i,ww}$, $K_{B,i,w\beta}$, and $K_{B,i,\beta\beta}$ as well as the vibration modes $\Omega_{i,r}$ together with $\nabla_{i,r}$ and $\Delta_{i,r}$ will be accurate.

***d.* Free vibration.** Table 3.13 shows the calculated values $p^2_{i,ww}$, $p^2_{i,\beta w}$, $p^2_{i,w\beta}$, and $p^2_{i,\beta\beta}$ obtained from Eq. (3.4.56). The natural frequencies $f_{i,\mathrm{I}}$ and $f_{i,\mathrm{II}}$ are estimated, as shown in Table 3.14.

The amplitudes $\bar{w}_i$ and $\bar{\beta}_i$ of free vibration can be obtained by substituting p_i^2 back into Eq. (3.4.57). Accordingly, the patterns of the free vibration modes are illustrated in Fig. 3.61, where the skeleton of the bridge is sketched in a straight line for the sake of clarity. We see from this figure that the first, second, and fourth modes are predominant in

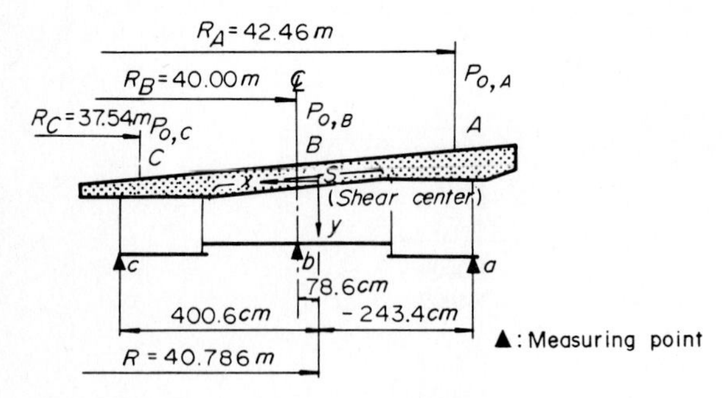

Figure 3.59 Load and measuring points.

TABLE 3.12 Static Displacements w_{st} and β_{st}

	Load point A		Load point C	
i	$w_{st,i,j1}$ ($\times 10^{-5}$ cm)	$\beta_{st,i,j1}$ ($\times 10^{-8}$ rad)	$w_{st,i,j1}$ ($\times 10^{-5}$ cm)	$\beta_{st,i,j1}$ ($\times 10^{-9}$ rad)
1	1.7517	−2.1102	0.9781	1.2835
2	0.5259	−0.7965	0.3085	4.3052
4	0.0081	−0.1269	0.0052	0.2057
5	0.0159	−0.0364	0.0108	0.4215
$\sum_{i=1}^{5}$	2.3016×10^{-5}	-3.0700×10^{-8}	1.3026×10^{-5}	6.2159×10^{-9}

TABLE 3.13 Values of p_i^2

i	$p^2_{i,ww}$	$p^2_{i,\beta w}$	$p^2_{i,w\beta}$	$p^2_{i,\beta\beta}$
1	3.2816×10^2	2.1539×10^3	2.4455×10^3	3.8029×10^3
2	8.0735×10^2	4.6528×10^3	5.2542×10^3	7.6008×10^3

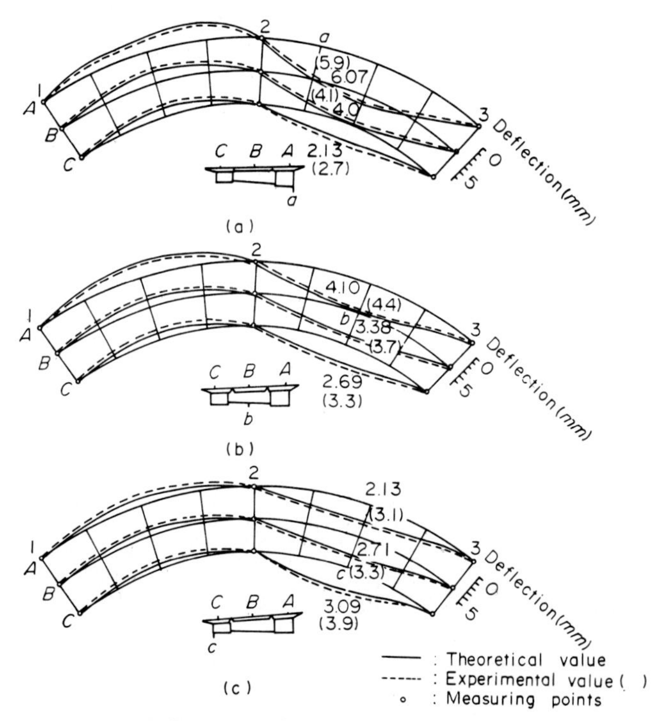

Figure 3.60 Influence surfaces of deflection at (*a*) point *a*, (*b*) point *b*, and (*c*) point *c*.

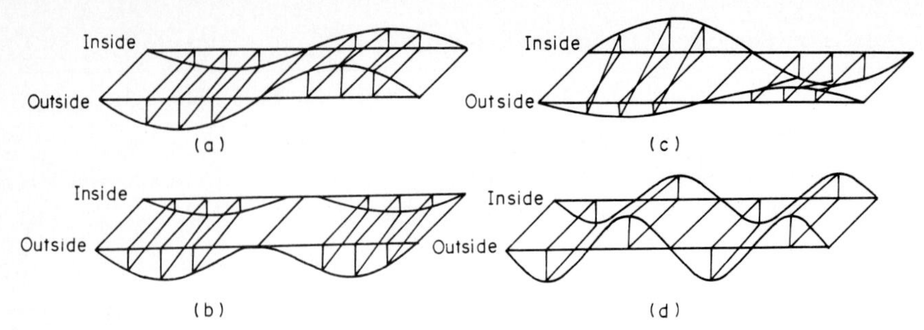

Figure 3.61 Free vibration modes: (*a*) first mode, (*b*) second mode, (*c*) third mode, (*d*) fourth mode.

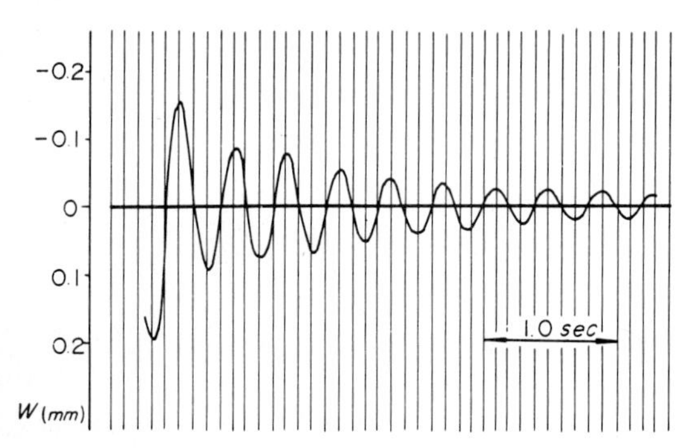

Figure 3.62 Free vibration record.

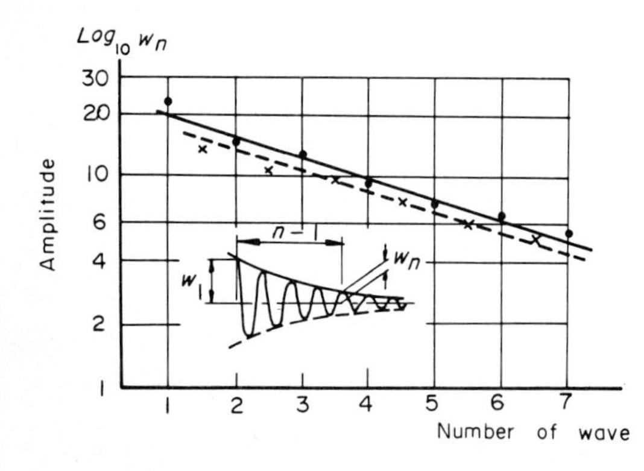

Figure 3.63 Examination of damping effect.

TABLE 3.14 Natural Frequency f_i (Hz)

i	$f_{i,\mathrm{I}}$	$f_{i,\mathrm{II}}$
1	2.655	9.492
2	4.195	12.735

TABLE 3.15 Damping Coefficient d_i

i	$d_{i,w}$	$d_{i,\beta}$
1	1.227	4.385
2	1.933	5.884

vertical bending vibration, and the second mode is predominant in torsional vibration.

Free vibration tests have been carried out by recording the deflections w after a test truck has passed over the bridge. Typical results are plotted in Fig. 3.62.

From this record, the measured frequency is equal to 2.60 Hz, and this corresponds to the calculated lowest frequency $f_{1,\mathrm{I}} = 2.655$ Hz in Table 3.14. The ratio between $(f_{1,\mathrm{I}})_{\mathrm{cal}}$ and $(f_{1,\mathrm{I}})_{\mathrm{test}}$ is 1.02, indicating quite good agreement between the values. Accordingly, the validity of the evaluation of the p_i^2 values has been confirmed by these tests.

Let us now investigate the damping effects of this curved girder bridge. Figure 3.63 show the envelope curves of the free vibration record of Fig. 3.62 from which the damping effects will be considered as viscous damping, because the amplitudes of free vibration decrease linearly. The corresponding logarithmic damping coefficient can be evaluated by

$$D_{i,\mathrm{I}} = \frac{1}{n-1} \ln 10 \log \frac{w_i}{w_n} \qquad (3.4.111)$$

Although the damping coefficient for only the lowest mode of vibration can be measured by the experiments, the remaining damping coefficients other than $i = 1$, $d_{i,w}$, and $d_{i,\beta}$ can be deduced by assuming $D_{i,\mathrm{I}} = D_{i,\mathrm{II}}$, as in Table 3.15.

e. Response due to the vehicle

(i) Dynamic properties of the vehicle. For a test vehicle, a truck weighing 7.2 tf was loaded with steel plates weighing 12.7 tf. The total weight was thus 19.9 tf (the weights of the front and rear wheels were 3.7 and 16.2 tf, respectively) with a wheel distance of 4.82 m.

The stiffness of the rear-wheel springs was measured by dial gages during the loading of the steel plates and was

$$K_V = 1.49 \times 10^3 \text{ kgf/cm} \qquad (3.4.112a)$$

Next, the acceleration of the sprung mass of the vehicle at the center of gravity of the vehicle was measured by driving the vehicle across the curved girder bridge. Typical data are plotted in Fig. 3.64. From these data, the natural frequency of the vehicle is found to be

$$f_v = 2.15 \text{ Hz} \qquad (3.4.112b)$$

and the damping coefficient is

$$D_V = 0.2 \qquad (3.4.112c)$$

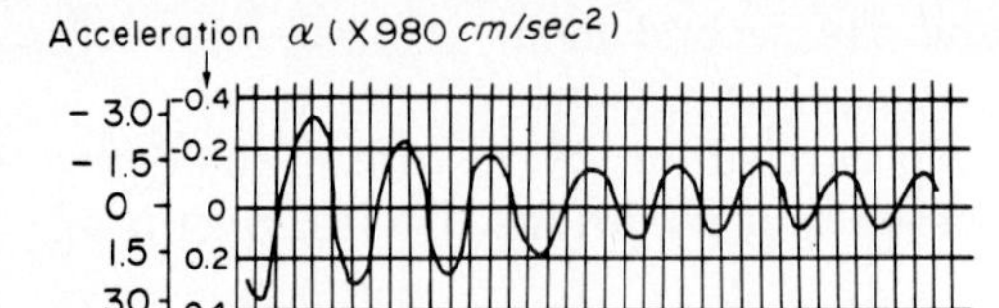

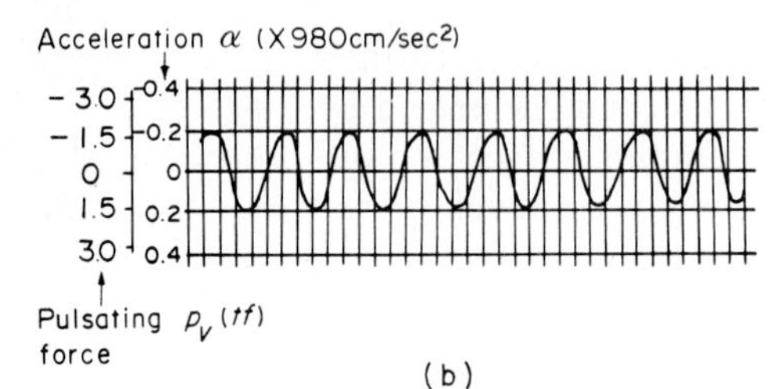

Figure 3.64 Vibration records of vehicle: (*a*) damping type and (*b*) harmonic type.

Combining these data, we can estimate the sprung mass of the vehicle $M_{V,u}$ as

$$M_{V,u} = 8.164 \text{ kgf/(cm·s}^2) \tag{3.4.112d}$$

It follows that $M_{V,u}/M_V = 0.4$, namely, the sprung mass of the vehicle is 40 percent of the total mass M_V of the vehicle.

(ii) Forcing terms. The exact response of the curved girder bridge could be obtained by considering all the vibration modes $i = 1, 2, 3, \ldots$. However, the following approximations were made because of the limitations of the elements in the analog computer.[57,58]

First, it is clear that the influence of vibration modes higher than the second mode is negligible when the fundamental frequency of the bridge is the same order of magnitude as the natural frequency of the test vehicle. In this case, the response of the curved girder bridge due to the test vehicle can be approximately determined by taking only the first mode of vibration.

However, for a continuous curved girder bridge, the static deflections do not coincide with the predicted values when the vibration modes higher than the second mode are ignored. So the forcing terms q_{V,i,j_1} and τ_{V,i,j_1} can be approximated by Eq. (3.4.88) by considering $w_{st,1,j_1}$ and $\beta_{st,1,j_1}$ equal to the sum of $\Sigma_{i=1}^{5} w_{st,1,j_1}$ and $\Sigma_{i=1}^{5} \beta_{st,1,j_1}$ in Table 3.12. These results are summarized in Table 3.16.

(iii) Initial conditions. There are four initial conditions—$w(0)$, $\dot{w}(0)$, $\beta(0)$, and $\dot{\beta}(0)$—for the curved girder bridge. They are assumed to be initially at rest for $t = 0$, or

$$w_{1,j_1}(0) = \dot{w}_{1,j_1}(0) = 0 \tag{3.4.113a}$$

$$\beta_{1,j_1}(0) = \dot{\beta}_{1,j_1}(0) = 0 \tag{3.4.113b}$$

TABLE 3.16 Forcing Terms due to Test Vehicle

Load line	q_{V,i,j_1}	τ_{V,i,j_1}
A	1.4229×10^2	-1.0832
C	5.3123×10^1	1.2071

But two initial conditions $y_{V,1}(0)$ and $\dot{y}_{V,1}(0)$ actually exist for the vehicle. An initial condition of vehicle deflection $y_{V,1}(0) = \delta_1(0)$ due to the shock at the expansion joint of the curved girder bridge is important in estimating the dynamic response of the curved girder bridge. The deflection is determined through the use of an accelometer attached to the vehicle to measure $\alpha(0)$. The deflection $\delta_1(0)$ is

$$\delta_1(0) = \alpha_1(0)\frac{M_{V,u}}{K_V} \tag{3.4.114}$$

For example, setting $\alpha_1(0) = 0.3g$ ($g = 980$ cm/s^2), as measured and indicated in Fig. 3.64, we obtain $\delta_1(0) = 1.62$ cm. The corresponding pulsating force $P_v(t) = K_V\delta_1$ is shown in Fig. 3.64.

The remaining initial condition $\dot{y}_{V,1}(0)$ is a parameter that is governed by the absolute displacement $y_{V,1}(t)$ of the sprung mass $M_{V,u}$ of the vehicle and the phase angle of the displacement $\delta_1(t)$ for the spring K_V of the vehicle. So $y_{V,1}(0)$ falls in the following range:

$$-p_u y_{V,1}(0) \leqq \dot{y}_{V,1}(0) \leqq p_v y_{V,1}(0) \tag{3.4.115}$$

Figure 3.65 illustrates the results w_{dy} of simulations obtained from an analog computer.[57,58]

(iv) Comparisons with experimental results. The dynamic load test of a vehicle on the curved girder bridge has been carried out by driving a vehicle from span 1 to span 2 of the bridge (see Fig. 3.58). The driving positions of the vehicle were chosen as the same lines A, B, and C in Fig. 3.59 with speeds of the vehicle V_o equal to 10, 20, and 30 km/h, respectively. The deflection of the curved girder bridge was measured by vibrometers attached at points a and c, shown in Fig. 3.59 at the outer and inner main girders, respectively. At the same time, the acceleration of the vehicle was recorded by oscillographs. Typical data plotted are the dimensionless values w_{dy}/w_{st} and s_1/L_1 versus the static deflection w_{st} and span length L_1, as shown in Fig. 3.65 (the measured values on the second span are omitted for the sake of simplicity).

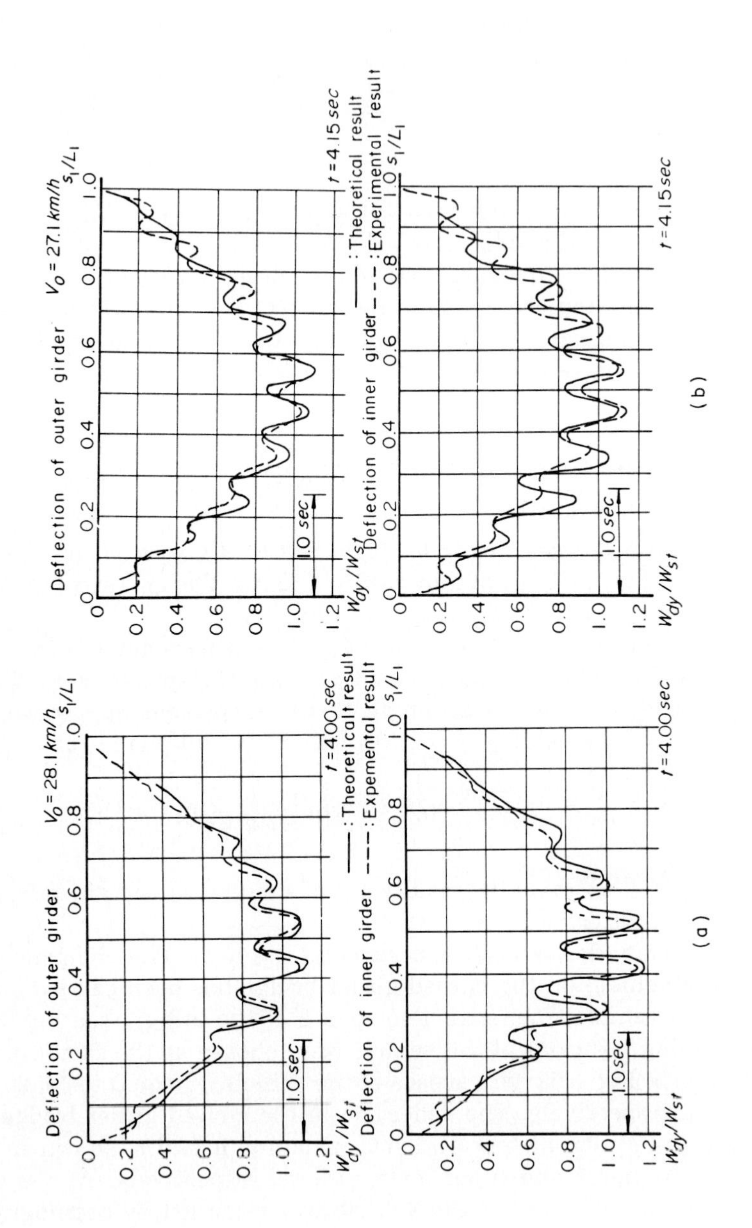
Deflection of outer girder
V_o = 28.1 km/h
s_1/L_1
t = 4.00 sec
W_dy/W_st
1.0 sec
——— : Theoretical result
– – – : Expemental result
Deflection of inner girder
t = 4.00 sec
(a)
Deflection of outer girder
V_o = 27.1 km/h
t = 4.15 sec
——— : Theoretical result
– – – : Experimental result
Deflection of inner girder
t = 4.15 sec
(b)

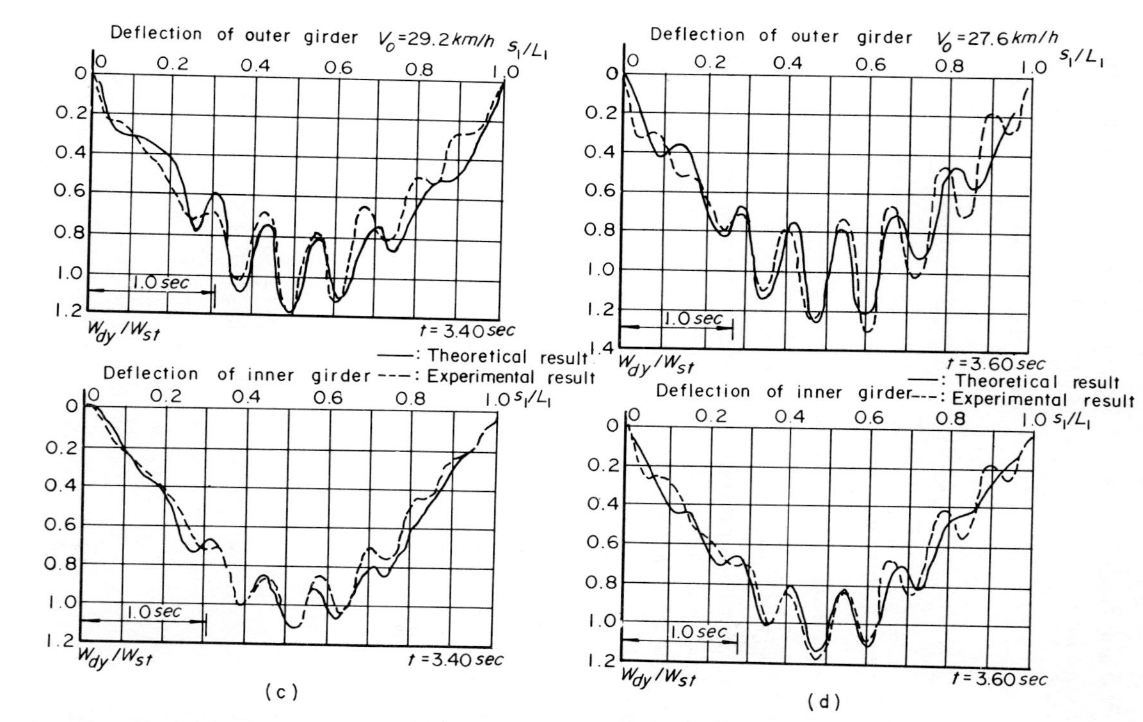

Figure 3.65 Response due to moving vehicle. Moving across the outer girder: (*a*) damping type and (*b*) harmonic type. Moving across the inner girder: (*c*) damping type and (*d*) harmonic type.

3.5 Dynamic Behavior of Curved Girders

3.5.1 Natural frequencies and damping coefficients of curved girder bridges

(1) Natural frequencies.[62] Let us consider the natural frequencies of simply supported curved girder bridges having

$$L = R_s \Phi \qquad \text{span length} \tag{3.5.1a}$$

$$\Phi = \frac{L}{R_s} \qquad \text{central angle} \tag{3.5.1b}$$

with vibration modes

$$\Omega_i = \sin \frac{i\pi s}{L} \qquad i = 1, 2, \ldots \tag{3.5.2}$$

The two natural circular frequencies $p_{\mathrm{I}} < p_{\mathrm{II}}$ rad/s of these curved girder bridges can be estimated from Eqs. (3.4.56) through (3.4.59) by taking the first mode of vibration $i = 1$ and considering Eq. (3.5.1) as follows:

$$p_{\mathrm{I,II}} = \sqrt{b \mp \sqrt{\frac{b^2 - 4ac}{2a}}} \tag{3.5.3}$$

where $a = A_y I_S$ (3.5.4a)

$$b = A_y I_s (p_{ww}^2 + p_{\beta\beta}^2) - S_y S_y' (p_{\beta w}^2 + p_{w\beta}^2) \tag{3.5.4b}$$

$$c = A_y I_s p_{ww}^2 p_{\beta\beta}^2 - S_y S_y' p_{w\beta}^2 \tag{3.5.4c}$$

and

$$p_{ww}^2 = E_s I_x' \frac{\pi^2(\pi^2 - \Phi^2)}{m A_y L^4} \tag{3.5.5a}$$

$$p_{\beta w}^2 = E_s I_x' \frac{\Phi(\pi^2 - \Phi^2)}{m S_y' L^3} \tag{3.5.5b}$$

$$p_{w\beta}^2 = E_s I_x' \frac{\pi^2 \Phi(\gamma + 1)}{m S_y L^3} \tag{3.5.5c}$$

$$p_{\beta\beta}^2 = E_s I_x' \frac{\gamma\pi^2 + \Phi^2}{m I_s L^2} \tag{3.5.5d}$$

in which γ is the equivalent torsional-flexural rigidity ratio designated by [see Eq. (3.3.69)]

$$\gamma = \frac{G_s K + E_s I_\omega (\pi / L)^2}{E_s I'_x} \tag{3.5.6}$$

To determine these parameters and classify bridge types, numerous parametric surveys were conducted by studying the design data of 21 actual curved girder bridges with a span length L of 20 to 50 m and radius of curvature R_s of 30 to 90 m, which have already been constructed in Japan.[65]

Table 3.17 shows the geometry and cross-sectional quantities of these bridges, where the cross-sectional shape has been categorized according to the types illustrated in Fig. 3.32.

From Eqs. (3.5.3) to (3.5.5) and Table 3.17, the influence of γ and Φ on the lowest natural frequency

$$f_B = \frac{p_\mathrm{I}}{2\pi} \qquad \mathrm{Hz} \tag{3.5.7}$$

of the curved girder bridges is plotted in Fig. 3.66, where f_B is dimensionless with respect to the lowest natural frequency $f_\Phi = 0$ with central angle $\Phi = 0$.

Figure 3.66 is similar in form and consequence to the static deflections shown in Fig. 3.49. In the regions where the error is less than 2.5 percent, the dynamic behavior of curved girder bridges may be analyzed according to the following criteria:

$$\Phi \geqq 0.2 \qquad \text{when } \gamma \leqq 0.1 \tag{3.5.8a}$$

$$\Phi \geqq 0.2 + 0.5(\gamma - 0.1) \qquad \text{when } 0.1 < \gamma < 0.5 \tag{3.5.8b}$$

$$\Phi \geqq 0.4 \qquad \text{when } \gamma \geqq 0.5 \tag{3.5.8c}$$

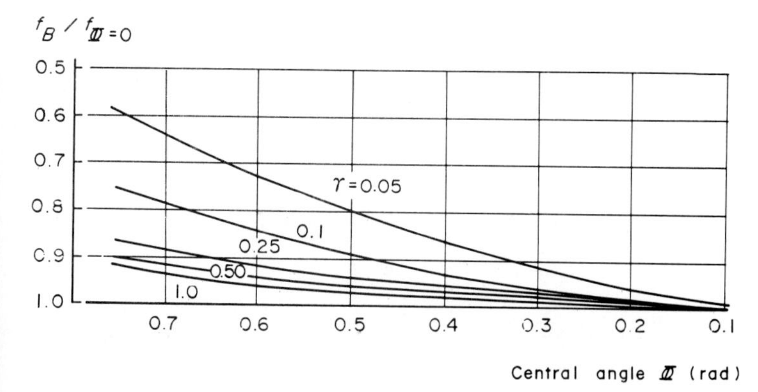

Figure 3.66 Variations of lowest natural frequency due to parameters γ and Φ.

TABLE 3.17 **Cross-Sectional Quantities of Curved Girder Bridges**[65]

Bridge no.	Span L, m	Radius of curvature R, m	Central angle Φ, rad	Width of roadway B, m	Mass m, 10^{-5} kgf·s^3/cm^4	Cross-sectional area A_s, $\times 10^3$ cm^2	Static moment S_y, $\times 10^5$ cm^3	Moment of inertia I'_x, $\times 10^6$ cm^6	Polar moment of inertia I_s, $\times 10^4$ cm^4	Reduced torsional stiffness γ	Cross-sectional shape
1	19.5	57.0	0.342	6.5	2.013	2.667	0.774	8.210	1.321	0.114	
2*	20.0	50.0	0.400	9.2	2.473	3.478	4.452	10.570	2.086	0.156	
3	25.0	51.0	0.490	6.5	2.733	2.076	1.714	8.181	1.705	0.104	
4	27.0	31.0	0.871	7.5	2.138	3.070	1.909	14.219	2.248	0.088	Fig. 3.32*a*
5	30.0	60.0	0.500	7.0	2.070	3.117	0.505	20.436	1.749	0.055	
6	33.4	90.0	0.371	8.2	2.027	3.692	0.872	22.830	1.976	0.059	
7	40.0	74.0	0.528	8.5	2.240	3.154	0.700	20.443	2.818	0.050	
8	22.0	61.0	0.361	9.7	2.659	3.352	1.615	6.416	2.698	0.470	
9	25.0	40.7	0.614	7.8	2.427	3.098	0.748	8.436	2.050	0.377	
10	25.0	40.9	0.611	8.5	2.591	3.441	3.264	7.442	2.073	0.304	
11*	30.0	40.0	0.750	7.8	2.427	3.098	5.353	13.830	4.220	0.344	
12	40.0	64.2	0.623	8.0	1.846	4.584	0.280	33.586	3.144	0.262	Fig. 3.32*b*
13	45.0	80.0	0.563	12.7	2.038	4.642	1.450	39.124	6.703	0.271	
14	45.0	80.5	0.559	12.7	2.233	5.581	2.860	41.815	8.918	0.328	
15	45.0	85.5	0.528	13.1	2.294	5.583	1.615	46.595	9.246	0.304	
16	50.0	88.7	0.564	12.6	2.273	5.434	2.763	44.305	8.839	0.341	
17	50.0	84.9	0.589	13.1	2.810	5.617	3.014	50.162	9.348	0.322	
18	22.0	60.0	0.340	7.9	2.302	3.406	0.725	7.067	2.142	0.890	
19	30.0	79.0	0.380	7.9	2.302	3.406	0.728	11.064	2.302	0.890	
20*	32.0	50.0	0.640	9.2	2.291	3.382	1.407	15.430	1.147	0.686	Fig. 3.32*c*
21	37.4	54.2	0.690	8.2	1.872	4.472	0.653	17.355	2.102	0.785	

*Complete dynamic analyses have been made of these bridges.

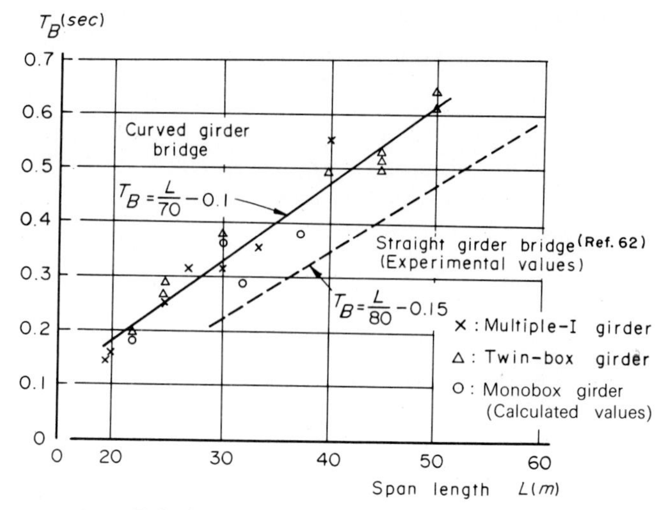

Figure 3.67 Relationships between T_B and L.

Next the relationships between the period of the natural frequency T_B

$$T_B = 1/f_B \qquad \text{s} \tag{3.5.9}$$

and the span length L (m) are illustrated in Fig. 3.67. In this figure, the experimental results from 21 straight girder bridges are plotted on a dotted line on the basis of the least-squares method. The period of calculated natural frequency T_B of curved girder bridges can be found from

$$T_B = \frac{L}{70} - 0.1 \qquad 20\ \text{m} < L < 50\ \text{m} \tag{3.5.10}$$

For example, the periods of natural frequency are $T_B = 0.47$ s and $T_B = 0.35$ s for curved and straight girder bridges with span $L = 40$ m, respectively, and T_B of the curved girder bridge with span length $L = 40$ m is almost equivalent to that of a straight girder bridge with span length $L = 50$ m. These differences are thought to be caused mainly by the influence of the curvature of the bridge axis.

(2) Damping coefficients.[62,63] The relationship between the logarithmic damping coefficient D_I and span length L of curved girder bridges is plotted in Fig. 3.68 on the basis of experimental studies. In this figure, the logarithmic damping coefficients of the straight girder bridges are shown to supply the lack of data for the curved girder bridges. Although it is, of course, desirable to collect lots more data,

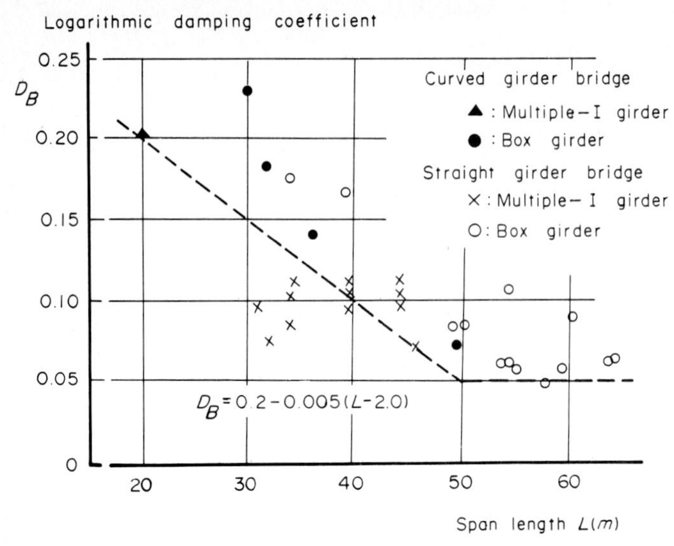

Figure 3.68 Relationship between D_B (measured values) and L.

the logarithmic damping coefficients D_B of the curved girder bridges may be assumed conservatively to be

$$D_B = D_{\mathrm{I}} = D_{\mathrm{II}} = 0.2 - 0.005(L - 20) \qquad 20\ \mathrm{m} < L < 50\ \mathrm{m} \qquad (3.5.11)$$

3.5.2 Dynamic response due to a single vehicle

(1) Dynamic properties of vehicles. The vehicle moving across a curved girder bridge will first undergo a shock at the expansion joint of the bridge, and then the vehicle will oscillate at various frequencies owing to the roughness of the roadway surface. Since these random vibrations should be taken into account, we assumed in this deterministic simulation method that the vehicle vibrates nearly equal to its own natural frequency $f_V = p_v/(2\pi)$(Hz) with a vertical acceleration αg (cm/s^2).

We see from Harris and Grede[64] that the relationships between f_V and α fall within the ranges shown in Fig. 3.69 when the vehicle is moving across the highway. These facts are supported by experimental results, which are plotted in this figure.

Therefore, two parameters f_V and α for a vehicle can be classified in the following regions

$$f_V = 2\text{–}6\ \mathrm{Hz} \qquad (3.5.12a)$$

$$\alpha = 0.1\text{–}0.4\ \mathrm{g} \qquad (3.5.12b)$$

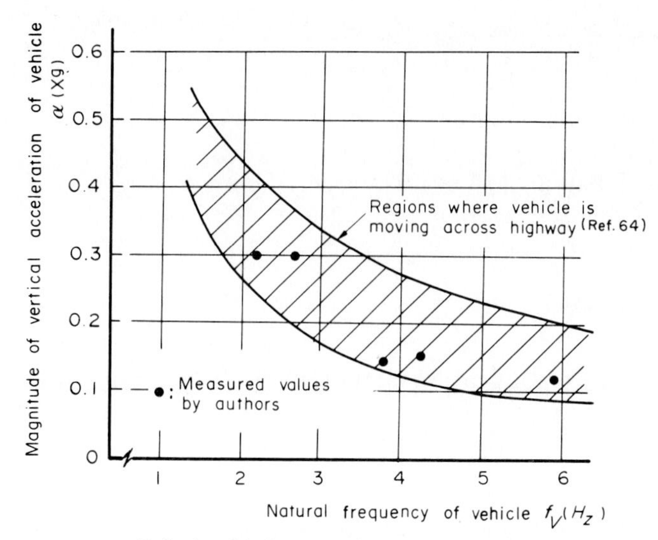

Figure 3.69 Relationship between f_V and α.

From this equation, the initial displacement of a vehicle reduces to

$$y_V(0) = \frac{M_{V,u}}{K_V}\alpha = \frac{\alpha}{(2\pi f_v)^2} \tag{3.5.13}$$

and the corresponding dynamic force can be represented by

$$F = M_{V,u}\alpha = K_V y_V(0) \tag{3.5.14}$$

With regard to the other parameters for the vehicles, the speed V_o, mass ratio $M_{V,u}/M_V$, and logarithmic damping coefficient D_V remain, and these values will vary with the type of vehicle. However, the running speed V_o on the curved girder bridges with a sharp radius of curvature is usually limited to less than 40 km/h, and our experimental results show that $M_{V,u}/M_V \geqq 0.5$ and $D_V = 0.2$ for heavy trucks, so that simulations can be conducted within these ranges.

(2) Definition of the dynamic amplification factor.[57] The analytical method for determining the dynamic response of a curved girder bridge under moving vehicles on a digital computer is explained in detail in the next section. Now we denote the maximum dynamic deflection by $w_{d,\max}$ and the corresponding static one by $w_{st,\max}$, as shown in Fig. 3.70, so that

$$w_{st,\max} = w_{st,\max} + x_p \beta_{st,\max} \tag{3.5.15}$$

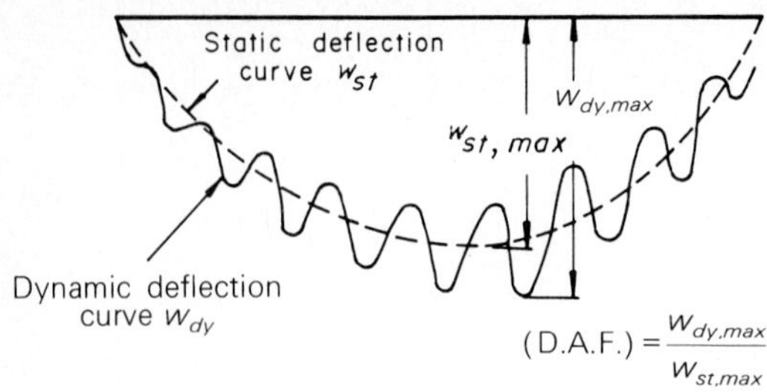

Figure 3.70 Definition of DAF.

Therefore, the dynamic amplification factor (DAF) can be evaluated by

$$\mathrm{DAF} = \frac{w_{dy,\max}}{w_{st,\max}} \tag{3.5.16}$$

(3) Variations in the DAF due to a single vehicle.[65] To investigate how the values of the DAF for curved girder bridges are altered by various parameters mentioned in the previous section, the dynamic response of curved girder bridges due to a single vehicle is computed by the deterministic simulation method described in Sec. 3.4.4(3).

For these analyses, the adopted parameters are summarized in Table 3.18, where the reference values correspond to the values of curved girder bridges numbers 2, 11, and 20, as shown in Table 3.17.

TABLE 3.18 Ranges of Parameters

Parameter	Range	Reference value	Remarks
(1) Frequency ratio f_B/f_V	0.3–3.0	1.0	See Eqs. (3.5.10) and (3.5.12*a*) and Figs. 3.67 and 3.68.
(2) Rigidity parameter γ	0.05–1.0	0.3	See Eqs. (3.5.6) and (3.5.8) and Fig. 3.66.
(3) Moving position x_p (m)	± 3.0	0	
(4) Central angle Φ	0–1.0	0.75	Reference values are $L = 30.0$ m and $R_s = 40.0$ m.
(5) Mass ratio $\zeta = M_{V,u}/M_B$	0.1–1.0	0.2	Assume $p_o = 20$ tf
(6) Damping coefficient D_B	0–0.2	0.1	See Eq. (3.5.11) and Fig. 3.68.
(7) Moving speed V_o (km/h)	10–50	40	
(8) Vehicle initial condition α	0–0.5	0.3	See Eq. (3.5.12*b*) and Fig. 3.69.
(9) Stiffness of vehicle K_V (tf/m)	0.5–5.0	2.5	
(10) Damping coefficient D_V	0.2		Deduced from test results
(11) Mass ratio $M_{V,u}/M_V$	Decide from (1), (8), and (9)		

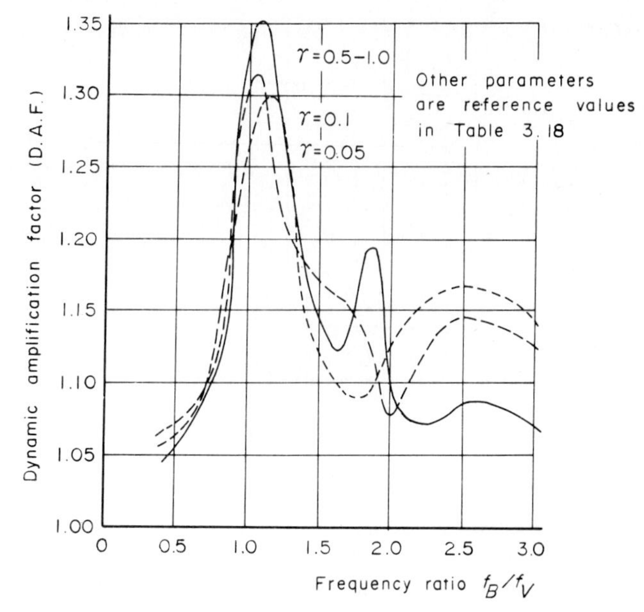

Figure 3.71 Variations of DAF due to f_B/f_V and γ.

a. **Effects of the frequency ratio f_B/f_V.** The variations in the DAF due to the frequency ratio f_B/f_V and γ can be plotted in Fig. 3.71. We see in this figure that the effect of f_B/f_V is a remarkable one, since the DAF does not change greatly over a wide range of γ, as can be seen from Fig. 3.71. Note that the values of the DAF always peak near $f_B/f_V = 1$. Strictly speaking, the maximum DAF does not coincide with $f_B = f_V$, because the natural frequency of the girder bridge is somewhat decreased by the effect of the unsprung mass of the vehicle. However, the maximum DAF can be obtained at $f_B/f_V \cong 1$ for practical purposes.

b. **Effects of γ.** Figure 3.72 shows the variations of DAF with γ by setting f_B/f_V equal to 0.5, 1, and 2.

In general, the DAF becomes large as γ increases for $f_B/f_V = 1.0$. However, changes in the DAF are almost constant for f_B/f_V equal to 0.5 and 2.0.

Accordingly, these phenomena should be considered in determining a rational impact coefficient for curved girder bridges.

c. **Effects of moving the position R_p.** In the previous analyses, a vehicle is assumed to drive along the shear center axis of the curved girder bridge for the sake of convenience. However, let us now investigate the influence of the torsional vibration induced by a vehicle moving along with an eccentricity from the shear center axis. For these analyses, curved girder bridges 1 and 11 were selected as representative of

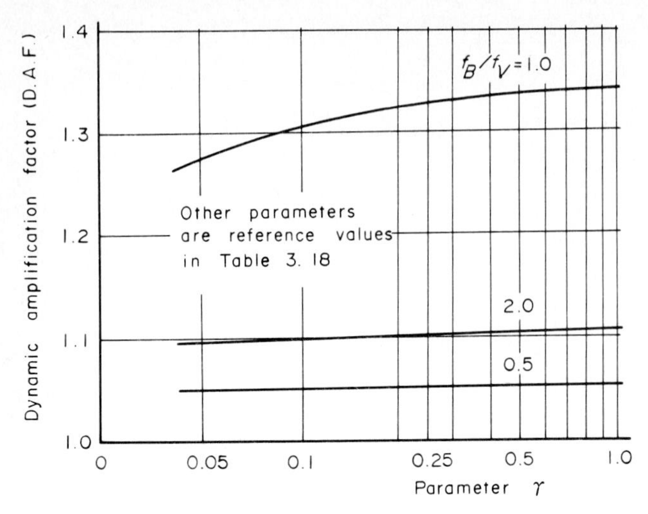

Figure 3.72 Variations of DAF due to γ and f_B/f_V.

curved girder bridges, and their cross-sectional quantities are used (see Table 3.18).

Figures 3.73 and 3.74 illustrate the variation in the midspan static and dynamic deflection curves along the bridge axis as well as the maximum deflections in the cross section at midspan. Clearly we need to take into account the effect of the vehicle changing positions, because this greatly affects the torsional vibration. The design live load shall be imposed so as to make the stress resultants and displacements maximum at the point of interest. Curved girder bridges are analyzed according to these load conditions in Chap. 4, and an impact coefficient based on these load conditions is developed.

To investigate the effects of the changing positions of the vehicle as simply as possible, we focus on the values of the DAF of the outer (inner) girder when a vehicle moves along the outer (inner) girder.

Observing the static and dynamic deflection curves in the cross section of Figs. 3.73(*b*) and 3.74(*b*) shows that the values of the DAF of the inner girder are greater than those of the outer girder. This is caused by the fact that the increments of dynamic deflection $\Delta w = w_{dy} - w_{st}$ are not affected by the changing position of the vehicle. The static deflections w_{st} have entirely different values as the position of the vehicle changes, and those of the outer girder are always greater than those of the inner girder. Thus the resulting DAF ($= \Delta w / w_{st} + 1$) of the outer girder reduces to a smaller DAF than that of the inner girder. This tendency is clearly observed in multiple-I-girder curved bridges.

***d.* Effects of the central angle Φ.** The effect of the radius of curvature of curved girder bridges on the DAF was investigated by altering the

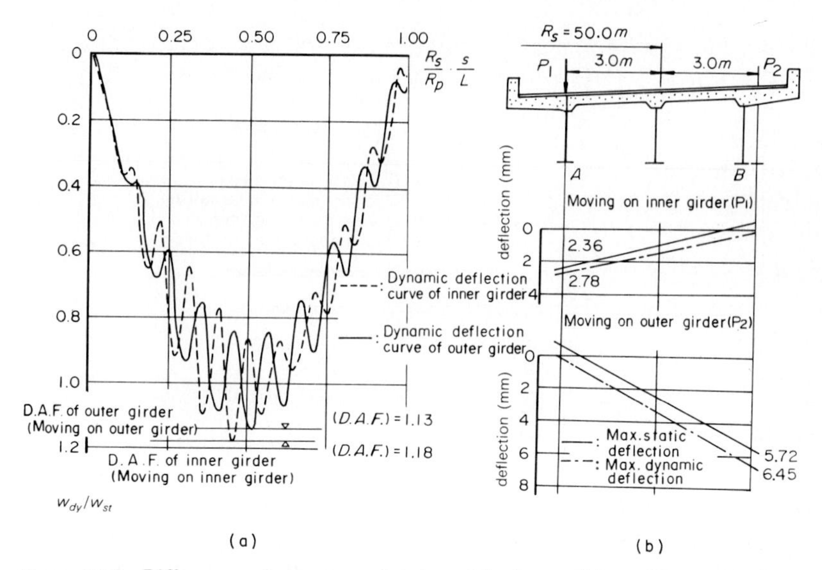

Figure 3.73 Difference of responses due to vehicular position change: (*a*) dynamic deflection curves at midspan and (*b*) maximum static deflection curve and maximum dynamic deflection curve (curved multiple-I-girder bridge).

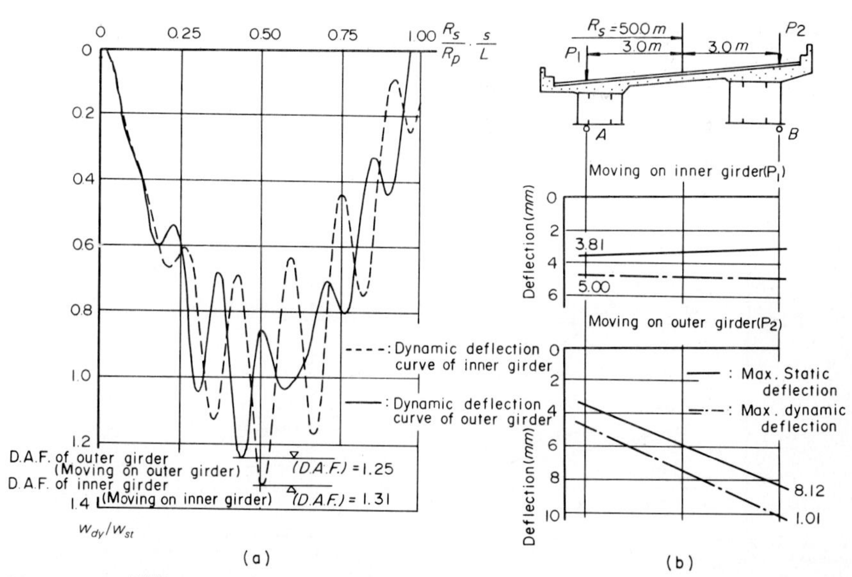

Figure 3.74 Difference of responses due to vehicular position: (*a*) dynamic deflection curves at midspan and (*b*) maximum static deflection curve and maximum dynamic deflection curve change (curved twin-box-girder bridge).

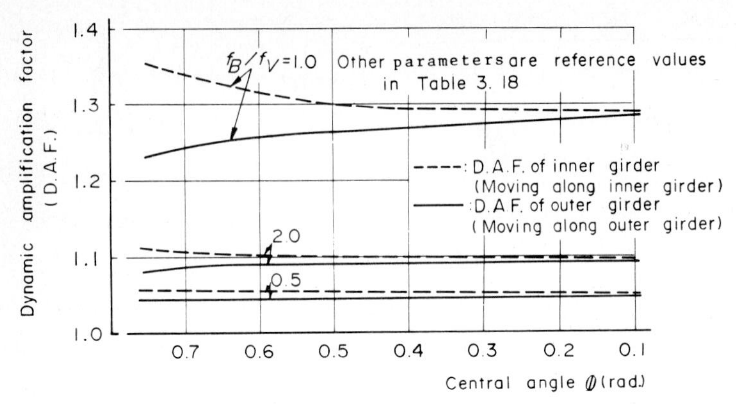

Figure 3.75 Variations of DAF due to Φ.

central angle Φ, as shown in Fig. 3.75. The influence of Φ is greatest in the case when $f_B = f_V$, and the DAF values become larger as the central angle Φ increases. Therefore, this effect should also be taken into consideration in our analyses.

***e.* Effect of the mass ratio $M_{V,u}/M_B$.** The variations in the DAF with the mass ratio $\zeta = M_{V,u}/M$ are very significant (see Fig. 3.76). In general,

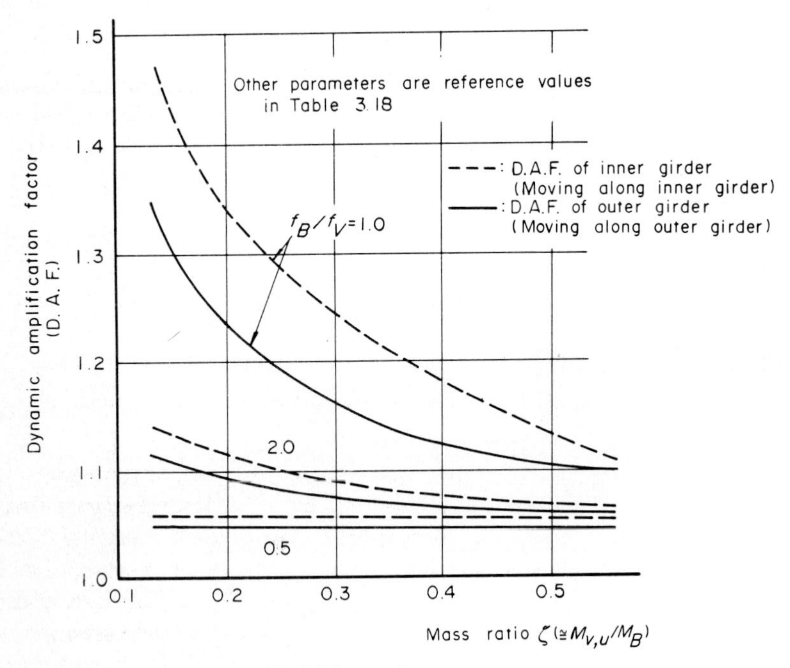

Figure 3.76 Variations of DAF due to ζ.

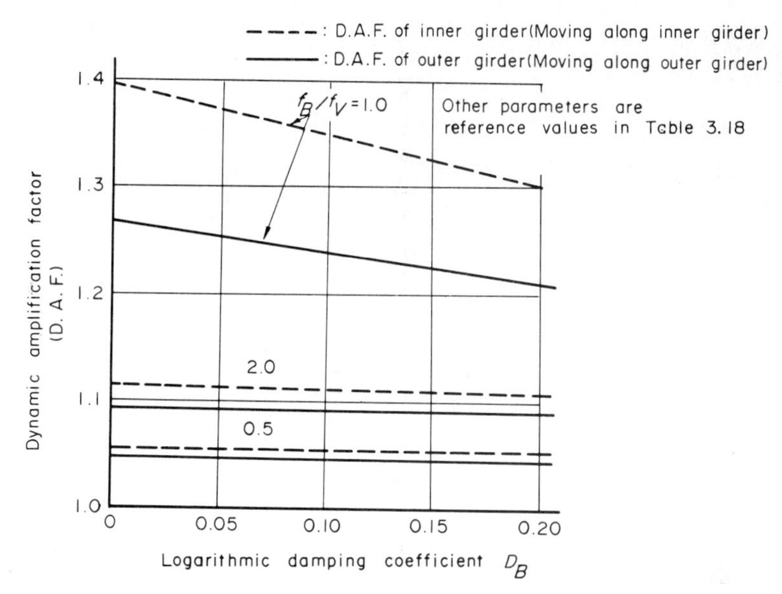

Figure 3.77 Variations of DAF due to D_B.

the DAF values become greater as the mass ratio ζ decreases, and the DAF values decrease when the value of ζ increases.

f. **Effects of the damping coefficient D_B.** Figure 3.77 shows the variation in the logarithmic damping coefficient D_B with the DAF. Small variations due to damping are noticeable when $f_B/f_V = 1$. It is, however, reasonable to approximate the logarithmic damping coefficient D_B with Eq. (3.5.11).

Furthermore, the DAF values are constant when $D_B = 0$ and $f_B/f_V = 1$, because the dynamic response of a curved girder bridge is a transcendental phenomenon due to the forced vibration of the vehicle over a comparatively short time. This is the main reason why the curved girder bridge does not resonate with the vibration of the vehicle, and therefore the values of the DAF are not infinite.

g. **Effects of the moving velocity V_o.** The final results of the DAF due to the variations in moving velocity V_o of a vehicle can be plotted as an envelope of curves by omitting the unevenness of the DAF curves (see Fig. 3.78). We see that the DAF increases gradually with an increase in speed V_o of a vehicle and is almost reduced to a constant value around $V_o = 40$ km/h.

h. **Effects of the vertical acceleration on V_o.** The variations in the DAF due to the vehicle's acceleration α were estimated by the initial condition $y_V(0)$, of Eq. (3.5.13), and these results are plotted in Fig. 3.79. In

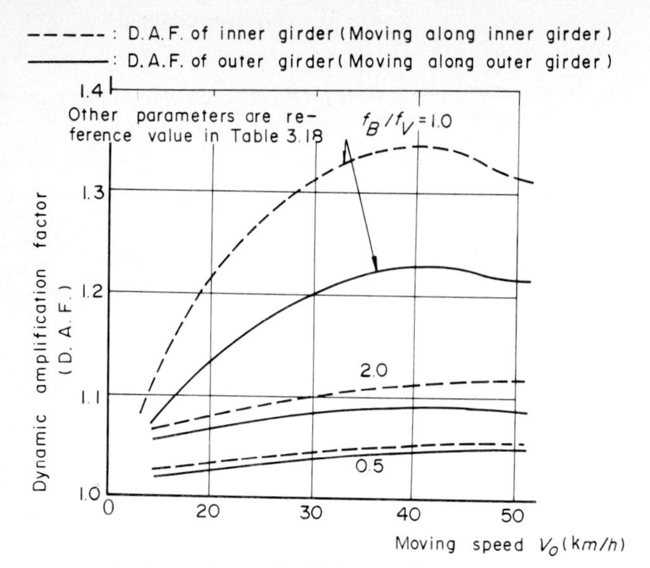

Figure 3.78 Variations of DAF due to V_o.

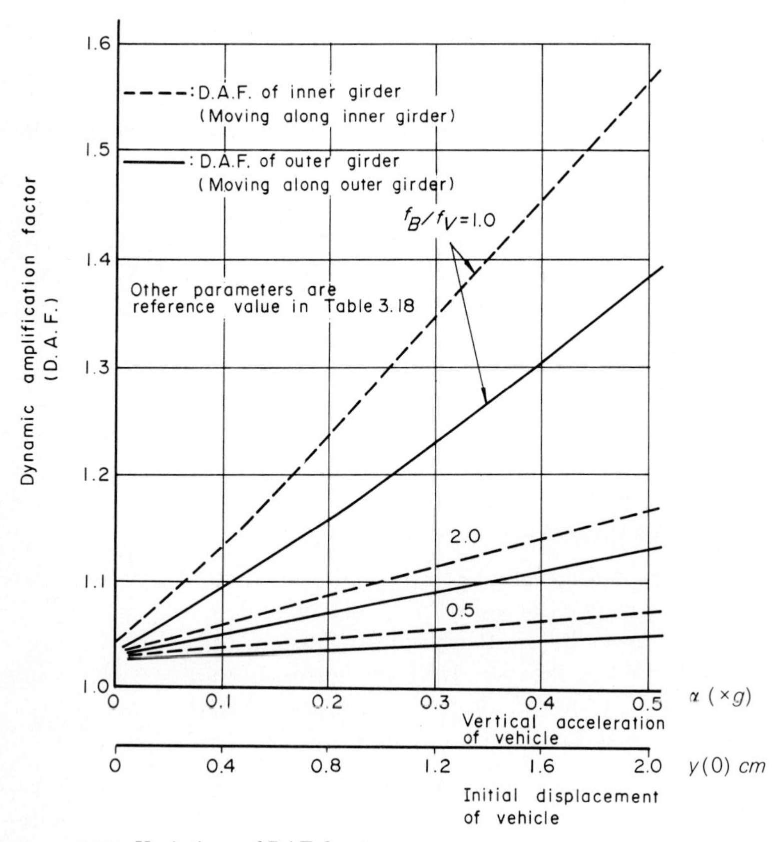

Figure 3.79 Variations of DAF due to α.

general, the DAF seems to have a linear relationship with the magnitude of the vehicle acceleration, since the forcing term is directly proportional to α, as seen in Eq. (3.5.14). However, the DAF can be approximated as 1.05, even in the initial condition where $\dot{y}_V(0) = 0$ (i.e., a vehicle merely moves without vibrations).

Furthermore, we conclude that the remaining conditions, $\dot{y}(0)$, influences only the phase angle of the response of the curved girder bridge. Therefore, this effect is completely ignored in the foregoing analyses.

i. **Effects of the stiffness K_V of the vehicle.** The changes in the DAF with changes in the stiffness K_V of the vehicle are different when the vehicle force is pulsating and when it is constant [see Eq. (3.5.14)]. Therefore, the relationships between the stiffness K_V of the vehicle and the DAF were evaluated by setting $F(0) = 3.0$ tf (tf = tons of force), as shown in Fig. 3.80.

From this figure, we see that the DAF increases when $f_B/f_V = 1$, and the stiffness K_V of the vehicle decreases, although the pulsating force is constant. Since the displacement of a vehicle spring δ can be expressed by the difference of the displacement y_V of the sprung mass of the vehicle and the curved girder bridge w_d, this means that $y_V \gg w_{dy}$ and then $\delta \cong y_V$ for a small value of K_V. In such a situation, the vibration of the vehicle does not couple with that of the curved girder bridge, and

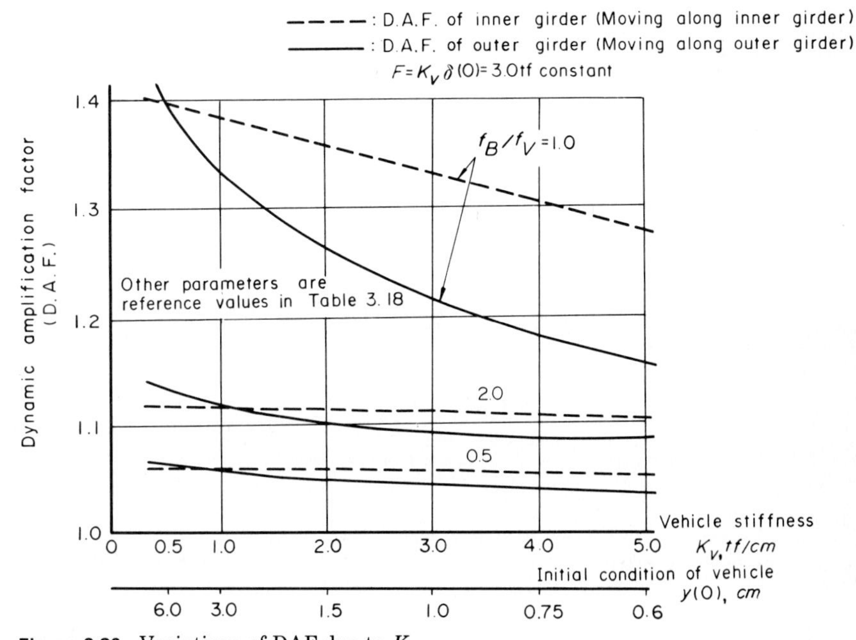

Figure 3.80 Variations of DAF due to K_V.

so the induced response can be analyzed by assuming that the motion of the vehicle induces a sinusoidal force.

j. **Other effects.** The mass ratio $M_{V,u}/M_V$ of a vehicle does not affect the DAF and the damping coefficient D_V of a vehicle has little affect on the DAF; therefore the details of these influences are not explained here.

3.5.3 Dynamic amplification factor due to traffic vehicles

(1) Simulation method of DAF

a. **Modeling of traffic vehicles.**[60,61] A vehicle that moves on the curved girder bridge is idealized as a mechanical vibration system with one degree of freedom on the basis of the analysis in Sec. 3.4.4(3). Therefore, the moving situations of these modified vehicles, as observed in actual traffic flow, can be illustrated as in Fig. 3.81, where

$M_{V,i,k}$, $M_{V,l,k}$ = sprung and unsprung masses, respectively, for kth vehicle

$D_{v,k}$ = damping coefficient for kth vehicle

y_k = absolute displacement of sprung mass $M_{v,k}$ for kth vehicle

δ_k = displacements of spring K_v and dashpot $D_{v,k}$ for kth vehicle

V_o = moving speed of vehicles

λ_k = load point of kth vehicle from left end of curved girder bridge

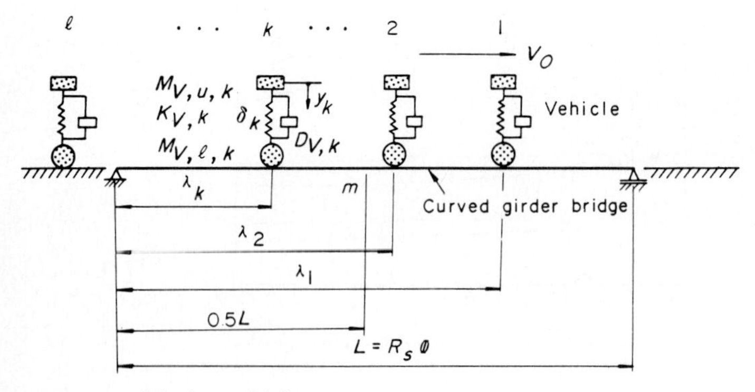

Figure 3.81 Moving vehicles.

Thus
$$M_{V,k} = M_{V,u,k} + M_{V,l,k} \tag{3.5.17a}$$

$$P_{o,k} = M_{V,k}g \tag{3.5.17b}$$

***b.* Mass, damping, and stiffness of curved girder bridges.**[65] The dynamic response of a curved girder bridge can be treated as a vibration system with two degrees of freedom, shown in Fig. 3.82, and described in Sec. 3.4.4(4). In the case of a simply supported curved girder bridge, we conclude that the analytical results for displacements w and β at the midspan of the curved girder bridge have sufficiently high accuracy, even if the vibration modes $\Omega_i(s)$ are limited to only the first mode of vibration $i = 1$. Thus,

$$\Omega(s) = \sin\frac{\pi s}{L} \tag{3.5.18}$$

The mass of the bridge will be affected by the unsprung mass $M_{V,1}$, $M_{V,l,1}, \ldots, M_{V,l,n}$ of the vehicles in Fig. 3.81. The resulting effective mass (M_B, I_B, and N_B) can be written as

$$M_B = \frac{mA_sL}{2} + \sum_{k=1}^{l} M_{V,l,k}\Upsilon^2(v_ot)U(\lambda_k) \tag{3.5.19a}$$

$$I_B = \frac{mI_sL}{2} + \sum_{k=1}^{l} M_{V,l,k}x_p^2\Upsilon^2(v_ot)U(\lambda_k) \tag{3.5.19b}$$

$$N_B = \frac{mS_yL}{2} + \sum_{k=1}^{l} M_{V,l,k}x_p\Upsilon^2(v_ot)U(\lambda_k) \tag{3.5.19c}$$

where $\Upsilon(v_ot)$ is a function of running speed V_o and the position of the vehicle, given by

$$\Upsilon(v_ot) = \sin\left(\frac{R_s}{R_p}\frac{\pi V_o}{L}t\right) \tag{3.5.20}$$

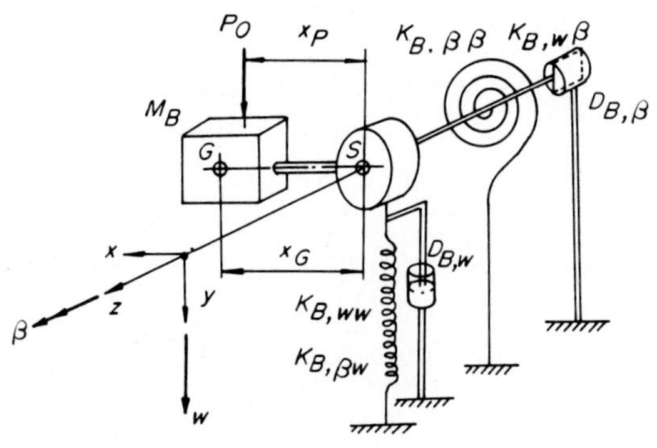

Figure 3.82 Equivalent vibration system for a curved girder bridge.

and $U(\lambda_k)$ is the unit step function defined by

$$U(\lambda_k) = 1 \qquad 0 < \lambda_k < L \tag{3.5.21a}$$

$$U(\lambda_k) = 0, \qquad 0 \geqq \lambda_k \geqq L \tag{3.5.21b}$$

The stiffnesses $K_{B,ww}$, $K_{B,w\beta}$, and $K_{B,\beta\beta}$ for a simply supported curved girder bridge are represented by

$$K_{B,ww} = E_s I'_x \frac{\pi^2(\pi^2 + \Phi^2\gamma)}{2L^3} \tag{3.5.22a}$$

$$K_{B,w\beta} = K_{B,\beta w} = E_s I'_x \frac{\pi^2\Phi(\gamma + 1)}{2L^2} \tag{3.5.22b}$$

$$K_{B,\beta\beta} = E_s I'_x \frac{\pi^2\gamma + \Phi^2}{2L} \tag{3.5.22c}$$

where γ is the equivalent torsional-flexural rigidities ratio designated by Eq. (3.5.6).

The final expression for the damping coefficients $D_{B,w}$ and $D_{B,\beta}$ for bending and torsional vibrations of a curved girder bridge can, respectively, be obtained as follows:

$$D_{B,w} = \frac{M_B D_{\mathrm{I}} p_{\mathrm{I}}}{\pi} \tag{3.5.23a}$$

$$D_{B,\beta} = \frac{M_B D_{\mathrm{II}} p_{\mathrm{II}}}{\pi} \tag{3.5.23b}$$

c. Coupled vibrations of curved girder bridges with traffic vehicles.[65] The dynamic response of a curved girder bridge due to several vehicles can be analyzed by expanding the equations of motion given in Eqs. (3.4.105) through (3.4.107) in Sec. 3.4.4(3). To evaluate the influence of various parameters as exact as possible, all the basic equations are nondimensionalized as follows:

First, the real time t is converted to

$$\tau = \Lambda t \tag{3.5.24}$$

where

$$\Lambda = \frac{R_s}{R_p}\frac{V_o}{L} \tag{3.5.25}$$

The derivative with respect to computational time τ is denoted

$$d/d\tau = {}' \tag{3.5.26}$$

Next, all the displacements w and β of the curved girder bridge including that of the kth vehicle (y_k and δ_k) are nondimensionalized as follows:

$$W = \frac{w}{w_{st}} \tag{3.5.27a}$$

$$B = \frac{\beta}{\beta_{st}} \tag{3.5.27b}$$

$$Y_k = \frac{y_k}{w_{st}} \tag{3.5.27c}$$

$$\Delta_k = \frac{\delta_k}{w_{st}} \tag{3.5.27d}$$

where w_{st} and β_{st} are the maximum static displacements of the curved girder bridge at midspan m under the set of traffic vehicles shown in Fig. 3.81. These values can be estimated by imposing the static displacements $\tilde{w}_{st,k}$ and $\tilde{\beta}_{st,k}$ due to the kth vehicle at point m of a simply supported curved girder bridge, which gives

$$\tilde{w}_{st,k} = \frac{R_p}{R_s} \frac{M_{V,k}g}{K_{B,ww}} \frac{1 - x_p k_\beta}{1 - k_w k_\beta} \tag{3.5.28a}$$

$$\tilde{\beta}_{st,k} = \frac{R_p}{R_s} \frac{M_{V,k}g}{K_{B,\beta\beta}} \frac{x_p - k_w}{1 - k_w k_\beta} \tag{3.5.28b}$$

The maximum displacements w_{st} and β_{st} are given by the maximum values of

$$w_{st} = \sum_{k=1}^{l} w_{st,k} \sum_{k=1}^{l} \left[\tilde{w}_{st,k} U(\lambda_k) \sin \frac{\pi \lambda_k}{L} \right] \tag{3.5.29a}$$

$$\beta_{st} = \sum_{k=1}^{l} \beta_{st,k} \sum_{k=1}^{l} \left[\tilde{\beta}_{st,k} U(\lambda_k) \sin \frac{\pi \lambda_k}{L} \right] \tag{3.5.29b}$$

where

$$k_w = L\Phi \frac{1 + \gamma}{\pi^2 + \kappa \Phi^2} \tag{3.5.30a}$$

$$k_\beta = \frac{\pi^2 \Phi}{L} \frac{1 + \gamma}{\gamma \pi^2 + \Phi^2} \tag{3.5.30b}$$

Note that $k_w = k_\beta = 0$ for a straight girder bridge.

Through this nondimensionalized process, a set of simultaneous differential equations of motion with respect to the vibration of the curved girder bridge $W(\tau)$ and $B(\tau)$ and vehicles $Z_k(\tau)$ and $\Delta_k(\tau)$ can be expressed as

$$\ddot{W} = -\rho_w(W - \alpha_w B) - \mu_w(\dot{W} - \varepsilon_w \dot{B}) + \sum_{k=1}^{l} [\rho_w \psi_{w,k} + \zeta_k(\rho_k \Delta_k + \mu_k \dot{\Delta}_k]U(\lambda_k) \sin(\pi\tau - \varphi_k) \tag{3.5.31a}$$

$$\ddot{B} = -\rho_\beta(B + \alpha_\beta W) - \mu_\beta(\dot{B} - \varepsilon_\beta \dot{W}) + \sum_{k=1}^{l} [\rho_\beta \psi_{\beta,k} + \chi_k(\rho_k \Delta_k + \mu_k \dot{\Delta}_k)]U(\lambda_k) \sin(\pi\tau - \varphi_k) \tag{3.5.31b}$$

$$\ddot{Z}_k = -\rho_k \Delta_k - \mu_k \dot{\Delta}_k \tag{3.5.31c}$$

$$\dot{\Delta}_k = \dot{Z}_k - (\dot{W} + \eta_p \dot{B}) \sin(\pi\tau - \varphi_k) - (W + \eta_p B)\pi \cos(\pi\tau - \varphi_k) \tag{3.5.31d}$$

$$\Delta_k = Z_k - (W + \eta_p B) \sin(\pi\tau - \varphi_k) \qquad k = 1, 2, \ldots, l) \tag{3.5.31e}$$

where the newly introduced parameters are expressed by the following formulas:

For a curved girder bridge,

$$\rho_w = \frac{1}{\Lambda^2} \frac{K_{B,ww}(I_B - N_B k_w)}{M_B I_B - N_B^2} \tag{3.5.32a}$$

$$\rho_\beta = \frac{1}{\Lambda^2} \frac{K_{B,\beta\beta}(M_B - N_B k_\beta)}{M_B I_B - N_B^2} \tag{3.5.32b}$$

$$\mu_w = \frac{1}{\Lambda} \frac{I_B D_{B,w}}{M_B I_B - N_B^2} \tag{3.5.33a}$$

$$\mu_\beta = \frac{1}{\Lambda} \frac{M_B D_{B,\beta}}{M_B I_B - N_B^2} \tag{3.5.33b}$$

$$\alpha_w = \frac{N_B - I_B k_\beta}{I_B - N_B k_w} \frac{x_p - k_w}{1 - x_p k_\beta} \tag{3.5.34a}$$

$$\alpha_\beta = \frac{N_B k_w - N_B}{M_B - N_B k_\beta} \frac{1 - x_p k_\beta}{x_p - k_w} \tag{3.5.34b}$$

$$\varepsilon_w = \frac{N_B K_{B,ww}}{I_B K_{B,\beta\beta}} \frac{D_{B,\beta}}{D_{B,w}} \frac{x_p - k_w}{1 - x_p k_\beta} \tag{3.5.35a}$$

$$\varepsilon_\beta = \frac{N_B K_{B,\beta\beta}}{M_B K_{B,ww}} \frac{D_{B,w}}{D_{B,\beta}} \frac{1 - x_p k_\beta}{x_p - k_w} \tag{3.5.35b}$$

Note that $Z_B = 0$; therefore $\alpha_w = \alpha_\beta = \varepsilon_w = \varepsilon_\beta = 0$ in the case of a straight girder bridge having a symmetric cross section.

For vehicles,

$$\rho_k = \frac{1}{\Lambda^2} \frac{K_{V,k}}{M_{V,u,k}} \tag{3.5.36a}$$

$$\mu_k = \frac{1}{\Lambda} \frac{D_{V,k}}{M_{V,u,k}} \tag{3.5.36b}$$

$$\eta_p = \frac{K_{B,ww}}{K_{B,\beta\beta}} \frac{x_p(x_p - k_w)}{1 - x_p k_\beta} \tag{3.5.37}$$

$$\varphi_k = \frac{\Lambda\pi}{V_o} \sum_{k=1}^{l} \lambda_k \tag{3.5.38}$$

$$\psi_{w,k} = \frac{w_{st,k}}{w_{st}} \frac{I_B - N_B x_p}{I_B - N_B k_w} \frac{1 - k_w k_\beta}{1 - x_p k_\beta} \tag{3.5.39a}$$

$$\psi_{\beta,k} = \frac{\beta_{st,k}}{\beta_{st}} \frac{M_B y_p - N_B}{M_B - N_B k_\beta} \frac{1 - k_w k_\beta}{y_p - k_w} \tag{3.5.39b}$$

$$\zeta_k = M_{V,u,k} \frac{R_p}{R_s} \frac{I_B - N_B x_p}{M_B I_B - N_B^2} \tag{3.5.40a}$$

$$\chi_k = M_{V,u,k} \frac{R_p}{R_s} \frac{M_B x_p - N_B}{M_B I_B - N_B^2} \frac{K_{B,\beta\beta}}{K_{B,ww}} \frac{1 - x_p k_\beta}{x_p - k_w} \tag{3.5.40b}$$

Equation (3.5.31) can be solved easily by using the parameters given above, provided that the initial conditions imposed on the curved girder bridge are

$$W(0) = B(0) = \dot{W}(0) = \dot{B}(0) = 0 \tag{3.5.41a–d}$$

The initial conditions of the vehicles can approximately be described by

$$Y_k(0) = \frac{y_k(0)}{w_{st}} \tag{3.5.42a}$$

$$\dot{Y}_k(0) = 0 \tag{3.5.42b}$$

by neglecting the influences of their phase angles.

***d.* Simulation by the Runge-Kutta-Gill method.**[65] The numerical integration of Eq. (3.5.31) can be effectively conducted by virtue of the Runge-Kutta-Gill method or the Newmark β method by taking into

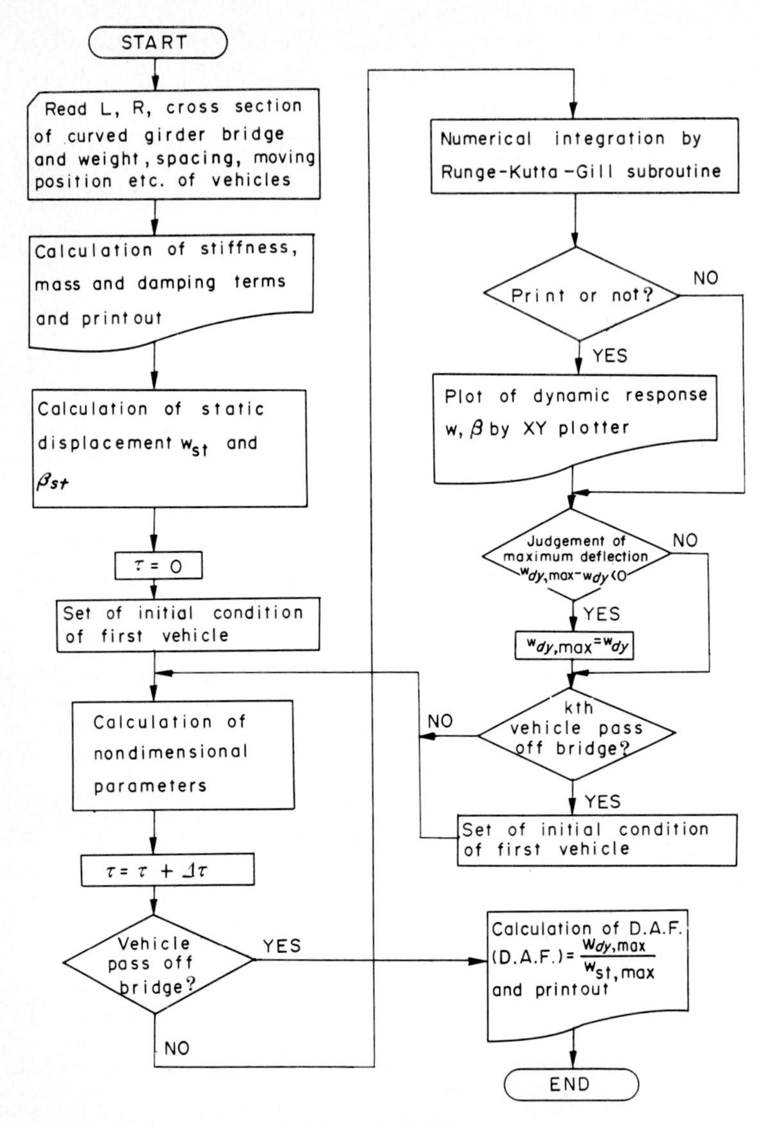

Figure 3.83 Flowchart to estimate the DAF.

consideration the above initial conditions. This section adopts the former method by setting the computational time interval to

$$\Delta\tau = \frac{\Lambda}{30}\frac{1}{f_B} \tag{3.5.43}$$

where

$$f_B = \frac{p_1}{2\pi} \tag{3.5.44}$$

Here f_B is the fundamental natural frequency of the curved girder bridge, in hertz. The computational time interval was determined through various preliminary calculations.

Final results of Eq. (3.5.31) can now be obtained by reconverting the dimension through Eqs. (3.5.27) and (3.5.42). Then the desired deflection w_{dy} at a section in the midspan of a curved girder bridge with a radius of curvature R_P can be estimated by

$$w_{dy} = w + x_p \beta \tag{3.5.45}$$

Figure 3.83 shows a flowchart to simulate the dynamic and static deflection as well as the DAF due to the traffic vehicles by a deterministic simulation method through the use of Runge-Kutta-Gill's subroutine installed in digital computer software.

(2) Parameters of traffic vehicles.[65] The values of the DAF due to several vehicles, as is observed in actual traffic flows, are analyzed by the deterministic simulation method. For these analyses, the parameters for a set of traffic vehicles are summarized as shown in Table 3.19 by referring to the parametric analyses of the DAF mentioned earlier. Thus the assumed traffic vehicles are not random and have constant dynamic properties, but the DAF of the curved girder evaluated from these parameters will be the maximum values.

(3) Load conditions.[65] The load conditions, i.e., especially the running position of the vehicles, are an important factor, so that the values of the DAF for the outer and inner main girders relative to the

TABLE 3.19 Parameters for Traffic Vehicles

Bridge and vehicle	Item		Values of parameter
Curved girder bridge	Span L, radius of curvature R_s, cross-sectional value, logarithmic damping coefficient D_B		See Table 3.17. See Eq. (3.5.11)
Vehicle	Weight	(tf)	$P_{o,k} = 20 \quad k = 1, 2, \ldots, l$
	Spacing	(m)	$\lambda_{k+1} - \lambda_k = 10 \quad k = 1, 2, \ldots, l$
	Number		$l = L/10$
	Speed	(km/h)	$V_{o,k} = 40 \quad k = 1, 2, \ldots, l$
	Position	(m)	$R_p = R_s$, $R_p = R_s \pm 3.0$ (see Fig. 3.85)
	Frequency	(Hz)	$f_{V,k} = f_B \quad k = 1, 2, \ldots, l$
	Force	(tf)	$F_k = 3.0 \quad k = 1, 2, \ldots, l$
	Others		$D_{V,k} = 0.2 \quad k = 1, 2, \ldots, l$

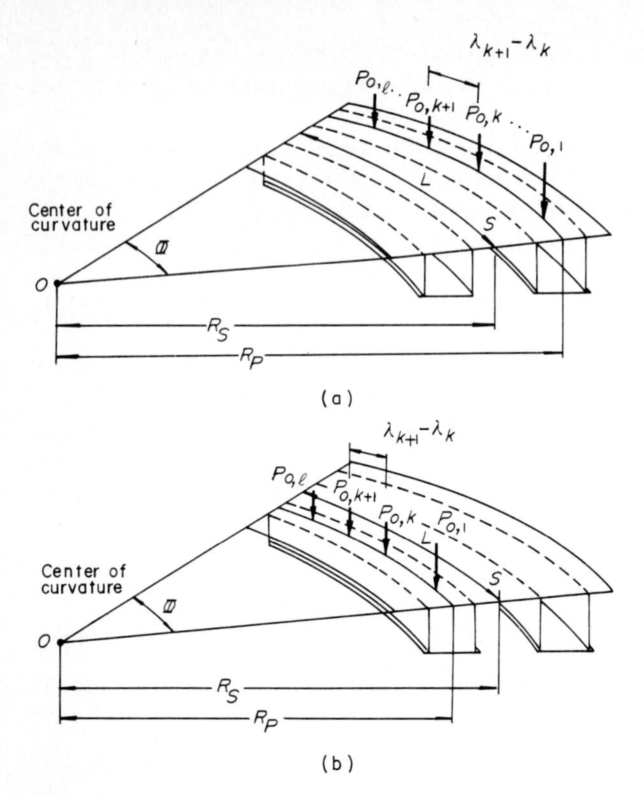

Figure 3.84 Load condition of vehicles. Evaluation of DAF of (*a*) outer girder and (*b*) inner girder.

center of curvature are evaluated by the load conditions, as illustrated in Fig. 3.84*a* and *b*, respectively.

(4) DAF of curved girder bridges under traffic vehicles.[65] The simulation results of the DAF for the curved girder bridges under a set of traffic vehicles are plotted as a function of the span length of the curved girder bridges and are shown in Fig. 3.85.

From these figures, we see that the values of the DAF are more or less different from each other, even when the curved girder bridges have the same span length. This may be caused by the differences in cross-sectional shape of the curved girders and various irregularities due to the forcing terms of the vehicles.

However, the overall tendency of the DAF of both the curved box-girder and multiple-I-girder bridges is to decrease as the span length increases. The envelope of curves of these values of the DAF is shown

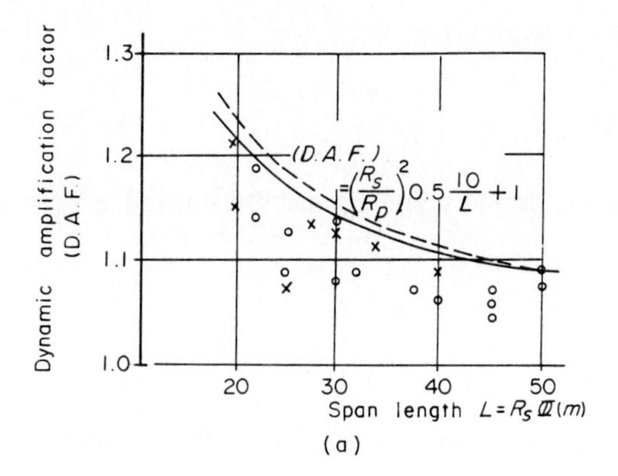

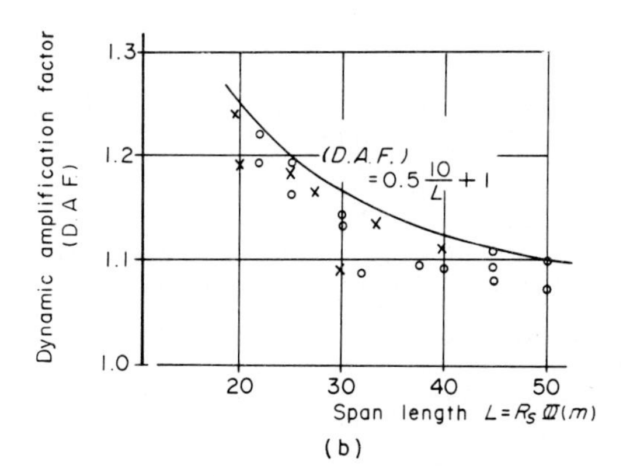

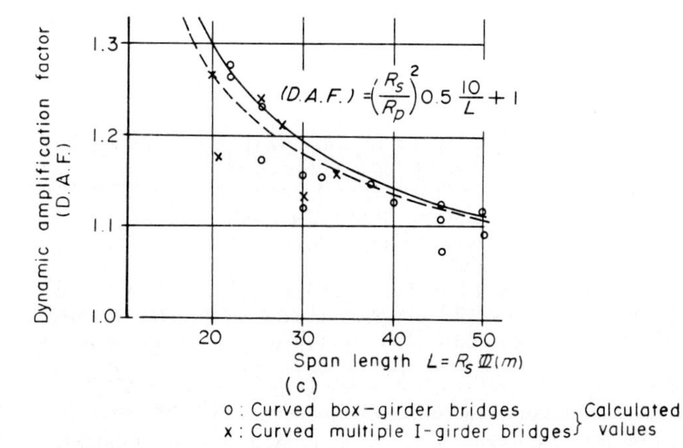

Figure 3.85 Relationship between DAF and L: (*a*) DAF of outer girder, (*b*) DAF of midgirder (shear center), and (*c*) DAF of inner girder.

by the solid lines in Fig. 3.85, which leads to

$$\mathrm{DAF} = \left(\frac{R_s}{R_p}\right)^2 (0.5)\left(\frac{10}{L}\right) + 1 \qquad (3.5.46)$$

From this equation, a rational impact coefficient for curved girder bridges can be determined.

References

1. Love, A. E. H.: *Mathematical Theory of Elasticity*, 4th ed., McGraw-Hill, New York, 1927.
2. Vlasov, V. Z.: *Thin-Walled Elastic Beams*, National Science Foundation, Washington, D.C., 1961.
3. Kuranishi, M.: *Theory of Elasticity*, Japan Society of Mechanics, Tokyo, 1943 (in Japanese).
4. Konishi, I., and S. Komatsu: "On Fundamental Theory of Thin-Walled Curved Girder," *Transactions of the Japanese Society of Civil Engineers*, no. 87, pp. 35–46, Nov. 1962 (in Japanese).
5. Kuranishi, S.: "Analysis of Thin-Walled Curved Beam," *Transactions of the Japanese Society of Civil Engineers*, no. 108, pp. 7–12, Aug. 1964 (in Japanese).
6. Fukazawa, Y.: "Fundamental Theory on Statical Analysis of Thin-Walled Curved Bars," *Transactions of the Japanese Society of Civil Engineers*, no. 110, pp. 30–51, Oct. 1964 (in Japanese).
7. Becker, G.: Ein Beitrag zur statischen Berechnung beliebige galarter ebener gekrümmter Stäbe mit einfach symmetrischen dünnwadige offen Profilen von Stabachse veranderlischen Querschnitt unter Berücksichtigen der Wölbkragttorsion, *Der Stahlbau*, 34, s.334–346 und 368–377, Nov. und Dez., 1965 (in German).
8. Dabrowski, R.: *Gekrümmte dumwandige Träger*, Springer-Verlag, Berlin, 1968 (in German).
9. Ochi, Y.: *Structural Analysis by Computer*, Bridge Engineering Association, Tokyo, 1969 (in Japanese).
10. Yoo, C. H.: "Flexural-Torsional Stability of Curved Beams," *Proceedings of the American Society of Civil Engineers*, vol. 108, EM-6, pp. 1361–1369, Dec. 1982.
11. Task Committee on Curved Box Girder of ASCE-ASSHTO Committee of Metal of ASCE Structural Division: "Curved Steel Box Girder Bridges—State-of-the-Art," *Proceedings of the American Society of Civil Engineers*, vol. 103, no. ST-11, pp. 1719–1739, Nov. 1978.
12. Galambos, T. V.: *Structural Members and Frames*, Prentice-Hall, Englewood Cliffs, N.J., 1968.
13. Kollbrunner-Basler: *Torsion*, Springer-Verlag, Berlin, 1972 (in German).
14. Kollbrunner-Hajdin: *Dünnwandige Stäbe*, Springer-Verlag, Berlin, 1972 (in German).
15. Reissner, E.: "Analysis of Shear Lag in Box Beams by Principle of Minimum Potential Energy," *Quarterly of Applied Mathematics*, 4(3):268–278, 1946.
16. Chwalla, E.: *Über das Problem der voll mittrangender Breite von Gurt und Rippen Platten*, Alfons-Leon-Gederskchrift, Wein, 1952.
17. Kuhn, P.: *Stress in Aircraft and Shell Structures*, McGraw-Hill, New York, 1956.
18. Kondo, K., S. Komatsu, and H. Nakai: "Theoretical and Experimental Researches on Effective Width of Girder Bridges with Steel Deck Plate," *Transactions of the Japanese Society of Civil Engineers*, no. 86, pp. 1–17, Oct. 1962 (in Japanese).
19. Komatsu, S., H. Nakai, and T. Kitada: "Study on Shear Lag and Effective Width of Curved Girder Bridges," *Proceedings of the Japanese Society of Civil Engineers*, no. 191, pp. 1–14, July 1971 (in Japanese).
20. Yoshimura, T., and N. Nirasawa: "On the Stress Distribution and Effective Width of Curved Girder Bridges by the Folded Plate Theory," *Proceedings of the Japanese Society of Civil Engineers*, no. 233, pp. 45–54, Jan. 1975 (in Japanese).

21. Moffatt, K. R., and P. J. Dowling: "Shear Lag in Steel Box Girder Bridges," *The Structural Engineers*, 53:439–448, Oct. 1975.
22. Wright, R. N., S. R. Abdel-Samad, and A. R. Robinson: "BEF Analogy for Analysis of Box Girders," *Proceedings of the American Society of Civil Engineers*, vol. 94, no. ST-7, pp. 1719–1743, July 1968.
23. Oleinik, J. C., and C. P. Heins: "Diaphragms for Curved Box Beam Bridges," *Proceedings of American Society of Civil Engineers*, vol. 101, no. ST-10, pp. 2161–2178, Oct. 1975.
24. Sakai, F., and M. Nagai: "A Recommendation on the Design of Intermediate Diaphragms in Steel Box Girder Bridges," *Proceedings of the Japanese Society of Civil Engineers*, no. 261, pp. 21–34, May 1977 (in Japanese).
25. Nomachi, S.: "On Torsion Bending of Thin-Walled Rectangular Beams with Equal-Distant Rigid Diaphragms," *Transactions of the Japanese Society of Civil Engineers*, no. 146, pp. 13–21, May 1967 (in Japanese).
26. Nakai, H., and Y. Murayama: "Distortional Stress Analysis and Design Aid for Horizontally Curved Box Girder Bridges with Diaphragms," *Proceedings of the Japanese Society of Civil Engineers*, no. 309, pp. 25–39, May 1981 (in Japanese).
27. Timoshenko, S. P., and S. Woinowsky-Krieger: *Theory of Plates and Shells*, McGraw-Hill, New York, 1969.
28. Jaeger, L. G.: *Elementary Theory of Elastic Plates*, Pergamon Press, New York, 1964.
29. Lekhnitski, S. G.: *Anisotropic Plates*, Gordon Breach, New York, 1968.
30. Cusens, A. R., and R. P. Pama: *Bridge Deck Analysis*, Wiley, New York, 1975.
31. Heins, C. P.: *Applied Plate Theory for Engineers*, Lexington Books, Lexington, Mass., 1976.
32. Hambly, E. C.: *Bridge Deck Behavior*, Wiley, New York, 1976.
33. Komatsu, S.: "Practical Formulas for the Curved Bridge with Multiple Plate Girders," *Transactions of the Japanese Society of Civil Engineers*, no. 93, pp. 1–9, May 1963 (in Japanese).
34. Konishi, I., and S. Komatsu: *Three-Dimensional Analysis of Curved Girders with Thin-Walled Cross-Sections*, Publication of IABSE, Zürich, 1965.
35. McMacus, P. F., G. A. Nasir, and C. G. Culver: "Horizontally Curved Girder—State-of-the-Art," *Proceedings of the American Society of Civil Engineers*, vol. 91, no. ST-5, pp. 835–870, May 1965.
36. Simada, S., and S. Kuranishi: *Formulas and Tables for Curved Beams*, Gihodo, Tokyo, 1966 (in Japanese).
37. Watanabe, N.: *Theory and Calculation of Curved Girder*, Gihodo, Tokyo, 1967 (in Japanese).
38. Heins, C. P., and K. R. Spates: "Behavior of Single Horizontally Curved Girder," *Proceedings of the American Society of Civil Engineers*, vol. 99, no. ST-7, pp. 1511–1524, July 1970.
39. Heins, C. P.: "Behavior and Design of Curved Girder Bridges," Developments in Bridge Design and Construction, *Conference at University College*, Cardiff, Wales, March/April 1971.
40. Buchanan, J. D., Yoo C. H., and Heins, C. P.: Field Study of a Curved Box Beam Bridge, *Civil Engineering Report No. 89, University of Maryland*, Sept. 1974.
41. Komatsu, S., H. Nakai, and Y. Taido: "A Proposition for Designing the Horizontally Curved Girder in Connection with Ratio between Torsional and Flexural Rigidities," *Proceedings of the Japanese Society of Civil Engineers*, no. 224, pp. 55–66, April 1974 (in Japanese).
42. Bell, L. C., and C. P. Heins: "Analysis of Curved Girder Bridges," *Proceedings of the American Society of Civil Engineers*, vol. 96, no. ST-8, pp. 1657–1673, Aug. 1970.
43. Nakai, H., and T. Tani: "An Approximate Method for the Evaluation of Torsional and Warping Stress in Box Girder Bridges," *Proceedings of the Japanese Society of Civil Engineers*, no. 227, pp. 41–55, Sept. 1978 (in Japanese).
44. Nakai, H., and C. P. Heins: "Analysis Criterion for Curved Bridges," *Proceedings of the American Society of Civil Engineers*, vol. 103, no. ST-7, pp. 1419–1427, July 1977.
45. British Standard Institution: BS 5400, Part 3, *Code of Practice for Design of Steel Bridges*, April 1982.

46. Japanese Road Association: *Japanese Specification for Designing of Highway Bridges*, Maruzen, Tokyo, 1980 (in Japanese).
47. Timoshenko, S. P.: *Vibration Problems in Engineering*, 3d ed., McGraw-Hill, New York, 1955.
48. Schumpich, G.: "Beitrag zur Kinetic und Statik ebener Stabwerke mit gekrümten Stäben," *Österreiches Ingenieur Archiv.*, Bd. 3, s. 194–225, 1957 (in German).
49. Gere, J. M., and Y. K. Lin: "Coupled Vibrations of Thin-Walled Beams of Open Cross-section," *Journal of Applied Mechanics*, Sept. 35, 1958, pp. 373–378.
50. Yonezawa, H.: "Moments and Free Vibration in Curved Girder Bridges," *Proceedings of the American Society of Civil Engineers*, vol. 88, no. EM-1, pp. 1–21, Feb. 1962.
51. Komatsu, S., and H. Nakai: "Study of Free Vibration of Curved Girder Bridges," *Transactions of the Japanese Society of Civil Engineers*, no. 136, pp. 27–38, Dec. 1966.
52. Yamazaki, T., and K. Sakiyama: "Dynamic Response of Circular Beam to Moving Loads," *Memoirs of Faculty of Engineering*, Kyushu University, vol. 41, Reprint no. 2, 1968.
53. Tan, C. P., and S. Shore: "Dynamic Response of Horizontally Curved Bridges," *Proceedings of the American Society of Civil Engineers*, vol. 94, no. ST-3, pp. 761–781, March 1968.
54. Tan, C. P., and S. Shore: "Response of Horizontally Curved Bridges to Moving Load," *Proceedings of the American Society of Civil Engineers*, vol. 94, no. ST-9, pp. 2135–2151, Sept. 1968.
55. Komatsu, S., and H. Nakai: "Fundamental Study on Forced Vibration of Curved Girder Bridges," *Proceedings of the Japanese Society of Civil Engineers*, no. 174, pp. 27–38, Feb. 1970.
56. Yoo, C. H. and Fehrenbach, J. P.: Natural Frequencies of Curved Girders, *Proceedings of the American Society of Civil Engineers*, vol. 107, no. EM-2, pp. 339–354, April 1981.
57. Norris, C. H., R. J. Hansen, M. J. Holley, J. M. Biggs, S. Namyet, and J. K. Minami: *Structural Design for Dynamic Load*, McGraw-Hill, New York, 1963.
58. Fifer, S.: *Analogue Computation*, vols. 1–3, McGraw-Hill, New York, 1961.
59. Komatsu, S., and H. Nakai: "Application of Analogue Computer to the Analysis of Dynamical Response of Curved Girder Bridges," *Proceedings of the Japanese Society of Civil Engineers*, no. 178, pp. 11–26, June 1970 (in Japanese).
60. Hirai, I.: "Fundamental Equations of Beam Structure under Moving Loads and Its Application," *Transactions of the Japanese Society of Civil Engineers*, no. 90, pp. 29–36, Feb. 1963 (in Japanese).
61. Yamada, Y., and T. Kobori: "Studies on Highway Bridge Impact due to Random Moving Vehicles," *Transactions of the Japanese Society of Civil Engineers*, no. 119, pp. 1–9, July 1965.
62. Matsunaga, Y., H. Nakai, H. Kotoguchi, and R. Ohminami: *Experimental Study on Dynamic Properties of Straight Girder Bridges*, 24th Annual Conference of JSCE, Paper no. I-82, Nov. 1969 (in Japanese).
63. Ito, M., and T. Katayama: "Damping of Bridge Structures," *Transactions of the Japanese Society of Civil Engineers*, no. 117, pp. 12–22, May 1965 (in Japanese).
64. Harris, C. M., and C. E. Crede: *Shock and Vibration Handbook*, vol. 3, McGraw-Hill, New York, 1961.
65. Komatsu, S., H. Nakai, and H. Kotoguchi: "Study on Dynamic Response and Impact of Horizontally Curved Girder Bridges under Moving Vehicles," *Proceedings of the Japanese Society of Civil Engineers*, no. 192, pp. 55–68, Aug. 1971 (in Japanese).

Chapter

4

Analysis of Flexural and Torsional Stress Resultants and Displacements in Curved Girders

4.1 Introduction

This chapter analyzes flexural and torsional stress resultants and displacements in a curved girder based on the fundamental theory presented in Chap. 3.

First, the force method for analyzing stress resultants and displacements in a curved girder is introduced on the basis of pure torsional theory. The statically determinate and indeterminate systems for these analyses are shown, and the resulting elastic equations are derived from the compatibility condition of the torsional angle. For practical design purposes, the influence lines of stress resultants are shown for a curved girder with single span. Furthermore, this method is developed for the analysis of stress resultants in a continuous curved girder.

Second, the stress resultants and displacements in a curved girder, where the torsional warping behavior is not negligible, are determined from their governing differential equations. All the solutions of these stress resultants and displacements are summarized in tables for a curved girder with a single span. The analytical procedures used to apply this method to a continuous curved girder are also shown by using the energy method.

Third, we discuss curved girder analysis by using transfer matrix methods. For these analyses, a curved girder is treated with or without torsional warping rigidities. The field transfer matrices are derived on the basis of the fundamental differential equations of the curved girders. The point transfer matrices are found from the equilibrium conditions of the stress resultants and the compatibility conditions of the displacements. The initial conditions, the boundary conditions at the intermediate supports of a continuous curved girder, and the end conditions are categorized according to their geometric and static boundary conditions. Moreover, the analytical procedure for computer use is detailed with flowcharts.

4.2 Force Method (Pure Torsional Theory)[1–3]

4.2.1 Statically determinate systems

Figure 4.1 shows a structural model of a curved girder, where

$$R = \text{radius of curvature (constant)}$$

$$\Phi = \text{central angle}$$

$$EI\,(=EI_X) = \text{flexural rigidity of main girder}$$

$$GK = \text{pure torsional rigidity of main girder}$$

And the end floor beams are assumed to have infinite flexural and torsional rigidities. Four vertical reactions V_1 to V_4 are produced at each support under the application of the load, as illustrated in Fig. 4.1, and these reactions cannot be ascertained by the equilibrium conditions of the forces, that is, $\Sigma\, V = 0$, $\Sigma\, M = 0$, and $\Sigma\, T = 0$, alone. Thus this curved girder is a statically indeterminate structure with one degree of redundancy.

To determine the stress resultants and displacements in this curved girder as simply as possible, we idealize it as a statically determinate system by releasing the reaction against the torsional angle and introducing the statically indeterminate torque X_1 at the left end of the support, as shown in Fig. 4.2. Thus, the complete solution can be

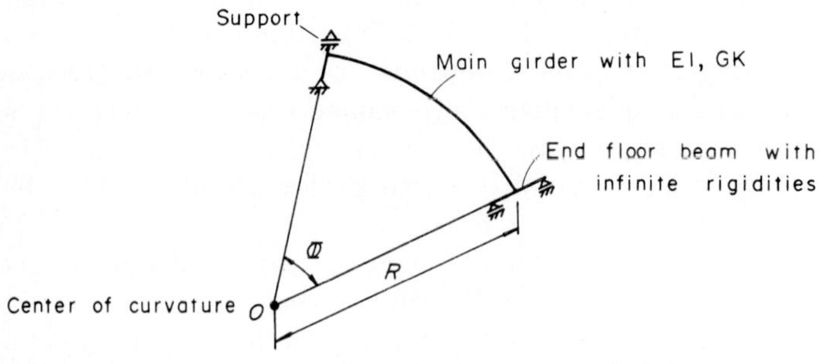

Figure 4.1 Structural model for a simple span curved girder.

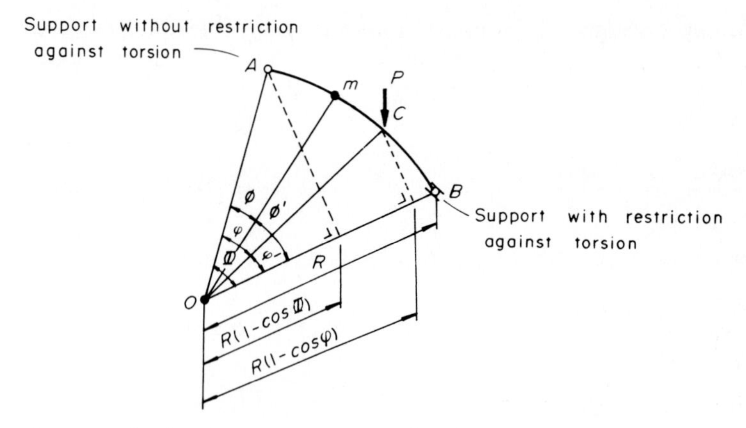

Figure 4.2 Statically determinate curved girder.

obtained from the compatibility condition of torsional angle at the left end support.

4.2.2 Stress resultants in a statically determinate system

When a vertical concentrated load P is applied at a point C with the radius of curvature R and the central angle φ or φ' or a statically determinate curved girder, as shown in Fig. 4.2, the corresponding stress resultants can be estimated as follows:

(1) Reaction. The reactions V_A^o and V_B^o at ends A and B, respectively, of the idealized curved girder in Fig. 4.2 can be easily obtained by considering the equilibrium conditions $\Sigma\, M = 0$ with respect to the OB axis and $\Sigma\, V = 0$, which gives

$$V_A^o = P\frac{\sin\varphi'}{\sin\Phi} \tag{4.2.1a}$$

$$V_B^o = P\left(1 - \frac{\sin\varphi'}{\sin\Phi}\right) \tag{4.2.1b}$$

(2) Bending moment. The bending moment M_o at point m with angular coordinate ϕ, as shown in Fig. 4.2, is given by

$$M_o = V_A^o R\sin\phi = PR\frac{\sin\varphi'}{\sin\Phi}\sin\phi \qquad 0 \leqq \phi \leqq \varphi \tag{4.2.2a}$$

$$\begin{aligned} M_o &= V_A^o R\sin\phi - PR\sin(\phi - \varphi) \\ &= PR\left[\frac{\sin\varphi'}{\sin\Phi}\sin\phi - \sin(\phi-\varphi)\right] \qquad \varphi \leqq \phi \leqq \Phi \end{aligned} \tag{4.2.2b}$$

(3) Torsional moment. In a similar manner, the torsional moment T_o is

$$T_o = V_A^o R(1 - \cos\phi)$$

$$= PR\frac{\sin\varphi'(1-\cos\phi)}{\sin\Phi} \qquad 0 \leqq \phi \leqq \varphi$$

$$T_o = V_A^o R(1 - \cos\phi) - PR[1 - \cos(\phi - \varphi)] \qquad (4.2.3a)$$

$$= PR\left\{\frac{\sin\varphi'(1-\cos\phi)}{\sin\Phi} - [1 - \cos(\phi - \varphi)]\right\} \qquad \varphi \leqq \phi \leqq \Phi \quad (4.2.3b)$$

where ϕ is the angular coordinate of point m in Fig. 4.2.

4.2.3 Stress resultants due to the statically indeterminate torque

The stress resultants due to a unit statically indeterminate torque $X_1 = 1$ which is applied to the left end support A, as shown in Fig. 4.3, can be determined directly in the manner listed above as follows:

(1) Reactions V'_A and V'_B:

$$V'_A = \frac{1}{R} \qquad (4.2.4a)$$

$$V'_B = -\frac{1}{R} \qquad (4.2.4b)$$

(2) Bending moment M_1:

$$M_1 = V'_A R \sin\phi - 1 \cdot \sin\phi \equiv 0 \qquad (4.2.5)$$

This equation shows that the bending moment M_1 is always equal to

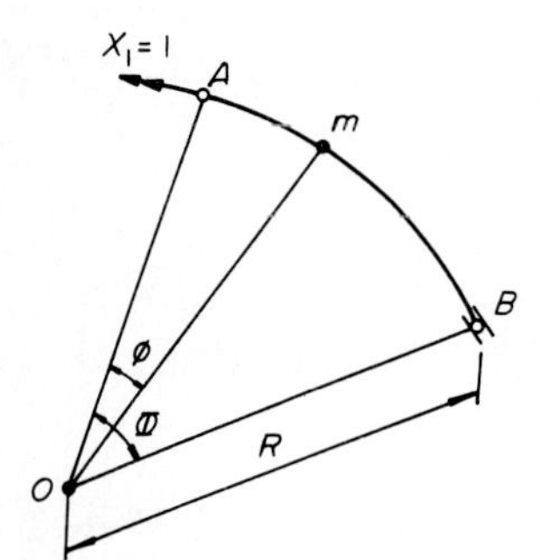

Figure 4.3 Application of unit statically indeterminate torque $X_1 = 1$.

zero, even when the statically indeterminate torque $X_1 = 1$ is applied to a statically determinate curved girder.

(3) Torsional moment T_1. It follows that

$$T_1 = V'_A R(1 - \cos\phi) + 1 \cdot \cos\phi \equiv 1 \tag{4.2.6}$$

4.2.4 Stress resultants and the compatibility condition for the torsional angle in a curved girder

(1) Stress resultants in a curved girder. The stress resultants in a curved girder, in which the torsional angle is restrained at both ends of the girder, can be found from Secs. 4.2.2 and 4.2.3. Thus

$$M = M_o + M_1 X_1 = M_o \tag{4.2.7a}$$

$$T = T_o + T_1 X_1 = T_o + X_1 \tag{4.2.7b}$$

(2) Compatibility condition for the torsional angle. The strain energy U stored in the curved girder can then be estimated by

$$U = \int_0^{\Phi} \frac{M^2}{2EI} R\, d\phi + \int_0^{\Phi} \frac{T^2}{2GK} R\, d\phi \tag{4.2.8}$$

To make the torsional angle equal to zero at the left end A of the curved girder, the following equation, given in Castigliano's theorem, must be satisfied:

$$\frac{\partial U}{\partial X_1} = 0 \tag{4.2.9}$$

From this condition, the statically indeterminate torque X_1 can be found from the following elastic equation:

$$X_1 = -\frac{\delta_{1o}}{\delta_{11}} \tag{4.2.10}$$

where $\delta_{11} = \int_0^{\Phi} \frac{T_1^2}{GK} R\, d\phi = \frac{R\Phi}{GK}$ (4.2.11a)

and
$$\delta_{1o} = \int_0^{\Phi} \frac{T_o T_1}{GK} R\, d\phi$$

$$= \frac{PR^2}{GK} \left\{ \int_0^{\Phi} \frac{\sin\varphi'}{\sin\Phi} (1 - \cos\phi)\, d\phi - \int_{\varphi}^{\Phi} [1 - \cos(\phi - \varphi)]\, d\phi \right\}$$

$$= \frac{PR^2\Phi}{GK} \left(\frac{\sin\varphi'}{\sin\Phi} - 1 + \frac{\varphi}{\Phi} \right) \tag{4.2.11b}$$

Therefore, Eq. (4.2.10) reduces to

$$X_1 = -\frac{\delta_{1o}}{\delta_{11}} = -PR\left(\frac{\sin\varphi'}{\sin\Phi} - 1 + \frac{\varphi}{\Phi}\right) \tag{4.2.12}$$

4.2.5 Influence lines of stress resultants for a single-span curved girder

(1) Vertical unit load. The influence line (see Fig. 4.2) for stress resultants in a curved girder with a single span subject to a vertical unit load $P = 1$ can be derived from Eqs. (4.2.1) through (4.2.12) as follows: Reactions V_A and V_B:

$$V_A = \frac{\varphi'}{\Phi} \tag{4.2.13a}$$

$$V_B = \frac{\varphi}{\Phi} \tag{4.2.13b}$$

Shearing force $Q\ (= Q_z)$:

$$Q = \begin{cases} \dfrac{\varphi}{\Phi} & 0 \leqq \varphi \leqq \phi \qquad (4.2.14a) \\[2ex] -\dfrac{\varphi'}{\Phi} & \phi \leqq \varphi \leqq \Phi \qquad (4.2.14b) \end{cases}$$

Bending moment $M\ (= M_X)$:

$$M = \begin{cases} R\dfrac{\sin\varphi \sin\phi'}{\sin\Phi} & 0 \leqq \varphi \leqq \phi \qquad (4.2.15a) \\[2ex] R\dfrac{\sin\varphi \sin\phi}{\sin\Phi} & \phi \leqq \varphi \leqq \Phi \qquad (4.2.15b) \end{cases}$$

Torsional moment $T\ (= T_z)$

$$T = \begin{cases} R\left(-\dfrac{\sin\varphi \cos\phi'}{\sin\Phi} + \dfrac{\varphi}{\Phi}\right) & 0 \leqq \varphi \leqq \phi \qquad (4.2.16a) \\[2ex] R\left(\dfrac{\sin\varphi' \cos\phi}{\sin\Phi} - \dfrac{\varphi'}{\Phi}\right) & \phi \leqq \varphi \leqq \Phi \qquad (4.2.16b) \end{cases}$$

In a similar manner, the influence line for a unit torque $M_T = 1 \cdot e$, shown in Fig. 4.4, where e is the eccentricity of a vertical unit load $P = 1$

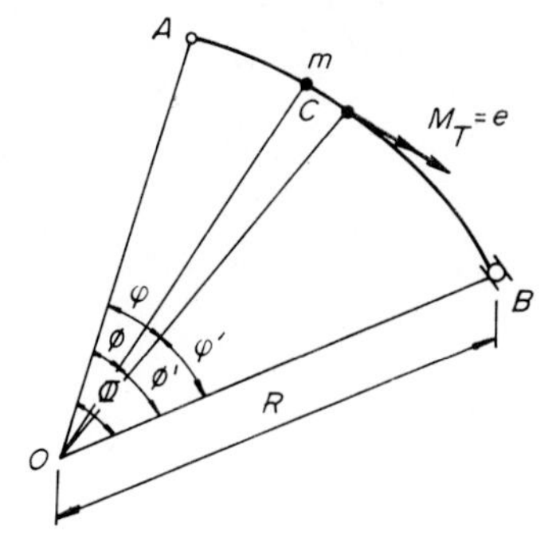

Figure 4.4 Application of unit torque $M_T = e$.

from the shear center of a curved girder, can be summarized as follows:
Bending moment $M\ (=M_X)$:

$$M = \begin{cases} (R-e)\dfrac{\sin\varphi \sin\varphi'}{\sin\Phi} & 0 \leqq \varphi \leqq \phi \qquad (4.2.17a) \\[2ex] (R-e)\dfrac{\sin\varphi' \sin\phi}{\sin\Phi} & \phi \leqq \varphi \leqq \Phi \qquad (4.2.17b) \end{cases}$$

Torsional moment $T\ (=T_z)$:

$$T = \begin{cases} -(R-e)\dfrac{\sin\varphi \cos\phi'}{\sin\Phi} + R\dfrac{\varphi}{\Phi} & 0 \leqq \varphi \leqq \phi \qquad (4.2.18a) \\[2ex] (R-e)\dfrac{\sin\varphi' \cos\phi}{\sin\Phi} - R\dfrac{\varphi'}{\Phi} & \phi \leqq \varphi \leqq \Phi \qquad (4.2.18b) \end{cases}$$

4.2.6 Application of analysis to a continuous curved girder[4]

Figure 4.5*a* illustrates a three-span continuous curved girder. For this curved girder, the same analytical method can be adopted by separating each span into a single span and introducing the statically indeterminate torques X_1, X_3, and X_5, to eliminate the torsional angles at the left end of each single curved girder (see Fig. 4.5*b*). However, the additional statically indeterminate bending moments X_2 and X_4 should also be applied at each intermediate support, to fulfill the continuity conditions for the deflection angle, as shown in Fig. 4.5*b*. The stress resultants at an arbitrary point *m* caused by these statically

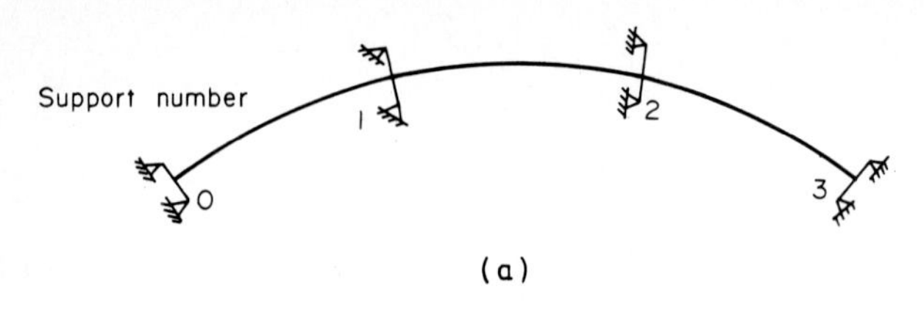

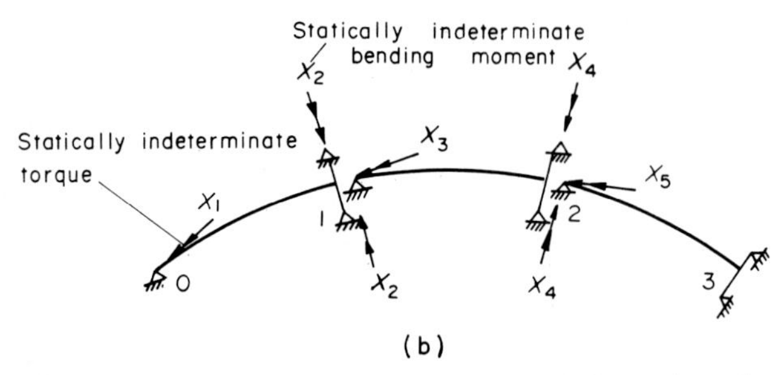

Figure 4.5 Three-span continuous curved girder and statically determinate system: (*a*) three-span continuous girder and (*b*) statically determinate system and corresponding statically indeterminate forces.

indeterminate forces with suffixes *i*, *j*, and *k* can be written by referring to Fig. 4.6 as follows:
For Fig. 4.6*a*,

$$\bar{M}_i = 0 \tag{4.2.19a}$$

$$\bar{T}_i = 1 \tag{4.2.19b}$$

For Fig. 4.6*b*,

$$\bar{M}_j = \frac{\sin \phi'}{\sin \Phi} \qquad \bar{M}_k = \cos \phi' - \frac{\sin \phi'}{\tan \Phi} \tag{4.2.20a}$$

$$\bar{T}_j = \frac{1 - \cos \phi'}{\sin \Phi} \qquad \bar{T}_k = \sin \phi' - \frac{\cos \phi'}{\tan \Phi} \tag{4.2.20b}$$

Therefore, the expansions of Eq. (4.2.12) give the following set of elastic equations for these statically indeterminate forces:

$$\delta_{11}X_1 + \delta_{12}X_2 = -\delta_{1o} \tag{4.2.21a}$$

$$\delta_{21}X_1 + \delta_{22}X_2 + \delta_{23}X_3 + \delta_{24}X_4 = -\delta_{2o} \tag{4.2.21b}$$

$$\delta_{32}X_2 + \delta_{33}X_3 + \delta_{34}X_4 = -\delta_{3o} \tag{4.2.21c}$$

$$\delta_{42}X_2 + \delta_{43}X_3 + \delta_{44}X_4 + \delta_{45}X_5 = -\delta_{4o} \tag{4.2.21d}$$

$$\delta_{54}X_4 + \delta_{55}X_5 = -\delta_{5o} \tag{4.2.21e}$$

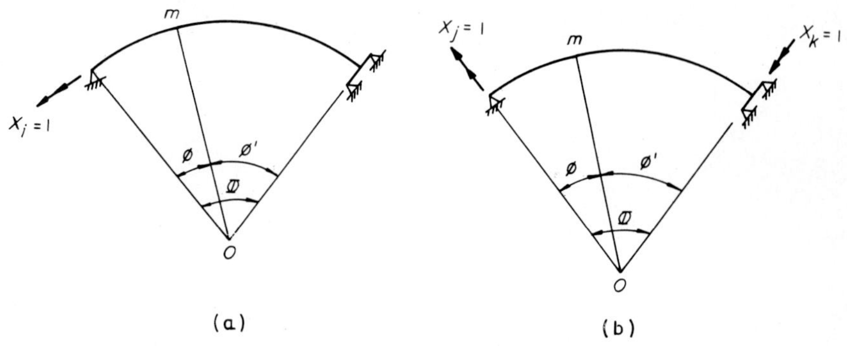

Figure 4.6 Unit statically indeterminate forces: (*a*) statically indeterminate torque $X_i = 1$, (*b*) statically indeterminate bending moments $X_j = 1$ and $X_k = 1$.

where δ_{ij} and δ_{io} can be estimated by

$$\delta_{ij} = \int_0^L \frac{\bar{M}_i \bar{M}_j}{EI} ds + \int_0^L \frac{\bar{T}_i \bar{T}_j}{GK} ds \tag{4.2.22a}$$

$$\delta_{io} = \int_0^L \frac{M_o \bar{M}_i}{EI} ds + \int_0^L \frac{T_o \bar{T}_i}{GK} ds \tag{4.2.22b}$$

in which L is the total span length of the continuous curved girder. Variables M_o and T_o are calculated based on the stress resultants due to the applied loads [see Eqs. (4.2.2) and (4.2.3)].

Thus, the final solutions can be obtained easily by superimposing these stress resultants. For example, bending and torsional moments for the second span can be expressed as

$$M_2 = M_{o,2} + X_2 \frac{\sin \phi'}{\sin \Phi} + X_4\left(\cos \phi' - \frac{\sin \phi'}{\tan \Phi}\right) \tag{4.2.23a}$$

and

$$T_2 = T_{o,2} + X_2 \frac{1 - \cos \phi'}{\sin \Phi} + X_3 + X_4\left(\sin \phi' - \frac{\cos \phi'}{\tan \Phi}\right) \tag{4.2.23b}$$

4.3 Analytical Method Based on Differential Equations (Torsional Warping Theory)

The closed-form solutions of stress resultants and displacements of a curved girder on the basis of torsional warping theory are very complicated. The solutions obtained by Konishi and Komatsu[5] are introduced here. Note that all these stress resultants and displacements are those of transferred values with respect to the centroidal axis.

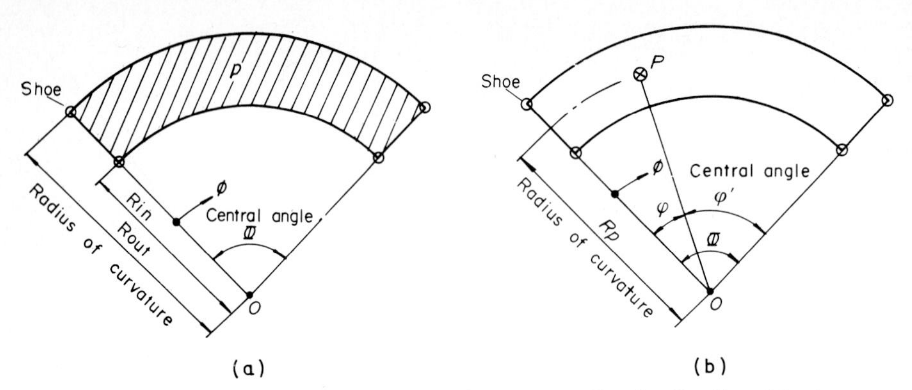

Figure 4.7 Load conditions (plane configuration): (*a*) uniformly distributed load *p*, (*b*) concentrated load *P*.

4.3.1 Load conditions

As for the load conditions imposed on a simply supported curved girder, only two typical loads are considered, for the sake of convenience, as shown in Fig. 4.7; i.e., one is a uniformly distributed load p over the girder surface, and the other is a concentrated load P at $\phi = \varphi$, where the radius of curvature is R_P.

In deriving the solutions of stress resultants and displacements, the following notation is introduced for the uniformly distributed load p;

$$L_1 = \tfrac{1}{3}(R_{\text{out}}^3 - R_{\text{in}}^3) \tag{4.3.1a}$$

$$L_2 = \frac{R_s}{2}(R_{\text{out}}^2 - R_{\text{in}}^2) \tag{4.3.1b}$$

In these equations, R_{out} and R_{in} are the radii of curvature of the outside and inside edges, respectively, of the uniformly distributed load p, as illustrated in Fig. 4.7*a*.

4.3.2 Solutions of stress resultants and displacements for single-span curved girders

(1) Bending moment M_X. The bending moment $M_X(\phi)$ can be obtained from Eq. (3.2.21):

$$\frac{d^2M_X}{d\phi^2} + M_X = pL_1 \tag{4.3.2a}$$

$$\frac{d^2M_X}{d\phi^2} + M_X = PR_P\delta(\phi - \varphi) \tag{4.3.2b}$$

where $\delta(\phi - \varphi)$ is Dirac's delta function. The boundary conditions for

TABLE 4.1 Solutions of the Bending Moment

Load	Solutions for M_X	
p	$pL_1\left[\dfrac{\sin\phi + \sin(\Phi - \phi)}{\sin\Phi} - 1\right]$	
P	$PR_P\dfrac{\sin\varphi'}{\sin\Phi}\sin\phi$	$0 \leqq \phi \leqq \varphi$
	$PR_P\dfrac{\sin(\Phi - \phi)}{\sin\Phi}\sin\varphi$	$\varphi \leqq \phi \leqq \Phi$

M_X at both ends $\phi = 0$ and $\phi = \Phi$ of a simply supported curved girder are

$$[M_X]_{\phi=0} = 0 \tag{4.3.3a}$$

$$[M_X]_{\phi=\Phi} = 0 \tag{4.3.3b}$$

The solutions of M_X are listed in Table 4.1.

(2) Shearing force due to bending $\bar{Q}_Y$. The shearing force $\bar{Q}_Y(\phi)$ due to bending can be estimated from Eq. (3.2.47) as

$$\bar{Q}_Y = \frac{1}{R_o}\frac{dM_X}{d\phi} \tag{4.3.4}$$

and the solutions to this equation are given in Table 4.2.

(3) Warping moment (bimoment) M_ω. The differential equation for the warping moment $M_\omega(\phi)$ is given by Eqs. (3.2.25), (3.2.103), and (3.2.104):

$$\frac{d^2M_\omega}{d\phi^2} - \alpha^2 M_\omega = R_o[q(L_2 - L_1) - M_X] \tag{4.3.5}$$

where

$$\alpha = R_o\sqrt{\frac{G_sK_o}{E_sI_{\omega o}}} \tag{4.3.6}$$

TABLE 4.2 Solutions of Shearing Force due to Bending

Load	Solutions for $\bar{Q}_Y$	
p	$p\dfrac{L_2}{R_s}\left(\dfrac{\Phi}{2} - \phi\right)$	
P	$P\dfrac{\varphi'}{\phi}$	$0 \leqq \phi \leqq \varphi$
	$-P\dfrac{\varphi}{\Phi}$	$\varphi \leqq \phi \leqq \Phi$

The boundary values of M_ω for a simply supported curved girder at the ends $\phi = 0$ and $\phi = \Phi$ are

$$[M_\omega]_{\phi=0} = 0 \tag{4.3.7a}$$

$$[M_\omega]_{\phi=\Phi} = 0 \tag{4.3.7b}$$

The solutions of M_ω are summarized in Table 4.3.

(4) Torsional angle θ. Once the warping moment M_ω has been determined, the torsional angle $\theta(\phi)$ can be easily obtained from Eq. (3.2.27) as

$$\theta = \int_0^\Phi \left(\int \frac{M_\omega}{E_s I_{\omega o}} R_o \, d\phi \right) R_o \, d\phi + C_1 R_o \phi + C_2 \tag{4.3.8}$$

For the boundary conditions of $\theta(\phi)$ for a supported curved girder at both ends $\phi = 0$ and $\phi = \Phi$, integration constants C_1 and C_2 can be determined by

$$[\theta]_{\phi=0} = [\theta]_{\phi=\Phi} \tag{4.3.9}$$

The solution to Eq. (4.3.8) is listed in Table 4.4.

(5) Pure torsional moment T_s. The pure torsional moment $T_s(\phi)$ can be easily obtained from Eq. (3.2.28):

$$T_s = G_s K_o \frac{d\theta}{R_o \, d\phi} \tag{4.3.10}$$

The solutions for T_s are given in Table 4.5.

TABLE 4.3 Solutions for the Warping Moment (Bimoment)

Load	Solutions for M_ω	
p	$pR_o\left[\left(\frac{L_2}{\alpha_2} - \frac{L_1}{\alpha^2+1}\right)\frac{\sinh \alpha\phi + \sinh \alpha(\Phi-\phi)}{\sinh \alpha\Phi} + \frac{L_1}{\alpha^2+1}\frac{\sin\phi + \sin(\Phi-\phi)}{\sin\Phi} - \frac{L_2}{\alpha_2}\right]$	
P	$PR_o\left[\left(\frac{R_P}{\alpha^2+1} - \frac{R_S}{\alpha^2}\right)\alpha\frac{\sinh \alpha\varphi'}{\sinh \alpha\Phi}\sinh \alpha\phi + \frac{R_P}{\alpha^2+1}\frac{\sin\varphi'}{\sin\Phi}\sin\phi\right]$	$0 \leqq \phi \leqq \varphi$
	$PR_o\left[\left(\frac{R_P}{\alpha^2+1} - \frac{R_S}{\alpha^2}\right)\alpha\frac{\sinh \alpha(\Phi-\phi)}{\sinh \alpha\Phi}\sinh \alpha\varphi + \frac{R_P}{\alpha^2+1}\frac{\sin(\Phi-\phi)}{\sin\Phi}\sin\varphi\right]$	$\varphi \leqq \phi \leqq \Phi$

TABLE 4.4 Solutions of the Torsional Angle

Load	Solutions for θ
p	$\frac{pR_o}{G_sK}\left[\left(\frac{L_2}{\alpha^2}-\frac{L_1}{\alpha^2+1}\right)\frac{\sinh\alpha\phi+\sinh\alpha(\Phi-\phi)}{\sinh\alpha\Phi}-\frac{L_1}{\alpha^2+1}\alpha^2\frac{\sin\phi+\sin(\Phi-\phi)}{\sin\Phi}+\frac{L_2}{\alpha^2}\left(\frac{\alpha^2\phi\Phi}{2}-\frac{\alpha^2\phi^2}{2}-1\right)+L_1\right]$
P	$\frac{PR_o}{G_sK}\left[\left(\frac{R_P}{\alpha^2+1}-\frac{R_S}{\alpha^2}\right)\alpha\frac{\sinh\alpha\varphi'}{\sinh\alpha\Phi}\sinh\alpha\phi-\frac{R_P}{\alpha^2+1}\alpha^2\frac{\sin\varphi'}{\sin\Phi}\sin\phi+R_S\frac{\varphi'}{\Phi}\phi\right]$ $\quad 0\leqq\phi\leqq\varphi$ $\frac{PR_o}{G_sK}\left[\left(\frac{R_P}{\alpha^2+1}-\frac{R_S}{\alpha^2}\right)\alpha\frac{\sinh\alpha(\Phi-\phi)}{\sinh\alpha\Phi}\sinh\alpha\varphi-\frac{R_P}{\alpha^2+1}\alpha^2\frac{\sin(\Phi-\phi)}{\sin\Phi}\sin\varphi+R_S\frac{\Phi-\phi}{\Phi}\right]$ $\quad \varphi\leqq\phi\leqq\Phi$

(6) Secondary torsional moment T_ω. The remaining secondary torsional moment $T_\omega(\phi)$ can also be evaluated from Eq. (3.2.29):

$$T_\omega = G_sK_o\frac{d^3\theta}{R_o^3\,d\phi^3} \tag{4.3.11}$$

or

$$T_\omega = T_X - T_s \tag{4.3.12}$$

where $T_X(\phi)$ is the total torsional moment, shown in Table 4.6.

TABLE 4.5 Solutions of the Pure Torsional Moment

Load	Solutions for T_s
p	$p\left[\left(\frac{L_1}{\alpha^2+1}-\frac{L_2}{\alpha^2}\right)\alpha\frac{\cosh\alpha(\Phi-\phi)-\cosh\alpha\phi}{\sinh\alpha\Phi}+\frac{L_1\alpha^2}{\alpha^2+1}\frac{\cos(\Phi-\phi)-\cos\phi}{\sin\Phi}+L_2\left(\frac{\Phi}{2}-\phi\right)\right]$
P	$P\left[\left(\frac{R_P}{\alpha^2+1}-\frac{R_S}{\alpha^2}\right)\alpha^2\frac{\sinh\alpha\varphi'}{\sinh\alpha\Phi}\cosh\alpha\phi-\frac{R_P\alpha^2}{\alpha^2+1}\frac{\sin\varphi'}{\sin\Phi}\cos\phi+R_S\frac{\varphi'}{\Phi}\right]$ $\quad 0\leqq\phi\leqq\varphi$ $P\left[\left(\frac{R_S}{\alpha^2}-\frac{R_P}{\alpha^2+1}\right)\alpha^2\frac{\cosh\alpha(\Phi-\phi)}{\sinh\alpha\Phi}\sinh\alpha\varphi+\frac{R_P\alpha^2}{\alpha^2+1}\frac{\cos(\Phi-\phi)}{\sin\Phi}\sin\varphi-R_S\frac{\varphi}{\Phi}\right]$ $\quad \varphi\leqq\phi\leqq\Phi$

TABLE 4.6 Solutions of the Total Torsional Moment

Load	Solutions of $T_X = T_s + T_\omega$	
p	$p\left[L_1 \dfrac{\cos(\Phi-\phi)-\cos\phi}{\sin\Phi} + L_2\left(\dfrac{\Phi}{2}-\phi\right)\right]$	
P	$-P\left[R_P \dfrac{\sin\varphi'}{\sin\Phi}\cos\phi - R_s\dfrac{\varphi'}{\Phi}\right]$	$0 \leqq \phi \leqq \varphi$
	$P\left[R_P \dfrac{\cos(\Phi-\phi)}{\sin\Phi}\sin\varphi - R_s\dfrac{\varphi}{\Phi}\right]$	$\varphi \leqq \phi \leqq \Phi$

(7) Deflection w. The deflection $w(\phi)$ of a simply supported curved girder with the boundary conditions

$$[w]_{\phi=0} = 0 \tag{4.3.13a}$$

$$[w]_{\phi=\Phi} = 0 \tag{4.3.13b}$$

at both ends $\phi = 0$ and $\phi = \Phi$ can be estimated from Eq. (3.2.30):

$$\frac{d^2w}{d\phi^2} + w = -R_o^2\left(\frac{M_X}{E_sI_X} - \frac{\theta}{R_o}\right) \tag{4.3.14}$$

These results are summarized in Table 4.7, where the following notation

TABLE 4.7 Solutions of Deflection

Load	Solutions for w
p	$pR_oR_S\left\{-\dfrac{\kappa_2}{\alpha^2}\dfrac{\sinh\alpha\phi+\sinh\alpha(\Phi-\phi)}{\sinh\alpha\Phi} - \omega_1\dfrac{\sin\phi+\sin(\Phi-\phi)}{\sin\Phi} + \dfrac{1}{GK}\left[\dfrac{L_2}{\alpha^2}\left(\dfrac{\alpha^2\Phi\phi}{2}-\dfrac{\alpha^2\phi^2}{2}-1\right)+L_1\right] - \kappa_1\left[\dfrac{1-\cos\Phi}{\sin\Phi}\left(\Phi\cos\Phi\dfrac{\sin\phi}{\sin\Phi}-\phi\cos\phi\right)+(\phi-\Phi)\sin\phi\right]+\kappa_3\right\}$
P	$P\dfrac{R_oR_S}{R_P}\left\{\dfrac{\mu_2}{\alpha}\dfrac{\sinh\alpha\varphi'}{\sinh\alpha\Phi}\sinh\alpha\phi + \alpha^2\left[\mu_2 - \dfrac{R_P^2}{GK(\alpha^2+1)}\right]\dfrac{\sin\varphi'}{\sin\Phi}\sin\varphi - \mu_1\left[\dfrac{\sin\varphi'}{\sin\Phi}\left(\Phi\cos\Phi\dfrac{\sin\varphi}{\sin\Phi}-\Phi\cos\phi\right)+(\sin\varphi'-\varphi'\cos\varphi')\dfrac{\sin\varphi}{\sin\Phi}\right] + \dfrac{R_oR_P}{GK}\dfrac{\varphi'}{\Phi}\phi\right\}$ $\quad 0 \leqq \phi \leqq \varphi$ For $\varphi < \phi < \Phi$, set $\varphi' = \Phi - \phi$ and $\phi = \varphi$

is introduced:

$$\omega_1 = \frac{L_1 \alpha^2}{G_s K_o(\alpha^2+1)} + \kappa_2 + \kappa_3 \tag{4.3.15a}$$

$$\omega_2 = \frac{R_p^2 \alpha^2}{G_s K_o(\alpha^2+1)} + \mu_2 + \mu_3 \tag{4.3.15b}$$

where

$$\kappa_1 = \left[\frac{R_o^2}{E_s I_{\omega o}(\alpha^2+1)} + \frac{1}{E_s I'_X}\right]\frac{L_1}{2} \tag{4.3.16a}$$

$$\kappa_2 = \frac{R_o^2}{E_s I_{\omega o}(\alpha^2+1)}\left(\frac{L_1}{\alpha^2+1} - \frac{L_2}{\alpha^2}\right) \tag{4.3.16b}$$

$$\kappa_3 = \frac{R_o^2 L_2}{E_s I_{\omega o}\alpha^2} + \frac{L_1}{E_s I'_X} \tag{4.3.16c}$$

and

$$\mu_1 = \frac{R_p^2}{2}\left[\frac{R_o^2}{E_s I_{\omega o}(\sigma^2+1)} + \frac{1}{E_s I'_X}\right] \tag{4.3.17a}$$

$$\mu_2 = \frac{R_o^2 R_p}{E_s I_{\omega o}(\alpha^2+1)}\left(\frac{R_p}{\alpha^2+1} - \frac{R_s}{\alpha^2}\right) \tag{4.3.17b}$$

$$\mu_3 = R_p\left(\frac{R_s}{G_s K_o} + \frac{R_p}{E_s I'_X}\right) \tag{4.3.17c}$$

in which

$$I'_X = I_X - \frac{I_{XY}^2}{I_Y} \tag{4.3.18}$$

(8) Rotational angle β. The rotational angle $\beta(\phi)$ can be directly calculated from Eq. (3.2.14a):

$$\beta = \theta - \frac{w}{R_o} \tag{4.3.19}$$

4.3.3 Application to continuous curved girders[5]

Figure 4.8 shows the side elevation of a continuous curved girder bridge having supports $m = 1, 2, \ldots, n$, central angle Φ_m, and radius of

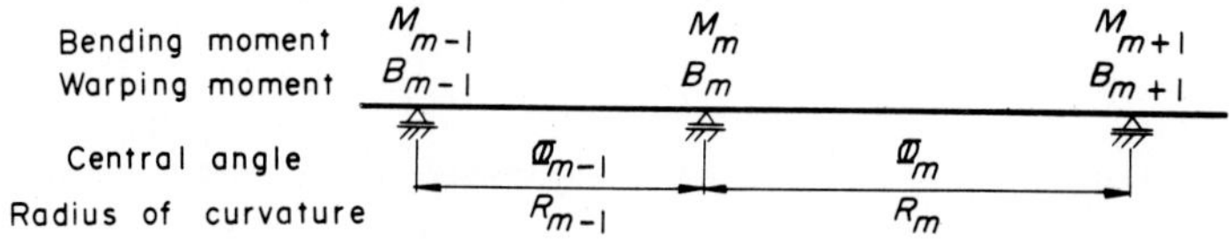

Bending moment M_{m-1} M_m M_{m+1}
Warping moment B_{m-1} B_m B_{m+1}
Central angle Φ_{m-1} Φ_m
Radius of curvature R_{m-1} R_m

Figure 4.8 Side elevation of continuous curved girder and statically indeterminate bending moment M_m and warping moment B_m.

curvature R_m. This continuous curved girder is separated into simply supported spans by cutting the girder at each intermediate support. The statically indeterminate bending moment M_m and warping moment B_m are introduced, to maintain the continuity conditions for the deflectional angle and the rotational angles. Now, by assigning the stress resultants for these simple curved girders (see Sec. 4.3.2) with the suffix o and superimposing the statically indeterminate stress resultants, the following stress resultants can be obtained, provided that

$$\phi'_m = \Phi_m - \phi_m \tag{4.3.20}$$

Bending moment:

$$M_{X,m} = M_{m,o} + M_m \frac{\sin \phi'_m}{\sin \Phi_m} + M_{m+1} \frac{\sin \phi_m}{\sin \Phi_m} \tag{4.3.21}$$

Warping moment:

$$\begin{aligned} M_{\omega,m} = M_{\omega,o} &+ B_m \frac{\sinh \alpha_m \phi'_m}{\sinh \alpha_m \Phi_m} + B_{m+1} \frac{\sinh \alpha_m \phi_m}{\sinh \alpha_m \Phi_m} \\ &+ \frac{R_{o,m}}{\alpha^2_{m+1}} \left[M_m \left(\frac{\sin \phi'_m}{\sin \Phi_m} - \frac{\sinh \alpha_m \phi'_m}{\sinh \alpha_m \Phi_m} \right) \right. \\ &\left. + M_{m+1} \left(\frac{\sin \phi_m}{\sin \Phi_m} - \frac{\sinh \alpha_m \phi_m}{\sinh \alpha_m \Phi_m} \right) \right] \end{aligned} \tag{4.3.22}$$

Pure torsional moment:

$$\begin{aligned} T_{s,m} = T_{s,mo} &+ \frac{B_m}{R_{o,m}} \left(\frac{1}{\Phi_m} - \frac{\alpha_m \cosh \alpha_m \phi'_m}{\sinh \alpha_m \Phi_m} \right) \\ &+ \frac{B_{m+1}}{R_{o,m}} \left(\frac{\alpha_m \cosh \alpha_m \phi_m}{\sinh \alpha_m \Phi_m} - \frac{1}{\Phi_m} \right) \\ &+ M_m \left[\frac{1}{\alpha^2_m + 1} \left(\alpha^2_m \frac{\cos \phi'_m}{\sin \Phi_m} + \alpha_m \frac{\cosh \alpha_m \phi'_m}{\sinh \alpha_m \Phi_m} \right) - \frac{1}{\Phi_m} \right] \\ &+ M_{m+1} \left[\frac{1}{\Phi_m} - \frac{1}{\alpha^2_m + 1} \left(\alpha^2_m \frac{\cos \phi_m}{\sin \Phi_m} + \alpha_m \frac{\cosh \alpha_m \phi_m}{\alpha_m \Phi_m} \right) \right] \end{aligned} \tag{4.3.23}$$

where

$$\alpha_m = R_{o,m} \sqrt{\frac{G_s K_{o,m}}{E_s I_{\omega o,m}}} \tag{4.3.24}$$

There will, of course, be secondary torsional moments $T_{\omega,m}$ and shearing forces $Q_{y,m}$ due to bending. However, the strain energy stored in a continuous curved girder can be approximated through the use of the above stress resultants $M_{X,m}$, $M_{\omega,m}$, and $T_{s,m}$ with high accuracy. Thus

$$U = \sum_{m=1}^{n} \left(\int_0^{\Phi_m} \frac{M_{X,m}^2}{2E_s I_{X,m}} R_{o,m}\, d\phi_m + \int_0^{\Phi_m} \frac{M_{\omega,m}^2}{2E_s I_{\omega,o}} R_{o,m}\, d\phi_m + \int_0^{\Phi_m} \frac{T_{s,m}^2}{2G_s K_{o,m}} R_{o,m}\, d\phi_m \right) \qquad (4.3.25)$$

According to the principle of least work, the statically indeterminate bending moment M_m and warping moment B_m are

$$\frac{\partial U}{\partial M_m} = 0 \qquad (4.3.26a)$$

$$\frac{\partial U}{\partial B_m} = 0 \qquad (4.3.26b)$$

From these equations, the following elastic equations (three moment equations) can be obtained:

$$a_{m,m-1} M_{m-1} + a_{m,m} M_m + a_{m,m+1} M_{m+1} + b_{m,m-1} B_{m-1} + b_{m,m} B_m + b_{m,m+1} B_{m+1} = -d_m \qquad (4.3.27a)$$

$$b_{m,m-1} M_{m-1} + b_{m,m} M_m + b_{m,m+1} M_{m+1} + c_{m,m-1} B_{m-1} + c_{m,m} B_m + c_{m,m+1} B_{m+1} = -e_m \qquad m = 1, 2, \ldots, n-1 \qquad (4.3.27b)$$

A continuous curved girder which is simply supported at both ends $m = 0$ and $m = n$ yields

$$M_o = M_n = 0 \qquad (4.3.28a)$$

and

$$B_o = B_n = 0 \qquad (4.3.28b)$$

For further details on coefficients a, b, and c and load terms d and e in Eqs. (4.3.27a) and (4.3.27b), the interested reader is referred to Konishi and Komatsu.[5]

4.4 Transfer Matrix Method

The determination of stress resultants and displacements in a curved girder with arbitrary conditions is very complicated and takes great effort to handle with either the force method or the analytical method, based on the differential equations given in Sec. 4.3. To avoid complex and time-consuming procedures, the transfer matrix method[6–9] is introduced, which is a powerful tool that can be applied with a computer.

This method is essentially based on the general solutions of the differential equations governing the behavior of curved girders. These solutions are obtained through the use of a matrix calculus detailed in the following sections.

4.4.1 Pure torsional theory

The following transfer matrix method, based on pure torsional theory, was developed by Becker.[10] Therefore, the notations and signs of stress resultants and displacements are in accordance with his work.[11,12]

(1) Notation and signs. Figure 4.9*a* shows the general plan of a curved girder with variable cross section and arbitrary boundary conditions, where the curved girder is subdivided into small segments by the panel points $k = 1, 2, \ldots, k, \ldots, n$ with radii of curvature $R_{s,1}, R_{s,2}, \ldots, R_{s,k}, \ldots, R_{s,n}$ and central angles $\Phi_1, \Phi_2, \ldots, \Phi_k, \ldots, \Phi_n$. Thus the panel length of the kth panel is

$$l_k = R_{s,k}\Phi_k \tag{4.4.1}$$

In the coordinate system of these segments, the right-hand-side coordinate system (x_k, y_k, z_k) is taken as the shear central axis of the kth panel. In addition, the curvilinear coordinate system (ρ_k, ϕ_k) is taken as the subcoordinate system shown in Fig. 4.9*b*.

Thus, the stress resultants and displacements acting on these coordinate axes can be defined as

w = deflection in direction of y axis

θ = rotational angle with respect to z axis

φ = deflection angle with respect to x axis

M = bending moment with respect to x axis

T = torsional moment with respect to z axis

Q = shearing force in direction of y axis

Note that the signs of the bending moment and shearing force are different from the definitions given in ordinary strength of materials.

The applied loads are also considered:

P = intensity of concentrated load

q = intensity of uniformly distributed load

R_p = radius of curvature at loading point of concentrated load

R_{out}, R_{in} = radius of curvature at outer and inner edges of uniformly distributed load, respectively

ψ = central angle between left end and loading edge of uniformly distributed load

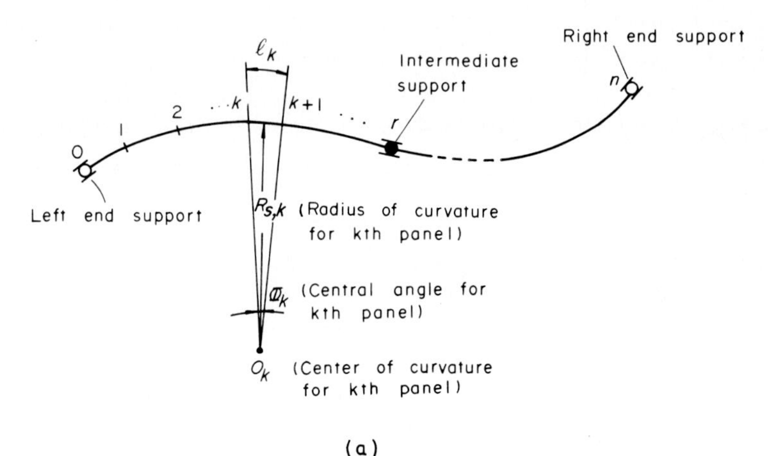

(b)

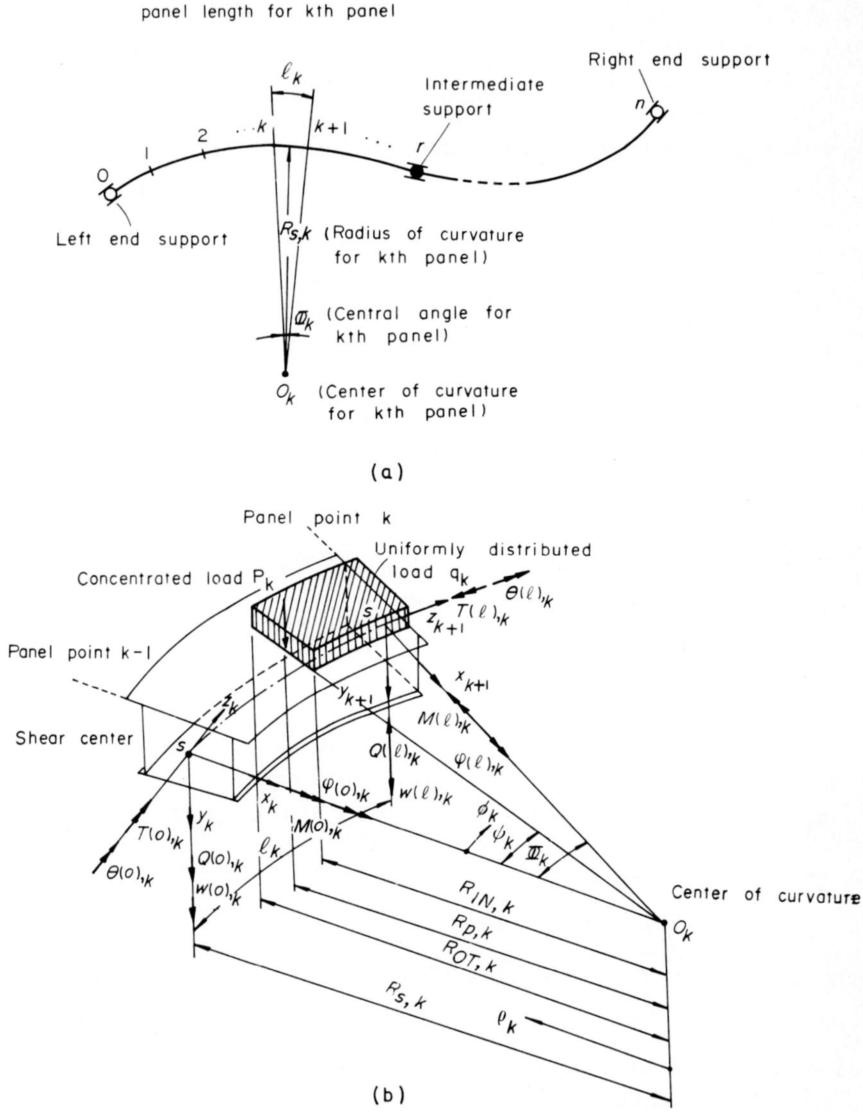

Figure 4.9 Curved box girder: (*a*) general plan, (b) detail of geometry, loads, stress resultants, and displacements for *k*th panel.

The properties of material are represented by

E = Young's modulus

G = shear modulus of elasticity

I = geometric moment of inertia

K = pure torsional constant

Strictly speaking, the subscript k must be added to all the above quantities to express the quantities of the kth panel. Moreover, the following subscripts are attached for the stress resultants and displacements:

0 = values at left end panel

Φ = values at right end panel

Finally, the following quantities are introduced to nondimensionalize the stress resultants and displacements:

I_c = specific geometric moment of inertia

R_c = specific radius of curvature

(2) Derivation of the field transfer matrix. The general solutions of the differential equations for the stress resultants and displacements consist of homogeneous solutions and particular solutions. These solutions are treated separately in the following sections.

a. **Homogeneous solution.** Figure 4.10 shows the stress resultants acting on a small element $dl_k = R_{s,k}\, d\phi_k$ of the kth panel of a curved girder. From this figure, the equilibrium conditions of the stress

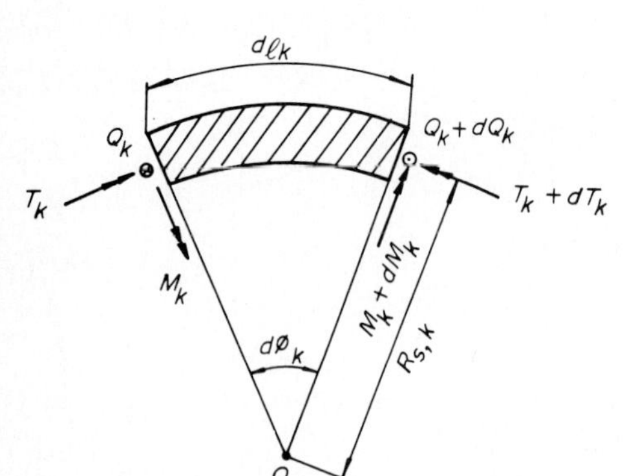

Figure 4.10 Equilibrium of stress resultants.

resultants can be reduced to

$$\frac{dM_k}{d\phi_k} + T_k = Q_k R_{s,k} \tag{4.4.2a}$$

$$\frac{dT_k}{d\phi_k} = M_k \tag{4.4.2b}$$

$$\frac{dQ_k}{d\phi_k} = 0 \tag{4.4.2c}$$

The above equations give the following differential equation for the bending moment:

$$\frac{d^2 M_k}{d\phi_k^2} + M_k = 0 \tag{4.4.3}$$

Solving this equation for the bending moment M_k and expressing the bending moment $M(\Phi)_{,k}$ at the right-end side of panel k by the bending moment $M(0)_{,k}$, torsional moment $T(0)_{,k}$, and shearing force $Q(0)_{,k}$ at the left-end side of panel k, we get

$$M(\Phi)_{,k} = M(0)_{,k} \cos \Phi_k - T(0)_{,k} \sin \Phi_k + Q(0)_{,k} R_{s,k} \sin \Phi_k \tag{4.4.4}$$

The shearing force obviously can be represented by Eq. (4.4.2c) as

$$Q(\Phi)_{,k} = Q(0)_{,k} \tag{4.4.5}$$

Therefore, the torsional moment can be obtained from

$$T_k = Q_{s,k} R_{s,k} - \frac{dM_k}{d\phi_k} \tag{4.4.6}$$

This results in

$$T(\Phi)_{,k} = M(0)_{,k} \sin \Phi_k + T(0)_{,k} \cos \Phi_k + Q(0)_{,k} R_{s,k}(1 - \cos \Phi_k) \tag{4.4.7}$$

However, the relationships between stress resultants and displacements can be determined by referring to Fig. 4.11:

$$E_k I_k \left(\frac{d\varphi_k}{dl_k} + \frac{\theta_k}{R_{s,k}} \right) = -M_k \tag{4.4.8a}$$

$$G_k K_k \left(\frac{d\theta_k}{dl_k} - \frac{\varphi_k}{R_{s,k}} \right) = -T_k \tag{4.4.8b}$$

where φ_k is the deflection angle designated by

$$\varphi_k = -\frac{dw_k}{R_{s,k}\, d\phi_k} \tag{4.4.9}$$

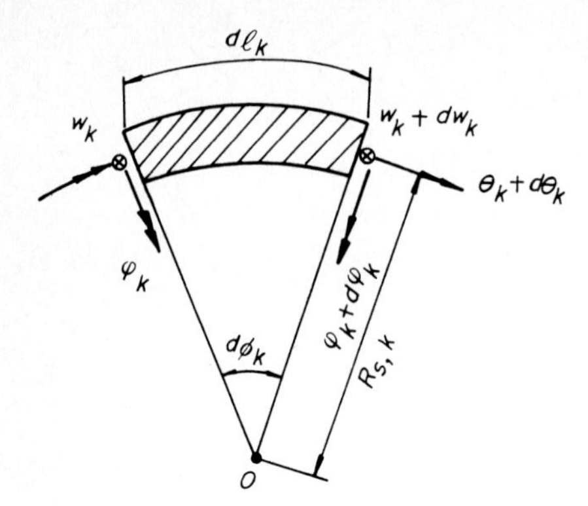

Figure 4.11 Displacements.

The differential equation for the deflection angle is [see Eq. (4.4.8*a*)]

$$\frac{d^2\phi_k}{R_{s,k}^2\,d\phi_k^2} + \frac{\varphi_k}{R_{s,k}^2} = -\frac{1}{E_kI_k}\frac{dM_k}{R_{s,k}\,d\phi_k} + \frac{1}{R_{s,k}}\frac{T_k}{G_kK_k} \tag{4.4.10}$$

The solution of this equation results in

$$\begin{aligned}\varphi(\Phi)_{,k} &= -\theta(0)_{,k}\sin\Phi_k + \varphi(0)_{,k}\cos\Phi_k\\ &+ M(0)_{,k}R_{s,k}\left[\frac{1}{2}\left(\frac{1}{E_kI_k}+\frac{1}{G_kK_k}\right)(\sin\Phi_k - \Phi_k\cos\Phi_k) - \frac{1}{E_kI_k}\sin\Phi_k\right]\\ &+ T(0)_{,k}\left[\frac{R_{s,k}}{2}\left(\frac{1}{E_kI_k}+\frac{1}{G_kK_k}\right)\Phi_k\sin\Phi_k\right]\\ &+ Q(0)_{,k}R_{s,k}^2\left[\frac{1}{G_kK_k}(1-\cos\Phi_k) - \frac{1}{2}\left(\frac{1}{E_kI_k}+\frac{1}{G_kK_k}\right)\Phi_k\sin\Phi_k\right]\end{aligned} \tag{4.4.11}$$

Next, the rotational angle θ_k is given by [see Eq. (4.4.8*b*)]

$$\theta(\Phi)_{,k} = \int_0^{\Phi_k} R_{s,k}\left(-\frac{T_k}{G_kK_k} + \frac{\varphi_k}{R_{s,k}}\right)d\phi_k + C_1 \tag{4.4.12}$$

from Eq. (4.4.8*b*) in which the integration constant C_1

$$C_1 = \theta(0)_{,k}\cos\Phi_k \tag{4.4.13}$$

Equation (4.4.12) has the following solution:

$$\begin{aligned}\theta(\Phi)_{,k} &= \theta(0)_{,k}\cos\Phi_k + \varphi(0)_{,k}\sin\Phi_k\\ &- M(0)_{,k}R_{s,k}\left[\frac{\Phi_k}{2}\left(\frac{1}{E_kI_k}+\frac{1}{G_kK_k}\right)\Phi_k\sin\Phi_k\right]\end{aligned}$$

$$+ T(0)_{,k} R_{s,k} \left[\frac{1}{2} \left(\frac{1}{E_k I_k} + \frac{1}{G_k K_k} \right) (\sin \Phi_k - \Phi_k \cos \Phi_k) - \frac{1}{G_k K_k} \sin \Phi_k \right]$$

$$- Q(0)_{,k} R^2_{s,k} \left[\frac{1}{2} \left(\frac{1}{E_k I_k} + \frac{1}{G_k K_k} \right) (\sin \Phi_k - \Phi_k \cos \Phi_k) \right] \quad (4.4.14)$$

Finally, the deflection w_k can be obtained from Eq. (4.4.9):

$$w(\Phi)_{,k} = - \int_0^{\Phi_k} R_{s,k} \varphi_k d\phi_k + C_2 \quad (4.4.15)$$

where the integration constant C_2 is

$$C_2 = w(0)_{,k} \quad (4.4.16)$$

and the solution of Eq. (4.4.15) is

$$w(\Phi)_{,k} = w(0)_{,k} - \theta(0)_{,k} R_{s,k} (\cos \Phi_k - 1) - \varphi(0)_{,k} R_{s,k} \sin \Phi_k$$

$$+ M(0)_{,k} R^2_{s,k} \left[\left(\frac{1}{E_k I_k} + \frac{1}{G_k K_k} \right) \left(\cos \Phi_k + \frac{\Phi_k}{2} \sin \Phi_k \right) \right.$$

$$\left. - \frac{1}{E_k I_k} \cos \Phi_k - \frac{1}{G_k K_k} \right] - T(0)_{,k} \left[\frac{R^2_{s,k}}{2} \left(\frac{1}{E_k I_k} + \frac{1}{G_k K_k} \right) \right.$$

$$\left. \times (\sin \Phi_k - \Phi_k \cos \Phi_k) \right] - Q(0)_{,k} R^3_{s,k} \left[\frac{1}{2} \left(\frac{1}{E_k I_k} + \frac{1}{G_k K_k} \right) \right.$$

$$\left. \times (\Phi_k \cos \Phi_k - \sin \Phi_k) + \frac{1}{G_k K_k} (\Phi_k - \sin \Phi_k) \right] \quad (4.4.17)$$

Now, all the above stress resultants and displacements are nondimensionalized (indicated by asterisk) as follows:

$$w = \frac{P_c R_c^3}{E_c I_c} w^* \quad (4.4.18a)$$

$$\theta = \frac{P_c P_c^2}{E_c I_c} \theta^* \quad (4.4.18b)$$

$$\varphi = \frac{P_c R_c^2}{E_c I_c} \varphi^* \quad (4.4.18c)$$

$$M = P_c R_c M^* \quad (4.4.18d)$$

$$T = P_c R_c T^* \quad (4.4.18e)$$

$$Q = P_c Q^* \quad (4.4.18f)$$

In addition, the following nondimensionalized parameters are introduced:

$$\alpha_k = \frac{G_k K_k}{E_k I_k} \tag{4.4.19a}$$

$$g_k = \frac{E_c I_c}{E_k I_k} \tag{4.4.19b}$$

Now, the state vectors $\mathbf{x}^*_{(0),k}$ and $\mathbf{x}^*_{(\Phi),k}$ at the left and right ends of panel k, which are respectively represented by

$$\mathbf{x}^*_{(0),k} = [w(0)^*_{,k}\ \theta(0)^*_{,k}\ \varphi(0)^*_{,k}\ M(0)^*_{,k}\ T(0)^*_{,k}\ Q(0)^*_{,k}\ 1] \tag{4.4.20a}$$

$$\mathbf{x}^*_{(\Phi),k} = [w(\Phi)^*_{,k}\ \theta(\Phi)^*_{,k}\ \varphi(\Phi)^*_{,k}\ M(\Phi)^*_{,k}\ T(\Phi)^*_{,k}\ Q(\Phi)^*_{,k}\ 1] \tag{4.4.20b}$$

and may be written as

$$\mathbf{x}^*_{(\Phi),k} = \mathbf{F}^*_k \mathbf{x}^*_{(0),k} \tag{4.4.21}$$

where

$$\mathbf{F}^*_k = \begin{array}{c} \begin{matrix} w^* & \theta^* & \varphi^* & M^* & T^* & Q^* & 1 \end{matrix} \\ \begin{bmatrix} 1 & a^*_{12} & a^*_{13} & a^*_{14} & a^*_{15} & a^*_{16} & \tilde{w}^*_k \\ 0 & \cos\Phi_k & \sin\Phi_k & a^*_{24} & a^*_{25} & a^*_{26} & \tilde{\theta}^*_k \\ 0 & -\sin\Phi_k & \cos\Phi_k & a^*_{34} & a^*_{35} & a^*_{36} & \tilde{\varphi}^*_k \\ 0 & 0 & 0 & \cos\Phi_k & -\sin\Phi_k & a^*_{46} & \tilde{M}^*_k \\ 0 & 0 & 0 & \sin\Phi_k & \cos\Phi_k & a^*_{56} & \tilde{T}^*_k \\ 0 & 0 & 0 & 0 & 0 & 1 & \tilde{Q}^*_k \\ 0 & 0 & 0 & 0 & 0 & 0 & 1 \end{bmatrix} \end{array} \tag{4.4.22}$$

Here $\mathbf{F}^*_k$ is the field transfer matrix, and the remaining undefined elements a^*_{ij} in the above equation can be estimated as follows:

$$a^*_{12} = \frac{R_{s,k}}{R_c}(1 - \cos\Phi_k) = a^*_{56} \tag{4.4.23a}$$

$$a^*_{13} = -\frac{R_{s,k}}{R_c}\sin\Phi_k = -a^*_{46} \tag{4.4.23b}$$

$$a^*_{14} = \frac{g_k}{2\alpha_k}\left(\frac{R_{s,k}}{R_c}\right)^2[(1+\alpha_k)\Phi_k\sin\Phi_k + 2\cos\Phi_k - 2] = -a^*_{36} \tag{4.4.23c}$$

$$a^*_{15} = \frac{1+\alpha_k}{2\alpha_k}g_k\left(\frac{R_{s,k}}{R_c}\right)^2(\Phi_k\cos\Phi_k - \sin\Phi_k) = a^*_{26} \tag{4.4.23d}$$

$$a_{16}^{*} = \frac{g_k}{2\alpha_k}\left(\frac{R_{s,k}}{R_c}\right)^3 [(3+\alpha_k)\sin\Phi_k - (1+\alpha_k)\Phi_k\cos\Phi_k - 2\Phi_k] \tag{4.4.23e}$$

$$a_{24}^{*} = -\frac{1+\alpha_k}{2\alpha_k} g_k \frac{R_{s,k}}{R_c}\Phi_k \sin\Phi_k = -a_{35}^{*} \tag{4.4.23f}$$

$$a_{25}^{*} = \frac{g_k}{2\sigma_k}\left(\frac{R_{s,k}}{R_c}\right)[(\alpha_k - 1)\sin\Phi_k - (\alpha_k+1)\Phi_k\cos\Phi_k] \tag{4.4.23g}$$

$$a_{34}^{*} = \frac{g_k}{2\alpha_k}\left(\frac{R_{s,k}}{R_c}\right)[(1-\alpha_k)\sin\Phi_k - (\alpha_k+1)\Phi_k\cos\Phi_k] \tag{4.4.23h}$$

***b.* Particular solutions.** The loading terms $\tilde{w}_k^*$, $\tilde{\theta}_k^*$, $\tilde{\varphi}_k^*$, $\tilde{M}_k^*$, $\tilde{T}_k^*$, and $\tilde{Q}_k^*$ in Eq. (4.4.22) are obtained from the particular solutions (indicated by tilde) of the differential equations (4.4.2) and (4.4.8). These solutions can be found easily by using the following equations for the loading conditions illustrated in Figs. 4.12 and 4.13:

$$\frac{d\tilde{Q}_k}{d\phi_k} = -\int_{R_{\text{in}}}^{R_{\text{out}}} P_o\delta(\Phi_k - \Psi_k)\,d\rho_k \qquad \text{for Fig. 4.12} \tag{4.4.24a}$$

$$\frac{d\tilde{Q}_k}{d\phi_k} = -\int_{R_{\text{in}}}^{R_{\text{out}}} q\rho_k U(\Phi_k - \Psi_k)\,d\rho_k \qquad \text{for Fig. 4.13} \tag{4.4.24b}$$

$$\frac{d^2\tilde{M}_k}{d\phi_k^2} + \tilde{M}_k = \frac{P_o}{R_{s,k}} L_{1,k}\delta(\Phi_k - \Psi_k) \qquad \text{for Fig. 4.12} \tag{4.4.25a}$$

$$\frac{d^2\tilde{M}_k}{d\phi_k^2} + \tilde{M}_k = qL_{2,k}U(\Phi_k - \Psi_k) \qquad \text{for Fig. 4.13} \tag{4.4.25b}$$

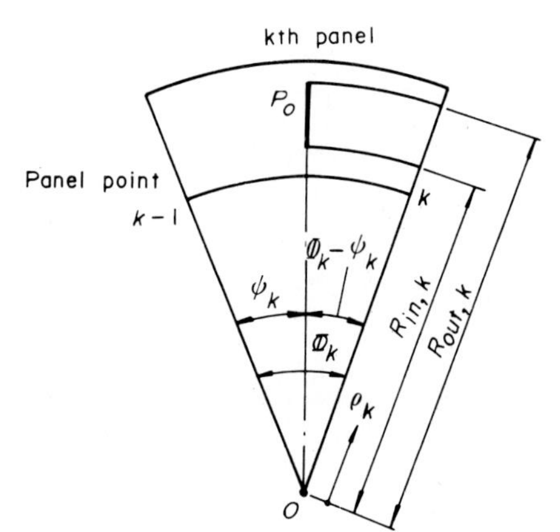

Figure 4.12 Line load P_o in radial direction.

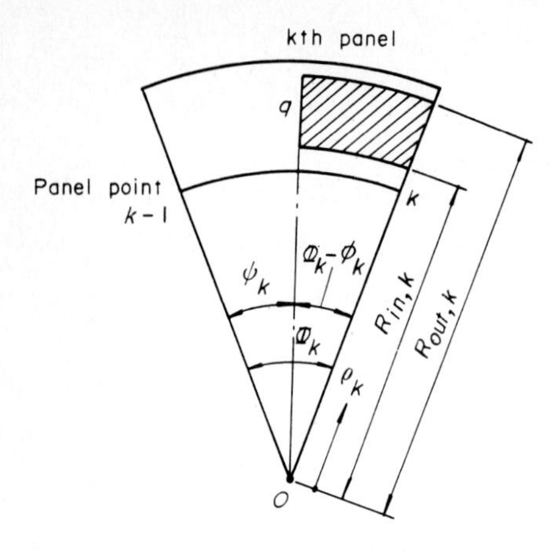

Figure 4.13 Partially distributed load q.

where
$$L_{1,k} = \frac{R_{s,k}}{2}(R^2_{\text{out},k} - R^2_{\text{in},k}) \tag{4.4.26a}$$

$$L_{2,k} = \tfrac{1}{3}(R^3_{\text{out},k} - R^3_{\text{in},k}) \tag{4.4.26b}$$

$$L_{3,k} = R^2_{s,k}(R_{\text{out},k} - R_{\text{in},k}) \tag{4.4.26c}$$

with δ and U being, respectively, defined as Dirac's delta function and the unit step function. These results are listed in Tables 4.8 and 4.9.

TABLE 4.8 Load Terms in Eq. (4.4.22) for Radial Line Force of Fig. 4.12

Load term	Setting formulas
$\tilde{w}^*_k$	$\frac{P_o R_{s,k} g_k}{2P_c R_c^3 \alpha_k}\{L_{1,k}(\alpha_k+1)[\sin(\Phi_k-\Psi_k)-(\Phi_k-\Psi_k) \times\cos(\Phi_k-\Psi_k)-2L_{3,k}(\Phi_k-\Psi_k)-\sin(\Phi_k-\Psi_k)]\}$
$\tilde{\theta}^*_k$	$\frac{P_o g_k}{2P_c R_c^2 \alpha_k}\{-L_{1,k}(\alpha_k+1)[\sin(\Phi_k-\Psi_k)-(\Phi_k-\Psi_k) \times\cos(\Phi_k-\Psi_k)]+2(L_{1,k}-L_{3,k})\sin(\Phi_k-\Psi_k)\}$
$\tilde{\varphi}^*_k$	$\frac{P_o g_k}{2P_c R_c^2 \alpha_k}\{-L_{1,k}(\alpha_k+1)(\Phi_k-\Psi_k) \times\sin(\Phi_k-\Psi_k)+2L_{3,k}[1-\cos(\Phi_k-\Psi_k)]\}$
$\tilde{M}^*_k$	$\frac{P_o L_{1,k}}{P_c R_c R_{s,k}}\sin(\Phi_k-\Psi_k)$
$\tilde{T}^*_k$	$\frac{P_o}{P_c R_c R_{s,k}}[L_{3,k}-L_{1,k}\cos(\Phi_k-\Psi_k)]$
$\tilde{Q}^*_k$	$\frac{P_o L_{3,k}}{P_c R^2_{s,k}}$

TABLE 4.9 Load Terms in Eq. (4.4.22) for Partially Distributed Force of Fig. 4.13

Load term	Setting formulas
$\tilde{w}_k^*$	$\frac{qR_{s,k}g_k}{2P_cR_c^2\alpha_k}\{L_{2,k}(\alpha_k+1)[2-(\Phi_k-\Psi_k)\sin(\Phi_k-\Psi_k)-2\cos(\Phi_k-\Psi_k)]+L_{1,k}[2-(\Phi_k-\Psi_k)^2-2\cos(\Phi_k-\Psi_k)]\}$
$\tilde{\theta}_k^*$	$\frac{qR_{s,k}g_k}{2P_cR_c^2\alpha_k}\{l_{2,k}(\alpha_k+1)[-2+(\Phi_k-\Psi_k)\sin(\Phi_k-\Psi_k)+2\cos(\Phi_k-\Psi_k)]+2(L_{2,k}-L_{1,k})[1-\cos(\Phi_k-\Psi_k)]\}$
$\tilde{\varphi}_k^*$	$\frac{qR_{s,k}g_k}{2P_cR_c^2\alpha_k}\{L_{2,k}(\alpha_k+1)[(\Phi_k-\Psi_k)\cos(\Phi_k-\Psi_k)-\sin(\Phi_k-\Psi_k)]+2L_{1,k}[-\sin(\Phi_k-\Psi_k)+(\Phi_k-\Psi_k)]\}$
$\tilde{M}_k^*$	$\frac{qL_{2,k}}{P_cR_c}[1-\cos(\Phi_k-\Psi_k)]$
$\tilde{T}_k^*$	$\frac{q}{P_cR_c}[L_{1,k}(\Phi_k-\Psi_k)-L_{2,k}\sin(\Phi_k-\Psi_k)]$
$\tilde{Q}_k^*$	$\frac{qL_{1,k}}{P_cR_{s,k}}(\Phi_k-\Psi_k)$

These formulas can also be simplified for the forces illustrated in Figs. 4.14 through 4.16 as follows: For concentrated load P (Fig. 4.14),

$$P_o = P \qquad L_{1,k} = R_{s,k}R_{p,k} \qquad L_{3,k}^2 = R_{s,k}^2 \tag{4.4.27}$$

in Table 4.8. For the uniformly distributed load p (Fig. 4.15),

$$\Psi_k = 0 \tag{4.4.28}$$

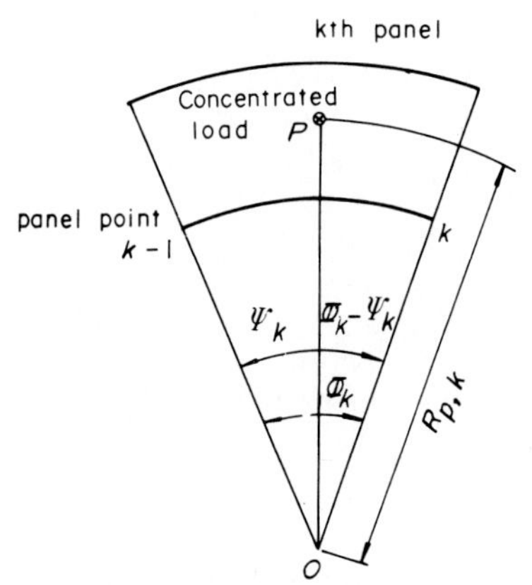

Figure 4.14 Concentrated load P.

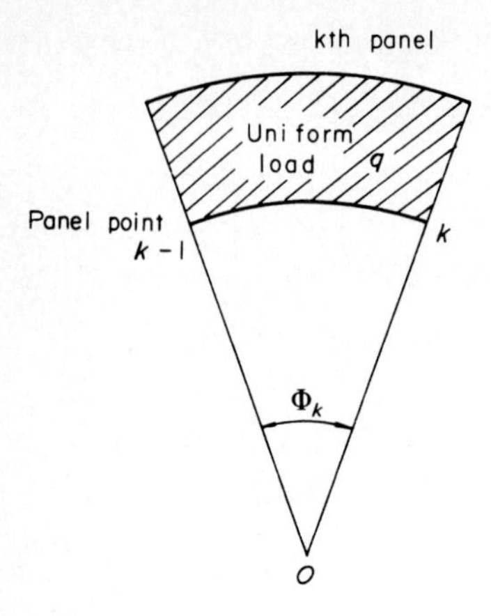

Figure 4.15 Uniformly distributed load p.

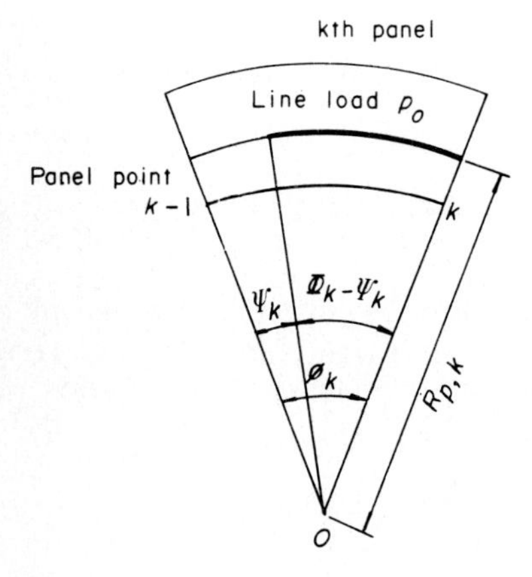

Figure 4.16 Longitudinal line load P_o.

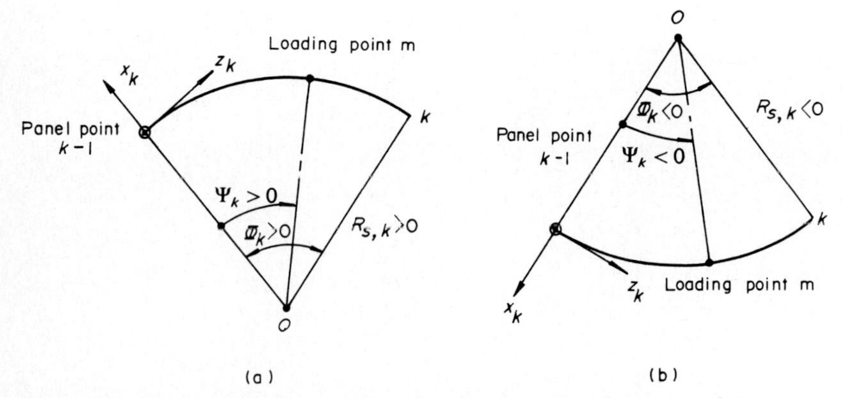

Figure 4.17 Two examples (*a*) and (*b*) of definitions of signs $R_{s,k}$, Φ_k, and Ψ_k.

in Table 4.9. For the longitudinal line load P_o (Fig. 4.16),

$$q = P_o \qquad L_{1,k} = R_{s,k}R_{p,k} \qquad L_{2,k} = R_{p,k}^2 \tag{4.4.29}$$

in Table 4.9.

Consequently, the signs of $R_{s,k}$, Φ_k, and Ψ_k to derive Eq. (4.4.22) can be set in accordance with the signs indicated in Fig. 4.17.

(3) Derivation of the point transfer matrix. Let us now consider the relationships between the displacements at panel points k and $k+1$ by referring to Fig. 4.18. Obviously the deflection w and rotational angle θ can be represented as

$$\varphi(0)_{,k+1} = \varphi(\Phi)_{,k} \tag{4.4.30a}$$

$$\theta(0)_{,k+1} = \theta(\Phi)_{,k} \tag{4.4.30b}$$

Therefore, the deflection must be given by the following compatibility condition:

$$w(0)_{,k+1} = w(\Phi)_{,k} - (R_{s,k+1} - R_{s,k})\theta(\Phi)_{,k} \tag{4.4.31}$$

While the situations of stress resultants acting on panels k and $k+1$ can be sketched as shown in Fig. 4.19, the following equations should be satisfied. For the shearing force Q, the equilibrium forces in the vertical direction give

$$Q(0)_{,k+1} = Q(\Phi)_{,k} + P_k - \Lambda_k \tag{4.4.32}$$

where P_k is a concentrated load at panel point k, and Λ_k is a statically indeterminate reaction at the intermediate support k.

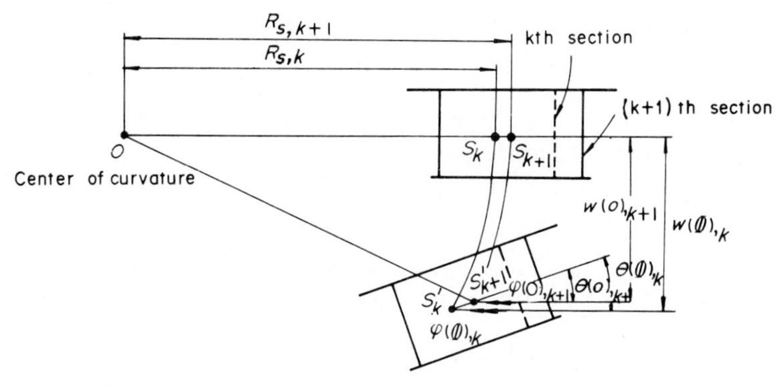

Figure 4.18 Displacements at panel point k.

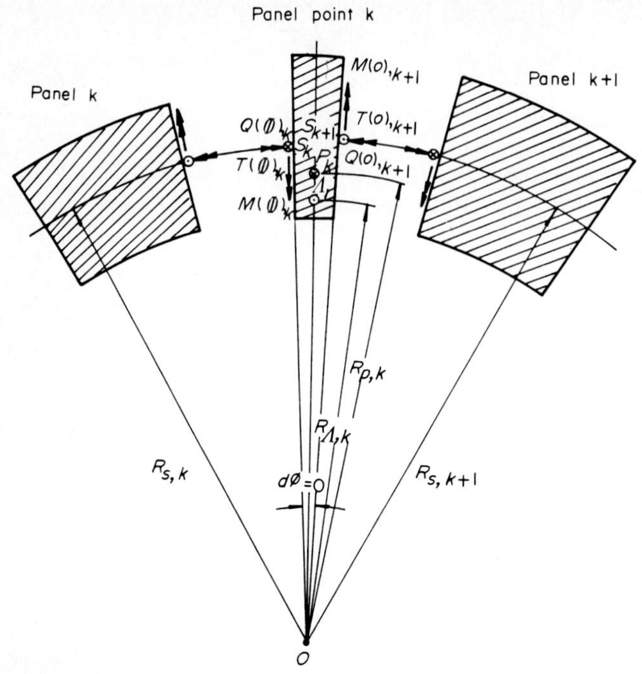

Figure 4.19 Stress resultants acting on panel point k.

The bending moment is no other than

$$M(0)_{,k+1} = M(\Phi)_{,k} \tag{4.4.33}$$

The remaining condition for the torsional moment T can be expressed by the equilibrium condition for torque:

$$T(0)_{,k+1} = T(\Phi)_{,k} + (R_{s,k+1} - R_{s,k})Q(\Phi)_{,k} + \Lambda_{k=r}(R_{s,k+1} - R_{\Lambda,k}) + P_k(R_{s,k+1} - R_{p,k}) \tag{4.4.34}$$

All the above equations can be written as follows by introducing an equation of the nondimensional parameters:

$$\mathbf{x}^*_{(0),k+1} = \mathbf{P}^*_k \mathbf{x}^*_{(\Phi),k} \tag{4.4.35}$$

where $\mathbf{P}^*_k$ is the point transfer matrix given by

$$\mathbf{P}^*_k = [\mathbf{P}^*_{s,k} \,|\, \mathbf{P}^*_{\Lambda,k}] \tag{4.4.36}$$

in which

$$
\mathbf{P}^*_{s,k} = \begin{array}{c} \begin{matrix} w^* & \theta^* & \varphi^* & M^* & T^* & Q^* & 1 \end{matrix} \\ \begin{bmatrix} 1 & b^*_{12} & 0 & 0 & 0 & 0 & 0 \\ 0 & 1 & 0 & 0 & 0 & 0 & 0 \\ 0 & 0 & 1 & 0 & 0 & 0 & 0 \\ 0 & 0 & 0 & 1 & 0 & 0 & 0 \\ 0 & 0 & 0 & 0 & 1 & b^*_{56} & b^*_{57} \\ 0 & 0 & 0 & 0 & 0 & 1 & b^*_{67} \\ 0 & 0 & 0 & 0 & 0 & 0 & 1 \end{bmatrix} \end{array} \tag{4.4.37}
$$

(the original point transfer matrix with seven rows and seven columns),

$$b^*_{12} = -\frac{R_{s,k+1} - R_{s,k}}{R_c} \tag{4.4.38a}$$

$$b^*_{56} = \frac{R_{s,k+1} - R_{s,k}}{R_c} \tag{4.4.38b}$$

$$b^*_{57} = \frac{P_k(R_{s,k+1} - R_{p,k})}{R_c P_c} \tag{4.4.38c}$$

$$b^*_{67} = \frac{P_k}{P_c} \tag{4.4.38d}$$

and

$$
\mathbf{P}^*_{\Lambda,k} = \begin{array}{c} \Lambda^*_r \\ \begin{bmatrix} 0 \cdots 0 & \cdots 0 \\ 0 \cdots 0 & \cdots 0 \\ 0 \cdots 0 & \cdots 0 \\ 0 \cdots 0 & \cdots 0 \\ 0 \cdots c^*_{5r} & \cdots 0 \\ 0 \cdots 1 & \cdots 0 \\ 0 \cdots 0 & \cdots 0 \end{bmatrix} \end{array} \tag{4.4.39}
$$

[this is the point transfer matrix considering unknown reaction at the rth intermediate support with seven rows and r_n columns (the total number of unknown reactions)],

and

$$c^*_{5r} = -\frac{R_{s,k+1} - R_{\Lambda,k}}{R_c} \tag{4.4.40}$$

Accordingly, the state vectors $\mathbf{x}^*_{(\Phi),k}$ and $\mathbf{x}^*_{(0),k+1}$ can be represented as

$$\mathbf{x}^*_{(\Phi),k} = [\mathbf{X}^*_{(\Phi),k} \,\vdots\, \mathbf{Y}^*_{(\Phi),k}] \begin{bmatrix} \mathbf{x}^*_{(0),1} \\ \cdots \\ \mathbf{y}^* \end{bmatrix} \tag{4.4.41a}$$

and

$$\mathbf{x}^*_{(0),k+1} = [\mathbf{X}^*_{(0),k+1} \,\vdots\, \mathbf{Y}^*_{(0),k+1}] \begin{bmatrix} \bar{\mathbf{x}}^*_{(0),1} \\ \cdots \\ \mathbf{y}^* \end{bmatrix} \tag{4.4.41b}$$

where $\mathbf{x}^*_{(0),1}$ is the initial state vector (unknown integration constants at the left end), as mentioned in the preceding section and $\mathbf{y}^*$ consists of the unknown intermediate reactions

$$\mathbf{y}^* = [\Lambda_1^* \ \Lambda_2^* \cdots \Lambda_r^* \cdots \Lambda_{rn}^*] \tag{4.4.42}$$

where r_n is the total number of intermediate reactions. Thus, matrices $\mathbf{Y}^*_{(\Phi),k}$, $\mathbf{Z}^*_{(\Phi),k}$, $\mathbf{Y}^*_{(0),k+1}$, and $\mathbf{Z}^*_{(0),k+1}$ can be estimated by

$$\mathbf{X}^*_{(\Phi),k} = \mathbf{F}_k^* \mathbf{X}^*_{(0),k} \tag{4.4.43a}$$

$$\mathbf{Y}^*_{(\Phi),k} = \mathbf{F}_k^* \mathbf{Y}^*_{(0),k} \tag{4.4.43b}$$

$$\mathbf{X}^*_{(0),k+1} = \mathbf{P}^*_{s,k} \mathbf{X}^*_{(\Phi),k} \tag{4.4.44a}$$

$$\mathbf{Y}^*_{(0),k+1} = \mathbf{P}^*_{s,k} \mathbf{Y}^*_{(\Phi),k} + \mathbf{P}^*_{\Lambda,k} \tag{4.4.44b}$$

(4) Boundary conditions

***a.* Initial conditions.** When a curved girder is simply supported at panel point $k = 0$, then $w(0)_{,1} = M(0)_{,1} = 0$ and the initial vector at the left end of a curved girder reduces to

$$\bar{\mathbf{x}}^*_{(0),1} = [\theta(0)^*_{,1} \ \varphi(0)^*_{,1} \ Q(0)^*_{,1} \ 1] \tag{4.4.45}$$

From the above equation, the initial vector can be represented by the following matrix:

$$\mathbf{X}_{(0),1} = \begin{array}{c} \begin{array}{cccc} \theta(0)^*_{,1} & \varphi(0)^*_{,1} & Q(0)^*_{,1} & 1 \end{array} \\ \begin{bmatrix} 0 & 0 & 0 & 1 \\ 1 & 0 & 0 & 0 \\ 0 & 1 & 0 & 0 \\ 0 & 0 & 0 & 0 \\ 0 & 0 & 0 & 0 \\ 0 & 0 & 1 & 0 \\ 0 & 0 & 0 & 1 \end{bmatrix} \end{array} \tag{4.4.46}$$

(with seven rows and four columns)

and

$$\mathbf{Y}^*_{(0),1} = 0 \tag{4.4.47}$$

(with seven rows and r_n columns).

***b.* Boundary conditions at an intermediate support.** If an intermediate support at panel point k is a rigid one for the deflection w, then

$$\mathbf{w}^*_{(\Phi),k} = 0 \tag{4.4.48}$$

c. **Boundary conditions at the right end.** The calculations of the state vector can be successfully performed by combining Eqs. (4.4.21) and (4.4.35) throughout a curved girder. Then the state vector at the right end n can be written as

$$\mathbf{x}^*_{(\Phi),n} = \mathbf{F}^*_n \mathbf{P}^*_{n-1} \cdots \mathbf{F}^*_2 \mathbf{P}^*_1 \mathbf{F}^*_1 \mathbf{x}^*_{(0),1} \tag{4.4.49}$$

For the right end n, the following boundary conditions are valid provided that the curved girder is also simply supported:

$$\mathbf{w}^*_{(\Phi),n} = 0 \tag{4.4.50a}$$

$$\mathbf{M}^*_{(\Phi),n} = 0 \tag{4.4.50b}$$

$$\mathbf{T}^*_{(\Phi),n} = 0 \tag{4.4.50c}$$

(5) Determination of unknown values. There are $3 + r_n$ unknown values in the above transfer matrix method. The unknown values $Q(0)^*_{,1}$, $\varphi(0)^*_{,1}$, $Q(0)^*_{,1}$ in Eq. (4.4.45) and intermediate reactions Λ^*_1, $\Lambda^*_2, \ldots, \Lambda^*_{rn}$ in Eq. (4.4.42) can be determined from the corresponding boundary condition given in Eqs. (4.4.48) and (4.4.50). Thus, the following set of simultaneous equations can be obtained:

$$\mathbf{A}\mathbf{a} = -\mathbf{b} \tag{4.4.51}$$

where $$\mathbf{a} = [\theta(0)^*_{,1}\ \varphi(0)^*_{,1}\ Q(0)^*_{,1}\ \Lambda^*_1\ \Lambda^*_2 \cdots \Lambda^*_{rn}] \tag{4.4.52}$$

a = unknown values

$\mathbf{A}$ = coefficient matrix of unknown values

$\mathbf{b}$ = load terms from last column of boundary conditions for unknown values

From this equation, the unknown values can be easily determined by

$$\mathbf{a} = -\mathbf{A}^{-1}\mathbf{b} \tag{4.4.53}$$

Therefore, the initial vector $\mathbf{x}^*_{(0),1}$ can be found by inserting the load term, 1.0, into the fourth term and separating the intermediate reaction $\mathbf{y}^*$ from the above vector as follows:

$$\mathbf{x}^*_{(0),1} = [\theta(0)^*_{,1}\ \varphi(0)^*_{,1}\ Q(0)^*_{,1}\ 1] \tag{4.4.54a}$$

$$\mathbf{y}^* = [\Lambda_1 \Lambda_2 \cdots \Lambda_{rn}] \qquad r = 1, 2, \ldots, r_n \tag{4.4.54b}$$

Thus, the complete solution of stress resultants and displacements can be obtained by substituting these values again into Eqs. (4.4.44).

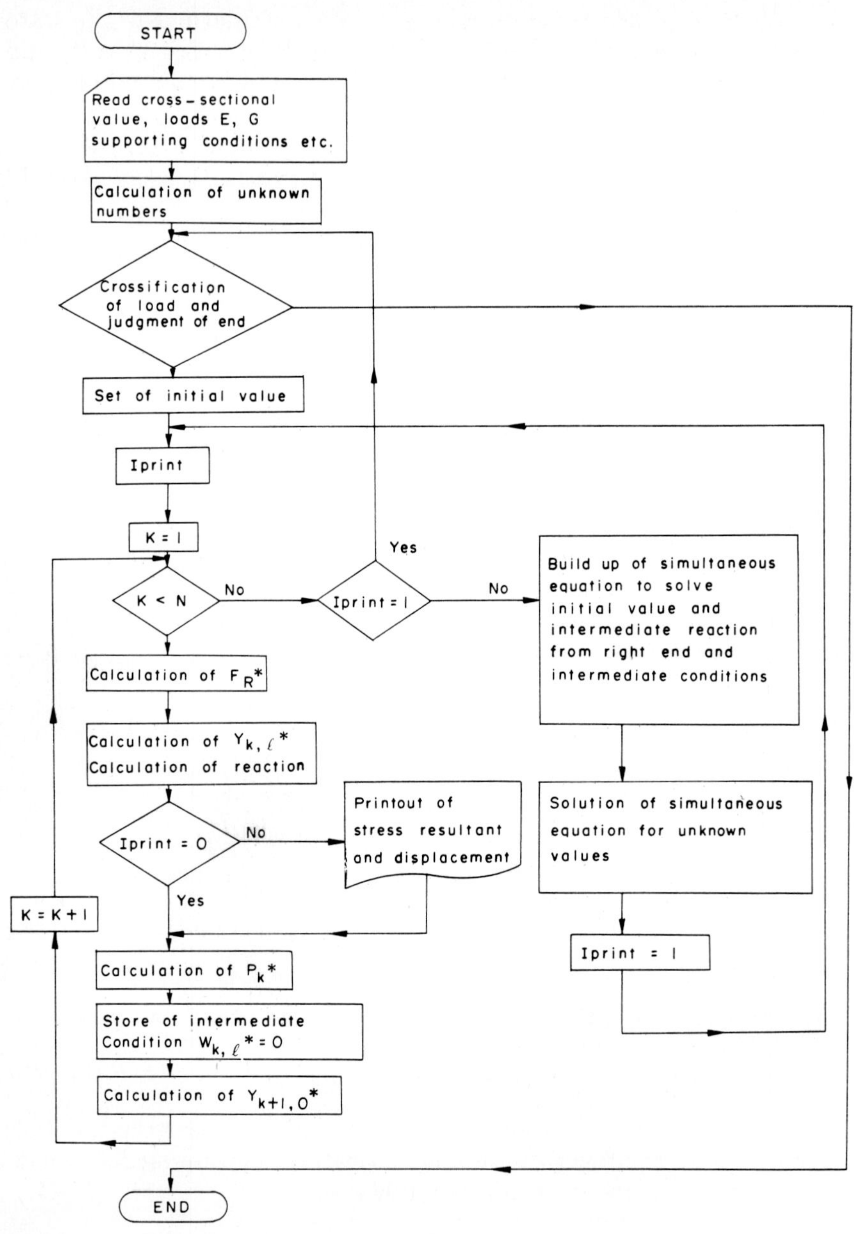

Figure 4.20 Flowchart of transfer matrix method: K = panel number, N = last number of panel, and Iprint = switch for printout of stress resultant and displacement.

(6) Flowchart for computer use. The analytical procedure for estimating the stress resultants and displacements in a curved girder by the transfer matrix method can be performed by a computer program, as illustrated by the flowchart in Fig. 4.20.

(7) Application to analysis of curved grillage.[13] By applying the above transfer matrix, a curved girder with twin-box girders, illustrated in Fig. 4.21*a*, can be easily treated in the following manner. To begin, this girder is idealized as a grillage curved girder with main girders $i = 1$ and $i = 2$ and floor beams as shown in Fig. 4.21*b*. The corresponding cross-sectional quantities are EI_1, GK_1, EI_2, GK_2, and $E\bar{I}_Q$.

Next, the stress resultants and displacements for each main girder and floor beam can be treated by the transfer matrix method in almost the same way as mentioned above—through the introduction of the statically indeterminate shear forces Q^s, torque T^s, and bending moment M^s to the main girders and floor beam between junction points i and j, so as to satisfy the following equation, shown in Fig. 4.22:

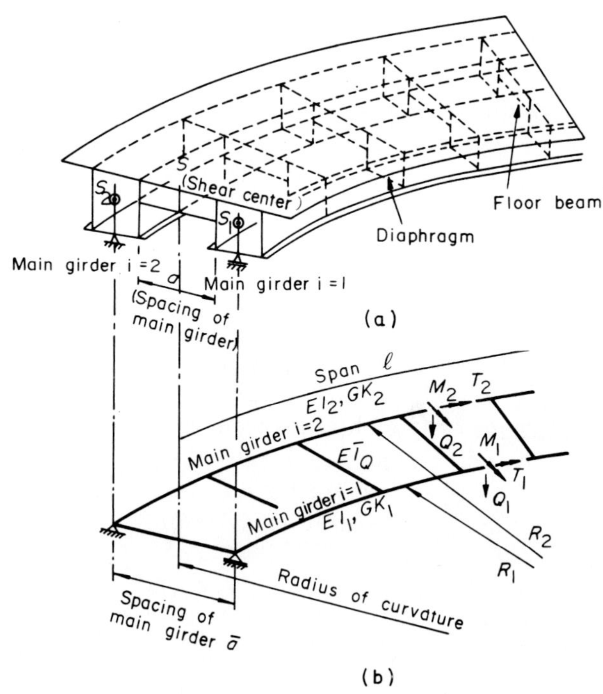

Figure 4.21 Grillage curved box girder: (*a*) Twin-box girder, and (*b*) idealization of curved grillage girder.

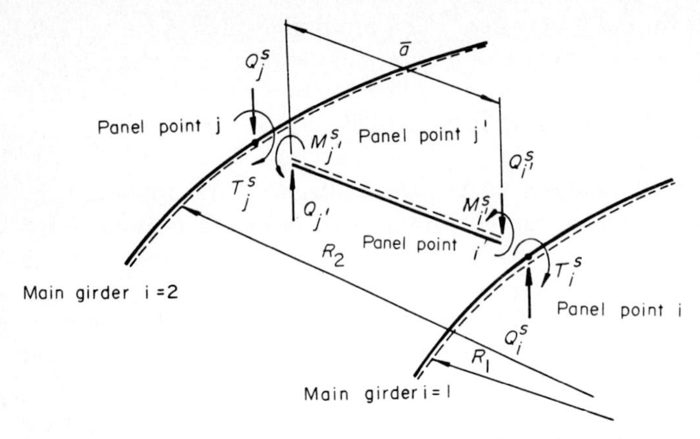

Figure 4.22 Introductions of statically indeterminate forces.

$$T_i^s{}^* = M_{i'}^s{}^* \tag{4.4.55a}$$

$$Q_i^s{}^* = Q_{i'}^s{}^* \tag{4.4.55b}$$

and

$$T_j^s{}^* = M_{j'}^s{}^* \tag{4.4.56a}$$

$$Q_j^s{}^* = Q_{j'}^s{}^* \tag{4.4.56b}$$

Finally, the following compatibility conditions should be fulfilled for the deflection w, torsional angle θ, and deflection angle φ:

$$w^*(\Phi)_{,i} = -w^*(0)_{,i'} \tag{4.4.57a}$$

$$\theta^*(\Phi)_{,i} = -\varphi^*(0)_{,i'} \tag{4.4.57b}$$

and

$$w^*(\Phi)_{,j} = -w^*(\bar{a})_{,j'} \tag{4.4.58a}$$

$$\theta^*(\Phi)_{,j} = -\varphi^*(\bar{a})_{,j'} \tag{4.4.58b}$$

where $\bar{a}$ is the spacing of the main girder, i.e., the span of floor beams (see Fig. 4.21*b*).

Thus, the complete solutions can be obtained from these additional calculations. In general, the floor beams are usually made of straight girders without the applied loads, so that the field transfer matrix can be simplified by taking into account flexural behavior only:

$$\mathbf{F}_k^* = \begin{array}{c} \begin{matrix} w^* & \varphi^* & M^* & Q^* & 1^* \end{matrix} \\ \begin{bmatrix} 1 & f_{12} & f_{13} & f_{14} & 0 \\ 0 & 1 & f_{23} & f_{24} & 0 \\ 0 & 0 & 1 & f_{34} & 0 \\ 0 & 0 & 0 & 0 & 0 \\ 0 & 0 & 0 & 0 & 1 \end{bmatrix}_k \end{array} \tag{4.4.59}$$

Here
$$f^*_{12} = -f^*_{34} = -\frac{l_k}{l_c} \tag{4.4.60a}$$

$$f^*_{13} = -f^*_{24} = -\frac{l_k^3 I_c}{6 l_c^3 I_k} \tag{4.4.60b}$$

$$f^*_{14} = \frac{l_k^3}{6 l_c^3 I_k} \tag{4.4.60c}$$

$$f^*_{23} = -\frac{l_k I_c}{l_c I_k} \tag{4.4.60d}$$

and I_k = geometric moment of inertia of kth panel of floor beam

l_k = panel length of kth panel of floor beam

I_c = reference geometric moment of inertia

l_c = reference panel length

4.4.2 Torsional warping theory

Let us now analyze a curved girder by using the torsional warping theory. Strictly speaking, coupled behavior of out-of-plane deformation and in-plane deformation occurs in thin-walled curved girders. However, these phenomena can be neglected, since they produce small effects, as described by Nakai, Kotoguchi, and Tani.[14] The uncoupled contributions are analyzed simultaneously by the transfer matrix method.

(1) Coordinate system, displacements, and stress resultants. Figure 4.23 shows the coordinates axes of a curved girder bridge where (X, Y, Z) is taken at centroid C and (x, y, z) is at the shear center S. At centroid C, there are three conjugate pairs of deformations and stress resultants (u, N_Z), (φ_X, M_X), and (φ_Y, M_Y), defined as

u = axial displacement in longitudinal direction of Z girder axis

N_Z = axial force in that Z axis

φ_X, φ_Y = deflection angles with respect to X and Y axes, respectively

M_X, M_Y = bending moments with respect to X and Y axes, respectively

Applying the shearing force along the centroidal axis (X, Y), the girder will be subjected to torsional moments. Therefore, a shearing force should be imposed on the shear center S. Thus a new set of forces

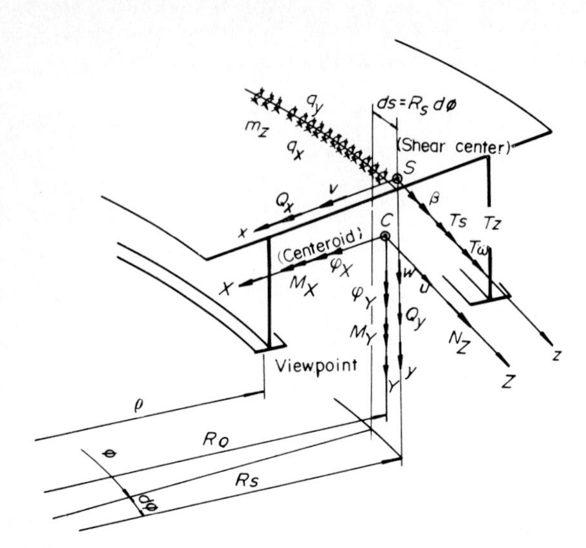

Figure 4.23 Coordinate axes, corresponding displacements, and stress resultants.

and displacements (β, T_z), (v, Q_x), and (w, Q_y) are defined as in Fig. 4.23, where

β = angle of rotation of cross section with respect to z axis

T_z = torsional moment with respect to z axis

v = deflection in direction of x axis

w = deflection in direction of y axis

Q_x, Q_y = shearing forces in direction of x and y axes, respectively

The radii of curvature and central angle are defined as

R_o = radius of curvature at centroid C

R_s = radius of curvature at shear center S

and ϕ = central angle of curved axis

Therefore, the length of a small segment cut off by a differential angle $d\phi$ is designated

$$ds = R_s\, d\phi \tag{4.4.61}$$

along the shear center axis. The equilibrium conditions of all the above stress resultants and the relationships between stress resultants and displacements are estimated by Eq. (4.4.61) for the sake of simplicity.

Finally, the applied loads are

q_x, q_y = uniformly distributed loads acting on x and y axes, respectively

m_z = uniformly distributed torque with respect to z axis

The magnitude of these loads are shown in Table 4.10 for actual load conditions.

(2) Differential equation for stress resultants and displacements

a. Out-of-plane deformation. The following linear equations are given by Nakai, Kotoguchi, and Tani.[12]

Equilibrium conditions:

$$\frac{dQ_y}{ds} = -q_y \tag{4.4.62a}$$

$$\frac{dM_X}{ds} - Q_y + \frac{T_z}{R_s} = 0 \tag{4.4.62b}$$

$$\frac{dT_z}{ds} - \frac{M_X}{R_s} = -m_z \tag{4.4.62c}$$

$$T_z = T_s + T_\omega \tag{4.4.62d}$$

$$T_\omega = -\frac{dM_\omega}{ds} \tag{4.4.62e}$$

Relationships among displacements and stress resultants:

$$E_s I_X\left(\frac{d^2w}{ds^2} - \frac{\beta}{R_s}\right) = -\frac{R_0}{R_s} M_X \tag{4.4.63a}$$

$$E_s I_\omega\left(\frac{d^2\beta}{ds^2} + \frac{1}{R_s}\frac{d^2w}{ds^2}\right) = M_\omega \tag{4.4.63b}$$

$$G_s K\left(\frac{d\beta}{ds} + \frac{1}{R_s}\frac{dw}{ds}\right) = T_s \tag{4.4.63c}$$

b. In-plane deformation. In a similar manner, the following basic equations can be derived from Nakai, Kotoguchi, and Tani.[12]

TABLE 4.10 Loading Conditions for Practical Designs

	Line load	Area load	Concentrated load	Radial load
Actual loading condition	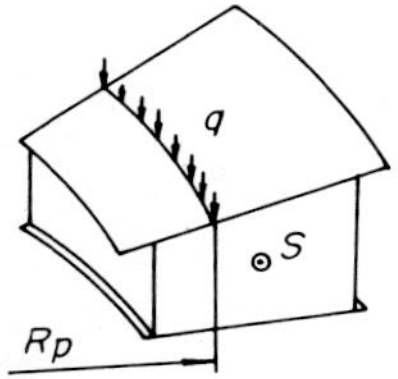 q, S, Rp	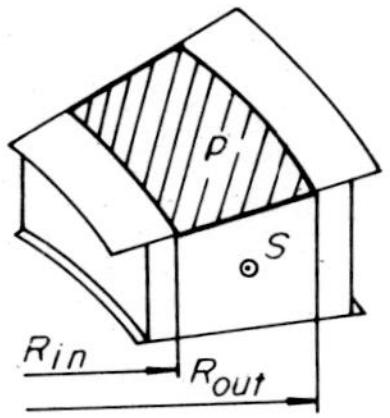 p, S, Rin, Rout	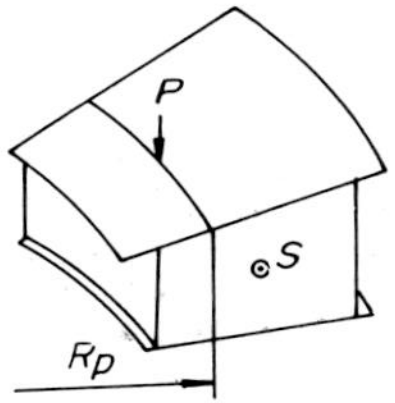 P, S, Rp	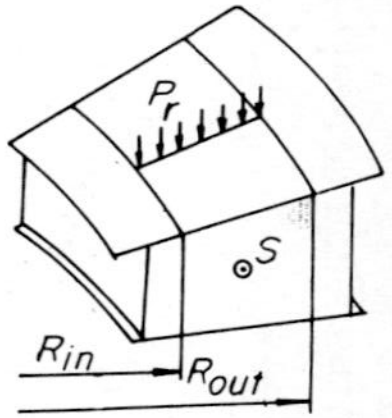 Pr, S, Rin, Rout
Idealized loading condition	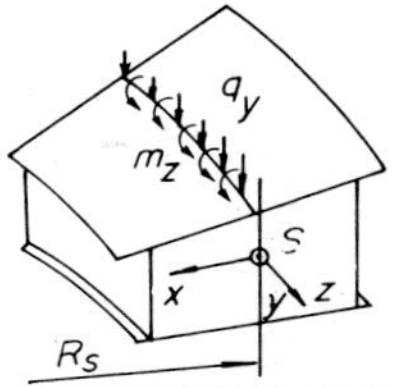 qy, mz, S, x, y, z, Rs		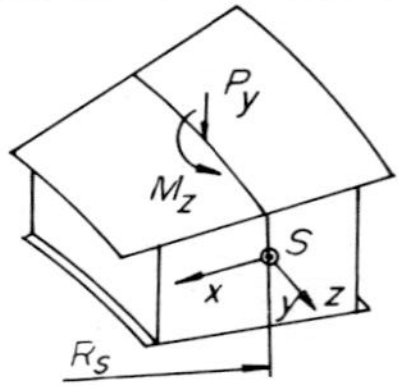 Py, Mz, S, x, y, z, Rs	
Equivalent load	q_y $\quad q\frac{R_p}{R_s}$	q_y $\quad \frac{P}{2}\frac{R_{\text{out}}^2 - R_{\text{in}}^2}{R_s}$	P_y $\quad P$	P_y $\quad Pr(R_{\text{out}} - R_{\text{in}})$
	m_z $\quad q(R_s - R_p)\frac{R_p}{R_s}$	m_z $\quad \frac{P}{2}(R_{\text{out}}^2 - R_{\text{in}}^2)\frac{R_s - R_p}{R_s}$ where $R_p = \frac{2}{3}\frac{R_{\text{out}}^3 - R_{\text{in}}^3}{R_{\text{out}}^2 - R_{\text{in}}^2}$	M_x $\quad P(R_s - R_p)$	M_z $\quad Pr(R_{\text{out}} - R_{\text{in}})(R_s - R_p)$ where $R_p = \frac{R_{\text{out}} + R_{\text{in}}}{2}$

Stress resultants:

$$\frac{dN_Z}{ds} - \frac{Q_x}{R_s} = 0 \tag{4.4.64a}$$

$$\frac{dQ_x}{ds} + \frac{N_Z}{R_s} = -q_x \tag{4.4.64b}$$

$$\frac{dM_Y}{ds} + Q_x = 0 \tag{4.4.64c}$$

Relationships between deformations and stress resultants:

$$E_s A_X\left(\frac{du}{ds} - \frac{v}{R_s}\right) = \frac{R_0}{R_s} N_Z \tag{4.4.65a}$$

$$E_s I_Y\left(\frac{d^2v}{ds^2} + \frac{1}{R_s}\frac{du}{ds}\right) = \frac{R_0}{R_s} M_Y \tag{4.4.65b}$$

In Eqs. (4.4.63) and (4.4.65), the cross-sectional quantities are defined as

$E_s A_Z$ = axial stiffness

$E_s I_X$, $E_s I_Y$ = flexural rigidities of X and Y axes, respectively

$G_s K$ = torsional rigidity

$E_s I_\omega$ = warping rigidity

(3) Solution of stress resultants and displacements

***a.* Out-of-plane deformation.** To solve the simultaneous differential equations (4.4.62*a*) to (4.4.62*c*) and (4.4.63) as simply as possible, these equations are rewritten as follows. First, the shearing force $Q_y(\phi)$ at section Φ can be directly solved by Eq. (4.4.62*a*):

$$Q_y(\Phi) = -R_s \int_0^{\Phi} q_y\, d\phi + Q_y(0) \tag{4.4.66}$$

Here $Q_y(0)$ is the integration constant, and the shearing forces are at $\phi = 0$. Second, the differential equation for the bending moment $M_X(\phi)$ can be written from Eqs. (4.4.62*a*) to (4.4.62*c*) as

$$\frac{d^2M_X(\phi)}{d\phi^2} + M_X(\phi) = -q_y R_s^2 + m_z R_s \tag{4.4.67}$$

Third, the torsional moment $T_z(\Phi)$ can be obtained from

$$T_z(\Phi) = \int_0^{\Phi} M_X(\phi)\, d\phi - R_s \int_0^{\Phi} m_z\, d\phi + C_1 \tag{4.4.68}$$

where C_1 is an integral constant.

Also the differential equation for the warping moment $M_\omega(\phi)$ can be developed from Eqs. (4.4.62*c*) to (4.4.62*e*) and (4.4.63*b*) to (4.4.63*c*) as follows:

$$\frac{d^2 M_\omega(\phi)}{d\phi^2} - \alpha^2 M_\omega(\phi) = -R_s M_X(\phi) + R_s^2 m_z \tag{4.4.69}$$

where

$$\alpha = R_s \sqrt{\frac{G_s K}{E_s I_\omega}} \tag{4.4.70}$$

The basic equation for the rotation angle $\beta(\phi)$ can be expressed by Eqs. (4.4.63*a*) and (4.4.63*b*) as

$$\frac{d^2 \beta(\phi)}{d\phi^2} + \beta(\phi) = \frac{R_s^2}{E_s I_\omega} M_\omega(\phi) + \frac{R_s}{E_s I_x} M_X(\phi) \tag{4.4.71}$$

where

$$I_x = \frac{R_s}{R_o} I_X \tag{4.4.72}$$

The deflection $w(\Phi)$ can then be found from Eq. (4.4.63*a*):

$$w(\Phi) = -R_s \int_0^{\Phi} \left[\int \frac{M_Y(\phi)}{EI_x}\, d\phi + \int \beta(\phi)\, d\phi \right] d\phi - \varphi_X(0) R_s \Phi + w(0) \tag{4.4.73}$$

where

$$\varphi_X(\phi) = -\frac{1}{R_s} \frac{dw(\phi)}{d(\phi)} \tag{4.4.74}$$

which is the deflection angle.

Finally, St. Venant's torsional moment $T_s(\phi)$ can be estimated from Eq. (4.4.63*c*):

$$T_s(\phi) = G_s K \left[\frac{1}{R_s} \frac{d\beta(\phi)}{d\phi} + \frac{1}{R_s^2} \frac{dw(\phi)}{d\phi} \right] \tag{4.4.75}$$

and Wagner's torsional moment $T_\omega(\phi)$ can be found from Eq. (4.4.62*e*).

The differential equations (4.4.66) through (4.4.75) can be easily solved by using a Laplace transformation:

$$\begin{aligned} w(\Phi) = {} & w(0) + \beta(0) R_s (1 - \cos\Phi) - \varphi_X(0) R_s \sin\Phi \\ & + T_s(0) f_{14} + M_\omega(0) f_{15} + M_X(0) f_{16} \\ & + T_z(0) f_{17} + Q_y(0) f_{18} + f_{19} \end{aligned} \tag{4.4.76a}$$

$$\beta(\Phi) = \beta(0)\cos\Phi + \varphi_X(0)\sin\Phi + T_s(0)f_{24}$$
$$+ M_\omega(0)f_{25} + M_X(0)f_{26} + T_z(0)f_{27} + Q_X(0)f_{28} + f_{29} \qquad (4.4.76b)$$

$$\varphi_X(\Phi) = -\beta(0)\sin\Phi + \varphi_X(0)\cos\Phi + T_s(0)f_{34}$$
$$+ M_\omega(0)f_{35} + M_X(0)f_{36} + T_z(0)f_{37} + Q_y(0)f_{38} + f_{39} \qquad (4.4.76c)$$

$$T_s(\Phi) = T_s(0)\cosh\alpha\Phi + M_\omega(0)\frac{\alpha}{R_s}\sinh\alpha\Phi$$
$$+ M_X(0)f_{46} + T_z(0)f_{47} + Q_y(0)f_{48} + f_{49} \qquad (4.4.76d)$$

$$M_\omega(\Phi) = T_s(0)\frac{R_s}{\alpha}\sinh\alpha\Phi + M_\omega(0)\cosh\alpha\Phi$$
$$+ M_X(0)f_{56} + T_z(0)f_{57} + Q_y(0)f_{58} + f_{59} \qquad (4.4.76e)$$

$$M_X(\Phi) = M_X(0)\cos\Phi - T_z(0)\sin\Phi + Q_y(0)R_s\sin\Phi + f_{69} \qquad (4.4.76f)$$

$$T_z(\Phi) = M_X(0)\sin\Phi + T_z(0)\cos\Phi$$
$$+ Q_y(0)R_s(1-\cos\Phi) + f_{79} \qquad (4.4.76g)$$

$$Q_y(\Phi) = Q_y(0) + f_{89} \qquad (4.4.76h)$$

where

$$f_{14} = -\frac{R_s^2}{E_sI_x}\frac{1}{\lambda\alpha^2(1+\alpha^2)}\left(\sin\Phi - \frac{1}{\alpha}\sin\alpha\Phi\right)$$
$$= -\frac{R_s^2}{E_sI_x}\frac{1}{\lambda\alpha^2}f_{58} \qquad (4.4.77a)$$

$$f_{15} = -\frac{R_s}{E_sI_x}\frac{1+\alpha^2-\alpha^2\cos\Phi-\cosh\alpha\Phi}{\lambda\alpha^2(1+\alpha^2)}$$
$$= -\frac{R_s}{E_sI_x}\frac{1}{\lambda\alpha^2}f_{48} \qquad (4.4.77b)$$

$$f_{16} = -\frac{R_s^2}{E_sI_x}\left\{\frac{1}{2}\left[1+\frac{1}{\lambda(1+\alpha^2)}\right]\Phi\sin\Phi\right.$$
$$\left. - \frac{(1+\alpha^2)^2-\alpha^2(2+\alpha^2)\cos\Phi-\cosh\alpha\Phi}{\lambda\alpha^2(1+\alpha^2)^2}\right\}$$
$$= -f_{38} \qquad (4.4.77c)$$

$$f_{17} = \frac{R_s^2}{E_s I_x} \left\{ \frac{1}{2} \left[1 + \frac{1}{\lambda(1+\alpha^2)} \right] (\sin \Phi - \Phi \cos \Phi) + \frac{\sin \Phi - (1/\alpha) \sinh \alpha\Phi}{\lambda(1+\alpha^2)^2} \right\}$$

$$= f_{28} \tag{4.4.77d}$$

$$f_{18} = -\frac{R_s^3}{E_s I_x} \left\{ \frac{1}{2} \left[1 + \frac{1}{\lambda(1+\alpha^2)} \right] (\sin \Phi - \Phi \cos \Phi) - \frac{(1+\alpha^2)^2 \Phi - \alpha^2(2+\alpha^2) \sin \Phi - (1/\alpha) \sinh \alpha\Phi}{\lambda\alpha^2(1+\alpha^2)^2} \right\} \tag{4.4.77e}$$

$$f_{24} = \frac{R_s}{E_s I_x} \frac{1}{\lambda\alpha^2(1+\alpha^2)} (\sin \Phi + \alpha \sinh \alpha\Phi)$$

$$= -\frac{R_s}{E_s I_x} \frac{1}{\lambda\alpha^2} f_{57} \tag{4.4.77f}$$

$$f_{25} = -\frac{1}{E_s I_x} \frac{\cos \Phi - \cosh \alpha\Phi}{\lambda(1+\alpha^2)}$$

$$= -\frac{1}{E_s I_x} \frac{1}{\lambda\alpha^2} f_{47} \tag{4.4.77g}$$

$$f_{26} = \frac{R_s}{E_s I_x} \left\{ \frac{1}{2} \left[1 + \frac{1}{\lambda(1+\alpha^2)} \right] \Phi \sin \Phi + \frac{\cos \Phi - \cosh \alpha\Phi}{\lambda(1+\alpha^2)^2} \right\}$$

$$= -f_{37} \tag{4.4.77h}$$

$$f_{27} = -\frac{R_s}{E_s I_x} \left\{ \frac{1}{2} \left[1 + \frac{1}{\lambda(1+\alpha^2)} \right] (\sin \Phi - \Phi \cos \Phi) - \frac{\alpha^2[\sin \Phi - (1/\alpha) \sinh \alpha\Phi]}{\lambda(1+\alpha^2)^2} \right\} \tag{4.4.77i}$$

$$f_{34} = \frac{R_s}{E_s I_x} \frac{1}{\lambda\alpha^2(1+\alpha^2)} (\cos \Phi - \cosh \alpha\Phi)$$

$$= \frac{R_s}{E_s I_x} \frac{1}{\lambda\alpha^2} f_{56} \tag{4.4.77j}$$

$$f_{35} = \frac{1}{E_s I_x} \frac{\alpha \sin \Phi - \sinh \alpha\Phi}{\lambda\alpha(1+\alpha^2)}$$

$$= \frac{1}{E_s I_x} \frac{1}{\lambda\alpha^2} f_{46} \tag{4.4.77k}$$

$$f_{36} = \frac{R_s}{E_s I_x} \left\{ \frac{1}{2} \left[1 + \frac{1}{\lambda(1+\alpha^2)} \right] (\sin \Phi + \Phi \cos \Phi) - \frac{\alpha(2+\alpha^2) \sin \Phi - \sinh \alpha\Phi}{\lambda\alpha(1+\alpha^2)^2} \right\} \quad (4.4.77l)$$

and

$$f_{19} = -q_z \frac{R_s^4}{E_s I_x} \left\{ \frac{1}{\lambda(1+\alpha^2)^2} \left[\frac{(1+\alpha^2)^2}{\alpha^2} \frac{\Phi}{2} - \frac{\cosh \alpha\Phi - 1}{\alpha^4} \right] - \left[1 + \frac{(3+2\alpha^2)}{\lambda(1+\alpha^2)^2} \right] (1 - \cos \Phi) + \left[1 + \frac{1}{\lambda(1+\alpha^2)} \right] \frac{\Phi}{2} \sin \Phi \right\}$$
$$+ m_z \frac{R_s^3}{E_s I_x} \left\{ \frac{1}{\lambda(1+\alpha^2)^2} \frac{1}{\alpha^2} (\cosh \alpha\Phi - 1) + \left[1 + \frac{(2+\alpha^2)}{\lambda(1+\alpha^2)^2} \right] (\cos \Phi - 1) + \left[1 + \frac{1}{\lambda(1+\alpha^2)} \right] \frac{\Phi}{2} \sin \Phi \right\} \quad (4.4.78a)$$

$$f_{29} = q_y \frac{R_s^3}{E_s I_x} \left\{ \frac{1}{\lambda(1+\alpha^2)^2} \frac{1}{\alpha^2} (\cosh \alpha\Phi - 1) + \left[1 + \frac{(2+\alpha^2)}{\lambda(1+\alpha^2)^2} \right] (\cos \Phi - 1) + \left[1 + \frac{1}{\lambda(1+\alpha^2)} \right] \frac{\Phi}{2} \sin \Phi \right\}$$
$$+ m_x \frac{R_s^2}{E_s I_x} \left\{ \frac{1}{\lambda(1+\alpha^2)^2} (\cosh \alpha\Phi - \cos \Phi) + 1 - \cos \Phi - \left[1 + \frac{1}{\lambda(1+\alpha^2)} \right] \frac{\Phi}{2} \sin \Phi \right\} \quad (4.4.78b)$$

$$f_{39} = -q_y \frac{R_s^3}{E_s I_x} \left\{ \frac{1}{\lambda(1+\alpha^2)^2} \left[\frac{\sinh \alpha\Phi}{\alpha^3} - \frac{(1+\alpha^2)^2}{\alpha^2} \Phi \right] + \left[1 + \frac{(5+3\alpha^2)}{\lambda(1+\alpha^2)^2} \right] \frac{1}{2} \sin \Phi - \left[1 + \frac{1}{\lambda(1+\alpha^2)} \right] \frac{\Phi}{2} \cos \Phi \right\}$$
$$- m_z \frac{R_s^2}{E_s I_x} \left\{ \frac{1}{\lambda(1+\alpha^2)^2} \frac{1}{\alpha} \sinh \alpha\Phi - \left[1 + \frac{(3+\alpha^2)}{\lambda(1+\alpha^2)^2} \right] \frac{1}{2} \sin \Phi + \left[1 + \frac{1}{\lambda(1+\alpha^2)} \right] \frac{\Phi}{2} \cos \Phi \right\} \quad (4.4.78c)$$

$$f_{49} = q_y \frac{R_s^2}{1+\alpha^2} [\alpha \sinh \alpha\Phi - (1+\alpha^2)\Phi + \alpha^2 \sin \Phi] + m_z \frac{R_s}{1+\alpha^2} (\alpha \sinh \alpha\Phi - \alpha^2 \sin \Phi) \quad (4.4.78d)$$

$$f_{59} = -q_y \frac{R_s^3}{1+\alpha^2}\left(\frac{1+\alpha^2}{\alpha^2} - \frac{\cosh \alpha\Phi}{\alpha^2} - \cos \Phi\right)$$

$$+ m_z \frac{R_s^2}{1+\alpha^2}(\cosh \alpha\Phi - \cos \Phi) \quad (4.4.78e)$$

$$f_{69} = -q_y R_s^2(1 - \cos \Phi) + m_z R_s(1 - \cos \Phi) \quad (4.4.78f)$$

$$f_{79} = -q_y R_s^2(\Phi - \sin \Phi) - m_z R_s \sin \Phi \quad (4.4.78g)$$

$$f_{89} = -q_y R_s \Phi \quad (4.4.78h)$$

in which the new parameter λ is defined as

$$\lambda = \frac{I_\omega}{R_s^2 I_x} \quad (4.4.79)$$

***b.* In-plane deformation.** The differential equation for the axial force $N_Z(\phi)$ can be written from Eqs. (4.4.64*a*) and (4.3.64*b*) as

$$\frac{d^2 N_Z}{d\phi^2} + N_Z = R_s q_x \quad (4.4.80)$$

Then the shearing force $Q_x(\phi)$ is [see Eq. (4.4.64*a*)]

$$Q_x(\phi) = \frac{dN_Z}{d\phi} \quad (4.4.81)$$

Also, the bending moment $M_Y(\Phi)$ is [see Eq. (4.4.64*c*]

$$M_Y(\Phi) = -R_s \int_0^\Phi Q_x(\phi)\, d\phi + M_Y(0) \quad (4.4.82)$$

Next the differential equation relative to the vertical displacement $v(\phi)$ can be written from Eqs. (4.4.65*a*) and (4.4.65*b*) as

$$\frac{d^2 v(\phi)}{d\phi^2} + v(\phi) = R_s^2\left[\frac{M_Y(\phi)}{E_s I_y} - \frac{1}{R}\frac{N_Z(\phi)}{E_s A_z}\right] \quad (4.4.83)$$

where

$$I_y = \frac{R_s}{R_0} I_Y \quad (4.4.84a)$$

$$A_z = \frac{R_s}{R_0} A_Z \quad (4.4.84b)$$

Now, the axial displacement $u(\phi)$ can be obtained from Eq. (4.4.65*a*) as

$$u(\phi) = \int_0^\Phi v(\phi)\, d\phi + R_s \int_0^\Phi \frac{N_Z(\phi)}{E_s A_z}\, d\phi + u(0) \quad (4.4.85)$$

and the deflection angle $\varphi_Y(\Phi)$ can be found from Eq. (4.4.65b):

$$\varphi_Y(\Phi) = R_s \int_0^{\Phi} \frac{M_Y(\phi)}{EI_y} d\phi + \varphi_Y(0) \tag{4.4.86}$$

Solving differential equations (4.4.80) through (4.4.86) by means of a Laplace transformation, we find the following solutions:

$$u(\Phi) = u(\Phi) \cos \Phi + v(0) \sin \Phi + \varphi_Y(0) R_s(1 - \cos \Phi) + M_Y(0) g_{14} + Q_x(0) g_{15} + N_Z(0) g_{16} + g_{17} \tag{4.4.87a}$$

$$v(\Phi) = -u(0) \sin \Phi + v(0) \cos \Phi + \varphi_Y(0) g_{23} + M_Y(0) g_{24} + Q_x(0) g_{25} + N_Z(0) g_{26} + g_{27} \tag{4.4.87b}$$

$$\varphi_Y(\Phi) = \varphi_Y(0) + M_Y(0) g_{34} + Q_x(0) g_{35} + N_Z(0) g_{36} + g_{37} \tag{4.4.87c}$$

$$M_Y(\Phi) = M_Y(0) - Q_x(0) R_s \sin \Phi + N_Z(0) R_s(1 - \cos \Phi) + g_{47} \tag{4.4.87d}$$

$$Q_x(\Phi) = Q_x(0) \cos \Phi - N_Z(0) \sin \Phi + g_{57} \tag{4.4.87e}$$

$$N_Z(\Phi) = Q_x(0) \sin \Phi + N_Z(0) \cos \Phi + g_{67} \tag{4.4.87f}$$

where

$$g_{14} = \frac{R_s^2}{E_s I_y} (\Phi - \sin \Phi) = g_{36} \tag{4.4.88a}$$

$$g_{15} = \frac{R_s^3}{2E_s I_y} (\Phi \sin \Phi + 2 \cos \Phi - 2) + \frac{R_s}{2EA_z} \Phi \sin \Phi = -g_{26} \tag{4.4.88b}$$

$$g_{16} = \frac{R_s^3}{2E_s I_y} (\Phi \cos \Phi - 3 \sin \Phi + 2\Phi) + \frac{R_s}{2E_s A_z} (\sin \Phi + \Phi \cos \Phi) \tag{4.4.88c}$$

$$g_{24} = \frac{R_s^2}{E_s I_y} (1 - \cos \Phi) = -g_{35} \tag{4.4.88d}$$

$$g_{25} = -\frac{R_s(I_y + A_z R_s^2)}{2E_s A_z I_y} (\sin \Phi - \Phi \cos \Phi) \tag{4.4.88e}$$

$$g_{34} = \frac{R_s}{E_s I_y} \Phi \tag{4.4.88f}$$

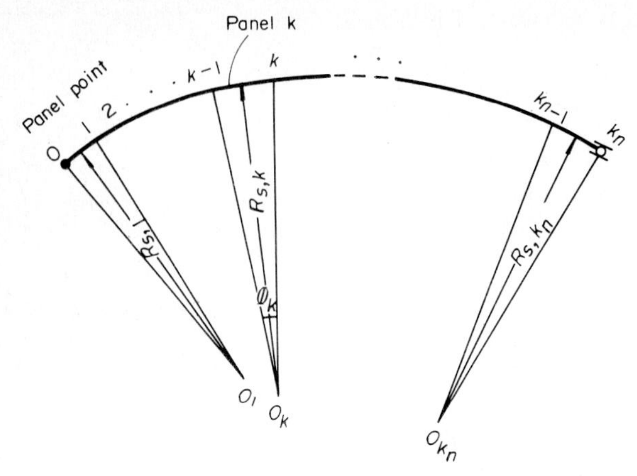

Figure 4.24 Plane view and panel point of curved girder.

and

$$g_{17} = q_x \left[\frac{R_s^4}{2E_s I_y} (\Phi \cos \Phi - 3 \sin \Phi + 2\Phi) - \frac{R_s^2}{2E_s A_z} (\sin \Phi - \Phi \cos \Phi) \right] \quad (4.4.89a)$$

$$g_{27} = -q_x \frac{R_s^2(I_y + A_z R_s^2)}{2E_s A_z I_y} (\Phi \sin \Phi + 2 \cos \Phi - 2) \quad (4.4.89b)$$

$$g_{37} = q_x \frac{R_s^3}{E_s I_y} (\Phi - \sin \Phi) \quad (4.4.89c)$$

$$g_{47} = q_x R_s^2 (1 - \cos \Phi) \quad (4.4.89d)$$

$$g_{57} = -q_x R_s \sin \Phi \quad (4.4.89e)$$

$$g_{67} = -q_x R_s (1 - \cos \Phi) \quad (4.4.89f)$$

(4) Field transfer matrix

***a.* Out-of-plane deformation.** The solutions of the stress resultants and displacements shown in Eqs. (4.4.76a) to (4.4.76h) can be rewritten in matrix form for a panel k with panel points k and $k-1$, radius of curvature $R_{s,k}$, and central angle Φ_k, as illustrated in Fig. 4.24, as follows:

$$
\begin{bmatrix} w(\Phi) \\ \beta(\Phi) \\ \varphi_X(\Phi) \\ T_s(\Phi) \\ M_\omega(\Phi) \\ M_X(\Phi) \\ T_z(\Phi) \\ Q_y(\Phi) \\ 1 \end{bmatrix}_k = \begin{bmatrix} 1 & R_s(1-\cos\Phi) & -R_s\sin\Phi & f_{14} & f_{15} & f_{16} & f_{17} & f_{18} & f_{19} \\ 0 & \cos\Phi & \sin\Phi & f_{24} & f_{25} & f_{26} & f_{27} & f_{28} & f_{29} \\ 0 & -\sin\Phi & \cos\Phi & f_{34} & f_{35} & f_{36} & f_{37} & f_{38} & f_{39} \\ 0 & 0 & 0 & \cosh\alpha\Phi & \dfrac{\alpha}{R_s}\sinh\alpha\Phi & f_{46} & f_{47} & f_{48} & f_{49} \\ 0 & 0 & 0 & \dfrac{R_s}{\alpha}\sinh\alpha\Phi & \cosh\alpha\Phi & f_{56} & f_{57} & f_{58} & f_{59} \\ 0 & 0 & 0 & 0 & 0 & \cos\Phi & -\sin\Phi & R_s\sin\Phi & f_{59} \\ 0 & 0 & 0 & 0 & 0 & \sin\Phi & \cos\Phi & R_s(1-\cos\Phi) & f_{79} \\ 0 & 0 & 0 & 0 & 0 & 0 & 0 & 0 & f_{89} \\ 0 & 0 & 0 & 0 & 0 & 0 & 0 & 0 & 0 \end{bmatrix}_k \begin{bmatrix} w(0) \\ \beta(0) \\ \varphi_X(0) \\ T_s(0) \\ M_\omega(0) \\ M_X(0) \\ T_z(0) \\ Q_y(0) \\ 1 \end{bmatrix}_k \tag{4.4.90}
$$

or

$$\mathbf{x}_{(\Phi),k} = \mathbf{F}_k \mathbf{x}_{(0),k} \tag{4.4.91}$$

where

$$\mathbf{x}_{(0),k} = [w(0)\ \beta(0)\ \varphi_X(0)\ T_s(0)\ M_\omega(0)\ M_X(0)\ T_z(0)\ Q_y(0)\ 1]_k \tag{4.4.92a}$$

$$\mathbf{x}_{(\Phi),k} = [w(\Phi)\ \beta(\Phi)\ \varphi_X(\Phi)\ T_s(\Phi)\ M_\omega(\Phi)\ M_X(\Phi)\ T_z(\Phi)\ Q_y(\Phi)\ 1]_k \tag{4.4.92b}$$

In Eqs. (4.4.92), $\mathbf{x}_{(0),k}$ and $\mathbf{x}_{(\Phi),k}$ are, respectively, defined by the state vector at the left and right sides of panel k. Also we have

$$\mathbf{F}_k = \begin{bmatrix} 1 & R_s(1-\cos\Phi) & -R_s\sin\Phi & f_{14} & f_{15} & f_{16} & f_{17} & f_{18} & f_{19} \\ 0 & \cos\Phi & \sin\Phi & f_{24} & f_{25} & f_{26} & f_{27} & f_{28} & f_{29} \\ 0 & -\sin\Phi & \cos\Phi & f_{34} & f_{35} & f_{36} & f_{37} & f_{38} & f_{39} \\ 0 & 0 & 0 & \cosh\alpha\Phi & \frac{\alpha}{R_s}\sin\alpha\Phi & f_{46} & f_{47} & f_{48} & f_{49} \\ 0 & 0 & 0 & \frac{R_s}{\alpha}\sinh\alpha\Phi & \cosh\alpha\Phi & f_{56} & f_{57} & f_{58} & f_{59} \\ 0 & 0 & 0 & 0 & 0 & \cos\Phi & -\sin\Phi & R_s\sin\Phi & f_{69} \\ 0 & 0 & 0 & 0 & 0 & \sin\Phi & \cos\Phi & R_s(1-\cos\Phi) & f_{79} \\ 0 & 0 & 0 & 0 & 0 & 0 & 0 & 1 & f_{89} \\ 0 & 0 & 0 & 0 & 0 & 0 & 0 & 0 & 1 \end{bmatrix}_k \tag{4.4.93}$$

which is defined by the field transfer matrix. The coefficients f_{ij} in the above equation have been given previously in Eqs. (4.4.77) and (4.4.78).

When the value of $\alpha\Phi$ becomes larger than 30, or

$$\kappa = \alpha\Phi = R_s\Phi\sqrt{\frac{G_sK}{E_sI_\omega}} \geqq 30 \tag{4.4.94}$$

the warping can be ignored. Therefore,

$$M_\omega = 0 \tag{4.4.95a}$$

$$T_z = T_s \tag{4.4.95b}$$

Omitting the terms corresponding to these stress resultants [shown in the dotted parts of Eq. (4.4.93)], we find the following simplified formula:

$$\mathbf{x}_{(0),k} = [w(0)\ \beta(0)\ \varphi_X(0)\ M_X(0)\ T_x(0)\ Q_y(0)\ 1]_k \qquad (4.4.96a)$$

$$\mathbf{x}_{(\Phi),k} = [w(\Phi)\ \beta(\Phi)\ \varphi_X(\Phi)\ M_X(\Phi)\ T_z(\Phi)\ Q_y(\Phi)\ 1]_k \qquad (4.4.96b)$$

and

$$\mathbf{F}_k = \begin{bmatrix} 1 & R_s(1-\cos\Phi) & R_s\sin\Phi & f_{14} & f_{15} & f_{16} & f_{17} \\ 0 & \cos\Phi & \sin\Phi & f_{24} & f_{25} & f_{26} & f_{27} \\ 0 & -\sin\Phi & \cos\Phi & f_{34} & f_{35} & f_{36} & f_{37} \\ 0 & 0 & 0 & \cos\Phi & -\sin\Phi & R_s\sin\Phi & f_{47} \\ 0 & 0 & 0 & \sin\Phi & \cos\Phi & R_s(1-\cos\Phi) & f_{57} \\ 0 & 0 & 0 & 0 & 0 & 1 & f_{67} \\ 0 & 0 & 0 & 0 & 0 & 0 & 1 \end{bmatrix}_k \qquad (4.4.97)$$

where

$$f_{14} = -\frac{R_s^2}{E_sI_x}\left[\frac{1}{2}\left(1+\frac{1}{\chi}\right)\Phi\sin\Phi + \frac{1}{\chi}(\cos\Phi - 1)\right] = -f_{26} \qquad (4.4.98a)$$

$$f_{15} = \frac{R_s^2}{E_sI_x}\left[\frac{1}{2}\left(1+\frac{1}{\chi}\right)(\sin\Phi - \Phi\cos\Phi)\right] = f_{26} \qquad (4.4.98b)$$

$$f_{16} = -\frac{R_s^3}{E_sI_x}\left[\frac{1}{2}\left(1+\frac{3}{\chi}\right)\sin\Phi - \frac{1}{2}\left(1+\frac{1}{\chi}\right)\Phi\cos\Phi - \frac{\Phi}{\chi}\right] \qquad (4.4.98c)$$

$$f_{24} = \frac{R_s}{E_sI_x}\left[\frac{1}{2}\left(1+\frac{1}{\chi}\right)\Phi\sin\Phi\right] = -f_{35} \qquad (4.4.98d)$$

$$f_{25} = \frac{R_s}{E_sI_x}\left[\frac{1}{2}\left(1+\frac{1}{\chi}\right)\Phi\cos\Phi - \frac{1}{2}\left(1-\frac{1}{\chi}\right)\sin\Phi\right] \qquad (4.4.98e)$$

$$f_{34} = \frac{R_s}{E_sI_x}\left[\frac{1}{2}\left(1+\frac{1}{\chi}\right)\Phi\cos\Phi + \frac{1}{2}\left(1-\frac{1}{\chi}\right)\sin\Phi\right] \qquad (4.4.98f)$$

and

$$f_{17} = q_x\frac{R_s^4}{E_sI_x}\left[\left(1-\frac{2}{\chi}\right)(1-\cos\Phi) - \frac{1}{2}\left(1+\frac{1}{\chi}\right)\Phi\sin\Phi - \frac{\Phi^2}{2\chi}\right] + m_z\frac{R_s^3}{E_sI_x}\left[\left(1+\frac{1}{\chi}\right)(\cos\Phi - 1) + \frac{1}{2}\left(1+\frac{1}{\chi}\right)\Phi\sin\Phi\right] \qquad (4.4.99a)$$

$$f_{27} = q_x \frac{R_s^3}{E_s I_x}\left[\left(1+\frac{1}{\chi}\right)(\cos\Phi - 1) + \frac{1}{2}\left(1+\frac{1}{\chi}\right)\Phi\sin\Phi\right] + m_z \frac{R_s^2}{E_s I_x}\left[(1-\cos\Phi) - \frac{1}{2}\left(1+\frac{1}{\chi}\right)\Phi\sin\Phi\right] \quad (4.4.99b)$$

$$f_{37} = q_x \frac{R_s^3}{E_s I_x}\left[\frac{1}{2}\left(1+\frac{1}{\chi}\right)\Phi\cos\Phi - \frac{1}{2}\left(1+\frac{3}{\chi}\right)\sin\Phi + \frac{\Phi}{\chi}\right] + m_z \frac{R_s^2}{E_s I_x}\left[\frac{1}{2}\left(1+\frac{1}{\chi}\right)(\sin\Phi - \Phi\cos\Phi)\right] \quad (4.4.99c)$$

$$f_{47} = q_x R_s^2(\cos\Phi - 1) + m_z R_s(1-\cos\Phi) \quad (4.4.99d)$$

$$f_{57} = q_x R_s^2(\sin\Phi - \Phi) - m_z R_s \sin\Phi \quad (4.4.99e)$$

$$f_{67} = -q_x R_s \Phi \quad (4.4.99f)$$

where

$$\chi = \frac{G_s K}{E_s I_z} \quad (4.4.100)$$

***b.* In-plane deformation.** In a similar manner, Eq. (4.4.87) can be expressed in matrix form:

$$\mathbf{y}_{(\Phi),k} = \mathbf{G}_k \mathbf{y}_{(0),k} \quad (4.4.101)$$

where

$$\mathbf{y}_{(0),k} = [u(0)\ v(0)\ \varphi_y(0)\ M_y(0)\ Q_x(0)\ N_Z(0)\ 1]_k \quad (4.4.102a)$$

$$\mathbf{y}_{(\Phi),k} = [u(\Phi)\ v(\Phi)\ \varphi_Y(\Phi)\ M_Y(f)\ Q_x(\Phi)\ Q_x(\Phi)\ N_Z(\Phi)\ 1]_k \quad (4.4.102b)$$

and

$$\mathbf{G}_k = \begin{bmatrix} \cos\Phi & \sin\Phi & R_s(1-\cos\Phi) & g_{14} & g_{15} & g_{16} & g_{17} \\ -\sin\Phi & \cos\Phi & R_s\sin\Phi & g_{24} & g_{25} & g_{26} & g_{27} \\ 0 & 0 & 1 & g_{34} & g_{35} & g_{36} & g_{37} \\ 0 & 0 & 0 & 1 & -R_s\sin\Phi & R_s(1-\cos\Phi) & g_{47} \\ 0 & 0 & 0 & 0 & \cos\Phi & -\sin\Phi & g_{57} \\ 0 & 0 & 0 & 0 & \sin\Phi & \cos\Phi & g_{67} \\ 0 & 0 & 0 & 0 & 0 & 0 & 1 \end{bmatrix}_k \quad (4.4.103)$$

The coefficients g_{ij} have been given by Eqs. (4.4.88) and (4.4.89).

(5) Point transfer matrix

***a.* Out-of-plane deformation.** The continuity conditions for the displacements and the equilibrium conditions for the stress resultants at panel point k can be expressed as

$$w(0)_{,k+1} = w(\Phi)_{,k} - (R_{s,k+1} - R_{s,k})\beta(\Phi)_{,k} \tag{4.4.104a}$$

$$\beta(0)_{,k+1} = \beta(\Phi)_{,k} \tag{4.4.104b}$$

$$\varphi_X(0)_{,k+1} = \varphi_X(\Phi)_{,k} \tag{4.4.104c}$$

$$T_s(0)_{,k+1} = T_s(\Phi)_{,k} \tag{4.4.104d}$$

$$M_\omega(0)_{,k+1} = M_\omega(\Phi)_{,k} \tag{4.4.104e}$$

$$M_X(0)_{,k+1} = M_X(\Phi)_{,k} \tag{4.4.104f}$$

$$T_z(0)_{,k+1} = T_z(\Phi)_{,k} + (R_{s,k+1} - R_{s,k})Q_y(\Phi)_{,k} - T_{p,k} \tag{4.4.104g}$$

$$Q_y(0)_{,k+1} = Q_y(\Phi)_{,k} - P_{y,k} \tag{4.4.104h}$$

Note that $R_{s,k+1}$ and $R_{s,k}$ are the radii of curvature at panels $k+1$ and k, respectively, and $T_{z,k}$ and $P_{y,k}$ are the concentrated external torque and vertical load, respectively, which act on panel point k.

Therefore, this condition can be written in matrix form as

$$\mathbf{x}_{(0),k+1} = \mathbf{P}_k \mathbf{x}_{(\Phi),k} \tag{4.4.105}$$

where the state vectors at the left and right sides of panel k are, respectively, given by

$$\mathbf{x}_{(0),k+1} = [w(0)\ \beta(0)\ \varphi_X(0)\ T_s(0)\ M_\omega(0)\ M_X(0)\ T_y(0)\ Q_y(0)\ 1]_{k+1} \tag{4.4.106a}$$

$$\mathbf{x}_{(\Phi),k} = [w(\Phi)\ \beta(\Phi)\ \varphi_X(\Phi)\ T_s(\Phi)\ M_\omega(\Phi)\ M_X(\Phi)\ T_z(\Phi)\ Q_y(\Phi)\ 1]_k \tag{4.4.106b}$$

The point transfer matrix is

$$\mathbf{P}_k = \begin{bmatrix} 1 & -(R_{s,k+1}-R_{s,k}) & 0 & 0 & 0 & 0 & 0 & 0 & 0 \\ 0 & 1 & 0 & 0 & 0 & 0 & 0 & 0 & 0 \\ 0 & 0 & 1 & 0 & 0 & 0 & 0 & 0 & 0 \\ 0 & 0 & 0 & 1 & 0 & 0 & 0 & 0 & 0 \\ 0 & 0 & 0 & 0 & 1 & 0 & 0 & 0 & 0 \\ 0 & 0 & 0 & 0 & 0 & 1 & 0 & 0 & 0 \\ 0 & 0 & 0 & 0 & 0 & 0 & 1 & R_{s,k+1}-R_{s,k} & -T_{p,k} \\ 0 & 0 & 0 & 0 & 0 & 0 & 0 & 1 & -P_{y,k} \\ 0 & 0 & 0 & 0 & 0 & 0 & 0 & 0 & 1 \end{bmatrix} \tag{4.4.107}$$

In the case of pure torsion, shown in Eq. (4.4.97), the dotted parts of Eq. (4.4.107) will vanish.

By combining Eqs. (4.4.91) and (4.4.105), the following formulas can be obtained:

$$\mathbf{x}_{(\Phi),k} = \mathbf{F}_k\mathbf{P}_{k-1}\mathbf{F}_{k-2}\mathbf{P}_{k-2}\cdots\mathbf{F}_2\mathbf{P}_1\mathbf{F}_1\mathbf{x}_{(0),1} \tag{4.4.108a}$$

$$\mathbf{x}_{(0),k+1} = \mathbf{P}_k\mathbf{F}_k\mathbf{P}_{k-1}\mathbf{F}_{k-2}\mathbf{P}_{k-2}\cdots\mathbf{F}_2\mathbf{P}_1\mathbf{F}_1\mathbf{x}_{(0),1} \tag{4.4.108b}$$

where

$$\mathbf{x}_{(0),1} = [w(0)\ \beta(0)\ \Phi_X(0)\ T_s(0)\ M_\omega(0)\ M_X(0)\ T_z(0)\ Q_y(0)\ 1]_1 \tag{4.4.109}$$

is defined as the initial vector at the left end of panel 1 (see Fig. 4.24).

***b.* In-plane deformation.** In a manner similar to that used to describe out-of-plane deformation, the state vector

$$\mathbf{y}_{(0),k+1} = [u(0)\ v(0)\ \varphi_Y(0)\ M_Y(0)\ Q_x(0)\ N_Z(0)\ 1]_{k+1} \tag{4.4.110}$$

at the left side of panel $k+1$ can be expressed by using the state vector

$$\mathbf{y}_{(\Phi),k} = [u(\Phi)\ v(\Phi)\ \varphi_Y(\Phi)\ M_Y(\Phi)\ Q_x(\Phi)\ N_Z(\Phi)\ 1]_k \tag{4.4.111}$$

at the right-hand side of panel k as follows:

$$\mathbf{y}_{(0),k+1} = \mathbf{Q}_k\mathbf{y}_{(\Phi),k} \tag{4.4.112}$$

Here the point transfer matrix $\mathbf{Q}_k$ is given by

$$\mathbf{Q}_k = \begin{bmatrix} 1 & 0 & 0 & 0 & 0 & 0 & 0 \\ 0 & 1 & 0 & 0 & 0 & 0 & 0 \\ 0 & 0 & 1 & 0 & 0 & 0 & 0 \\ 0 & 0 & 0 & 1 & 0 & 0 & 0 \\ 0 & 0 & 0 & 0 & 1 & 0 & P_{x,k} \\ 0 & 0 & 0 & 0 & 0 & 1 & 0 \\ 0 & 0 & 0 & 0 & 0 & 0 & 1 \end{bmatrix} \tag{4.4.113}$$

and $P_{x,k}$ is a concentrated load acting in the direction of the x axis at panel point k.

The state vectors at the right-hand side of panel k and left-hand side of panel $k+1$ can also be written from Eqs. (4.4.101) and (4.4.112) as

$$\mathbf{y}_{(\Phi),k} = \mathbf{G}_k\mathbf{Q}_{k-1}\mathbf{G}_{k-1}\mathbf{Q}_{k-2}\cdots\mathbf{G}_2\mathbf{Q}_1\mathbf{G}_1\mathbf{y}_{(0),1} \tag{4.4.114a}$$

$$\mathbf{y}_{(0),k+1} = \mathbf{Q}_k\mathbf{G}_k\mathbf{Q}_{k-1}\mathbf{G}_{k-1}\mathbf{Q}_{k-2}\cdots\mathbf{G}_2\mathbf{Q}_1\mathbf{G}_1\mathbf{y}_{(0),1} \tag{4.4.114b}$$

where

$$y_{(0),1} = [u(0)\ v(0)\ \varphi_Y(0)\ M_Y(0)\ Q_x(0)\ N_Z(0)\ 1]_1 \tag{4.4.115}$$

is the initial vector at the left end of panel 1.

(6) Boundary conditions

a. **Initial conditions.** Figure 4.25 shows the boundary conditions at the left end of typical curved girder bridges. From this figure, the initial conditions can be determined for out-of-plane deformation. Since the displacements $w(0)_{,1}$ and $\beta(0)_{,1}$ and stress resultants $M_\omega(0)_{,1}$ and $M_x(0)_{,1}$ must vanish at the left end of panel 1, that is,

$$w(0)_{,1} = 0 \tag{4.4.116a}$$

$$\beta(0)_{,1} = 0 \tag{4.4.116b}$$

and

$$M_\omega(0)_{,1} = 0 \tag{4.4.117a}$$

$$M_X(0)_{,1} = 0 \tag{4.4.117b}$$

we see that the state vector $\mathbf{x}_{(0),1}$ can be taken into consideration by the vector

$$\mathbf{x}_{(0),1} = [\varphi_X(0)_{,1}\ T_s(0)_{,1}\ T_z(0)_{,1}\ Q_y(0)_{,1} \cdots T^s_{z,k}\ Q^s_{y,k} \cdots 1] \tag{4.4.118}$$

where $T^s_{z,k}$ and $Q^s_{y,k}$ are the reactions at the internal support k, which is discussed in detail in the next section.

This equation can also be expressed by the following state matrix $\mathbf{x}_{(0),1}$ which would be utilized by a computer program:

$$\mathbf{X}_{(0),1} = \begin{array}{c} \begin{matrix} \varphi_Y(0)_{,1} & T_s(0)_{,1} & T_z(0)_{,1} & Q_y(0)_{,1} & \cdots & T^s_{y,k} & Q^s_{y,k} & \cdots & 1 \end{matrix} \\ \begin{bmatrix} 0 & 0 & 0 & 0 & & 0 & 0 & & 0 \\ 0 & 0 & 0 & 0 & & 0 & 0 & & 0 \\ 1 & 0 & 0 & 0 & & 0 & 0 & & 0 \\ 0 & 1 & 0 & 0 & & 0 & 0 & & 0 \\ 0 & 0 & 0 & 0 & & 0 & 0 & & 0 \\ 0 & 0 & 0 & 0 & & 0 & 0 & & 0 \\ 0 & 0 & 1 & 0 & & 0 & 0 & & 0 \\ 0 & 0 & 0 & 1 & & 0 & 0 & & 0 \\ 0 & 0 & 0 & 0 & & 0 & 0 & & 1 \end{bmatrix} \end{array} \tag{4.4.119}$$

The dotted elements can be ignored in the case of pure torsion.

In a similar manner, the following initial vector can be obtained from Fig. 4.25 for in-plane deformation:

For type A,

$$\mathbf{y}_{(0),1} = [\varphi_Y(0)_{,1}\ Q_x(0)_{,1}\ N_Z(0)_{,1} \cdots Q^s_{y,k}\ N^s_{z,k} \cdots 1] \tag{4.4.120}$$

For type B,

$$\mathbf{y}_{(0),1} = [u(0)_{,1}\ \varphi_Y(0)_{,1}\ Q_x(0)_{,1} \cdots Q^s_{y,k} \cdots 1] \tag{4.4.121}$$

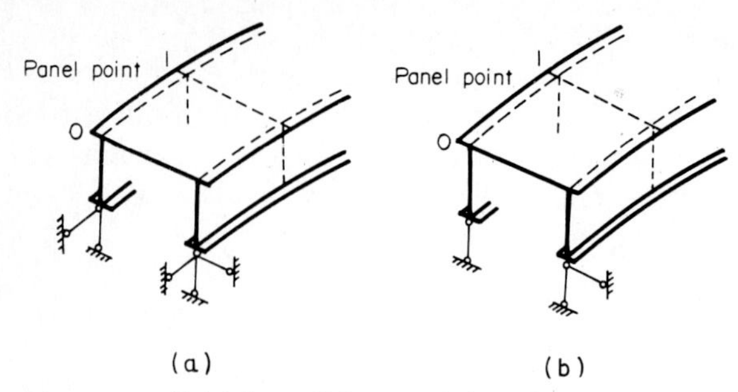

Figure 4.25 Initial conditions at left end for (*a*) type-A girder and (*b*) type-B girder.

Here $Q^s_{y,k}$ and $N^s_{Z,k}$ are the internal reactions at the support k. These vectors can also be written by the state matrix as follows:
For type A,

$$
\mathbf{Y}_{(0),1} = \begin{array}{c} \begin{matrix} \varphi_Y(0)_{,1} & Q_x(0)_{,1} & N_Z(0)_{,1} & & N^s_{Z,k} & Q^s_{y,k} & & 1 \end{matrix} \\ \begin{bmatrix} 0 & 0 & 0 & & 0 & 0 & & 0 \\ 0 & 0 & 0 & & 0 & 0 & & 0 \\ 1 & 0 & 0 & & 0 & 0 & & 0 \\ 0 & 0 & 0 & \cdots & 0 & 0 & \cdots & 0 \\ 0 & 1 & 0 & & 0 & 0 & & 0 \\ 0 & 0 & 1 & & 0 & 0 & & 0 \\ 0 & 0 & 0 & & 0 & 0 & & 1 \end{bmatrix} \end{array} \tag{4.4.122}
$$

For type B,

$$
\mathbf{Y}_{(0),1} = \begin{array}{c} \begin{matrix} u(0)_{,1} & \varphi_Y(0)_{,1} & Q_x(0)_{,1} & & Q^s_{x,k} & & 1 \end{matrix} \\ \begin{bmatrix} 1 & 0 & 0 & & 0 & & 0 \\ 0 & 0 & 0 & & 0 & & 0 \\ 0 & 1 & 0 & & 0 & & 0 \\ 0 & 0 & 0 & \cdots & 0 & \cdots & 0 \\ 0 & 0 & 1 & & 0 & & 0 \\ 0 & 0 & 0 & & 0 & & 0 \\ 0 & 0 & 0 & & 0 & & 1 \end{bmatrix} \end{array} \tag{4.4.123}
$$

***b.* Boundary conditions at the intermediate support.** Two unknown internal reactions, $T^s_{z,k}$ and $Q^s_{y,k}$, occur at the intermediate support k for both type A and B supports, as shown in Fig. 4.26. Since these reactions have been considered in Eqs. (4.4.118) and (4.4.119), the following

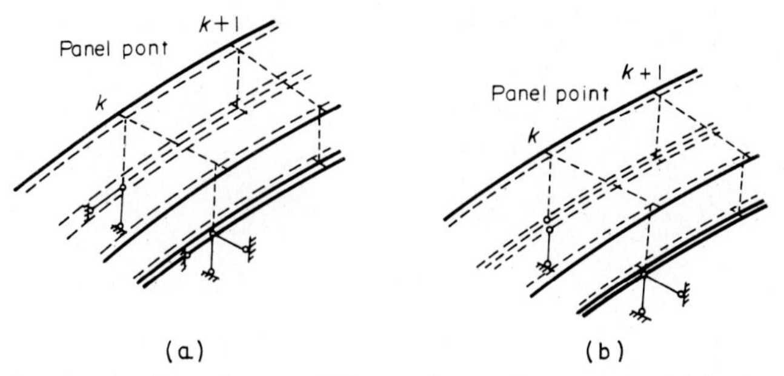

Figure 4.26 Boundary conditions at internal support for (*a*) type-A girder and (*b*) type-B girder.

boundary conditions must be satisfied at point k to determine these unknown values:

$$\mathbf{w}_{(\Phi),k} = 0 \tag{4.4.124a}$$

$$\boldsymbol{\beta}_{(\Phi),k} = 0 \tag{4.4.124b}$$

Here $\mathbf{w}_{(\Phi),k}$ and $\boldsymbol{\beta}_{(\Phi,k}$ are defined by the first and second rows, respectively, of the matrix:

$$\mathbf{X}_{(\Phi),k} = \mathbf{F}_k\mathbf{P}_{k-1}\mathbf{F}_{k-1}\mathbf{P}_{k-2}\cdots\mathbf{F}_2\mathbf{P}_1\mathbf{F}_1\mathbf{X}_{(0),1} \tag{4.4.125}$$

Corresponding to these procedures, the state matrix

$$\mathbf{X}_{(0),k+1} = \mathbf{P}_k\mathbf{F}_k\mathbf{P}_{k-1}\mathbf{F}_{k-2}\cdots\mathbf{F}_2\mathbf{P}_1\mathbf{F}_1\mathbf{X}_{(0),1} \tag{4.4.126}$$

must be modified as follows:

$$\mathbf{X}_{(0),k+1} = \begin{bmatrix} \varphi_X(0)_{,1} & T_s(0)_{,1} & T_z(0)_{,1} & Q_y(0)_{,1} & T^s_{z,k} & Q^s_{y,k} & 1 \\ & & & & 0 & 0 & 0 \\ & \text{nonzero element} & & & 0 & 0 & 0 \\ & & & & 0 & 0 & 0 \\ & & & \cdots & 0 & 0 \quad \cdots & 0 \\ & & & & 0 & 0 & 0 \\ & & & & 0 & 0 & 0 \\ & & & & 1 & R_{s,k+1}-R_{s,k} & 0 \\ & & & & 0 & 1 & 0 \\ & & & & 0 & 0 & 1 \end{bmatrix} \tag{4.4.127}$$

The dashed parts can be omitted in the case of pure torsion.

The boundary conditions for the in-plane deformation can be written as follows:
For type A,

$$\mathbf{u}_{(\Phi),k} = 0 \tag{4.4.128a}$$

$$\mathbf{v}_{(\Phi),k} = 0 \tag{4.4.128b}$$

For type B,

$$\mathbf{v}_{(\Phi),k} = 0 \tag{4.4.129}$$

where $\mathbf{u}_{(\Phi),k}$ and $\mathbf{v}_{(\Phi),k}$, respectively, designate the first and second rows of the state matrix, calculated by

$$\mathbf{Y}_{(\Phi),k} = \mathbf{G}_k \mathbf{Q}_{k-1} \mathbf{G}_{k-1} \mathbf{Q}_{k-2} \cdots \mathbf{G}_2 \mathbf{Q}_1 \mathbf{G}_1 \mathbf{Y}_{(0),1} \tag{4.4.130}$$

On the other hand, the matrix

$$\mathbf{Y}_{(0),k+1} = \mathbf{Q}_k \mathbf{G}_k \mathbf{Q}_{k-1} \mathbf{G}_{k-2} \cdots \mathbf{G}_2 \mathbf{Q}_1 \mathbf{G}_1 \mathbf{Y}_{(0),1} \tag{4.4.131}$$

must be modified as follows:
For type A,

$$\begin{array}{cccccccc} & \varphi_Y(0)_{,1} \quad Q_x(0)_{,1} \quad N_Z(0)_{,1} & \cdots & Q^s_{x,k} & N^s_{Z,k} & \cdots & 1 \end{array}$$

$$\mathbf{Y}_{(0),k+1} = \left[\begin{array}{c:cccccc} & & & 0 & 0 & & 0 \\ & & & 0 & 0 & & 0 \\ & & & 0 & 0 & & 0 \\ & \text{nonzero element} & \cdots & 0 & 0 & \cdots & 0 \\ & & & 1 & 0 & & 0 \\ & & & 0 & 1 & & 0 \\ & & & 0 & 0 & & 1 \end{array}\right] \tag{4.4.132}$$

For type B,

$$\begin{array}{cccccc} & u(0)_{,1} \quad \varphi_Y(0)_{,1} \quad Q_x(0)_{,1} & \cdots & Q^s_{x,k} & \cdots & 1 \end{array}$$

$$\mathbf{Y}_{(0),k+1} = \left[\begin{array}{c:ccccc} & & & 0 & & 0 \\ & & & 0 & & 0 \\ & & & 0 & & 0 \\ & \text{nonzero element} & \cdots & 0 & \cdots & 0 \\ & & & 1 & & 0 \\ & & & 0 & & 0 \\ & & & 0 & & 1 \end{array}\right] \tag{4.4.133}$$

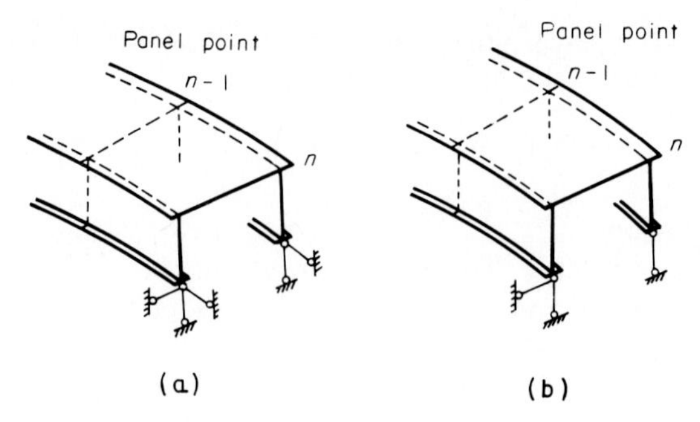

Figure 4.27 Boundary conditions at right end for (*a*) type-A girder and (*b*) type-B girder.

c. **Boundary conditions at the right end.** Figure 4.27 shows the boundary conditions at the right end of panel n of a curved girder bridge. In out-of-plane deformation, for the boundary conditions shown at the right end of panel n, the following equations must be satisfied:

$$\mathbf{w}_{(\Phi),n} = 0 \tag{4.4.134a}$$

$$\boldsymbol{\beta}_{(\Phi),n} = 0 \tag{4.4.134b}$$

$$\mathbf{M}_{\omega(\Phi),n} = 0 \tag{4.4.134c}$$

$$\mathbf{M}_{X(\Phi),n} = 0 \tag{4.4.134d}$$

where $\mathbf{w}_{(\Phi),n}$, $\boldsymbol{\beta}_{(\Phi),n}$, $\mathbf{M}_{\omega(\Phi),n}$, and $\mathbf{M}_{X(\Phi),n}$, respectively, represent the first, second, fifth, and sixth rows of the state matrix

$$\mathbf{X}_{(\Phi),n} = \mathbf{F}_n\mathbf{P}_{n-1} \cdots \mathbf{P}_k\mathbf{F}_k \cdots \mathbf{P}_2\mathbf{F}_2\mathbf{P}_1\mathbf{F}_1\mathbf{X}_{(0),1} \tag{4.4.135}$$

For in-plane deformation:
For type A,

$$\mathbf{u}_{(\Phi),n} = 0 \tag{4.4.136a}$$

$$\mathbf{v}_{(\Phi),n} = 0 \tag{4.4.136b}$$

$$\mathbf{M}_{Y(\Phi),n} = 0 \tag{4.4.136c}$$

For type B,

$$\mathbf{v}_{(\Phi),n} = 0 \tag{4.4.137a}$$

$$\mathbf{M}_{Y(\Phi),n} = 0 \tag{4.4.137b}$$

$$\mathbf{N}_{Z(\Phi),n} = 0 \tag{4.4.137c}$$

where $\mathbf{u}_{(\Phi),n}$, $\mathbf{v}_{(\Phi),n}$, $\mathbf{M}_{Y(\Phi),n}$, and $\mathbf{N}_{Z(\Phi),n}$ designate, respectively, the first, second, fourth, and sixth rows of the state matrix

$$\mathbf{Y}_{(\Phi),n} = \mathbf{G}_n \mathbf{Q}_{n-1} \cdots \mathbf{Q}_k \mathbf{G}_k \cdots \mathbf{Q}_2 \mathbf{F}_2 \mathbf{Q}_1 \mathbf{F}_1 \mathbf{Y}_{(0),1} \tag{4.4.138}$$

(7) Procedure of computation

***a.* Determination of unknown values.** The unknown values of $\varphi_X(0)_{,1}$, $T_s(0)_{,1}$, $T_z(0)_{,1}$, $Q_y(0)_{,1}, \ldots,$ $T^s_{z,k}$, and $Q^s_{y,k}$ can be estimated by the boundary conditions in Eqs. (4.4.134*a*) to (4.4.134*d*) and (4.4.124) in the case of out-of-plane deformation. These equations yield a set of simultaneous equations:

$$\mathbf{Aa} = -\mathbf{b} \tag{4.4.139}$$

where
$$\mathbf{a} = [\varphi_X(0)_{,1} \;\; T_s(0)_{,1} \;\; T_z(0)_{,1} \;\; \cdots \;\; T^s_{z,k} \;\; Q^s_{y,k} \;\; \ldots] \tag{4.4.140}$$

with $\mathbf{A}$ being the coefficient matrix made by unknown values and $\mathbf{b}$ the load-terms mode by the last column of the boundary conditions for the unknown values. From this equation, the unknown values can be easily solved:

$$\mathbf{a} = -\mathbf{A}^{-1}\mathbf{b} \tag{4.4.141}$$

Thus, the initial vector can be determined by adding the load term, 1.0, to vector $\mathbf{a}$:

$$\mathbf{y}_{(0),1} = [\mathbf{a} \,\vdots\, 1] \tag{4.4.142}$$

The unknown values for the in-plane deformations can also be estimated similarly.

***b.* Flowchart for computer use.** The analysis of deformations and stress-resultants of curved girder bridges have been performed with the computer program flowchart shown in Fig. 4.28. Note, however, that numerical errors will occur in the computations, since the elements of the transfer matrix have remarkably different dimensions. To avoid such errors, the matrix is nondimensionalized as follows.

For out-of-plane deformations:

Displacement terms,

$$w = w^* \frac{P_c R_c^3}{E_s I_c} \tag{4.4.143a}$$

$$\beta = \beta^* \frac{P_c R_c^2}{E_s I_c} \tag{4.4.143b}$$

$$\varphi_X = \varphi_X^* \frac{P_c R_c^2}{E_s I_c} \tag{4.4.143c}$$

Stress resultant terms,

$$T_s = T_s^* P_c R_c \qquad (4.4.144a)$$

$$M_\omega = M_\omega^* P_c R_c \qquad (4.4.144b)$$

$$M_X = M_X^* P_c R_c \qquad (4.4.144c)$$

$$T_z = T_z^* P_c R_c \qquad (4.4.144d)$$

$$T_z^s = T_z^{s*} P_c R_c \qquad (4.4.144e)$$

$$Q_y = Q_y^* P_c \qquad (4.4.144f)$$

$$Q_y^s = Q_y^{s*} P_c \qquad (4.4.144g)$$

Load terms,

$$q_y = q_y^* \frac{P_c}{R_c} \qquad (4.4.145a)$$

$$m_z = m_z^* P_c \qquad (4.4.145b)$$

$$P_y = P_y^* P_c \qquad (4.4.145c)$$

$$T_z = T_z^* P_c R_c \qquad (4.4.145d)$$

In Eqs. (4.3.143) to (4.3.145), R_c, I_c, and P_c can be chosen arbitrarily to make the nondimensionalization as simple as possible.

Thus, all the vectors are transformed to nondimensional values and are defined by the superscript asterisk, such as w^*, β^*, T_s^*, M_ω^*, and Q_x^*. The elements of the field transfer matrix and point transfer matrix should, of course, be similarly changed to $\mathbf{F}_k^*$ and $\mathbf{P}_k^*$, as shown in Fig. 4.20. Accordingly

$$\mathbf{X}^*(\Phi)_{,k} = \mathbf{F}_k^* \mathbf{Y}(0)_{,b} \qquad (4.4.146a)$$

$$\mathbf{X}(0)_{,k+1} = \mathbf{P}_k^* \mathbf{X}(\Phi)_{,k} \qquad (4.4.146b)$$

Another source of error occurs when the parameter $\alpha_k \phi_k = R_{s,k}\sqrt{G_s K_k/(E_s I_{\omega,k})}$ has a larger value than 4.0 for the kth panel—then the values given by the hyperbolic function sinh $\alpha_k \phi_k$ or cosh $\alpha_k \phi_k$, as seen in Eqs. (4.4.76) and (4.4.77), become greater than 1.0. Therefore,

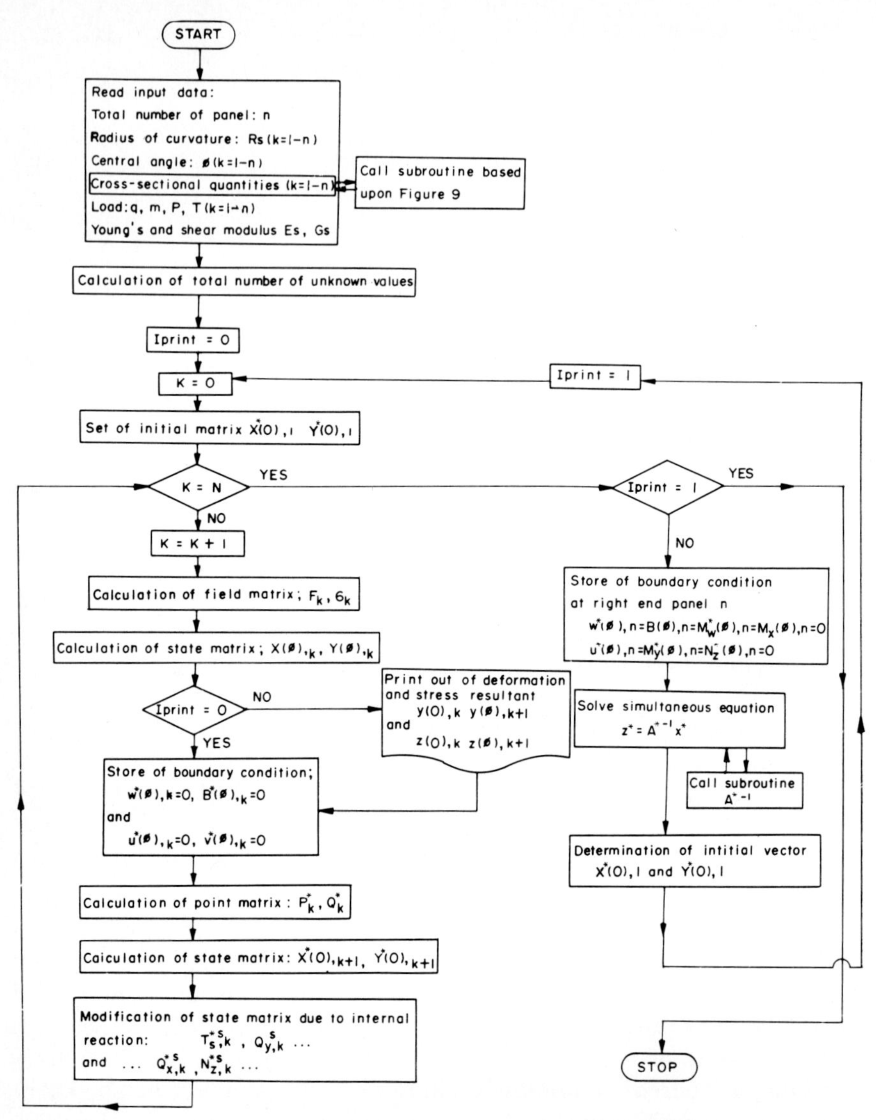

Figure 4.28 Flowchart for analyzing the displacements and stress resultants of a curved girder bridge by the transfer matrix method.

the elements other than the hyperbolic function may be reduced to meaningless values, and the results will contain numerical errors.

To avoid this problem in using the transfer matrix method, the following special method has been developed by Nakai, Kotoguchi, and Tani.[12]

For in-plane deformation, to obtain accurate results, the deformations and the stress resultants should also be nondimensionalized as follows:
Displacements,

$$u = u^* \frac{P_c R_c^3}{E_s I_c} \tag{4.4.147a}$$

$$v = v^* \frac{P_c R_c^3}{E_s I_c} \tag{4.4.147b}$$

$$\varphi_Y = \varphi_Y^* \frac{P_c R_c^2}{E_s I_c} \tag{4.4.147c}$$

Stress resultants,

$$M_Y = M_Y^* P_c R_c \tag{4.4.148a}$$

$$Q_x = Q_x^* P_c \tag{4.4.148b}$$

$$Q_x^s = Q_x^{s*} P_c \tag{4.4.148c}$$

$$N_Z = N_Z^* P_c \tag{4.4.148d}$$

$$N_Z^s = N_Z^{s*} P_c \tag{4.4.148e}$$

Forces,

$$q_x = q_x^* \frac{P_c}{R_c} \tag{4.4.149a}$$

$$P_x = P_x^* P_c \tag{4.4.149b}$$

For this case, the calculations for the curved girder bridge can be performed by the following method

$$\mathbf{y}_{(\Phi),k}^* = \mathbf{G}_k^* \mathbf{y}_{(0),k}^* \tag{4.4.150a}$$

$$\mathbf{y}_{(0),k+1}^* = \mathbf{Q}_k^* \mathbf{y}_{(\Phi),k}^* \tag{4.4.150b}$$

where the elements of the field and point transfer matrices $\mathbf{G}_k^*$ and $\mathbf{Q}_k^*$ are all nondimensional values obtained from Eqs. (4.4.147) to (4.4.149).

References

1. Wittfoht, H.: *Kreisforming gekrümmte Träger*, Springer-Verlag, Berlin, 1964 (in German).
2. Shimada, S., and S. Kuranishi: *Formulas for Calculation of Curved Beam*, Gihodo, Tokyo, 1966 (in Japanese).

3. Watanabe, N.: *Theory and Calculation of Curved Girder*, Gihodo, Tokyo, 1967 (in Japanese).
4. Vreden, W.: *Die Berechnung des gekrümmten Durchlaufträgers*, Verlag von Wilhelm Ernst & Sohn, Berlin, 1969 (in German).
5. Konishi, I., and S. Komatsu: "*Three-Dimensional Analysis of Curved Girder with Thin-Walled Cross-Sections*," Publication of IABSE (International Association for Bridge and Structural Engineering), Zurich, 1965.
6. Kersten, R.: *Das Reduktion Verfahren in Baustatik*, Springer-Verlag, Berlin, 1962 (in German).
7. Pestel, C. E., and A. F. Leckie: *Matrix Method in Elastomechanics*, McGraw-Hill, New York, 1963.
8. Dabrowski, R.: *Gekrümmte dünnwandige Träger*, Springer-Verlag, Berlin, 1968 (in German).
9. Ochi, Y.: *Structural Analysis by Computer*, Bridge Engineering Association, Tokyo, April 1969 (in Japanese).
10. Becker, G.: "Ein Beitrag zur statischen Berechnung beliebge gelarter ebener gekrümmter einfach symmetrischen dünnwadige offen Profilen von Stabachse veranderlischen Querchnitt unter Berucksichtigen der Wölbkrafttorsion," Der Stahlbau, 34 Jahrgang, s.334–369, Nov. 1965 (in German).
11. Komatsu, S., H. Nakai, and M. Nakanishi: "Statical Analysis of Horizontally Curved Skew Box Girder Bridges," *Proceedings of the Japanese Society of Civil Engineers*, no. 193, pp. 1–12. Sept. 1971 (in Japanese).
12. Nakai, H., T. Kotoguchi, and T. Tani: "Matrix Structural Analysis of Thin-Walled Curved Girder Bridges Subjected to Arbitrary Loads," *Proceedings of the Japanese Society of Civil Engineers*, no. 255, pp. 1–15, Nov. 1976 (in Japanese).
13 Nakai, H., T. Okumura, and R. Tanida: "Structural Analysis of Arbitrary Grillage Girder Bridges by Transfer Matrix Method Sustained with Model Tests," *Proceedings of the Japanese Society of Civil Engineers*, no. 243, pp. 1–18, Nov. 1975 (in Japanese).
14. Nakai, H., and T. Tani: "An Approximate Method for Evaluation of Torsional and Warping Stresses in Box Girder Bridges," *Proceedings of the Japanese Society of Civil Engineers*, no. 277, pp. 41–55, Sept. 1978 (in Japanese).

Chapter

5

Buckling Stability and Strength of Curved Girders

5.1 Introduction

This chapter deals with the buckling and ultimate strength of curved girders, and we develop a design method for curved girders and compare the equations and methods developed for straight girders.

In Sec. 5.2, the lateral buckling behavior of multiple curved I girders is determined through numerous model girders derived from dimensions and configurations of actual curved girder bridges. The overall lateral buckling behavior of such girders is not an important problem, when the ratio between the span and spacing of the main girder is small. However, the local lateral buckling of curved I girders at the supported portions between floor beams or sway and lateral bracings is important. These buckling behaviors are also investigated through studies of many model girders. Based on these experimental studies, an analytical method for evaluating the lateral buckling strength of curved I girders is developed in accordance with the second-order buckling analysis. Furthermore, the lateral buckling strength and the corresponding effective buckling length to be applied to the design of multiple curved I girders are proposed on the basis of these experimental and theoretical studies.

In Sec. 5.3, the problems of buckling stability and the ultimate strength of the web plate of a curved girder are presented in comparison with those of straight girders. However, there are many unclarified points concerning the buckling and collapse behavior of web plates in the curved girders. Therefore, the buckling and ultimate strength of curved girders for bending, shear, and their combinations are investigated through a series of experimental studies on numerous model I girders having different curvatures, aspect ratios, slenderness of web plates, rigidity of stiffeners, and slenderness of flange plates. Thus, we show how much the buckling and ultimate strength are affected by curvature parameters of curved girders. Moreover, these behaviors are examined by selecting the various inherent parameters of curved girders.

In Sec. 5.4, a method for evaluating the web slenderness requirements of curved girders under bending moment action is presented. The stress and displacement in the web plate of a curved girder within the elastic region are analyzed by using the isoparametric finite-element method on the basis of large displacement theory, since the clear buckling phenomena are not observed in the curved web plates according to the experimental studies in Sec. 5.2. These results are shown in the form of many graphs. Then, after a serviceability limit state for stresses and out-of-plane deflections of a curved web plate is set, a requirement for the slenderness of web plates is proposed by referring to the design criteria of straight girders. The application of this proposal to the design of curved girders is also discussed in detail.

In Sec. 5.5, a design method for longitudinal stiffeners to prevent the overall buckling of curved web plates is presented. To design the curved girder stiffened by the longitudinal stiffeners without collapse of the longitudinal stiffeners prior to the yielding of flange plates, a beam-column model of the longitudinal stiffener subjected to an axial compressive force and lateral load due to the curvature of the web plate is proposed, and their strength is analyzed through the limit state analysis. By comparing the rigidity of longitudinal stiffeners of curved girders with that of straight girders, the required rigidity of the longitudinal stiffener in the curved girder is revealed.

In Sec. 5.6, a method of designing the transverse stiffeners in curved girders is presented. First, the shear buckling strength of a web plate in a curved girder is determined according to the elastic buckling theory of shallow shells, and the variations in shear buckling strength due to the curvature of the web plates are investigated numerically. Second, the ultimate strength of curved girders for shear is determined, and the induced forces in transverse stiffeners are clarified, when the curved web plate collapses after the propagation of the diagonal tension field. Thereafter, a beam-column model of the transverse stiffener subjected to axial compressive force and lateral force due to the curvature of the web plate is proposed, and the yielding and ultimate strengths are determined corresponding to the induced forces in the beam-column model. Then we show that the web plates of curved girders do not reach the overall shear buckling, if the transverse stiffeners have enough strength and rigidity against the initial yield state even when the curved girder results in collapse due to shear. From these facts, a design method for estimating the rigidity of transverse stiffeners is proposed.

In Sec. 5.7, the local buckling strength and the corresponding ultimate strength of flange plates in curved girders are predicted. In an elastic buckling analysis, the influence of the curvature of flange plates in ordinary curved girders is generally so small that the ultimate strength of flange plates without curvature is analyzed by taking into account only the warping stress in the flange plates. According to these analyses, the requirements for the slenderness of flange plates in curved girders are discussed in detail.

5.2. Lateral Buckling of Multiple Curved I Girders

5.2.1 Lateral buckling of I girders

When an I girder with a bisymmetric cross section is subjected to equal bending moments M at both ends, the I girder initially deflects a distance w in the vertical direction, as seen in Fig. 5.1. However, the centroid C of this I girder suddenly deflects a distance v in the horizontal direction and rotates an angle θ around the z axis simultaneously in accordance with the increase in bending moment M. Thus, M is always governed by a critical bending moment M_{cr}. This phenomenon is referred to as the *lateral buckling*, and it must be considered in designing the bridges composed of I girders.[1-4]

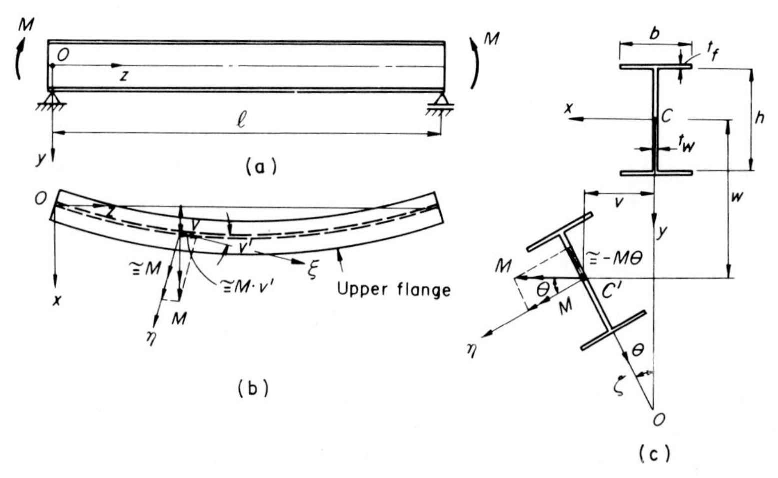

Figure 5.1 Lateral buckling of I girders: (*a*) span and load condition, (*b*) displacement of upper flange, (*c*) displacement of cross section.

Since the original coordinate axes (z, x, y) will change to new coordinate axes (ξ, η, ζ) after buckling, this I girder undergoes bending moment $-M\theta$ with respect to the ζ axis and torsional moment Mv' with respect to the ξ axis. Therefore, the governing differential equation of lateral buckling is represented according to the fundamental equations of flexure and torsional warping of a beam given by Eqs. (2.3.81) and (2.3.117a):

$$EI_y v^{\mathrm{IV}} + M\theta'' = 0 \tag{5.2.1a}$$

$$EI_\omega \theta^{\mathrm{IV}} - GK\theta'' + Mv'' = 0 \tag{5.2.1b}$$

where I_y = geometric moment of inertia with respect to y axis
I_ω = warping constant
K = torsional constant

When both ends of the I girder are simply supported, the solutions of displacements v and θ, which satisfy all the boundary conditions imposed on the I girder, are given by

$$v = a \sin \frac{\pi z}{l} \tag{5.2.2a}$$

$$\theta = b \sin \frac{\pi z}{l} \tag{5.2.2b}$$

The substitution of these equations into Eq. (5.2.1) gives the following homogeneous and simultaneous equations concerning the unknown coefficients a and b:

$$\left(EI_y \frac{\pi^4}{l^4}\right)a - \left(M\frac{\pi^2}{l^2}\right)b = 0 \tag{5.2.3a}$$

$$-\left(M\frac{\pi^2}{l^2}\right)a + \left(EI_\omega \frac{\pi^4}{l^4} + GK\frac{\pi^2}{l^2}\right)b = 0 \tag{5.2.3b}$$

To obtain a nontrivial solution, the derminant built by coefficients a and b should equal zero:

$$\begin{vmatrix} EI_y \dfrac{\pi^4}{l^4} & -M\dfrac{\pi^2}{l^2} \\ -M\dfrac{\pi^2}{l^2} & EI_\omega \dfrac{\pi^4}{l^4} + GK\dfrac{\pi^2}{l^2} \end{vmatrix} = 0 \tag{5.2.4}$$

Thus, the lateral buckling moment, i.e., the critical moment $M = M_{\mathrm{cr}}$, can be found from

$$M_{\mathrm{cr}} = \frac{\pi}{l}\sqrt{EI_y GK\left[1 + \frac{EI_\omega}{GK}\left(\frac{\pi}{l}\right)^2\right]} \tag{5.2.5}$$

With regard to the cross-sectional quantities, setting

$$I_y \cong 2\frac{b^3 t_f}{12} = \frac{A_f}{6} b^2 \tag{5.2.6a}$$

$$I_\omega \cong 2\frac{I_y}{2}\left(\frac{h}{2}\right)^2 = \frac{I_y}{4} h^2 \tag{5.2.6b}$$

$$A = 2bt_f + t_w h = 2A_f + A_w \tag{5.2.6c}$$

and neglecting the contribution of K in Eq. (5.2.5), we can write the critical stress $\sigma_{\rm cr}$ as

$$\sigma_{\rm cr} \cong \frac{M_{\rm cr}}{I_x}\frac{h}{2} = \frac{\pi^2 E\sqrt{I_y I_\omega}}{l^2 I_x}\frac{h}{2}$$

$$= \frac{\pi^2 E}{4(I_x/I_y)(l/h)^2} \tag{5.2.7}$$

Hence

$$I_x \cong 2bt_f\left(\frac{h}{2}\right)^2 + \frac{t_w h^3}{12}$$

$$= \frac{A_f}{2} h^2\left(1 + \frac{1}{6}\frac{A_w}{A_f}\right) \tag{5.2.8}$$

The term $(I_x/I_y)(l/h)^2$ can be simplified as follows:

$$\frac{I_x}{I_y}\left(\frac{l}{h}\right)^2 = \frac{(A_f/2)h^2}{(A_f/6)b^2}\left(1 + \frac{1}{6}\frac{A_w}{A_f}\right)\left(\frac{l}{h}\right)^2$$

$$= \left(3 + \frac{A_w}{2A_f}\right)\left(\frac{l}{b}\right)^2 \tag{5.2.9}$$

Letting A_f and A_w equal the cross-sectional area of the compression flange and web plates, respectively, and introducing a parameter k given by

$$k = \sqrt{3 + \frac{A_w}{2A_f}} \tag{5.2.10}$$

we can write Eq. (5.2.7) as

$$\sigma_{\rm cr} = \frac{\pi^2 E}{4(kl/b)^2} \tag{5.2.11}$$

For a noncomposite girder with a bisymmetric cross section, where $A_w/A_f \leqq 2$,

$$k \cong 2 \tag{5.2.12}$$

Now, dividing both sides of Eq. (5.2.11) by the yield stress σ_y, we can write σ_{cr}/σ_y as

$$\frac{\sigma_{cr}}{\sigma_y} = \frac{1}{\alpha^2} \tag{5.2.13}$$

where α is the nondimensionalized buckling parameter, which gives

$$\alpha = \frac{2k}{\pi}\frac{l}{b}\sqrt{\frac{\sigma_y}{E}} \tag{5.2.14}$$

The relationship between σ_{cr}/σ_y and α can be plotted as shown in Fig. 5.2. In this figure, the region where $\alpha < \sqrt{2}$ corresponds to nonelastic buckling due to the residual stresses.[5]

Therefore, a practical formula to evaluate the ultimate stress σ_u is[6]

$$\sigma_u = \sigma_y[1 - 0.412(\alpha - 0.2)] \tag{5.2.15}$$

In fact, the Japanese Specification for Highway Bridges[6] provides the following allowable compressive bending stresses σ_{ba}, by dividing by the factor of safety ν ($\cong 1.7$):

$$\sigma_{ba} = \begin{cases} \dfrac{\sigma_u}{\nu} & \dfrac{l}{b} \leqq c \quad (5.2.16a) \\[2ex] \dfrac{\sigma_u}{\nu} - a\left(\dfrac{l}{b} - c\right) & c < \dfrac{l}{b} \leqq c_o \quad (5.2.16b) \end{cases}$$

where a, c, and c_o are coefficients determined by the steel grade as shown in Table 5.1 and l is the length of the unsupported portion of the compression flange, as illustrated in Fig. 5.3.

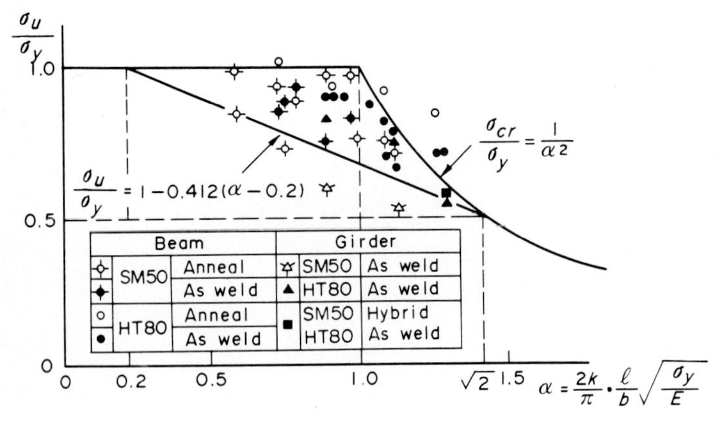

Figure 5.2 Lateral buckling curve $\sigma_u/\sigma_y \sim \alpha$.

TABLE 5.1 Coefficients a, c, and c_o

Steel grade	σ_y, N/mm^2	a	c	c_o
SS 41	235	24	4.5	30
SM 50	314	38	4.0	30
SM 53	353	44	3.5	27
SM 58	451	66	5.0	25

5.2.2 Experiments on lateral buckling of multiple curved I girders

The lateral buckling behavior of multiple curved I-girder bridges can be classified into two distinct categories: overall buckling and local buckling of the main girders between the support points, such as floor beams or sway and lateral bracings. To clarify these phenomena, a series of experimental studies was conducted prior to discussing the lateral buckling strength of curved I girders.[7]

(1) Overall lateral buckling tests of multiple curved I girders

a. **Model girders.** Three model girder bridges MG-1, MG-2, and MG-3 with two main I girders having floor beams and three different sizes of lateral bracings were built on the scale of 1 : 3.8 based on a dimensional analysis of actual multiple curved I-girder bridges. Figures 5.4 and 5.5 show the plan and the cross section of model girders, respectively. The ratio between the span length L and girder spacing B was $L/B = 5.56$. The local buckling of flange and web plates could effectively be prevented by making their thicknesses as large as possible.

The cross sections of lateral bracings in model girders MG-1, MG-2, and MG-3 were, respectively, chosen to have 2, 1, and 0.5 times the cross-sectional area of lateral bracings based on the dimensional analysis of actual bridges, as shown in Fig. 5.6.

All these members were made of structural steel SS-41, with Young's

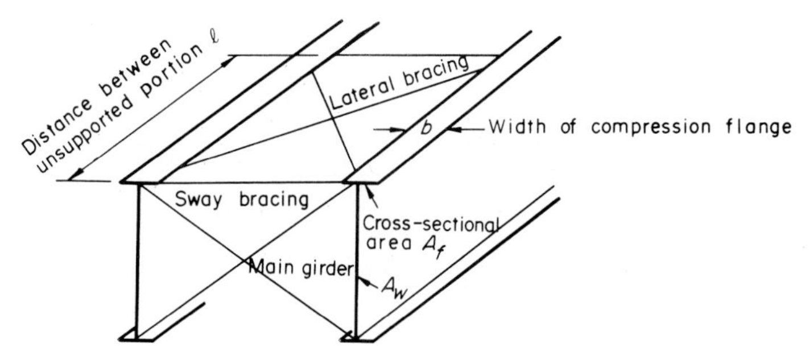

Figure 5.3 Distance between unsupported portion of compression flange.

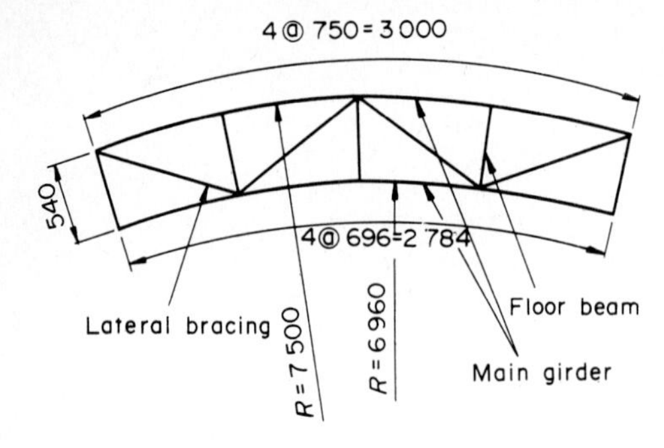

Figure 5.4 Plan of model girder (in millimeters).

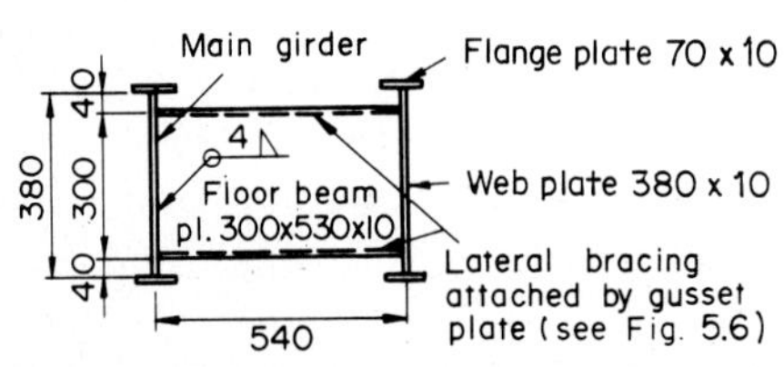

Figure 5.5 Cross section of model girder (in millimeters).

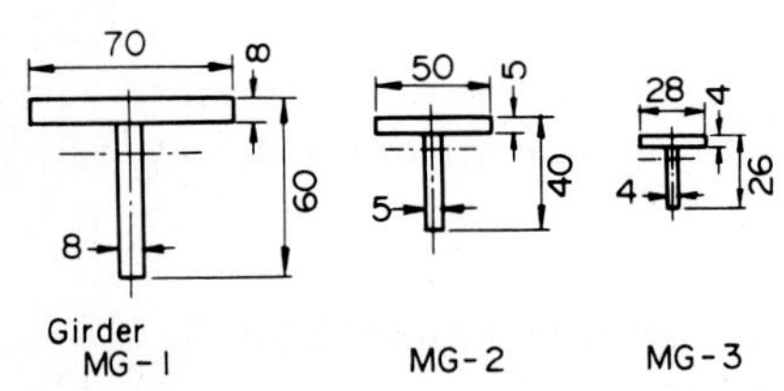

Figure 5.6 Cross section of lateral bracings (in millimeters).

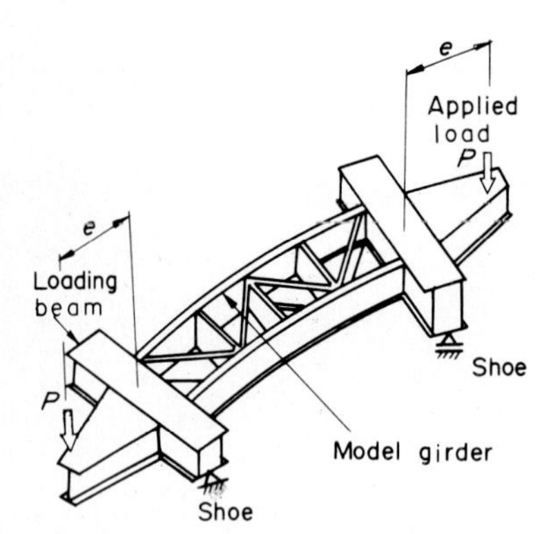

Figure 5.7 Load conditions.

modules $E = 2.1 \times 10^6$ kgf/cm^2 (205.8 GPa), Poisson's ratio, $\mu = 0.3$, and yield stress $\sigma_y = 3100$ kgf/cm^3 (303.8 MPa).

***b.* Loading conditions.** It is preferable that the lateral buckling tests be performed under a condition where the shearing force will not appear as in the pure bending of a straight beam. To satisfy this condition as exactly as possible, two loading beams with cantilever tongues were bolted to the model girders. The upper shoes using the load cell transducers were also bolted to these loading beams, and the lower shoes were set so that the model girder was supported on a roller at one end and a pin at the other. And then the same vertical loads P were applied to the ends of cantilevers by hydraulic jacks, as seen in Fig. 5.7.

Thus, the model girder can be approximately subjected to the pure bending moment $M = Pe$, where e is the eccentricity of load P and the shoe in the direction of girder axis.

***c.* Measurements of displacements and strains.** The lateral displacements were measured by special instruments which consist of steel wire ($\phi = 0.3$ mm), a roller with ball bearings, counterweight (300 gr), and dial gages with accuracy 1/100 mm, as illustrated in Fig. 5.8. However, the strains in main girders and lateral bracings were measured with the ordinary electric wire strain gages with gage lengths 5 and 3 mm for the main girder and lateral bracings, respectively.

***d.* Test results.** Figure 5.9 is an example of load-strain curves for the main girders. The corresponding lateral displacements of main girders can be plotted as shown in Fig. 5.10.

From Fig. 5.9 it is clear that the strain at point 1 in the cross section changes from compression to tension when the applied moment reaches a certain value. This bending moment can be considered as the lateral buckling moment. Then this moment can be determined by obtaining an intersection of tangents to the load-strain curve at two points A and B, as shown in Fig. 5.9. Thus, the lateral buckling

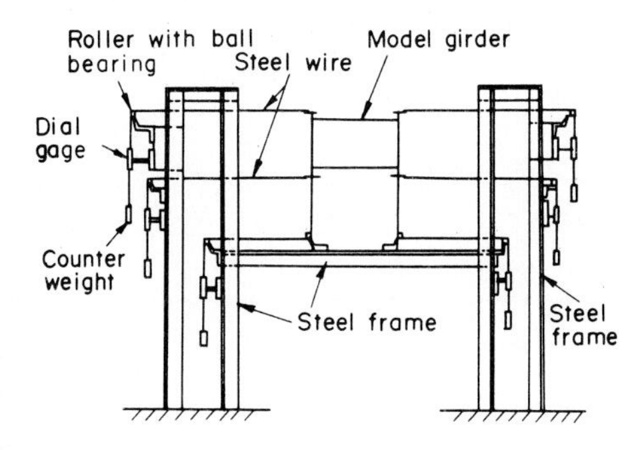

Figure 5.8 Measurement of displacements.

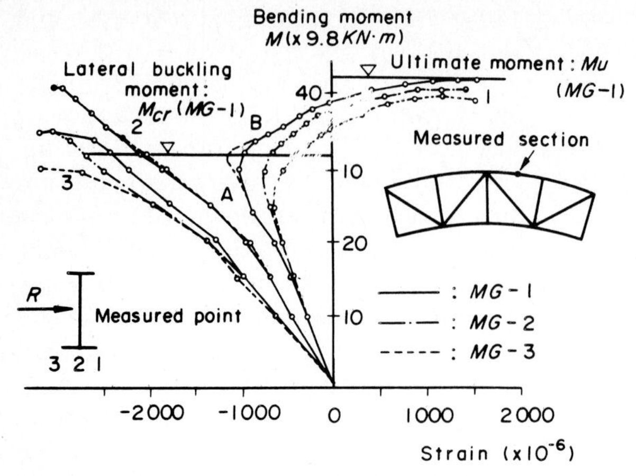

Figure 5.9 Load–strain curves (main girder).

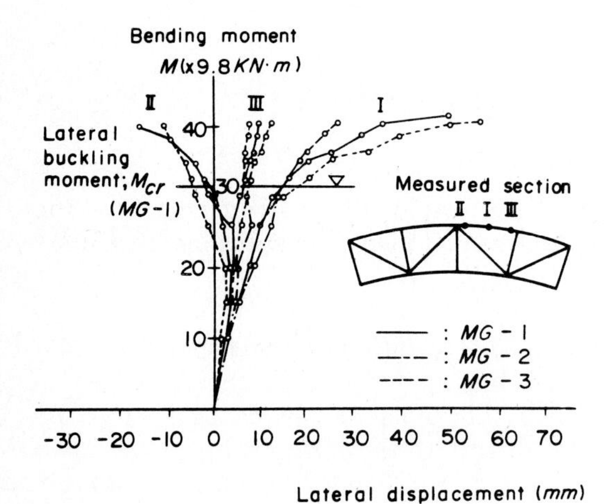

Figure 5.10 Load–lateral-displacement curves (main girder).

TABLE 5.2 M_{cr} **and** M_u**, kN·m**

Girder	M_{cr}	M_u
MG-1	303.8	416.5
MG-2	294.0	401.8
MG-3	284.2	392.0

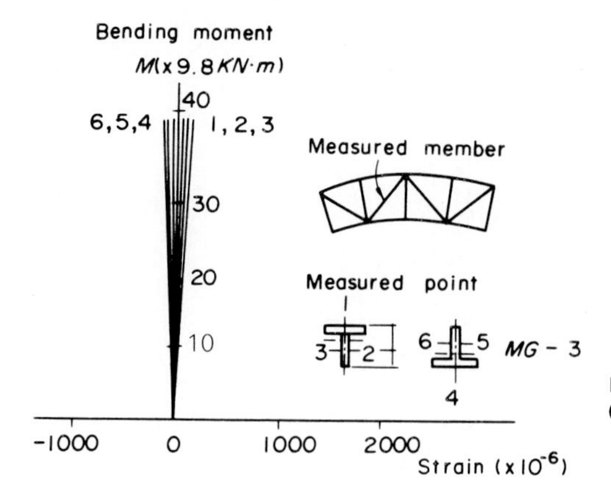

Figure 5.11 Load–strain curves (lateral bracing).

moments M_{cr} and the ultimate moments M_u of girders MG-1 through MG-3 can be clearly obtained, and these values are summarized in Table 5.2.

From this table, it seems that neither M_{cr} nor M_u is significantly affected by the differences in lateral bracings, because the strains of lateral bracing remain within elastic ranges even in the case of the minimum cross-sectional area, shown in Fig. 5.11.

The relative lateral buckling displacements at the compression flanges of the outer girder of MG-3 for each loading step can be plotted as in Fig. 5.12 by subtracting the absolute displacements at the junction points of floor beams and lateral bracings. Observing this figure,

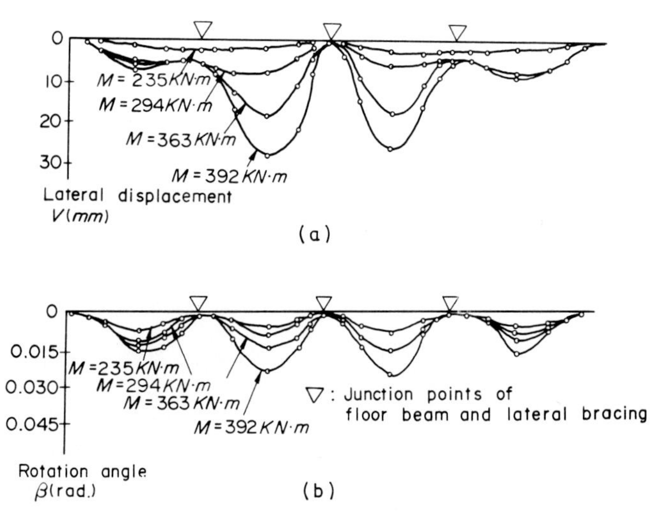

Figure 5.12 Lateral buckling modes of outer girder of MG-3: (*a*) lateral displacement v and (*b*) rotation angle β.

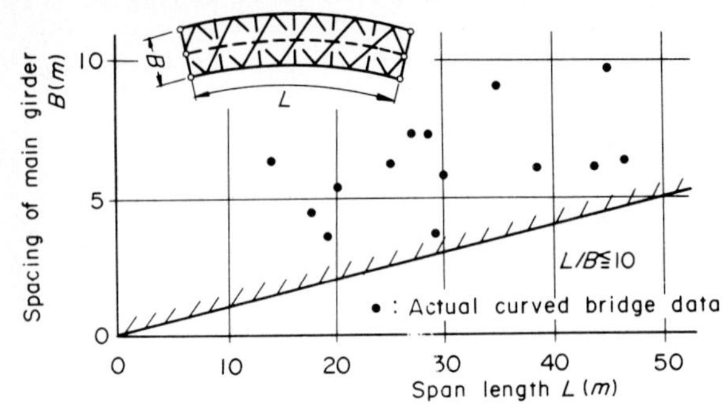

Figure 5.13 Survey of L/B in actual curved bridges.

we see that the local lateral buckling modes of main girders between floor beams and lateral bracings predominate rather than the overall buckling.

So we conclude from these facts that the overall lateral buckling in multiple curved I-girder bridges is not as critical as the local buckling of the main girder between the portions supported by floor beams or sway and lateral bracings if the ratio L/B of span length to girder spacing has a small value, as illustrated by the survey of the multiple curved girder bridge in Fig. 5.13. The floor beams or sway and lateral bracings are designed to have enough strength and rigidity against the lateral loads, as described in Chap. 6.

(2) Local buckling tests of curved I girders. The buckling tests for a supported portion of main girders among the junction points of floor beams and sway or lateral bracings, which is idealized in Fig. 5.14, were undertaken in a procedure similar to that mentioned above, in order to evaluate the local buckling strength of main girders. Twenty-seven model girders, IG-1 through IG-27 (see Table 5.3), were tested under loading conditions illustrated in Fig. 5.15.

Typical load-displacement curves and lateral buckling modes are shown in Figs. 5.16 and 5.17, respectively. However, the material nonlinearity predominates over the geometric nonlinearity. This tendency

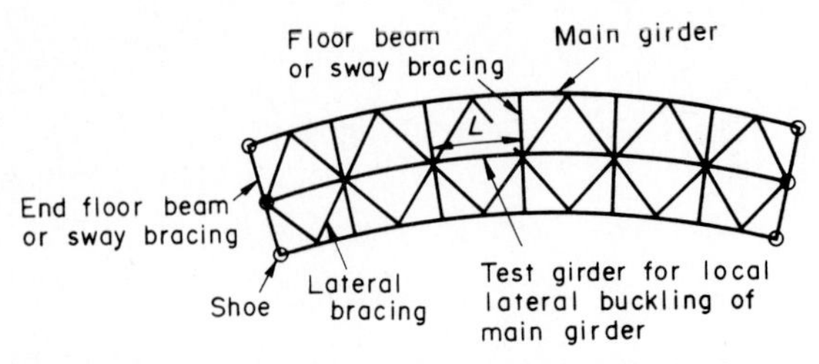

Figure 5.14 Detail of multiple curved I-girder bridges.

TABLE 5.3 Comparison of Ultimate Moments M_u

Girder	Span L, m	Radius R_s, m	Central angle, rad	Flange area A_f, cm^2	Web area A_w, cm^2	Ultimate moment M_u, kN·m		Experiment / Theory
						Experiment	Theory	
IG-1	2.5	5	0.5	8	38	52.9	58.8	0.90
IG-2	2.5	15	0.17	8	38	86.2	91.1	0.95
IG-3	2.5	30	0.08	8	38	109.8	113.7	0.97
IG-4	2.3	4.6	0.5	18	38	135.2	178.4	0.76
IG-5	2.3	13.8	0.17	18	38	202.9	209.8	0.97
IG-6	2.3	27.6	0.08	18	38	202.9	223.4	0.91
IG-7	1.8	3.6	0.5	20	38	161.7	192.1	0.84
IG-8	1.8	10.8	0.17	20	38	261.7	256.8	1.02
IG-9	1.8	21.6	0.08	20	38	274.4	263.6	1.04
IG-10	0.9	1.8	0.5	20	38	201.9	251.9	0.80
IG-11	0.9	5.4	0.17	20	38	271.5	280.3	0.97
IG-12	0.9	10.8	0.08	20	38	301.8	314.6	0.97
IG-13	2.5	5	0.5	6	38	45.1	35.3	1.28
IG-14	2.5	15	0.17	6	38	65.7	66.6	0.99
IG-15	2.5	30	0.08	6	38	80.4	80.4	1.00
IG-16	2.5	5	0.5	10	38	61.7	89.2	0.69
IG-17	2.5	15	0.17	10	38	113.7	109.8	1.04
IG-18	2.5	30	0.08	10	38	136.2	120.5	1.13
IG-19	2.5	5	0.5	12	38	76.4	115.6	0.66
IG-20	2.5	15	0.17	12	38	139.2	135.2	1.03
IG-21	2.5	30	0.08	12	38	176.4	150.9	1.17
IG-22	2-5	15	0.17	14	38	117.4	156.8	1.13
IG-23	2.5	30	0.08	14	38	193.1	162.7	1.19
IG-24	2.3	4.6	0.5	7	38	51.0	56.8	0.90
IG-25	2.3	13.8	0.17	7	38	80.4	79.4	1.01
IG-26	2.3	27.6	0.08	7	38	100.9	91.1	1.11
IG-27	2.3	∞	0	7	38	146.0	113.7	1.28

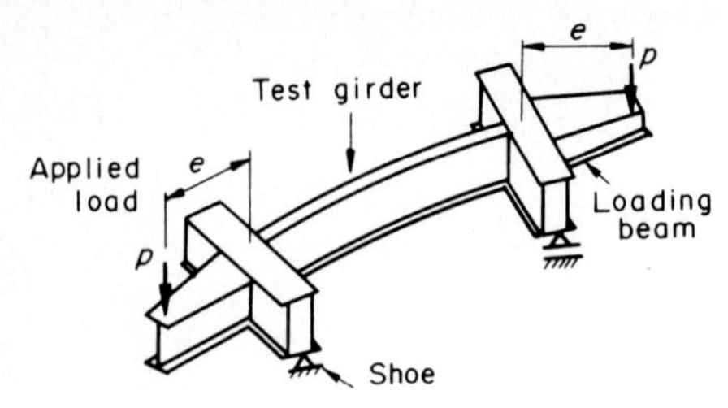

Figure 5.15 Load conditions.

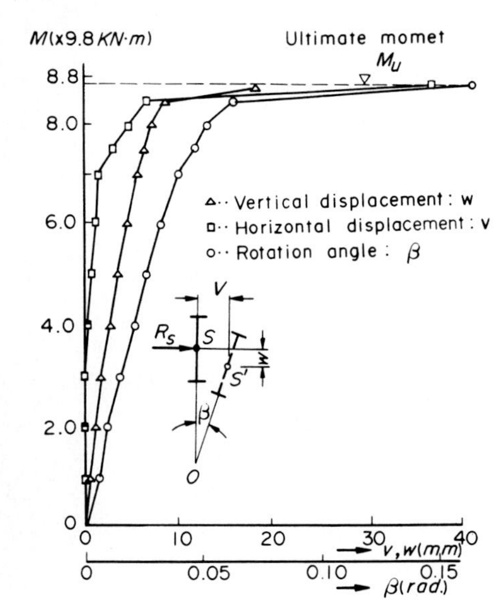

Figure 5.16 Load-displacement curves (IG-2).

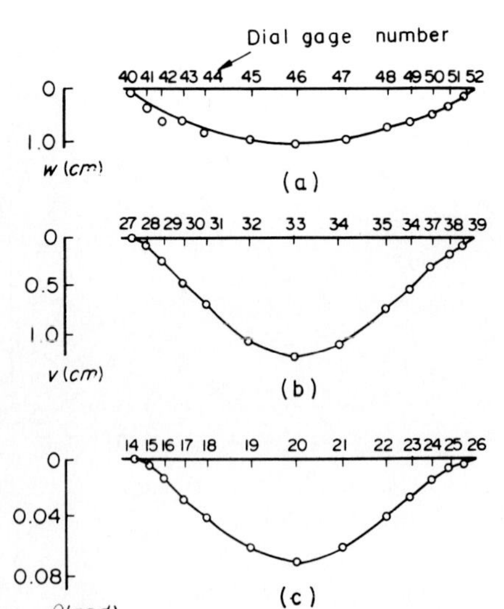

Figure 5.17 Buckling modes (IG-2): (*a*) vertical displacement w, (*b*) horizontal displacement v, and (*c*) rotation angle β.

obviously can be observed in a survey of actual curved bridges, as shown in Fig. 5.18, and almost all the lateral buckling of I girders falls in the elastoplastic regions, since the buckling parameters [cf. Eq. (5.2.14)] have $\alpha = 0.3$ to 1.2.[8]

5.2.3 Analysis of lateral buckling for curved I girders

(1) Analytical study. The horizontal and vertical displacements and rotation angle of a curved I beam will always occur in a beam even though the applied moment is relatively small. However, once a certain critical moment is reached, the curved beam bows out sideways. This bending gives the large displacements which cause a curved I beam to buckle. A few finite displacement theories, which have been developed by many researchers,[9–19] can be used to investigate this complex phenomenon.

One can deal with lateral buckling behavior with a theory based on a second-order analysis.[17–19] For this analysis, if the buckling displacements at state 2 are assumed in reference to the undeformed state 1 in which the displacements in a beam are assumed to be negligibly small until the applied moment reaches a lateral buckling moment, then the following relationships between displacements and stress resultants can be obtained by referring to Fig. 5.19 with an asymmetric curved I girder.[7,20]

$$EI_X\left(\frac{d^2w}{ds^2} - \frac{\beta}{R_s}\right) + EI_{XY}\left(\frac{d^2y}{ds^2} + \frac{1}{R_s}\frac{du}{ds}\right) = -\frac{R_o}{R_s}M_{\bar{\eta}} \tag{5.2.17a}$$

$$EI_Y\left(\frac{d^2v}{ds^2} + \frac{1}{R_s}\frac{du}{ds}\right) + EI_{XY}\left(\frac{d^2w}{ds^2} - \frac{\beta}{R_s}\right) = \frac{R_o}{R_s}M_{\bar{\zeta}} \tag{5.2.17b}$$

$$EA_Z\left[\frac{du}{ds} - \frac{v}{R_s} - \left(\frac{d^2w}{ds^2} - \frac{\beta}{R_s}\right)y_s - \left(\frac{d^2v}{ds^2} + \frac{1}{R_s}\frac{du}{ds}\right)x_s\right] = \frac{R_o}{R_s}N_{\bar{\xi}} \tag{5.2.17c}$$

$$EI_\omega\frac{d^2\theta}{ds^2} - \overline{GK}\frac{d\theta}{ds} = -T_{\bar{\xi}} \tag{5.2.17d}$$

where

s = curvilinear coordinate at shear center

x, y, z = horizontal, vertical, and axial coordinate axes, respectively, at shear center

u, v, w = displacement in the direction of z, x, and y coordinate axes, respectively

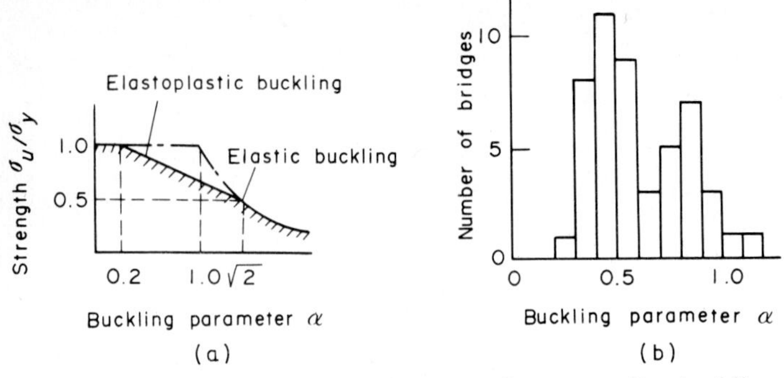

Figure 5.18 Lateral buckling parameter α and corresponding buckling behaviors: (*a*) definition of α and (*b*) histogram of α.

$$\theta = \beta + \frac{w}{R_s} = \text{torsional angle} \tag{5.2.18}$$

θ = torsional angle

β = rotational angle

EI_X, EI_Y, EI_{XY} = flexural rigidity with respect to centroidal X, Y, and XY axes, respectively

EA_Z = elongation rigidity

EI_ω = warping rigidity

$$\overline{GK} = GK + \int_A (x^2 + y^2)\sigma \, dA \tag{5.2.19}$$

in which

GK = torsional rigidity

$\int_A (x^2 + y^2) \, dA$ = additional torsional rigidity due to normal stress[3]

R_o, R_s = radius of curvature at centroid C' and shear center S', respectively

x_s, y_s = eccentricity between C' and S' in the direction of x and y coordinate axes, respectively

Now, the additional stress resultants in the right-hand side of Eq. (5.2.17) at state 2 through state 1 can be written as

$$N_\xi = 0 \tag{5.2.20a}$$

$$M_\eta = -T_z^o \varphi_y \tag{5.2.20b}$$

$$M_\xi = -M_X^o \beta + T_z^o \varphi_x \tag{5.2.20c}$$

$$T_\xi = M_X^o \varphi_y \tag{5.2.20d}$$

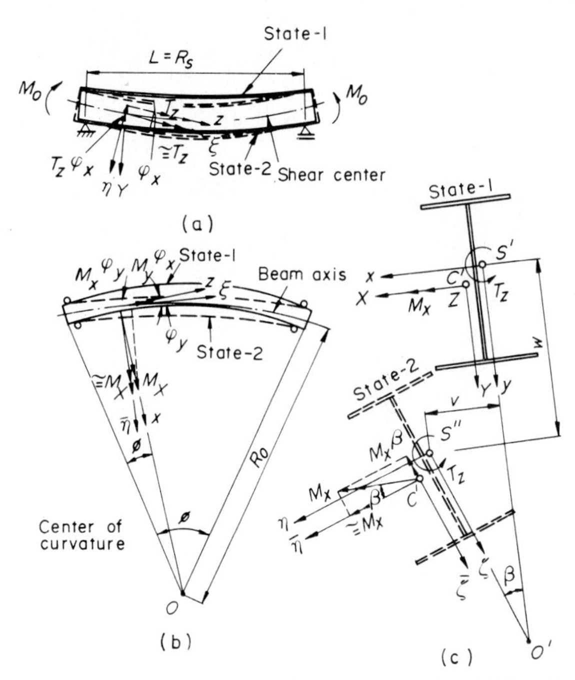

Figure 5.19 Buckling displacements and additional stress resultants: (*a*) side elevation, (*b*) plan, (*c*) cross section.

When a girder is subjected to only a bending moment about the strong axis at the ends of the girder $\phi = 0$ and $\phi = \Phi$, the bending moment M^o_Y and torsional moments T^o_x at an arbitrary section ϕ will be reduced to

$$M^o_X = \frac{M_o \cos(\phi - \Phi/2)}{\cos \Phi/2} \tag{5.2.21a}$$

$$T^o_z = \frac{M_o \sin(\phi - \Phi/2)}{\cos \Phi/2} \tag{5.2.21b}$$

Moreover, the warping moment M^o_ω can be determined by solving the following equation:

$$\frac{d^2 M^o_\omega}{ds^2} - \frac{GK}{EI_\omega} M^o_\omega = -\frac{M_o}{R_o} \tag{5.2.22}$$

The deflection angles φ_x and φ_y due to bending can also be found from

$$\varphi_x = -\frac{dw}{ds} \tag{5.2.23a}$$

$$\varphi_y = \frac{dv}{ds} + \frac{u}{R_s} \tag{5.2.23b}$$

Accordingly, the substitution of Eqs. (5.2.18)–(5.2.23) into Eq. (5.2.17) gives a set of simultaneous differential equations for the lateral buckling displacements u, v, w, β, and θ as follows:

$$EI_x\left(\frac{d^4w}{ds^4}-\frac{1}{R_s}\frac{d^2\beta}{ds^2}\right)-\frac{d^2}{ds^2}\left[T_z^o\left(\frac{dv}{ds}+\frac{u}{R_s}\right)\right]$$
$$-\frac{d^2}{ds^2}\left(M_X^o\beta+T_z^o\frac{dw}{ds}\right)\frac{I_{XY}}{I_Y}=0 \quad (5.2.24a)$$

$$EI_y\left[\frac{R_s}{R_o}\left(\frac{d^4v}{ds^4}+\frac{1}{R_s^2}\frac{d^2v}{ds^2}\right)+\frac{y_s}{R_o}\left(\frac{d^4w}{ds^4}-\frac{1}{R_s}\frac{d^2\beta}{ds^2}\right)\right]$$
$$+\frac{d^2}{ds^2}\left(M_X^o\beta+T_y^o\frac{dw}{ds}\right)+\frac{d^2}{ds^2}(T_y^o\varphi_y)\frac{I_{XY}}{I_X}=0 \quad (5.2.24b)$$

$$EI_\omega\frac{d^4\theta}{ds^4}-\overline{GK}\frac{d^2\theta}{ds^2}+\frac{d}{ds}\left[M_X^o\left(\frac{dv}{ds}+\frac{u}{R_x}\right)\right]=0 \quad (5.2.24c)$$

where
$$u=\int_0^s\frac{v}{R_s}\,ds+u_o \quad (5.2.25)$$

u_o = axial displacement at the origin $s=0$

and I_x and I_y are defined as

$$I_x=\frac{R_s}{R_o}\left(I_X-\frac{I_{XY}^2}{I_Y}\right) \quad (5.2.26a)$$

$$I_y=\frac{R_s}{R_o}\left(I_Y-\frac{I_{XY}^2}{I_X}\right) \quad (5.2.26b)$$

The boundary conditions for displacements and stress resultants for a simply supported curved I girder can be set as

$$v=w=\beta=\varphi_z=T_s=0 \qquad M_X=M_o \quad (5.2.27a,b)$$

at both ends $\phi=0$ and $\phi=\Phi$ of the girder, and

$$u=0 \qquad N_Y=0 \quad (5.2.27c,d)$$

for $\phi=0$ and $\phi=\Phi$ of the girder, respectively. Incidentally, these equations perfectly coincide with Eq. (5.2.1), provided that the girder is straight with $R_s=\infty$ having a symmetric cross section.

For the analyses of the material nonlinearity, a cross section of the girder is divided into small elements, as shown in Fig. 5.20, and the

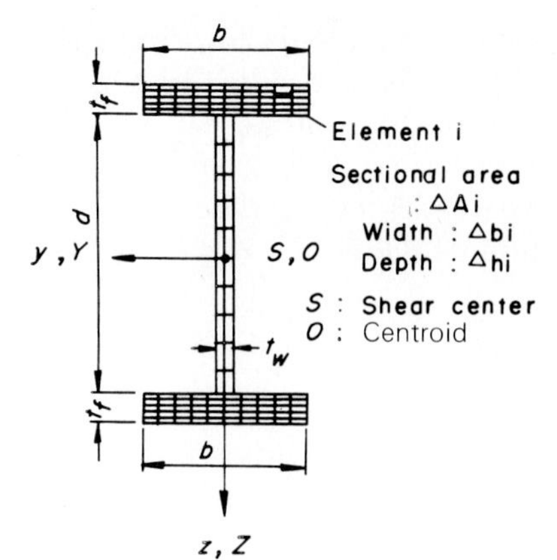

Figure 5.20 Segmental method of cross section.

following assumptions are made to estimate the elastoplastic behavior for each element:

1. The material behaves as an ideal elastoplastic.
2. The Bernoulli-Euler hypothesis can be applied to both the bending strain ε_b $[=M_X^o/(EI_Y Z)]$ and warping strain ε_ω $[=M_\omega^o/(EI_\omega \omega)]$, as illustrated in Fig. 5.21*a* and *b*, respectively.
3. The residual stress distributions are shown in Fig. 5.21*c*, in which the residual strain is defined as $\varepsilon_r = \sigma_r/E$.
4. The stiffness of the element vanishes when the sum of the strain exceeds the yield strain $\varepsilon_y = \sigma_y/E$ of the material.

Thus, Eq. (5.2.24) can be successively solved by determining the rigidities EI_X, EI_Y, EI_{XY}, and EI_ω on the basis of the above assumptions by regarding this problem as an eigenvalue problem.[21]

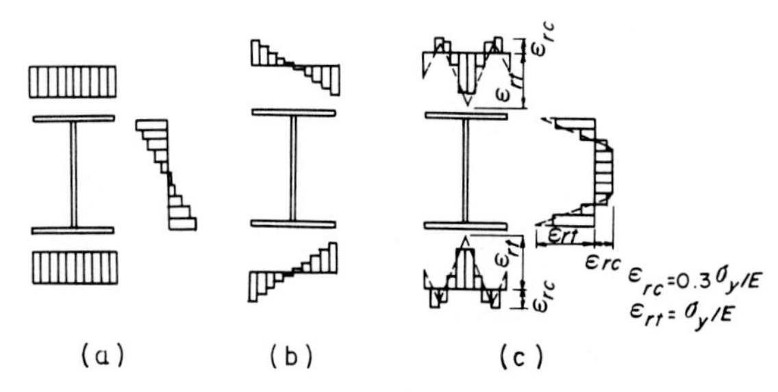

Figure 5.21 Strain distribution in I girder: (*a*) bending strain ε_b, (*b*) warping strain ε_ω, and (*c*) residual strain ε_r.

(2) Comparison with test results. Numerical calculations were performed for test girders IG-1 through IG-27. These results are summarized in Table 5.3 together with the experimental ones.

From this table, it seems that the theoretical values coincide well with the experimental ones (90 to 110 percent) for the case where the central angle Φ of the girder is less than 0.2 rad.

5.2.4 Lateral buckling strength of curved I girders

(1) Lateral buckling strength. The lateral buckling stress σ_u of a curved I girder can be obtained by dividing the ultimate moment M_u by the corresponding section modulus W_x:

$$\sigma_u = \frac{M_u}{W_x} \tag{5.2.28}$$

To simplify the expression for σ_u, let us now arrange σ_u by means of Eq. (5.2.14) and the effective buckling length given by

$$l = \lambda L \tag{5.2.29}$$

In Eq. (5.2.29), L ($= R_s\Phi$) is the unsupported length of the curved I girder, and the coefficient λ can be approximated by

$$\lambda = 1 - 1.97\Phi^{1/3} + 4.25\Phi - 26.3\Phi^3 \tag{5.2.30}$$

This is based on parametric studies of elastic buckling analysis, in which Φ = central angle between the unsupported portion of I girders (rad) and Φ can be adopted in the following ranges:

$$0.02\alpha \le \Phi \le 0.2 \text{ rad} \tag{5.2.31}$$

The lower bound of 0.02α is determined by considering the differences in buckling modes between curved and straight multiple-I-girder bridges, illustrated in Fig. 5.22. Note that Eq. (5.2.15) can be applied to curved I girders with $\Phi < 0.02\alpha$.

Figure 5.23 shows the relationships among σ_u/σ_y, α, and Φ. The buckling stress σ_u is entirely dependent on the buckling parameter α and the central angle Φ of the curved I girder.

(2) Allowable bending compressive stress for curved I girders. The allowable bending compressive stress $(\sigma_{ba})_c$ can be obtained by dividing the lateral buckling stress by the safety factor ν:

$$(\sigma_{ba})_c = \frac{\sigma_u}{\nu} \tag{5.2.32}$$

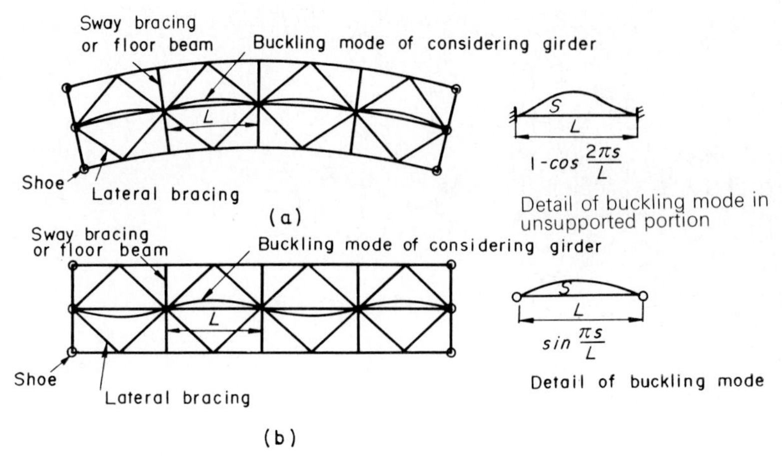

Figure 5.22 Difference of buckling modes between (*a*) curved and (*b*) straight I-girder bridges.

To express this stress in terms of that of a straight I girder $(\sigma_{ba})_s$, the reduction factor $\psi_1(\alpha, \Phi)$, which is a function of the buckling parameter α and the central angle Φ of the curved I girder, must be multiplied by $(\sigma_{ba})_s$. Thus,

$$(\sigma_{ba})_c = \psi_1(\sigma_{ba})_s \tag{5.2.33}$$

This reduction factor, $\psi_1(\alpha, \Phi)$, is plotted in Fig. 5.24. Moreover, ψ_1 can be rearranged by the following formula through a least-squares analysis by using the lower bound of the test results within the ranges

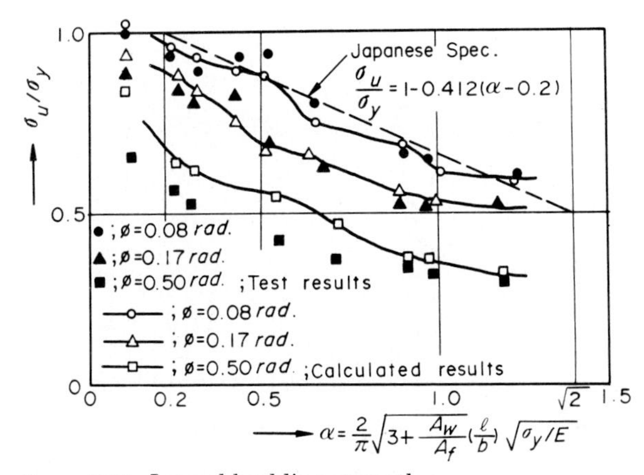

Figure 5.23 Lateral buckling strength σ_u.

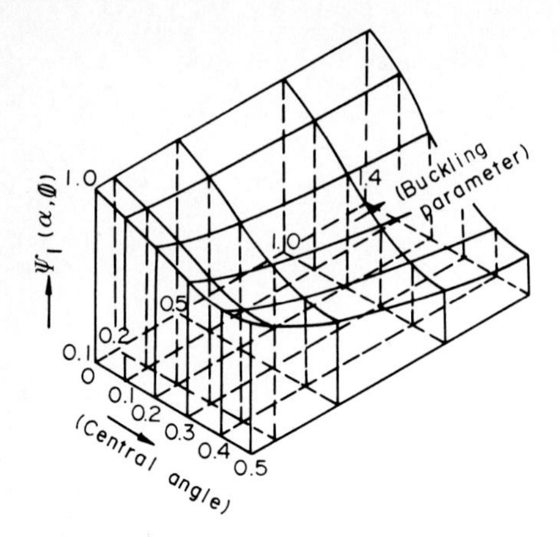

Figure 5.24 Variations of ψ_1 (α, Φ).

of $0.1 \le \alpha \le \sqrt{2}$ and $\Phi < 0.2$, as shown in Fig. 5.25[22]

$$\psi_1 = 1 - 1.05\sqrt{\alpha}(\Phi + 4.52\Phi^2) \tag{5.2.34}$$

5.2.5 Application to design of multiple curved I-girder bridges

(1) Stresses in main girders.[23] The bending stress in each main girder of a multiple curved I-girder bridge is

$$\sigma_b = \frac{M_x}{I_x} y \tag{5.2.35}$$

The additional warping stress σ_ω due to the out-of-plane bending of flange plates can be estimated by referring to Fig. 5.26*a* to *c* as follows:

$$\sigma_\omega = \pm \frac{\chi \sigma_b}{R W_f} \left(A_f + \frac{A_s}{3} \right) L^2 \tag{5.2.36}$$

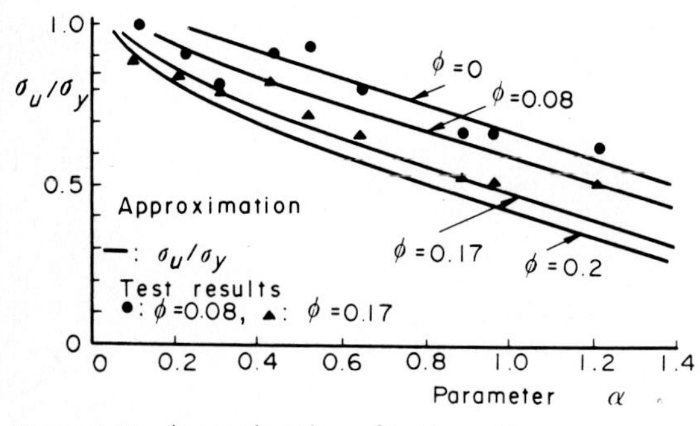

Figure 5.25 Approximation of test results.

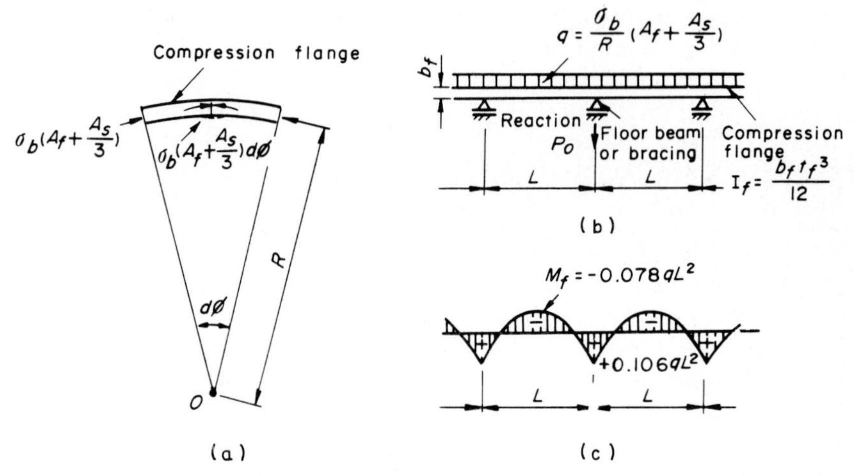

Figure 5.26 Additional warping stress σ_ω: (*a*) out-of-plane bending of compression flange, (*b*) equivalent model, and (*c*) additional bending moment.

provided that rigid supports are assumed at the junction points of floor beams or sway and lateral bracings, and where

A_f = cross-sectional area of flange plate

A_s = cross-sectional area of web plate from flange to neutral axis

L = longitudinal spacing of floor beams or sway and lateral bracings

R = radius of curvature of main girder

W_f = section modulus of flange plate ($=t_f b_f^2/6$, where t_f is flange thickness and b_f is flange width)

χ = bending moment coefficient as a continuous beam with $\chi = 0.106$ as maximum value

(2) Check for stress and lateral buckling strength. The total compressive stresses at the compression flange of main girders should satisfy the following criteria:

$$\sigma_{bc} + \sigma_{\omega c} \leqq b_{bao} \qquad \text{for stress check} \tag{5.2.37a}$$

$$\sigma_{bc} \leqq (\sigma_{ba})_c \qquad \text{for buckling check} \tag{5.2.37b}$$

where σ_{bc} = bending compressive stress
$\sigma_{\omega c}$ = additional warping compressive stress
σ_{bao} = upper limit of allowable compressive stress
$(\sigma_{ba})_c$ = allowable compressive stress for lateral buckling

It is, however, inconvenient to solve simultaneously Eqs. (5.2.37*a*) and (5.2.37*b*). Dividing both sides of Eq. (5.2.37*a*) by σ_{bao} and approximating the denominator of the first term in the left-hand side of this equation with $(\sigma_{ba})_c$, we find this simplified interaction formula:

$$\frac{\sigma_{bc}}{(\sigma_{ba})_c} + \frac{\sigma_{\omega c}}{\sigma_{bao}} \leqq 1.0 \tag{5.2.38}$$

Note that Eq. (5.2.37*b*) holds when $\sigma_{\omega c} = 0$ and gives conservative answers when $\sigma_{\omega c} = 0$.

Figure 5.27 shows the interaction curves for σ_{bc}/σ_{bao} versus $\sigma_{\omega c}/\sigma_{bao}$. Note that the buckling modes of curved I girders between the unsupported portions between the floor beams or sway and lateral bracings are the same as for the straight I girder when the central angle Φ is sufficiently small (see Fig. 5.22). This condition has been derived from various parametric analyses, which give $\Phi \leqq 0.02\alpha$ [see Eq. (5.2.31)]. For these curved girders, the criterion relative to the straight girder can be applied by setting $\lambda = 1$.

For the composite girder, the stress check can be determined from Eq. (5.2.37*a*) alone. Moreover, the stress check for the tension flange can be found from

$$\sigma_{bt} + \sigma_{\omega t} \leqq \sigma_{ta} \tag{5.2.39}$$

Now, Eq. (5.2.38) can be rewritten as

$$\sigma_{bc} \leqq (\sigma_{ba})_c \left(1 - \frac{\sigma_{\omega c}}{\sigma_{bao}}\right) \tag{5.2.40}$$

The substitution of Eq. (5.2.33) into the above equation gives

$$\sigma_{bc} \leqq (\sigma_{ba})_s \psi_1 \psi_2 \tag{5.2.41}$$

where

$$\psi_2 = 1 - \frac{\sigma_{\omega c}}{\sigma_{bao}} = 1 - \left(\frac{\sigma_{bc}}{\sigma_{bao}}\right)\left(\frac{\sigma_{\omega c}}{\sigma_{bc}}\right) \tag{5.2.42}$$

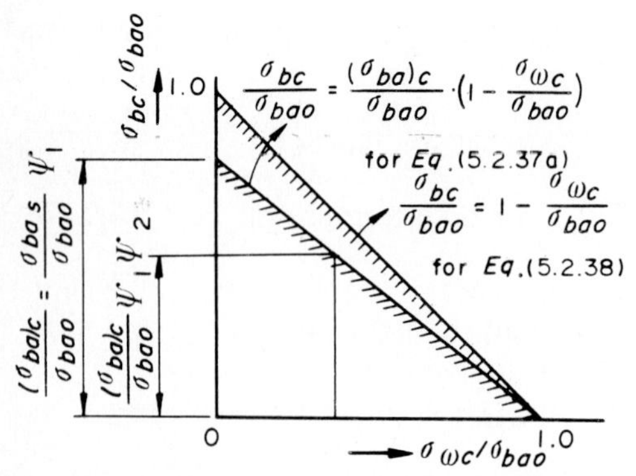

Figure 5.27 Interaction curves for compression flange due to bending and additional warping stresses.

This equation is similar in form to the recommendation of the AASHTO specification (AASHTO = American Association of State Highway and Transportation Officials).[24]

5.2.6 Design of floor beams and sway and lateral bracings for multiple curved I-girder bridges

The multiple curved I girder bridges, which are stiffened by the floor beams or sway and lateral bracings, always undergo additional stresses due to the curvature of the bridge axis. These stresses are thought to be primary ones rather than the secondary stresses in the straight girder bridges. Therefore, the floor beams or sway and lateral bracings should be designed to have enough strength and rigidity as prescribed in Chap. 6.[25–28]

5.3 Buckling Stability and Strength of Web Plates in Curved Girders

5.3.1 Ultimate and buckling strengths of plate girders

To build a plate girder having enough strength against bending and shear, one must know the true ultimate and buckling strengths of the girder, since the elements of the plate girder such as the web and flange plates should be designed to have a strength corresponding to that of the girder. This section discusses the ultimate and buckling strengths of plate girders.

(1) Ultimate strength of a plate girder

a. **Ultimate strength for bending.** Let us now consider an I girder with a bisymmetric cross section, as shown in Fig. 5.28*a*. We do, of course, assume that the flange and web plates do not buckle before the ultimate strength of the girder is reached.

The yielding moment M_y in which the stress in the flange plate reduces to the yield stress σ_y, shown in Fig. 5.28*b*, can be estimated by

$$M_y = A_f \sigma_y h\left(1 + \frac{1}{6}\frac{A_w}{A_f}\right) \tag{5.3.1}$$

where h = girder depth

$A_f = b_{tf}$, $A_w = h_{tw}$ = cross-sectional areas of flange and web plate, respectively

The fully plastic moment M_p where the entire cross section reaches the

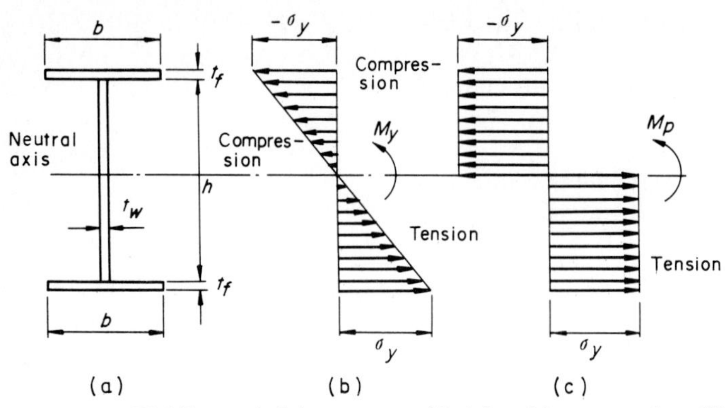

Figure 5.28 Yielding and ultimate state of I girder: (*a*) cross section, (*b*) yielding state, and (*c*) ultimate state.

yield stress σ_y and reaches the ultimate state, shown in Fig. 5.28*c*, can also be evaluated as

$$M_p = A_f \sigma_y h \left(1 + \frac{1}{4} \frac{A_w}{A_f} \right) \tag{5.3.2a}$$

The fully plastic moment M_p of an I girder without the buckling can therefore be written as

$$M_p = f M_y = \frac{1 + \frac{1}{4} A_w / A_f}{1 + \frac{1}{6} A_w / A_f} M_y \tag{5.3.2b}$$

In Eq. (5.3.2*b*), f is the shape factor and falls between 1.10 and 1.23 for ordinary I girders.

The ultimate strength, however, decreases, because the flange and web plates undergo considerable compressive forces at the upper side of the neutral axis, as seen in Fig. 5.28*b*, and they might buckle in this region. Basler and Thürliman[29] studied the influence of buckling of the web plate on the basis of experimental and theoretical research. They suggest that the ultimate strength M_u can be determined according to Fig. 5.29, when the lateral buckling (see Galambos[1]) and torsional buckling (see Galambos[3]) of the flange plates are prevented. In this figure, point A with $\beta = 53$ corresponds to the fully plastic moment M_p. When β is less than 53, stress hardening occurs and M_u is greater than M_p. When the thickness of the web decreases and β becomes large, one part of the girder reaches the plastic state and the other part remains in the elastic state. For this elastoplastic state, M_u is always greater than M_y. In this case, elastic buckling occurs at point B, shown in Fig. 5.29; that is, $\beta_o = 170$, and M_u is equal to the yield moment M_y for this situation. Furthermore, when β is larger than 170, a part of the web

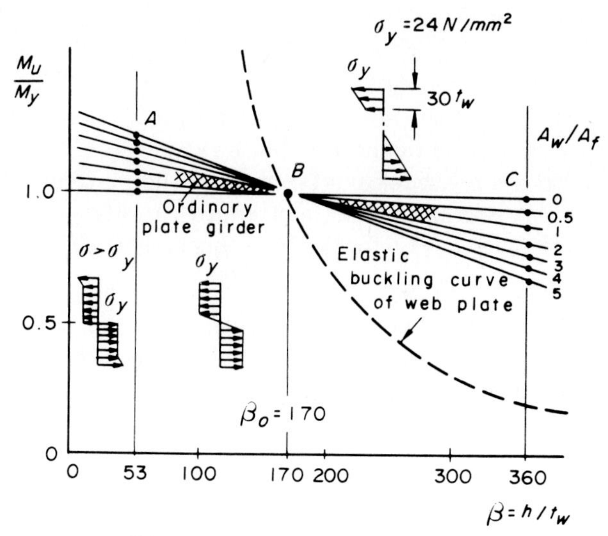

Figure 5.29 Ultimate strength M_u/M_y of plate girder due to web slenderness β.

plate has already buckled, but the reduction of M_u is not extreme, because a T section is composed of the flange plate and conjuncting web plate with width $30t_w$, which are still effective as compression members, and considerable post buckling strength can be expected. This behavior continues up to point C in Fig. 5.29 with $\beta = 360$. For $\beta > 360$, however, the ultimate strength suddenly drops due to the vertical downward buckling of the web plate.

The approximate formula to evaluate the ultimate strength M_u was also proposed by Basler and Thürliman[29] for $\beta = 53$ to 360:

$$M_u = M_y\left[1 - 0.0005\,\frac{A_w}{A_f}\left(\frac{h}{t_w} - \beta_o\right)\right] \tag{5.3.3}$$

where $\beta_o = 5.7\sqrt{\dfrac{E}{\sigma_y}}$ (5.3.4)

σ_y = yield stress
E = Young's modulus

Equation (5.3.4) derived from Eq. (5.3.27) by setting $b = h$, $t = t_w$, $\beta_o = h/t_w$, $k_\sigma = 36.0$ [cf. Eqs. (5.3.37) and (5.3.38)], and $\sigma_{cr} = \sigma_y$. For $E = 2.1 \times 10^6$ kgf·cm² and $\sigma_y = 2400$ kgf/cm², β_o is about 170.

It is obvious from Fig. 5.29 and Eq. (5.3.3) that the ultimate strength is not greatly affected by the buckling strength of the web plate, but is entirely governed by the postbuckling strength of the girders. The design of plate girders can, therefore, be based on this ultimate strength. However, the actual design method of web plates is fundamentally based on the buckling strength of the web plate from the view of the serviceability limit state of the girder. In this design method, the factor of safety ν against buckling of the web plate is not $\nu = 1.7$, but $\nu = 1.4$ for bending due to the postbuckling strength of the plate girders; i.e., the factor of safety against ultimate strength can always be ensured to be $\nu \cong 1.7$ if one uses $\nu = 1.4$ for the buckling strength of the web plate.

***b.* Ultimate strength for shear.** Let us investigate the ultimate strength of plate girders for shear such as in the end panels of simply supported girders or intermediate supports of continuous girders. In the I girder, shown in Fig. 5.30*a*, having a thick web plate without shear buckling, the ultimate strength can be evaluated by

$$Q_p = \tau_y A_w \tag{5.3.5}$$

where

$$\tau_y = \frac{\sigma_y}{\sqrt{3}} \tag{5.3.6}$$

But if buckling is impending, the above equation must be modified as follows:

$$Q_{\text{cr}} = Q_p \frac{\tau_{\text{cr}}}{\tau_y} \tag{5.3.7}$$

where τ_{cr} is the buckling stress for shear, as detailed in the next section.

Figure 5.30*b* shows this buckling state, where the portions of the web plate subjected to compression are buckled and they do not cooperate with the cross section of the I girder. Tensile forces are induced in the diagonal direction of the web plate, as shown in Fig. 5.30*b*, and the girder is in the so-called diagonal tension field. In this situation,

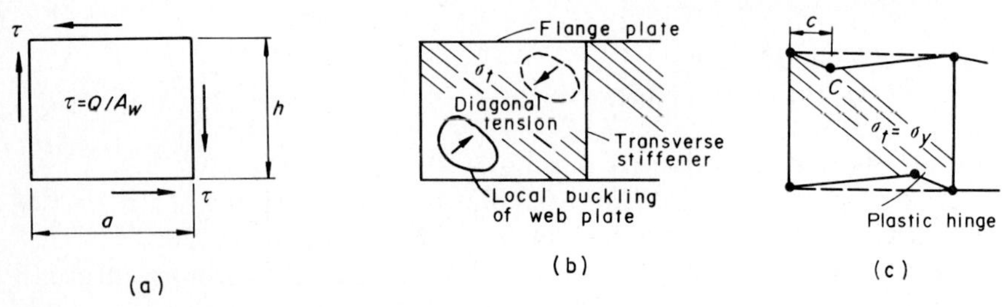

Figure 5.30 Ultimate state of web plate for shear: (*a*) elastic beam behavior, (*b*) buckling state, and (*c*) ultimate state.

the girder can be treated as a sort of truss structure composed of flange plates, transverse stiffeners, and diagonal elements of the web plate. The girder will then be able to support a greater shear force than Q_{cr}.

Finally, the diagonal tensile stress σ_t reaches the yield stress σ_y, and the plastic hinges are propagated in the flange plates at points C apart from the transverse stiffeners c, as shown in Fig. 5.30c. Thus, the mechanism is formed, and the girder reduces to the ultimate limit state.

The ultimate strength Q_u for shear is also derived by Basler[30] as

$$Q_u = Q_p\left[\frac{\tau_{cr}}{\tau_y} + \frac{1 - \tau_{cr}/\tau_y}{1.15\sqrt{1 + (a/h)^2}}\right] \tag{5.3.8}$$

where a is the spacing of the transverse stiffeners.

From these facts, the plate girder has enough strength against the shearing forces, and the factor of safety ν against the buckling of the web plate is $\nu \cong 1.25$, to ensure the ultimate strength for shear by a safety factor of $\nu = 1.7$ as determined by much experimental and theoretical research.[4]

(2) Buckling strength of a plate girder[31–33]

a. Fundamental equation for buckling of a plate. The fundamental equation of buckling for a plate is derived here. Figure 5.31 shows a plate element $t\,dx\,dy$, where n_x and n_y are the in-plane resultant forces of normal stresses and t_x and t_y are those of shearing stresses. The induced stress resultants are

$$m_x, m_y = \text{bending moments}$$

$$m_{xy}, m_{yx} = \text{torsional moments}$$

$$q_x, q_y = \text{shearing forces}$$

for the axes x and y, respectively.

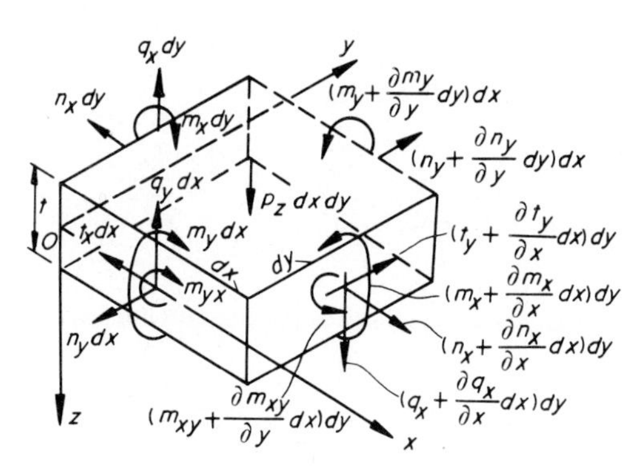

Figure 5.31 Applied load and induced stress resultants in plate.

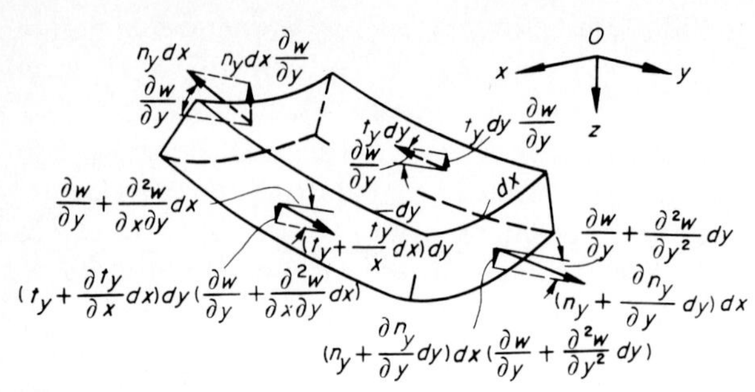

Figure 5.32 Components of applied forces after deformation.

From the equilibrium conditions of the torsional moment with respect to the axis parallel to the z axis, the following equations can be obtained:

$$t_x = t_y = t_{x,y} \tag{5.3.9}$$

The plate element after deformation with the deflection surface $w(x, y)$ can be sketched as in Fig. 5.32. Then the components of applied in-plane forces in the direction of the z axis parallel to the xz plane are

$$-n_y\, dx \frac{\partial w}{\partial y} + \left(n_y + \frac{\partial n_y}{\partial y} dy\right) dx \left(\frac{\partial w}{\partial y} + \frac{\partial^2 w}{\partial y^2} dy\right) \tag{5.3.10a}$$

$$-t_y\, dy \frac{\partial w}{\partial y} + \left(t_y + \frac{\partial t_y}{\partial y} dy\right) dy \left(\frac{\partial w}{\partial y} + \frac{\partial^2 w}{\partial x\, \partial y} dx\right) \tag{5.3.10b}$$

Similar equations can be derived for the components of the applied in-plane forces in the direction of the z axis parallel to the yz plane by changing the index y to x. The summation of these components of applied in-plane forces in the direction of the z axis can be written by omitting the smaller terms such as $\partial n_y / \partial y$ and $\partial t_y / \partial y$ as follows:

$$n_x \frac{\partial^2 w}{\partial x^2} dx\, dy + n_y \frac{\partial^2 w}{\partial y^2} dx\, dy + 2t_{x,y} \frac{\partial^2 w}{\partial x\, \partial y} dx\, dy \tag{5.3.11}$$

In addition, the equilibrium conditions for shearing forces q_x and q_y and applied vertical force p_z can be represented as

$$\frac{\partial q_x}{\partial x} dx\, dy + \frac{\partial q_y}{\partial y} dx\, dy + p_z\, dx\, dy = 0 \tag{5.3.12}$$

Adding the additional terms of Eq. (5.3.11) to the above equation, we get

$$n_x \frac{\partial^2 w}{\partial x^2} + n_y \frac{\partial^2 w}{\partial y^2} + 2t_{x,y} \frac{\partial^2 w}{\partial x\, \partial y} + \frac{\partial q_x}{\partial y} + \frac{\partial q_y}{\partial y} + p_z = 0 \tag{5.3.13}$$

The equilibrium condition for bending and torsional moment m_x, m_y, and m_{xy} is

$$\frac{\partial^2 m_x}{\partial x^2} + 2\frac{\partial^2 m_{xy}}{\partial x\,\partial y} + \frac{\partial^2 m_y}{\partial y^2} = -p_z \tag{5.3.14}$$

where the relationships among the stress resultants q_x, q_y, m_x, m_y, and m_{xy} reduce to

$$q_x = -D\left(\frac{\partial^3 w}{\partial x^3} + \frac{\partial^3 w}{\partial x\,\partial y^2}\right) \tag{5.3.15a}$$

$$q_y = -D\left(\frac{\partial^3 w}{\partial y^3} + \frac{\partial^3 w}{\partial x^2\,\partial y}\right) \tag{5.3.15b}$$

$$m_x = -D\left(\frac{\partial^2 w}{\partial x^2} + \mu\frac{\partial^2 w}{\partial y^2}\right) \tag{5.3.16a}$$

$$m_y = -D\left(\frac{\partial^2 w}{\partial y^2} + \mu\frac{\partial^2 w}{\partial x^2}\right) \tag{5.3.16b}$$

$$m_{xy} = -2C\frac{\partial^2 w}{\partial x\,\partial y} \tag{5.3.16c}$$

in which D is the flexural rigidity of the plate, i.e.,

$$D = \frac{Et^3}{12(1-\mu^2)} \tag{5.3.17}$$

C is the torsional rigidity of the plate given by

$$C = \frac{Gt^3}{12} = \frac{1-\mu}{2}D \tag{5.3.18}$$

and μ is Poisson's ratio.

Therefore, by substituting p_z of Eq. (5.3.13), q_x and q_y of Eq. (5.3.15), and m_x, m_y and m_{xy} of Eq. (5.3.16), the following fundamental equation describing a plate subject to in-plane stresses can be obtained:

$$D\left(\frac{\partial^4 w}{\partial x^4} + 2\frac{\partial^4 w}{\partial x^2\,\partial y^2} + \frac{\partial^4 w}{\partial y^4}\right) - n_x\frac{\partial^2 w}{\partial x^2} - 2t_{x,y}\frac{\partial^2 w}{\partial x\,\partial y} - n_y\frac{\partial^2 w}{\partial y^2} = p_z \tag{5.3.19}$$

***b.* Buckling of a plate subjected to compression.** As the most simple example, let us analyze the buckling strength of a simply supported rectangular plate subjected to uniaxial compression p_x along the x axis, as shown in Fig. 5.33. The fundamental equation for this plate can be

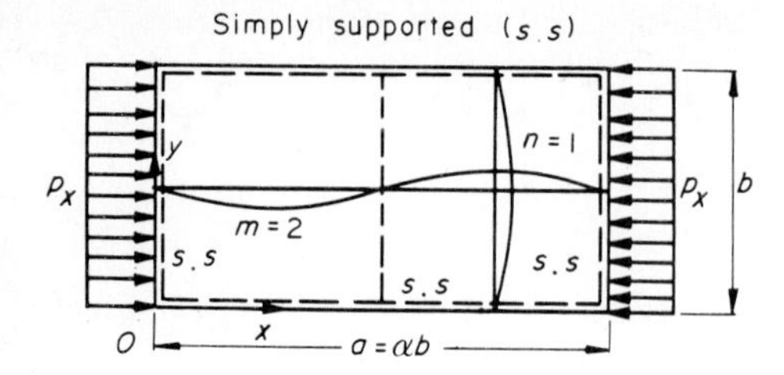

Figure 5.33 Buckling of a plate under uniaxial compression.

written by setting $n_x = -p_x$, $t_{x,y} = 0$, $n_y = 0$, and $p_z = 0$ in Eq. (5.3.19):

$$\frac{\partial^4 w}{\partial x^4} + 2\frac{\partial^4 w}{\partial x^2\,\partial y^2} + \frac{\partial^4 w}{\partial y^4} + \frac{p_x}{D}\frac{\partial^2 w}{\partial x^2} = 0 \tag{5.3.20}$$

In this case, the deflection surface $w(x, y)$ can be exactly expressed by

$$w = A_{mn} \sin\frac{m\pi x}{a} \sin\frac{n\pi y}{b} \qquad m = 1, 2, \ldots;\ n = 1, 2, 3, \ldots \tag{5.3.21}$$

This equation fulfills all the boundary conditions concerning w, m_x, and m_y of this plate.

The substitution of Eq. (5.3.21) into Eq. (5.3.20) leads to using the elastic buckling load $p_{cr} = p_x$ as an eigenvalue problem, which gives

$$p_{cr} = \frac{D\pi^2[(m/a)^2 + (n/b)^2]^2}{(m/a)^2} \tag{5.3.22}$$

In this equation, m and n designate the number of half wavelengths of sinusoidal curves, i.e., the buckling modes. The elastic buckling stress σ_{cr} can then be obtained by taking $n = 1$ in the direction of the y axis and dividing by the cross-sectional area $1 \cdot t$. Thus,

$$\sigma_{cr} = \frac{D\pi^2}{b^2 t}\left(\frac{m}{\alpha} + \frac{\alpha}{m}\right)^2 = k_\sigma \sigma_e \tag{5.3.23}$$

where $$\alpha = \frac{a}{b} = \text{aspect ratio of plate} \tag{5.3.24}$$

$$k_\sigma = \text{buckling coefficient}$$

and $$\sigma_e = \frac{D\pi^2}{b^2 t} = \frac{E\pi^2}{12(1-\mu^2)}\left(\frac{t}{b}\right)^2 \tag{5.3.25}$$

with σ_e being Euler's buckling stress.

The variations of the buckling coefficient k_σ due to the aspect ratio are plotted in Fig. 5.34. The corresponding buckling modes for $m = 1$,

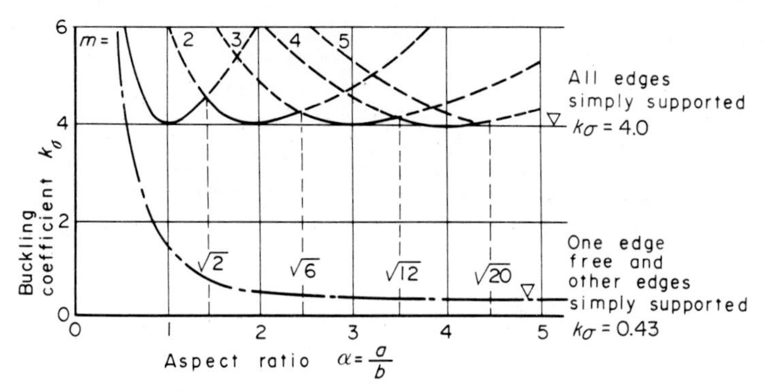

Figure 5.34 Variation of k_σ due to α.

2, and 3 are illustrated in Fig. 5.35. From this figure, the minimum value of k_σ results in

$$k_\sigma = 4.0 \tag{5.3.26}$$

and the buckling stress σ_{cr} can be written by setting Poisson's ratio as $\mu = 0.3$ as follows:

$$\sigma_{cr} \cong 0.904 k_\sigma E \left(\frac{t}{b}\right)^2 \tag{5.3.27}$$

c. **Buckling of a plate subjected to bending.** Let us consider the buckling of a rectangular plate subjected to an arbitrary normal stress σ_x, which varies linearly along the y axis as shown in Fig. 5.36, i.e.,

$$\sigma_x = \sigma_o \left(1 - \varphi \frac{b-y}{b}\right) \tag{5.3.28}$$

In Eq. (5.3.28), φ is a coefficient that describes the distributions of the normal stresses σ_o and $\sigma_o(1-\varphi)$ at the upper and lower sides of the plate, respectively. For example, $\varphi = 2$ corresponds to pure bending. Therefore, the fundamental equation of buckling can be written as

$$D\left(\frac{\partial^4 w}{\partial x^4} + 2\frac{\partial^2 w}{\partial x^2\,\partial y^2} + \frac{\partial^4 w}{\partial y^4}\right) + \sigma_o t\left(1 - \varphi\frac{b-y}{b}\right) = 0 \tag{5.3.29}$$

The exact solution of the deflection surface $w(x, y)$ is so complicated

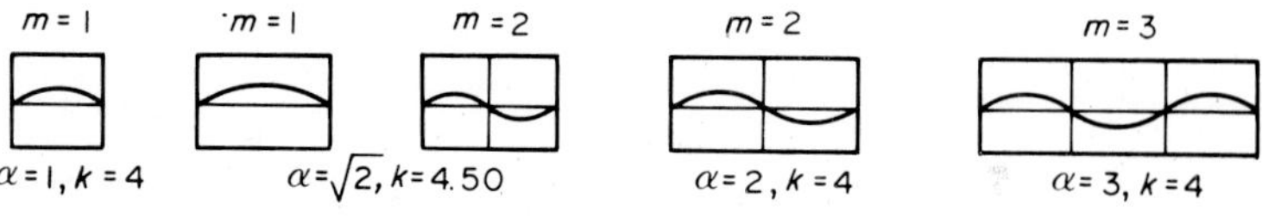

Figure 5.35 Buckling modes.

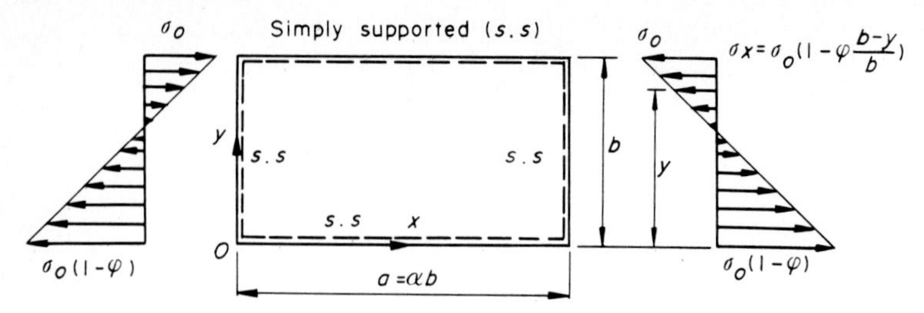

Figure 5.36 Buckling of plate under bending.

that the following approximate solution with unknown buckling amplitude A_n is used:

$$w = \sin \frac{m\pi x}{a} \sum_{n=1}^{\infty} A_n \sin \frac{n\pi y}{b} \tag{5.3.30}$$

However, it is impossible to determine a particular eigenvalue as above, even when Eq. (5.3.30) is substituted into Eq. (5.3.29). In general, the energy method is used in this case. The strain energy U according to Timoshenko and Gere[34] can be estimated as

$$U = \frac{D}{2} \int_0^a \int_0^b \left\{ \left(\frac{\partial^2 w}{\partial x^2} + \frac{\partial^2 w}{\partial y^2} \right)^2 - 2(1-\mu) \left[\frac{\partial^2 w}{\partial x^2} \frac{\partial^2 w}{\partial y^2} - \left(\frac{\partial^2 w}{\partial x \, \partial y} \right)^2 \right] \right\} dx\, dy \tag{5.3.31}$$

The energy expended by the external force is

$$V = -\frac{t}{2} \int_0^a \int_0^b \sigma_x \left(\frac{\partial w}{\partial x} \right)^2 dx\, dy \tag{5.3.32}$$

Accordingly, a condition to determine the most adequate unknown buckling amplitudes A_n ($n = 1, 2, 3, \ldots$) can be found from the principle of least work:

$$\frac{\partial}{\partial A_n} (U + V) = 0 \tag{5.3.33}$$

This condition gives a set of homogenous equations for A_n:

$$A_n \left[\left(1 + \frac{n^2 \lambda^2}{b^2} \right) - \frac{\sigma_o t \lambda^2}{D\pi^2} \left(1 - \frac{\varphi}{2} \right) \right] - \frac{8\varphi \sigma_o t \lambda^2}{D\pi^2} \sum_{i=1}^{\infty} \frac{ni}{(n^2 - i^2)^2} A_i = 0 \qquad n = 1, 2, \ldots \tag{5.3.34}$$

where

$$\lambda = \frac{a}{m} \tag{5.3.35}$$

and Σ_i^{∞} is expanded for i of the odd numbers of $n + i$.

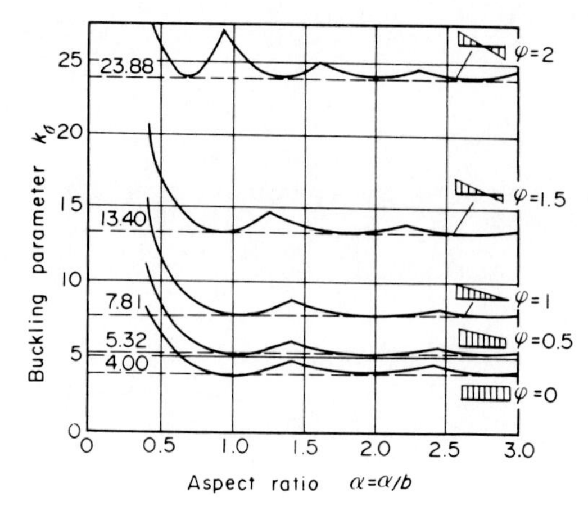

Figure 5.37 Variations of k_σ due to φ and α.

The determinant built by the coefficient matrix of Eq. (5.3.34) must equal zero to have a nontrivial solution. Thus, the buckling stress $\sigma_{cr} = \sigma_o$ can be obtained and expressed as

$$\sigma_{cr} = k_\sigma \sigma_e \tag{5.3.36}$$

which is similar to Eq. (5.3.23). Figure 5.37 shows the variations of k_σ due to the stress gradient factor φ and aspect ratio α. Specifically, the buckling coefficient k_σ for pure bending, that is, $\varphi = 2$, can be written

$$k_\sigma = \begin{cases} 23.9 & \alpha \geqq 2/3 \quad (5.3.37a) \\ 15.87 + \dfrac{1.87}{\alpha^2} + 8.6\alpha^2 & \alpha < 2/3 \quad (5.3.37b) \end{cases}$$

When the web plate is clamped at the upper and lower flange plates and is simply supported at the transverse stiffeners,

$$k_\sigma = 39.7 \tag{5.3.38}$$

***d.* Buckling of a plate subjected to shear.** Let us analyze the shear buckling strength of a simply supported rectangular plate, shown in Fig. 5.38. The deflection surface $w(x, y)$ can be approximated by

$$w = \sum_{m=1}^{\infty} \sum_{n=1}^{\infty} A_{mn} \sin \frac{m\pi x}{a} \sin \frac{n\pi y}{b} \tag{5.3.39}$$

For this case, the work done by the external force should be calculated by[34]

$$V = -\tau_{xy} t \int_0^a \int_0^b \frac{\partial w}{\partial x} \frac{\partial w}{\partial y} \, dx \, dy \tag{5.3.40}$$

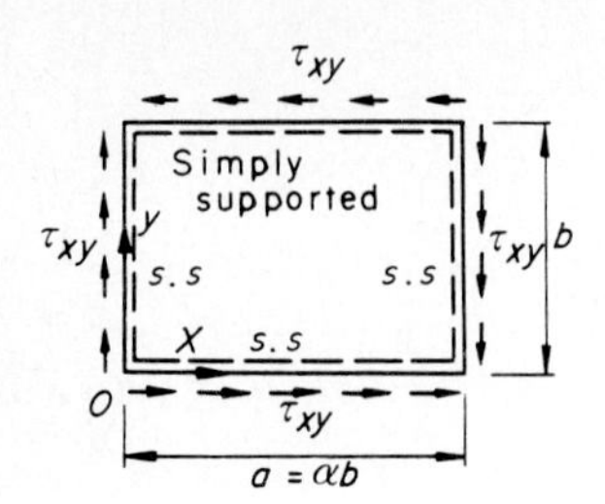

Figure 5.38 Buckling of a plate under shear.

and the simultaneous equations for unknown amplitude A_{mn} give

$$\lambda(m^2 + n^2\alpha^2)A_{mn} - mn \sum_i^\infty \sum_j^\infty \frac{ijA_{ij}}{(i^2 - m^2)(j^2 - n^2)} = 0 \tag{5.3.41}$$

where

$$\lambda = \frac{\pi^4 D}{32\alpha^3 b^2 t\tau_{xy}} \tag{5.3.42}$$

and Σ_i^∞ is expanded for i of the odd numbers of $m + j$ and Σ_i^∞ is expanded for j of the odd numbers of $n + j$.

The buckling strength is

$$\tau_{cr} = k\sigma_e \tag{5.3.43}$$

Figure 5.39 shows the variation of k_τ due to $1/\alpha$. The coefficient k_τ can be approximated as

$$k_\tau = \begin{cases} 5.34 + \dfrac{4.00}{\alpha^2} & \alpha \geqq 1 \quad (5.3.44a) \\[2ex] 4.00 + \dfrac{5.34}{\alpha^2} & \alpha < 1 \quad (5.3.44b) \end{cases}$$

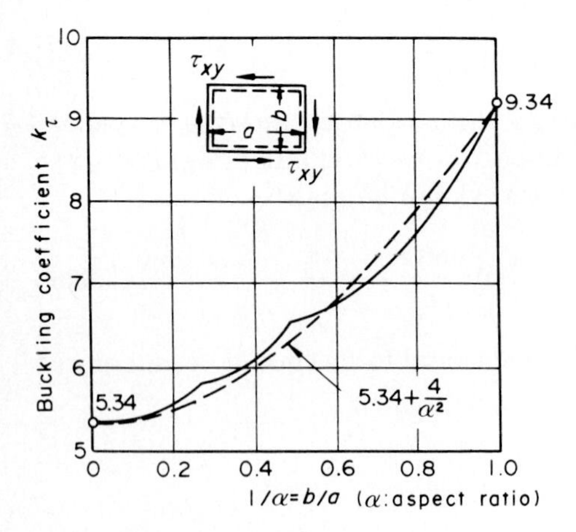

Figure 5.39 Variations of k_τ due to $1/\alpha$.

5.3.2 Buckling and ultimate strengths of a curved girder

To find the buckling and ultimate strengths of a curved girder, the following series of experimental studies were carried out.[35] Through these tests, the properties of buckling strength and ultimate strength of curved girders under bending, shear, and their combinations are analyzed and discussed in detail.

(1) Buckling and ultimate strengths for bending[36]

a. Test method for bending. Nine test specimens made of the curved I girders listed in Table 5.4 were tested for bending. In Table 5.4,

R = radius of curvature of web plate

h_w = depth of web plate (= 80 cm)

t_w = thickness of web plate (= 4.5 or 3.2 mm)

$\alpha = \dfrac{a}{h_w}$ = aspect ratio of test web plate panel

a = spacing of transverse stiffener

b_f = width of flange plate

t_f = thickness of flange plate (= 12 mm)

γ_l = relative stiffness of longitudinal stiffener

$\gamma_{l,\mathrm{req}}$ = required relative stiffness of longitudinal stiffener according to Japanese Specification for Highway Bridges [see Eq. (5.5.13)]

TABLE 5.4 Test Girders under Bending

Test panel	Specimen number	R, m	$\dfrac{h_w}{t_w}$	α	b_f, mm	$\dfrac{\gamma_l}{\gamma_l}$
	2	10, 30	178	1.0	180	—
	6	10, 30 ∞	178	0.5	90, 180	—
	1	10	250	0.5	180	1.0

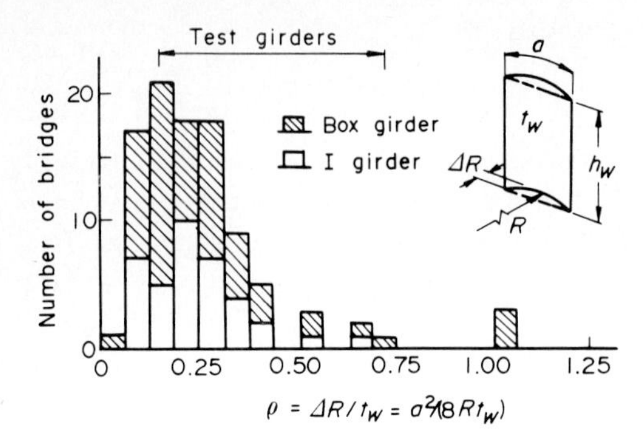

Figure 5.40 Distribution of curvature parameter ρ in test girders and actual bridges.

The dimensions of these test I girders were determined on the basis of actual data from curved girder bridges by referring to the results of a survey[8,37] of 125 bridges in Japan and the United States. A curvature parameter ρ

$$\rho = \frac{\Delta R}{t_w} = \frac{a^2}{8Rt_w} \tag{5.3.45}$$

is used as a representative index of curvature for a web panel between transverse stiffeners, as shown in Fig. 5.40, where ΔR is the eccentricity of the web plate due to the radius of curvature R. We can see from this figure that ρ is less than 1.0 in the case of actual bridges. Thus, the values of ρ of the test I girders obviously cover almost all the realistic ranges.

The aspect ratio α of the test panel was chosen as 0.5 or 1.0. For instance, the detail of a test girder is illustrated in Fig. 5.41.

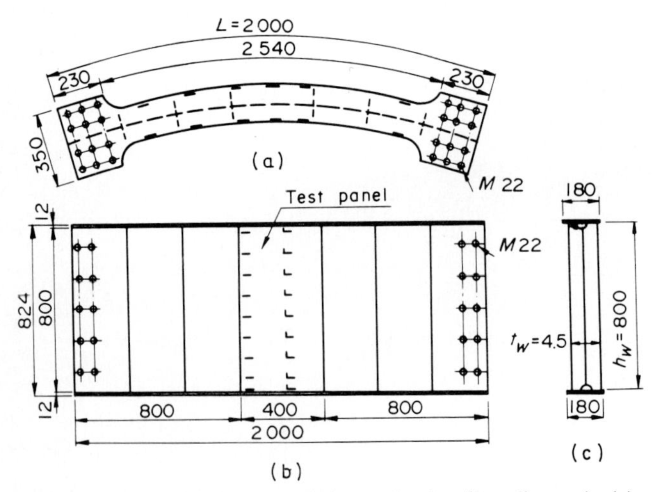

Figure 5.41 Detail of a test girder under bending (in mm): (*a*) plan, (*b*) side elevation, and (*c*) cross section.

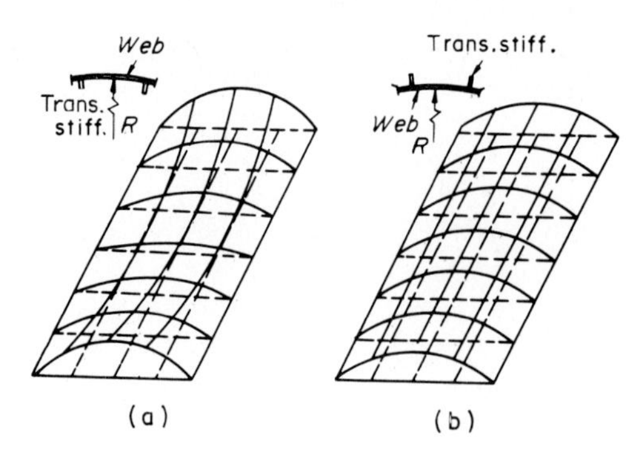

Figure 5.42 Typical initial deformations observed in curved web panels: (*a*) saddle type and (*b*) cylindrical type.

The yield points of steel grade SS-41 of the test girders are as follows: For flange plates,

$$\sigma_{yf} = 339\text{–}387 \text{ MPa} \tag{5.3.46a}$$

For web plates,

$$\sigma_{yw} = 254\text{–}319 \text{ MPa} \tag{5.3.46b}$$

To investigate the effects of initial deformations on the strength of the test web plate panels, their values were measured with a transit and categorized into two types due to the locations of transverse stiffeners, as shown in Fig. 5.42.

The loading and supporting devices for bending tests are sketched in Fig. 5.43, where the loads P are applied to end cantilevers with span e and then the end moments $M_o = Pe$ are introduced to the test girders. The lateral buckling of the test girder is prevented at the ends by the floor beams, which is typical of the loading conditions of actual bridges (see Fig. 5.14).

The induced stress resultants at the test panel can be analyzed on a

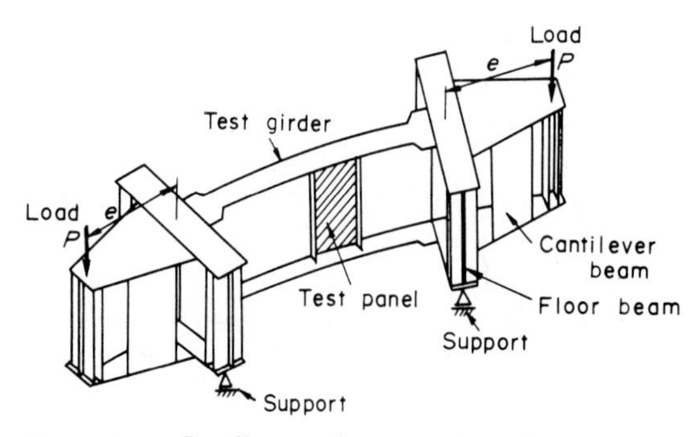

Figure 5.43 Loading and supporting device for bending test.

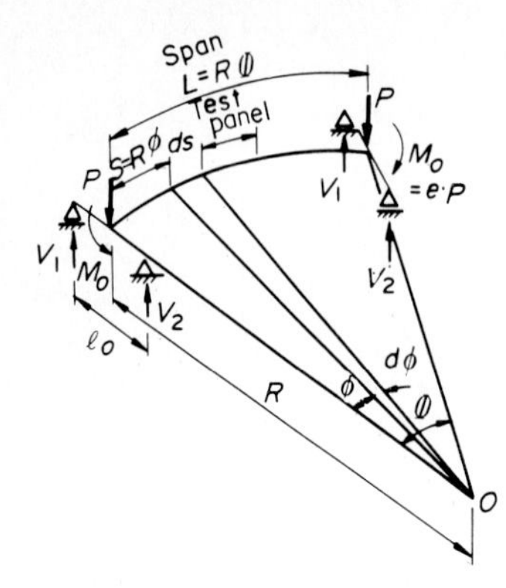

Figure 5.44 Analytical model of stress resultants for bending test.

model, shown in Fig. 5.44, as follows:

$$\text{Shearing force: } Q = 0 \tag{5.3.47a}$$

$$\text{Bending moment: } M = -\frac{M_o \cos(\phi - \Phi/2)}{\cos \Phi/2} \tag{5.3.47b}$$

$$\text{Torsional moment: } T = -\frac{M_o \sin(\phi - \Phi/2)}{\cos \Phi/2} \tag{5.3.47c}$$

$$\text{Reactions: } V_{1,2} = \frac{P}{2[1 \pm (e/l_o) \tan(\Phi/2)]} \tag{5.3.48}$$

For the central angle $\phi = \Phi/2$, then, $M \cong -M_o = -Pe$, $Q = 0$, and $T \cong 0$. Thus, the web plate panels can be tested very nearly under pure bending.

***b.* Buckling strength for bending.** For a web plate panel with curvature, the out-of-plane displacement of the web plate δ_w gradually increases in accordance with the increment of the applied bending moment. Note that the ideal buckling phenomenon does not occur in the curved web panels. But a critical bending moment, which makes the deflection δ_w of the web panel largely bow out, can clearly be observed, as shown in Fig. 5.45. From these treatments, the experimental buckling moments $M^*_{cr} = P^*_{cr}e$ are determined.

The theoretical buckling moments are determined on the assumption that a curved web plate panel can be approximately treated as a flat plate with two edges simply supported at the transverse stiffeners and the other edges fixed at the upper and lower flange plates (with the exception of the web panel having a longitudinal stiffener). Then, by

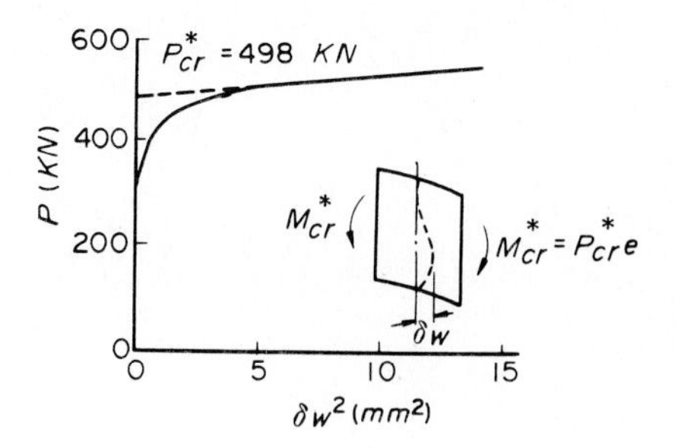

Figure 5.45 Load and square deflection curve of a web panel.

using $k_\sigma = 39.7$ in Eq. (5.3.38) and Eq. (5.3.27), the theoretical buckling moment M_{cr} can be estimated by

$$M_{cr} = \sigma_{cr} W_w \tag{5.3.49}$$

where W_w is the section modulus of the test girders at the top of the web plate.

The ratio between the test and calculated buckling moments M^*_{cr}/M_{cr}, is plotted in Fig. 5.46. This figure shows that M^*_{cr}/M_{cr} decreases in accordance with the increase in the curvature parameter ρ. A solid line in this figure is an empirical one found by means of the least-squares method.

To derive an alternative empirical curve of M^e_{cr}/M_{cr}, a new parameter $(a/R)(a/b_f)(a/t_w)$ is introduced by combining the lateral buckling

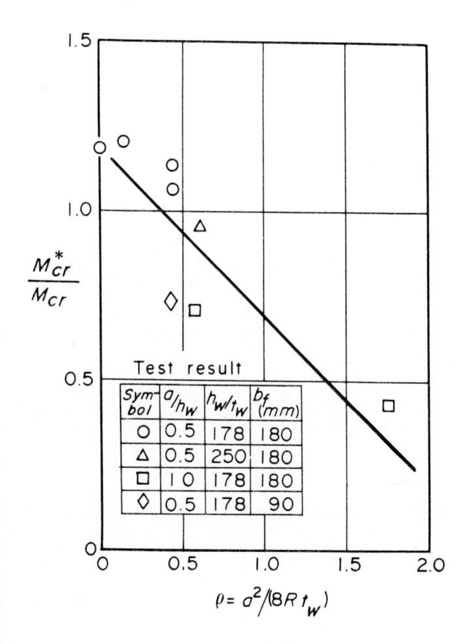

Figure 5.46 Variation of M^*_{cr}/M_{cr} with curvature parameter ρ (with longitudinal stiffener).

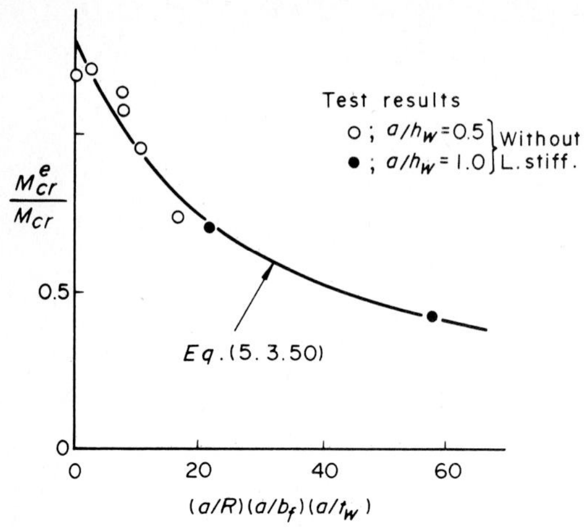

Figure 5.47 Relationships between M^e_{cr}/M_{cr} and $(a/R)(a/b_f)(a/t_w)$.

parameter a/b_f [cf. Eq. (5.2.14)] and a/t_w from Eq. (5.3.45). The relationship between M^*_{cr}/M_{cr} and $(a/R)(a/b_f)(a/t_w)$ is plotted in Fig. 5.47. Then a formula for predicting M^e_{cr}/M_{cr} can be approximated by the least-square method as

$$\frac{M^e_{cr}}{M_{cro}} = \frac{35.32}{(a/R)(a/b_f)(a/t_w) + 27.52} \tag{5.3.50}$$

c. Ultimate strength for bending. The fully plastic state of a curved I girder is generally considered to be as shown in Fig. 5.48 by taking into account the out-of-plane bending of flange plates due to curvature. From this figure, the fully plastic moment M^c_p is

$$M^c_p = \sigma_{yf} A_f (1 - 2\xi) h_w + \frac{\sigma_{yw} A_w h_w}{4} \tag{5.3.51}$$

where σ_{yf}, σ_{yw} = yield points of flange and web plate, respectively

$A_f = b_f t_f$, $A_w = t_w h_w$ = cross-sectional areas of flange and web plate, respectively

ξ = nondimensional parameter for location of neutral axis for in-plane bending of flange plate

The above equation can be simplified for a straight girder by setting $\xi = 0$ to get M^s_p:

$$M^s_p = \sigma_{yf} A_f h_w + \frac{\sigma_{yw} A_w h_w}{4} \tag{5.3.52}$$

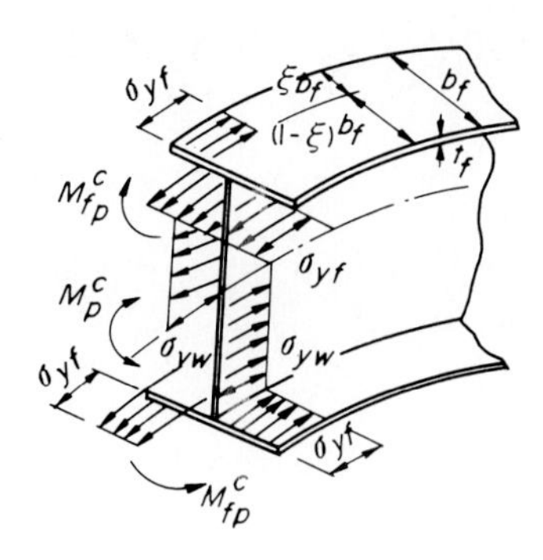

Figure 5.48 Fully plastic moment of a curved I girder under bending.

In addition, the out-of-plane fully plastic moment M_{fp}^c of the flange plate is given by

$$M_{fp}^c = \sigma_{yf} A_f (\xi - \xi^2) b_f \tag{5.3.53}$$

When there is only an in-plane bending moment in the flange plates, this equation gives the fully plastic moment M_{fp}, by setting $\xi = 1/2$, as

$$M_{fp} = \frac{\sigma_{yf} A_f b_f}{4} \tag{5.3.54}$$

The elimination of ξ from Eqs. (5.3.51) and (5.3.53) leads to the interaction curve for M_p^c/M_p^s and M_{fp}^c/M_{fp}, which gives

$$\frac{M_p^c}{M_p^s} = 1 - \frac{1}{\rho} \pm \frac{1}{\rho\sqrt{1 - M_{fp}^c/M_{fp}}} \tag{5.3.55}$$

where ρ is a constant given by

$$\rho = 1 + \frac{1}{4(A_w/A_f)(\sigma_{yw}/\sigma_{yf})} \tag{5.3.56}$$

The interaction curve of M_p^c/M_p^s versus M_{fp}^c/M_{fp} in Eq. (5.3.55) is shown in Fig. 5.49, where the gradient m of a line with an arbitrary combination of bending moment M_p^c of the curved I girder and in-plane bending moment M_{fp}^c is defined as

$$m = \frac{M_p^c}{M_{fp}^c} \frac{M_{fp}}{M_p^s} \tag{5.3.57}$$

The ordinate of this line, M_p^c/M_{fp}^c, is the value corresponding to the fully plastic state. Suppose that the ratio M_p^c/M_{fp}^c does not change in both the elastic and elastoplastic ranges and can be approximated as the ratio between the applied bending moment M^c of curved I girders

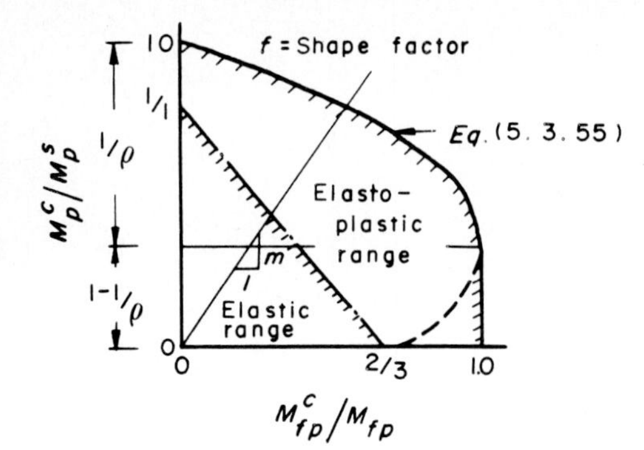

Figure 5.49 Interaction curve of M_p^c/M_p^s–M_{fp}^c/M_{fp}.

and the induced in-plane bending moment M_f^c of the flange plates in elastic ranges, so that

$$\frac{M_p^c}{M_{fp}^c} = \frac{M^c}{M_f^c} \tag{5.3.58}$$

Then the value of M_f^c/M_c in the above equation can also be related to the following formula, as explained in Eqs. (5.2.35) and (5.2.36):

$$\frac{M_f^c/W_f}{M_c/W} = \frac{\sigma_\omega}{\sigma_b} = \pm\chi\left(\frac{L}{\sqrt{Rb_f}}\right)^2\left(1+\frac{A_w}{6A_f}\right) \tag{5.3.59}$$

where W, W_f = section modulus of I girder with respect to strong axis and of flange plate with respect to weak axis, respectively

L = span length of test girder

Thus, the fully plastic moments M_p^c and M_{fp}^c can be obtained from Eq. (5.3.58) by substitutions of Eqs. (5.3.51) and (5.3.53).

Now, the relationships between test results of the ultimate moment $M_u^* = P_u^* e$, and calculated moment M_p^c are shown in Fig. 5.50. From this figure, M_u^* falls between $0.95M_p^c$ and $1.2M_p^c$. Moreover, these results are plotted in Fig. 5.51 as an interaction curve

$$\rho\left[\frac{M_p^c}{M_p^s} - \left(1-\frac{1}{\rho}\right)\right] = \sqrt{1-\frac{M_{fp}^c}{M_{fp}}} \tag{5.3.60}$$

instead of as Eq. (5.3.55), to avoid the errors due to ρ, which contains the variations due to yield points and cross-sectional areas of the flange and web plates. It is obvious from this figure that the test results coincide well with the calculated ones.

The fully plastic moment can, furthermore, be determined by the following alternative method. Let the allowable bending moment M_a of

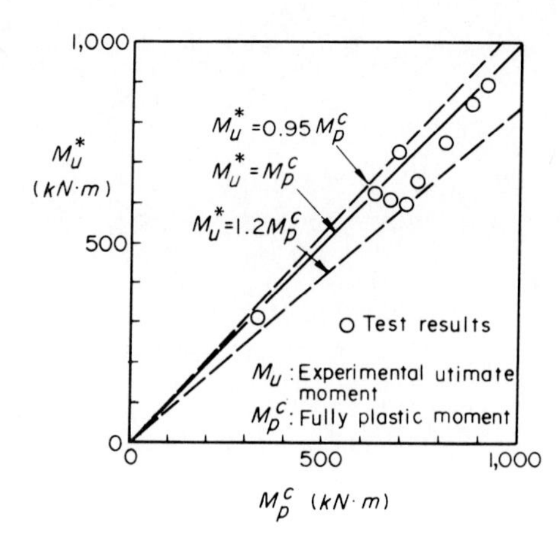

Figure 5.50 Relationships between M_u^* and M_p^c.

a curved I girder be calculated from the minimum values of the following equations, as shown in Fig. 5.52:

$$M_a = \min\{M_{wa}, M_{fa}\} \tag{5.3.61}$$

Here M_{wa} corresponds to the allowable bending moment $M_{wa} = \sigma_{yw} W_w$ in which $\nu = 1.7$ is the safety factor as provided by the Japanese Specifications for Highway Bridges, and M_{fa} is the allowable bending moment $M_{fa} = \sigma_{ba} W_f$ at which the compressive stress at an edge of the compression flange plate reaches the allowable bending stress σ_{ba}, as is detailed in Sec. 5.2.4(2).

By using this allowable bending moment M_a, the ratio of test results of ultimate moment M_u^* to M_a, that is, M_u^*/M_a, can be plotted versus a parameter $L^2/(Rb_f)$, as shown in Fig. 5.53. The figure includes 19 test results of the lateral buckling for curved I girders with compact cross section, from Nakai and Kotoguchi.[7] From this figure, a linear

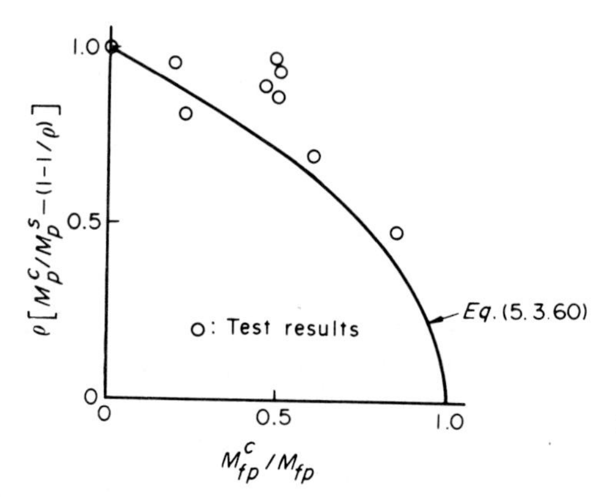

Figure 5.51 Plot of test results in interaction curve.

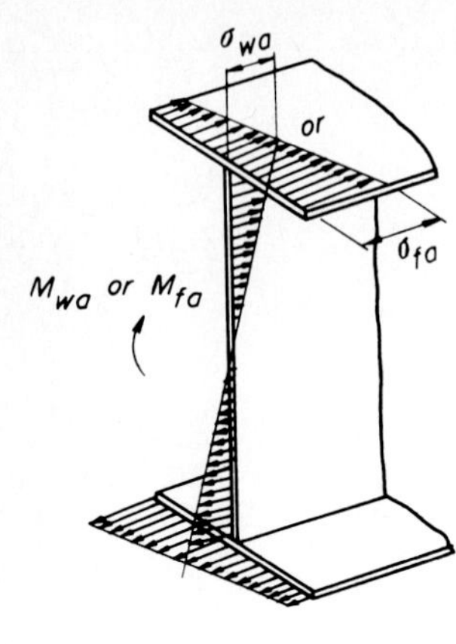

Figure 5.52 Normal stress distribution of curved I girder. $M_{wa} = \sigma_{wa} W_w$ or $M_{ta} = \sigma_{ba} W_f$.

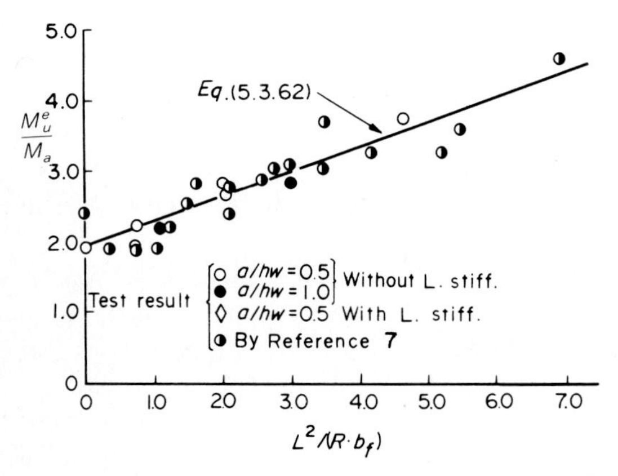

Figure 5.53 Relationships between M_u^e/M_a and $L^2/(Rb_f)$.

TABLE 5.5 Test Girder under Shear

Test panel	Specimen number	R, m	$\frac{h_w}{t_w}$	Aspect ratio α	$\frac{\gamma_l}{\gamma_{l,\text{req}}}$	$\frac{\gamma_t}{\gamma_{t,\text{req}}}$
	2	10, ∞	178	1.0	—	—
	4	10, ∞	178	0.5	—	1, 5
	4	10	250	0.5	1, 5	1, 5

relationship is recognized, so that the following empirical formula based on the least-squares method holds

$$\frac{M_u^e}{M_a} = 1.92 + \frac{0.357L^2}{Rb_f} \tag{5.3.62}$$

(2) Buckling and ultimate strengths for shear[38]

a. Test method for shear. Similar to the approach used in bending, 10 test specimens were built, as shown in Table 5.5, on the basis of the dimensional analysis of actual curved girder bridges. Hence the symbols are the same as for bending (see Table 5.1), and these notations are introduced:

γ_t = relative stiffness of transverse stiffener

$\gamma_{t,\text{req}}$ = required relative stiffness of transverse stiffener according to Japanese Specification for Highway Bridges [see Eq. (5.6.16)].

The details of the test girder are illustrated in Fig. 5.54. These model girders were tested with the loading and supporting devices shown in Fig. 5.55.

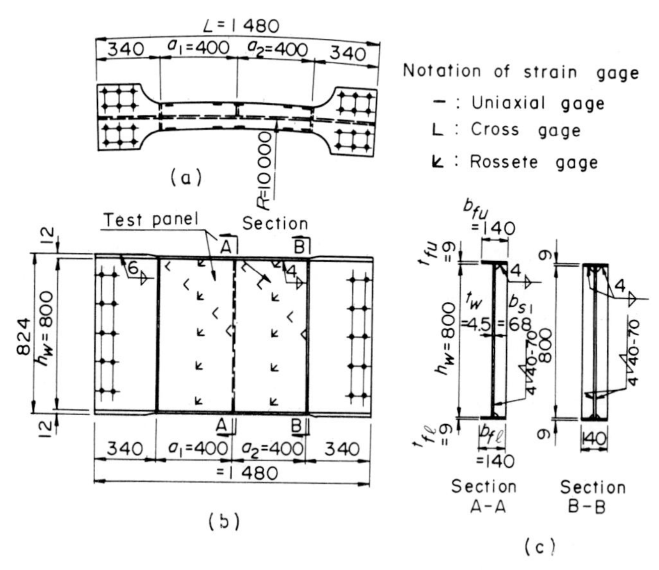

Figure 5.54 Detail of a test girder under shear (in mm): (*a*) plan, (*b*) side elevation, and (*c*) cross section.

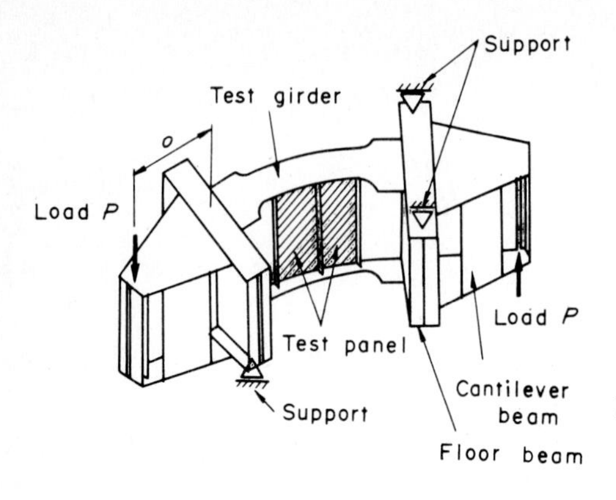

Figure 5.55 Loading and supporting device for shear.

The stress resultants in the test panel are analyzed based on the model shown in Fig. 5.56:
Shearing force:

$$Q = \frac{2M_o}{R\Phi} \tag{5.3.63a}$$

Bending moment:

$$M = M_o\left[\frac{(1+\cos\Phi)(1-\cos\phi)}{\sin\Phi}\sin\phi\right] \tag{5.3.63b}$$

Torsional moment:

$$T = -M_o\left[\frac{(1+\cos\Phi)\cos\phi}{\sin\Phi} + \sin\phi - 2\Phi\right] \tag{5.3.63c}$$

Reaction:

$$V_{1,2} = \frac{P+Q}{2} \pm \frac{[T]_{\phi=0}}{l_o} \tag{5.3.64}$$

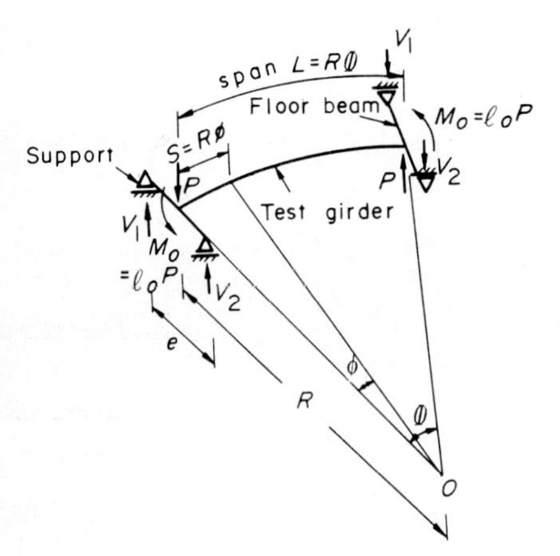

Figure 5.56 Analytical model of test girder under shear.

Therefore, if $\phi = \Phi/2$ and Φ is small, then $M = 0$ and $T = 0$. Thus, the test girders are subjected to very nearly pure shear.

***b.* Buckling strength for shear.** A typical load-deflection curve of Q versus δ_w and load-strain curve of Q versus ε_w in the web plate panels are plotted in Fig. 5.57. From these test data, the buckling loads Q^*_{cr} are determined and compared with the calculated ones by assuming that the curved subweb panel adjacent to a tension flange and a longitudinal stiffener behave as a flat and simply supported plate. Equations (5.3.43) and (5.3.44) were used in the calculations. In this case, if the elastic critical buckling stress τ_{cr} is greater than $0.8\tau_w$ where

$$\tau_{yw} = \frac{\sigma_{yw}}{\sqrt{3}} \tag{5.3.65}$$

then the elastoplastic buckling stress $\tau_{\text{cr}i}$ is estimated by

$$\frac{\tau_{\text{cr}i}}{\tau_{yw}} = \begin{cases} 1 & \tau_{\text{cr}} \leqq 0.8\tau_{yw} \quad (5.3.66a) \\[2ex] 1 - \dfrac{0.16\tau_{yw}}{\tau_{\text{cr}}} & \tau_{\text{cr}} > 0.8\tau_{yw} \quad (5.3.66b) \end{cases}$$

according to Johnston.[2] Therefore, the buckling load $Q_{\text{cr}o}$ can be obtained by

$$Q_{\text{cr}o} = \tau_{\text{cr}i} A_w \tag{5.3.67}$$

Figure 5.58 shows the comparisons of Q^*_{cr} and $Q_{\text{cr}o}$. Observing this figure, we see that $0.9Q_{\text{cr}} < Q^*_{\text{cr}} < 1.1Q_{\text{cr}o}$, and the test results almost

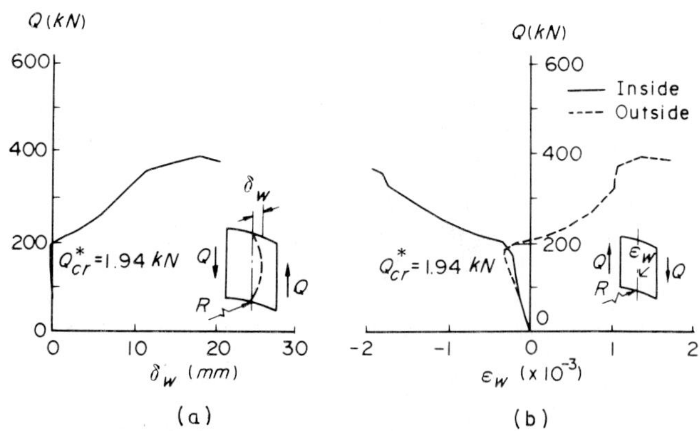

Figure 5.57 Determination of shear buckling load Q^*_{cr}: (*a*) Q–δ_w curves and (*b*) Q–ε_w curves.

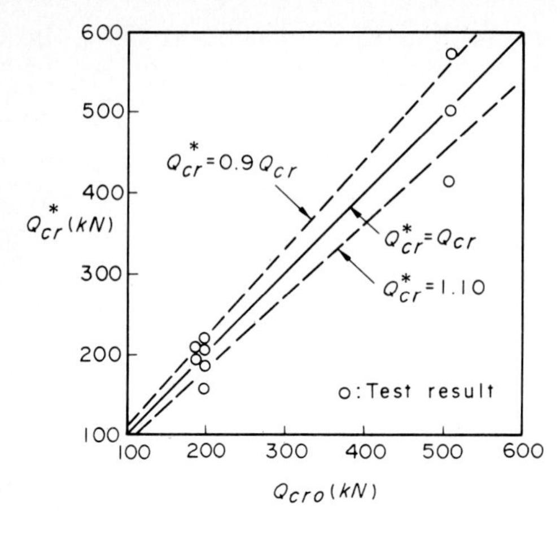

Figure 5.58 Experimental and calculated shear buckling forces.

coincide with the shear buckling load of a flat plate with simply supported edges.

The shear buckling strength of a curved web plate panel has been analyzed by Batdorf, Stein, and Shildcrout.[39] They suggest that the shear buckling strength Q_{cro} is affected by both

$$\lambda = \left(\frac{a}{R}\right)\left(\frac{a}{t_w}\right)\sqrt{1-\mu^2} \tag{5.3.68}$$

and the buckling coefficient k_τ

$$k_\tau = \frac{12(1-\mu^2)\tau_{cr}a^2}{Et_w^2} \tag{5.3.69}$$

which can be plotted as in Fig. 5.59. Additional results are plotted in

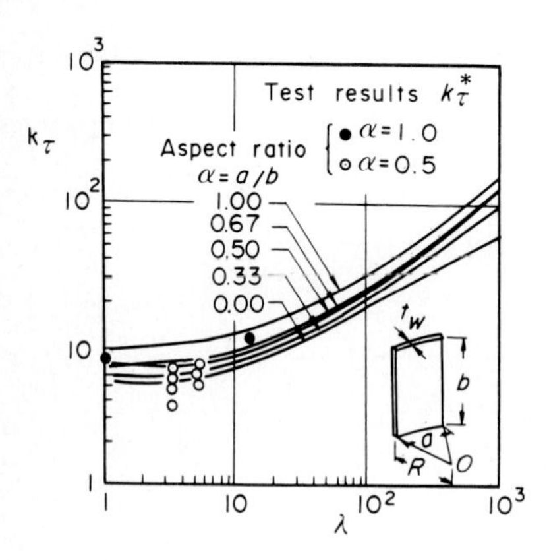

Figure 5.59 Variations of shear buckling coefficient k_τ due to curvature parameter λ.

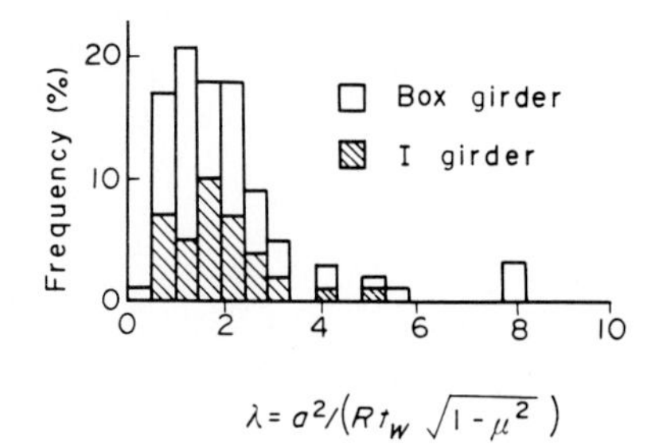

Figure 5.60 Distribution of curvature parameter λ.

this figure by replacing τ_{cr} with $\tau^*_{cr} = Q^*_{cr}/A_w$ and calculating k^*_τ from Eq. (5.3.69).

It is clear that k^*_{cr} increases in accordance with an increase in the curvature parameter λ when the aspect ratio $\alpha = a/b = 1.0$. However, this tendency is not as evident for $\alpha = 0.5$. In general, the value of λ is less than 10 from the survey of actual bridges, as shown in Fig. 5.60.[8]

Judging from Figs. 5.58 and 5.60, we see that the shear buckling strength of the web plate in the curved girders can be rationally evaluated by modeling the web plate as a flat plate with simply supported edges at flange plates and transverse stiffeners.

c. **Ultimate strength for shear.** The test results of ultimate strength for shear Q^*_u are compared with the theoretical values shown in Fig. 5.61, where Q^B_u is obtained by Basler and Thürliman[29] as predicted in Eq. (5.3.8) and Q^R_u is from Rockey and Skaloud.[40]

To explain the ultimate shear strength Q^R_u, Fig. 5.62 shows the diagonal tension field propagated after the buckling of the web plate. For this situation, the stress situation can be illustrated as in Fig. 5.63, where τ_{cr} is the shear buckling stress and σ_t is the diagonal tensile

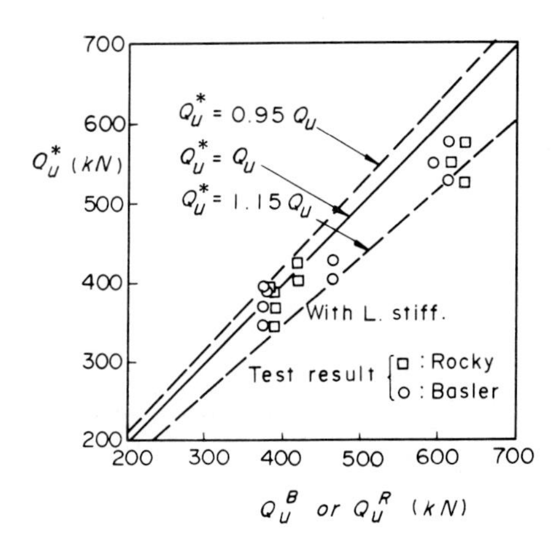

Figure 5.61 Comparisons of test results of Q^*_u with calculated ones of Q^B_u and Q^R_u.

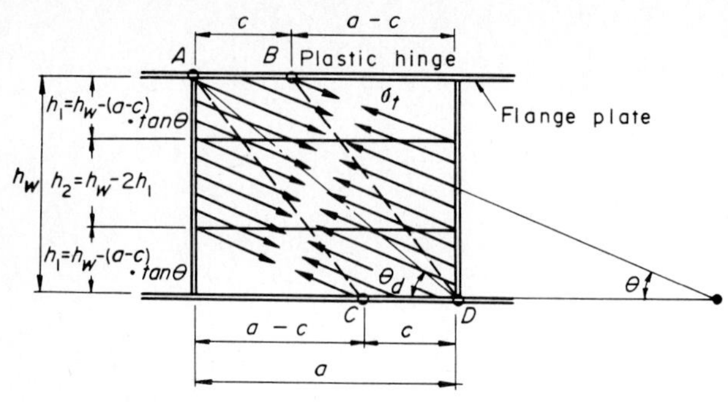

Figure 5.62 Diagonal tension field proposed by Rockey.[40]

stress in the web plate panel, with θ being the inclination angle of the diagonal tension field.

Therefore, the stresses acting in the direction of the x and y coordinate axes, shown in Fig. 5.63*b*, are

$$\sigma_x = \tau_{\text{cr}} \sin 2\theta + \sigma_t \tag{5.3.70a}$$

$$\sigma_y = -\tau_{\text{cr}} \sin 2\theta \tag{5.3.70b}$$

$$\tau_{xy} = \tau_{\text{cr}} \cos 2\theta \tag{5.3.70c}$$

Substituting this equation into von Mises' equation [cf. Eq. (2.5.6)] and solving for σ_t by setting $\sigma_y = \sigma_{yw}$, we find that

$$\sigma_t = -\tfrac{3}{2}\tau_{\text{cr}} \sin 2\theta + \sqrt{\sigma_{yw}^2 + \tau_{\text{cr}}^2(\tfrac{9}{4} \sin^2 2\theta - 3)} \tag{5.3.71}$$

Now, let us consider the postbuckling shear force Q_u^p according to the collapse mechanism shown in Fig. 5.64. From the strain energy (see

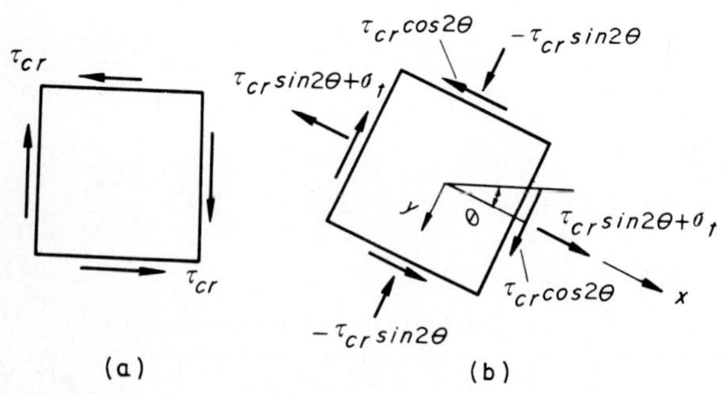

Figure 5.63 Stress situation in web plate: (*a*) in shear buckling state and (*b*) in postbuckling state.

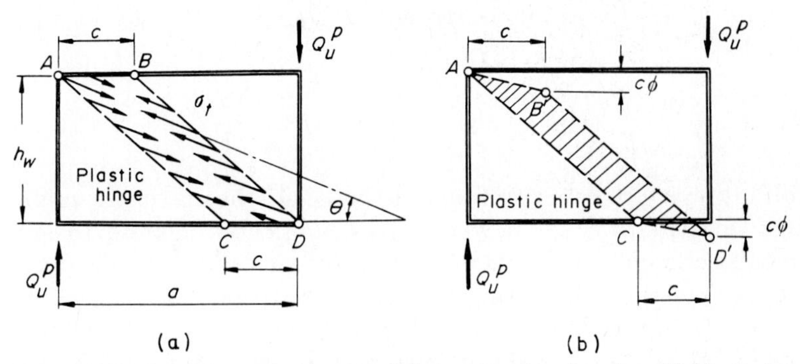

Figure 5.64 Collapse mechanism of girder under shearing force:[40] (*a*) strain energy of girder and (*b*) work due to external force.

Fig. 5.64*a*) and the work done by Q_u^P (see Fig. 5.64*b*), it follows that

$$Q_u^P = \frac{4M_{fp}}{c} + \sigma_t t_w c \sin^2\theta + \sigma_t A_w\left(\cos\theta - \frac{a}{b}\right)\sin^2\theta \qquad (5.3.72)$$

where M_{fp} is the plastic moment of the flange plate given by

$$M_{fp} = \frac{t_f}{4}\sigma_{yf}A_f \qquad (5.3.73)$$

The location of plastic hinges c from the corners of the web plate panel can be evaluated from Fig. 5.65 as

$$c = \frac{\sin\theta}{2}\sqrt{\frac{M_{fp}}{\sigma_t t_w}} \qquad (5.3.74)$$

Consequently, the substitution of Eqs. (5.3.73) and (5.3.74) into Eq.

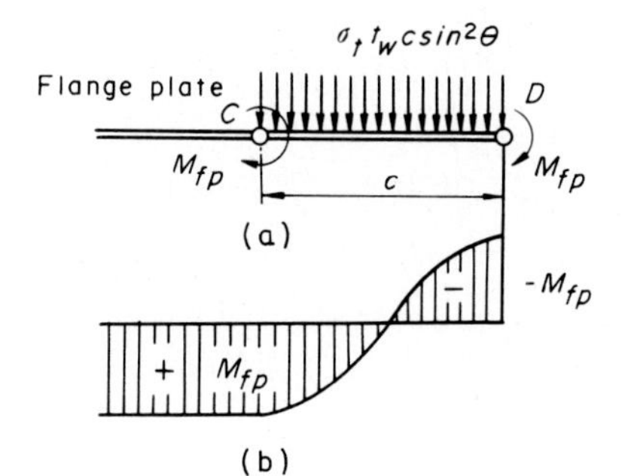

Figure 5.65 (*a*) Applied force and (*b*) induced bending moment in flange plate.

(5.3.72) determines the ultimate strength for shear $Q_u^R = Q_u^P + \sigma_{cr} A_w$, which yields

$$Q_u^R = Q_p \left[\frac{\tau_{cr}}{\tau_{yw}} + \frac{\sigma_t}{\sigma_{yw}} \sqrt{3} \sin^2\theta \left(\cos\theta - \frac{b}{a} \right) + 4\sqrt{3} \frac{\sigma_t}{\sigma_{yw}} \frac{M_{pf}}{bA_w \sigma_{yw}} \right] \qquad (5.3.75)$$

The inclination angle θ is chosen so as to make Q_u^R the maximum value. Although this equation is somewhat more complicated than Eq. (5.3.8), these results coincide well with the test results as shown in Fig. 5.61.

Finally, the test results Q_u^*/Q_p are rearranged through the use of the parameter $\sqrt{Q_p/Q_{cr}}$ and plotted together with the experimental results obtained by Mozer, Ohlson, and Culver,[41] as shown in Fig. 5.66. Recognizing the good correlation between them, we get the following empirical formulas:
For a web panel without longitudinal stiffeners,

$$\frac{Q_u^e}{Q_p} = \frac{0.55}{\sqrt{Q_p/Q_{cr}}} + 0.34 \qquad (5.3.76a)$$

For a web panel with a longitudinal stiffener,

$$\frac{Q_u^e}{Q_p} = \frac{0.27}{\sqrt{Q_p/Q_{cr}}} + 0.68 \qquad (5.3.76b)$$

(3) Buckling and ultimate strengths for combined actions of bending and shear[42]

***a.* Test method for bending and shear.** The buckling and ultimate strengths of a curved girder were investigated through tests on 12 girders, shown in Table 5.6. The symbol definitions are similar to those

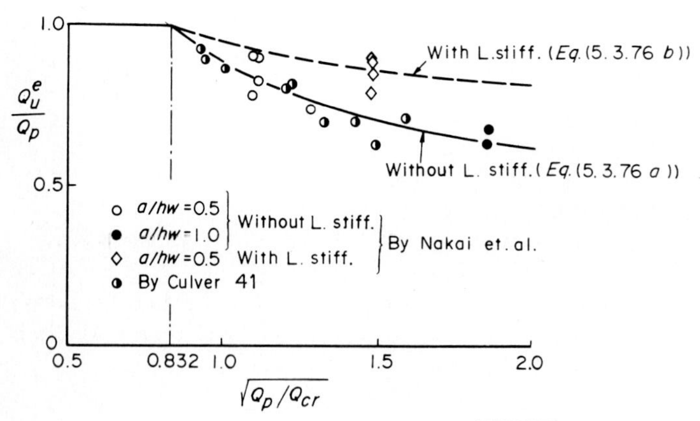

Figure 5.66 Relationship between Q_u^e/Q_p and $\sqrt{Q_p/Q_{cro}}$.

TABLE 5.6 Test Girder under Bending and Shear

Test panel	Specimen number	R, m	$\frac{h_w}{t_w}$	Aspect ratio α	$\frac{\gamma_l}{\gamma_{l,\mathrm{req}}}$	Q/Q_p (M/M_p)
	2	10	178	1.0	—	0.85 (1.65)
	4	10, ∞	178	0.5	—	0.85 (1.65)
	6	10, ∞	250	0.5	1, 5	0.85 (1.65)

in Tables 5.4 and 5.5, and the detail of a test specimen is illustrated in Fig. 5.67.

A load P was applied to the test specimen through the loading and supporting devices shown in Fig. 5.68 to produce the desired shear and bending ratio $(Q/Q_p)/(M/M_p)$ given in Table 5.6. Then the induced stress resultants were analyzed for the model shown in Fig. 5.69, which gives

$$M = M_o\left(\frac{\sin \phi}{\tan \Phi} - \cos \phi\right) \tag{5.3.77a}$$

$$Q = \frac{M_o}{R\Phi} \tag{5.3.77b}$$

$$T = -M_o\left(\sin \phi + \frac{\cos \phi}{\tan \Phi} - \frac{1}{\Phi}\right) \tag{5.3.77c}$$

where M_o is the end moment of the cantilever beam with length l_o (see Fig. 5.68), i.e.,

$$M_o = Pl_o \tag{5.3.78}$$

b. **Buckling strength for bending and shear.** The test buckling load P^*_{cr} was similarly determined from load-deflection and load-strain curves in the web plate of test specimens. Then the corresponding measured bending moment M^*_{cr} and shearing force Q^*_{cr} were determined from Eqs. (5.3.77) and (5.3.78).

The buckling moment M^e_{cr} was estimated by Eq. (5.3.50) and the shear buckling force Q_{cro} from Eq. (5.3.67). By combining these

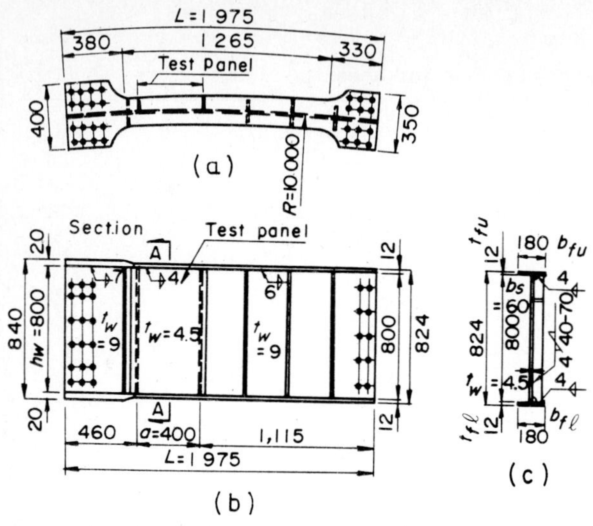

Figure 5.67 Detail of a test girder (in mm): (*a*) plan, (*b*) side elevation, and (*c*) cross section (*A–A*).

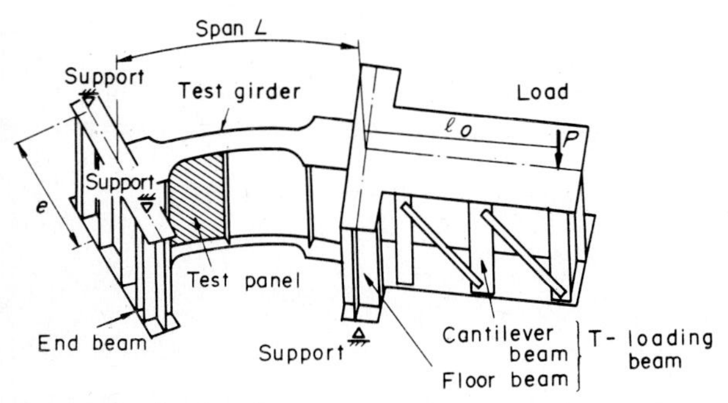

Figure 5.68 Load and support conditions of test girder for bending and shear.

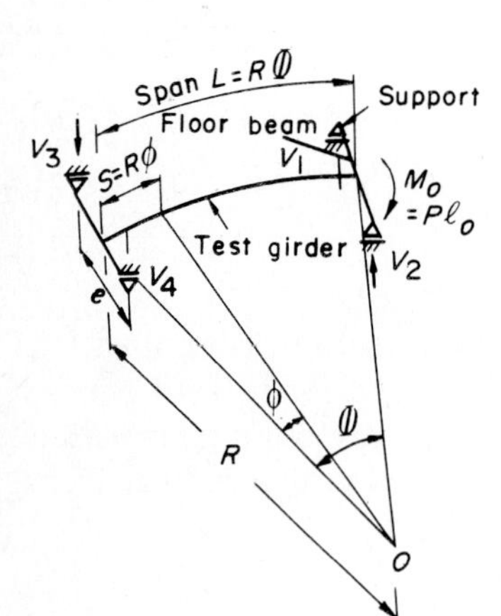

Figure 5.69 Analytical model of stress resultant for bending and shear test.

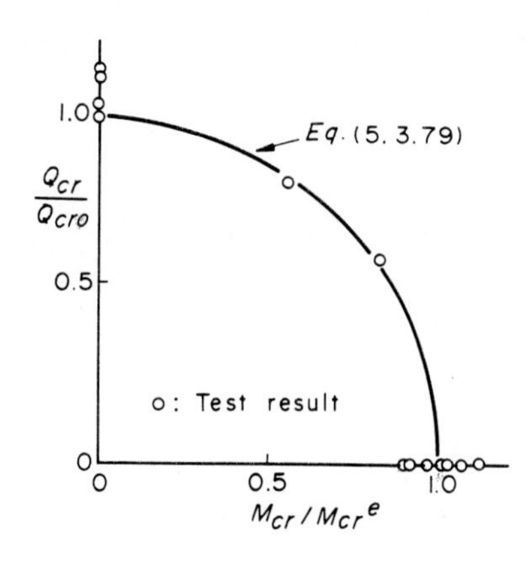

Figure 5.70 Interaction curve of web buckling.

buckling forces, the relationship between $Q^*_{\mathrm{cr}}/Q_{\mathrm{cr}o}$ and $M^*_{\mathrm{cr}}/M^{\mathrm{e}}_{\mathrm{cr}}$ was determined and plotted, as shown in Fig. 5.70.

From this figure, the following empirical interaction curve for the combined buckling of bending and shear for the curved web plate panel was obtained:

$$\left(\frac{M_{\mathrm{cr}}}{M^e_{\mathrm{cr}}}\right)^2 + \left(\frac{Q_{\mathrm{cr}}}{Q_{\mathrm{cr}\,o}}\right)^2 = 1 \tag{5.3.79}$$

c. **Ultimate strength for bending and shear.** The interaction curves for the ultimate strength of test girders was determined by investigating the relationships between Q^*_u/Q^R_u and M^*_u/M^c_p, where Q^R_u and M^c_p are theoretical strengths derived from Eqs. (5.3.75) and (5.3.55), respectively. Figure 5.71 shows these results, and the following interaction curve is proposed:

$$\frac{Q_u}{Q^R_u} = 0.9 \qquad \frac{M_u}{M^c_p} \leqq 0.5 \tag{5.3.80a}$$

$$\frac{Q_u}{Q^R_u} = 1.25 - 0.70\,\frac{M_u}{M^c_p} \qquad 0.5 < \frac{M_u}{M^c_p} < 1.0 \tag{5.3.80b}$$

$$\frac{M_u}{M^c_p} = 1.0 \qquad \frac{Q_u}{Q_R} \leqq 0.55 \tag{5.3.80c}$$

An alternative interaction is also proposed by using the empirical strengths Q^e_u [see Eq. (5.3.76)] and M^e_u [see Eq. (5.3.62)] instead of the theoretical strengths Q^R_u and M^c_p. These results are also shown in

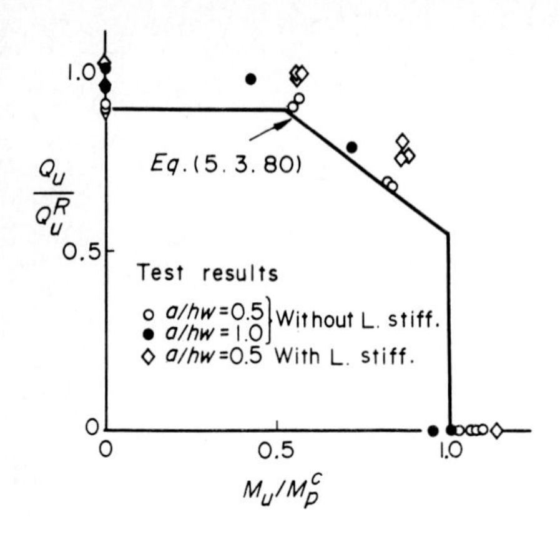

Figure 5.71 Interaction curves between Q_u/Q_u^R and M_u/M_p^c.

Fig. 5.72, which gives

$$\frac{Q_u}{Q_u^e} = 1.0 \qquad \frac{M_u}{M_u^e} \leqq 0.69 \tag{5.3.81a}$$

$$\frac{Q_u}{Q_u^e} = 1.44 - 0.62\frac{M_u}{M_u^e} \qquad 0.69 < \frac{M_u}{M_u^e} < 1.0 \tag{5.3.81b}$$

$$\frac{M_u}{M_u^e} = 1.0 \qquad \frac{Q_u}{Q_u^e} \leqq 0.81 \tag{5.3.81c}$$

5.4 Design of Web Plates in Curved Girders

5.4.1 Web slenderness of plate girders

The relationships between the depth and the thickness of the web plate, shown in Fig. 5.73, can be determined from Eqs. (5.3.23) and (5.3.25) by setting $b = h_w$ and $t = t_w$:

$$\frac{t_w}{h_w} = \frac{1}{\pi}\sqrt{\frac{12(1-\mu^2)}{k}}\sqrt{\frac{\sigma_{cr}}{E}} \tag{5.4.1}$$

where k is the buckling coefficient and can be taken as $k = k_\sigma = 23.9$ [cf. Eq. (5.3.37a)] for the most severe situation, t_w/h_w is the web slenderness, and σ_{cr} is the buckling stress. For this stress, a certain factor of safety ν must be provided against the yield stress σ_{yw} of the web plate in practical design. Thus we can set

$$\sigma_{cr} \leqq \frac{\sigma_{yw}}{\nu} \tag{5.4.2}$$

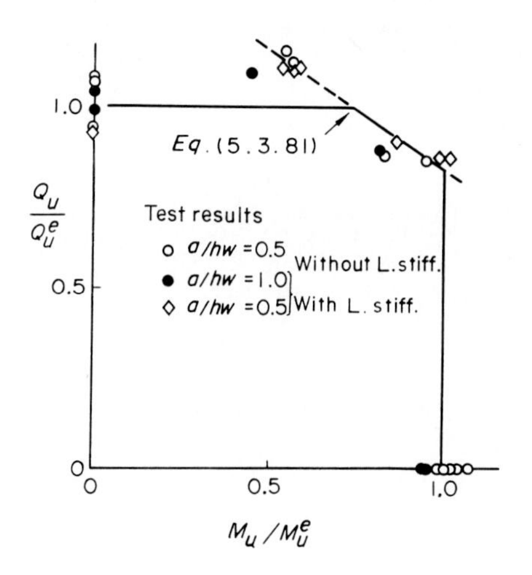

Figure 5.72 Interaction curves between Q_u/Q_u^e and M_u/M_u^e.

In determining the factor of safety ν, the postbuckling behavior is expected to be as predicted in Sec. 5.3. For example, this value is taken as $\nu = 1.4$ in the Japanese Specification for Highway Bridges (JSHB)[6] and $\nu = 1.2$ in AASHTO.[56] Accordingly, the substitution of Eq. (5.4.2) into Eq. (5.4.1) with $E = 2.1 \times 10^6$ kgf/cm^2 and $\sigma_{yw} = 2400$ kgf/cm^2 (mild steel) gives

$$\frac{t_w}{h_w} \geqq \frac{1}{\beta_o} \tag{5.4.3a}$$

where

$$\beta_o = 152 \qquad \text{JSHB} \tag{5.4.3b}$$

$$\beta_o = 165 \qquad \text{AASHTO} \tag{5.4.3c}$$

When longitudinal stiffeners are installed in the web plate as illustrated in Fig. 5.74, JSHB[6] codified the criteria for β_o, as shown in Table 5.7.

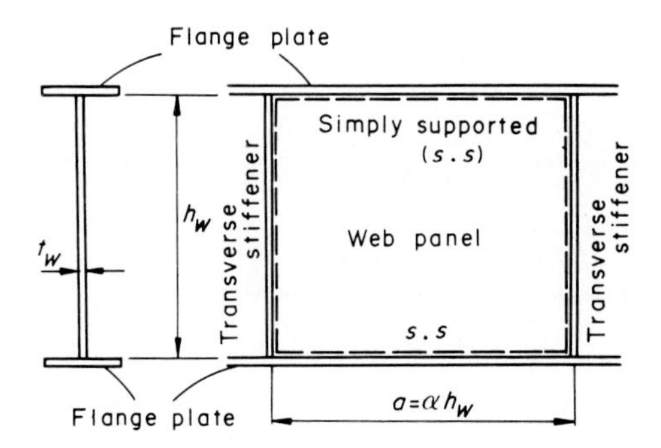

Figure 5.73 Details of a web plate.

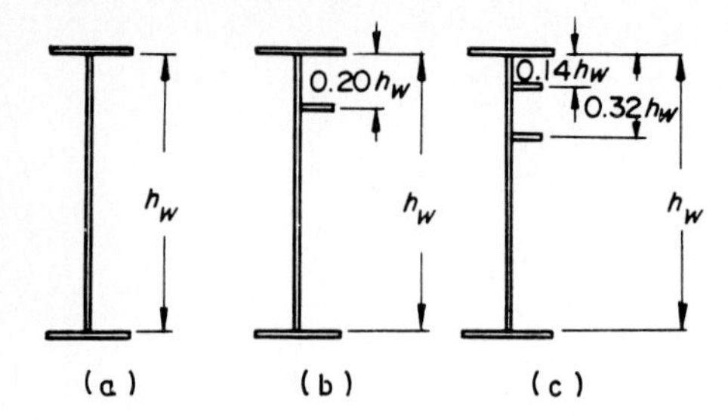

Figure 5.74 Location of longitudinal stiffener: (*a*) without longitudinal stiffener, (*b*) with one longitudinal stiffener, and (*c*) with two longitudinal stiffeners.

It is, however, difficult to select the web slenderness for a curved girder in a manner similar to that above, because the out-of-plane displacement of the web plate gradually increases in accordance with the increase in the applied load due to the curvature of the web plate, and then the clear buckling phenomenon cannot be observed in the curved girder.

These behaviors of the curved web plate have been pointed out and analyzed by Culver, Dym, and Brogan;[43] Dabrowski and Wachowiak;[44] Abdel-Sayed;[45] Mikami and Furunishi;[46] and Hiwatashi and Kuranishi.[47] It is necessary to consider the out-of-plane displacement of the curved web plate in designing curved girder bridges.

5.4.2 Parametric analysis of stress and displacement in curved web panels

(1) Analytical method. To determine the out-of-plane deformation of a curved web panel, the finite-element method (FEM) based on the large-displacement theory was adopted.[48,49]

a. Finite displacement according to incremental theory. For this problem, the configuration C_{n+1} following the increment of load can be defined by using the configuration C_n prior to the increment of load on

TABLE 5.7 Codified Web Slenderness Parameter β_o in JSHB

Longitudinal stiffener	σ_{yw}, kgf/cm^2 (steel material) 2400 (SS 41, SM 41)	3200 (SM 50)	3600 (SM 53, SM 50Y)	4600 (SM 58)
Without longitudinal stiffener	152	130	123	110
With one longitudinal stiffener	256	220	209	188
With two longitudinal stiffeners	310	310	294	262

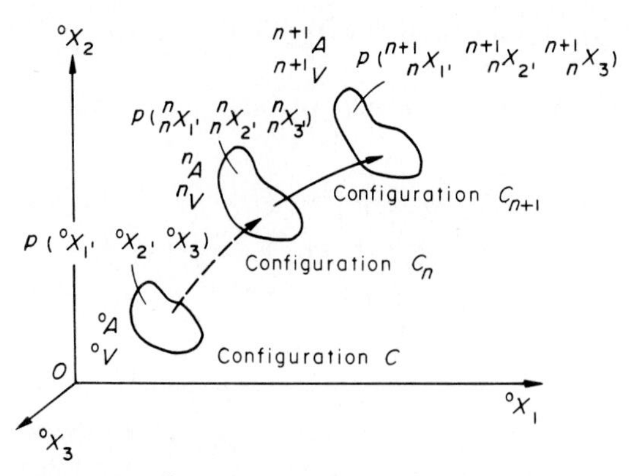

Figure 5.75 Cartesian coordinate system of UL configurations.

the basis of the *updated lagrangian method* (henceforth referred to as the UL method), as shown in Fig. 5.75. The virtual work based on the incremental theory can be written as

$$\int_{^nV} {}^{n+1}_{\ n}s_{ij}\, \delta(\Delta\varepsilon_{ij})\, d\,{}^nV = {}^{n+1}r \tag{5.4.4}$$

where ${}^{n+1}_{\ n}s_{ij}$ = Kirchhoff's stress estimated at configurations C_{n+1} through C_n

$\Delta\varepsilon_{ij}$ = Green's incremental strain from configuration C_{n+1} to C_n

${}^nV$ = volume at configuration C_n

${}^{n+1}r$ = virtual work of applied force at configuration C_{n+1}

The Kirchhoff's stress ${}^{n+1}_{\ n}s_{ij}$ in Eq. (5.4.4) can be related to Cauchy's stress ${}^n\sigma_{ij}$ at configuration C_n and Kirchhoff's stress increment Δs_{ij}, which gives

$${}^{n+1}_{\ n}s_{ij} = {}^n\sigma_{ij} + \Delta s_{ij} \tag{5.4.5}$$

However, the incremental strains $\Delta\varepsilon_{ij}$ consist of the linear strain Δe_{ij} and nonlinear strain $\Delta\eta_{ij}$, which can be expressed as

$$\Delta\varepsilon_{ij} = \Delta e_{ij} + \Delta\eta_{ij} \tag{5.4.6}$$

where

$$\Delta e_{ij} = \frac{\Delta u_{i,j} + \Delta u_{j,i}}{2} \tag{5.4.7a}$$

$$\Delta\eta_{ij} = \frac{\Delta u_{k,i}\, \Delta u_{k,j}}{2} \tag{5.4.7b}$$

and Δu_i signifies the incremental displacement from configurations C_n to C_{n+1}.

Furthermore, Δs_{ij} and $\Delta \varepsilon_{ij}$ can be combined through the material rigidity tensor D_{ijrs} as follows:

$$\Delta s_{ij} = D_{ijrs}\, \Delta \varepsilon_{rs} \tag{5.4.8}$$

Substitution of Eqs. (5.4.5), (5.4.6), and (5.4.8) into Eq. (5.4.4) gives

$$\int_{^nV} D_{ijrs}\, \Delta\varepsilon_{rs}\, \delta(\Delta e_{ij})\, d^nV + \int_{^nV} {}^n\sigma_{ij}\, \delta(\Delta \eta_{ij})\, d^nV = {}^{n+1}r - \int_{^nV} {}^n\sigma_{ij}\, \delta(\Delta e_{ij})\, d^nV \tag{5.4.9}$$

The solution of this equation cannot be obtained directly, since nonlinear terms in the right-hand side of Eq. (5.4.9) are involved. Accordingly, it is necessary to formulate Eq. (5.4.4) through the linearization by setting $\Delta\varepsilon_{ij} = \Delta e_{ij}$.

***b.* Formulation by using isoparametric shell elements.** The isoparametric shell element having eight nodal points, shown in Fig. 5.76, is adopted here in order to analyze the stress and displacement in the curved web panel with complex shapes such as curvature and initial imperfections.

Details of the element are summarized as follows:

(*i*) *Geometry of element.* The x, y, and z coordinate axes at an arbitrary point in the shell element can be related to the nodal coordinate axes of x_i, y_i, and z_i by using the shape function $N_i(\xi, \eta)$, which is expressed by the curvilinear coordinate axes (ξ, η, ζ) as

$$\begin{bmatrix} x \\ y \\ z \end{bmatrix} = \sum_{i=1}^{8} N_i(\xi,\eta) \begin{bmatrix} x_i \\ y_i \\ z_i \end{bmatrix} + \frac{1}{2} \sum_{i=1}^{8} N_i(\xi,\eta)\zeta t_i \mathbf{v}_{3i} \tag{5.4.10}$$

where t_i = thickness at each nodal point
$\mathbf{v}_{3i}$ = unit normal vector at midplane of each nodal point

(*ii*) *Displacement in element.* The displacement (u, v, w) at an arbitrary point in the element can be written by using displacement (u_i, v_i, w_i) for each nodal point represented by the global coordinates and the rotation angles β_i and α_i with respect to the coordinate axes (x', y', z') of the element, as shown in Fig. 5.77. The corresponding displacement function N_i can be written as

$$\begin{bmatrix} u \\ v \\ w \end{bmatrix} = \sum_{i=1}^{8} N_i(\xi,\eta) \begin{bmatrix} u_i \\ v_i \\ w_i \end{bmatrix} + \frac{1}{2} \sum_{i=1}^{8} N_i(\xi,\eta)\zeta t_i \boldsymbol{\Phi}_i \begin{bmatrix} \alpha_i \\ \beta_i \end{bmatrix} \tag{5.4.11}$$

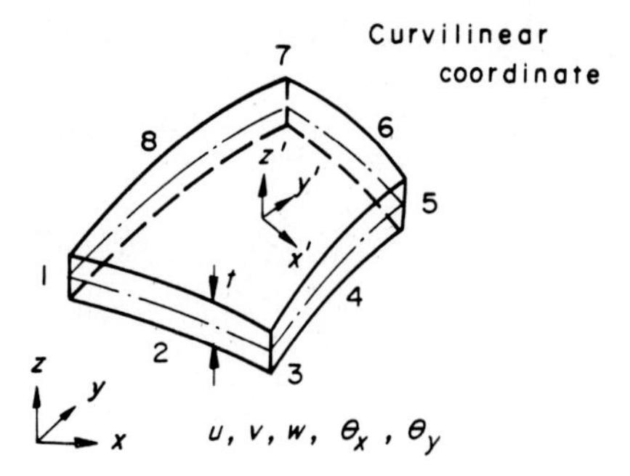

Figure 5.76 Isoparametric shell element.

where $\boldsymbol{\phi}_i$ is a matrix consisting of the unit vector in the direction of the x' and y' coordinate axes so that

$$\boldsymbol{\phi}_i = [\nu_{1i} - \nu_{2i}] \tag{5.4.12}$$

(*iii*) *Strain.* In general, the strain at an arbitrary point in the element can be represented by the terms of the partial derivative of the displacement (u', v', w') of the local coordinate system of the element, which leads to

$$[\mathbf{e}_{u'}, \mathbf{e}_{v'}\ \mathbf{e}_{w'}] = \begin{bmatrix} \dfrac{\partial u'}{\partial x'} & \dfrac{\partial v'}{\partial x'} & \dfrac{\partial w'}{\partial x'} \\ \dfrac{\partial u'}{\partial y'} & \dfrac{\partial v'}{\partial y'} & \dfrac{\partial w'}{\partial y'} \\ \dfrac{\partial u'}{\partial z'} & \dfrac{\partial v'}{\partial z'} & \dfrac{\partial w'}{\partial z'} \end{bmatrix}$$

$$= \boldsymbol{\theta}^T \mathbf{J}^{-1} \begin{bmatrix} \dfrac{\partial u}{\partial \xi} & \dfrac{\partial v}{\partial \xi} & \dfrac{\partial w}{\partial \xi} \\ \dfrac{\partial u}{\partial \eta} & \dfrac{\partial v}{\partial \eta} & \dfrac{\partial w}{\partial \eta} \\ \dfrac{\partial u}{\partial \zeta} & \dfrac{\partial v}{\partial \zeta} & \dfrac{\partial w}{\partial \zeta} \end{bmatrix} \boldsymbol{\theta} \tag{5.4.13}$$

where $\mathbf{J}$ = Jacobian matrix
$\boldsymbol{\theta}$ = transformation matrix from local to global coordinate system

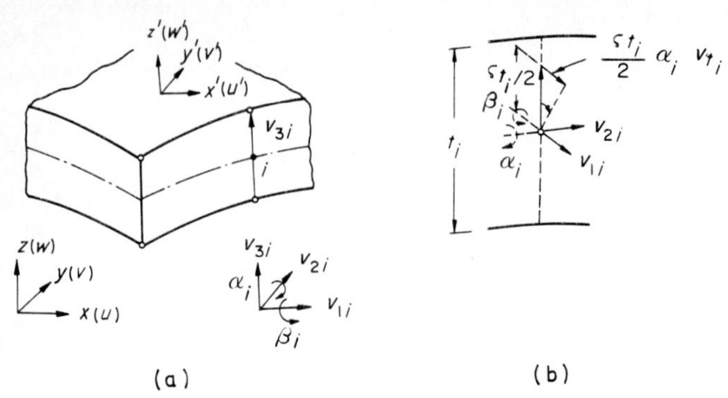

Figure 5.77 Displacement field: (*a*) displacement of local cartesian coordinate and (*b*) rotational displacement α_i and β_i.

The term $\boldsymbol{\theta}^T\mathbf{J}^{-1}$ in the above equation can be expressed by

$$\mathbf{A} = \boldsymbol{\theta}^T\mathbf{J}^{-1} = \begin{bmatrix} a_{11} & a_{12} & 0 \\ a_{21} & a_{22} & 0 \\ 0 & 0 & a_{33} \end{bmatrix} \tag{5.4.14}$$

Then Eq. (5.4.13) can be rewritten as

$$\begin{bmatrix} \mathbf{e}_{u'} \\ \mathbf{e}_{v'} \\ \mathbf{e}_{w'} \end{bmatrix} = \sum_{i=1}^{8} [\mathbf{B}_i\boldsymbol{\theta}^T \tfrac{1}{2}t_i(\zeta\mathbf{B}_i + \mathbf{C}_i)\boldsymbol{\theta}^T\boldsymbol{\phi}_i] \begin{bmatrix} u_i \\ v_i \\ w_i \\ \alpha_i \\ \beta_i \end{bmatrix} \tag{5.4.15}$$

where matrices $\mathbf{B}_i$ and $\mathbf{C}_i$ are

$$\mathbf{B}_i = \begin{bmatrix} b_1 & 0 & 0 \\ b_2 & 0 & 0 \\ 0 & 0 & 0 \\ 0 & b_1 & 0 \\ 0 & b_2 & 0 \\ 0 & 0 & 0 \\ 0 & 0 & b_1 \\ 0 & 0 & b_2 \\ 0 & 0 & 0 \end{bmatrix}_i \tag{5.4.16}$$

$$\mathbf{C}_i = \begin{bmatrix} 0 & 0 & 0 \\ 0 & 0 & 0 \\ c_1 & 0 & 0 \\ 0 & 0 & 0 \\ 0 & 0 & 0 \\ 0 & c_1 & 0 \\ 0 & 0 & 0 \\ 0 & 0 & 0 \\ 0 & 0 & c_1 \end{bmatrix}_i \tag{5.4.17}$$

in which

$$b_{1i} = a_{11}N_{i,\xi} + a_{12}N_{i,\eta} \tag{5.4.18a}$$

$$b_{2i} = a_{21}N_{i,\xi} + a_{22}N_{i,\eta} \tag{5.4.18b}$$

$$c_{1i} = a_{33}N_i \tag{5.4.18c}$$

(*iv*) *Stiffness matrix of an element.* The stiffness matrix of an element for the small displacement $\mathbf{K}_{Lij}$ can be obtained by setting $\Delta\varepsilon_{ij} \cong \Delta e_{ij}$ in the first term of the left-hand side of Eq. (5.4.9):

$$\mathbf{K}_{Lij} = \int\int\int_{-1}^{1} \begin{bmatrix} \boldsymbol{\theta} & 0 \\ 0 & \frac{1}{2}t_i\boldsymbol{\Phi}_i^T\boldsymbol{\theta} \end{bmatrix} \mathbf{K}_{Lij} \begin{bmatrix} \boldsymbol{\theta}^T & 0 \\ 0 & \frac{1}{2}t_j\boldsymbol{\theta}^T\boldsymbol{\Phi}_j \end{bmatrix} |\mathbf{J}|\, d\xi\, d\eta\, d\zeta \tag{5.4.19}$$

where

$$\mathbf{K}_{Lij} = \begin{bmatrix} \mathbf{G}_i^T \\ \zeta\mathbf{G}_i^T + \mathbf{H}_i^T \end{bmatrix} \mathbf{D}[\mathbf{G}_j \quad \zeta\mathbf{G}_j + \mathbf{H}_j] \tag{5.4.20}$$

$$\mathbf{G}_i = \begin{bmatrix} b_1 & 0 & 0 \\ 0 & b_1 & 0 \\ b_2 & b_1 & 0 \\ 0 & 0 & b_1 \\ 0 & 0 & b_2 \end{bmatrix}_i \tag{5.4.21}$$

and

$$\mathbf{H}_i = \begin{bmatrix} 0 & 0 & 0 \\ 0 & 0 & 0 \\ 0 & 0 & 0 \\ c_1 & 0 & 0 \\ 0 & c_1 & 0 \end{bmatrix}_i \tag{5.4.22}$$

The geometrical stiffness matrix $\mathbf{K}_{Gij}$ can be determined from the second term of Eq. (5.4.9) and can be written in a form similar to

Eq. (5.4.20):

$$\mathbf{K}_{Gij} = \begin{bmatrix} \mathbf{B}_i^T \\ \zeta \mathbf{B}_i^T + \mathbf{C}_i^T \end{bmatrix} \begin{bmatrix} \boldsymbol{\sigma} & 0 & 0 \\ 0 & \boldsymbol{\sigma} & 0 \\ 0 & 0 & \boldsymbol{\sigma} \end{bmatrix} [\mathbf{B}_j \quad \zeta \mathbf{B}_j + \mathbf{C}_j] \tag{5.4.23}$$

where
$$\boldsymbol{\sigma} = \begin{bmatrix} \sigma_{x'} & \tau_{x'y'} & \tau_{x'z'} \\ & \tau_{y'} & \tau_{y'z'} \\ \text{Sym.} & & \sigma_{z'} \end{bmatrix} \tag{5.4.24}$$

(Sym. = symmetry.)

(*v*) *Equivalent nodal force.* The equivalent nodal force due to the stress resultants on an element can be estimated from the second term of the right-hand side of Eq. (5.4.9):

$$\begin{bmatrix} f_x \\ f_y \\ f_z \\ m_1 \\ m_2 \end{bmatrix}_i = \int\int\int_{-1}^{1} \begin{bmatrix} \theta & 0 \\ 0 & \tfrac{1}{2} t_i \Phi_i^T \boldsymbol{\theta} \end{bmatrix} \begin{bmatrix} \mathbf{G}_i^T \\ \zeta \mathbf{G}_i^T + \mathbf{H}_i^T \end{bmatrix} \begin{bmatrix} \sigma_{y'} \\ \tau_{x'y'} \\ \tau_{x'z'} \\ \tau_{y'z'} \end{bmatrix} |\mathbf{J}| \, d\xi \, d\eta \, d\zeta \tag{5.4.25}$$

where f_{xi}, f_{yi}, f_{zi} = nodal forces in the global x, y, and z coordinate axes, respectively

m_{1i}, m_{2i} = bending moments with respect to local x' and y' coordinate axes, respectively

(*vi*) *Analysis of nonlinearity.* The stiffness matrix and equivalent nodal force in terms of the local coordinate system are transferred and assembled into the global coordinate system. Then the following equilibrium condition in terms of the global coordinate system can be obtained:

$$(\mathbf{K}_L^i + \mathbf{K}_G^i)\, \Delta\mathbf{u} = {}^{n+1}\mathbf{r} - \mathbf{f}^i = \Delta\mathbf{e}^i \tag{5.4.26}$$

where i = number of iterations between configurations C_n and C_{n+1}

$\mathbf{K}_i$ = stiffness matrix for small displacement in configuration C_i

$\mathbf{K}_G^i$ = geometric stiffness matrix in configuration C_i

${}^{n+1}\mathbf{r}$ = force vector in configuration C_{n+1}

$\mathbf{f}^i$ = equivalent nodal force in configuration C_i

Equation (5.4.9) can successively be iterated by using the Newton-Raphson method, as shown in Fig. 5.78, where the convergence of the

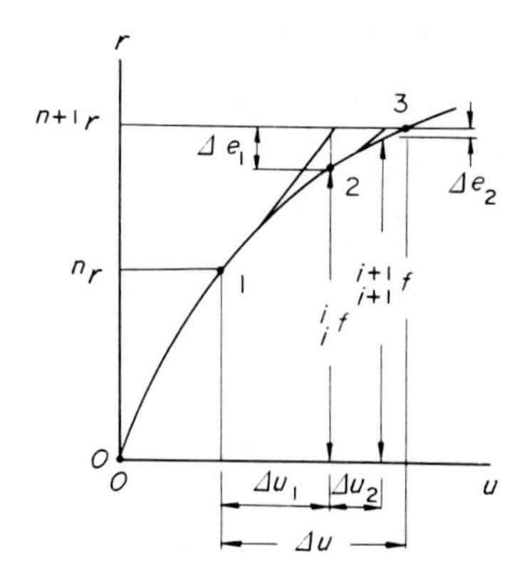

Figure 5.78 Process of iteration.

solution is checked by the norm given by the ratio between the unbalanced force $\Delta\mathbf{e}^i$ and applied force ${}^{n+1}\mathbf{r}$, or

$$|\Delta\mathbf{e}^i|/|{}^{n+1}\mathbf{r}| < \varepsilon \tag{5.4.27}$$

in which ε is the zero estimation value set by $\varepsilon \ll 1$. The flowchart for programming these procedures is shown in Fig. 5.79.

(2) Effects of boundary and other conditions on curved web panels

***a.* Effects of boundary conditions.** The curved web panel, which consists of flange plates and longitudinal stiffeners, is idealized as shown

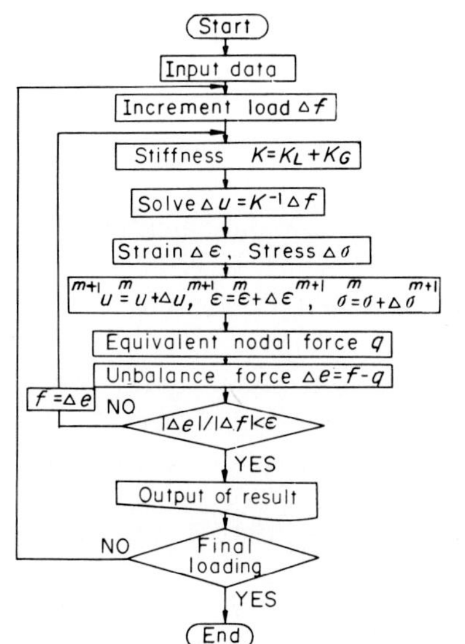

Figure 5.79 Flowchart for computer program. K = total stiffness matrix; K_L = stiffness matrix of small displacement; K_G = geometrical stiffness matrix; Δu = incremental displacement vector; Δf = incremental load vector; ε = index of desired accuracy; m = dummy counter.

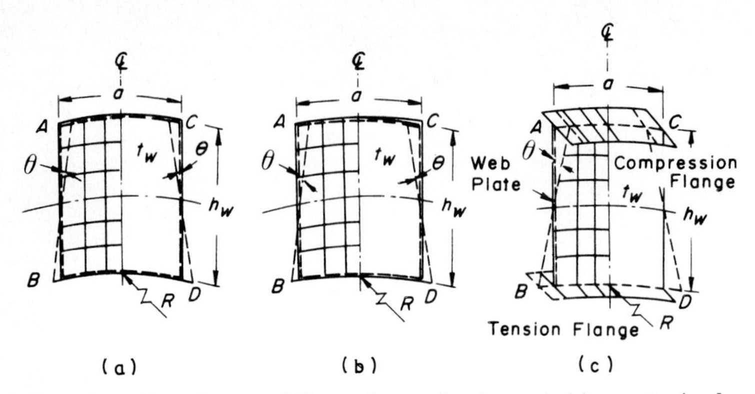

Figure 5.80 Boundary conditions of curved web panel: (*a*) case 1: simple supported model, (*b*) case 2: fixed supported model, (*c*) case 3: model with flange.

in Fig. 5.80, where the boundary conditions of the curved web panel are categorized into three cases:

1. Simply supported at flange plates (SS)
2. Fixed support at flange plates (FS)
3. Elastically supported by flange plate (ES)

The boundary conditions at the longitudinal stiffeners are assumed to be simply supported in all the cases. The bending moment is applied to the web plate panel by introducing the deflection angle θ through the rigid body attached along the loading edges $\overline{AB}$ and $\overline{CD}$.

Figure 5.81 shows the out-of-plane deflection δ_{max}/t_w of the web panel with thickness t_w, in accordance with the applied bending moment M, which is nondimensionalized by the initial yield moment M_{yw} of the web plate as a girder. And A_f/A_w is the ratio of the cross-sectional area of the flange to the web plates.

The longitudinal normal stress σ_{ln} and out-of-plane bending stress

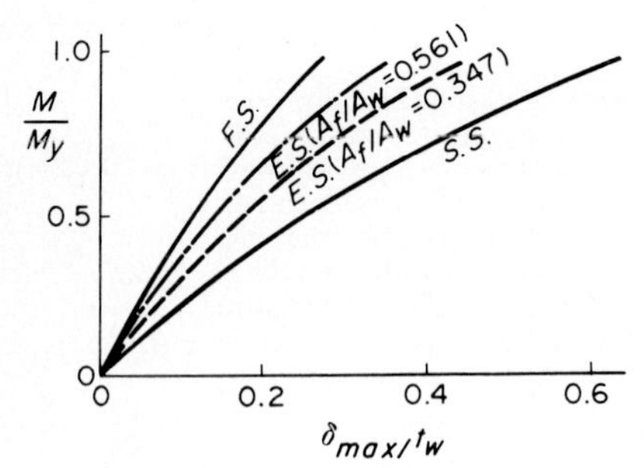

Figure 5.81 Relationships between M/M_{yw} and δ_{max}/t_w. $a/R = 0.0245$; $a/h_w = 0.7$; $h_w/t_w = 125$; $w_o/t_w = 0.5$.

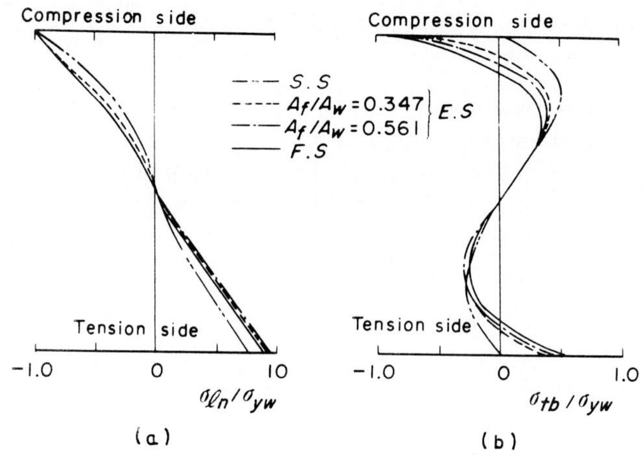

Figure 5.82 Stress distribution in web plate. (*a*) σ_{ln} and (*b*) σ_{tb}. $a/R = 0.0245$; $a/h_w = 0.7$; $h_w/t_w = 125$; $w_o/t_w = 0.5$.

σ_{tb} in the transverse direction, corresponding to the situation where σ_{ln} reaches the nominal yield stress $\sigma_{yw} = 2400$ kgf/cm^2 = 235.2 MPa at the junction point of the flange and web plates, can be plotted by nondimensionalizing and dividing through by σ_{yw}, as shown in Fig. 5.82. For these situations, let M_r denote the resistant moment of the girder. This was designated M_{wr} for the web plate in previous discussions.

By taking the values of M_{wr} as well as σ_{tb} and $\delta_{\max}$, which also correspond to the resisting bending moment M_{wr} of the web plate, equal to unity for case 2 (FS), the variations of these values due to boundary conditions can be summarized as in Table 5.8.

From this table note that $\delta_{\max}/t_w = 2$, $M_{wr}/M_{yw} \cong 0.8$, and $\sigma_{tb}/\sigma_{yw} = 0$ for case 1 (SS) and that this boundary condition (SS) may not describe the behavior of the curved web panel exactly. But the values of $\delta_{\max}/t_w$, M_{wr}/M_{yw}, and σ_{tb}/σ_{yw} for case 2 (FS) are close to those of case 3 (ES). Therefore, we conclude that case 3 (ES) is the most suitable boundary

TABLE 5.8 Effect of Boundary Conditions

	Case 1 (SS)	Case 2 (FS)	Case 3 (ES)	
			$A_f/A_w = 0.347$	$A_f/A_w = 0.561$
Maximum out-of-plane deflection $\delta_{\max}/t_w$	1.907	1.0	1.583	1.322
Resisting bending moment M_{wr}/M_{wy}	0.833	1.0	0.919	0.906
Maximum bending stress σ_{tb}/σ_y	0	1.0	0.825	0.699

condition for evaluating the actual stress and deflection of curved web panels as exactly as possible.

***b.* Effects of warping stresses in the flange plate.** The flange plates of curved girders usually undergo not only longitudinal normal stress σ_{ln} but also warping stress σ_ω due to the out-of-plane bending of the flange plates by the curvature of the girder axis. The ratio ζ between σ_ω and σ_{ln} at the flange plate can be obtained from Eqs. (5.2.35) and (5.2.36):

$$\zeta = \pm\chi \frac{L^2}{RW_f}\left(A_f + \frac{A_w}{6}\right) \tag{5.4.28}$$

where L = supported length of flange plate
R = radius of curvature of girder
W_f = section modulus of flange plate with respect to vertical axis
A_f, A_w = cross-sectional area of flange and web plate, respectively
$\chi = 0.106$ = coefficient to determine out-of-plane bending of flange plate.

According to the survey of actual curved girder bridges, ζ varies from 0.5 to -0.5. Therefore, the analysis of curved web panels is conducted for these conditions, and the distributions of σ_{ln} and the variations

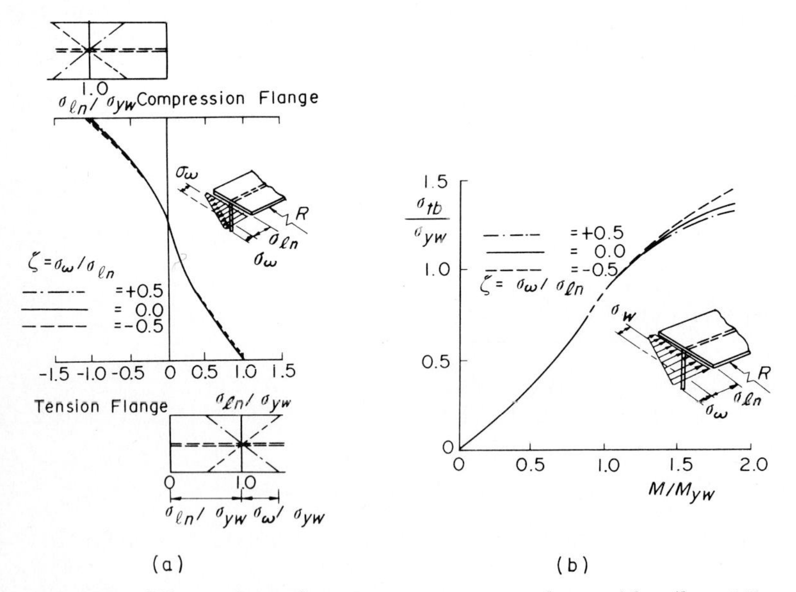

Figure 5.83 Effects of warping stress σ_ω on σ_{ln} and σ_{tb} with $a/h_w = 0.7$, $h_w/t_w = 156$, $w_o/h_w = 0.004$, $A_f/A_w = 0.432$, and $\sigma_{yw} = 235.2$ MPa: (*a*) longitudinal normal stress distribution and (*b*) variation of transverse bending stress.

of σ_{tb} due to M/M_{yw} are plotted in Fig. 5.83. From this figure it is evident that the warping stress does not influence σ_{ln} and σ_{tb} for the elastic range $M/M_{yw} < 1$. Consequently, the effect of warping can be disregarded in the preceding parametric analysis.

c. Effects of initial deflections of the curved web plate panels. It has been shown that the modes of initial deflections in the curved web plates clearly differ from the arrangements of welding surfaces to which the transverse and longitudinal stiffeners are attached (see Fig. 5.42). These initial deflection modes will affect the resisting moment M_r of curved girders.

For these initial deflection modes, two types of modes are assumed; one is a saddle type toward the center of curvature, and the other is a cylindrical type outward from the center of curvature. These modes are assumed to be double sinusoidal functions with a half wave, and their maximum amplitudes are taken equal to $w_o = h_w/250$ according to the fabrication tolerance of JSHB.[6]

Figure 5.84 shows the variations in the resisting moment M_r/M_y due to initial deflection modes. From this figure, we observe that the cylindrical type is more sensitive than the saddle type.

Similar analyses were conducted on the web plate panel with one and two longitudinal stiffeners. It was found that the cylindrical modes, shown in Fig. 5.85, give the most severe resisting moment.

(3) Results of parametric analysis.[50] Parametric analyses were made for the curved web plate panels with less than two longitudinal stiffeners, where the boundary conditions and the other conditions

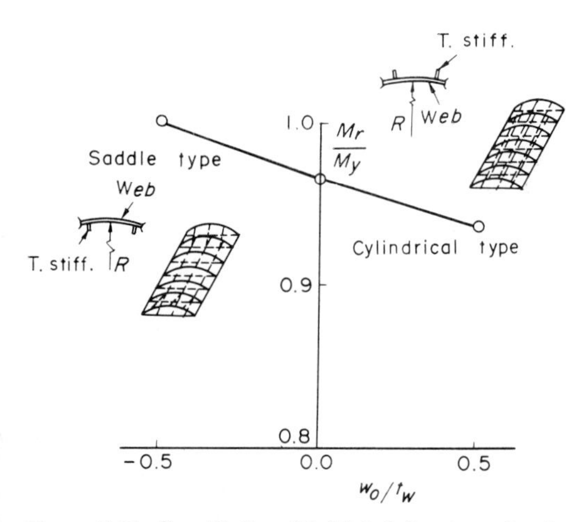

Figure 5.84 Sensitivity of initial deflection of web plate to resistance bending moment. $a/R = 0.0245$, $a/h_w = 0.7$, $h_w/t_w = 156$, $A_f/A_w = 0.347$.

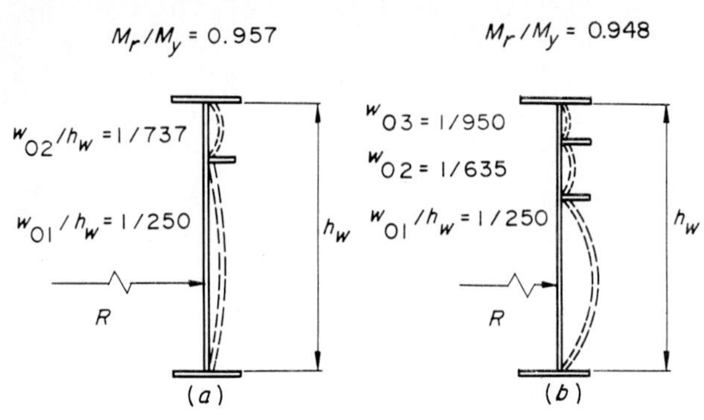

Figure 5.85 Assumed initial deflection modes for web plate panel with longitudinal stiffeners: (*a*) with one stiffener and (*b*) with two stiffeners.

were chosen to make their effects as large as possible, as mentioned above. These analytical parameters are listed in Table 5.9, where a/R, h_w/t_w, and A_f/A_w are determined by referring to a parametric survey in Nakai et al.[8] and JSHB.[6] The nominal yield stress is assumed to be $\sigma_y = 2400\ \text{kgf/cm}^2 = 235.2\ \text{MPa}$ for SS 41 and $\sigma_y = 3600\ \text{kgf/cm}^2 = 352.8\ \text{MPa}$ for SM 50.

***a.* Resisting moment.** Figure 5.86 shows the distribution of longitudinal normal stress σ_{ln} in the web plate when σ_{ln} reaches the yield stress σ_{yw}. The corresponding variations of resisting moment M_r nondimensionalized by the yield moment M_y, that is, M_r/M_y, due to the curvature parameter a/R and web slenderness h_w/t_w are plotted in Fig. 5.87.

TABLE 5.9 Analysis Parameters

Number of longitudinal stiffeners	Combinations of curvature parameter a/R and slenderness ratio h_w/t_w		$\dfrac{A_f}{A_w}$
0	$\dfrac{a}{R}$	0, 0.0098, 0.0245, 0.049	0.499–0.698
	$\dfrac{h_w}{t_w}$	100, 112, 125, 140, 156	
1	$\dfrac{a}{R}$	0, 0.0098, 0.0245, 0.049	0.250–0.722
	$\dfrac{h_w}{t_w}$	90, 120, 150, 210, 233, 260	
2	$\dfrac{a}{R}$	0, 0.0098, 0.0245, 0.049	0.467–1.113
	$\dfrac{h_w}{t_w}$	130, 190, 250, 310	

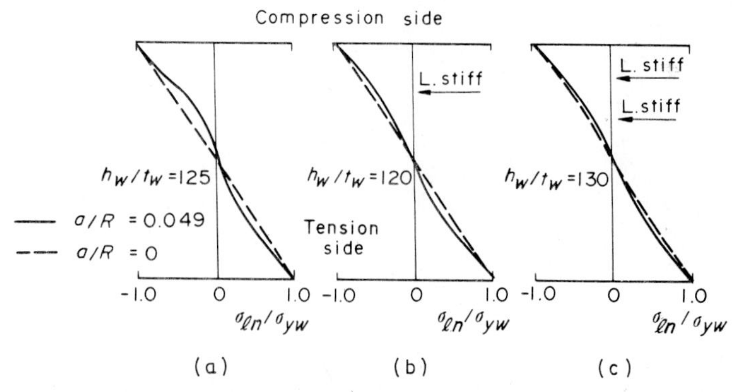

Figure 5.86 Longitudinal normal stress distribution ($a/h_w = 0.7$, $w_o/h_w = 0.004$, $\sigma_{yw} = 235.2$ MPa): (*a*) without longitudinal stiffeners, (*b*) with one longitudinal stiffener, (*c*) with two longitudinal stiffeners.

Clearly the values of M_r/M_y are not affected by h_w/t_w, but they become almost constant in spite of the existence of longitudinal stiffeners. Yet M_r/M_y decreases in proportion to the increase in the curvature parameter a/R, and the values of M_r/M_y are not reduced to less than 0.9 for the values a/R of actual curved girder bridges. These facts suggest that almost all the applied moments can be taken by the flange plates and that the contributions of the web plate are small enough to be ignored.

***b.* Out-of-plane deflection of web plates.** The out-of-plane deflection curve of the web plate, in the case where the applied moment equals the resisting moment M_r, is plotted in Fig. 5.88. We observe from this figure that the maximum out-of-plane deflection δ_{max} occurs at the location nearly equal to $0.20h_w$ or $0.14h_w$ and $0.36h_w$ from the compression flange. Accordingly, the optimum locations $b = 0.20h_w$ or $b_1 = 0.14h_w$ and $b_2 = 0.32h_w$ in JSHB,[6] shown in Fig. 5.74, for one or two longitudinal stiffeners can also be applied to curved girder bridges.

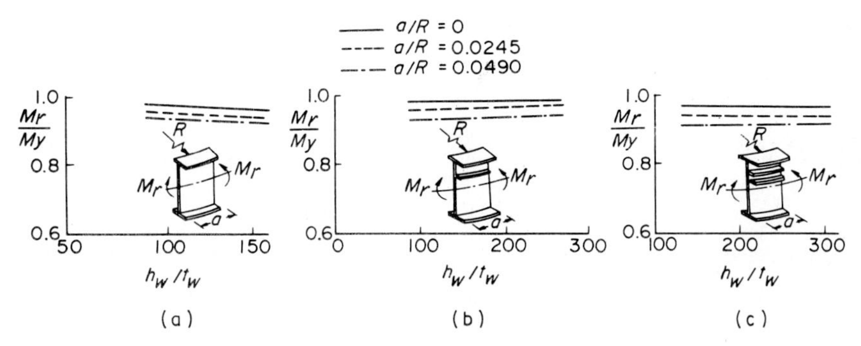

Figure 5.87 Variations of M_r/M_y due to h_w/t_w and a/R: (*a*) without longitudinal stiffeners, (*b*) with one longitudinal stiffener, (*c*) with two longitudinal stiffeners.

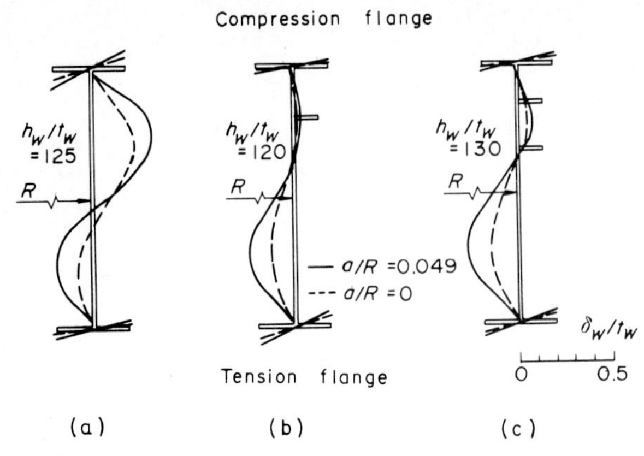

Figure 5.88 Out-of-plane deflection of web plate ($a/h_w = 0.7$, $w_o/h_w = 0.004$, $\sigma_{yw} = 235.2$ MPa): (*a*) without longitudinal stiffeners, (*b*) with one longitudinal stiffener, (*c*) with two longitudinal stiffeners.

Figure 5.89 illustrates the variations of δ_{max} due to the curvature parameter a/R and web slenderness h_w/t_w. In general, δ_{max} has a tendency to increase in accordance with the increase in h_w/t_w and a/R.

c. **Out-of-plane bending stress of web plates.** When the applied moment equals the resistance moment M_r, the patterns of out-of-plane bending stress of the web plate, σ_{tb}, in the direction of the depth are shown in Fig. 5.90. From the figure it seems that the distributions of σ_{tb} are analogous to the bending stress distribution near the intermediate support of a continuous girder at the longitudinal stiffeners, and the maximum bending stress occurs at the tension flange plates.

These maximum bending stresses σ_{tb} are nondimensionalized by the yield stress σ_{yw} and are plotted as variations with the curvature parameter a/R and the web slenderness h_w/t_w, as shown in Fig. 5.91. In all cases, σ_{tb}/σ_y increases in accordance with an increase in a/R and h_w/t_w.

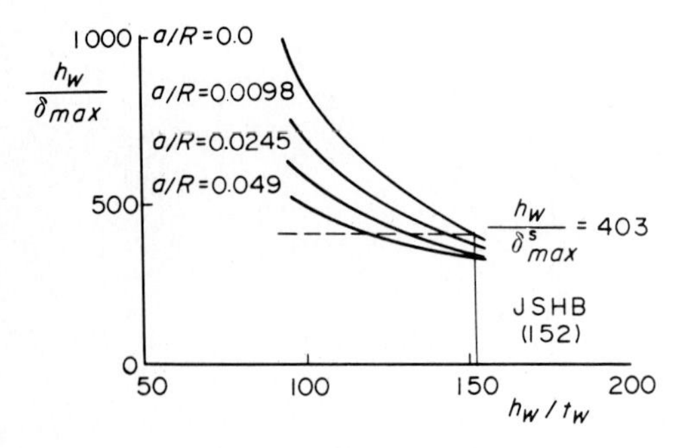

Figure 5.89 Curve of h_w/δ_{max} versus h_w/t_w for curvature parameter a/R (without longitudinal stiffener, $\sigma_{yw} = 235.2$ MPa).

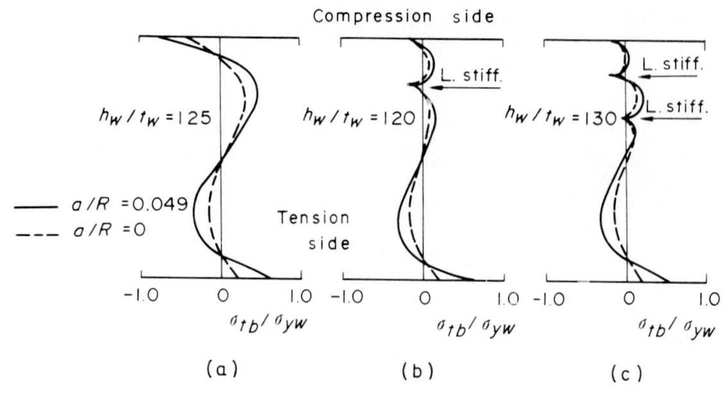

Figure 5.90 Transverse out-of-plane bending stress distribution ($a/h_w = 0.7$, $w_0/h_w = 0.004$, $\sigma_y = 235.2$ MPa): (*a*) without longitudinal stiffeners, (*b*) with one longitudinal stiffener, (*c*) with two longitudinal stiffeners.

5.4.3 Required web slenderness of curved girders[24,50,53]

(1) Criteria for limit state. The web slenderness requirements for the curved girder cannot be decided by a method similar to that of ordinary straight girders, since the out-of-plane deflections of curved web panels gradually increase in accordance with the load increments and a clear buckling phenomenon is not presented. Therefore, the requirements for the web slenderness of curved web panels are determined by devoting attention to the out-of-plane deflection and bending stress.

Given these criteria, the requirements for the web slenderness of straight girders specified in JSHB[6] with curvature parameter $a/R = 0$

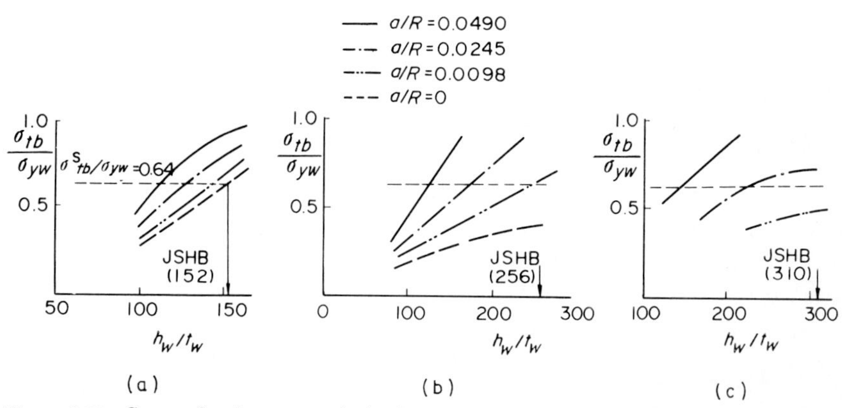

Figure 5.91 Curve of σ_{tb}/σ_{yw} versus h_w/t_w for curvature parameter a/R ($\sigma_{yw} = 235.2$ MPa): (*a*) without longitudinal stiffeners, (*b*) with one longitudinal stiffener, (*c*) with two longitudinal stiffeners.

are adopted. The out-of-plane deflection δ^s_{max} and stress σ^s_{max} in the web plate panel of straight girders are, of course, analyzed by regarding them as shells with a maximum initial deflection $\delta_o = b/250$. By comparing the out-of-plane deflections δ^c_{max} and δ^s_{max} and stresses σ^c_{max} and σ^s_{max}, the following criteria can be established as the limit state:

$$\delta^c_{max} \leqq \delta^s_{max} \tag{5.4.29a}$$

$$\sigma^c_{max} \leqq \sigma^s_{max} \tag{5.4.29b}$$

According to recent research,[51,52] the fatigue strength should be adopted as the limit state for the out-of-plane bending stress σ^s_{max}. This point should be investigated soon.

(2) Relationships between the required web slenderness and curvature parameter. In Fig. 5.92, the required web slenderness of straight girders specified in JSHB[6] is indicated by the solid line, and the dotted line indicates the same stress level as the straight girders. The required web slenderness h_w/t_w for the curved girders can be determined from Eq. (5.4.29b) as the intersecting points of this dotted line and the σ_{tb}–a/R curves for each case.

The required slenderness h_w/t_w for the curved girders can also be found by Eq. (5.4.29a) from a point of view of the limit state for the out-of-plane deflection curve $\delta - a/R$. These results, plotted in Fig. 5.92, indicate the relationships between the web slenderness h_w/t_w and the curvature parameter a/R.

5.4.4 Design recommendation[50]

The requirement for the web slenderness h_w/t_w for curved girders can be evaluated safely by the out-of-plane bending stress σ^c_{max} rather than

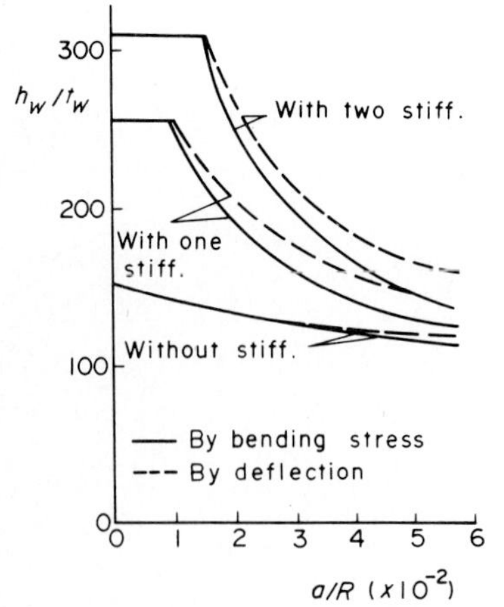

Figure 5.92 Curve of h_w/t_w versus a/R for curved web panel (SS 41).

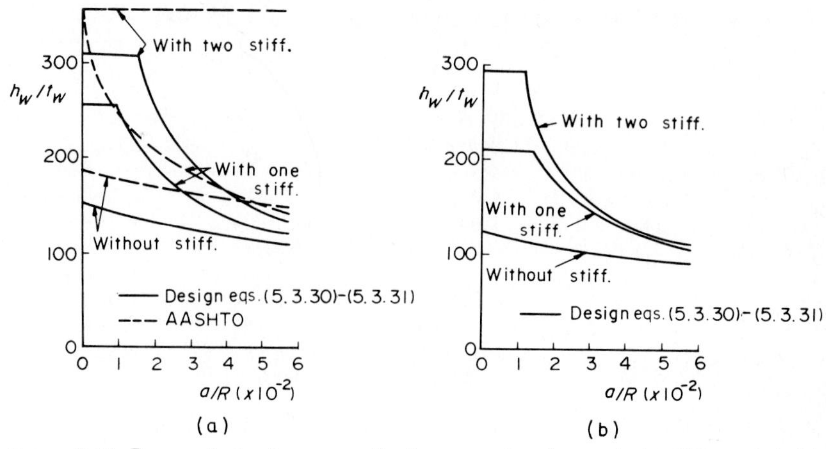

Figure 5.93 Proposed slenderness ratio for curved web panel: (*a*) SM 41 and (*b*) SM 50Y.

the deflection δ^c_{max}, as can be seen from Fig. 5.92. Therefore, the relationship between h_w/t_w and a/R can be plotted as in Fig. 5.93, where the specified values by AASHTO[54] are also plotted for the sake of comparison.

The arrangements of these curves through the least-squares method yields the following design recommendations:

In the case of nonlongitudinal stiffeners,

$$\frac{t_w}{h_w} \geqq \frac{1 + \alpha_o(a/R)}{\beta_o} \tag{5.4.30}$$

where β_o (see Table 5.7) and α_o are given in Table 5.10.

In the case of longitudinal stiffeners,

$$\frac{t_w}{h_w} \geqq \frac{1}{\beta_o} \qquad \frac{a}{R} \leqq \eta_o \tag{5.4.31a}$$

$$\frac{t_w}{h_w} \geqq \frac{1}{\beta_o} \frac{1}{\gamma_o - \delta_o a/R + \varepsilon_o(a/R)^2} \qquad \frac{a}{R} > \eta_o \tag{5.4.31b}$$

TABLE 5.10 Parameters β_o and α_o

	Steel material yield point	
Parameter	SS 41, SM 41 ($\sigma_{yw} = 2400$ kgf/cm^2)	SM 53, SM 50Y ($\sigma_{yw} = 3600$ kgf/cm^2)
β_o	152	123
α_o	6.623	4.587

TABLE 5.11 Parameters β_o, γ_o, δ_o, ε_o, and η_o

	Steel material yield point			
	SS 41, SM 41 (σ_{yw} = 2400 kgf/cm²)		SM 53, SM 50Y (σ_{yw} = 3600 kgf/cm²)	
Parameter	One longitudinal stiffener	Two longitudinal stiffeners	One longitudinal stiffener	Two longitudinal stiffeners
β_o	256	310	209	294
γ_o	1.232	1.643	1.748	1.510
δ_o	29.92	51.19	55.17	53.20
ε_o	303.7	556.8	631.0	625.6
η_o	0.090	0.015	0.014	0.011

where β_o is also given in Table 5.7 and the other parameters are listed in Table 5.11 corresponding to the appropriate grade of steel and the number of longitudinal stiffeners.

Note that the locations of the longitudinal stiffeners can be taken as the same as those in the straight plate girder bridges, as illustrated in Fig. 5.74.

5.5 Design of Longitudinal Stiffeners in Curved Girders

To limit the buckling of web plate panels locally up to the ultimate state of the plate girder, it is necessary to install longitudinal and transverse stiffeners in the web plate. For these plate girders, the web plate panel is considered to be simply supported along the flange plates and stiffeners, and they can be analyzed on the basis of the buckling theory of plates. In this case, the appropriate strength and rigidity should be provided with the stiffeners. This section and the next section illustrate the design of longitudinal and transverse stiffeners, respectively.

5.5.1 Design method for longitudinal stiffeners in straight girders

(1) Behavior of longitudinal stiffeners. Let us now consider a simple case of a rectangular plate with a longitudinal stiffener at the middle

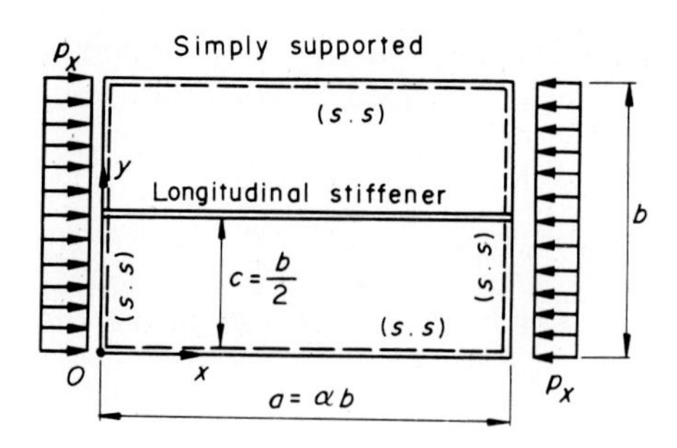

Figure 5.94 Buckling of rectangular plate with longitudinal stiffener.

and subjected to uniaxial compression p_x, as shown in Fig. 5.94, where

F = cross-sectional area of longitudinal stiffener

I = geometric moment of inertia of longitudinal stiffener with respect to web surface

$P = p_x F$ = applied axial force to longitudinal stiffener

The deflection surface $w(x, y)$ of this stiffened plate can be approximated in the same manner as given by Eq. (5.3.39):

$$w = \sum_m \sum_n A_{mn} \sin \frac{m\pi x}{a} \sin \frac{n\pi y}{b} \tag{5.5.1}$$

where A_{mn} is an unknown coefficient.

To determine this coefficient, the energy method can be employed. Since the strain energy stored in plate element U_P is given by Eq. (5.3.31), the total strain energy U can be obtained by adding the strain energy U_L stored in the longitudinal stiffener, i.e.,

$$U_L = \frac{EI}{2} \int_0^a \left[\frac{\partial^2 w}{\partial x^2} \right]_{y=c}^2 dx \tag{5.5.2}$$

Thus

$$U = U_P + U_L \tag{5.5.3}$$

However, the potential energy of plate element V_P is also given by Eq. (5.3.32), and the potential energy V_L of the longitudinal stiffener is

$$V_L = -\frac{P}{2} \int_0^a \left[\frac{\partial w}{\partial x} \right]_{y=c}^2 dx \tag{5.5.4}$$

This should be added to V_P; namely, the total potential energy V is

$$V = V_P + V_L \tag{5.5.5}$$

A condition which minimizes the energy $U + V$, explained in Eq.

(5.3.33), can be written for a case where the buckling mode is limited to the half-wave mode ($m = 1$) in the direction of the x axis as follows:

$$\sigma_e[(1+\alpha^2)^2 A_{11} + 2\gamma_l(A_{11} - A_{13} + A_{15} - \cdots)] \\ - \alpha^2\sigma_{cr}[A_{11} + 2\delta(A_{11} - A_{13} + A_{15} - \cdots)] = 0 \quad (5.5.6)$$

(the remaining terms A_{12}, A_{13}, A_{14}, . . . are omitted here) in which the parameters γ_l and δ are

$$\gamma_l = \frac{EI}{bD} \quad (5.5.7)$$

$$\delta = \frac{F}{bt} \quad (5.5.8)$$

Now, the eigenvalue σ_{cr} of the homogeneous equation involving coefficient A_{11} can be approximated from Eq. (5.5.6):

$$\sigma_{cr} = \frac{(1+\alpha^2)^2 + 2\gamma_l}{\alpha^2(1+2\delta)}\sigma_e = k_\sigma \sigma_e \quad (5.5.9)$$

In Fig. 5.95, the variations in the buckling coefficient k_σ due to the changes in γ_l and m are plotted. When there is no longitudinal stiffener ($\gamma_l = 0$), then $k_\sigma = 4$ for $\alpha > 1$ and $m = 1, 2, 3 \ldots$. For the case where the longitudinal stiffener is installed, the buckling coefficient k_σ is, however, increased in accordance with the increase in γ_l, provided that the aspect ratio $\alpha = a/b$ is kept constant. Accordingly, the value $k_\sigma = 16$ in Fig. 5.95 has the same effect where the width of the plate reduces to $b/2$. Therefore, if the longitudinal stiffener provides a parameter γ_l

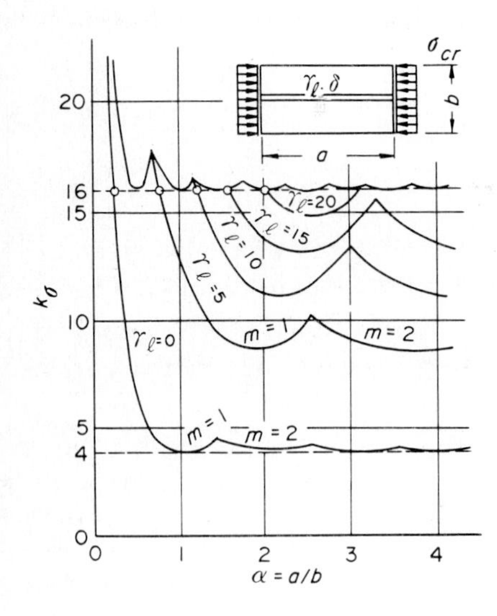

Figure 5.95 Relationships among k, α, and γ_l.

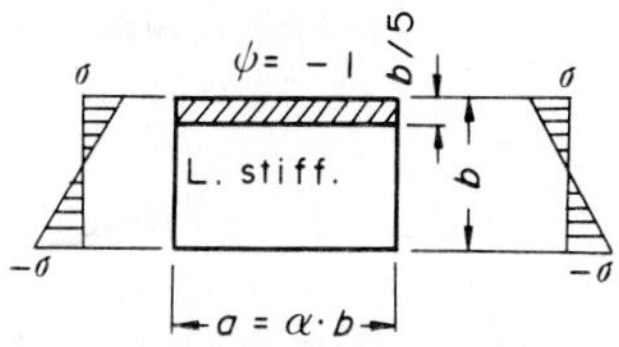

Figure 5.96 Web plate stiffened with a longitudinal stiffener at location $b/5$.

with a buckling coefficient $k_\sigma = 16$, then the edge $y = b/2$ along the longitudinal stiffener is thought to be the node of the buckling mode. Thus, γ_l is referred to as the *minimum required relative rigidity* γ_l^*.

(2) Required relative rigidity of longitudinal stiffeners. It is codified in DIN 4114[55] that γ_l^* depends on the aspect ratio $\alpha = a/b$ for the case shown in Fig. 5.96 as follows:

$$\gamma_l^* = \begin{cases} (21.3 + 112.6\delta)(\alpha - 0.1) & 0.5 \leqq \alpha \leqq 1.0 \qquad (5.5.10a) \\ (32.0 + 168.9\delta)(\alpha - 0.4) & \alpha > 1.0 \qquad (5.5.10b) \end{cases}$$

where $\quad \gamma_l^* \leqq 50 + 200\delta \qquad (5.5.11)$

Therefore, it is necessary to check the geometric moment of inertia I_l of longitudinal stiffeners in designing them, since γ_l is given by Eq. (5.5.7). Namely, the required geometric moment of inertia I_l^* is gotten by setting Poisson's ratio $\mu = 0.3$ as follows:

$$I_l^* = \frac{bD}{E}\gamma_l^* = \frac{b}{E}\frac{Et^3}{12(1-\mu^2)}\gamma_l^* \cong \frac{bt^3}{11}\gamma_l^* \qquad (5.5.12)$$

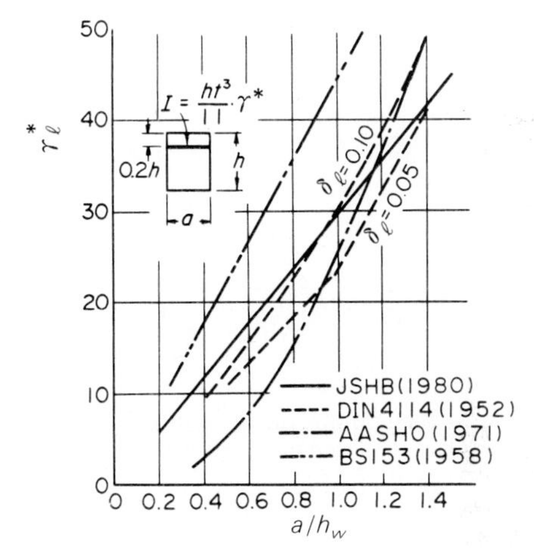

Figure 5.97 Variation of γ^* due to a/h in specifications.

In the design code of JSHB,[6] $\gamma_{l,\text{req}}$ is determined by the following convenient formula:

$$\gamma_{l,\text{req}} = 30.0\left(\frac{a}{h_w}\right) \tag{5.5.13}$$

where h_w is the web depth. The variations in γ_l^* due to a/h_w are plotted in Fig. 5.97 together with some specification such as DIN 4114,[55] AASHTO,[56] and BS 153.[57]

5.5.2 Ultimate strength of longitudinal stiffeners in curved girders[50]

Longitudinal stiffeners in curved girders undergo somewhat different forces from straight girders. Therefore, the longitudinal stiffeners should be designed to have enough strength that yielding of the flange plates occurs before the stiffeners fail.

(1) Model for analyzing the ultimate strength of longitudinal stiffeners. The ultimate strength of longitudinal stiffeners can be analyzed by modeling them as a curved beam-column model with a T cross section having an effective width

$$b_e = t_l + 0.181a \tag{5.5.14}$$

as shown in Fig. 5.98, according to Timoshenko and Goodier,[58] where

t_l = thickness of longitudinal stiffener

a = spacing of transverse stiffener

The corresponding applied forces to this model at yielding σ_y of the

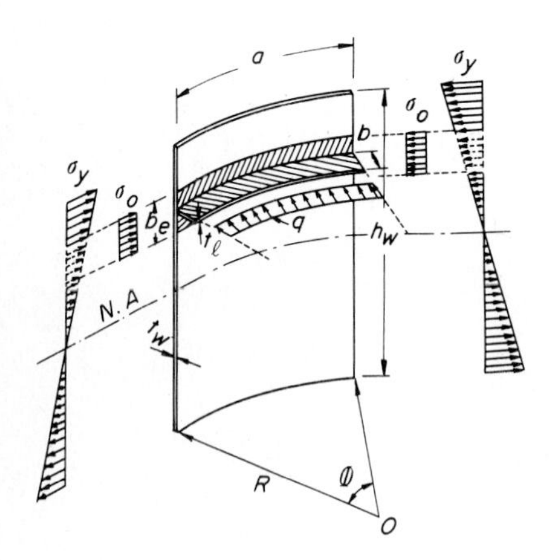

Figure 5.98 Model for analyzing the ultimate strength of a longitudinal stiffener.

flange plate are evaluated from the stress level of the web plate where the longitudinal stiffener is located. A curved beam-column model is subjected to an axial compressive force P and a uniformly distributed load q acting in the radial direction due to the radius of curvature R of the curved web plate panel

$$P = \psi \sigma_y A_{we} \tag{5.5.15a}$$

and

$$q = \frac{P}{R} \tag{5.5.15b}$$

where $A_{we} = b_e t_w$ (5.5.16)

ψ = reduction factor of σ_y (0.6 in the case of one longitudinal stiffener located at $0.2h_w$ from the compression flange)

t_w = thickness of web plate

Thus, the longitudinal stiffener can be transformed to a beam-column model with span a and curvature R, as shown in Fig. 5.99. When the boundary conditions of this model are assumed to be fixed at both ends, as is observed in experimental studies,[38] the end moment M_o is

$$M_o = qR^2\left(\frac{2}{\Phi}\tan\frac{\Phi}{2} - 1\right) \tag{5.5.17}$$

where Φ is the central angle of the curved web plate panel.

(2) Stress resultants in beam-column models. The stress resultants in a curved beam column can be easily analyzed by considering the effects of the additional deflection. Let us first assume that the initial deflection v_o is

$$v_o = g_o \sin\frac{\pi\phi}{\Phi} \tag{5.5.18}$$

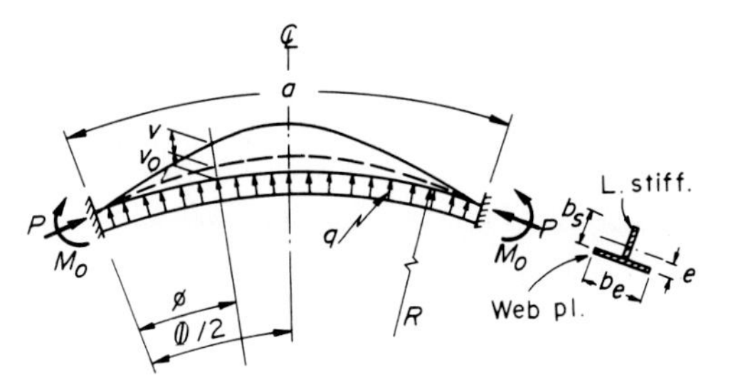

Figure 5.99 Beam-column model with curvature.

where the maximum value g_o and the induced deflection v can be approximated as

$$v = g\left(1 - \cos\frac{2\pi\phi}{\Phi}\right) \tag{5.5.19}$$

where $\bar{v}$ is the maximum value of deflection and ϕ is the angular coordinate of this model.

The following error function $\varepsilon(\phi)$ is obtained from the equilibrium condition of bending moment:

$$\varepsilon(\phi) = EI\frac{d^2v}{R^2\,d\phi^2} + qR^2\tan\frac{\Phi}{2}\sin\phi - qR^2(1-\cos\phi) + P(v + v_o + e) - M_o \tag{5.5.20}$$

where EI = flexural rigidity of beam-column model
e = eccentricity from neutral axis of beam-column model to midplane of web plate

By applying Galerkin's method, the most appropriate value for the coefficient g can be determined by

$$\int_0^{\Phi} \varepsilon(\phi)\left(1 - \cos\frac{2\pi\phi}{\Phi}\right)R\,d\phi = 0 \tag{5.5.21}$$

so that the following equation is obtained:

$$g = \frac{\dfrac{4qR^2\Phi}{4\pi^2 - \Phi^2}\tan\dfrac{\Phi}{2} + P\left(\dfrac{8g_o}{3\pi} + \dfrac{e}{\Phi}\right) - \dfrac{M_o}{\Phi}}{EI\left(\dfrac{2\pi}{R\Phi}\right)^2 - 3P} \tag{5.5.22}$$

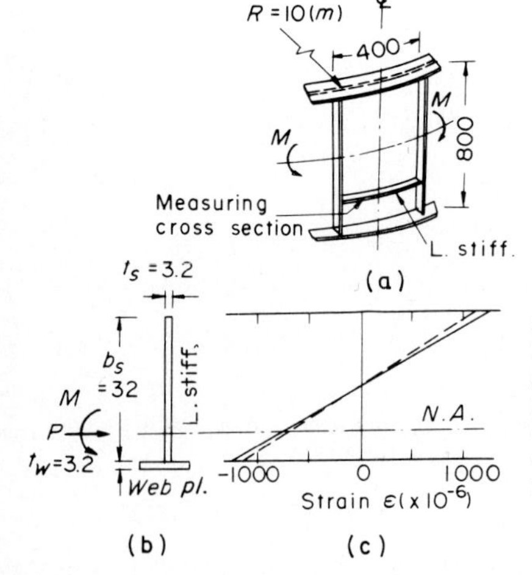

Figure 5.100 Comparison of strain distribution in longitudinal stiffeners by beam-column model (in mm):[38] (*a*) test specimen, (*b*) cross section of beam-column model, and (*c*) strain distribution (—— = analytical result; - - - = test result).

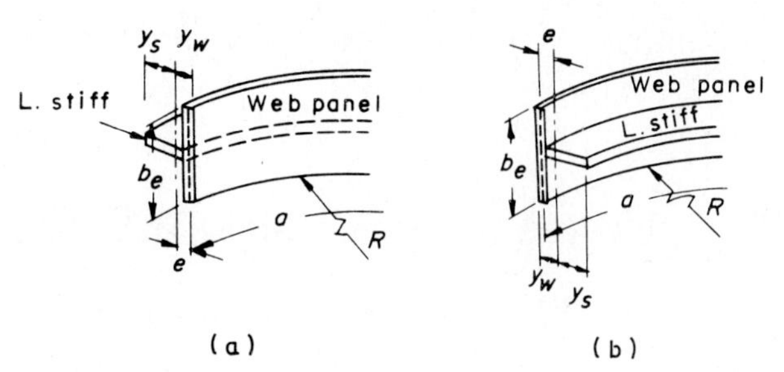

Figure 5.101 Locations of longitudinal stiffeners: (*a*) outside and (*b*) inside.

Therefore, the bending moment M at an arbitrary location on the beam-column model can readily be estimated as

$$M = -EI \frac{d^2 v}{R^2\, d\phi^2} \tag{5.5.23}$$

The validity of the above analysis was investigated through a comparison with an experimental study[38] on test curved girders, as illustrated in Fig. 5.100. The strain distribution, where the axial strain of the compression flange reached the yield point, coincided well with the analytical one. Accordingly, the beam-column method is useful in evaluating the ultimate strength of longitudinal stiffeners.

(3) Ultimate strength of beam-column models. In ordinary curved girders, the longitudinal stiffeners are located either inside or outside the surface of the web plate toward the center of curvature of the curved girders, so that their ultimate strength differs from each other as shown in Fig. 5.101. These ultimate strengths are analyzed on the basis of the Perry-Robertson method.[59] Then the initial yield line of the beam-column model subjected to a compressive force P and bending moment M, calculated from Eqs. (5.5.15a) and (5.5.23), is plotted as shown in Fig. 5.102, where their interaction curves can be categorized

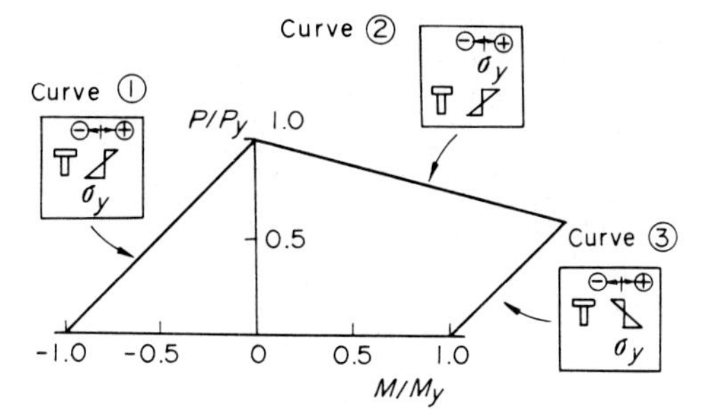

Figure 5.102 Initial yield line of beam-column model of longitudinal stiffener.

as follows:

Curve 1,
$$\frac{p}{p_y} - \frac{M}{M_y} \leqq 1 \qquad (5.5.24a)$$

Curve 2,
$$\frac{p}{p_y} + \left(\frac{x_w}{x_s}\right)\left(\frac{M}{M_y}\right) \leqq 1 \qquad (5.5.24b)$$

Curve 3,
$$-\frac{p}{p_y} + \frac{M}{M_y} \leqq 1 \qquad (5.5.24c)$$

where M_y = yield moment of beam-column model
P_y = squash load of beam-column model

5.5.3 Proposition for designing longitudinal stiffeners of curved girders

The above analytical method is so complicated to apply to the actual design of curved girders that the minimum relative required rigidity of longitudinal stiffeners γ_L^c can be obtained from the condition where the beam-column model does not yield by using the above interaction curve, which leads to

$$\gamma_L^c = \frac{EI}{hD} \qquad (5.5.25)$$

where I is the geometric moment of inertia of a longitudinal stiffener with respect to the surface of the web plate and

$$D = \frac{Et_w^3}{12(1-\mu^2)} \qquad (5.5.26)$$

$= $ flexural rigidity of web plate

t_w = thickness of web plate

E = Young's modulus of web plate

μ = Poisson's ratio of web plate

The relationships between the rigidity parameter β_L

$$\beta_L = \frac{\gamma_L^c}{\gamma_L^{\mathrm{JSHB}}} \qquad (5.5.27)$$

and the curvature parameter Z

$$Z = \frac{a^2}{Rt_w}\sqrt{1-\mu^2} \qquad (5.5.28)$$

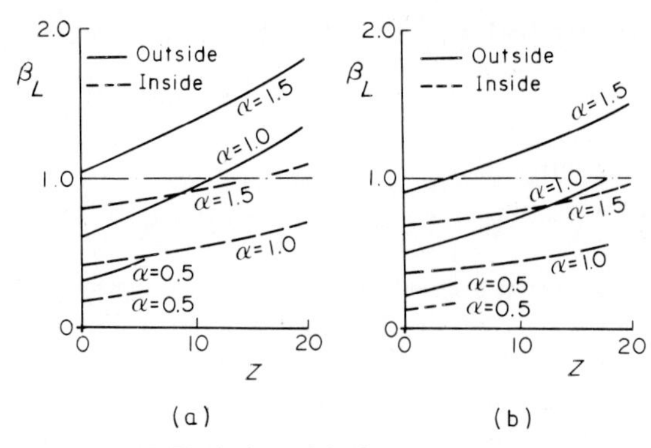

Figure 5.103 Variation of β_L due to curvature parameter Z: (*a*) SS 41 and (*b*) SM 50Y.

are shown in Fig. 5.103, where γ_L^{JSHB} is the specified minimum relative required rigidity of longitudinal stiffeners in the Japanese Specification for Highway Bridges[6] (JSHB), as indicated in Eq. (5.5.13). In this case, the relationship between the width b_l and thickness t_l of longitudinal stiffeners is, of course, determined from JSHB (see Table 5.14), and the stiffeners are attached to either the outer or the inner surfaces of the web plate located $0.2h_w$ from the compression flange of the curved girder.

We observe from this figure that greater rigidity should be required when the longitudinal stiffener is installed on the outer surface of the web plate toward the center of curvature than when it is installed on the inner surface. Some combinations of the aspect ratio α and curvature parameter Z also show $\beta_L < 1.0$. It is, however, thought that $\beta_L = 1.0$ for the sake of safety for design purposes.

By applying the least-squares method, the following design formula can be proposed for the rigidity parameter β_L:

$$\beta_L = (c_1 Z + c_2)Z + c_3\alpha - c_4 \geqq 1 \tag{5.5.29}$$

Coefficients c_1 through c_4 are listed in Table 5.12, and they vary with the location and the grade of steel of the longitudinal stiffeners.

TABLE 5.12 Coefficients c_1 to c_4 in Eq. (5.5.29)

Steel grade	Location of stiffener	c_1, $\times 10^{-4}$	c_2, $\times 10^{-2}$	c_3	c_4
SS 41	Outside	3.766	3.226	0.739	0.108
	Inside	2.838	0.163	0.775	0.163
SM 50Y	Outside	5.362	1.549	0.818	0.227
	Inside	3.286	−0.464	0.800	0.200

5.6 Design of Transverse Stiffeners in Curved Girders

5.6.1 Design method for transverse stiffeners in straight girders

(1) Buckling of a web plate under the combined action of normal and shearing stress. In the general case where the web plate is subjected to flexural compressive normal stress σ^M and the shearing stress τ_{xy}, the interaction curve can be approximated by the following circular formula for the combined actions of σ^M and τ_{xy}, as shown in Fig. 5.104.[32]

$$\left(\frac{\sigma^M}{\sigma_{cr}}\right)^2 + \left(\frac{\tau_{xy}}{\tau_{cr}}\right)^2 \leqq 1 \tag{5.6.1}$$

Here σ_{cr} is the buckling stress due to the bending moment, and τ_{cr} is the buckling stress due to the shearing force.

Taking an appropriate factor of safety v against the buckling stress yields

$$\sigma^M = v\sigma_w \tag{5.6.2a}$$

$$\tau_{xy} = v\tau_w \tag{5.6.2b}$$

where

$$\sigma_w = \frac{M}{W_w} \tag{5.6.3}$$

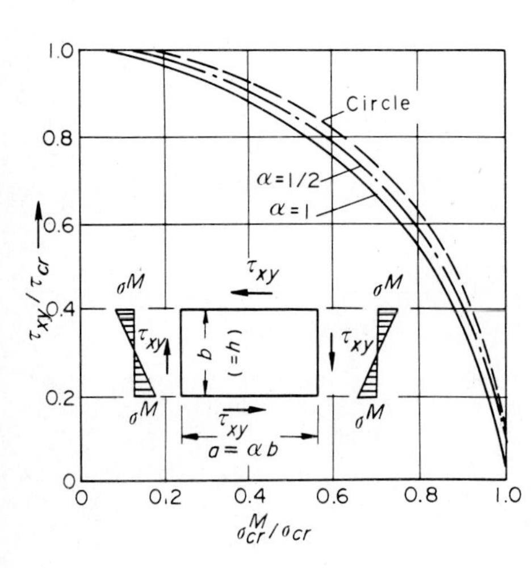

Figure 5.104 Interaction curve of τ_{xy}/τ_{cr} versus $\sigma^M\sigma_{cr}$.

with σ_w being the compressive stress of the web plate (in kgf/cm^2) and

$$\tau_w = \frac{Q}{A_w} \tag{5.6.4}$$

where
τ_w = shearing stress of web plate (kgf/cm^2)
M = applied bending moment
Q = applied shearing force
W_w = section modulus of web plate
$A_w = h_w t_w$ = cross-sectional area of web plate

Moreover, the buckling stresses σ_{cr} (kgf/cm^2) and τ_{cr} (kgf/cm^2) are

$$\sigma_{cr} = k_\sigma \frac{E\pi^2}{12(1-\mu^2)} \left(\frac{t_w}{h_w}\right)^2 \tag{5.6.5a}$$

$$\tau_{cr} = k_\tau \frac{E\pi^2}{12(1-\mu^2)} \left(\frac{t_w}{h_w}\right)^2 \tag{5.6.5b}$$

according to Eqs. (5.3.23), (5.3.25), and (5.3.43) by setting $b = h_w$.
The substitution of Eqs. (5.6.2) through (5.6.5) into Eq. (5.6.1) and setting of $E = 2.1 \times 10^6$ kgf/cm^2 and $\mu = 0.3$ lead to

$$\left(\frac{h_w}{100t_w}\right)^4 \left[\left(\frac{\sigma_w}{190k_\sigma}\right)^2 + \left(\frac{\tau_w}{190k_\tau}\right)^2\right] \leqq \frac{1}{\nu^2} \tag{5.6.6}$$

in which the factor of safety ν is

$$\nu = 1.25 \tag{5.6.7}$$

for a condition where

$$\eta = \frac{\tau_w}{\sigma_w} \leqq 0.8 \tag{5.6.8}$$

as provided by JSHB.[6]

(2) Spacing of transverse stiffeners. The spacing of the transverse stiffeners a (cm) can be checked by Eqs. (5.6.6) and (5.6.7) when the buckling coefficients k_σ and k_τ are known.

***a.* Without longitudinal stiffeners.** From Eqs. (5.3.37a) and (5.3.44), $k_\sigma = 23.9$, $k_\tau = 5.34 + 4.0/\alpha^2$ for $\alpha \geqq 1$, and $k_\tau = 4.0 + 5.34/\alpha^2$ for $\alpha < 1$.

Thus the following criteria are obtained from Eqs. (5.6.6) and (5.6.7):

$$\left(\frac{h_w}{100t_w}\right)^4 \left\{\left(\frac{\sigma_c}{3650}\right)^2 + \left[\frac{\tau_w}{810 + 610(h_w/a)^2}\right]^2\right\} \leqq 1 \qquad \frac{a}{h_w} > 1 \qquad (5.6.9a)$$

$$\left(\frac{h_w}{100t_w}\right)^4 \left\{\left(\frac{\sigma_c}{3650}\right)^2 + \left[\frac{\tau_w}{610 + 810(h_w/a)^2}\right]^2\right\} \leqq 1 \qquad \frac{a}{h_w} \leqq 1 \qquad (5.6.9b)$$

where σ_c is the maximum compressive stress in the web plate.

***b*. With one longitudinal stiffener.** The buckling coefficient for the panel between the tension flange and a longitudinal stiffener is approximated by

$$k_\sigma = 23.9 \qquad \text{safety assumption} \qquad (5.6.10a)$$

$$k_\tau = 8.34 + 4.00\left(\frac{a}{h_w}\right)^2 \qquad \frac{a}{h_w} > 0.8 \qquad (5.6.10b)$$

$$= 6.25 + 5.34\left(\frac{a}{h_w}\right)^2 \qquad \frac{a}{h_w} \leqq 0.8 \qquad (5.6.10c)$$

Then setting the compressive normal stress in this plate to $\sigma_w = 0.6\sigma_c$, we have

$$\left(\frac{h_w}{100t_w}\right)^4 \left\{\left(\frac{\sigma_c}{9500}\right)^2 + \left[\frac{\tau_w}{1270 + 610(h_w/a)^2}\right]^2\right\} \leqq 1 \qquad \frac{a}{h_w} > 0.80 \qquad (5.6.11a)$$

$$\left(\frac{h_w}{100t_w}\right)^4 \left\{\left(\frac{\sigma_c}{9500}\right)^2 + \left[\frac{\tau_w}{950 + 810(h_w/a)^2}\right]^2\right\} \leqq 1 \qquad \frac{a}{h_w} \leqq 0.80 \qquad (5.6.11b)$$

***c*. With two longitudinal stiffeners.** Similarly, setting $\sigma_w = 0.28\sigma_c$ for the panel between the tension flange and a lower longitudinal stiffener, we get

$$k_\sigma = 23.9 \qquad \text{safety assumption} \qquad (5.6.12a)$$

$$k_\tau = 13.0 + 4.0\left(\frac{a}{h_w}\right)^2 \qquad \frac{a}{h_w} > 0.64 \qquad (5.6.12b)$$

$$= 9.75 + 5.34\left(\frac{a}{h_w}\right)^2 \qquad \frac{a}{h_w} \leqq 0.64 \qquad (5.6.12c)$$

and the following formula results:

$$\left(\frac{h_w}{100t_w}\right)^4 \left\{\left(\frac{\sigma_c}{31{,}500}\right)^2 + \left[\frac{\tau_w}{1970 + 610(h_w/a)^2}\right]^2\right\} \leqq 1 \qquad \frac{a}{h_w} > 0.64 \qquad (5.6.13a)$$

$$\left(\frac{h_w}{100t_w}\right)^4 \left\{\left(\frac{\sigma_c}{31{,}500}\right)^2 + \left[\frac{\tau_w}{1480 + 810(h_w/a)^2}\right]^2\right\} \leqq 1 \qquad \frac{a}{h_w} \leqq 0.64 \qquad (5.6.13b)$$

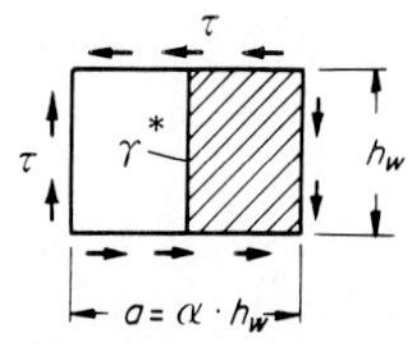

Figure 5.105 Plate with a transverse stiffener at middle.

However, note that

$$\frac{a}{h_w} \leqq 1.5 \tag{5.6.14}$$

in all cases.

(3) Required relative rigidity of transverse stiffeners. The transverse stiffener is an important member in ensuring sufficient strength against shearing forces. The overall shear buckling of the web plate must be limited within the web plate panel to local buckling having the nodal line at the transverse stiffener.

To enforce such a condition, the transverse stiffener must be designed for sufficient rigidity. For example, the minimum required rigidity γ_t^* of the transverse stiffener in a plate subjected to shearing stress τ, shown in Fig. 5.105, is

$$\gamma_t^* = \frac{5.4}{\alpha}\left(\frac{2}{\alpha} + \frac{2.5}{\alpha^2} - \frac{1}{\alpha^3} - 1\right) \qquad 0.5 \leqq \alpha \leqq 2.0 \tag{5.6.15}$$

according to DIN 4114.[55]

The flexural rigidity of the transverse stiffener, I_t^*, can also be determined from Eq. (5.5.12) by using the above γ_t^*. However, the factor of safety $\nu = 1.25$ in Eq. (5.6.7) is used to find the spacing of the transverse stiffener against the shear buckling of the web plate panel and underestimates the rigidity. Accordingly, the required relative rigidity of transverse stiffeners is provided by JSHB[6] as follows:

$$\gamma_{t,\text{req}} = 8.0\left(\frac{h_w}{a}\right)^2 \tag{5.6.16}$$

according to the study of Rockey, Valtinat, and Tang.[60] Figure 5.106 shows the variations of $\gamma_{t,\text{req}}^*$ with a/h_w for different specifications.

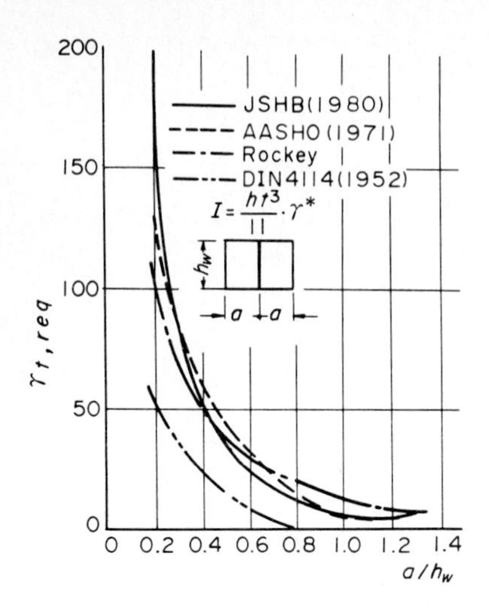

Figure 5.106 Variation of $\gamma_{t,\text{req}}$ with a/h_w for transverse stiffeners.

5.6.2 Ultimate strength of transverse stiffeners in curved girders[61]

(1) Analysis of the shear buckling strength of curved web plates with many transverse stiffeners. The elastic shear buckling strength of a curved web plate with a comparatively large radius of curvature, as is true in ordinary curved girder bridges, is analyzed on the basis of the theory of shallow shells. For this analysis, the fundamental equations for stress and strain are approximated by the well-known formula given by Donnell,[62] and the shear buckling strength is computed by means of the Rayleigh-Ritz method as an eigenvalue problem.

Figure 5.107 shows an analytical model under consideration,[63] where the web plate is stiffened by transverse stiffeners with constant spacing a and the basic dimensions of the web plate and transverse stiffeners are specified. The flange plates are neglected in this model, and the web plate is assumed to be simply supported along all the edges of the web plate. The following notation is used:

R = radius of curvature at midplane of web plate

b = depth of web plate

t_w = thickness of web plate

d = height of transverse stiffener

b_s = width of transverse stiffener

t_s = thickness of transverse stiffener

s = total longitudinal length of web plate

a = spacing of transverse stiffeners

n = total number of stiffeners

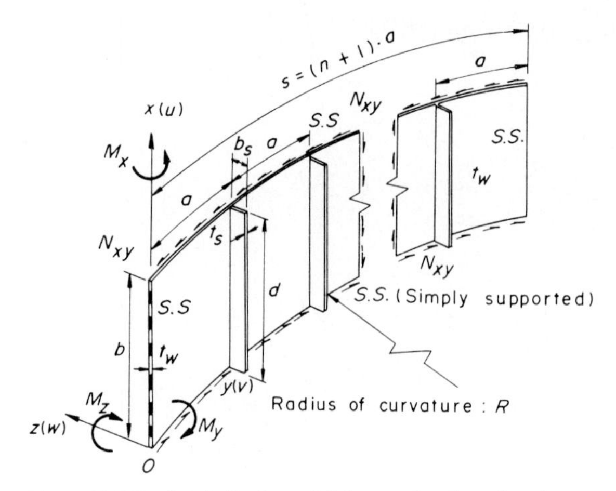

Figure 5.107 Analytical model for evaluating the shear buckling strength of a curved web plate with transverse stiffeners.

a. **Stress-strain relationships.** According to Donnell,[62] the relationships between stress and strain in a shallow shell can be approximated as follows:

$$\sigma_x = \frac{E}{1-\mu^2}(\varepsilon_x + \mu\varepsilon_y) \tag{5.6.17a}$$

$$\sigma_y = \frac{E}{1-\mu^2}(\varepsilon_y + \mu\varepsilon_x) \tag{5.6.17b}$$

$$\tau_{xy} = G\gamma_{xy} \tag{5.6.17c}$$

where σ_x, σ_y = normal stresses in direction of depth x and girder axis of web plate y, respectively

τ_{xy} = shearing stress

ε_x, ε_y = normal strain in direction of depth x and girder axis of web plate y, respectively

γ_{xy} = shear strain

E = Young's modulus

G = shear modulus of elasticity

μ = Poisson's ratio

b. **Strain-displacement relationships.** The relationships among displacements u, v, and w in the directions of the x, y, and z coordinate axes, shown in Fig. 5.107, and the above strains at arbitrary points of the web plate can also be approximated by

$$\varepsilon_x = \frac{\partial u}{\partial x} - z\frac{\partial^2 w}{\partial x^2} \tag{5.6.18a}$$

$$\varepsilon_y = \frac{\partial v}{\partial y} + \frac{w}{R} - z\frac{\partial^2 w}{\partial y^2} \tag{5.6.18b}$$

$$\gamma_{xy} = \frac{\partial u}{\partial y} + \frac{\partial v}{\partial x} - 2z\frac{\partial^2 w}{\partial x\,\partial y} \tag{5.6.18c}$$

For the transverse stiffeners with flexural rigidity alone, the corresponding normal stress σ_s reduces to

$$\sigma_s = E\varepsilon_s \tag{5.6.19}$$

where

$$\varepsilon_s = -z\frac{\partial^2 w}{\partial x^2} \tag{5.6.20}$$

By combining these equations for a thin cylindrical and shallow shell, as in the analysis of horizontally curved girders, the displacements u and v can be represented by the out-of-plane deflection w of the web plate:

$$-R\nabla^4 u = \mu\frac{\partial^3 w}{\partial x^3} - \frac{\partial^3 w}{\partial x\,\partial y^2} \tag{5.6.21a}$$

$$-R\nabla^4 v = (2+\mu)\frac{\partial^3 y}{\partial x^2\,\partial y} + \frac{\partial^3 w}{\partial y^3} \tag{5.6.21b}$$

where

$$\nabla^4 = \left(\frac{\partial^4}{\partial x^4} + 2\frac{\partial^4}{\partial x^2\,\partial y^2} + \frac{\partial^4}{\partial y^4}\right) \tag{5.6.22}$$

c. Buckling analysis by the energy method. It is only necessary to assume the out-of-plane deflection w of the web plate, which can be set so as to satisfy the boundary conditions of simple supports along all the edges of the web plate, as

$$w - \sum_{i=1}^{m}\sum_{j=1}^{n} A_{ij}\sin\frac{i\pi x}{b}\sin\frac{j\pi y}{s} \tag{5.6.23}$$

where A_{ij} = undetermined coefficient to determine out-of-plane deflection for each wave mode $i = 1, 2, \ldots$ and $j = 1, 2, \ldots$

i, j = number of wave modes for x and y axes, respectively

The total potential energy Π is then found by summing the strain energy U and potential energy V done by the external force N_{xy}:

$$\Pi = U + V \tag{5.6.24}$$

where

$$U = \tfrac{1}{2}\iiint_{\text{(web plate)}} (\sigma_x\varepsilon_x + \sigma_y\varepsilon_y + \tau_{xy}\gamma_{xy})\, dx\, dy\, dz + \sum_{k=1}^{n} \iint_{\text{(stiffener)}} [\sigma_s\varepsilon_s]_{y=ka}\, dx\, dz \tag{5.6.25a}$$

$$V = -\frac{1}{2}\iint_{\text{(web plate)}} N_{xy}\frac{\partial w}{\partial x}\frac{\partial w}{\partial y}\, dx\, dy \tag{5.6.25b}$$

The substitution of Eqs. (5.6.17) through (5.6.23) into Eq. (5.6.25) gives the total potential energy Π, represented by the terms of a_{ij} in Eq. (5.6.23). Accordingly, these unknown coefficients a_{ij} can be determined through the Rayleigh-Ritz method:

$$\delta\Pi = \frac{\partial \Pi}{\partial A_{ij}}\,\delta A_{ij} = 0 \tag{5.6.26}$$

which reduces to the following eigenvalue equation:

$$|\mathbf{K} - \kappa\mathbf{L}| = 0 \tag{5.6.27}$$

where $\mathbf{K}$ = square matrix containing web and stiffener rigidity terms
$\mathbf{L}$ = square matrix containing shearing force terms
κ = eigenvalues

From this equation, the minimum eigenvalue $\kappa_{\min}$ is found, and the corresponding buckling forces N_{cr} can be estimated as

$$N_{cr} = \kappa_{\min} N_{xy} \tag{5.6.28a}$$

$$\tau_{cr} = \frac{N_{cr}}{t_w} \tag{5.6.28b}$$

***d.* Analytical results.** The buckling coefficients k_τ, shown in Eq. (5.3.69), have been determined according to the above equation for a single curved web plate panel by taking $m = 10$ and $n = 18$, and their variations due to κ in Eq. (5.3.68) can be plotted. These results also coincide well with Batdorf's results, as shown in Fig. 5.59.

Next, to clarify the effects of the rigidity

$$\gamma_t = \frac{EI}{Da} \tag{5.6.29}$$

of transverse stiffeners on the buckling coefficient k_τ, we do additional calculations for the web plate model, which consists of two web panels and an intermediate stiffener for each web panel, where the central angle is equal to $\Phi = a/R = 0.0304$ and $\alpha = a/b = 1.0$. The variations of the buckling coefficient k_τ due to the rigidity of the stiffener γ_t are plotted in Fig. 5.108.

We see from this figure that the buckling coefficients k_τ in the region where $\gamma_t > 8$ converge into those of a single web panel. Furthermore, the out-of-plane deflections w in the buckled situation for $\gamma_t = 11.9$ are effectively restrained at the location of the intermediate stiffener. Therefore, we conclude that the approximate method mentioned in this section can be applied for estimating the buckling strength of curved web plates with and without transverse stiffeners.

(2) Ultimate strength of transverse stiffeners

***a.* Beam-column approach.** In the ultimate limit state of curve girders subjected to shear, the web plates have already buckled and the diagonal tension field is propagated in the web plates, as shown in Fig. 5.64. By assuming that the curved girder behaves as a straight one, the intensity of the tensile membrane stress σ_t can be approximated,[40] as shown in Eq. (5.3.71)

$$\sigma_t = -\tfrac{3}{2}\tau_{cr} \sin 2\theta + \sqrt{\sigma_{yw}^2 + \tau_{cr}^2(\tfrac{9}{4}\sin^2 2\theta - 3)} \tag{5.6.30}$$

where τ_{cr} = shear buckling stress of web plate given by Eq. (5.6.28*b*)
σ_{yw} = yield stress of web plate
θ = inclination angle of diagonal tension field to flange plate

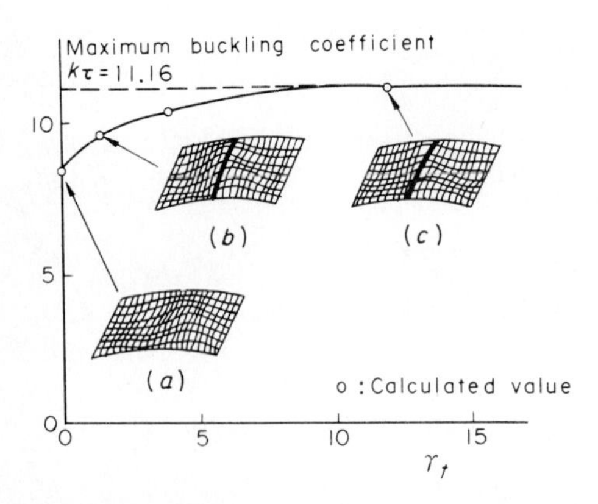

Figure 5.108 Relationships between buckling coefficient k_τ and rigidity of transverse stiffener γ. Buckling mode: (*a*) $\gamma_t = 0$, (*b*) $\gamma_t = 1.47$, (*c*) $\gamma_t = 11.9$.

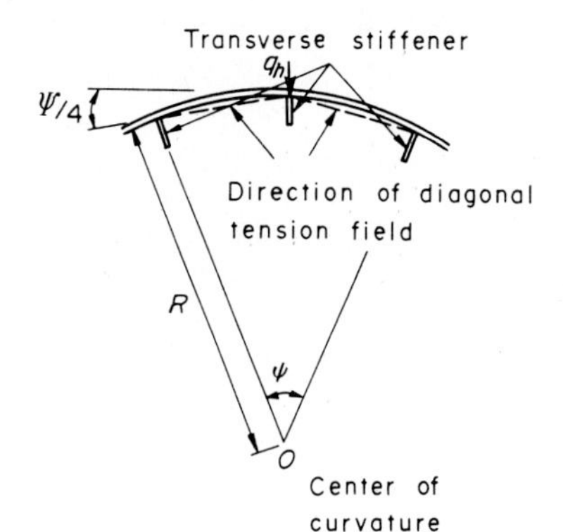

Figure 5.109 Diagonal tension field in curved web plate.

The components of the resulting forces due to σ_t act on the transverse stiffeners at the ultimate state. Note especially that the radial forces occur q_h in the transverse stiffener of the curved girder, as illustrated in Fig. 5.109. Thus, the induced forces in the transverse stiffeners can be summarized as in Fig. 5.110, which results in the following.

1. Axially concentrated force F_v:

$$F_v = \sigma_t t_w c \sin^2 \theta \tag{5.6.31}$$

2. Axially distributed forces q_v:

$$q_v = \sigma_t t_w \cos \theta \sin \theta \tag{5.6.32}$$

3. Radially distributed forces q_h:

$$q_h = \sigma_t t_w \cos^2 \theta \sin \frac{\Psi}{4} \tag{5.6.33}$$

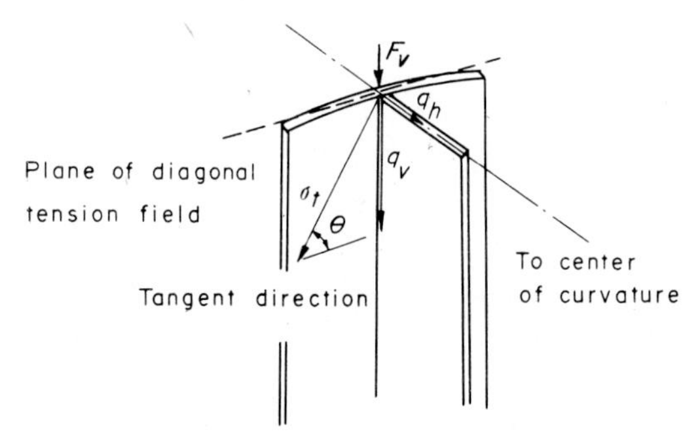

Figure 5.110 Forces acting on transverse stiffener.

$$\text{where } c = \frac{\sin \theta}{2} \sqrt{\frac{M_{pf}}{\sigma_t t_w}} \tag{5.6.34}$$

= distance from edge of web plate to location of plastic hinge in flange plate (see Fig. 5.64*b*)

$$M_{pf} = \tfrac{1}{4} A_f \sigma_{yf} t_f \tag{5.6.35}$$

= fully plastic moment of flange plate

A_f = cross-sectional area of flange plate

t_f = thickness of flange plate

σ_{yf} = yield stress of flange plate

Ψ = central angle of web plate panels between three adjacent transverse stiffeners

The transverse stiffeners can be idealized into a beam-column model with a T cross section, as illustrated in Fig. 5.111, by taking into account the effective width ηt_w of the web plate where $\eta = 40$, according to the study of Rockey, Valtinat, and Tang.[60]

***b.* Analyses of deflections and stress resultants in a beam column.** The deflections and stress resultants at any cross section of the beam column can be easily analyzed by considering the effects of additional deflection. For this analysis, the initial deflection w_o and additional deflection w can be assumed to be sinusoidal functions:

$$w_o = f_0 \sin \frac{\pi z}{d} \tag{5.6.36a}$$

$$w = f \sin \frac{\pi z}{d} \tag{5.6.36b}$$

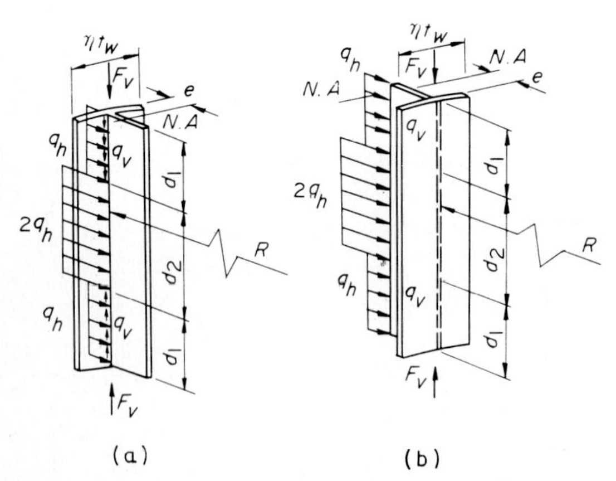

Figure 5.111 Beam-column model: (*a*) stiffener attached at inner side of curvature and (*b*) stiffener attached at outer side of curvature.

where f_o, f = maximum initial and additional deflections, respectively
d = length of beam column
z = distance from end of beam column to an arbitrary point along length of transverse stiffener

The governing differential equation with respect to the additional deflection w for this model is given by

$$\varepsilon(z) = EI \frac{d^2 w}{dz^2} + M_{Fv} + M_{qv} + M_{qh} \equiv 0 \qquad (5.6.37)$$

where
$$M_{Fv} = F_v \delta \qquad 0 \leqq z \leqq d \qquad (5.6.38a)$$

$$\begin{aligned} M_{qv} &= q_v z \delta & 0 &\leqq z \leqq d_1 \\ &= q_v d_1 \delta & d_1 &< z \leqq d_1 + d_2 \\ &= q_v (d - z) \delta & d_1 + d_2 &< z \leqq d \end{aligned} \qquad (5.6.38b)$$

$$\begin{aligned} M_{qh} &= \frac{qh}{2} [(d + d_2)z - z^2] & 0 &\leqq z \leqq d_1 \\ &= \frac{qh}{2} (-d_1^2 + 2dz - 2z^2) & d_1 &< z \leqq d_1 + d_2 \\ &= \frac{qh}{2} [d_2 d + (d - d_2)z - z^2] & d_1 + d_2 &< z \leqq d \end{aligned} \qquad (5.6.38c)$$

$$\delta = e + w_1 + w_o \qquad (5.6.39)$$

$$d_1 = d - (a - c) \tan \theta \qquad (5.6.40a)$$

$$d_2 = d - 2d_1 \qquad (5.6.40b)$$

and e = distance from midplane of web plate to neutral axis of beam column
a = spacing between transverse stiffeners

To determine the most suitable maximum deflection f in Eq. (5.6.36*b*), the following condition, which makes the potential energy a minimum, is a result of applying Galerkin's method:

$$\int_0^d \varepsilon(z) \sin \frac{\pi z}{d} \, dz = 0 \qquad (5.6.41)$$

where $\varepsilon(z)$ is the error function obtained from Eqs. (5.6.36) and (5.6.37). Substitution of Eq. (5.6.37) into Eq. (5.6.41) reveals the coefficient of additional deflection f. Then the stress resultant such as the axial force P and the bending moment M at any cross section of the beam column can be evaluated as

$$P = F_v + q_v z \qquad 0 \leqq z \leqq d_1 \tag{5.6.42a}$$

$$= F_v + q_v d_1 \qquad d_1 < z \leqq d_1 + d_2 \tag{5.6.42b}$$

$$= F_v - q_v(z - d_1 - d_2) \qquad d_1 + d_2 < z \leqq d \tag{5.6.42c}$$

$$M = P\delta + Mq_h \tag{5.6.43}$$

To examine the applicability of the above analysis, the strain distributions along the length of the transverse stiffener are calculated and plotted as in Fig. 5.112 together with the test results. The strain distributions of the test results clearly show that the radial force toward the outside direction against the center of curvature predominates, and the transverse stiffener behaves as a beam column, as assumed in this analysis. These analytical results also coincide well with the experimental ones, so that the transverse stiffeners can be treated by this beam-column approach.

c. Interaction curve for beam-column models. The interaction curve between M/M_p and P/P_s in the initial yield and fully plastic states for the beam column with T cross section and the yield stress is summarized in Table 5.13 corresponding to the stress situations, where

$$P_s = \sigma_y A_s \tag{5.6.44a}$$

$$M_p = \sigma_y(A_s y_1 + A_{s2} Y_2) \tag{5.6.44b}$$

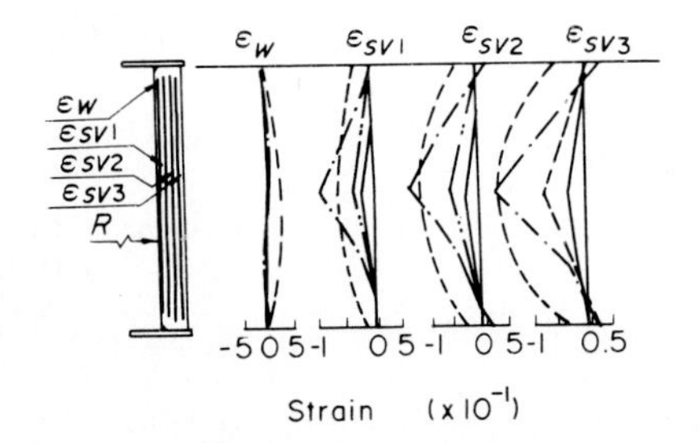

Figure 5.112 Strain distribution in transverse stiffener.

TABLE 5.13 Calculation Formulas for Interaction Curves

Curve	Equation number in Fig. 5.113	Stress situations	Calculation formulas
Initial yield line	1		$\frac{M}{M_p} = \frac{\sigma_y}{M_p}\frac{1}{y_s}\frac{P}{P_s} + \frac{\sigma_y}{M_p}\frac{1}{y_s}$
	2		$\frac{M}{M_p} = -\frac{\sigma_y}{M_p}\frac{1}{y_w}\frac{P}{P_s} + \frac{\sigma_y}{M_p}\frac{1}{y_s}$
	3		$\frac{M}{M_p} = \frac{\sigma_y}{M_p}\frac{1}{y_s}\frac{P}{P_s} - \frac{\sigma_y}{M_p}\frac{1}{y_s}$
Fully plastic curve	4		$\frac{M}{M_p} = \left\{b_w(y_1+\eta_1)[q_1 - \tfrac{1}{2}(y_1+\eta_1)] - b_w(t_w - y_1 - \eta_1)[q_1 - \tfrac{1}{2}(t_w + y_1 + \eta_1)] - \frac{t_s}{2}(b_s - 2b_s q_2)\right\}\frac{\sigma_y}{M_p}$
	5		$\frac{M}{M_p} = \left\{b_w t_w\left(q_1 - \frac{t_w}{2}\right) + t_s(y_1 + y_2 - t_w) \times [q_1 - \tfrac{1}{2}(y_1 + \eta_2 + t_w)] - \frac{t_s}{2}[(q_1 - y_1 - \eta_2)^2 - q_2^2]\right\}\frac{\sigma_y}{M_p}$
	6		$\frac{M}{M_p} = \left\{b_w(y_1 - \eta_1)[q_1 - \tfrac{1}{2}(y_1 - \eta_1)] - b_w[t_w - (y_1 - \eta_1)][q_1 - \tfrac{1}{2}(t_w + q_1 - \eta_1)] + \frac{t_s}{2}[q_2^2 - (q_1 - t_w)^2]\right\}\frac{\sigma_y}{M_p}$

Remarks: $y_1 = \frac{1}{2b_w}(b_w t_w + b_s t_s)$

$y_2 = b_s + t_w - y_1$

$q_1 = b_s + t_w - \frac{b_w t_w^2 + 2b_w b_s t_s + b_s^2 t_s}{b_w t_w + b_s t_s}$

$q_2 = \frac{b_w t_w^2 + 2b_w b_s t_s + b_s^2 t_s}{b_w t_w + b t_s}$

$\eta_1 = \frac{P_s}{2b_w \sigma_y}\frac{P}{P_s}$

$\eta_2 = y_2 - \left(1 - \frac{P}{P_s}\right)\frac{P_s}{2t_s \sigma_y}$

$x_w x_s$ = maximum distances from neutral axis to web and stiffener, respectively

I = geometric moment inertia of T cross section.

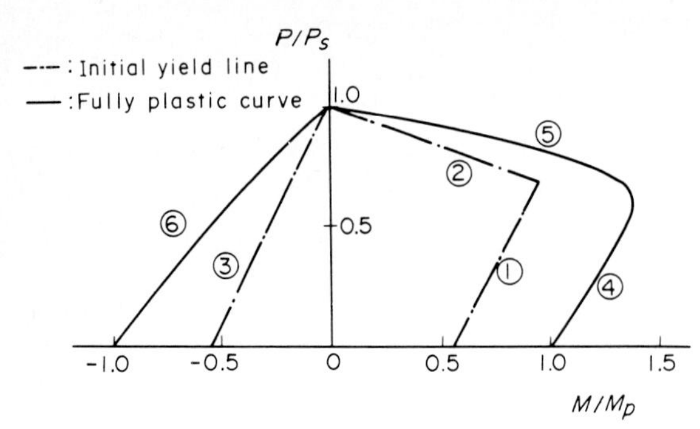

Figure 5.113 Interaction curves of M/M_p versus P/P_s for beam column.

and $A_s = A_{s1} + A_{s2}$ = cross-sectional area of beam column
A_{s1}, A_{s2} = cross-sectional areas of beam column divided by neutral axis for fully plastic moment, respectively
y_1, y_2 = distance from neutral axis to center of figure for A_{s1} and A_{s2} in fully plastic moment, respectively

From this table, the interaction curves can be plotted (see Fig. 5.113), where the interaction curves are categorized into two types corresponding to the sign of the bending moment M acting on the beam column.

In addition to the above analysis, it is confirmed that the collapse of the transverse stiffener in test girder is observed near the ultimate strength of the curved girders, and V_u/V_u^R is equal to 0.9, as shown by Nakai et al.[38]

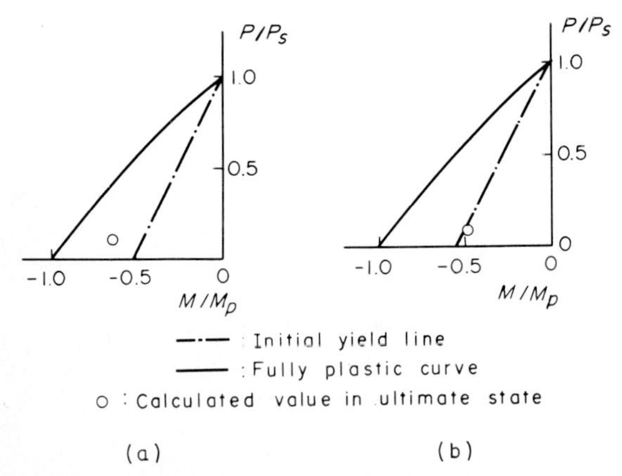

Figure 5.114 Examples of M/M_p–P/P_s interaction curves for beam-column model of transverse stiffener in test girders: (*a*) no. 4 [S10(0.5-178-12-0.5-0.0)] and (*b*) no. 5 [S10(0.5-178-12-1.0-0.0)].

To inquire as to the collapse loads of the transverse stiffeners, the interaction curves corresponding to the initial yield and fully plastic states of the stiffeners in test girders nos. 4 and 5 by Nakai et al.[38] are calculated as shown in Fig. 5.114. The calculated values from the induced forces on the transverse stiffeners in the ultimate state of test girders nos. 4 and 5, M/M_p and P/P_s, respectively, are situated very near to the initial yield line.[38]

Consequently, the required rigidity of the transverse stiffeners in the ultimate state of the curved girders can be evaluated from the condition corresponding to the initial yield state of the beam-column model.

5.6.3 Proposal for designing transverse stiffeners in curved girders

(1) Ultimate strength of transverse stiffeners designed by the elastic buckling theory. In Fig. 5.115*a*, the interaction curve of the beam column, in which the transverse stiffener is designed to have the required relative stiffness specified by JSHB[6] for the straight girders, is plotted together with the calculated values M/M_p and P/P_s at the ultimate state of this curved girder. In a similar manner, Fig. 5.115*b* illustrates a design example by applying the AASHTO guide specification.[54]

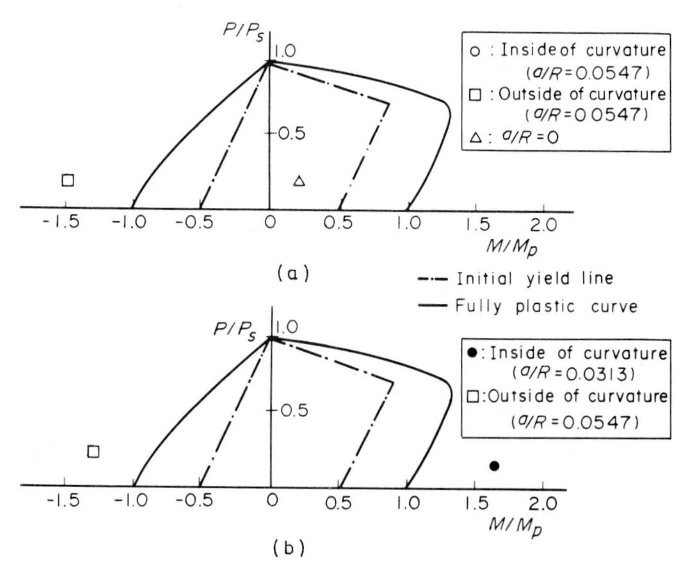

Figure 5.115 Strength of transverse stiffener designed by (*a*) JSHB and (*b*) AASHTO under $A_d/A_w = 0.2$ and $h_w/t_w = 152$.

This figure shows that the transverse stiffeners designed according to the elastic buckling theory are always unavailable at the ultimate state of the curved girders.

(2) Proposition for designing transverse stiffeners by the beam-column approach. In the ultimate state of the curved girders, the transverse stiffeners should have enough strength and rigidity against collapse, since the transverse stiffeners behave in a fashion similar to the vertical members of a Pratt truss in the postbuckling situation of the web plate.

From the analytical and experimental results, the ultimate strength of the transverse stiffeners can be reasonably estimated by the initial yield line of the interaction curve of the beam-column model.

***a.* Rigidity of transverse stiffeners.** Let us now evaluate the required rigidity γ^c_{req} of the transverse stiffeners in the ultimate state of the curved girders by using the initial yield line of the beam-column model and the calculated forces M and P from Eqs. (5.6.42) and (5.6.43). The ratio between $\gamma^c_{t,\text{req}}$ and $\gamma^{\text{JSHB}}_{t,\text{req}}$ is defined as

$$\beta_T = \frac{\gamma^c_{t,\text{req}}}{\gamma^{\text{JSHB}}_{t,\text{req}}} \qquad (5.6.45)$$

To find the magnification factor β_T, various numerical calculations are carried out by referring to dimensions of actual curved girder bridges.[8]

The variations of β_T due to the curvature parameter λ are shown in Fig. 5.116. Figure 5.117 shows the relationships between β_T and λ in which $\alpha = a/b$ is the aspect ratio of the web plate. From Figs. 5.116 and 5.117 we see that the magnification factor β_T increases in accordance with the increase in λ and α. Moreover, β_T becomes less than 1.0 in the region where $\alpha = 0.69$, so that the provision $\gamma^{\text{JSHB}}_{t,\text{req}}$ of the straight girders is effective for curved girders in this region.

From these figures, a design formula can be derived by fitting parabolic curves to the data as follows:

1. For stiffeners attached to one side of the web plate:

$$\begin{aligned}\beta_T &= 1.0 + (\alpha - 0.69)\lambda[9.38\alpha - 7.67 \\ &\quad -(1.49\alpha - 1.78)\lambda] \qquad 0.69 \leqq \alpha \leqq 1.0 \qquad (5.6.46a) \\ &= 1.0 \qquad \alpha < 0.69 \qquad (5.6.46b)\end{aligned}$$

2. For stiffeners attached to both sides:

$$\begin{aligned}\beta_T &= 1.0 + (\alpha - 0.65)\lambda(12.67\alpha - 10.42 \\ &\quad -(1.99\alpha - 2.49)\lambda] \qquad 0.65 \leqq \alpha \leqq 1.0 \qquad (5.6.47a) \\ &= 1.0 \qquad \alpha < 0.65 \qquad (5.6.47b)\end{aligned}$$

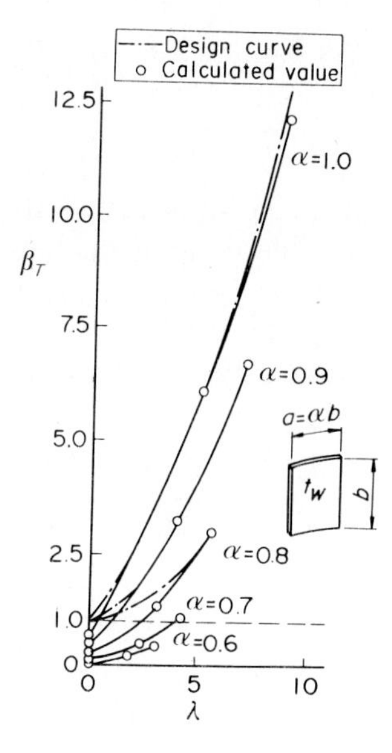

Figure 5.116 Variations of β_T due to curvature parameter λ.

where $$\lambda = \left(\frac{a}{R}\right)\left(\frac{a}{t_w}\right)\sqrt{1-\mu^2} \tag{5.6.48}$$

$$\mu = \text{Poisson's ratio } (=0.3)$$

***b*. Spacing of transverse stiffeners.** In deriving Eqs. (5.6.46) and (5.6.47), the aspect ratio α of a curved web plate panel is limited to less

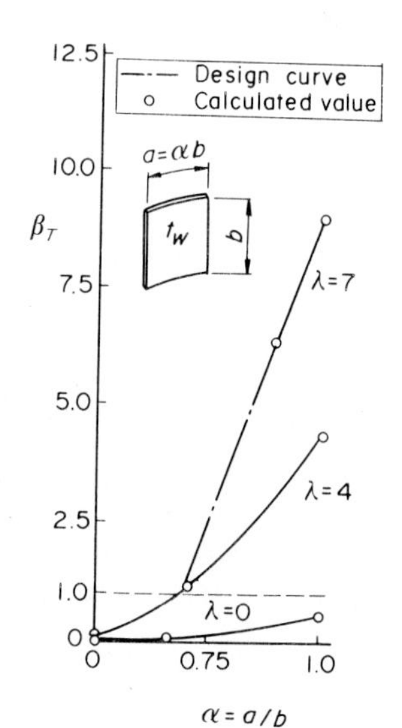

Figure 5.117 Variations of β_T due to aspect ratio α.

than unity. That is,

$$\frac{a}{h_w} \leqq 1.0 \tag{5.6.49}$$

instead of Eq. (5.6.14) holding, because the required rigidity of the transverse stiffener of the curved girder becomes too large to use in actual design unless this condition is imposed. This conclusion has also been reached by Mariani et al.[63]

Therefore, the spacing of transverse stiffeners in ordinary curved girders can be determined according to the design criteria in Eqs. (5.6.9), (5.6.11), and (5.6.13) within the limit of Eq. (5.6.49).

5.7 Design of Flange Plates in Curved Girders

5.7.1 Slenderness of flange plates according to buckling theory

(1) Flange slenderness and local buckling strength in straight girders

***a.* I girders.** Let us now consider the buckling of the compression flange by looking at a model of an outstanding plate with width b and thickness t, as shown in Fig. 5.118.[31,34,64] For this compressive plate, the nondimensionalized slenderness parameter R is available to determine the buckling strength, which can be evaluated[6] by

$$R = \sqrt{\frac{\sigma_y}{\sigma_{cr}}} \tag{5.7.1}$$

where σ_y is the yield stress and σ_{cr} is the elastic buckling stress given by Eqs. (5.3.23) and (5.3.25). Therefore, the above equation is equivalent to

$$R = \frac{b}{t}\sqrt{\frac{\sigma_y}{E}}\sqrt{\frac{12(1-\mu^2)}{\pi^2 k_\sigma}} \tag{5.7.2}$$

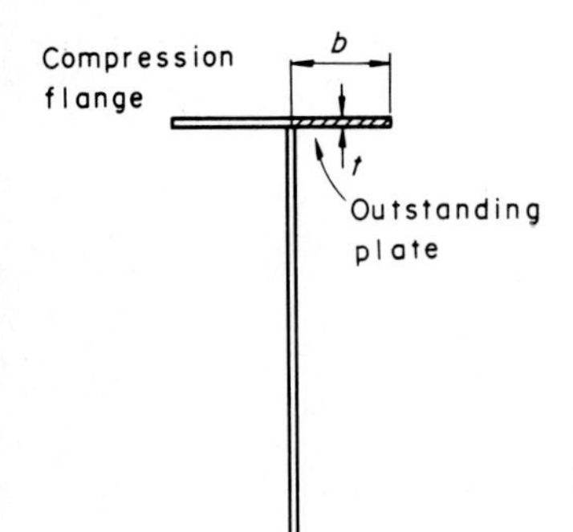

Figure 5.118 Dimension of compression flange in I girder.

where k_σ is the buckling coefficient for the outstanding plate with the boundary conditions given as simply supported along the web plate and transverse stiffeners as well as free along the outstanding edges. This results in

$$k_\sigma = 0.43 \tag{5.7.3}$$

as previously shown in Fig. 5.34.

However, the relationship between the ultimate strength σ_u/σ_y and R of the outstanding plate is plotted in Fig. 5.119. So local buckling does not occur, and the ultimate strength σ_u of the outstanding plate can be ensured up to the yield stress σ_y when

$$R \leqq 0.7 \tag{5.7.4}$$

according to various experimental and analytical studies.[6]

By combining Eqs. (5.7.2) and (5.7.4), the thickness t required for the outstanding plate without local buckling can be estimated from

$$t \geqq \frac{b}{632.4}\sqrt{\sigma_y} \tag{5.7.5}$$

Table 5.14 lists the thickness t for typical yield stresses σ_y (kgf/cm^2).

According to JSHB,[6] the thickness of the outstanding plate t can be reduced to the following minimum required thickness:

$$t_{\min} = \frac{b}{16} \tag{5.7.6}$$

For this case, local buckling occurs, so the ultimate stress σ_u should be

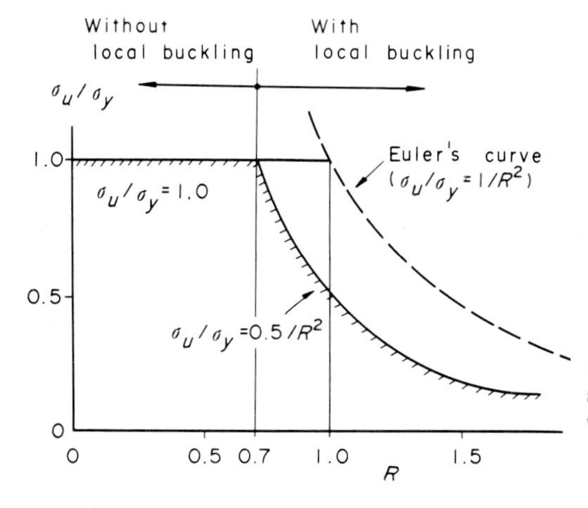

Figure 5.119 Curve of σ_u/σ_y versus R for outstanding plate (straight girder).

TABLE 5.14 Required Thickness for Outstanding Plate without Local Buckling (Straight Girders)

σ_y, kgf/cm^2	Thickness
2400	$t \geqq \dfrac{b}{13}$
3600	$t \geqq \dfrac{b}{11}$

calculated from

$$\frac{\sigma_u}{\sigma_y} = \begin{cases} 1.0 & R \leqq 0.7 \qquad (5.7.7a) \\ \dfrac{0.5}{R^2} & R > 0.7 \qquad (5.7.7b) \end{cases}$$

as plotted in Fig. 5.119.

Therefore, the allowable unit stress for local buckling σ_{cal} can be estimated by

$$\sigma_{cal} = \frac{\sigma_u}{\nu} \tag{5.7.8}$$

where ν (=1.7 in JSHB[6]) is the factor of safety.

***b.* Box girder with stiffened plates.** As an example, let us consider a box girder with a stiffened plate of width b and thickness t with the number of stiffeners s and panel number $n = s + 1$ as shown in Fig. 5.120. If the stiffened plate buckles with nodes at the transverse

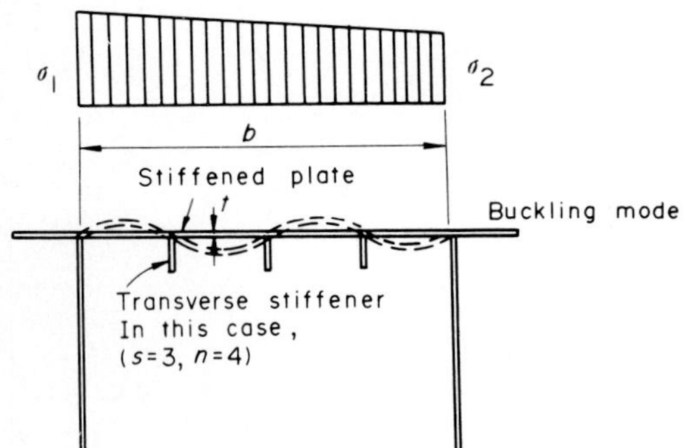

Figure 5.120 Dimensions of stiffened plate in box girder.

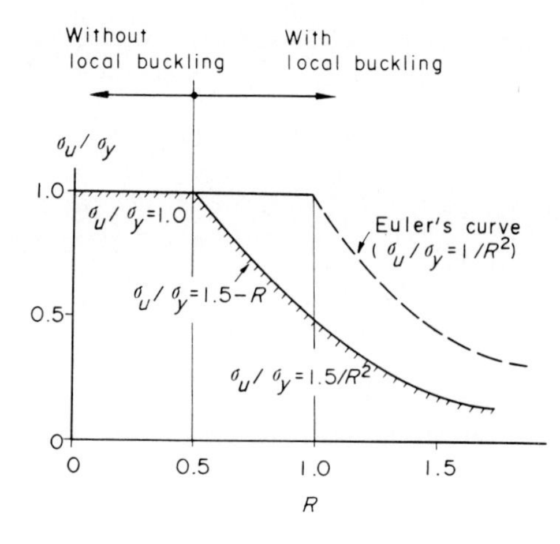

Figure 5.121 Curve of σ_u/σ_y versus R for stiffened plate.

stiffener, the buckling coefficient k_σ can be estimated[65] from Eq. (5.3.26) and Fig. 5.120 as

$$k_\sigma = 4n^2 \tag{5.7.9}$$

where the nondimensionalized parameter R is

$$R \leqq 0.5 \tag{5.7.10}$$

Experimental studies revealed that the local buckling of stiffened plates is completely prevented and the ultimate strength of the stiffened plate reduces to the yield stress σ_y, as shown in Fig. 5.121.

Thus, the required thickness t of the stiffened plates for such cases can be determined from Eq. (5.7.10) by using Eqs. (5.7.2) and (5.7.9):

$$t_o \geqq \frac{\mathrm{b}}{1378fn}\sqrt{\sigma_y} \tag{5.7.11}$$

For instance, JSHB[6] provides the thickness t_o of stiffened plates without local buckling, as shown in Table 5.15, where f is the factor for the

TABLE 5.15 Required Thickness t_o for Stiffened Plates without Local Buckling

σ_y, kgf/cm²	Thickness
2400	$t_o \geqq \dfrac{b}{28fn}$
3600	$t_o \geqq \dfrac{b}{22fn}$

stress gradient φ in the compression flange, as illustrated in Fig. 5.120, which gives

$$f = 0.65\varphi^2 + 0.13\varphi + 1.0 \tag{5.7.12}$$

where
$$\varphi = \frac{\sigma_1 - \sigma_2}{\sigma_1} \tag{5.7.13}$$

and
$$\sigma_1 \geqq \sigma_2 \tag{5.7.14}$$

(σ_1 and σ_2 are positive for compressive stress).

Yet the following minimum required thickness of the stiffened plate $t_{\min}$ is also specified in JSHB:

$$t_{\min} \geqq \frac{b}{80fn} \tag{5.7.15}$$

And the corresponding local buckling strength σ_u is determined from Fig. 5.121:

$$\frac{\sigma_u}{\sigma_y} = \begin{cases} 1.0 & R \leqq 0.5 & (5.7.16a) \\ 1.5 - R & 0.5 < R \leqq 1.0 & (5.7.16b) \\ \dfrac{0.5}{R^2} & R < 1.0 & (5.7.16c) \end{cases}$$

The allowable unit stress of local buckling σ_{cal} for stiffened plates can also be determined from Eq. (5.7.8).

The longitudinal stiffener of the stiffened plate should have a great enough geometric moment of inertia I_l and cross-sectional area A_l to make the buckling node occur at the longitudinal stiffeners, which can be written[65] as

$$I_l \geqq \frac{bt^3}{11}\gamma_{l,\text{req}} \tag{5.7.17}$$

$$A_l \geqq \frac{bt}{10} \tag{5.7.18}$$

The transverse stiffener of stiffened plates should also have

$$I_t \geqq \frac{bt^3}{11}\gamma_{t,\text{req}} \tag{5.7.19}$$

For example, the required relative stiffness $\gamma_{l,\text{req}}$ in Eq. (5.7.17) and $\gamma_{t,\text{req}}$ in Eq. (5.7.19) are codified in JSHB[6] as follows:

When $\alpha_o \leqq \alpha$ and $\gamma_{t,\text{req}}$ is as in Eq. (5.7.21),

$$\gamma_{l,\text{req}} = \begin{cases} 4\alpha^2 n\left(\dfrac{t_o}{t}\right)^2 (1 + n\delta_l) - \dfrac{(\alpha^2+1)^2}{n} & t \geqq t_o \qquad (5.7.20a) \\[2ex] 4\alpha^2 n(1 + n\delta_l) - \dfrac{(\alpha^2+1)^2}{n} & t < t_o \qquad (5.7.20b) \end{cases}$$

$$\gamma_{t,\text{req}} = \frac{1 + n\gamma_{l,\text{req}}}{4\alpha^2} \qquad (5.7.21)$$

In other cases,

$$\gamma_{l,\text{req}} = \begin{cases} \dfrac{1}{n}\left\{\left[2n^2\left(\dfrac{t_o}{t}\right)^2 (1 + n\delta_l) - 1\right]^2 - 1\right\} & t \geqq t_o \qquad (5.7.22a) \\[2ex] \dfrac{1}{n}\{[2n^2(1 + n\delta_l) - 1]^2 - 1\} & t < t_o \qquad (5.7.22b) \end{cases}$$

where the notation is defined by referring to Fig. 5.122:

$$\alpha = \frac{a}{b} = \text{aspect ratio of stiffened plate} \qquad (5.7.23)$$

$$\alpha_o = \sqrt[4]{1 + n\gamma_l} = \text{critical aspect ratio of stiffened plate} \qquad (5.7.24)$$

$$\delta_l = \frac{A_l}{bt} = \text{cross-sectional area ratio of longitudinal stiffener} \qquad (5.7.25)$$

$$\gamma_l = \frac{I_l}{bt^3/11} = \text{relative thickness of longitudinal stiffener} \qquad (5.7.26)$$

a = spacing of transverse stiffener

t_o = thickness of stiffened plate determined from Table 5.15

(2) Elastic buckling analysis of flange plates in curved I girders. Although the analysis of the ultimate strength of stiffened plates used in the compression flanges of curved box girders is a very difficult problem and an effective method to be applied to design is not yet established, an early study by Culver and Frampton[66] was founded on the elastic buckling analysis of compression flanges in curved I girders.

For this problem, Culver and Frampton treated a curved I girder as sketched in Fig. 5.123, where the out-of-plane bending moment M_f

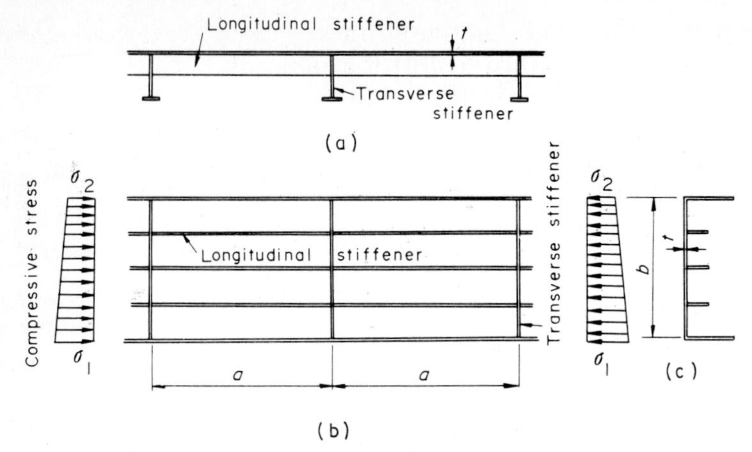

Figure 5.122 Detail of stiffened plate: (*a*) side elevation, (*b*) plane, (*c*) cross section.

acting on both the flange plates [a sort of bimoment $M_\omega = M_f h$—see Eq. (2.3.88)] is due to the curvature of the girder axis in conjunction with the bending moment M with respect to the major axis of the curved I girder. However, the pure torsional moment T_s and secondary torsional moment T_ω are omitted, since their effects are small enough to be ignored.

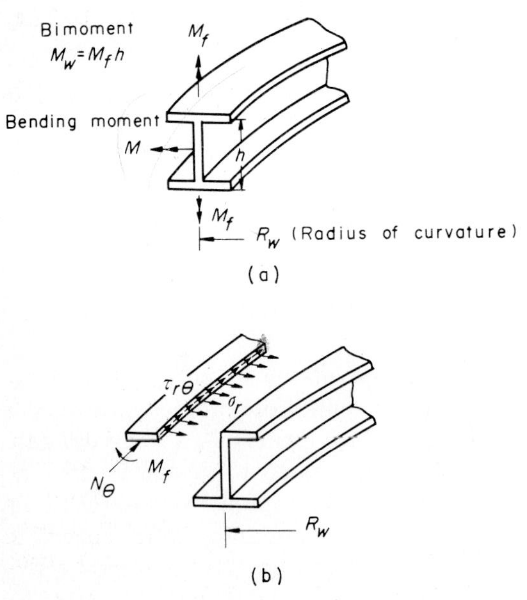

Figure 5.123 (*a*) Applied and (*b*) induced stress resultants in compressive flange plate of curved I girder.

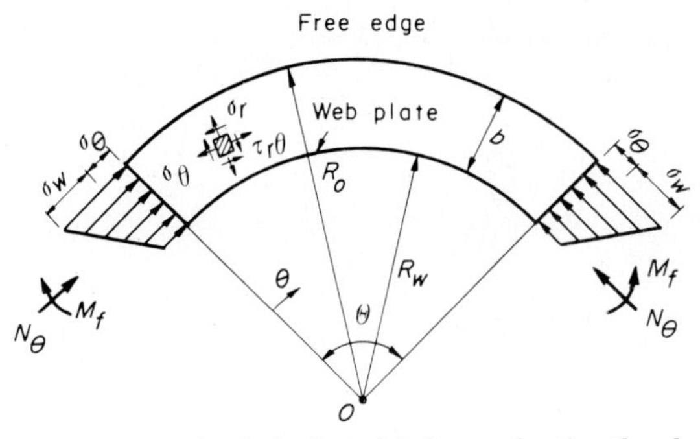

Figure 5.124 Analytical model for evaluating local buckling of flange plate in curved I girder.

Thus, the compressive flange plate of the curved I girder can be idealized analytically as in Fig. 5.124. In this sector plate, the fundamental equation of buckling can be written from Eq. (3.2.186) as

$$\nabla^2 w = \frac{t}{D}\left[\sigma_r \frac{\partial^2 w}{\partial r^2} + \sigma_\theta\left(\frac{1}{r}\frac{\partial w}{\partial r} + \frac{1}{r^2}\frac{\partial^2 w}{\partial \theta^2}\right) + 2\tau_{r\theta}\left(\frac{1}{r}\frac{\partial^2 w}{\partial r\,\partial \theta} - \frac{1}{r^2}\frac{\partial w}{\partial \theta}\right)\right] \tag{5.7.27}$$

where the stresses σ_θ, σ_r, and $\tau_{r\theta}$ can be estimated by a basic equation concerning the Airy stress function Φ as follows:

$$\nabla^2 \Phi = 0 \tag{5.7.28}$$

in which

$$\sigma_r = \frac{1}{r}\frac{\partial \Phi}{\partial r} + \frac{1}{r^2}\frac{\partial^2 \Phi}{\partial \theta^2} \tag{5.7.29a}$$

$$\sigma_\theta = \frac{\partial^2 \Phi}{\partial r^2} \tag{5.7.29b}$$

$$\tau_{r\theta} = \frac{1}{r^2}\frac{\partial \Phi}{\partial \theta} - \frac{1}{r}\frac{\partial^2 \Phi}{\partial r\,\partial \theta} \tag{5.7.29c}$$

and

$$-\int_{r=R_w}^{R_o} \sigma_\theta\left(r - \frac{R_w + R_o}{2}\right) t\,dr = \tfrac{1}{2} M_f \tag{5.7.30a}$$

$$-\int_{r=R_w}^{R_o} \sigma_\theta t\,dr = N_\theta \tag{5.7.30b}$$

The boundary conditions for Eqs. (5.7.27) and (5.7.28) can be ascertained by considering the torsional resistance of the web plate with a

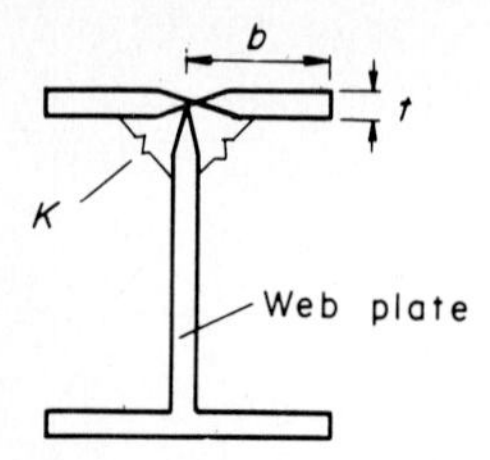

Figure 5.125 Torsional resistance of web plate and corresponding spring constant K.

spring constant K, as shown in Fig. 5.125, which leads to

$$[M_r]_{r=R_o} = 0 \tag{5.7.31a}$$

$$[V_r]_{r=R_o} = \left[Q_r - \frac{\partial M_{rt}}{r\,\partial\theta}\right]_{r=R_o} = 0 \tag{5.7.31b}$$

$$[w]_{r=R_w} = 0 \tag{5.7.31c}$$

$$\left[\frac{\partial w}{\partial r}\right]_{r=R_w} = \frac{D}{K}\left[\frac{\partial^2 w}{\partial r^2} + \frac{\mu}{r}\frac{\partial w}{\partial r}\right]_{r=R_w} \tag{5.7.31d}$$

These equations have been solved by means of the finite difference method, and

$$\sigma_{\rm cr} = k_\sigma \frac{\pi^2 E}{12(1-\mu^2)(b/t)^2} \tag{5.7.32}$$

where the buckling coefficient k_σ is summarized in Table 5.16 for when

TABLE 5.16 Buckling Coefficient k_σ for Flange Plate of Curved I Girders[66]

$\frac{b}{R_w}$	$\frac{\sigma_w}{\sigma_\theta}$	Outer flange plate toward center of curvature	Inner flange plate toward center of curvature
	0	0.431	0.431
0.0001	0.5	0.468	0.468
	1.0	0.489	0.489
	0	0.271	0.517
0.05	0.5	0.299	0.560
	1.0	0.315	0.587
	0	0.218	0.543
0.10	0.5	0.241	0.590
	1.0	0.255	0.616

the flange plate is simply supported along the junction edges of the web plate. Table 5.16 shows that the influence of the curvature parameter b/R_w is small when $b/R_w < 0.05$, but the stress gradient due to warping stress σ_w/σ_θ is predominant in curved I girders.

5.7.2 Ultimate-strength analysis of flange plates in I girders with stress gradients

As mentioned earlier, the influence of curvature on the buckling strength of flange plates of curved I girders is small enough to be neglected when the curvature parameter is

$$\frac{b}{R_w} \ll 0.05 \tag{5.7.33}$$

according to a survey on curved I-girder bridges.[8] However, the influence of the stress gradient due to the warping stress σ_w/σ_θ cannot be omitted in evaluating the buckling strength of the curved I girder, listed in Table 5.16 since warping stress due to the curvature of flange plate always occurs. Therefore, the ultimate strength of curved flange plates can be approximately analyzed on the basis of a flat plate analysis by taking the warping stress into account as follows:

(1) Analytical model and methods. The ultimate strength of outstanding plates without curvature is analyzed by applying an axial compressive displacement u or axial compressive force P with various eccentricities e from the junction point of the web plate, as shown in Fig. 5.126. The corresponding stress distributions in the outstanding plate are plotted in Fig. 5.127.

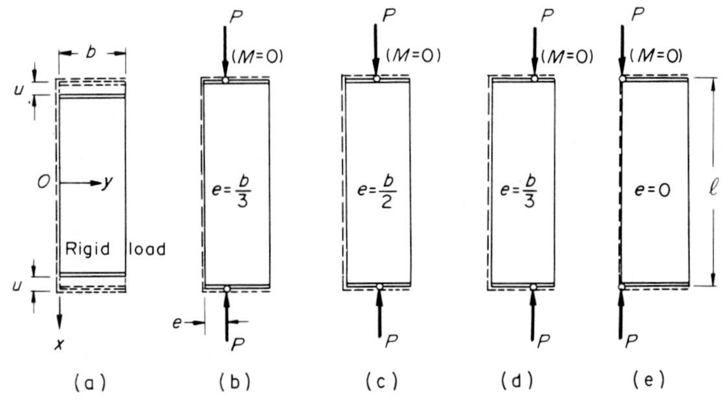

Figure 5.126 Applied axial displacement u or axial force P for (a) case 1, (b) case 2, (c) case 3, (d) case 4, and (e) case 5.

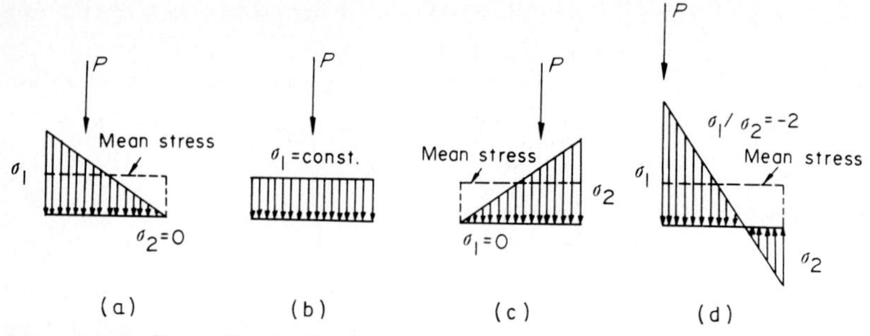

Figure 5.127 Stress distribution in the outstanding plate for (*a*) case 1, (*b*) cases 1 and 3, (*c*) case 4, and (*d*) case 5.

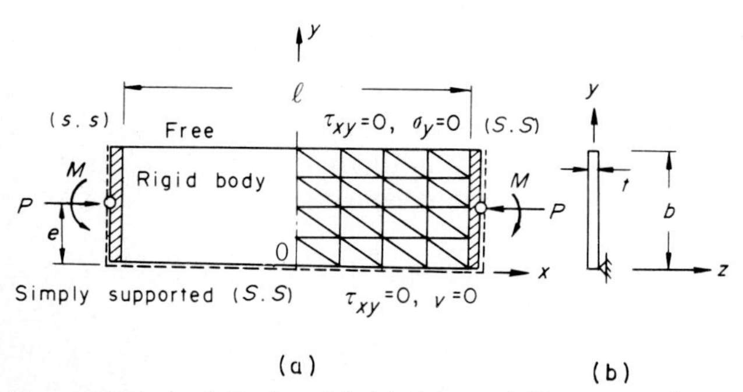

Figure 5.128 Analytical model: (*a*) plane and (*b*) cross section.

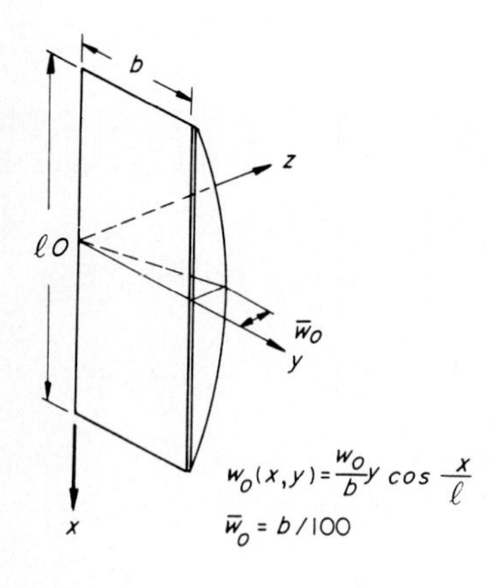

Figure 5.129 Initial deflection mode.

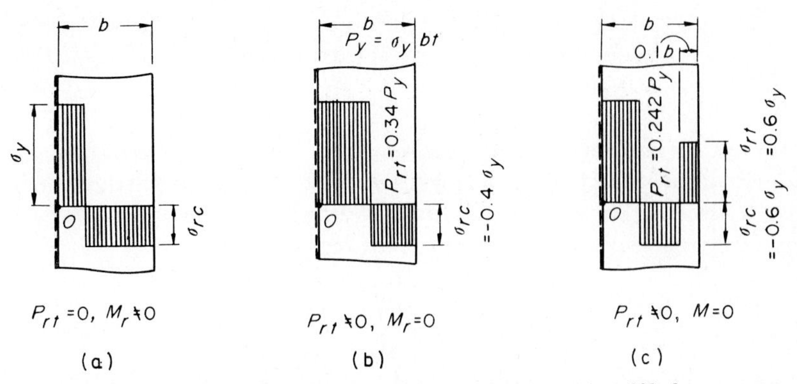

Figure 5.130 Patterns of residual stress distribution: (*a*) self-balance type, (*b*) machine cutting type, (*c*) gas cutting type.

Figure 5.128 shows the analytical model of the outstanding plate with width b and thickness t by using the finite-element method based on elastoplastic and large-deflection theory, as described by Komatsu, Kitada, and Miyazaki,[67] with an applied axial compressive force P and bending moment M as well as boundary conditions of simply supported and free edges.

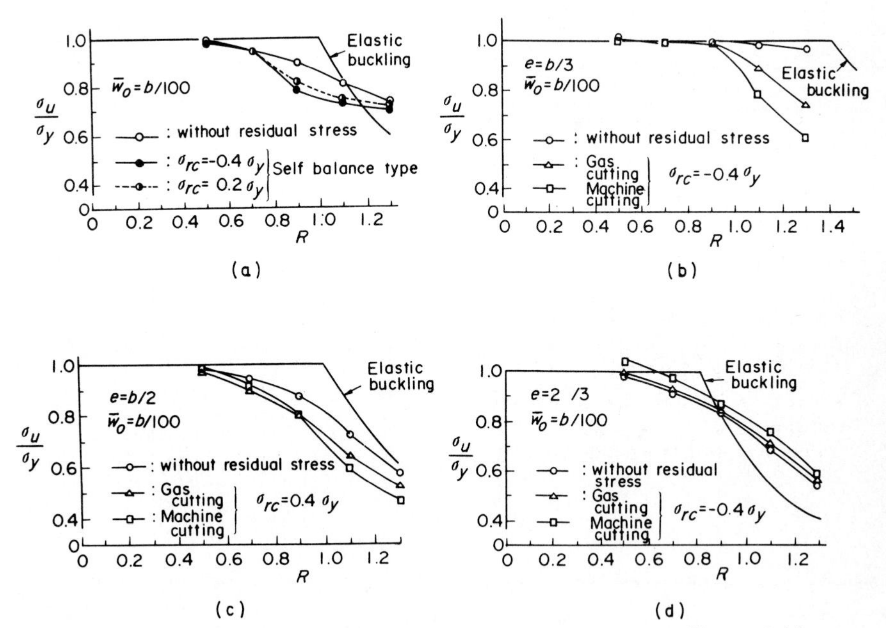

Figure 5.131 Ultimate strength of the outstanding plate for (*a*) case 1, (*b*) case 2, (*c*) case 3, and (*d*) case 4.

In this analysis, the initial deflection w_o and the residual compressive stress σ_{rc} are also taken into consideration, as shown in Figs. 5.129 and 5.130, respectively.

(2) Analytical results of ultimate strength. The ultimate strength σ_u, nondimensionalized by the yield stress σ_y, that is, σ_u/σ_y of the outstanding plate subjected to various loading conditions, can be plotted versus the nondimensionalized slenderness parameter R

$$R = \frac{b}{t}\sqrt{\frac{\sigma_y}{E}}\sqrt{\frac{12(1-\mu^2)}{0.43\,\pi^2}} \tag{5.7.34}$$

[see Eqs. (5.7.2) and (5.7.3)], as shown in Fig. 5.131.[68] This figure shows that the ultimate strength is $\sigma_u/\sigma_y = 1.0$ and that the local buckling of the outstanding plate does not occur when $R \leqq 0.5$ for all the loading conditions.

5.7.3 Design proposal for the flange plates of curved girders

When the warping stress is predominant in the compressive flange plate of curved I girders, the nondimensionalized slenderness parameter R should be designed by

$$R \leqq 0.5 \tag{5.7.35}$$

as mentioned above. For these conditions, the thickness t of the compressive flange plate can be determined according to the yield stress, as illustrated in Table 5.17.

If the thickness t is smaller than the value given in Table 5.17, the effect of local buckling should be considered. Although much more analytical and experimental studies should be conducted, the most conventional design may be carried out by assuming for $R > 0.5$, in a

TABLE 5.17 Required Thickness for Outstanding Plate for Curved I Girder without Local Buckling

σ_y, kgf/cm^2	Thickness
2400	$t \geqq \frac{b}{9.2}$
3600	$t \geqq \frac{b}{7.5}$

manner similar to Eq. (5.7.7*b*), that

$$\frac{\sigma_u}{\sigma_y} = \frac{0.5}{R^2} \tag{5.7.36}$$

For the tensile flange plate of curved I girders,

$$t_{\min} \geqq \frac{b}{16} \tag{5.7.37}$$

similar to the design code of JSHB[6] [see Eq. (5.7.6)].

However, there is such a lack of analytical and experimental studies on the ultimate strength of stiffened plates in curved box girders that these points should be investigated further.[69] But the effects of the stress gradient φ in stiffened plates have been taken into account by f in the design code of JSHB,[6] as shown in Table 5.15 and Eqs. (5.7.12) and (5.7.15). If the influence of the curvature can be ignored, the design of the stiffened plate in the curved box girder can follow that of straight box girders, discussed in Sec. 5.7.1(1)*b*.

References

1. Galambos, T. V.: "Inelastic Buckling of Beams," *Proceedings of the American Society of Civil Engineers*, vol. 89, no. ST-5, pp. 217–242, Oct. 1963.
2. Johnston, B. G.: *Guide to Design Criteria for Metal Compression Members*, Wiley, New York, 1966.
3. Galambos, T. V.: *Structural Members and Frames*, Prentice-Hall, New York, 1968.
4. Structural Stability Research Council: *Guide to Stability Design Criteria for Metal Structures*, 3d ed., Wiley, New York, 1976.
5. Fukumoto, Y., M. Fujiwara, and N. Watanabe: "Inelastic Lateral Buckling Tests on Welded Beams and Girders," *Proceedings of the Japanese Society of Civil Engineers*, no. 189, pp. 39–51, May 1971 (in Japanese).
6. The Japanese Road Association: *The Japanese Specification for Highway Bridges*, Maruzen, Tokyo, Feb. 1980.
7. Nakai, H., and H. Kotoguchi: "A Study on Lateral Buckling Strength and Design Aid for Horizontally Curved I-Girder Bridges," *Proceedings of the Japanese Society of Civil Engineers*, no. 339, pp. 195–204, Dec. 1983.
8. Nakai, H., S. Muramatsu, N. Yoshikawa, T. Kitada, and R. Ohminami: "A Survey for Web Plates of the Horizontally Curved Girder Bridges," *Bridge and Foundation Engineering*, 15, 38–45, May 1981 (in Japanese).
9. Love, A. E. H.: *Mathematical Theory of Elasticity*, 4th ed., Cambridge Press, London, 1952.
10. Namita, Y.: "Die Theorie II. Ordnung von Krümmten Stäben und ihre Anwendung auf das Kipp-Problem des Bogenträgers," *Transactions of the Japanese Society of Civil Engineers*, no. 155, s. 32–41, July 1970 (in German).
11. Ojalvo, M., E. Demutu, and F. Takarz: "Out-of-Plane Buckling of Curved Members," *Proceedings of the American Society of Civil Engineers*, vol. 95, no. ST-11, pp. 2305–2316, Nov. 1969.
12. Tameroglu, S., and I. Turkey: "Finite Theory of Thin Elastic Rods," *Acta Mechanica*, 11:271–282, Nov. 1971.
13. Schroeder, F. H.: Allegemine Stäbtheorie des raumlich Vorgekrümunten und Vorgewunden Trägers mit Grossen Verformungen, Ing-Archiv. 39, s. 87–103, 1970 (in German).

14. Enda, Y.: "Analysis of Thin-Walled Curved Beams with Open Cross-Section as Finite Displacement Theory by Transfer Matrix Method," *Proceedings of the Japanese Society of Civil Engineers*, no. 199, pp. 11–20, May 1972 (in Japanese).
15. Usuki, M., T. Kano, and N. Watanabe: "Analysis of Thin-Walled Curved Members in Account for Large Torsion," *Proceedings of the Japanese Society of Civil Engineers*, no. 290, pp. 1–12, Oct. 1979 (in Japanese).
16. Hirashima, M., M. Iura, and T. Yada: "Finite Displacement Theory of Naturally Curved and Twisted Thin-Walled Members," *Proceedings of the Japanese Society of Civil Engineers*, no. 292, pp. 13–25, Dec. 1979 (in Japanese).
17. Culver, C. G., and P. F. McManus: *Instability of Horizontally Curved Members—Lateral Buckling of Curved Plate Girder*, Carnegie-Mellon University Rep. P1, Pittsburgh, Sept. 1971.
18. McManus, P. F.: "Lateral Buckling of Curved Plate Girders," Ph.D. dissertation, Department of Civil Engineering, Carnegie-Mellon University, Pittsburgh, 1971.
19. Fukumoto, Y., and S. Nishida: "Ultimate Load Behavior of Curved I-Beams," *Proceedings of the American Society of Civil Engineers*, vol. 107, no. EM-2, pp. 367–388, Feb. 1981.
20. Nakai, H., H. Kotoguchi, and T. Tani: "Matrix Structural Analysis of Thin-Walled Curved Girder Bridges Subject to Arbitrary Loads," *Proceedings of the Japanese Society of Civil Engineers*, no. 225, pp. 1–15, Nov. 1976 (in Japanese).
21. Maegawa, K., and H. Yoshida: "Ultimate Strength Analysis of Curved I-Beams by Transfer Matrix Method," *Proceedings of the Japanese Society of Civil Engineers*, no. 312, pp. 27–42, Aug. 1981 (in Japanese).
22. The Hanshin Highway Public Corporation: *The Design Code, Part 2, Design Codes for Steel Bridge Structures*, Osaka, April 1980 (in Japanese).
23. Nakai, H., and C. P. Heins: "Analysis Criterion for Curved Bridges," *Proceedings of the American Society of Civil Engineers*, vol. 103, no. ST-7, pp. 1419–1427, July 1977.
24. The Task Committee on Curved Girders of the ASCE-AASHTO Committee on Flexural Member of the Committee on Metals of Structural Division: "Curved I-Girder Bridge Design Recommendations," *Proceedings of the American Society of Civil Engineers*, vol. 103, no. ST-5, pp. 1137–1167, May 1977.
25. Otsuka, H., and T. Yoshimura: "Studies on Additional Stresses of Main Girders and Member Forces of Lateral Bracing in Curved I- and Straight I-Girder Bridges," *Proceedings of the Japanese Society of Civil Engineers*, no. 290, pp. 17–29, Oct. 1979 (in Japanese).
26. Otsuka, H., T. Yoshimura, H. Hikosaka, and K. Hirata: "Analysis of Curved Girder Bridges Considering Eccentric Connection between a Deck Plate and Girders," *Proceedings of the Japanese Society of Civil Engineers*, no. 259, pp. 11–23, May 1977 (in Japanese).
27. Heins, C. P.: *Bending and Torsional Design in Structural Members*, Lexington Book, Lexington, Mass., 1975.
28. Komatsu, S., H. Nakai, and Y. Taido: "A Proposition for Designing the Horizontal Curved Bridges in Connection with Ratio between Torsional and Flexural Rigidities," *Proceedings of the Japanese Society of Civil Engineers*, no. 224, pp. 55–66, Apr. 1974 (in Japanese).
29. Basler, K., and B. Thürliman: "Strength of Plate Girder in Bending," *Transactions of the American Society of Civil Engineers*, vol. 128, pt. 2, p. 655, 1963.
30. Basler, K.: "Strength of Plate Girder in Shear," *Transactions of the American Society of Civil Engineers*, vol. 128, pt. 2, p. 783, 1963.
31. Bleich, F.: *Buckling Strength of Metal Structures*, McGraw-Hill, New York, 1952.
32. Kollbrunner, C. F., and M. Meister: *Ausbeulen*, Springer-Verlag, Berlin, 1958.
33. Hawranek/Steinhardt: *Theorie und Berechnung der Stahlbrücken*, Springer-Verlag, Berlin, 1958 (in German).
34. Timoshenko, S. P., and J. M. Gere: *Theory of Elastic Stability*, 2d ed., McGraw-Hill, New York, 1961.
35. Nakai, H., T. Kitada, and R. Ohminami: "Experimental Study on Ultimate Strength of Web Panels in Horizontally Curved Girder Bridges Subjected to Bending, Shear and Their Combinations," *Proceedings of SSRC 1984, Annual Technical Session and Meeting*, SSRC, San Francisco, pp. 91–102, April 1984.

36. Nakai, H., T. Kitada, and R. Ohminami: "Experimental Study on Bending Strength of Web Plate of Horizontally Curved Girder Bridges," *Proceedings of the Japanese Society of Civil Engineers*, no. 340, pp. 19–28, Dec. 1983 (in Japanese).
37. The Task Committee on Curved Girders of the ASCE-AASHTO Committee on Flexural Members of the Committee on Metals of Structural Division: "Curved Steel Box Girder Bridges: A Survey," *Proceedings of the American Society of Civil Engineers*, vol. 104, no. ST-11, pp. 1697–1718, Nov. 1978.
38. Nakai, H., T. Kitada, R. Ohminami, and K. Fukumoto: "Experimental Study on Shear Strength of Horizontally Curved Plate Girder," *Proceedings of the Japanese Society of Civil Engineers*, no. 350/I-2, pp. 281–290, Oct. 1984 (in Japanese).
39. Batdorf, S. B., M. Stein, and M. Shildcrout: "Critical Shear Stress of Curved Rectangular Panels," *NACA Technical Note*, no. 1348, May 1947.
40. Rockey, K. C., and M. Skaloud: "The Ultimate Load Behaviour of Plate Girder Loaded in Shear," *The Structural Engineer*, 53(8): 313–325, Aug. 1975.
41. Mozer, J., R. Ohlson, and C. G. Culver: *Horizontally Curved Highway Bridges—Stability of Curved Plate Girder*, Rep. P1–P3, Carnegie-Mellon University, Pittsburgh, Sept. 1970, Sept. 1971 and Nov. 1972.
42. Nakai, H., T. Kitada, and R. Ohminami: "Experimental Study on Buckling and Ultimate Strength of Curved Girder Subjected to Combined Loads of Bending and Shear," *Proceedings of the Japanese Society of Civil Engineers*, no. 356/I-3, pp. 445–454, Apr. 1985 (in Japanese).
43. Culver, C. G., C. L. Dym, and D. K. Brogan: "Bending Behavior of Cylindrical Web Panel," *Proceedings of the American Society of Civil Engineers*, vol. 98, no. ST-10, pp. 2201–2308, Oct. 1972.
44. Dabrowski, R., and J. Wachowiak: "Stress in Thin Cylindrical Web of Curved Plate Girder," *Colloquium on Design of Plate and Box Girder for Ultimate Strength*, IABSE Publ. vol. 11, pp. 352–377, 1972.
45. Abdel-Sayed, G.: "Curved Web under Combined Shear and Normal Stress," *Proceedings of the American Society of Civil Engineers*, vol. 99, no. ST-3, pp. 511–525, Mar. 1973.
46. Mikami, I., and K. Furunishi: "Nonlinear Behavior of Cylindrical Web Panels," *Proceedings of the American Society of Civil Engineers*, vol. 110, no. EM-2, pp. 230–251, Feb. 1984.
47. Hiwatashi, S., and S. Kuranishi: "The Finite Displacement Behavior of Horizontally Curved Elastic I-Section Plate Girder under Bending," *Proceedings of the Japanese Society of Civil Engineers* (Structural Engineering/Earthquake Engineering) 1(2): 59–69, Oct. 1984 (in Japanese).
48. Nakai, H., T. Kitada, and R. Ohminami: "An Elasto-Plastic and Finite Displacement Analysis of Web Plate for Curved Girder Bridges by Using Isoparametric Finite Element Method," *Memoirs of the Faculty of Engineering, Osaka City University*, 23: 191–204, Dec. 1982.
49. Bathe, K., E. Ramm, and E. L. Wilson: "Finite Element Formulations for Large Deformation Dynamic Analysis," *Journal of Numerical Methods in Engineering* 9: 353–386, 1972.
50. Nakai, H., T. Kitada, R. Ohminami, and T. Kawai: "A Study on Analysis and Design of Web Plate in Curved Girder Bridges Subjected to Bending," *Proceedings of the Japanese Society of Civil Engineers*, no. 368,I-5, pp. 23–31, Apr. 1986 (in Japanese).
51. Maeda, Y., and I. Okura: "Fatigue Stength of Plate Girder in Bending Considering Out-of-Plane Deformation," *Proceedings of the Japanese Society of Civil Engineers*, no. 350/I-2, pp. 35–45, Oct. 1984.
52. Daniels, J. H., J. W. Fisher, and B. T. Yen: *Fatigue of Curved Steel Bridge Elements: Design Recommendations for Fatigue of Curved Plate Girder and Box Girder Bridges*, Lehigh University, Bethlehem, Pa., Final Rep. 356, Apr. 1980.
53. Culver, C. G., C. L. Dym, and T. Uddin: "Web Slenderness Requirement for Curved Girders," *Proceedings of the American Society of Civil Engineers*, vol. 99, no. ST-3, pp. 417–430, Mar. 1973.
54. AASHTO: *Guide Specification for Horizontally Curved Highway Bridges*, Washington, D.C., 1980.

55. DIN 4114: Stabilitätfalle (Knickung, Kippung, Beulung), 1952 (in German).
56. AASHTO: *Standard Specification for Highway Bridges*, 12th ed., Washington, D.C., 1977.
57. British Standard Institution: BS153, London, 1958.
58. Timoshenko, S. P., and Goodier, J. N.: *Theory of Elasticity*, 3d ed., McGraw-Hill, New York, 1982.
59. Godfrey, G. B.: "The Allowable Stress in Axially-Loaded Steel Struts," *The Structural Engineer*, 40(3): 97–112, Mar. 1962.
60. Rockey, K. C., G. Valtinat, and K. H. Tang: "The Design of Transverse Stiffener on Webs Loaded in Shear—An Ultimate Approach," *Proceedings of the Institution of Civil Engineers*, vol. 71, pt. 2, 1069–1099, Dec. 1981.
61. Nakai, H., T. Kitada, and R. Ohminami: "A Proposition for Designing Intermediate Transverse Stiffeners in Web Plate of Horizontally Curved Girders, *Proceedings of the Japanese Society of Civil Engineers*, 362/I-4, pp. 249–257, Oct. 1985 (in Japanese).
62. Donnell, L. H.: *Stability of Thin-Walled Tubes under Torsion*, NACA Rep. 479, 1933.
63. Mariani, N., J. D. Moger, C. L. Dym, and C. G. Culver: "Transverse Stiffener Requirements for Curved Webs," *Proceedings of the American Society of Civil Engineers*, vol. 99, no. ST-4, pp. 757–771, Apr. 1973.
64. Wangner, H.: "Torsion and Buckling of Open Section," *NACA Technical Note* 807, 1936.
65. Giencke, E.: Uber die Berechnung regelmass iger Konstruktion aus Kontinium, *Der Stahlbau*, Heft 33, s. 39–48, Feb. 1964 (in German).
66. Culver, C. G., and R. E. Frampton: "Local Instability of Horizontally Curved Members," *Proceedings of the American Society of Civil Engineers*, vol. 96, no. ST-2, pp. 245–265, Feb. 1970.
67. Komatsu, S., T. Kitada, and S. Miyazaki: "Elastic-Plastic Analysis of Compressed Plate with Residual Stress and Initial Deflection," *Proceedings of the Japanese Society of Civil Engineers*, no. 244, pp. 1–14, Dec. 1975 (in Japanese).
68. Komatsu, S., and T. Kitada: "Ultimate Strength Characteristics of Outstanding Steel Plate with Initial Imperfection under Compression," *Proceedings of the Japanese Society of Civil Engineers*, no. 314, pp. 15–27, Oct. 1981 (in Japanese).
69. Yoo, C. H., and Heins, C. P.: "Plastic Collapse of Horizontally Curved Bridge Girders," *Proceedings of the American Society of Civil Engineers*, vol. 98, no. ST-4, pp. 899–914, Apr. 1972.

Chapter

6

Design Codes and Specifications

6.1 Introduction

This chapter presents the design codes and specifications for curved steel girder bridges. In Sec. 6.2, the working stress design method as it is used in the Japanese Specification for Highway Bridges (JSHB) is introduced. In connection with this specification, the design code of curved steel girder bridges established by the Hanshin Expressway Public Corporation (HEPC) is also discussed. Since there are so many opportunities for curved girder bridge construction in metropolitan districts at Osaka and Kobe City, the HEPC design code is much more complete than the JSHB.

Section 6.2 discusses general considerations in determining the radius of curvature and superelevation of bridge decks or slabs of curved girder bridges as well as how to select the structural types, such as composite and noncomposite curved girders and multiple-I girders, twin-box or monobox girders. The spacing of floor beams and lateral bracings is also indicated per HEPC code. The inherent loads acting on the curved girder bridges are explained on the basis of JSHB and HEPC code. Next, the steel materials used in Japan are discussed. With this information, we perform structural analyses of various types of curved girder bridges by developing the fundamental theory described in Chaps. 3 and 4. Then the design criteria for checking stress and deflection are presented in connection with JSHB. Furthermore, the design method of main girders is detailed according to JSHB and the analytical and experimental results of buckling strength obtained in Chap. 5. Finally, the methods of designing structural details such as shear connectors, floor beam or sway bracing, lateral bracing, intermediate diaphragms, and end transverse stiffeners are presented.

6.2 Working Stress Design Method of Curved Steel Bridges

At present, the design of highway bridges in Japan is based entirely on the allowable stress design method in which the working stress should be less than an allowable unit stress. However, an appropriate design method for curved steel bridges is not yet provided in the Japanese Specification for Highway Bridges[1] (hereafter referred to as JSHB). Therefore, a special design code has been established within the Hanshin Expressway Public Corporation[2] (henceforth referred to as HEPC), Nagoya Expressway Public Corporation, and so on to construct curved steel bridges in these districts. In this section we introduce the design method of curved girder bridges on the basis of the design specification or codes of JSHB and HEPC.

6.2.1 General considerations in determining the radius of curvature of curved girder bridges

(1) Radius of curvature and superelevation of roadway surface.[2] In planning a curved girder highway bridge, it is important to determine the radius of curvature with respect to the bridge axis and the suitable superelevation of the roadway surface corresponding to this radius of curvature. These geometries should be carefully determined with due consideration given to the various design conditions at a construction site.

Figure 6.1 shows a vehicle running on a curved girder bridge with radius of curvature R. Given that the weight of the vehicle is $W = Mg$ (M is the mass of the vehicle, and g is the acceleration of gravity) and the running speed is V, centrifugal force F will act on the vehicle:

$$F = \frac{MV^2}{R} \tag{6.2.1}$$

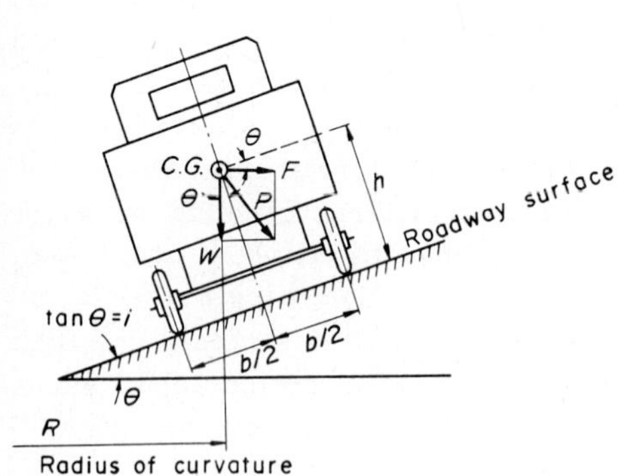

Figure 6.1 Centrifugal force of vehicle.

If we denote $i = \tan\theta$ as the superelevation of the roadway surface and define f as the frictional coefficient between the tires of the vehicle and the roadway surface, the following inequality is obtained for a condition where the vehicle does not slip sideways on the roadway:

$$F\cos\theta - W\sin\theta \leqq (W\cos\theta + F\sin\theta)f \tag{6.2.2}$$

From this equation and Eq. (5.2.1), the radius of curvature can be written as

$$R \geqq \frac{V^2}{g}\frac{1-fi}{i+f} \cong \frac{V^2}{127}\frac{1}{i+f} \tag{6.2.3}$$

In general, $f = 0.11$ to 0.16 and $i = 5$ to 8 percent, and an appropriate radius of curvature of the curved girder bridge R (m) can be determined by the corresponding moving speed V (km/h) of the vehicle. For instance, setting $i + f = 0.2$ and $V = 40$ km/h (25 mi/h), we find that $R > 63$ m.

(2) Sight distance and structural clearance for curved girder bridges.[2] In addition, consideration must be given to the sight distance $AB = S_d$ of the roadway surface, which is frequently inhibited by obstacles such as piers or illumination poles, as illustrated in Fig. 6.2*a*, where the ordinate m to the line-of-sight distance S_d in Fig. 6.2*b* can be expressed as

$$m = R\left(1 - \cos\frac{S_d}{2R}\right) \tag{6.2.4}$$

Furthermore, special care should be given to clearance for heavy trucks passing over the curved girder bridge, since there may be considerable superelevation on the slab surface, as shown in Fig. 6.3. If

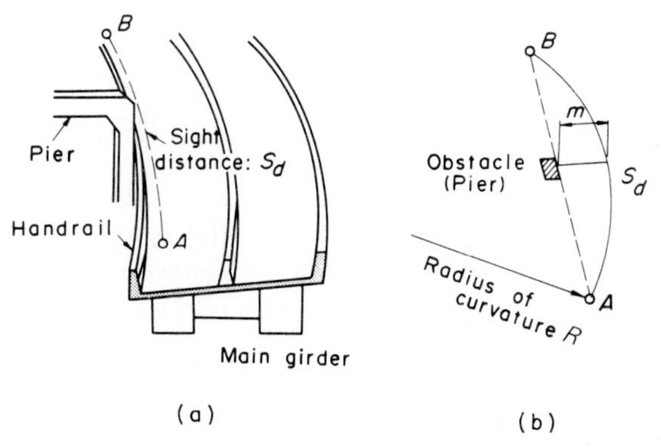

Figure 6.2 Sight distance for a curved girder bridge: (*a*) example of obstacle and (*b*) definition of sight distance S_d.

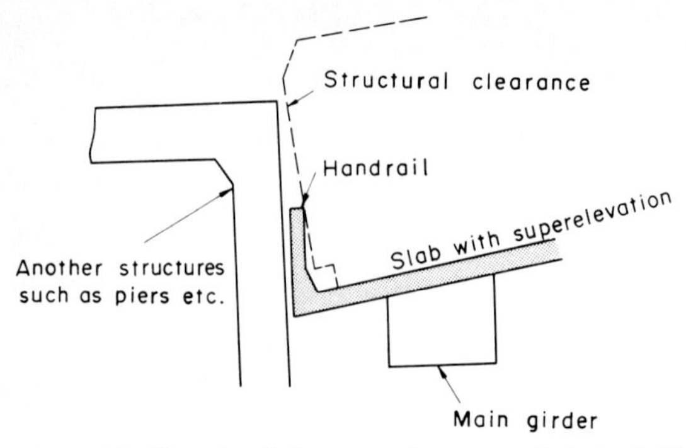

Figure 6.3 Structural clearance for a curved girder bridge.

this structural clearance cannot be met, the location of obstacles such as piers, handrails, and illumination poles should be shifted out of the clearance range in planning the curved girder bridges.

6.2.2 Selection of structural type of curved girder bridges

(1) Structural type corresponding to rigidity parameter γ.[3,4] As mentioned in Sec. 3.3, various types of cross-sectional shapes are used in planning horizontally curved girder bridges. The configuration of the girder is predetermined by the corresponding curvature R and span length L of the bridge, or angle Φ.

Applying Eq. (3.3.69) gives the relationship between Φ and γ, which is summarized in Table 6.1, where γ is

$$\gamma = \frac{GK + EI_{\omega}(\pi/L)^2}{EI_x} \tag{6.2.5}$$

and has the inherent values that correspond to the cross-sectional shape of the girders.

For a continuous curved girder bridge, the effective span length $L = L_{\text{eff}}$ in the above equation can be taken as the conventional values used in deflection analysis,[3] as shown in Fig. 6.4.

If the girder is continuous, then examination of the negative reactions at supports is necessary. When considerable uplift occurs, the designer should change the L/B ratio, where B is the girder width, as shown in Table 6.1, within the appropriate values shown in Eq. (3.3.23). That is,

$$\frac{L}{B} \leqq 10 \tag{6.2.6}$$

for a simple curved girder bridge.

TABLE 6.1 Appropriate Central Angle Φ due to Configuration and Cross-Sectional Shape of Curved Girder Bridges

Configuration	Cross section		γ	Appropriate central angle Φ less than, rad
$L = R_S \Phi$	B	$\gamma = 0.05$–0.20	0.05	0.33
			0.10	0.44
			0.20	0.60
$\Phi = L/R_S$, $R_S = L/\Phi$, O	B	$\gamma = 0.2$–0.5	0.2	0.60
			0.3	0.64
			0.4	0.68
			0.5	0.72
$\gamma = \dfrac{GK + EI_\omega(\pi/L)^2}{EI_X}$	B	$\gamma = 0.5$–1.0	$\geqq 0.5$	$\geqq 0.72$

Moreover, when the central angles Φ (rad) are small, as shown in Eq. (3.3.72), i.e.,

$$\Phi < \begin{cases} 0.09 & \text{multiple-I girder} \quad (6.2.7a) \\ 0.25 & \text{twin-box girder} \quad (6.2.7b) \\ 0.36 & \text{monobox girder} \quad (6.2.7c) \end{cases}$$

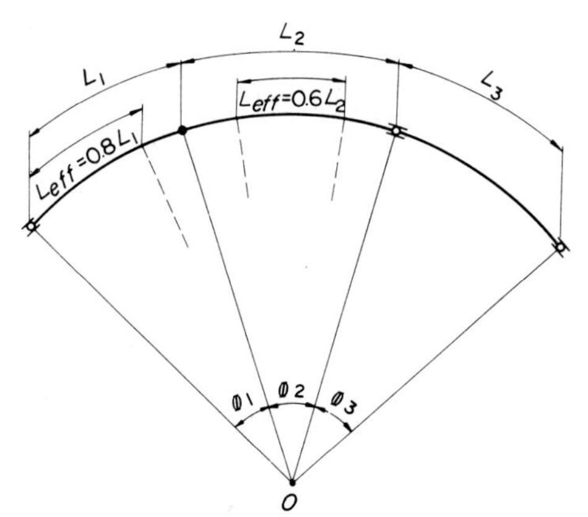

Figure 6.4 Effective span to estimate parameter γ for continuous curved girder bridges.

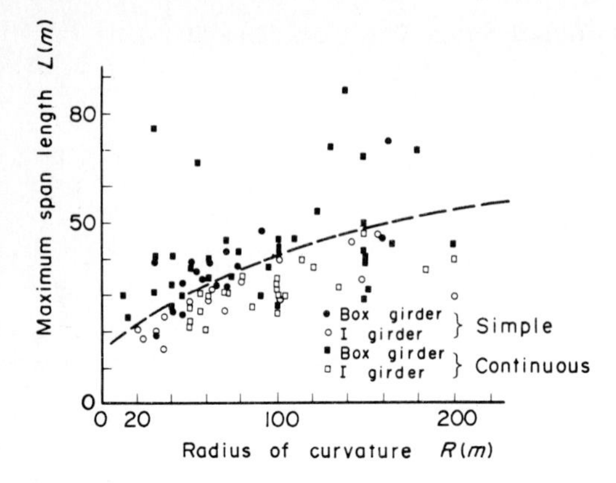

Figure 6.5 Relationship between L and R.

These curved girders may be analyzed by means of straight beam theory, except for the stress calculations of $\sigma_{\omega f}$ [see Eq. (6.2.104)] for multiple curved I-girder bridges.

The relationship between the maximum span length L and radius of curvature R has been investigated for actual curved girder bridges and is shown in Fig. 6.5.[5]

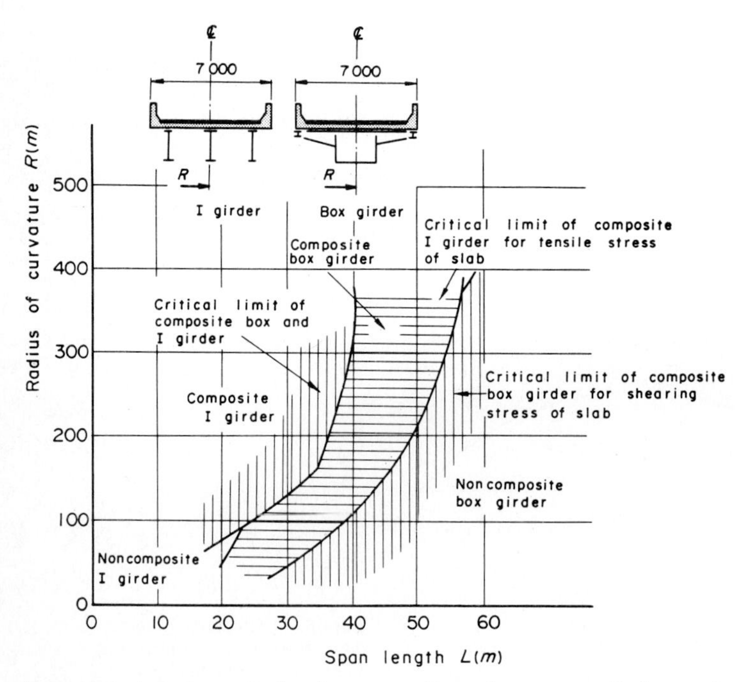

Figure 6.6 A design criterion for composite and noncomposite box or I girders.

(2) Composite and noncomposite curved girder bridges. In the HEPC[2] code, the trial designs of curved girder bridges having box girders and multiple-I girders with composite or noncomposite girders have been tested, and the design criteria for selecting the types of main girders are plotted in Fig. 6.6, where the criterion for a composite I girder is set by limiting the tensile stress in the concrete slab for the negative bending moment, and that of composite box girder is limited by the shearing stress due to flexure and torsion of the concrete slab.

Figure 6.6 is used in the determination of the type of main girders for curved girder bridges.

(3) Arrangement of bearing shoes in curved box-girder bridges.[2] The arrangement of bearing shoes in a curved box girder is an important factor in choosing the type of girder, i.e., monobox or twin-box girders. To explain this fact, let us consider a monobox girder, shown in Fig. 6.7, where two bearing shoes are located at supports 1 and 2 under each web plate. When the shearing force Q and torque T are applied to the end of this monobox girder, vertical reactions V_1 and V_2 can be estimated from

$$V_1 = \frac{Q}{2} + \frac{T}{B} \qquad (6.2.8a)$$

$$V_2 = \frac{Q}{2} - \frac{T}{B} \qquad (6.2.8b)$$

where B is the spacing of the bearing shoes.

The reaction V_1 will be negative; i.e., an uplift occurs at support 1 if the magnitude of the torque T is large or the spacing of the shoes B is small. Therefore, the ratio of span length L to spacing B should be limited to an appropriate value, as indicated in Eq. (6.2.6), to keep the reaction positive.

In a similar manner, uplift is introduced in each box girder when two shoes are attached to each box cell of a curved twin-box-girder bridge,

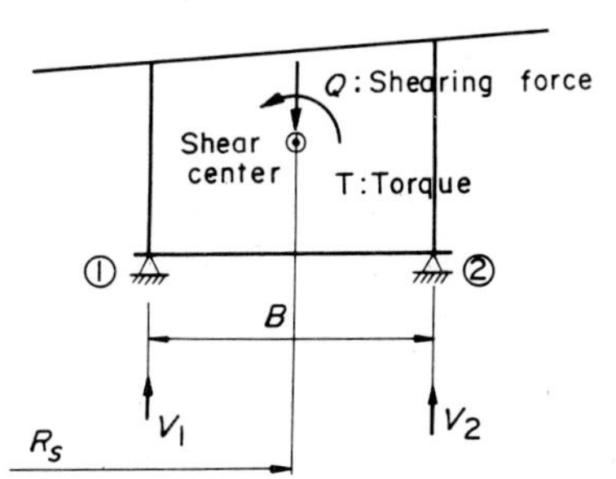

Figure 6.7 Reactions V_1 and V_2 in monobox girders.

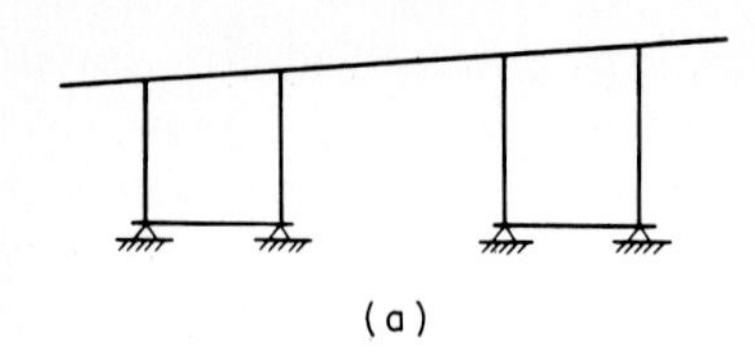

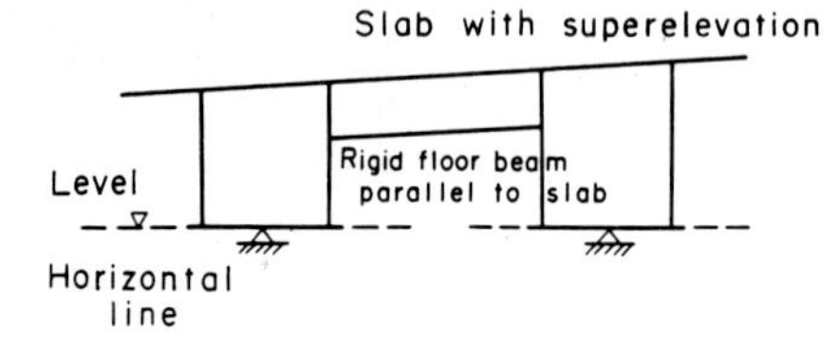

Figure 6.8 Arrangement of shoes in twin-box girders (*a*) with two shoes for each box cell and (*b*) with one shoe for each box cell.

illustrated in Fig. 6.8*a*. The details of designing shoes against uplift are very complicated for this case. For this case, the uplift is eliminated by using one shoe on each box cell, as shown in Fig. 6.8*b*. However, a rigid floor beam should be provided between each box girder to ensure the distribution of shearing forces as possible throughout the twin-box girder.

In determining the cross-sectional shape of a curved girder, whether monobox or twin-box, under the given clear width of roadway surface, the values of the central angle Φ, shown in Table 6.1, are given as reliable data. In addition, the static properties of monobox and twin-box girders are investigated from various points of view, as shown in Table 6.2.

(4) Spacing of floor beams.[2] In a multiple curved I-girder bridge, it is preferable to use floor beams instead of sway bracings because of their high flexural rigidity. These floor beams are usually located with spacing λ of 4 to 5 m and are connected to the main girders at right angles, as shown in Fig. 6.9. To ensure the static properties of floor

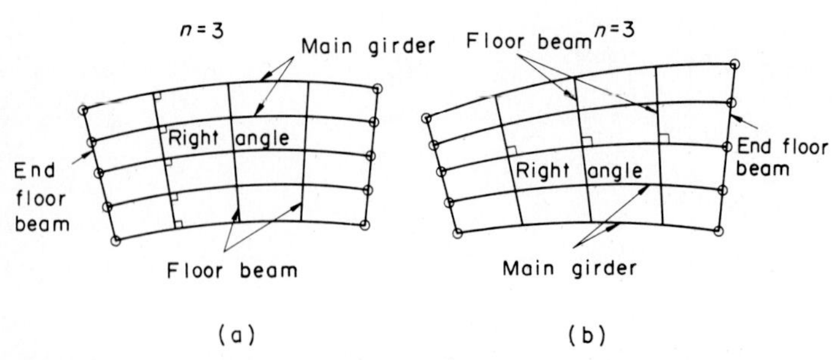

Figure 6.9 Spacing of floor beams: (*a*) standard part and (*b*) expanded part.

TABLE 6.2 Comparison of Monobox and Twin-Box Girders

Item	Monobox girder	Twin-box girder
Uplift of bearing shoes	Large.	Small.
Stress properties	Effective width is reduced in comparison with twin-box girder, and shearing stress in web plate is predominant.	Effective width is large in comparison with monobox. Shearing stress in web plate is not large.
Rigidity	Torsional rigidity is not small, but flexural rigidity is smaller than twin-box girder.	Torsional rigidity is large, but flexural rigidity is larger than that in monobox.
Distortion	Large; diaphragm should be carefully designed.	Small.
Fabrication, erection, and transportation	Very difficult to handle because cross section is large.	Comparatively easy to handle dividing into small segments.
Steel weight	Smaller than twin-box girder because there are only two web plates.	Larger than monobox girder.

beams, the web plates should be solid ones, and one should be located at the midspan of multiple curved I-girder bridges. The number of floor beams should be odd, such as $n = 1, 3, \ldots$.

The floor beams of twin-box girders, however, are usually located within spacing $\lambda = 20$ m. In this case, the relationships between spacing of intermediate diaphragms [see Sec. 6.2.8(4)] and spacings of transverse stiffeners [see Sec. 6.2.7(2)] in compression flange plates or web plates of main girders (see Sec. 6.2.7) should be taken into consideration.

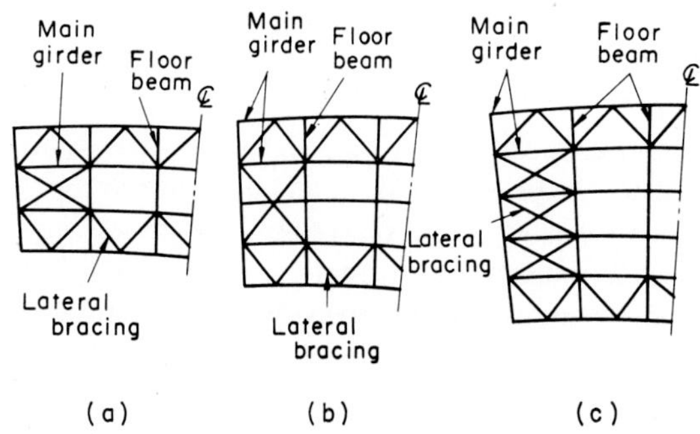

Figure 6.10 Spacing of lateral bracings with (*a*) four girders, (*b*) five girders, (*c*) six girders.

(5) Spacing of lateral bracing in multiple curved I-girder bridges.[2] The lateral bracing of a multiple curved I-girder bridge is not treated as a secondary member, but a main member of the bridge, as in Sec. 6.2.8(3). The spacing of the lateral bracing in multiple curved I-girder bridges with different numbers of main girders is specified in HEPC,[2] as shown in Fig. 6.10. These lateral bracings should, of course, be located at the top and bottom of the main girders, to decrease the secondary out-of-plane bending stress of the flange plate [see Sec. 6.2.5(3)] and to ensure stability during the erection of multiple curved I-girder bridges.

6.2.3 Loads

(1) Dead load ***D*****.** In designing a curved girder, it is important to know the exact intensity of the dead load. If this intensity is predicted from previous data, the trial design procedures will be significantly reduced and will consequently lead to an economical design of curved girder bridges.

For this information, the dead load intensity w_s (kgf/m^2) of the steel girder, i.e.,

$$w_s = \frac{W_s}{LB_c} \tag{6.2.9}$$

where W_s = total weight of steel materials of curved girder bridge (kgf)
L = span length of curved girders (m)
B_c = clear width of roadway (m)

was investigated by Konishi and Komatsu,[6] and these results are plotted in Fig. 6.11*a* to *c*. In Fig. 6.11*a*, the steel weight falls within the ranges *AOB* for noncomposite I girders and *BOC* for composite I girders. But the steel weights of both noncomposite and composite box girders fall within *AOB*, as shown in Fig. 6.11*b*. For the continuous curved girders, the steel weight varies from an upper bound *CAO* to a lower bound *BO* as in Fig. 6.11*c*.

(2) Live load L**.** In the design specification of JSHB, the live loads are categorized into two kinds: T load and L load. The former is a single-vehicle load that is used for the design of slabs, floor beams, and stringers, as shown in Fig. 6.12, where the intensity of wheel loads is detailed in Table 6.3. For some special highway bridges near the hinterland of important harbors of Japan, a heavy trailer, termed TT-43, is also specified, as in Fig. 6.13.[1]

The latter load, or the L load, is the live load for designing the main

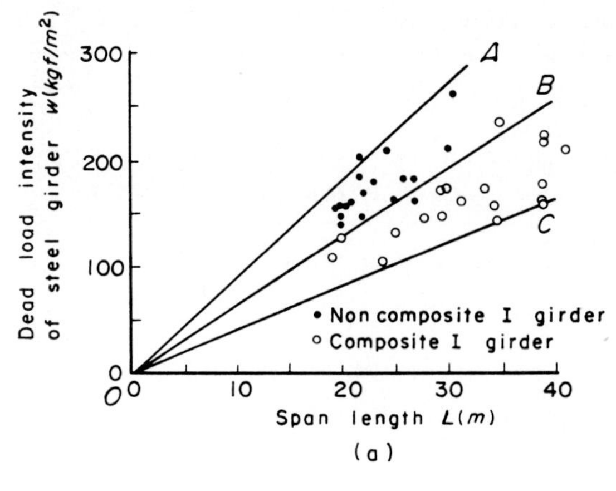

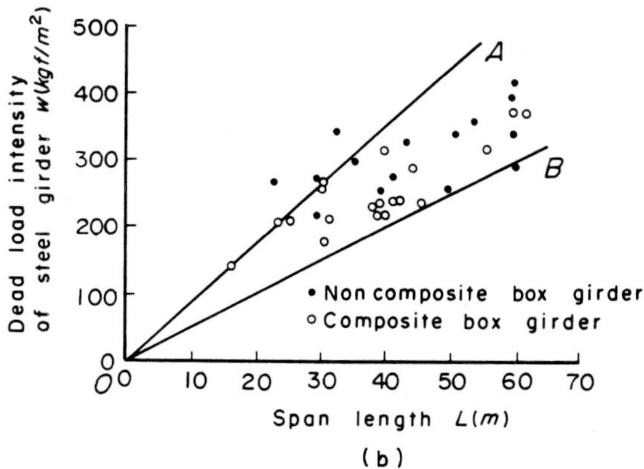

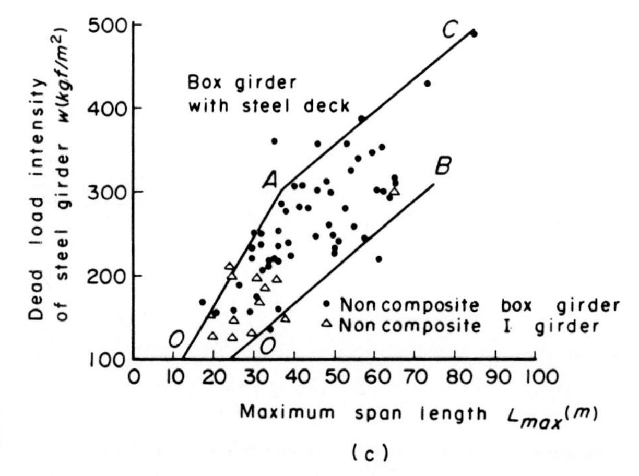

Figure 6.11 Dead load intensity of steel girders: (*a*) multiple curved I-girder bridges, (*b*) curved box-girder bridges, and (*c*) continuous curved girder bridges.

TABLE 6.3 Intensity of T Load

Class	Name of T load	Total wheel load W (tf)	Front-wheel load $0.1W$ (kgf)	Rear-wheel load $0.4W$ (kgf)
1st	T 20	20	2000	8000
2d	T 14	14	1400	5600

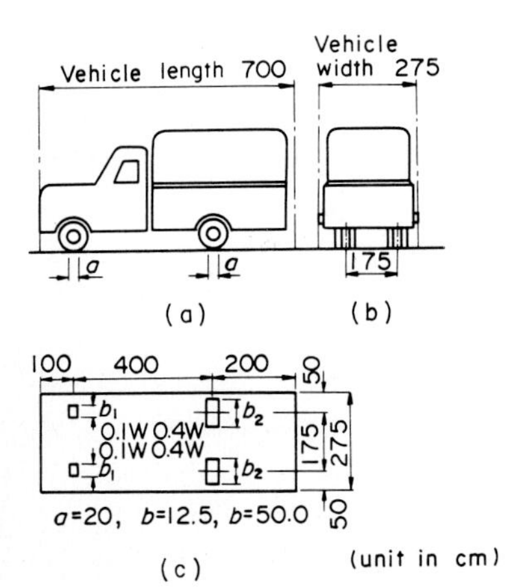

Figure 6.12 Dimension of T load: (*a*) side elevation, (*b*) cross section, and (*c*) plan.

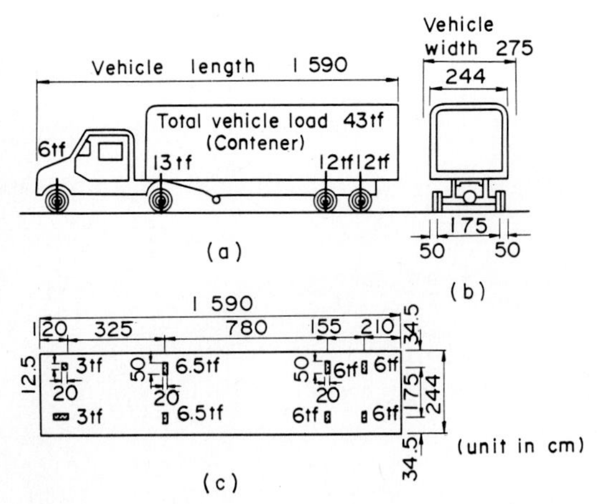

Figure 6.13 TT-43 load: (*a*) side elevation, (*b*) cross section, and (*c*) plan.

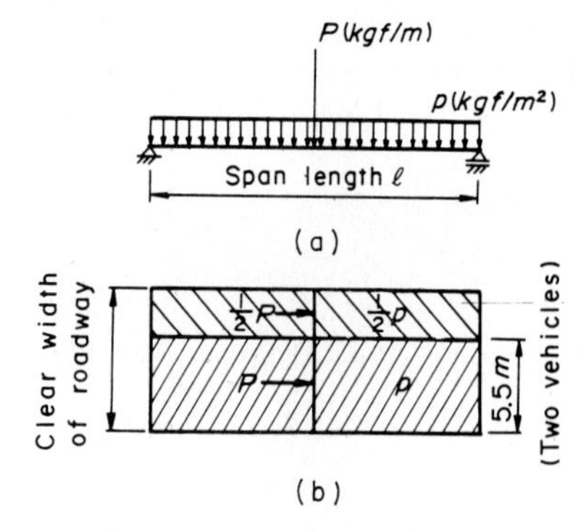

Figure 6.14 Illustration of L load: (a) side elevation and (b) plan.

members of bridges and is composed of a knife-edge load P and a uniformly distributed load p, as illustrated in Fig. 6.14.[1] The intensities of these loads are listed in Table 6.4.

(3) Impact I. The impact coefficient i of the live load is specified in JSHB as

$$i = \frac{20}{50 + l} \tag{6.2.10}$$

where l = span length of a straight girder (m). However, the dynamic behavior of curved girder bridges is somewhat different from that of the ordinary straight girder bridges, as detailed in Sec. 3.5. Thus the rational impact coefficient i_m for curved girder bridges is proposed on the basis of random vibration theory[7] as follows:

$$i_m = 0.5\left(\frac{R_c}{R_p}\right)\frac{10}{L} \leqq 0.4 \tag{6.2.11}$$

where R_c = radius of curvature at midline of roadway (m)
R_p = radius of curvature of main girder (m)
$L = R_c\Phi$ = span length of curved girder bridge (m)
Φ = central angle of curved girder bridge (rad)

TABLE 6.4 Intensity of L Load

Bridge class	Name of L load	Knife-edge load, P (kgf/m)	Uniformly distributed load p (kgf/m²), $l \leqq 80$ m	Uniformly distributed load p (kgf/m²), $l > 80$ m
1st	L 20	5000	350	$430 - l \geqq 300$
2d	L 14	70% of 1st-class bridge		

Equation (6.2.11) can be categorized into the following three cases corresponding to parameter γ, as in Eq. (6.2.5):
For multiple curved I-girder bridges,

$$\gamma \leqq 0.1 \quad \text{and} \quad \Phi \geqq 0.2 \tag{6.2.12a}$$

For curved twin-box-girder bridges,

$$0.1 < \gamma < 0.5 \quad \text{and} \quad \Phi \geqq 0.2 + 0.5(\gamma - 0.1) \tag{6.2.12b}$$

For curved monobox-girder bridges,

$$\gamma \geqq 0.5 \quad \text{and} \quad \Phi \geqq 0.4 \tag{6.2.12c}$$

Otherwise, i_m can be estimated as in straight girder bridges[8]

$$i_m = 0.6 \frac{10}{L} \leqq 0.4 \tag{6.2.13}$$

Figure 6.15 shows the variations in impact coefficient i_m with span length $L = R_c\Phi$ in comparison with some foreign specifications.

(4) Wind load W. In Japan, typhoons with high wind velocity occur frequently, so that the wind load should be carefully determined in designing highway bridges.

The wind pressure p (kgf/m²) under a constant wind velocity V (m/s) is generally determined from hydrodynamics as

$$p = \tfrac{1}{2}\rho V^2 C_D \tag{6.2.14}$$

where

$$\rho = 0.125 \text{ kgf·s}^2/\text{m}^4 \tag{6.2.15}$$

$$\rho = \text{density of air}$$

$$C_D = \text{drag coefficient}$$

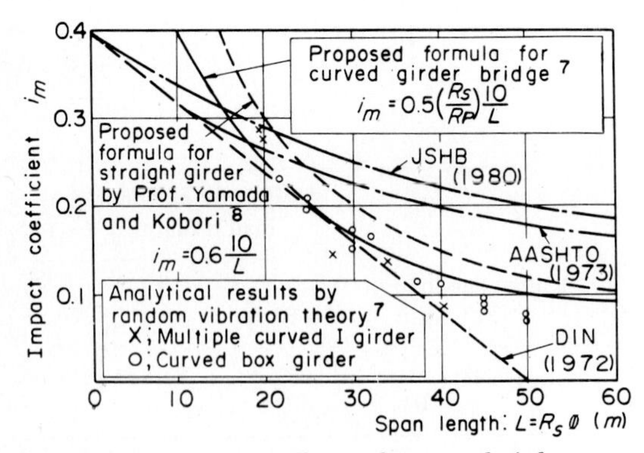

Figure 6.15 Impact coefficient for curved girder.

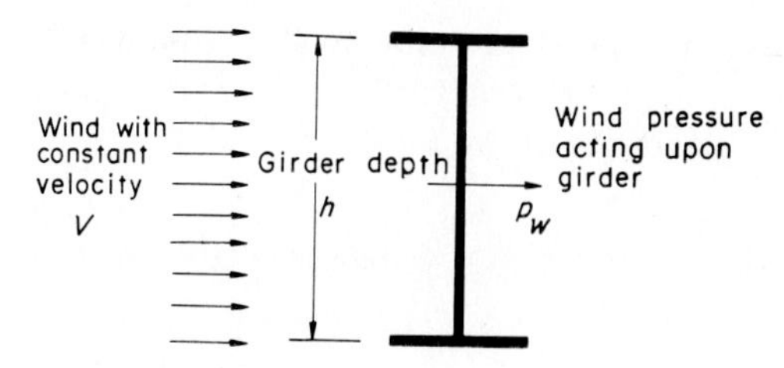

Figure 6.16 Wind pressure p_w.

Therefore, the wind pressure p_w (kgf/m) acting along the unit length of an I-girder with depth h (m), illustrated in Fig. 6.16, can be obtained by multiplying h by Eq. (6.2.14):

$$p_w = \tfrac{1}{2}\rho V^2 C_D h \tag{6.2.16}$$

The drag coefficient C_D for the usual I and box girder generally falls within[1]

$$C_D = 1.6\text{–}1.8 \tag{6.2.17}$$

For a special cross-sectional configuration of curved girders such as twin-, flat-, or trapezoidal-box girders, the drag coefficient C_D should be investigated through wind tunnel tests.

However, since the wind velocity V (m/s) always varies, as sketched in Fig. 6.17a, the speed V is defined as the mean value with respect to a time interval, 10 min, as illustrated in Fig. 6.17b. This wind velocity also varies according to the altitude of the structure; however, the

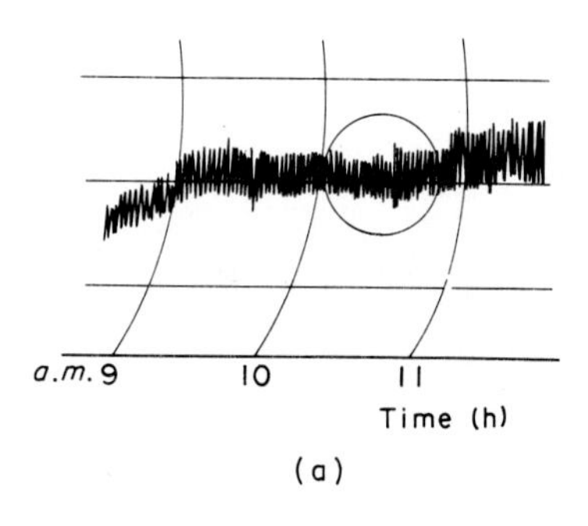

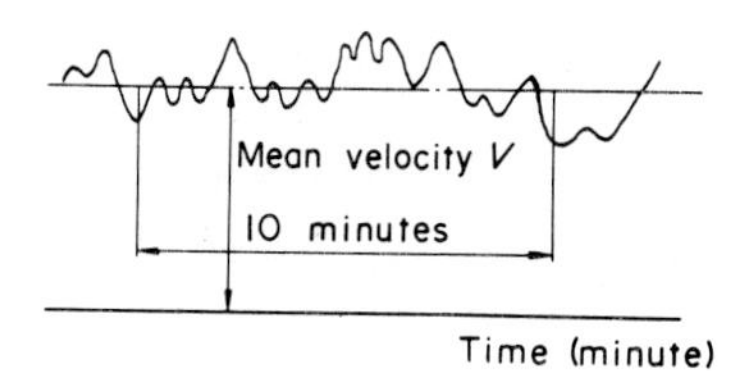

Figure 6.17 (a) Record and (b) detail of wind velocity (in a).

wind velocity can be related to V_{10}, the wind velocity 10 m above the ground. For the wind velocity V_y at an arbitrary altitude y other than V_{10}, the following exponential formula[2] can be used:

$$V_y = \begin{cases} V_{10}\left(\dfrac{y}{10}\right)^n & 10\text{ m} < y \leqq 240\text{ m} \quad (6.2.18a) \\ V_{10} & y \leqq 10\text{ m} \quad (6.2.18b) \end{cases}$$

where $$n = \frac{1}{2.5} - \frac{1}{7} \quad (6.2.19)$$

= roughness of ground surface

Moreover, this fundamental wind velocity (for example, $V = 40$ m/s in JSHB) should be modified by a certain factor (for example, 1.4 in JSHB) corresponding to the horizontal length of superstructures in the case of long-span bridges.

Synthesizing these facts, the JSHB has codified the wind pressure as $p = 300$ kgf/m^2 from Eq. (6.2.14) by using $V = 55$ m/s and $C_D = 1.6$ for ordinary plate girder bridges. The wind load p_w (kgf/m) is

$$p_w = 240 + 450h \leqq 600 \text{ kgf/m} \quad (6.2.20)$$

where h is the girder depth in meters.

In the case of curved girder bridges, the direction of the applied wind load p_w is taken as the corresponding data of wind rose. Otherwise, the maximum effect of wind loads p_w should be determined by changing their direction of wind angle φ with a small pitch, as illustrated in Fig. 6.18.[2]

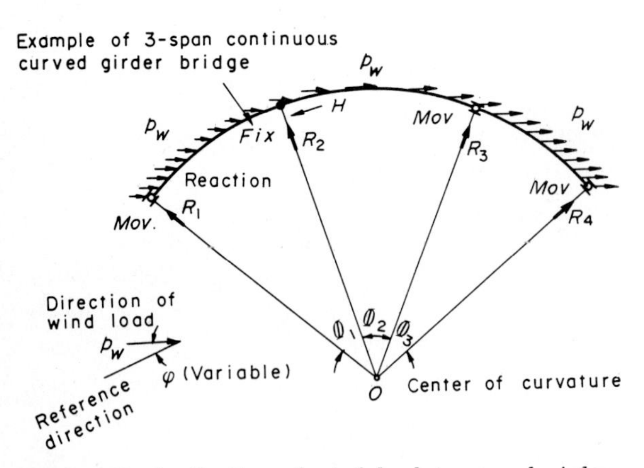

Figure 6.18 Application of wind load to curved girder bridge.

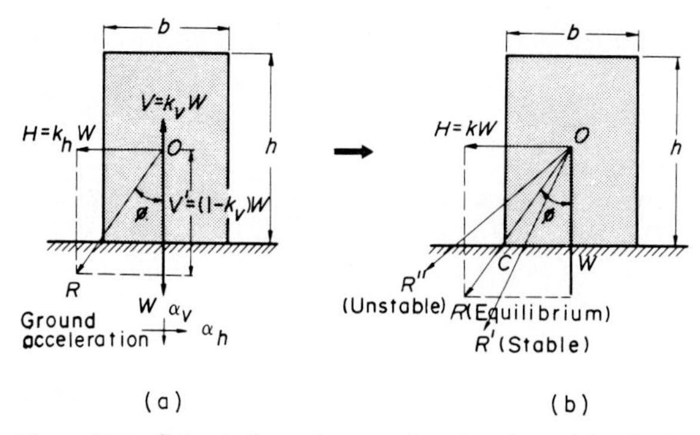

Figure 6.19 Seismic force for massive structure: (*a*) seismic forces H and V and (*b*) equivalent system.

(5) Seismic load E_Q. Japan is subject to strong earthquakes. The seismic design method of bridge structures is outlined in this section according to JSHB.[9]

***a.* Definition of seismic load.** When a massive structure having deadweight W is supported by the ground directly, as shown in Fig. 6.19*a*, and is subjected to an earthquake with horizontal and vertical accelerations α_h and α_v, respectively, the induced horizontal and vertical forces H and V follow from Newton's law:

$$H = k_h W \tag{6.2.21a}$$

$$V = k_v W \tag{6.2.21b}$$

where the seismic coefficients k_h and k_v are

$$k_h = \frac{\alpha_h}{g} \tag{6.2.22a}$$

$$k_v = \frac{\alpha_v}{g} \tag{6.2.22b}$$

and $\qquad g = \text{acceleration of gravity} = 980\ \text{cm/s}^2 \text{ or } 980\ \text{gal.}$

Since the most dangerous situation vis-à-vis the stability of this structure is caused by the horizontal force $H = k_h W$ and the resulting vertical force $V' = (1 - k_v)W$, the corresponding angle ϕ between these forces is determined by

$$\tan \phi = \frac{k_h W}{(1 - k_v)W} = \frac{k_h}{1 - k_v} = k \tag{6.2.23}$$

and k is called the resulting *seismic coefficient.*

This angle ϕ can be determined directly, as shown in Fig. 6.19*b*.

Therefore, the seismic design of structures should use

$$H \geqq kW \tag{6.2.24}$$

b. Seismic design according to JSHB[9]

(*i*) *Fundamental method.* In the design specification of JSHB,[9] the fundamental seismic coefficient k_{ho} to be applied to the value of k in Eq. (6.2.24) is determined by

$$k_{ho} = \nu_1 \nu_2 \nu_3 k_o \tag{6.2.25}$$

where $k_o = 0.2$ (200 cm/s² or 200 gal) (6.2.26)

= standard design seismic coefficient

and ν_1 = modification factor due to location of structure shown in Fig. 6.20

ν_2 = modification factor due to condition of ground, shown in Table 6.5

ν_3 = important factor due to category of bridge, shown in Table 6.6

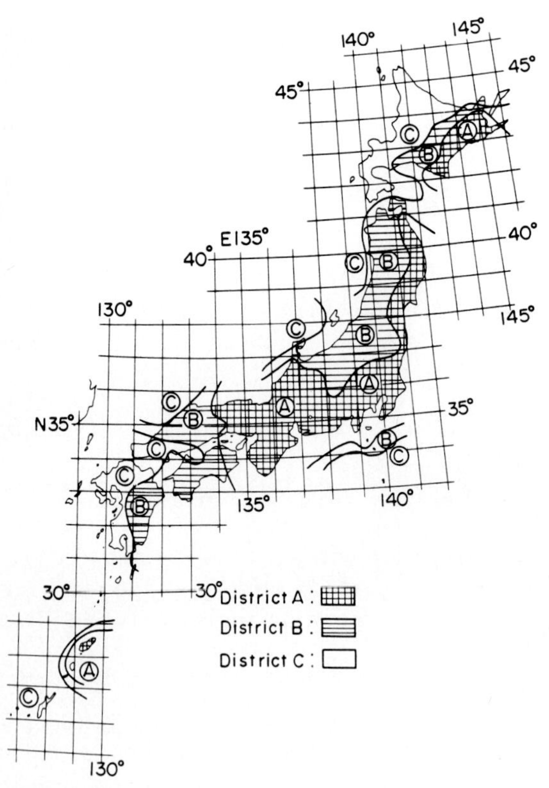

Figure 6.20 Modification factor ν_1 due to location of bridge structures.

TABLE 6.5 Modification Factor ν_2 due to Condition of Ground

Classification	Situations of ground	Modification factor due to ground condition ν_2
1st	1. Ground before tertiary period (thereafter referred to as rock bed) 2. Diluvial soil thinner than 10 m against rock bed	0.9
2d	1. Diluvial soil thicker than 10 m against rock bed 2. Alluvial soil thinner than 10 m against rock bed	1.0
3d	Alluvial soil thinner than 10 m and soft ground thinner than 5 m	1.1
4th	Other soils than the above (for example, artificial island by reclamation)	1.2

(ii) Modified method. If the natural period of vibration T of bridge structures is,

$$T > 0.5 \text{ s} \tag{6.2.27}$$

then JSHB[9] calls for the value of k_{ho} in Eq. (6.2.25) to be modified as

$$k_{km} = \beta k_{ho} \tag{6.2.28}$$

and this modified seismic coefficient k_{hm} must be used for the value of k in Eq. (6.2.24) to determine the seismic force H. The modification factor β can be determined from Fig. 6.21 according to the natural period of vibration of the bridge structures T and the ground classification, listed in Table 6.5.

c. Application to seismic design of curved girder bridges. When the dead load intensity of the curved girder bridge w_d along the bridge axis is known, the seismic force p_{eq} (kgf/m) can be determined from either of

TABLE 6.6 Importance Factor ν_3 due to Category of Bridge

Classification	Importance factor ν_3	Category of bridges
1st	1.0	Bridges for expressway, national highway, important local highway; bridges for important prefectural, municipal, town and village highway
2d	0.8	Other bridges than the above

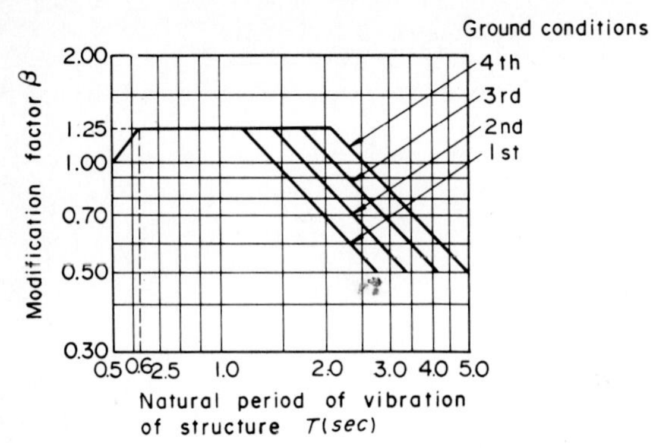

Figure 6.21 Modification factor β.

two methods as follows:

$$p_{\text{eq}} = k_{ho} w_d \qquad \text{fundamental method} \qquad (6.2.29a)$$

or

$$p_{\text{eq}} = k_{km} w_d \qquad \text{modified method} \qquad (6.2.29b)$$

The maximum stress resultants due to these loads can be determined in a manner similar to wind loads, as illustrated in Fig. 6.18. In general, the induced stress in the curved girders is usually found by using either the maximum value of the wind load p_w, derived in the previous section, or seismic load p_{eq}, derived in this section through consideration of the incremental factor for allowable stress under the combinations of various loads (see Table 6.9).

(6) Thermal load T. When the temperature of a straight girder bridge with both ends simply supported is suddenly changed by temperature T (°C), the length of span l will change by

$$\Delta l = \alpha l \, \Delta T \qquad (6.2.30)$$

and no thermal stress is produced for this case, where α = elongation coefficient per unit temperature.

In the case of a curved girder bridge with supporting conditions shown in Fig. 6.22*a*, elongations $\Delta l_i = \alpha l_i$ ($i = 1, 2, 3$) occur in the direction of the interconnecting lines i between a fixed support and the movable supports. For this case, the supports will move in the radial direction by δ_1, δ_2, and δ_3, as shown in Fig. 6.22*b*. If displacements are completely restrained by the bearing shoes, such as $\delta_1 = 0$, $\delta_2 = 0$, and $\delta_3 = 0$, then reactions H_0, H_1, H_2, and H_3 in the radial direction of the curved girder bridge are produced. In addition to a tangential reaction L_0, the frictional forces F_1, F_2, and F_3 are produced, and the thermal stress state can be analyzed by using these reactions.

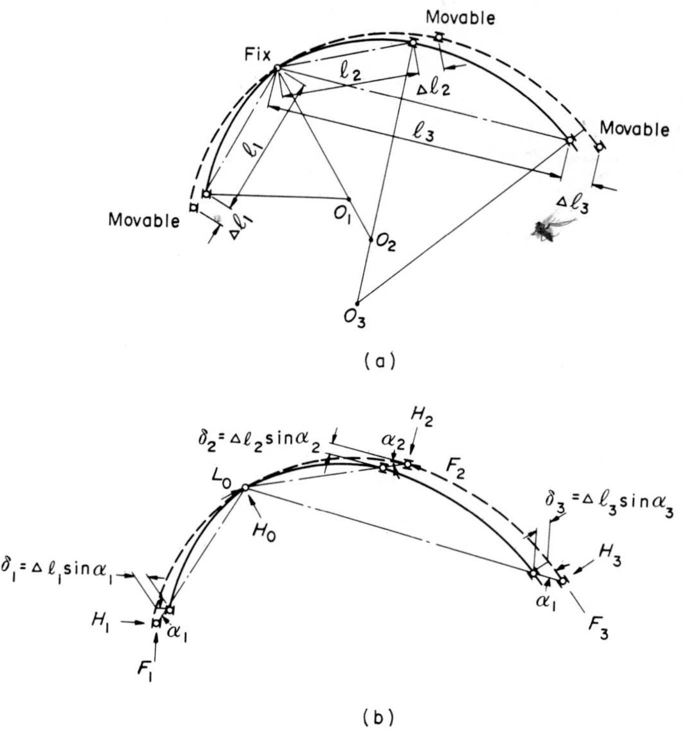

Figure 6.22 Thermal elongation of curved bridge and corresponding reaction: (*a*) movements of supports and (*b*) induced reactions L_0, H_0, H_1, H_2, H_3 to restrain radial movement δ_i and frictional forces F_1, F_2, F_3.

In JSHB[1] the elongation coefficient α is set by

$$\alpha = 10 \times 10^{-6} \tag{6.2.31}$$

for steel bridges, and the ranges of temperature change are given in Table 6.7.

(7) Collision load to handrail C_o. When a vehicle traveling on a highway bridge collides with handrails and causes failure in highway

TABLE 6.7 Range of Temperature Change

Type of bridge	Temperature change	
	Ordinary district	Cold district
Steel bridge with concrete slab	−10°C to +40°C	−20°C to +40°C
Steel bridge with steel deck	−10°C to +50°C	−20°C to +40°C

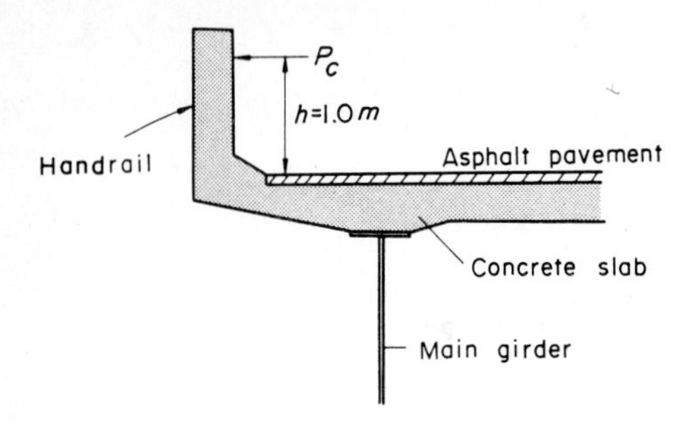

Figure 6.23 Collision load of vehicle.

bridges, the bridge structure and vehicle may be in danger. This collision force p_c (kgf/m) acting along the bridge axis at $h = 1.0$ m above the roadway surface, as shown in Fig. 6.23, is, according to HEPC,[2]

$$p_c = \left(\frac{V}{60}\right)^2 750 + 250 \qquad \text{kgf/m} \tag{6.2.32}$$

where V is the design speed of the vehicle in kilometers per hour. In this case, $p_c = 1000$ kg/m for the standard design speed $V = 60$ km/h.

For curved sections of roadways, the accidents due to these collisions are much more frequent than for straight roadways. Therefore, HEPC sets the collision load p_c at twice the value in Eq. (6.2.32) for sections where the radius of curvature of the highway bridge is smaller than 200 m, that is, $R \leqq 200$ m.

6.2.4 Steel material

In this section the steel materials used in curved girder bridges in Japan are discussed.

(1) Standard of Japanese steel materials. JSHB[1] provides four classes of steel materials, as shown in Table 6.8, where the yield point

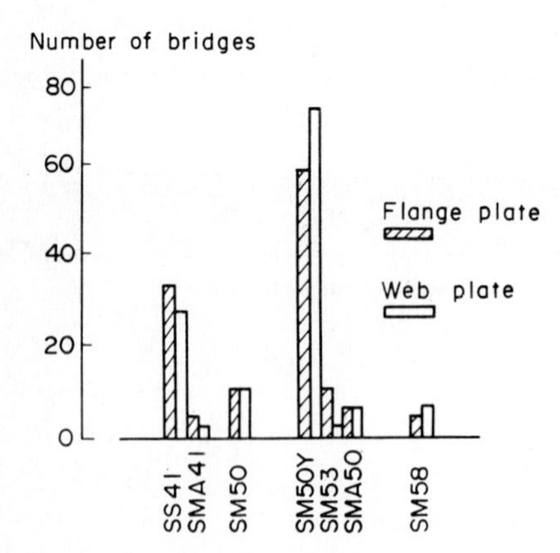

Figure 6.24 Histogram of steel material used in curved girder bridges in Japan.

TABLE 6.8 Japanese Steel Materials

Steel materials	Yield point σ_y, kgf/cm^2	Tensile strength (lower value) σ_u, kgf/cm^2	Allowable stress: Allowable tensile stress σ_{ta}, kgf/cm^2	Allowable stress: Allowable shearing stress τ_a, kgf/cm^2
SS 41 SM 41 SMA 41	2400	4100	1400	800
SM 50	3200	5000	1900	1100
SM 50Y SM 53 SMA 50	3600	5300	2100	1200
SM 58 SMA 58	4600	5800	2600	1500

TABLE 6.9 Combinations of Loads and Corresponding Incremental Factors for Allowable Stress

Cases	Combination of loads	Incremental factor (%)
1	D + L + I	0
2	D + L + I + T	15
3	D + L + I + W	25
4	D + L + I + T + W	35
5	D + L + I + C_o	70
6	W	20
7	D + E_Q + T	70
8	D + E_r	25

Remarks: D = dead load, L = live load, I = impact, T = thermal load, W = wind load, E_Q = seismic load, C_o = collision load, E_r = temporary load during erection.

σ_y, tensile strength σ_u, allowable tensile stress σ_{ta}, and shearing stress τ_a are also indicated. For reference, the histogram of steel materials used for curved girder bridges is plotted in Fig. 6.24.[5]

(2) Incremental factor of allowable stress for combinations of loads. In combining various load effects, as described in Sec. 6.2.3, the incremental factors for the allowable stresses permitted are shown in Table 6.9.[1]

6.2.5 Structural analysis of curved girder bridges

The stress analysis of curved girder bridges can be accomplished by combining four types of load, as illustrated in Table 4.10.[10,11] However, the basic load terms q_y and m_z, which act on a curved girder, are live loads per unit length about the shear center. Therefore, all the design loads should be transferred to the shear center and have the same intensity. The actual load conditions and the corresponding intensities have, therefore, been idealized as summarized in Table 4.10.

The dead load per unit area is $q_y = q_d$, and it is computed by using the relationship

$$R_{\text{out}} - R_{\text{in}} = W \tag{6.2.33}$$

where W is the clear width of roadway, as shown in Fig. 3.32.

In the design code of HEPC, the induced stress resultants and displacements due to these loads are outlined. These analytical methods are detailed now.

(1) Structural analysis of monobox curved girder bridges

***a.* Stress resultants and displacements.** The differential equation for the bending moment M_X can be written by using Eqs. (4.4.61) and (4.4.62*a*) to (4.4.62*c*) and Table 4.10, to give

$$\frac{d^2M_X}{d\phi^2} + M_X = R_s^2\left(-q_y + \frac{m_z}{R_s}\right) \tag{6.2.34}$$

This differential equation can be solved by considering the boundary conditions

$$[M_X]_{\phi=0} = [M_X]_{\phi=\Phi} = 0 \tag{6.2.35}$$

which corresponds to a simply supported curved girder bridge.

The differential equation of the twisting angle of rotation θ can be obtained from Eqs. (4.4.61), (4.4.62*c*), and (3.2.24*a*) as

$$\frac{d^2\theta}{d\phi^2} = \frac{R_s^2}{GK}\left(m_z - \frac{M_X}{R_s}\right) \tag{6.2.36}$$

where the contribution due to warping of the monobox section is neglected. Integrating the above equation twice with respect to $R_s\, d\phi$ gives

$$\theta = \int_0^{\phi}\left[\int \frac{R_s^2}{GK}\left(m_X - \frac{M_X}{R_s}\right) d\phi\right] d\phi + C_1\phi + C_2 \tag{6.2.37}$$

The integration constants C_1 and C_2 can be determined from the boundary conditions

$$[\theta]_{\phi=0} = [\theta]_{\phi=\Phi} = 0 \tag{6.2.38}$$

which corresponds to a simply supported curved girder bridge.

The torsional moment T_s $(=T_z)$ and the shearing force Q_y can be found from Eqs. (4.4.63c) and (4.4.62b):

$$T_z = \frac{GK}{R_s}\frac{d\theta}{d\phi} \tag{6.2.39}$$

$$Q_y = \frac{1}{R_s}\left(\frac{dM_X}{d\phi} + T_z\right) \tag{6.2.40}$$

The vertical deflection w can be determined readily by solving

$$\frac{d^3w}{d\phi^3} + \frac{dw}{d\phi} = R_s^2\left(\frac{T_z}{GK} - \frac{1}{EI_X}\frac{dM_X}{d\phi}\right) \tag{6.2.41}$$

where the boundary conditions imposed are

$$[w]_{\phi=0} = [w]_{\phi=\Phi} = 0 \tag{6.2.42}$$

These correspond to a simply supported curved girder bridge. In addition, the angle of rotation β can be found by solving Eq. (3.2.14a). All the above equations have been solved and reported in References 10 to 17.

***b.* Stresses in monobox curved girder bridges.** In general, the cross-sectional shape of curved girder bridges is always asymmetric owing to the inclination of the deck plate. Therefore, the centroid C does not coincide with the shear center S, as shown in Fig. 6.25. Thus, a new set of X and Y coordinate axes is chosen at centroid C, where the bending moment M_X is applied with respect to the horizontal X axis.

The normal stress σ_b due to bending is therefore given by

$$\sigma_b = \frac{R_s}{\rho}\frac{I_Y Y - I_{XY}X}{I_X I_Y - I_{XY}^2} M_X \tag{6.2.43}$$

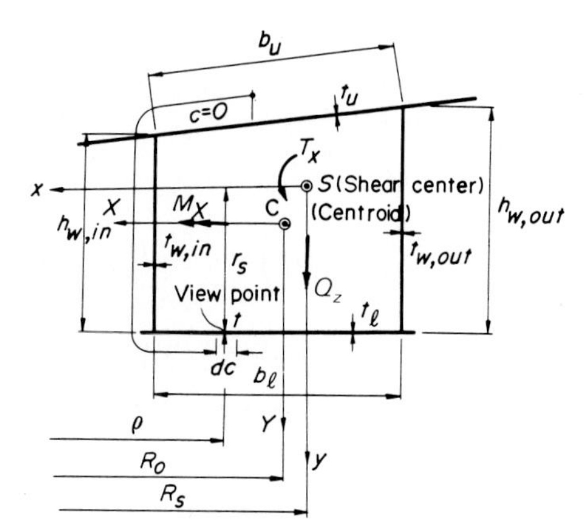

Figure 6.25 Cross section and coordinate axes (X, Y) for monobox curved girder bridges.

for an asymmetric cross section.[10] This expression is applicable provided the moments of inertia I_X, I_Y, and I_{XY} with respect to centroidal X and Y axes are given by

$$I_X = R_o \int_A \frac{Y^2}{\rho} t\,dc \tag{6.2.44a}$$

$$I_Y = R_o \int_A \frac{X^2}{\rho} t\,dc \tag{6.2.44b}$$

and

$$I_{XY} = R_o \int_A \frac{XY}{\rho} t\,dc \tag{6.2.44c}$$

where R_o and ρ are the radii to the centroid and an arbitrary point, respectively. The area of a small segment is denoted $dA = t\,dc$, where the curvilinear coordinate c and thickness t are as shown in Fig. 6.25.

The shearing stress τ_b is given by an expression similar to that used for straight girder bridges[10]

$$\tau_b = \frac{I_Y S_X - I_{XY} S_Y}{(I_X I_Y - I_{XY}^2)\,t} Q_y \tag{6.2.45}$$

where S_X and S_Y represent the static moments with respect to the centroidal axes and are given by

$$S_X = R_o \int_0^c \frac{Y}{\rho} t\,dc + S_{X,o} \tag{6.2.46a}$$

$$S_Y = R_o \int_0^c \frac{X}{\rho} t\,dc + S_{Y,o} \tag{6.2.46b}$$

with $S_{X,o}$ and $S_{Y,o}$ defined as

$$S_{X,o} = \oint \frac{R_o^3}{\rho^3 t} \left(\int_0^c \frac{R_o Y}{\rho} t\,dc \right) dc \left(\oint \frac{R_o^3}{\rho^3 t} dc \right)^{-1} \tag{6.2.47a}$$

$$S_{Y,o} = \oint \frac{R_o^3}{\rho^3 t} \left(\int_0^c \frac{R_o X}{\rho} t\,dc \right) dc \left(\oint \frac{R_o^3}{\rho^3 t} dc \right)^{-1} \tag{6.2.47b}$$

The final equation, required to estimate the shearing stress τ_s due to pure torsion, is given by[10]

$$\tau_s = \left(\frac{R_s}{\rho} \right)^2 \frac{T_s}{Kt} \tilde{q} \tag{6.2.48}$$

where q is the torsional function, defined as

$$\tilde{q} = \frac{\oint (R_s/\rho)^2 r_s \, dc}{\oint (R_s/\rho)^3 (dc/t)} \tag{6.2.49}$$

and the torsional constant K is

$$K = \oint \left(\frac{R_s}{\rho}\right)^3 \tilde{q}^2 \frac{dc}{t} \tag{6.2.50}$$

The term r_s given in Eq. (6.2.49) is the perpendicular distance from the shear center S to a tangent line of a segment dc (see Fig. 6.25).

If the effects of curvature ρ are small, then Eqs. (6.2.43) through (6.2.50) will coincide with well-known formulas for straight girder bridges.[18–20] For example, $\tilde{q}$ can be written by using the symbols shown in Fig. 6.25 as

$$\tilde{q} = \frac{2F}{b_u/t_u + h_{w,\mathrm{in}}/t_{w,\mathrm{in}} + h_{w,\mathrm{out}}/t_{w,\mathrm{out}} + b_l/t_l} \tag{6.2.51}$$

where
$$F = \frac{b_l}{2}(h_{w,\mathrm{in}} + h_{w,\mathrm{out}}) \tag{6.2.52}$$

The torsional constant K is therefore

$$K = \frac{4F^2}{b_u/t_u + h_{w,\mathrm{in}}/t_{w,\mathrm{in}} + h_{w,\mathrm{out}}/t_{w,\mathrm{out}} + b_l/t_l} \tag{6.2.53}$$

Consequently, Eq. (6.2.48) reduces to

$$\tau_s = \frac{T_s}{2Ft} \tag{6.2.54}$$

which is Bredt's formula.

(2) Structural analysis of twin-box curved girder bridges

a. Stress resultants and displacements

(i) Beam theory with warping. The analytical procedure for evaluating the bending moment is the same as that used in analyzing monobox girders. However, the basic equation for the bimoment M_ω, which must be taken into consideration in the case of twin-box girders, is

$$\frac{d^2 M_\omega}{d\phi^2} - \alpha^2 M_\omega = R^2\left(m_z - \frac{M_X}{R_s}\right) \tag{6.2.55}$$

where the boundary conditions applicable for simply supported curved girder bridges are

$$[M_\omega]_{\phi=0} = [M_\omega]_{\phi=\Phi} = 0 \tag{6.2.56}$$

Upon solving Eq. (6.2.55), the twisting angle θ can be determined by integrating Eq. (3.2.23) twice with respect to $R_s\, d\phi$:

$$\theta = R_s^2 \int_0^\phi \left(\int \frac{M_\omega}{EI_\omega}\, d\phi \right) d\phi + C_1\phi + C_2 \tag{6.2.57}$$

The integration constants C_1 and C_2 are determined by applying

$$[\theta]_{\phi=0} = [\theta]_{\phi=\Phi} = 0 \tag{6.2.58}$$

for simply supported curved girder bridges.

By using the solution for θ, St. Venant's torsional moment T_s and the warping torsional moment T_ω can be found by solving

$$T_s = \frac{GK}{R_s}\frac{d\theta}{d\phi} \tag{6.2.59a}$$

$$T_\omega = -\frac{EI_\omega}{R_s^3}\frac{d^3\theta}{d\phi^3} \tag{6.2.59b}$$

as described previously in Eqs. (3.2.28) and (3.2.29). The total torsional moment T_z is given by Eq. (3.2.24*b*), or $T_z = T_s + T_\omega$. The shearing force Q_y can be obtained by solving Eq. (6.2.40).

Finally, the deflection w can be ascertained by solving Eqs. (6.2.41) and (6.2.42), as described previously. These results are summarized in References 10 to 17.

(ii) Grillage theory. Twin-box curved girder bridges with floor beams may be analyzed by idealizing the system as a curved grillage, shown in Fig. 6.26. With this idealization, the complex procedures for analyzing the stresses due to warping are not necessary because the warping of the main girders that have monobox sections is small and so may be ignored. Thus, the structural system will consist of only two constants, $EI_{X,i}$ and GK_i, for each main girder with $i = 1$ and $i = 2$, where EI_Q represents the stiffness of the floor beam.

The analysis of these types of bridges has been studied by many researchers,[14,15,21] so the details of the analytical procedures for evaluating the stress resultants $M_{X,i}$, $Q_{y,i}$, and $T_{s,i}$ $(= T_{z,i})$ in each girder, $i = 1$ and $i = 2$, as shown in Fig. 6.26, are omitted. However, note that the increase in the span length of the floor beam from a to a', shown in Fig. 6.27, must be considered. An estimation of the grid parameter is

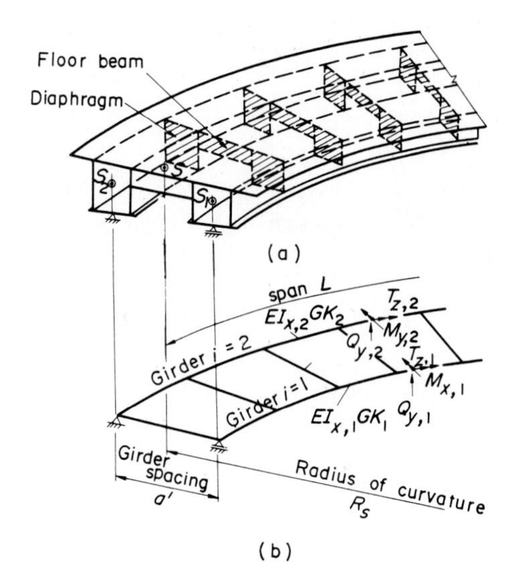

Figure 6.26 Structural system for analyzing twin-box curved girder bridges: (*a*) general view of twin box and (*b*) equivalent grillage system.

given as [22,23]

$$Z_g = \left(\frac{L}{2a}\right)^3 \frac{I_Q^*}{I_Y} \tag{6.2.60}$$

where

$$I_Q^* = \left(\frac{a'}{a}\right)^3 I_Q \tag{6.2.61}$$

***b*. Stresses in twin-box curved girder bridges**

(i) Beam theory with warping. The following stress analysis including warping is necessary in addition to the stress analysis for bending and pure torsion, described in Sec. 3.2.2. The general definition

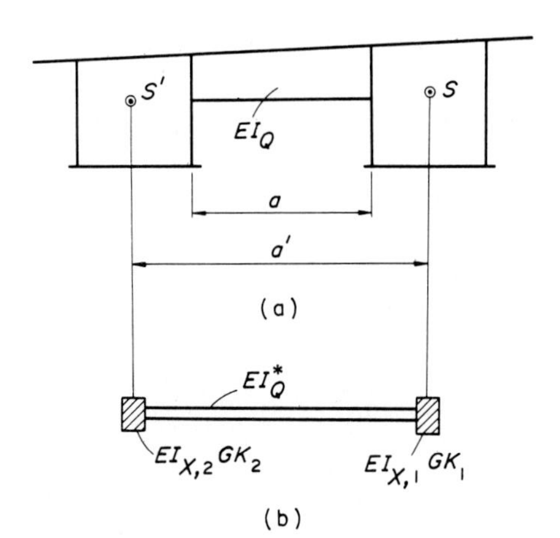

Figure 6.27 Replacing (*a*) twin box to (*b*) equivalent grillage system.

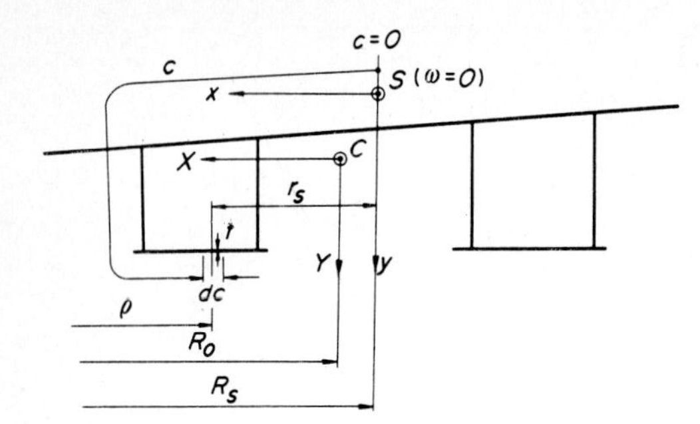

Figure 6.28 Definition of curvilinear coordinate c and arm r_s for twin-box girder.

of the normal warping stress σ_ω can be written as[10]

$$\sigma_\omega = \frac{M_\omega}{I_\omega}\,\omega \tag{6.2.62}$$

where ω is the warping function defined by Eq. (3.2.62), which can be estimated by using the torsional function $\tilde{q}_s$, given by Eq. (6.2.49), and c and r_s, given in Fig. 6.28:

$$\omega = \int_0^c \left(\frac{R_s}{\rho}\right)^3 \tilde{q}\,\frac{dc}{t} - \int_0^c \left(\frac{R_s}{\rho}\right)^2 r_s\,dc \tag{6.2.63}$$

Similarly, the shearing stress τ_ω due to warping can be evaluated by[10]

$$\tau_\omega = \frac{S_\omega}{I_\omega t}\,T_\omega \tag{6.2.64}$$

where

$$S_\omega = R_s \int_0^c \frac{\omega}{\rho}\,t\,dc + S_{\omega,o} \tag{6.2.65}$$

and $S_{\omega,o}$ is defined as

$$S_{\omega,o} = \frac{\oint (R_s/\rho)^3(\omega/t)\,dc}{\oint (R_s/\rho)^3\,(dc/t)} \tag{6.2.66}$$

for each box girder.

An approximate formula[14] that estimates the cross-sectional quantities by assuming the twin-box girder as two monobox girders can be obtained by taking the effective widths $b_{m,1}$ and $b_{m,2}$, as detailed

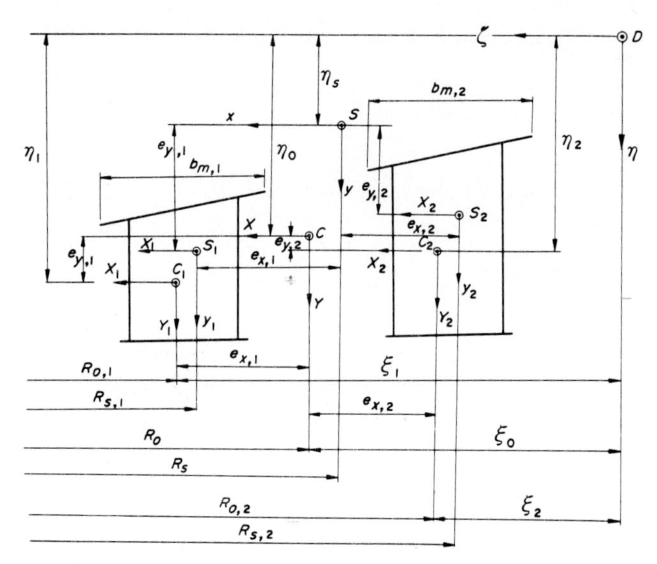

Figure 6.29 Calculation of cross-sectional quantities for twin-box girder.

previously,[1,24] for the deck plates of girders $i = 1$ and $i = 2$, as shown in Fig. 6.29.

The position of centroid C is

$$\xi_o = \frac{\sum_{i=1}^{2} A_i \xi_i / R_{o,i}}{\sum_{i=1}^{2} A_i / R_{o,i}} \tag{6.2.67a}$$

$$\eta_o = \frac{\sum_{i=1}^{2} A_i \eta_i / R_{o,i}}{\sum_{i=1}^{2} A_i / R_{o,i}} \tag{6.2.67b}$$

where A_i, the cross-sectional area for the ith girder, is

$$A_i = R_{o,i} \int_A \frac{1}{\rho} t \, dc \tag{6.2.68}$$

and

$R_{o,i}$ = radius of curvature at centroid C_i for ith girder

ξ_i, η_i = horizontal and vertical distances respectively, between C_i and an arbitrary chosen origin D.

The geometric moment of inertia and product of inertia are

$$I_X = \sum_{i=1}^{2} \frac{R_o}{R_{o,i}} (I_{X,i} + A_i e_{Y,i}^2) \tag{6.2.69a}$$

$$I_Y = \sum_{i=1}^{2} \frac{R_o}{R_{o,i}} (I_{Y,i} + A_i e_{X,i}^2) \tag{6.2.69b}$$

$$I_{XY} = \sum_{i=1}^{2} \frac{R_o}{R_{o,i}} (I_{XY,i} + A_i e_{X,i} e_{Y,i}) \tag{6.2.69c}$$

where $I_{X,i}$, $I_{Y,i}$, $I_{XY,i}$ = geometric moment of inertia and product of inertia, respectively, with respect to (Y_i, Z_i) of ith girder axes

R_o = radius of curvature at centroid C

$e_{X,i}$, $e_{Y,i}$ = horizontal and vertical distances between centroid C_i for ith girder and global centroid C, respectively

The location of the shear center S is

$$R_s = \frac{\sum_{i=1}^{2} I_{X,i}/R_{o,i}}{\sum_{i=1}^{2} I_{X,i}/(R_{s,i}R_{o,i})} \tag{6.2.70a}$$

$$\eta_s = \frac{\sum_{i=1}^{2} I_{Y,i}\eta_i}{\sum_{i=1}^{2} I_{Y,i}} \tag{6.2.70b}$$

where $R_{s,i}$ is the radius of curvature at shear center S_i for the ith girder. The warping function ω is

$$\omega = \left(\frac{R_s}{R_{s,i}}\right)^2 \left[\frac{R_s}{R_{s,i}} \omega_i + \frac{R_{s,i}}{\rho} (e_{x,i} y - e_{y,i} z)\right] \tag{6.2.71}$$

where ω_i = warping function for ith girder with respect to S_i

$e_{x,i}$ $e_{y,i}$ = horizontal and vertical distances, respectively, between S_i and S

The torsional and warping constants are

$$K = \sum_{i=1}^{2} \frac{R_s^4}{R_{s,i}R_{o,i}R_o^2} \left[K_i + \frac{E}{G} \frac{1}{R_{s,i}} (I_{X,i} e_{x,i} + I_{Y,i} e_{y,i})\right] \tag{6.2.72a}$$

and

$$I_\omega = \sum_{i=1}^{2} \frac{R_s^5}{R_{s,i}R_{o,i}R_o} \left[I_{\omega,i} + \frac{R_{o,i}^2}{R_{s,i}R_s} (I_{X,i} e_{x,i}^2 + I_{Y,i} e_{y,i}^2)\right] \tag{6.2.72b}$$

Equations (6.2.67) through (6.2.72) coincide with the formulas[18–20] for straight girder bridges, if the effects of curvature are ignored.

(ii) Grillage theory. The stress analysis of twin-box girder bridges that uses grillage theory results in a category described in Sec. 4.4.1. Therefore estimation of stresses $\sigma_{b,i}$, $\tau_{\omega,i}$, and $\tau_{s,i}$ is necessary only for individual monobox girders $i = 1$ and $i = 2$. Stresses $\sigma_{\omega,i}$ and $\tau_{\omega,i}$ may efficiently involve in those stresses. The validity of this fact is discussed minutely in the next section.

(3) Structural analysis of multiple curved I-girder bridges

a. Stress resultants and displacements

(i) Beam theory with warping. The stress resultants for multiple simple-span curved I-girder bridges can be analyzed by using the warping theory described in Sec. 3.2.2. Such analysis requires the cross-sectional constants of I girders, as shown in Fig. 6.30. These constants can be evaluated as follows, if the effects of curvature are neglected:

$$I_X \cong \left(\frac{h_w}{2}\right)^2 (b_u t_u + b_l t_l) + \frac{t_w h_w^3}{12} - A e_Y^2 \tag{6.2.73}$$

where

$$A \cong b_u t_u + t_w h + b_l t_l = A_u + A_w + A_l \tag{6.2.74}$$

$$e_Y = \frac{h}{2} \frac{A_u - A_l}{A} \tag{6.2.75}$$

$$K = \tfrac{1}{3}(b_u t_u^3 + t_w h^3 + b_l t_l^3) \tag{6.2.76}$$

$$I_\omega = I_{Y,u}(h_u - e_y)^2 + I_{Y,l}(h_l + e_y)^2 \tag{6.2.77}$$

with

$$h_u = \frac{h}{2} - e_Y \qquad h_l = \frac{h}{2} + e_Y \tag{6.2.78a,b}$$

$$I_{Y,u} = \frac{b_u^3 t_u}{12} \qquad I_{Y,l} = \frac{b_l^3 t_l}{12} \tag{6.2.79a,b}$$

$$e_y = \frac{I_{Y,u} h_u - I_{Y,l} h_l}{I_{Y,u} + I_{Y,l}} \tag{6.2.80}$$

If the bridge consists of multiple curved I girders, then Eqs. (6.2.67) through (6.2.72) should be applied. If the effects of curvature can be ignored, then Eqs. (3.3.6) through (3.3.10) should be applied.

However, when a multiple curved I-girder bridge has lateral bracing, as illustrated in Fig. 6.31*a*, the cross-sectional constants can be estimated on the basis of a quasi box girder, assuming there is a thin deck

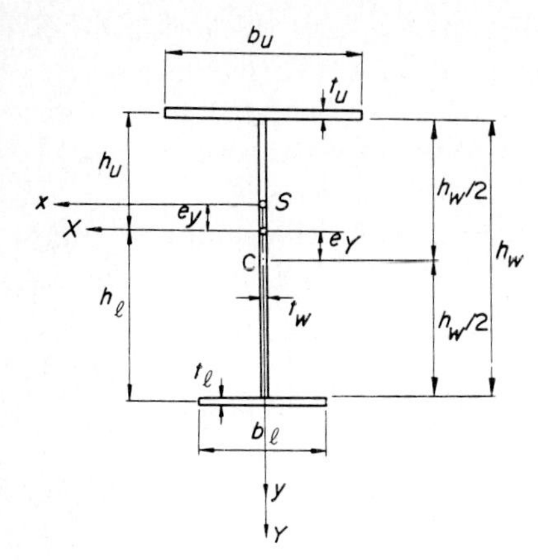

Figure 6.30 Cross section of I girder.

or bottom plate, as illustrated in Fig. 6.31*b*.[12,18,19] The equivalent thickness t_{eq} of such bracing is given by

$$t_{eq} = \frac{G}{E}\frac{a\lambda}{d^3/A_d + 2\lambda^3/(3A_f)} \tag{6.2.81}$$

where A_f, A_d = cross-sectional areas of flange and lateral bracing, respectively

a = spacing of girder

λ = pitch of lateral bracing

Thus, the multiple-I-girder bridge can be changed to a box section with

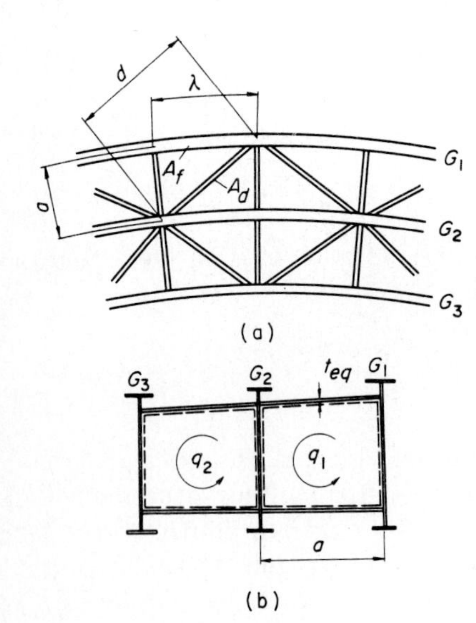

Figure 6.31 Quasi box-girder bridge having lateral bracing: (*a*) spacing of lateral bracing and (*b*) cross section of quasi box.

two cells. The cross-sectional constants can be estimated by expanding the general equations described in Sec. 6.2.8(3). For this cross section, a higher torsional rigidity than that given by Eq. (6.2.76) may be expected, because the shear flows q_1 and q_2 occur for a closed section, as shown in Fig. 6.31.

(ii) Grillage theory. The analytical procedures of multiple curved I-girder bridges are as described in the above section. However, the grid parameter Z_g, defined by Eq. (6.2.60), is different from that for twin-box-girder bridges. According to the definition of grillage theory given by Leonhardt and Andra,[22] the grid parameter is defined as

$$Z_g^* = \frac{\text{deflection of main girder}}{\text{deflection of floor beam}} \tag{6.2.82}$$

Hence the deflection of the main girder is a function of the flexural rigidity EI_X, the torsional rigidity GK, the warping rigidity EI_ω, the curvature R_s, and the central angle Φ. From Eq. (3.3.65), the deflection of the main girder is

$$w_H = \frac{2L^3}{EI_H \pi^4} \nu \tag{6.2.83}$$

However, the midspan floor beam deflection due to a point load $P = 1$ of span $2a$ and flexural rigidity EI_Q can be approximated as

$$w_Q = \frac{2(2a)^3}{EI_Q \pi^4} \tag{6.2.84}$$

The grid parameter Z_g^*, therefore, is

$$Z_g^* = \left(\frac{L}{2a}\right)^3 \frac{I_H}{I_Q} \nu \tag{6.2.85}$$

The parameter ν for W-shaped I girders, as shown in Fig. 6.32, can now be evaluated. First the cross-sectional constants can be determined from Eqs. (6.2.73) through (6.2.80):

$$I_X = \frac{A_f h^2}{2}\left(1 + \frac{1}{6}\frac{A_w}{A_f}\right) \tag{6.2.86}$$

$$K = \frac{A_f t_f^2}{3}\left[2 + \frac{A_w}{A_f}\left(\frac{t_w}{t_f}\right)^2\right] \tag{6.2.87}$$

$$I_\omega = \frac{A_f h^2}{24} b_f^2 \tag{6.2.88}$$

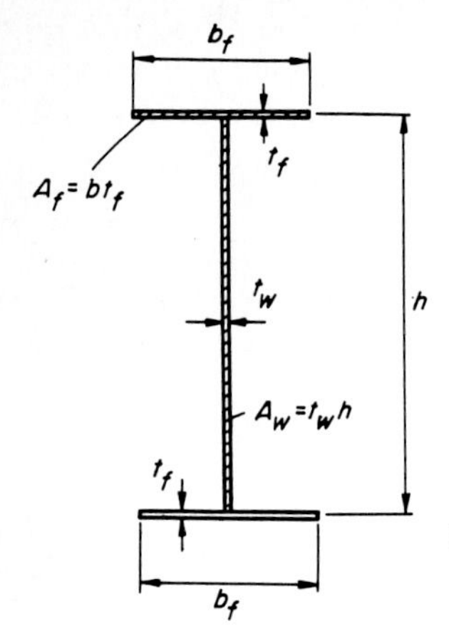

Figure 6.32 W-shaped I girder.

where $A_f = t_f b_f$ and $A_w = t_w h$. Then γ can be written as

$$\nu = \frac{GK}{EI_X} + \frac{I_w}{I_X}\left(\frac{\pi}{L}\right)^2$$
$$= \frac{1}{1 + \frac{1}{6}A_w/A_f}\left[\frac{2 + (A_w/A_f)(t_w/t_f)^2}{3(1+\mu)}\left(\frac{t_f}{h}\right)^2 + \frac{\pi^2}{12}\left(\frac{b_f}{L}\right)^2\right] \quad (6.2.89)$$

where Poisson's ratio $\mu = 0.3$. From this equation, the values of ν are approximately equal to 0.0003 for values of

$$\frac{h}{L} = 20 \quad (6.2.90a)$$

$$\frac{b_f}{h} = \frac{1}{3} \quad (6.2.90b)$$

$$\frac{t_w}{h} = 150 \quad (6.2.90c)$$

$$\frac{t_f}{b_f} = \frac{1}{25} \quad (6.2.90d)$$

Therefore, the curves for ν are plotted and shown in Fig. 6.33. This figure shows that changes in ν between 0.0001 and 0.001 create negligible changes and can be represented by a single curve. Therefore, it is obvious that for increases of values the central angle becomes large. For example, $\nu = 5$ for $\Phi = 0.2$ and $\nu = 17$ for $\Phi = 0.4$.

Accordingly, when the stiffness of the floor beam has the following condition

$$Z_g = \left(\frac{l}{2a}\right)^3 \frac{I_Q}{I_H} \geqq 10 \quad (6.2.91)$$

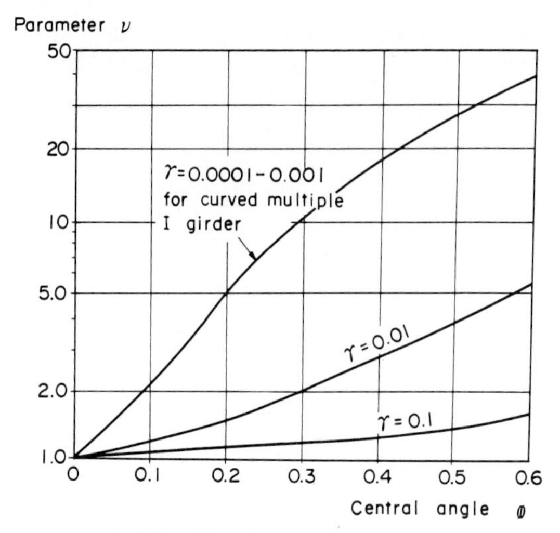

Figure 6.33 Magnification factor ν to estimate grid parameter $Z_g = (L/2a)^3(I_Q/I_H)$.

based on the design criteria for a straight girder bridge,[3] the grid parameter for multiple curved I-girder bridges may be expected to have a value of $Z_g^* = 50$ for $\Phi = 0.2$ and $Z_g^* = 170$ for $\Phi = 0.4$.

b. Stresses in multiple curved I-girder bridges

(i) Beam theory with warping. The stresses σ_b and τ_b due to bending can be evaluated as in Sec. 6.2.5(1) except that $S_{X,o} = S_{Y,o} = 0$, as given in Eq. (6.2.47), because the girder has an open cross section. However, the stress τ_s due to St. Venant's torsional moment T_s must be estimated by

$$\tau_s = \left(\frac{R_s}{\rho}\right)^2 \frac{T_s}{Kt} \tag{6.2.92}$$

rather than Eq. (6.2.48).

The warping stresses σ_ω and τ_ω, can also be evaluated from the general equations (6.2.62) and (6.2.64), respectively, provided that

$$\omega = -\int_0^c \left(\frac{R_s}{\rho}\right)^2 r_s\, dc \tag{6.2.93}$$

and $S_{\omega,o} = 0$, as in Eq. (6.2.65).

For a bridge with lateral bracing, illustrated in Fig. 6.31, the normal stress σ_d of the lateral bracing with length d can be determined by

$$\sigma_d = \frac{q_s d}{A_d} \tag{6.2.94}$$

where $q_s = \tau_s t_{eq}$ is the shear flow for a quasi box section with a thin cover plate t_{eq}, given by Eq. (6.2.81) and illustrated in Fig. 6.31.

(ii) Grillage theory. When the stress resultants $M_{X,i}$, $Q_{X,i}$, $M_{\omega,i}$, $T_{s,i}$, and $T_{\omega,i}$ for the ith girder are calculated by using grillage theory, the stresses can be estimated from the following:
Bending stress,

$$\sigma_{b,i} = \frac{M_{X,i}}{I_{X,i}} Y_i \tag{6.2.95}$$

Shearing stress,

$$\tau_b \cong \frac{Q_{y,i}}{A_{w,i}} \tag{6.2.96}$$

where $I_{X,i}$ is given by Eq. (6.2.73) or (6.2.86) and $A_w = t_w h_w$ is the cross-sectional area of the web plate.

The torsional warping stresses are

$$\tau_{s,i} = \frac{T_{s,i}}{K_i t} \tag{6.2.97a}$$

$$\sigma_{\omega,i} = \frac{M_{\omega,i}}{I_{\omega,i}} \omega_i \tag{6.2.97b}$$

$$\tau_{\omega,i} = \frac{T_{\omega,i}}{I_{\omega,i} t} S_{\omega,i} \tag{6.2.97c}$$

where K_i and $I_{\omega,i}$ are given by Eqs. (6.2.76) and (6.2.77) or (6.2.87) and (6.2.88).

The torsional stresses $\tau_{s,i}$ and $\tau_{\omega,i}$, given above, are generally small compared to the bending stresses. Therefore the proportioning of the I section may be carried out by applying Eqs. (6.2.95) and (6.2.96).

(iii) Local stresses due to curvature of the I girder. In multiple curved I-girder bridges, the out-of-plane bending of the flange due to curvature will occur at those locations where the floor beams or lateral bracing does not support the flanges. These additional stresses will cause yielding at the flanges.[25]

To examine this phenomenon, consider the W-shaped I girder shown in Fig. 6.34*a*. The stress distribution $\sigma_f = (M_X/I_X)(h/2)$ due to bending moment M_X is also given in this figure. The axial force that acts on the lower flange is therefore

$$F_z = \sigma_f A_f + \frac{\sigma_f A_w}{6} = \sigma_f A_f \left(1 + \frac{1}{6}\frac{A_w}{A_f}\right) \tag{6.2.98}$$

This force acts on a small element in the tangential direction, as shown in Fig. 6.35*a*; therefore the effects of curvature will produce the

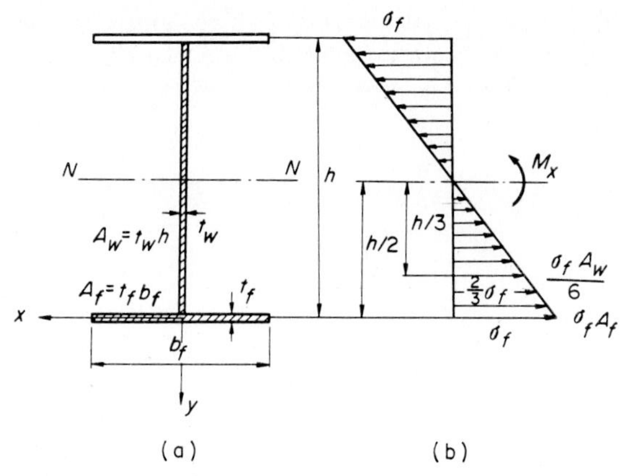

Figure 6.34 (*a*) Cross section of I girder and (*b*) stress distribution.

radial force

$$P_x = F_z\,d\phi = \sigma_f A_f\left(1 + \frac{1}{6}\frac{A_w}{A_f}\right)d\phi \tag{6.2.99}$$

Dividing this force by $ds = R_s\,d\phi$, we get for the intensity of the uniform load q_x

$$q_x = \frac{\sigma_f A_f}{R_s}\left(1 + \frac{1}{6}\frac{A_w}{A_f}\right) \tag{6.2.100}$$

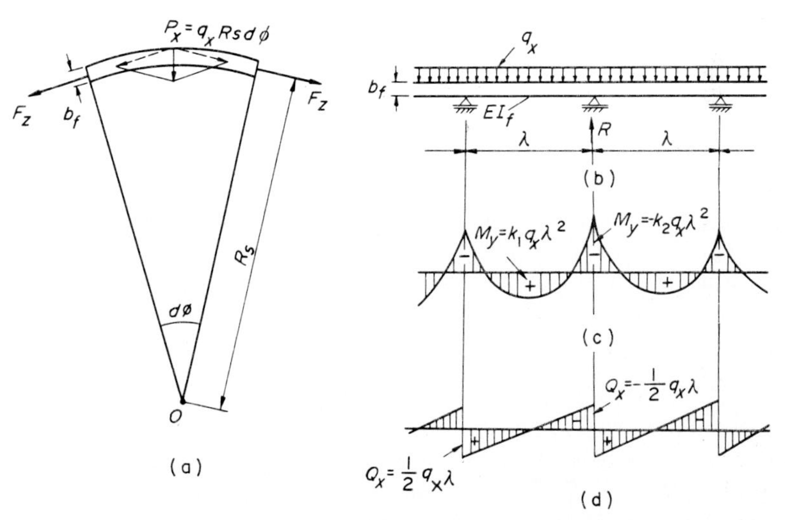

Figure 6.35 Out-of-plane bending of flange: (*a*) radial forces P_x and q_x, (*b*) equivalent continuous beam, (*c*) bending moment, and (*d*) shearing force.

For this horizontal load, the flange will behave as a straight continuous beam, supported by lateral bracing with pitch λ, as shown in Fig. 6.34b. Then the bending moment of the flange with respect to the x axis, given in Fig. 6.35c, is

$$M_y = \pm k_{1,2} q_x \lambda^2$$
$$= \pm k_{1,2} \frac{\sigma_f A_f}{R_s} \left(1 + \frac{1}{6}\frac{A_w}{A_f}\right) \lambda^2 \tag{6.2.101}$$

And the shearing force in the direction of the x axis, shown in Fig. 6.35d, is

$$Q_x = \pm \tfrac{1}{2} q_x \lambda$$
$$= \pm \frac{\sigma_f A_f}{2R_s} \left(1 + \frac{1}{6}\frac{A_w}{A_f}\right) \lambda \tag{6.2.102}$$

where $k_{1,2}$ in Eq. (6.2.101) is a constant approximated as

$$k_{1,2} \cong 0.106 \tag{6.2.103}$$

which is the maximum value required in a practical design.[14]

Accordingly, the additional normal stress is

$$\sigma_{\omega f} = \frac{M_z}{I_f} \frac{b_f}{2}$$
$$= \pm k_{1,2} \frac{\sigma_f A_f}{W_f R_s} \left(1 + \frac{1}{6}\frac{A_w}{A_f}\right) \lambda^2 \tag{6.2.104}$$

which will occur at the outer or inner side of the flanges, where W_f is the section modulus of the flange, or

$$W_f = \frac{I_f}{b_f/2} = \frac{t_f b_f^2}{6} \tag{6.2.105}$$

The additional maximum shearing stress $\tau_{\omega f}$ at the middle of the flange can be written as

$$\tau_{\omega f} = \frac{3}{2}\frac{Q_x}{A_f} = \pm \frac{3}{2}\frac{\sigma_f \lambda}{2R_s} \left(1 + \frac{1}{6}\frac{A_w}{A_f}\right) \tag{6.2.106}$$

Note that these stresses should be added to the results obtained from beam theory with warping or the grillage theory. The stress distribution of the lower flange at the junction point where the lateral bracing is attached to the girder is shown in Fig. 6.36.

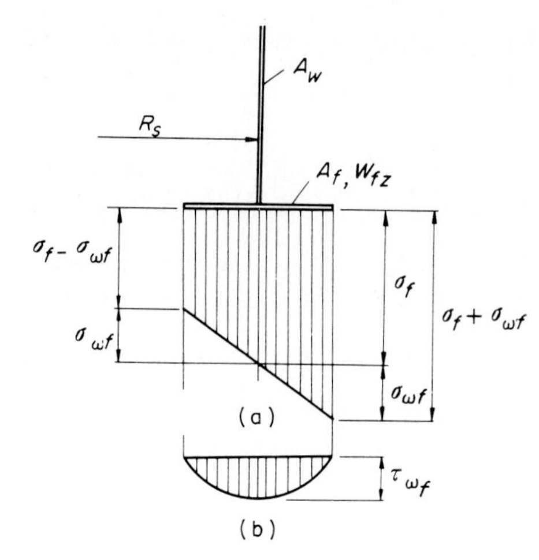

Figure 6.36 Stress distributions in lower flange: (*a*) normal and (*b*) shearing stress.

c. Difference between beam theory with warping and grillage theory.[3] To clarify the difference between beam theory with warping and grillage theory, consider a simple straight girder bridge consisting of two W-shaped I girders with span length l. At the midspan of the girder, a concentrated load P that acts eccentrically e from the centerline of the cross section is applied. The distortions of the cross section are effectively prevented by the rigid sway bracing, shown in Fig. 6.37*a*.

The forces that act on this bridge, assuming a single beam, are divided into a vertical force P and torque $T = Pe$, as shown in Fig. 6.37*b*. The bending stress is obtained from

$$\sigma_b = \frac{M_x}{I_x}\frac{h}{2} = \frac{M_x}{2I_H}\frac{h}{2} \tag{6.2.107}$$

as given by Eq. (6.2.95).

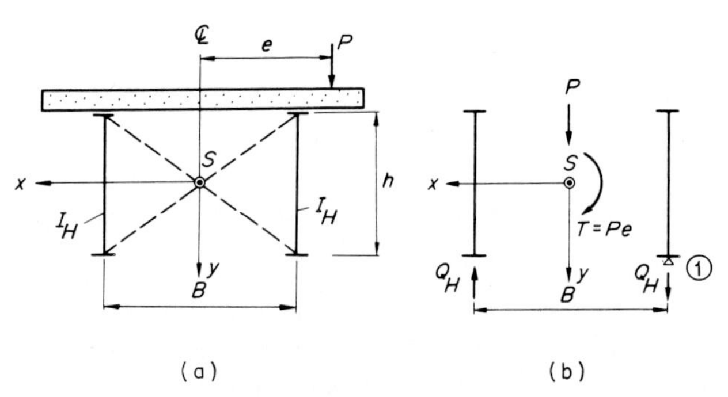

Figure 6.37 Multiple-I-girder bridge subjected to eccentric load: (*a*) cross-sectional dimension and (*b*) acting forces.

The shearing force Q_H is induced by the torque T; therefore an equilibrium condition for these forces gives

$$Q_H \cong \frac{T}{B} = \frac{Pe}{B} \tag{6.2.108}$$

if we neglect the small contribution of St. Venant's torsional moment T_s. This shearing force Q_H also produces an additional bending moment M_H, as shown in Fig. 6.38. Therefore,

$$M_H = \frac{Q_H l}{4} = \frac{Ple}{4B} = \frac{e}{B} M_x \tag{6.2.109}$$

The bimoment M_ω is the product of M_H and B:

$$M_\omega = M_H B = M_x e \tag{6.2.110}$$

The warping stress σ_ω can thus be estimated from Eqs. (6.2.97*b*) and (6.2.110). Referring to point 1, shown in Fig. 6.37*b*, we see that

$$\sigma_\omega = \frac{M_\omega}{I_\omega} \omega_1 = \frac{M_x e}{I_H B^2/2} \frac{Bh}{4} = \frac{M_x}{I_H} \frac{h}{2} \frac{e}{B} \tag{6.2.111}$$

The total stress σ is therefore the sum of Eqs. (6.2.107) and (6.2.111):

$$\sigma = \sigma_b + \sigma_\omega = \frac{M_x}{I_H} \left(\frac{1}{2}\right) \left[\frac{1}{2}\left(1 + \frac{2e}{B}\right)\right] \tag{6.2.112}$$

This equation can be used to estimate for various eccentricities e the

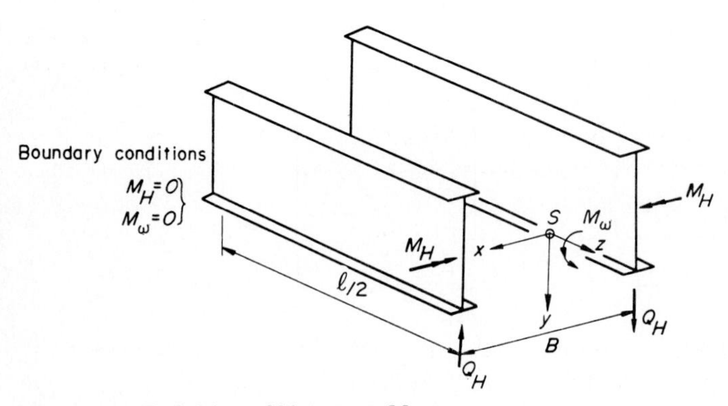

Figure 6.38 Definition of bimoment M_ω.

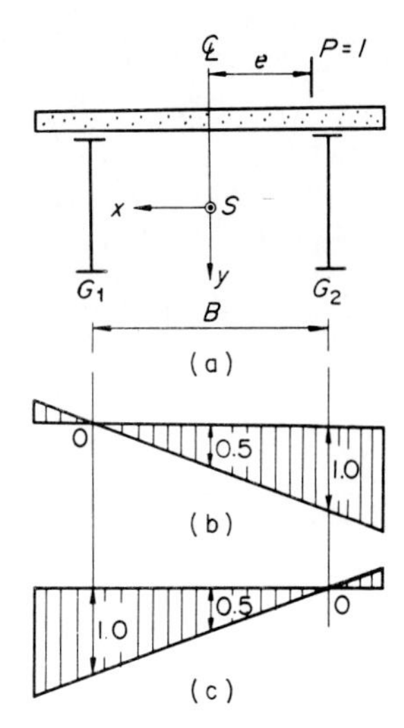

Figure 6.39 Load distribution factors for (*a*) twin I-girder bridge, (*b*) G_2, and (*c*) G_1 influence line.

following stress results:

$$\sigma = \begin{cases} \dfrac{M_x}{I_H}\dfrac{h}{2}(1.0) & \text{for } e = \dfrac{B}{2} \qquad (6.2.113a) \\[2ex] \dfrac{M_x}{I_H}\dfrac{h}{2}(0.5) & \text{for } e = 0 \qquad (6.2.113b) \\[2ex] \dfrac{M_x}{I_H}\dfrac{h}{2}(0) & \text{for } e = -\dfrac{B}{2} \qquad (6.2.113c) \end{cases}$$

The values enclosed in parentheses in the above equations agree with the conventional values of the load distribution factor for straight girder bridges, as shown in Fig. 6.39. The validity of beam theory with warping, therefore, can be seen from the following example.

Consider now the grillage girder shown in Fig. 6.40. In this bridge, $I_x = 3I_H$, so the normal stress due to bending is

$$\sigma_b = \frac{M_x}{I_H}\left(\frac{h}{2}\right)\left(\frac{1}{3}\right) \qquad (6.2.114)$$

However, the middle girder G_2 does not contribute to the warping, so the warping stress σ_ω is the same as that given by Eq. (6.2.111).

Therefore by adding Eqs. (6.2.114) and (6.2.111), the total stress σ can

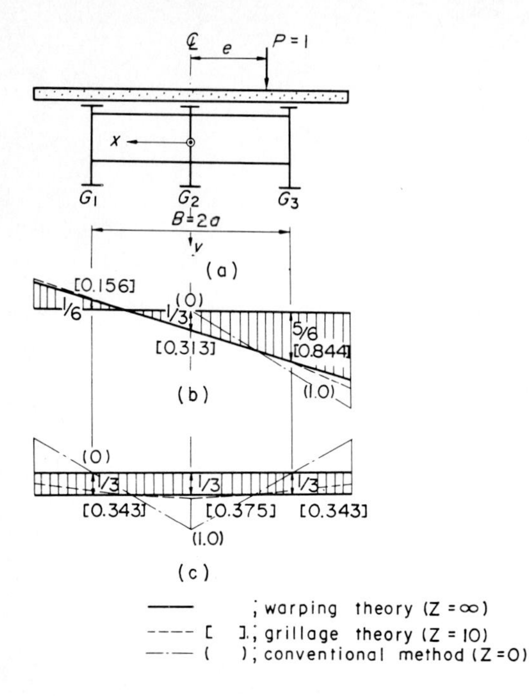

Figure 6.40 Load distribution factors for (*a*) grillage girder, (*b*) G_3, and (*c*) G_2 influence line.

be obtained:

$$\sigma = \begin{cases} \dfrac{M_x}{I_H}\dfrac{h}{2}\left(\dfrac{5}{6}\right) & \text{for } e = \dfrac{B}{2} \qquad (6.2.115a) \\ \dfrac{M_x}{I_H}\dfrac{h}{2}\left(\dfrac{1}{3}\right) & \text{for } e = 0 \qquad (6.2.115b) \\ \dfrac{M_x}{I_H}\dfrac{h}{2}\left(-\dfrac{1}{6}\right) & \text{for } e = -\dfrac{B}{2} \qquad (6.2.115c) \end{cases}$$

Figure 6.40 shows the corresponding load distribution factors, calculated from the above equations, which are shown by the solid line; the dotted line represents the conventional values.

Now, applying the grillage theory,[22,23] we obtain the following load distribution factor k_{ij}:
For girder G_1,

$$k_{11} = \frac{4 + 5Z_g}{4 + 6Z_g} \qquad (0.844 \text{ at } Z_g = 10) \qquad (6.2.116a)$$

$$k_{21} = \frac{Z_g}{2 + 3Z_g} \qquad (0.313 \text{ at } Z_g = 10) \qquad (6.2.116b)$$

$$k_{31} = -\frac{Z_g}{4 + 6Z_g} \qquad (-0.156 \text{ at } Z_g = 10) \qquad (6.2.116c)$$

For girder G_2,

$$k_{12} = k_{21} = k_{31} \tag{6.2.117a}$$

$$= \frac{Z_g}{2+3Z_g} \qquad (0.313 \text{ at } Z_g = 10) \tag{6.2.117b}$$

$$k_{22} = \frac{2+Z_g}{2+3Z_g} \qquad (0.375 \text{ at } Z_g = 10) \tag{6.2.117c}$$

Here Z_g is the grid parameter given by Eq. (6.2.91). For example, by substituting $Z_g = 10$, the load distribution factors can be calculated as indicated by the values in parentheses in Eqs. (6.2.116) and (6.2.117). Corresponding load distribution curves are shown by the dotted line in Fig. 6.40.

For large values of $Z_g^* \gg 10$, as in curved grillage girder bridges, shown in Sec. 6.2.8(2), a greater load distribution factor can be expected. Therefore, taking the limit of k_{ij} for an infinite value of Z_g, we get

$$\lim_{Z_g \to \infty} k_{ij} \tag{6.2.118}$$

This gives values that coincide with the terms in parentheses in Eq. (6.2.115).

This condition suggests that the analytical result obtained from beam theory with warping is similar to that obtained from grillage theory.

6.2.6 Check for stress and deflection

(1) Check for stress. The stress analysis of a curved girder bridge gives the following criteria[1]:

$$\sigma = \sigma_b + \sigma_\omega \leqq \sigma_a \tag{6.2.119a}$$

$$\tau = \tau_b + \tau_s + \tau_\omega \leqq \tau_a \tag{6.2.119b}$$

and

$$\sqrt{\sigma^2 + 3\tau^2} \leqq \sigma_{ay} \tag{6.2.120}$$

where σ_a, σ_{ay}, and τ_a are the allowable unit stresses. However, if the stress analysis is performed by using the methods described in Sec. 3.3.1, according to the design code of HEPC[2] the reduced stresses can be classified as in Table 6.10, corresponding to the cross-sectional shape of the curved girder bridge. In this table, the parameter κ is

$$\kappa = L\sqrt{\frac{GK}{EI_\omega}} \tag{6.2.121}$$

TABLE 6.10 Combinations of Stresses in Curved Girder Bridges

Condition for κ (cross section)	Beam theory with warping	Grillage theory
$\kappa > \kappa_{cr}$	$\sigma = \sigma_b$ $\tau = \tau_b + \tau_s$	—
$\kappa < \kappa_{cr}$ $i = 1$, 2	$\sigma = \sigma_b + \sigma_\omega$ $\tau = \tau_b + \tau_s + \tau_\omega$	$\sigma_i = \sigma_{b,i}$ $\tau = \tau_{b,i} + \tau_{s,i}$ $(i = 1, 2)$
$\kappa << \kappa_{cr}$ $i = 1$, 2 ⋯ n	$\sigma_i = \sigma_{b,i} + \sigma_{\omega,i} + \tau_{\omega,i}$ $\tau_i = \tau_{b,i} + \tau_{s,i} + \tau_\omega + \tau_{\omega f,i}$ $(i = 1, 2, \ldots, n)$	$\sigma_i = \sigma_{b,i} + \sigma_{\omega f.i}$ $\tau_i = \tau_{b,i} + \tau_{\omega f.i}$ $(i = 1, 2, \ldots, n)$

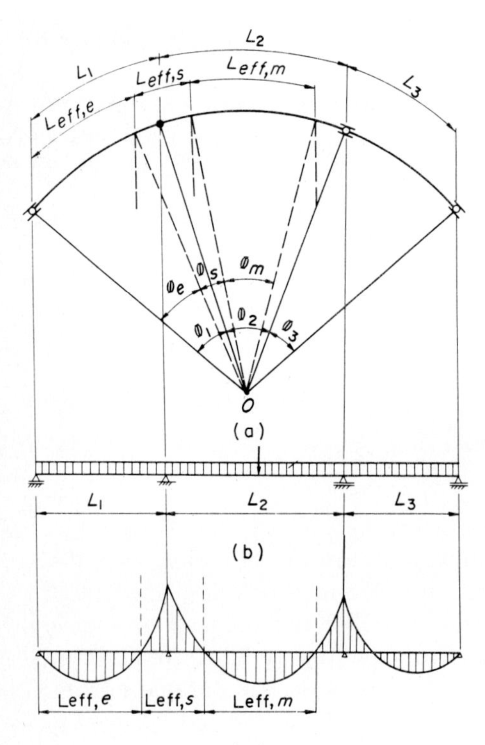

Figure 6.41 Effective span to estimate parameter for continuous curved girder bridge: (*a*) general plane, (*b*) load condition, and (*c*) bending moment diagram.

When the curved girder bridge is continuous, the effective span $L = L_{\text{eff}}$ can be determined as follows. First, assume that the maximum design loads act on the continuous girder bridge, as shown in Fig. 6.41*b*. The bending moment diagram is shown in Fig. 6.41*c* on the basis of straight beam theory.[3] The effective span L_{eff} for each span can be approximated, then, by taking the distance between the points where the moments are equal to zero, because the bimoment diagram will have a similar slope.[11]

The stresses σ_b, σ_ω, τ_b, τ_s, and τ_ω can be estimated as discussed in Sec. 6.2.5. The stress $\sigma_{\omega f}$, however, should be included in multiple curved I-girder bridges. In estimating the flexural stress σ_b, the effective width of the flange plate must, of course, be taken into account. For example, in the design code of HEPC the same effective width

$$\lambda = \lambda_1 = \lambda_2 \tag{6.2.122}$$

is specified for a curved box girder, illustrated in Fig. 6.42, based on the analytical results shown in Sec. 3.2.4. And the effective width λ can be evaluated according to the provision of JSHB[1] as follows:

$$\lambda = b \qquad \frac{b}{L} \leqq 0.05 \tag{6.2.123a}$$

$$\lambda = \left[1.1 - 2\left(\frac{b}{L}\right)\right]b \qquad 0.05 < \frac{b}{L} < 0.30 \tag{6.2.123b}$$

$$\lambda = 0.15L \qquad 0.30 \leqq \frac{b}{L} \tag{6.2.123c}$$

$$\lambda = b \qquad \frac{b}{L} \leqq 0.02 \tag{6.2.124a}$$

$$\lambda = \left[1.06 - 3.2\left(\frac{b}{L}\right) + 4.5\left(\frac{b}{L}\right)^2\right]b \qquad 0.02 < \frac{b}{L} < 0.30 \tag{6.2.124b}$$

$$\lambda = 0.15L \qquad 0.30 \leqq \frac{b}{L} \tag{6.2.124c}$$

where λ = effective width in Fig. 6.43
b = half of web plate spacing or width of outstanding flange plate in Fig. 6.43
L = equivalent span length in Table 6.11

Although there is another kind of stress—distortional warping due to the distortion of the cross section—these stresses can be effectively

TABLE 6.11 Formulas to Determine Effective Width in JSHB

Girder	Section	Effective width			Remarks
		Symbol	Equation	Equivalent span length	
Simple	1	λL	(6.2.123)	L	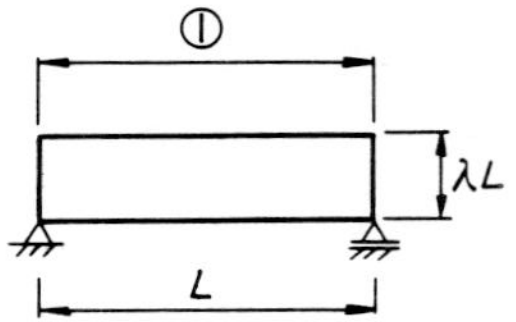 ① λL L
Continuous	1	λL_1	(6.2.123)	$0.8L_1$	① ② ③ ④ ⑤ ⑥ ⑦ ⑧
	5	λL_2		$0.6L_2$	λL_1 λS_1 λL_2 λS_2
	3	λS_1	(6.2.124)	$0.2(L_1 + L_2)$	$0.2L_1$ $0.2L_2$ $0.2L_2$ $0.2L_3$
	7	λS_2		$0.2(L_2 + L_3)$	L_1 L_2 L_3
	2 4 8 8	Linearly interpolate, using value of both sides			

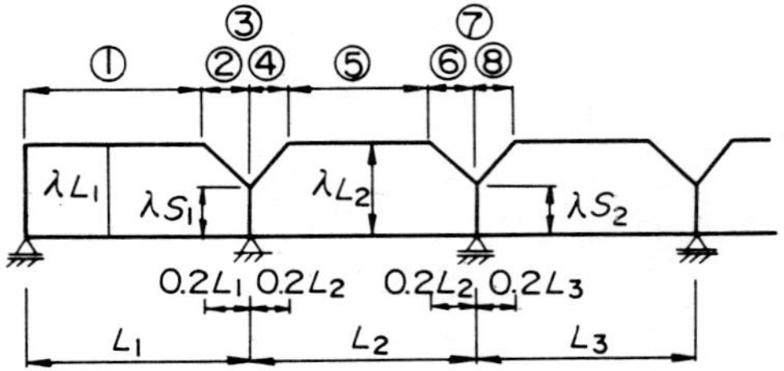

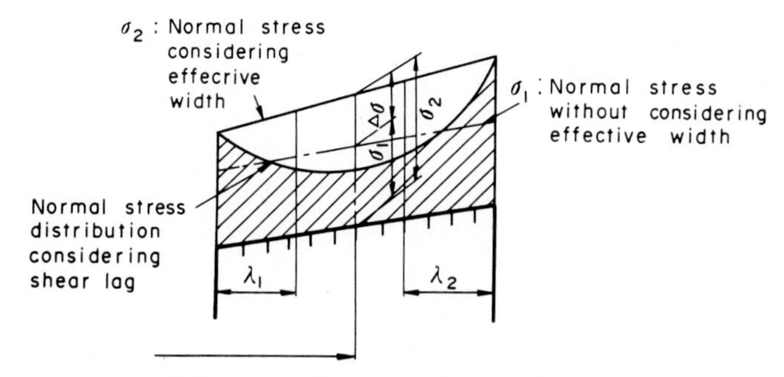

Figure 6.42 Effective widths λ_1 and λ_2 of flange plate of box girder.

prevented by the use of diaphragms with appropriate stiffness and spacing.[12,13,26] Further details are given in Sec. 6.2.8(4).

(2) Check for deflection. In JSHB,[1] the deflection δ_L due to the live load without impact of an ordinary straight girder bridge having span length l (m) should be limited by the following inequality:

$$\delta_L \leqq \begin{cases} \dfrac{l}{2000} & l \leqq 10\text{ m} \qquad (6.2.125a) \\ \dfrac{l}{20000/l} & 10\text{ m} < l \leqq 40\text{ m} \qquad (6.2.125b) \\ \dfrac{l}{500} & l > 40\text{ m} \qquad (6.2.125c) \end{cases}$$

For a curved girder bridge, the same criteria must be applied to the deflection δ_L of a curved girder bridge due to the live load without impact by taking

$$\delta_L = w + x\beta \qquad (6.2.126)$$

and span length L (m), or

$$L = R_s\Phi \qquad (6.2.127)$$

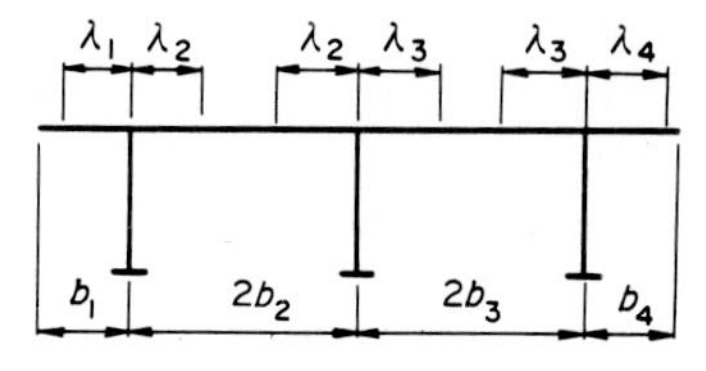

Figure 6.43 Effective widths λ_1 to λ_4.

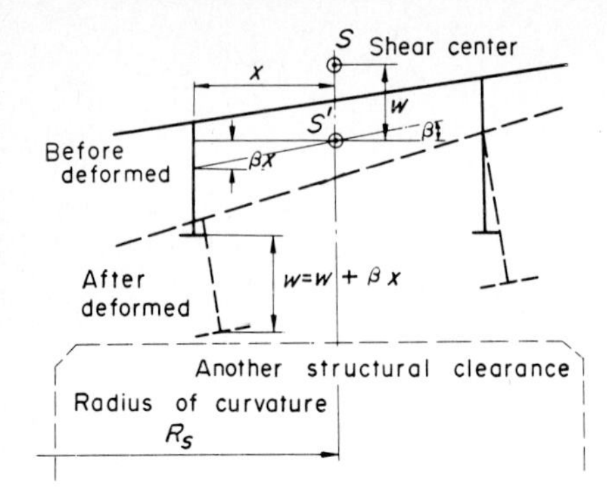

Figure 6.44 Deflection of curved girder bridge.

of the curved girder, to keep the structural clearance as shown in Fig. 6.44, where

w = deflection at shear center S

β = rotational angle of cross section around shear center S

x = distance between shear center S and considered point

R_s = radius of curvature of curved girder at shear center S (m)

Φ = central angle of curved girder (rad)

6.2.7 Design of main girders

The procedures for designing curved girders can be summarized by the flowchart in Fig. 6.45. The details of the design procedures of each girder element are explained later; note that there are two large feedback processes in this figure.

The first feedback is related to the check for the weight of the girders. To simplify this trial design procedure, the weight of the steel girder is predicted from Fig. 6.11*a* to *c*. Then the cross section of a curved girder, such as the web and flange plates, must be carefully proportioned according to the design method shown in previous sections.

The second large feedback procedure is concerned with the design criterion regarding the deflection. If the deflection of a curved girder bridge becomes significantly large, the structural clearance shown in Fig. 6.44 is not within acceptable limits. Moreover, the vibrational response due to various dynamic forces is predominant. To avoid such troublesome behavior, the depth of the girder must be designed within a certain limit, as shown in the next section.

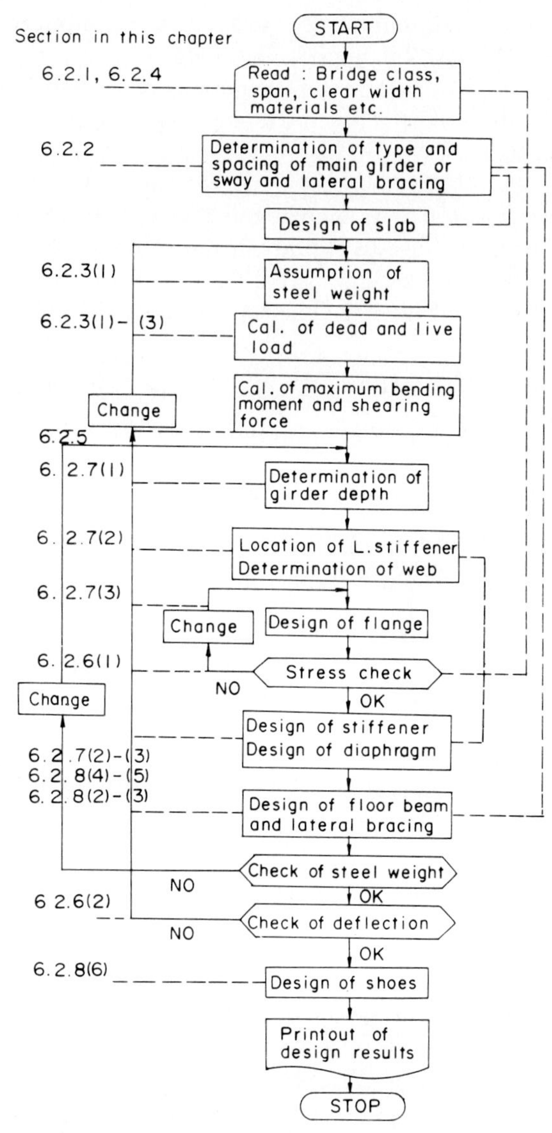

Figure 6.45 Flowchart for designing curved girder bridges.

In addition, there are so many relationships among the design details for each element of a curved girder bridge, as indicated by the dashed lines in Fig. 6.45, that we should change or adjust the details of other elements that may be affected.

Thus, the design of curved girder bridges will not be completed with a single design procedure; there will usually be a few trial and repeated processes. For the execution of trial-and-error methods, the computer is a powerful tool to reduce human labor and mistakes.

Accordingly, the design procedure of each element is discussed in the following sections together with flowcharts in cases where the design method is complicated.

(1) Determination of girder depth. As mentioned above, the girder depth is an important factor in the design of plate girder bridges. Thus, let us consider the simple case of a straight steel I girder with a symmetric cross section, as shown in Fig. 6.46. The total weight W of this steel girder is

$$W = \gamma(2A_f + 1.6A_w)l \tag{6.2.128}$$

where A_f = cross-sectional area of flange plate

$$1.6A_w = 1.6ht_w \tag{6.2.129}$$

= cross-sectional area of web plate including area of stiffeners increased by 60%

h = girder depth

t_w = thickness of web plate

γ = steel weight per unit volume (=7.85 tf/m^3 in JSHB[1])

l = span length of girder

The cross-sectional area A_f of the flange plate can also be related to the maximum bending moment M_{max} under the load condition illustrated in Fig. 6.46*b* and *c*:

$$A_f = \frac{0.8M_{max}}{\sigma_a h} - \frac{ht_w}{6} \tag{6.2.130}$$

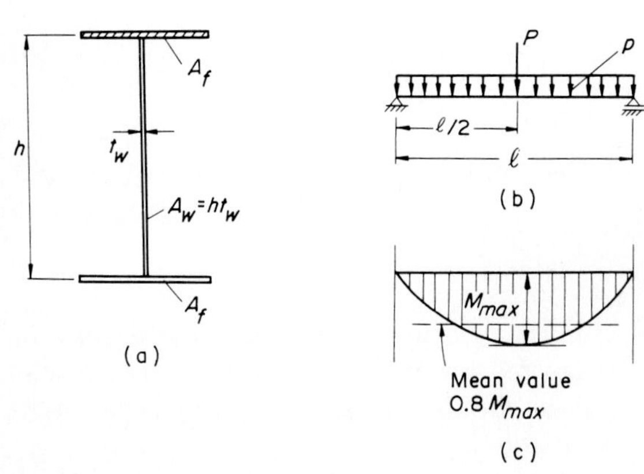

Figure 6.46 (*a*) Cross section of I girder, (*b*) load condition, and (*c*) bending moment to determine girder depth.

This is detailed later in Eq. (6.2.141), where σ_a is the allowable stress of the flange plate.

Substituting Eq. (6.2.130) into Eq. (6.2.128) and taking the derivative of W with respect to h give the following condition to minimize the steel weight:

$$\frac{dW}{dh} = \left[-1.6 \frac{M_{\max}}{\sigma_a h^2} + (1.6 - \tfrac{1}{3}) t_w \right] \gamma l = 0 \tag{6.2.131}$$

From this equation, the optimum girder depth h is represented by[27]

$$h = \sqrt{\frac{1.60}{1.27}} \sqrt{\frac{M_{\max}}{\sigma_a t_w}} \cong 1.1 \sqrt{\frac{M_{\max}}{\sigma_a t_w}} \tag{6.2.132}$$

The deflection w of the main girder under the load condition shown in Fig. 6.46*b*, with the exception of a concentrated load P whose influence is usually small, can be written as

$$\delta = \frac{5pl^4}{384EI} \tag{6.2.133}$$

where p is the intensity of the uniformly distributed load.

The corresponding induced flexural normal stress σ is given by

$$\sigma = \frac{M}{I} \frac{h}{2} = \frac{pl^2}{8I} \frac{h}{2} \tag{6.2.134}$$

where I is the geometric moment of inertia of the steel girder.

By eliminating the load term p from Eqs. (6.2.133) and (6.2.134), the following relationship between the girder depth h and span length l is

$$\frac{h}{l} = \frac{5}{24} \left(\frac{l}{\delta} \right) \left(\frac{\sigma}{E} \right) \tag{6.2.135}$$

Here, if we set $l/\delta = 500$ according to the criteria of JSHB,[1] as shown in Eq. (6.2.125*c*) with $E = 2.1 \times 10^6$ kgf/cm^2, and if we approximate the stress as the mean value throughout the girder $\sigma = 1000$ kgf/cm^2, then the girder depth h is[27]

$$h \cong \frac{l}{20} \tag{6.2.136}$$

As an example, the optimum girder depth h is indicated in HEPC[2] through trial design shown in Fig. 6.47.

For a curved girder bridge, the optimum girder depth h is determined as above.

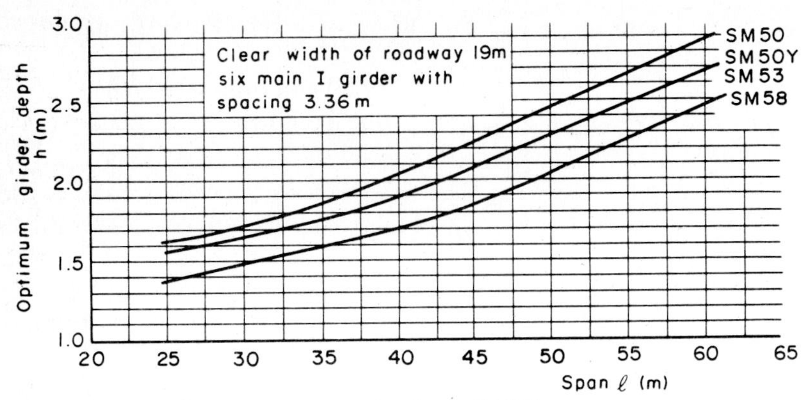

Figure 6.47 Example of optimum girder depth (HEPC).

(2) Design of web plate. Although the design method for the web plate of straight girders in JSHB is discussed in detail in Sec. 5.4.1, the maximum web depth h_w cm corresponding to thickness t_w (cm) of the web plate is listed in Table 6.12. For example, the web slenderness h/t_w of curved girder bridges is to be plotted from a survey of data[5] and is shown in Fig. 6.48.

Furthermore, the required relative stiffness $\gamma_{l,\text{req}}$ of stiffeners for a straight girder, i.e.,

$$\gamma_{l,\text{req}} = 30\left(\frac{a}{h_w}\right) \qquad \text{longitudinal stiffener} \tag{6.2.137}$$

$$\gamma_{t,\text{req}} = 8\left(\frac{h_w}{a}\right)^2 \qquad \text{transverse stiffener} \tag{6.2.138}$$

and the actual relative stiffnesses γ_l^c and γ_t^c, of curved girder bridges were also investigated[5] and are plotted in Figs. 6.49 and 6.50, respectively. From these figures, it seems that the design method of web

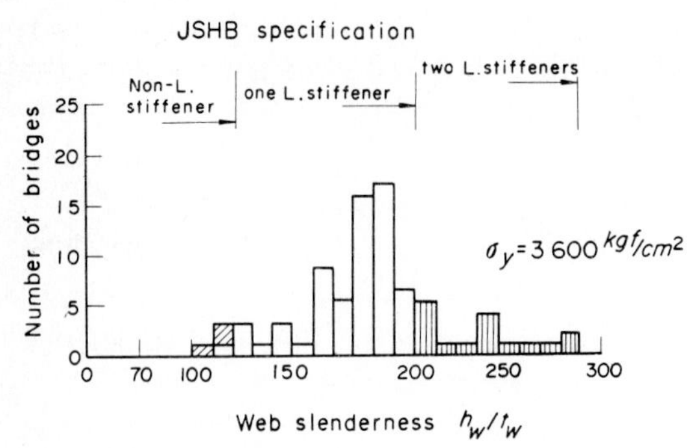

Figure 6.48 Histogram of h_w/t_w according to survey.[5]

TABLE 6.12 Web Depth h_w (cm) and Corresponding Location of Longitudinal Stiffeners

	Web thickness t_w (cm)	h_w (cm) Steel grade: SS 41, SM 41, SMA 41	SM 50	SM 50Y, SM 52, SMA 50	SM 58, SMA 58
No L stiffener	$t_w = 0.8$	121.6	104.0	98.4	88.0
	$t_w = 0.9$	136.8	117.0	110.7	99.0
	$t_w = 1.0$	152.0	130.0	123.0	110.0
	$t_w = 1.1$	167.2	143.0	135.3	121.0
One L stiffener	$t_w = 0.8$	204.8	176.0	167.2	150.4
	$t_w = 0.9$	230.4	198.0	188.1	169.2
	$t_w = 1.0$	256.0	220.0	209.0	188.0
	$t_w = 1.1$	281.6	242.0	229.9	206.8
Two L stiffeners	$t_w = 0.8$	248.0	248.0	235.2	209.6
	$t_w = 0.9$	279.0	279.0	264.6	235.8
	$t_w = 1.0$	310.0	310.0	294.0	262.0
	$t_w = 1.1$	341.0	341.0	323.4	288.2

t_w h_w — Non-L. stiffener
$0.2h_w$ t_w h_w — One L. stiffener
$0.14h_w$ $0.32h_w$ t_w h_w — Two L. stiffener

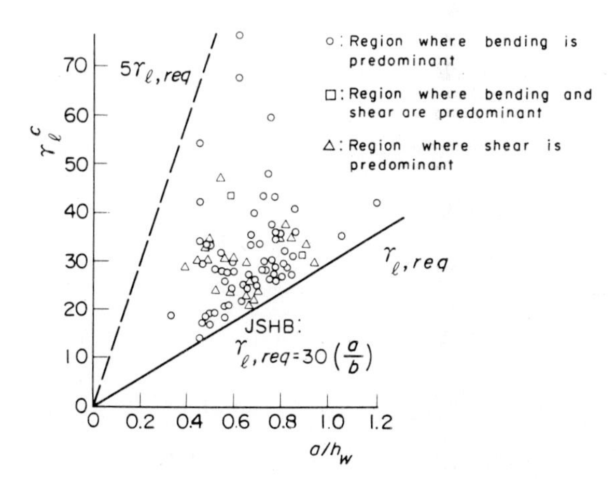

Figure 6.49 Relative stiffness γ_l^c of longitudinal stiffener in curved girder.[5]

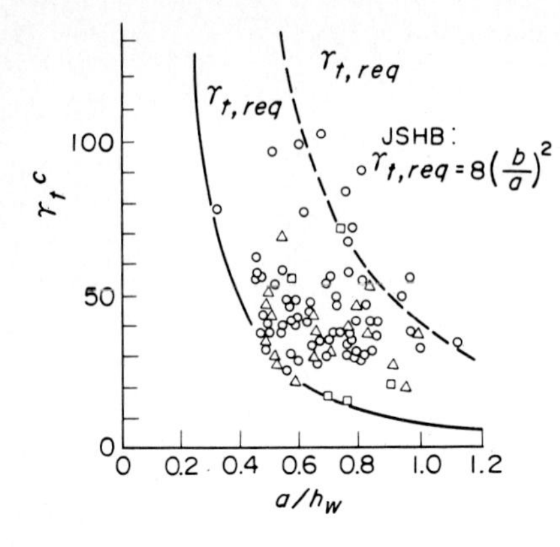

Figure 6.50 Relative stiffness γ_t^c of transverse stiffener in curved girder.[5]

plates of curved girder bridges is entirely based on the design criteria of JSHB.

However, the web plates of curved girders are situated in a severe condition when their radius of curvature is significantly small.[28] Then the design methods proposed in Secs. 5.5 and 5.6 are suitable. This approach is outlined in the flowchart shown in Fig. 6.51. In designing

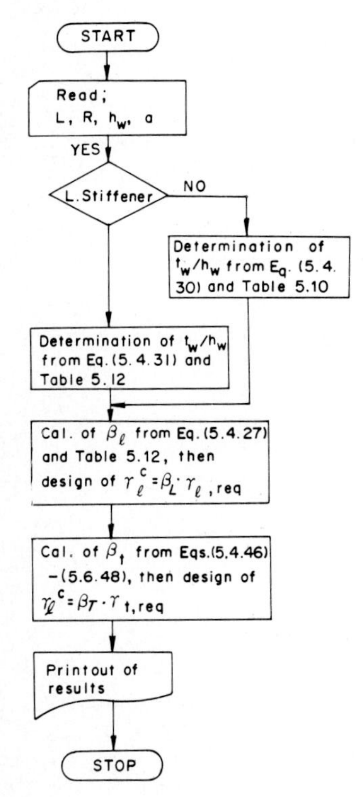

Figure 6.51 Flowchart for designing web plates.

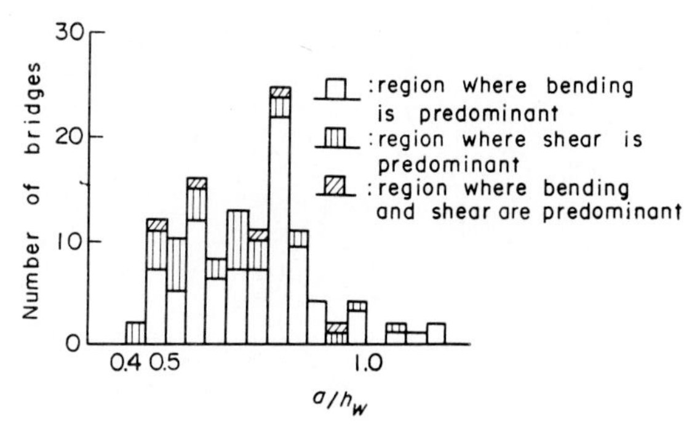

Figure 6.52 Histogram of aspect ratio a/h_w according to survey.

the web plates of curved girders, the aspect ratio $\alpha = a/h_w$ must be limited to the following range:

$$\alpha = \frac{a}{h_w} \leqq 1.0 \tag{6.2.139}$$

because of the shallow shell buckling behavior, discussed in Sec. 5.3.[29] Some of the curved girder bridges are over this limit according to the survey shown in Fig. 6.52.

(3) Design of flange plates

***a.* Cross section of flange plates.** Let us now consider a straight I girder with an asymmetric cross section subjected to bending moment M about the neutral N-N axis, as shown in Fig. 6.53, where A_c and A_t are the cross-sectional areas of the compression and tension flange plates, respectively. When the mean stresses σ_c and σ_t in the compression and tension flange plates, respectively, are known, the following equations[27] can be written:

$$S = A_c y_t - h t_w e + A_t y_t = 0 \tag{6.2.140a}$$

$$= \text{static moment about neutral axis}$$

$$M = A_c \sigma_c y_c + \frac{\sigma_c y_c t_w}{2} \frac{2}{3} y_c + A_t \sigma_t y_t + \frac{\sigma_t y_t t_w}{2} \frac{2}{3} y_t \tag{6.2.140b}$$

M = equilibrium condition of stress and applied moment M

The fiber distance y_c and y_t and eccentricity e of the neutral axis are also known from the geometric conditions shown in Fig. 6.53. The cross-sectional areas A_c and A_t can be found from Eqs. (6.2.140a) and

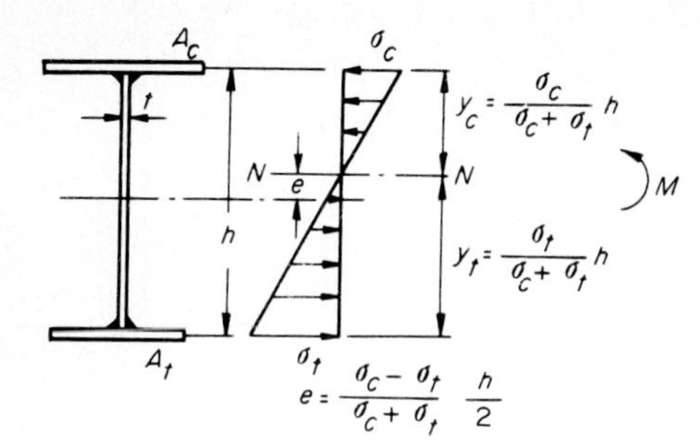

Figure 6.53 I girder with asymmetric cross section.

(6.2.140*b*) as follows (recall that h and t_w have already been determined as shown earlier):

$$A_c = \frac{M}{\sigma_c h} - \frac{ht_w}{6} \frac{2\sigma_c - \sigma_t}{\sigma_c} \tag{6.2.141a}$$

$$A_t = \frac{M}{\sigma_t h} - \frac{ht_w}{6} \frac{2\sigma_t - \sigma_c}{\sigma_t} \tag{6.2.141b}$$

These equations can be applied to a curved girder by taking σ_c and σ_t as the stresses at the junction points of the flange and web plates, respectively.

Once A_c has been determined, the corresponding width b and thickness t of the compression flange plate can be determined by considering the local buckling of the compression flange plate. For example, the thickness of the flange plate t can be determined from Table 6.13 (as explained in Sec. 5.7). The thickness t of the tension flange plate, however, can be taken as

$$t \geqq \frac{b}{16} \tag{6.2.142}$$

for both straight and curved girder bridges.

TABLE 6.13 Required Thickness *t* of Compressive Flange Plate

Yield stress σ_y (kgf/cm²)	Straight girder in JSHB	Curved girder proposed in Table 5.17	Remark
2400 (SS 41, SM 41)	$t \geqq \frac{b}{13}$	$t \geqq \frac{b}{9.2}$	(flange sketch with b and t)
3600 (SM 50Y, SM 53, SMA 50)	$t \geqq \frac{b}{11}$	$t \geqq \frac{b}{7.5}$	

***b.* Check for lateral buckling of curved I girders.** In addition to the local buckling of the compression flange plate, the lateral buckling strength of the curved I girder along the supported portion between floor beams or sway and lateral bracings should be carefully checked. For this criterion, HEPC gives the following formula:

$$\frac{\sigma_{bc}}{(\sigma_{ba})_c} + \frac{\sigma_{\omega c}}{\sigma_{bao}} \leqq 1.0 \tag{6.2.143}$$

for the following notation:

σ_{bc} = flexural compressive stress at junction point of flange and web plates

$\sigma_{\omega c}$ = additional compressive stress due to out-of-plane bending of flange plate [see Eq. (6.2.104)]

$$(\sigma_{ba})_c = (\sigma_{ba})_s \psi_1 \tag{6.2.144}$$

$(\sigma_{ba})_s$ = allowable flexural compressive stress of straight girder by JSHB in Eq. (5.2.16)

σ_{bao} = upper limit of allowable flexural compressive stress

$$\psi_1 = \begin{cases} 1.0 - 1.05\sqrt{\alpha_c}(\Phi + 4.52\Phi^2) & 0.02\alpha \leqq \Phi \leqq 0.2 \quad (6.2.145a) \\ 1.0 & \text{regard as straight I girder for } \Phi < 0.02\alpha \quad (6.2.145b) \end{cases}$$

$$\alpha_c = \gamma\alpha = \text{lateral buckling parameter of curved I girder} \tag{6.2.146}$$

$$\gamma = \begin{cases} 1.97\Phi^{1/3} + 4.25\Phi - 26.3\Phi^3 & 0.002\alpha \leqq \Phi \leqq 0.2 \quad (6.2.147a) \\ 1.0 & 0.02\alpha < \Phi \quad (6.2.147b) \end{cases}$$

$$\alpha = \frac{1}{\pi}\sqrt{3 + \frac{A_w}{2A_c}}\left(\frac{l}{b}\right)\sqrt{\frac{E}{\sigma_y}} \tag{6.2.148}$$

= lateral buckling parameter of straight I girder by JSHB

A_c = cross-sectional area of compression flange plate

A_w = cross-sectional area of web plate

b = full width of compression flange plate

E = Young's modulus

σ_y = yield stress

$$l = \gamma L \tag{6.2.149}$$

= effective lateral buckling length of curved I girder

$$L = R\Phi \tag{6.2.150}$$

= distance between supported portion of compression flange plate and floor beams

R = radius of curvature of curved I girder

Φ = central angle between supported portion of compression flange plate and floor beams

Figure 6.54 shows the flowchart for checking the lateral buckling

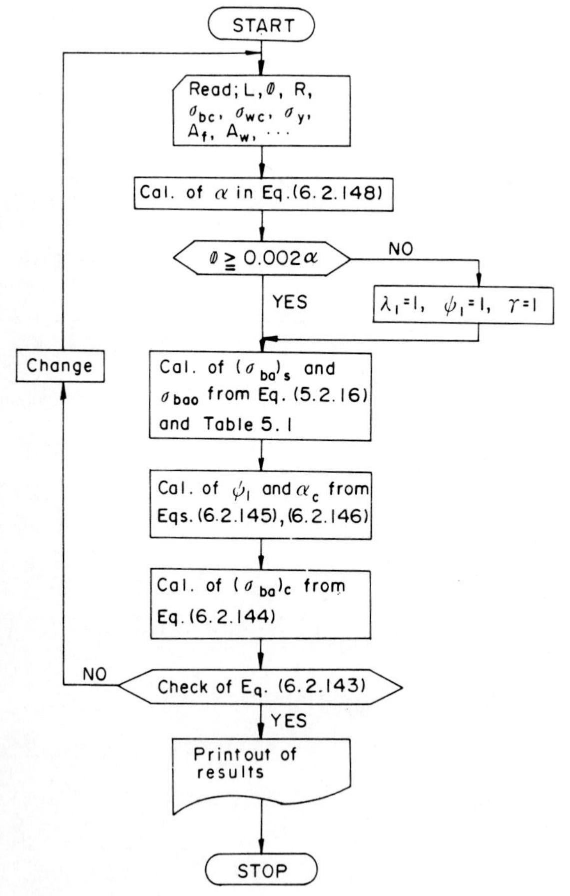

Figure 6.54 Flowchart checking lateral buckling strength.

strength. In addition, the following check should also be made[1] for the tension flange plate of the curved I girder:

$$\sigma_{bt} + \sigma_{\omega t} \leqq \sigma_{ta} \tag{6.2.151}$$

where σ_{bt} = flexural tensile stress of flange plate
$\sigma_{\omega t}$ = additional tensile stress due to out-of-plane bending of flange plate
σ_{ta} = allowable tensile stress

c. Design of stiffened compression flange plate in curved box girders. There are no design codes for proportioning the stiffened compressive flange plates of curved box girders in JSHB and HEPC. However, if the design method of JSHB is applied to curved girders by considering the stress gradient in the stiffened compression flange plate, as demonstrated in Sec. 5.7, the following design formulas can be adopted:

$$A_l \geqq \frac{bt}{10} \tag{6.2.152}$$

= cross-sectional area of longitudinal stiffener

$$I_l \geqq \frac{bt^3}{11} \gamma_{l,\mathrm{req}} \tag{6.2.153}$$

= geometric moment of inertia of longitudinal stiffener

and

$$I_t \geqq \frac{bt^3}{11} \gamma_{t,\mathrm{req}} \tag{6.2.154}$$

= geometric moment of inertia of transverse stiffener

where b = width of stiffened plate
t = thickness of stiffened plate
$\gamma_{l,\mathrm{req}}$ = required relative stiffness for longitudinal stiffener
$\gamma_{t,\mathrm{req}}$ = required relative stiffness for transverse stiffener

Figure 6.55 shows the flowchart for designing the stiffened compression flange plate of curved box-girder bridges.

6.2.8 Design of structural details

In this section, the design methods are presented for structural details such as shear connectors, floor beams or sway bracing, lateral bracing, intermediate diaphragms, stiffeners at supports, and reactions at bearing shoes of curved girder bridges.

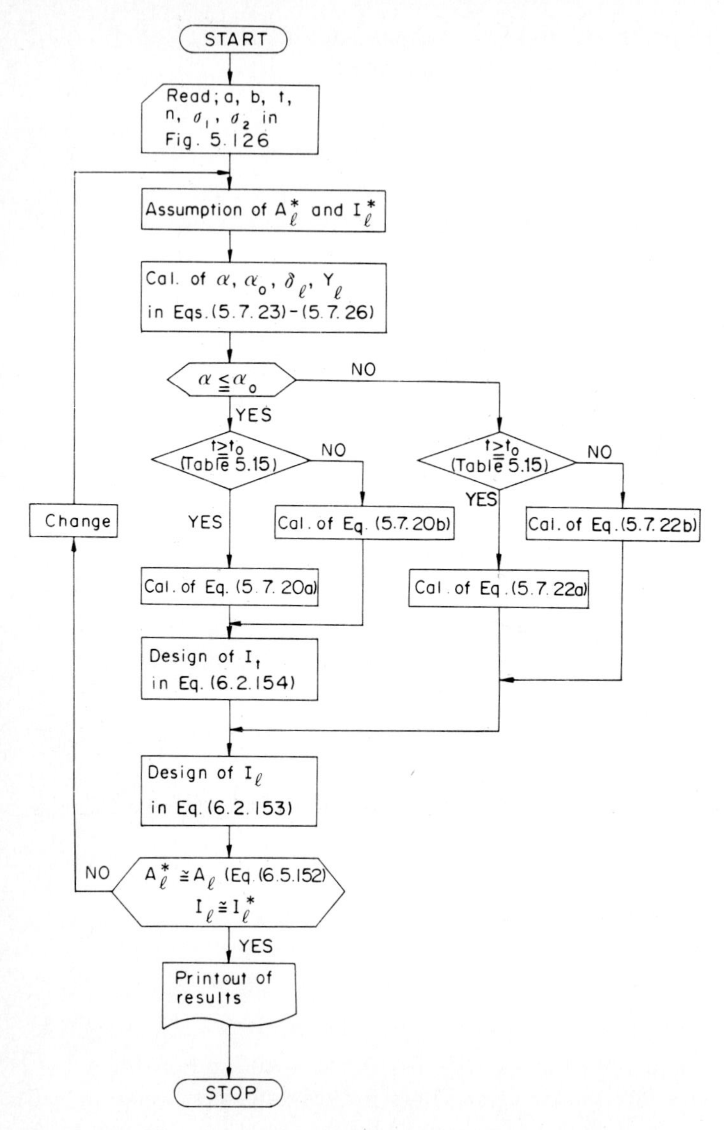

Figure 6.55 Flowchart to design stiffened compression flange plate.

(1) Design of shear connectors. In Japan, the shear studs, shown in Fig. 6.56, are generally adopted as the shear connectors for curved I-girder bridges. The shear flow q for these shear studs can be evaluated from the theory explained in Chaps. 2 and 3 as follows:

$$q = \frac{QA_c d_c}{nI_v} \tag{6.2.155}$$

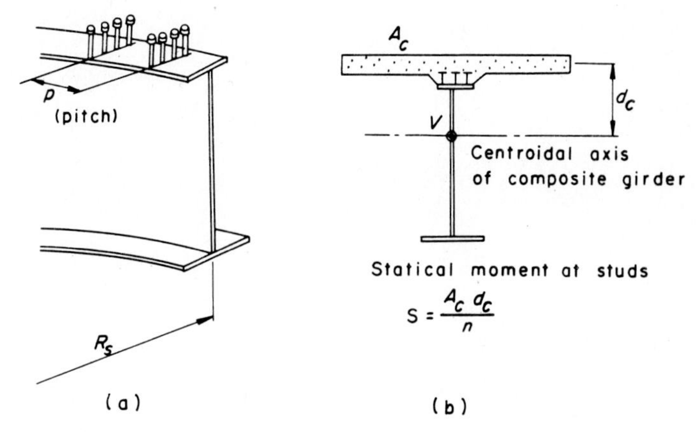

Figure 6.56 (*a*) Shear studs and (*b*) cross section of composite girder.

where Q = shearing force due to live load, shrinkage, and thermal change

A_c = cross-sectional area of concrete slab

d_c = distance between centroidal axis of concrete slab and composite girder

$n = E_s/E_c$ (=7 in JSHB)

= Young's modulus ratio between steel and concrete materials

I_v = geometric moment of inertia of composite girder with respect to the strong axis

Therefore, the pitch p of the shear studs with array m can be determined from

$$p \leqq \frac{mQ_a}{q} \tag{6.2.156}$$

where Q_a is the allowable strength of a shear stud as in JSHB

$$Q_a = \begin{cases} 30d^2\sqrt{\sigma_{ck}} & \text{for } \dfrac{H}{d} \geqq 5.5 \quad (6.2.157a) \\ 5.5dH\sqrt{\sigma_{ck}} & \text{for } \dfrac{H}{d} < 5.5 \quad (6.2.157b) \end{cases}$$

corresponding to the residual slip $\delta = 0.008$ cm (0.003 in), by the push-out

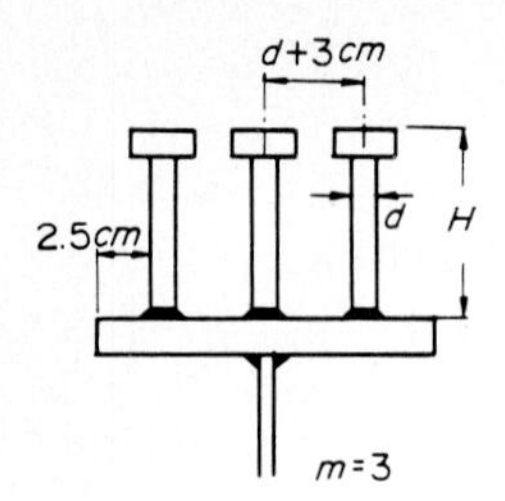

Figure 6.57 Spacing of shear studs.

test of shear stud specimens, in which

σ_{ck} = compressive strength of concrete (kgf/cm^2)

d = diameter of shear stud (cm)

H = height of shear stud (cm)

The spacing of shear studs with $m = 3$ is illustrated in Fig. 6.57.

(2) Design of floor beams or sway bracing. The necessary rigidity for the floor beams or sway bracing is estimated by a condition given in grillage theory, presented in Sec. 6.2.5(3), as follows:[2]

$$Z_g = \left(\frac{L}{2a}\right)^3 \left(\frac{I_Q}{I}\right) \geqq 10 \tag{6.2.158}$$

where L = span length of multiple curved I-girder bridge
a = spacing of main girder
I, I_Q = geometric moments of inertia of main girder and floor beam or sway bracing, respectively

The spacing λ of the floor beams or sway bracings is specified in JSHB as

$$\lambda \leqq \begin{cases} 6 \text{ m} & \text{for sway bracing} \quad (6.2.159a) \\ 20 \text{ m} & \text{for floor beam} \quad (6.2.159b) \end{cases}$$

However, the sway bracing is not adequate in multiple curved I-girder bridges because of their low flexural rigidity. Then, as in HEPC, a floor beam having a solid web should be provided to this type of curved girder bridge within the spacing

$$\lambda \leqq 6 \text{ m} \tag{6.2.159c}$$

(3) Design of lateral bracing. The lateral bracing should be attached to the upper and lower sides of the flange plates of multiple curved I-girder bridges to decrease additional stresses in the flange plates and to enhance stability against the overall lateral buckling of bridges during erection.[2,25] The strength of lateral bracing against horizontal forces such as wind or seismic load as well as the following check for multiple curved I-girder bridges, as outlined in HEPC, should be taken into consideration.

a. **Additional stress due to out-of-plane bending of flange plates.** The following reaction R must be applied to the lateral bracing at the junction points of main girders, as shown in Fig. 6.35*b*:

$$R = \frac{\sigma_b(A_f + A_s/3)}{R} \frac{l_1 + l_2}{2} \tag{6.2.160}$$

where l_1 and l_2 are the spacing of the lateral bracing, shown in Fig. 6.58.

The resulting stresses caused by this reaction R can be found by referring to Fig. 6.58:

$$\sigma_1 = \frac{R}{A_1} \frac{\sin \theta_2}{\sin (\theta_1 + \theta_2)} \tag{6.2.161a}$$

$$\sigma_2 = \frac{R}{A_2} \frac{\sin \theta_1}{\sin (\theta_1 + \theta_2)} \tag{6.2.161b}$$

where σ_1, σ_2 = normal stresses induced in lateral bracing (compression)

A_1, A_2 = cross-sectional areas of lateral bracing

θ_1, θ_2 = inclination angles of lateral bracing

b. **Additional stress due to torsion.** Multiple curved I-girder bridges are always subject to torsional moment. The stress due to torsion can be approximately evaluated by the following simplified method. First, let a multiple curved girder bridge be idealized as a quasi box bridge.[2]

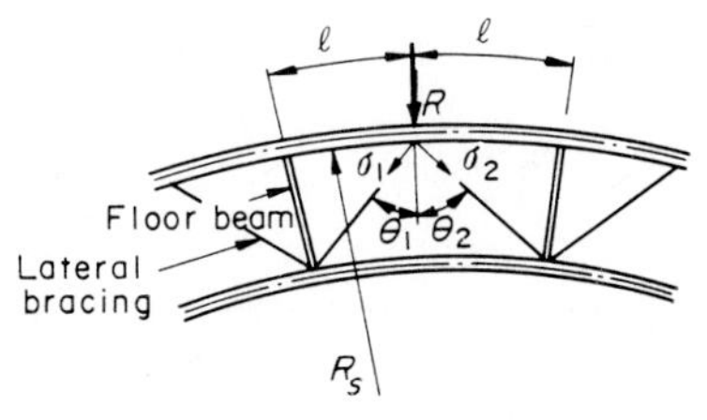

Figure 6.58 Stress caused by out-of-plane bending of flange plate.

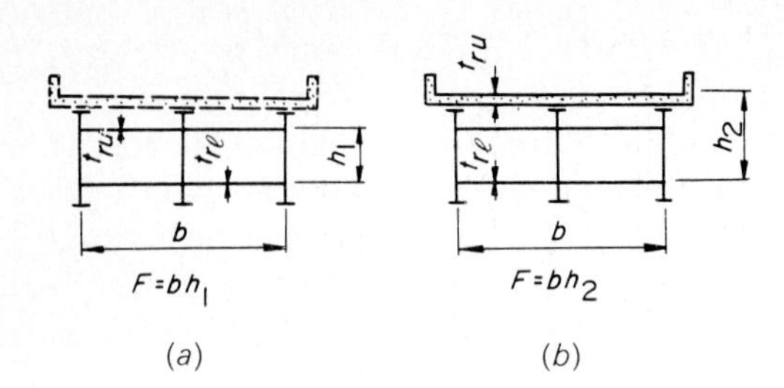

Figure 6.59 Quasi box-girder bridges: (*a*) noncomposite girder and (*b*) composite girder.

The equivalent thickness t_r of the top or bottom plates of these box girder bridges can be reduced as follows (see Fig. 6.61):

$$t_r = \frac{E}{G} \frac{al}{d^3/A_d + 2l^3/(3A_f)} \tag{6.2.162}$$

where A_d = cross-sectional area of lateral bracing
A_f = cross-sectional area of flange plate
a = spacing of main girder
d = length of lateral bracing
l = spacing of lateral bracing
E = Young's modulus
G = shear modulus of elasticity

Figure 6.59 illustrates the corresponding quasi box-girder bridges.

Next, the shear flow q due to the torsional moment T can be estimated by Bredt's formula:

$$q = \frac{T}{2F} \tag{6.2.163}$$

where the torsional moment can be determined approximately from Table 6.14 corresponding to the load conditions shown in Fig. 6.60.[2]

Finally, the force acting on the lateral bracing with length d is

$$Q = qd \tag{6.2.164}$$

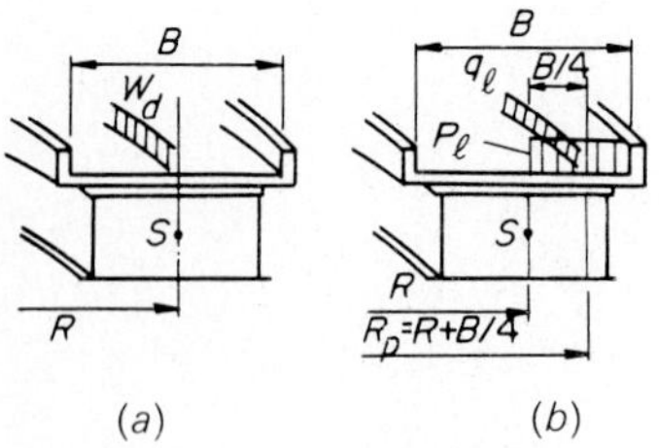

(*a*) (*b*)

Figure 6.60 Load conditions (*a*) and (*b*).

TABLE 6.14 Torsional Moment T

Load	Load position	Formulas for torsional moment	Interpolation at intermediate span
Dead load	Edge-span	$T_{d1} = w_d R^2 \left(\frac{1 - \cos \Phi}{\sin \Phi} - \frac{\Phi}{2} \right)$	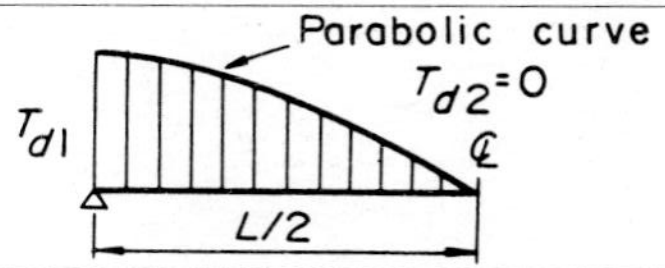 Parabolic curve; T_{d1}; $T_{d2} = 0$; ℄; $L/2$
	Midspan	$T_{d2} = 0$	
Live load	Edgespan (P_ℓ, q_ℓ, L)	$T_{l1} = P_l(R_p - R) + q_l R_p \times \left(R_p \frac{1 - \cos \Phi}{\sin \Phi} - \frac{R\Phi}{2} \right)$	Parabolic curve; $T_{\ell 1}$; $T_{\ell 2}$; ℄; $L/2$
	Midspan (q_ℓ, P_ℓ, $L/2$, L)	$T_{l2} = \frac{P_l}{2}(R_p - R) + \frac{q_l R_p}{4} \times \left(R_p \frac{1 - \cos \Phi}{\sin \Phi} - \frac{R\Phi}{2} \right)$	

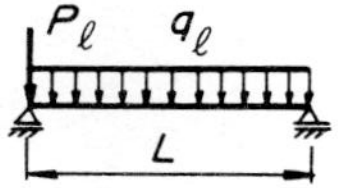

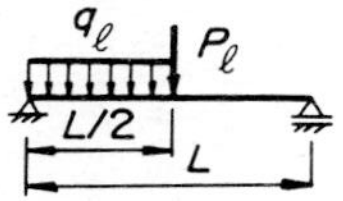

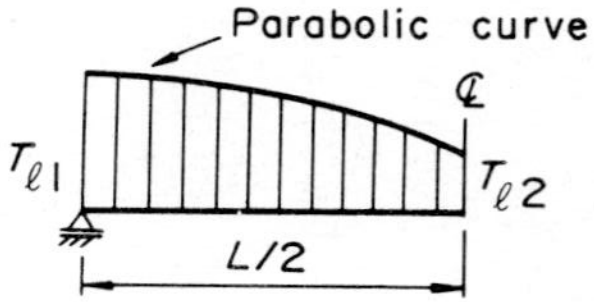

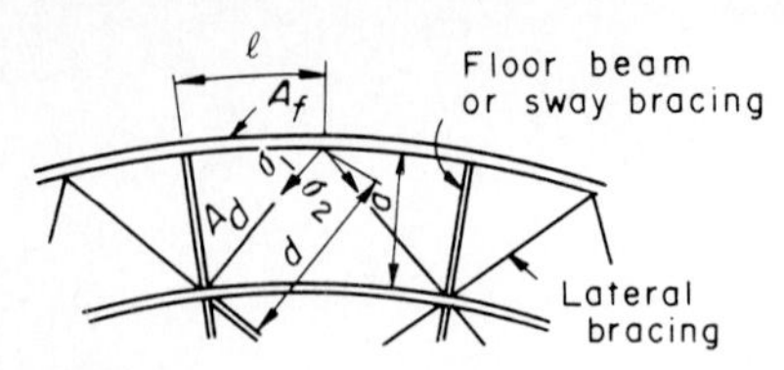

Figure 6.61 Additional stress caused by torsion.

The normal stresses due to torsion can be estimated by

$$\sigma_1 = -\frac{Q}{A_1} \qquad \text{compression} \tag{6.2.165a}$$

$$\sigma_2 = \frac{Q}{A_2} \qquad \text{tension} \tag{6.2.165b}$$

These are plotted in Fig. 6.61.

(4) Design of intermediate diaphragms. Although the design of intermediate diaphragms in curved box-girder bridges is important, there are no appropriate design criteria in either JSHB or HEPC. For the design of these diaphragms, the proposition described in Sec. 3.3.2 is a powerful approach and is outlined in the flowchart in Fig. 6.62.

In this flowchart, note that the distortional warping stress $\sigma_{D\omega}$ is taken into account in the stress combination shown in Table 6.10 when the ratio of $\sigma_{D\omega}$ to the flexural stress σ_b is large or $\sigma_{D\omega}/\sigma_b \geqq 0.05$.

(5) Design of end transverse stiffeners. Figure 6.63 shows the end transverse stiffener at a support with reaction R[30]. JSHB specifies that this end transverse stiffener should be designed as follows:

$$\sigma = \frac{R}{A_{\text{eff}}} \leqq \sigma_{ca} \tag{6.2.166}$$

where A_{eff} = effective cross-sectional area of column shown in Fig. 6.63*b*

σ_{ca} = allowable compressive stress (kgf/cm^2) of end transverse stiffener by taking effective buckling length

$$l = 0.5h \tag{6.2.167}$$

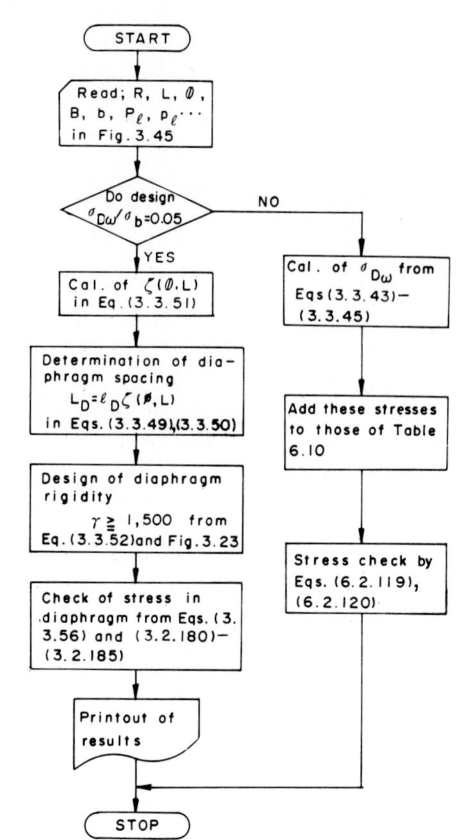

Figure 6.62 Flowchart for designing intermediate diaphragms.

for girder depth h and radius of gyration r of column in Fig. 6.63b, and

$$\sigma_{ca} = \begin{cases} \sigma_{cao} & \dfrac{l}{r} \leqq \beta & (6.2.168a) \\ \sigma_{cao} - \alpha\left(\dfrac{l}{r} - \beta\right) & \beta < \dfrac{l}{r} \leqq \gamma & (6.2.168b) \\ \dfrac{12{,}000{,}000}{\delta + (l/r)^2} & \dfrac{l}{r} > \gamma & (6.2.168c) \end{cases}$$

The values of σ_{cao}, α, β, γ, and δ are shown in Table 6.15.

(6) Determination of the reaction for designing bearing shoes. As an example, the reactions V_1, V_2, H_1 and H_2 induced in the curved box-girder bridge due to wind and seismic loads are sketched in Fig. 6.64.[2]

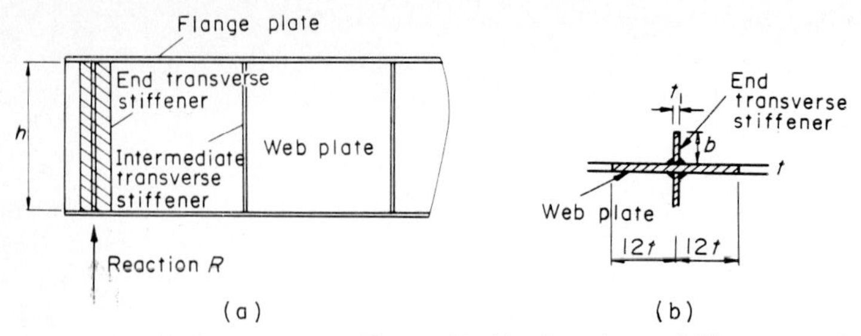

Figure 6.63 End transverse stiffener: (*a*) side elevation and (*b*) cross section.

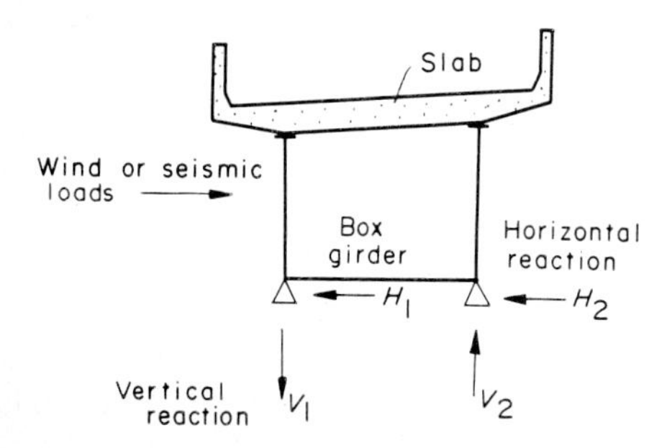

Figure 6.64 Reaction due to wind or seismic load.

These vertical and horizontal reactions that are caused by dead and live loads include impact, thermal change, wind load, and seismic load.

If the vertical reaction R (V_1 or V_2) of a support is negative (i.e., there is uplift), then by combining these loads effects the reaction R should be estimated as the most severe reaction of the following three

TABLE 6.15 Values of σ_{cao}, α, β, γ, and δ

Steel material	σ_{cao}, kgf/cm^2	α	β	γ	δ
SS 41 SM 41	1400	8.4	20	93	6700
SM 50	1900	13	15	80	5000
SM 50Y SM 53 SMA 50	2100	15	14	76	4500
SM 58	2600	12	14	67	3500

equations, specified in JSHB:[1]

$$R = 2R_{L+I} + R_{D1} + \frac{R_{D2}}{1.5} \quad (6.2.169a)$$

$$R = 2R_{L+I} + R_{D1} + \frac{R_{D2}}{1.5} + R_W \quad (6.2.169b)$$

$$R = R_{D1} + \frac{R_{D2}}{1.5} + R_{EQ} \quad (6.2.169c)$$

where R_{L+I} = maximum negative reaction due to live load including impact

R_{D1} = negative reaction due to dead load applied to region where negative reaction occurs

R_{D2} = positive reaction due to dead load applied to region where positive reaction occurs

R_W = maximum negative reaction due to wind load

R_{EQ} = maximum negative reaction due to seismic load

References

1. The Japanese Road Association: *The Japanese Specification for Highway Bridges*, pts. I and II, Maruzen, Tokyo, Feb. 1980 (in Japanese).
2. The Hanshin Expressway Public Corporation: *The Design Code*, *Part* 2, *Design Codes for Steel Bridge Structure*, Osaka, April 1980 (in Japanese).
3. Komatsu, S., H. Nakai, and Y. Taido: "A Proposition for Designing the Horizontally Curved Girder Bridges in Connection with Ratio between Torsional and Flexural Rigidities," *Proceedings of the Japanese Society of Civil Engineers*, no. 224, pp. 35–66, Apr. 1974 (in Japanese).
4. Nakai, H., and C. P. Heins: "Analysis Criterion for Curved Bridges," *Proceedings of the American Society of Civil Engineers*, vol. 103, no. ST-7, pp. 1419–1427, July 1977.
5. Nakai, H., S. Muramatsu, O. Yoshikawa, T. Kitada, and R. Ohminami: "A Survey for Web Plates of Horizontally Curved Girder Bridges," *Bridge and Foundation Engineering*, 81(5):38–45, May 1981 (in Japanese).
6. Konishi, I., and S. Komatsu: *Steel Bridges, Design Part II*, Maruzen, Tokyo, 1976 (in Japanese).
7. Nakai, H., and H. Kotoguchi: "Dynamic Response of Horizontally Curved Girder Bridges under Random Traffic Flow," *Proceedings of the Japanese Society of Civil Engineers*, no. 244, pp. 117–128, Dec. 1975.
8. Yamada, Y., and T. Kobori: "Dynamic Response of Highway Bridges due to Live Load by Spectral Analysis," *Transactions of the Japanese Society of Civil Engineers*, no. 148, pp. 40–50, Dec. 1967 (in Japanese).
9. The Japanese Road Association: *The Japanese Specification for Highway Bridges*, *Part V*, Maruzen, Tokyo, Feb. 1980 (in Japanese).
10. Konishi, I., and S. Komatsu: *Three-Dimensional Analysis of Curved Girder with Thin-Walled Cross-Section*, publication of IABSE, Zürich, 1965.
11. Nakai, H., H. Kotoguchi, and T. Tani: "Matrix Structural Analysis of Thin-Walled Curved Girder Bridges Subjected to Arbitrary Load," *Proceedings of the Japanese Society of Civil Engineers*, no. 255, pp. 1–15, Nov. 1976.
12. Heins, C. P.: *Bending and Torsional Design in Structural Members*, Lexington Book, Lexington, Mass., 1975.

13. Dabrowski, R.: *Gekrümmte dünnwandige Träger*, Springer-Verlag, Berlin, 1968 (in German).
14. Komatsu, S.: "Practical Formulas for the Curved Bridge with Multiple Plate Girders," *Transactions of the Japanese Society of Civil Engineers*, no. 93, pp. 1–9, May 1963 (in Japanese).
15. Simada, S., and S. Kuranishi: *Formula and Tables for Curved Beam*, Gihodo, Tokyo, 1966 (in Japanese).
16. Watanabe, N.: *Theory and Calculation of Curved Girder*, Gihodo, Tokyo, 1967 (in Japanese).
17. Takaba, K., and M. Naruoka: "An Analysis of Grillage Girder Bridges with Thin-Walled Cross-Section by the Deformation Method," *Transactions of the Japanese Society of Civil Engineers*, no. 178, pp. 1–9, June 1970 (in Japanese).
18. Kollbrunner-Basler: *Torsion*, Springer-Verlag, Berlin, 1972 (in German).
19. Kollbrunner-Hajdin: *Dünnwandige Stäbe*, Springer-Verlag, Berlin, 1972 (in German).
20. Galambos, T. V.: *Structural Members and Frames*, Prentice-Hall, New York, 1968.
21. Nakai, H., and T. Tani: "An Approximate Method for the Evaluation of Torsional Warping Stresses in Box Girder Bridges," *Proceedings of the Japanese Society of Civil Engineers*, no. 277, pp. 41–55, Sept. 1978 (in Japanese).
22. Leonhardt, H., and W. Andra: *Dei vereinfachte Träger—rostberechnung*, Julis Hoffman, Stuttgart, 1950 (in German).
23. Homberg, H.: *Kruzwerke, Forschungshefte aus dem Gebiete des Stahlbaues*, Heft 8, Springer-Verlag, Berlin 1951 (in German).
24. Komatsu, S., H. Nakai, and T. Kitada: "Study on Shear Lag and Effective Width of Horizontally Curved Girder Bridges," *Proceedings of the Japanese Society of Civil Engineers*, no. 191, pp. 1–14, July 1971 (in Japanese).
25. Nakai, H., and H. Kotoguchi: "A Study on Lateral Buckling Strength and Design Aid for Horizontally Curved I-Girder Bridges," *Proceedings of the Japanese Society of Civil Engineers*, no. 339, pp. 195–205, Nov. 1983.
26. Nakai, H., and Y. Murayama: "Distortional Stress Analysis and Design Aid for Horizontally Curved Box Girder Bridges with Diaphragm," *Proceedings of the Japanese Society of Civil Engineers*, no. 309, pp. 25–39, May 1981 (in Japanese).
27. Tachibana, Y., and H. Nakai: *Bridge Engineering*, 2d ed., Kyoritsu-shuppan, Tokyo, 1981 (in Japanese).
28. Nakai, H., T. Kitada, R. Ohminami, and T. Kawai: "A Study on Analysis and Design of Web Plate in Curved Girder Bridges Subjected to Bending," *Proceedings of the Japanese Society of Civil Engineers*, no. 368/I-5, pp. 235–244, Apr. 1986 (in Japanese).
29. Nakai, H., T. Kitada and R. Ohminami: "A Proposition for Designing Intermediate Transverse Stiffeners in Web Plate of Horizontally Curved Girders," *Proceedings of the Japanese Society of Civil Engineers*, no. 362/I-4, pp. 273–275, Oct. 1985 (in Japanese).
30. Nakai, H., and M. Sakano: "An Experimental and Analytical Study on Ultimate Strength of End Transverse Stiffener in Plate Girder Bridges," *Journal of Structural Engineering of the Japanese Society of Civil Engineers*, 32A:339–410, March 1986 (in Japanese).

Chapter

7

Fabrication, Details, Painting, and Erection of Curved Bridges

7.1 Introduction

In this chapter, fabrication methods and the corresponding details of curved girder bridges are explained in terms of the design criteria for straight girder bridges. Construction procedures for these types of bridges can generally be outlined in a flowchart, as shown in Fig. 7.1.[1] Fabrication methods are also detailed in Fig. 7.2.[1]

Section 7.2 presents the concept and criteria of initial imperfections such as initial deflections and residual stresses in steel girder bridges. In Sec. 7.3, the fabrication methods of the main girders using welding and high-strength bolted joints are discussed together with their design methods.

In Sec. 7.4, we investigate the design and fabrication methods of floor beams and sway and lateral bracings. Furthermore, the fabrication methods and details of intermediate, interior, and end diaphragms in box-girder bridges are outlined in Sec. 7.5.

In Sec. 7.6, the design and fabrication methods of bearing shoes are discussed in detail for various types of bearing. In connection with this, some attention has been given to constructing steel girder bridges in strong earthquake zones.

Finally, the painting and erection methods for the steel girder bridges are introduced through tables and figures in Secs. 7.7 and 7.8, respectively.

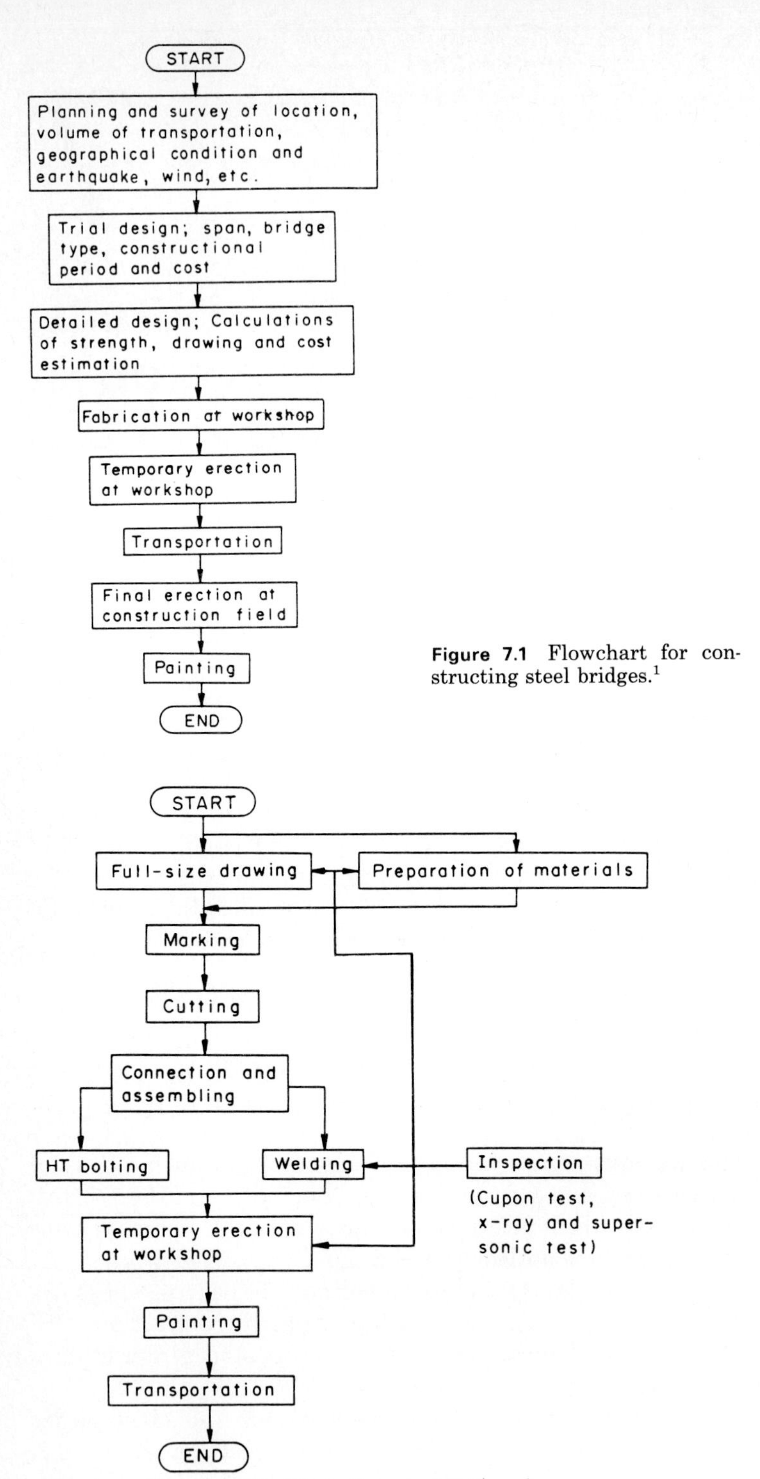

Figure 7.1 Flowchart for constructing steel bridges.[1]

Figure 7.2 Flowchart for fabricating steel bridges.[1]

7.2 Initial Imperfections in Curved Girder Bridges

7.2.1 Fabrication tolerances

To ensure some degree of quality in the bridges, the structural members should be fabricated within certain tolerances. For example, the Japanese Specification for Highway Bridges (JSHB)[2] code requires the following tolerances for the structural members in straight girder bridges (see Fig. 7.3):

For the length of a member (Fig. 7.3*a*),

$$\Delta l = \begin{cases} \pm 3\text{ mm} & l \leqq 10\text{ m} \qquad (7.2.1a) \\ \pm 4\text{ mm} & l > 10\text{ m} \qquad (7.2.1b) \end{cases}$$

For the cross-sectional dimensions (Fig. 7.3*b*),

$$\Delta d = \begin{cases} \pm 2\text{ mm} & d \leqq 0.5\text{ m} \qquad (7.2.2a) \\ \pm 3\text{ mm} & 0.5 < d \leqq 1.0\text{ m} \qquad (7.2.2b) \\ \pm 4\text{ mm} & 1.0 < d \leqq 2.0\text{ m} \qquad (7.2.2c) \\ \pm\left(3 + \dfrac{b}{2}\right)\text{ mm} & d > 2.0\text{ m} \qquad (7.2.2d) \end{cases}$$

where d is representative of flange width b, web depth h_w, and the spacing of web plates b' in Fig. 7.3*b*.

In addition, the final tolerance of a bridge should fall within the following limitations:

For total span length L in meters,

$$\Delta L = \pm\left(10 + \frac{L}{10}\right)\text{ mm} \qquad (7.2.3)$$

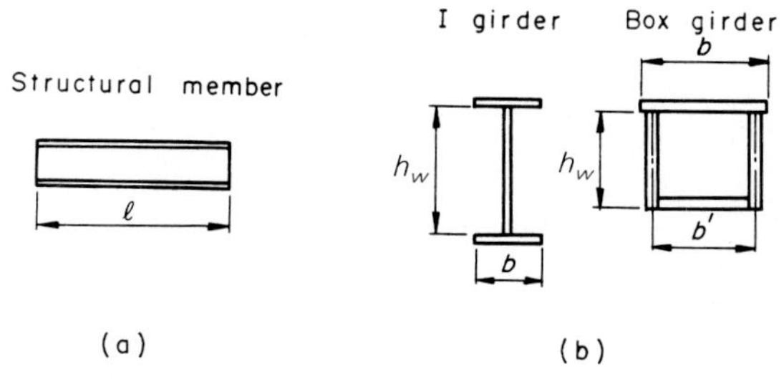

Figure 7.3 Dimensions of structural members: (*a*) side elevation and (*b*) cross section.

For the spacing of main girder B (m),

$$\Delta B = \begin{cases} \pm 4 \text{ mm} & B \leqq 2 \text{ m} \quad (7.2.4a) \\ \pm\left(3 + \dfrac{B}{2}\right) \text{ mm} & B > 2 \text{ m} \quad (7.2.4b) \end{cases}$$

These fabrication tolerances can also be applied to curved girder bridges.

7.2.2 Initial deflections

The structural members are usually built up by welding. During these welding procedures, considerable initial deflections occur in the structural members due to uneven heating and cooling cycles. Therefore, the structural members are usually welded at the fabrication shops through initial reverse-welding procedures. Thus, the flatness of the structural members can be ensured after welding, as shown in Fig. 7.4.[3]

Nevertheless, it is very difficult to manufacture all the members of a bridge in such a manner. Therefore, the following initial imperfections δ (mm) are allowed in JSHB:[1]

For the vertical deviation of a flange plate in an I girder (see Fig. 7.5*a*),

$$\delta = \frac{h}{200} \tag{7.2.5}$$

For the horizontal deviation of a flange plate in an I girder (see Fig. 7.5*b*),

$$\delta = 3 + \frac{h}{1000} \tag{7.2.6}$$

For the flatness of a web plate (see Fig. 7.3*b*),

$$\delta \leqq \frac{h_w}{250} \tag{7.2.7}$$

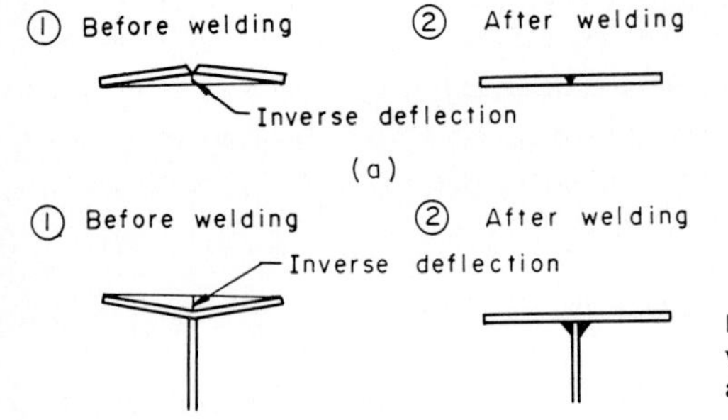

Figure 7.4 Initial deflections of weld joint: (*a*) groove weld joint and (*b*) fillet welding joint.

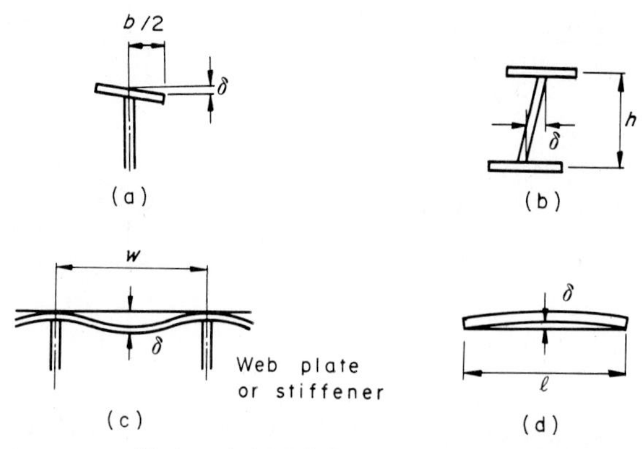

Figure 7.5 Various initial deflections: (*a*) vertical deviation of flange plate in I girder, (*b*) horizontal deviation of flange plate in I girder, (*c*) flatness of flange plate in box girder, and (*d*) eccentricity of compression member.

For the flatness of a flange plate (see Fig. 7.5*c*),

$$\delta \leqq \frac{w}{150} \tag{7.2.8}$$

For the eccentricity of a compression member (see Fig. 7.5*d*),

$$\delta \leqq \frac{l}{1000} \tag{7.2.9}$$

These criteria can also be adopted for curved girder bridges.

7.2.3 Residual stresses

Residual stresses are induced in the structural members of a bridge through welding. This stress is sort of a thermal stress,[4] as pointed out by Stüssi.[5]

Let us now consider a simple case of the groove welded joint illustrated in Fig. 7.6. The deformation at the joint is assumed to follow the pattern shown in the figure. This deformation can be removed if we apply tensile stress σ_{zo}. Corresponding to the deformation of this welded joint, the restraint force P of the stress resultant of tensile stress σ_{zo} can be derived from Fig. 7.7*a* as follows:

$$P = (\text{area of } \sigma_{zo}\text{-diagram})t \tag{7.2.10}$$

where t is the throat of the joint. Next, if we release this restraint force P, then the following compressive stress σ_m will occur throughout the

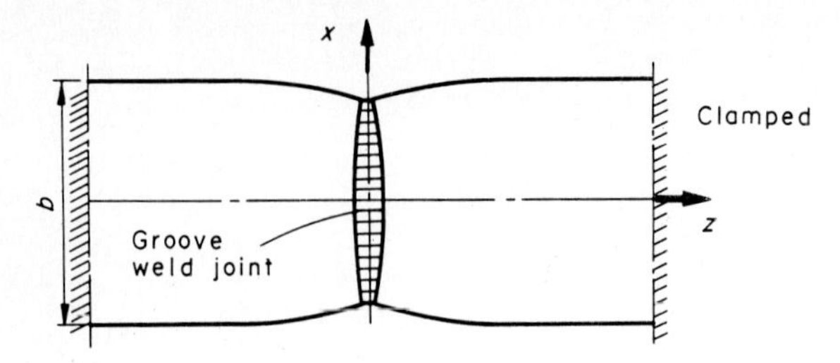

Figure 7.6 Deformation due to groove weld joint.

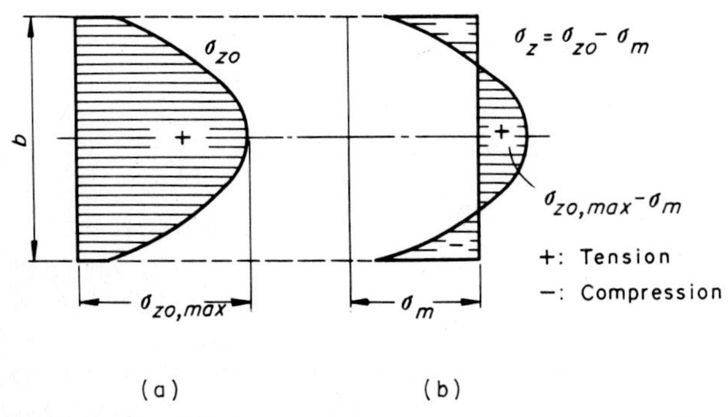

Figure 7.7 Residual stress distribution in simple groove weld joint of plates: (*a*) σ_{zo}-diagram and (*b*) final residue stress.

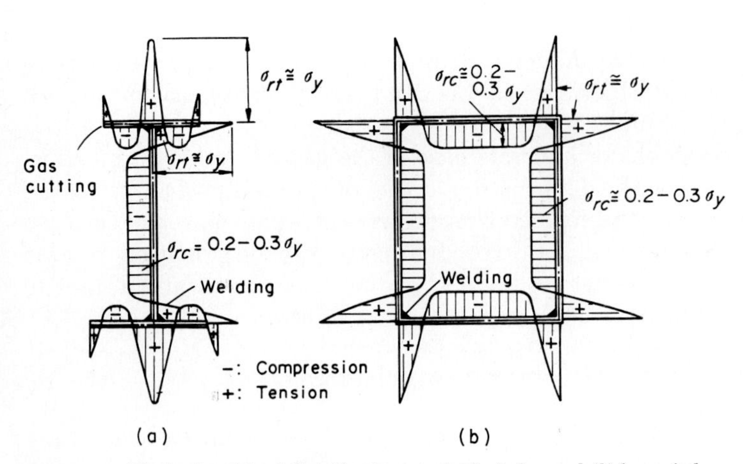

Figure 7.8 Typical residual distributions in (*a*) I girder and (*b*) box girder.

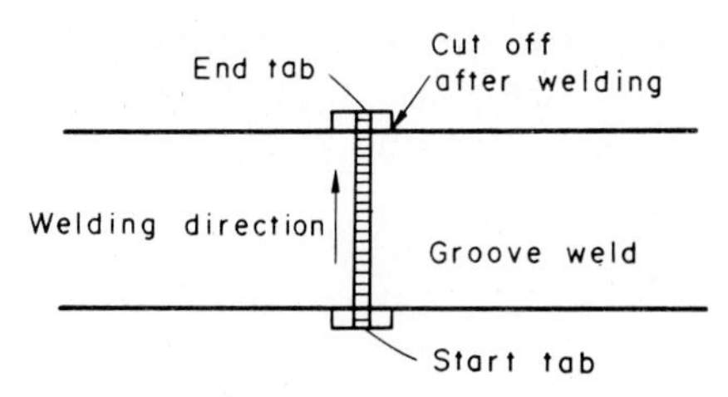

Figure 7.9 Groove weld joint with tabs.

groove weld joint:

$$\sigma_m = \frac{P}{A} \tag{7.2.11}$$

where A is the cross-sectional area of the groove joint. Consequently, the final residual stress σ_z can be represented by

$$\sigma_z = \sigma_{zo} - \sigma_m \tag{7.2.12}$$

The distribution is plotted in Fig. 7.7*b*.

The residual stress distributions in I girders and box girders are much more complicated than the above. However, the residual stress distributions and their magnitudes have been extensively investigated through research,[3,5] and they can be modified as in Fig. 7.8.

According to research, the residual tensile stress σ_{rt} occurs along the welded joints, and the residual compressive stress σ_{rc} is predominant at the portion somewhat apart from these welded joints. These magnitudes fall within the following ranges, even in the case of curved girder bridges:[6,7]

$$\sigma_{rt} \cong \sigma_y \tag{7.2.13}$$

$$\sigma_{rc} \cong (0.2\text{–}0.3)\sigma_y \tag{7.2.14}$$

where σ_y is the yield stress of steel.

These residual stresses affect the ultimate strength of the structural members of a curved girder bridge, as predicted in Chap. 5. In fabricating these structural members, it would be beneficial to reduce these

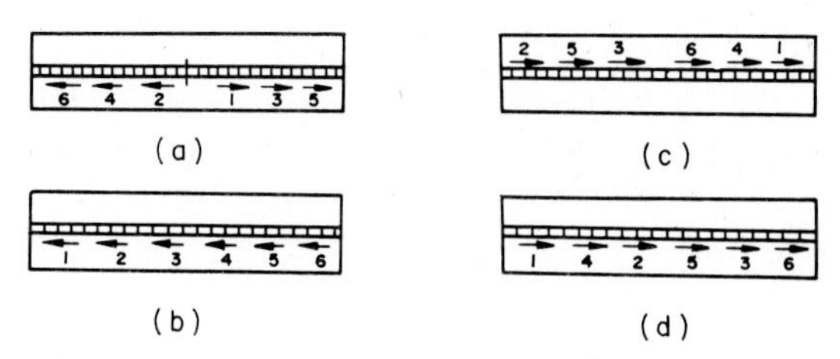

Figure 7.10 Welding procedures to reduce residual stresses:[3] (*a*) symmetric method, (*b*) backward method, (*c*) alternate method, and (*d*) stepping-stone method.

stresses as much as possible. An example is the groove welded joint with start and end tabs, as shown in Fig. 7.9.[5] Another example concerns the procedures for welding illustrated in Fig. 7.10.[3]

7.3 Fabrication of Main Girders

7.3.1 Flange plates

The flange plates of curved girder bridges are usually fabricated by using thin plates and an NC automatic gas cutting machine to obtain the desired degree of curvature. In general, the cross section of a flange plate must vary along the longitudinal direction of the curved girder to satisfy the following condition, illustrated in Fig. 7.11:

$$M_r > M_{\text{app}} \tag{7.3.1}$$

where M_{app} = applied bending moment

$$M_r = \min\left\{\frac{\sigma_{ba} I}{y_c}, \frac{\sigma_{ta} I}{y_t}\right\} \tag{7.3.2}$$

= allowable resisting moment of cross section

σ_{ba} = allowable flexural compressive stress

σ_{ta} = allowable tensile stress

I = geometric moment of inertia of girder

y_c, y_t = fiber distance at compression and tension flange plates, respectively

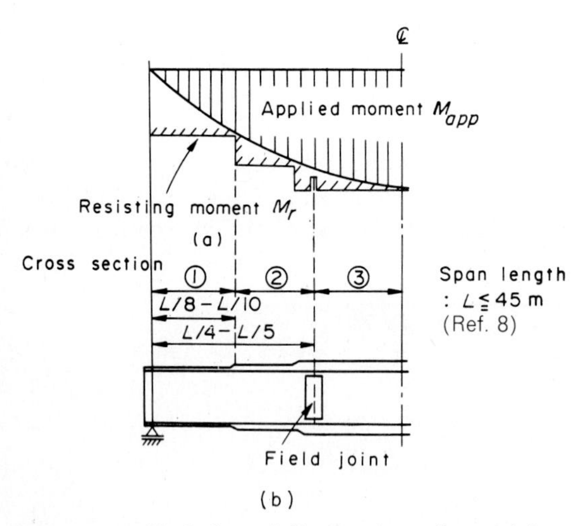

Figure 7.11 Variation of flange plate along girder axis. Span length $L \leqq 45$ m.[8] (*a*) M_r and M_{app}. (*b*) Side elevation.

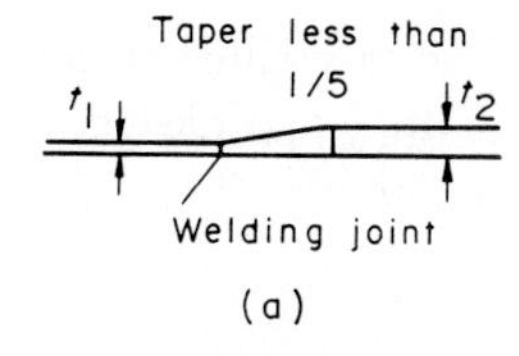

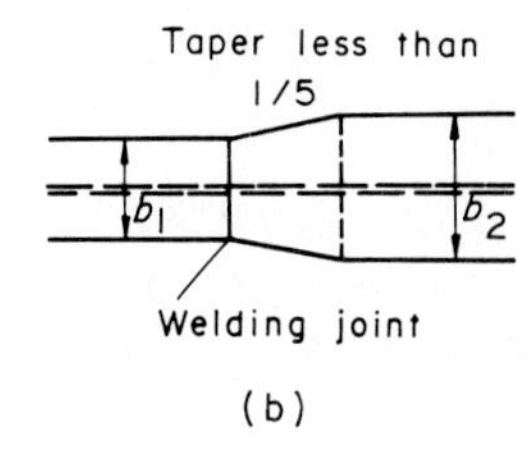

Figure 7.12 Change in (*a*) thickness and (*b*) width of flange plates.

For the welded joint at cross sections 1, 2, and 3 in Fig. 7.11, JSHB[2] specifies that either the thickness t or the width b of the flange plate can be altered as long as the taper is less than or equal to $\frac{1}{5}$, as shown in Fig. 7.12.

The groove welded joint can be designed according to Fig. 7.13 by

$$\sigma = \frac{M}{I} y \leqq \sigma_a \tag{7.3.3}$$

where M = applied bending moment at welding joint
y = fiber distance (in this case $y = y_t = y_c$)

$$I = \frac{a_w h_w^3}{12} + 2ba_f \left(\frac{h_w f + a_f}{2} \right)^2 \tag{7.3.4}$$

= geometric moment of inertia of welding joint
a_f, a_w = throat of welding in flange and web plate, respectively
h_w = depth of girder
b = width of flange plate
σ_a = allowable stress in groove welding (the same as in Table 6.8)

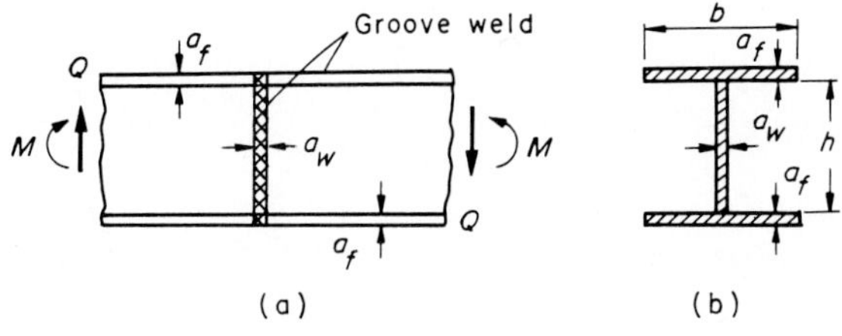

Figure 7.13 Example of groove weld joint: (*a*) side elevation and (*b*) cross section.

7.3.2 Web plates

Figure 7.14 shows the relationships between the median ordinate ΔR and the radius of curvature R in the web plate of a curved girder. This relationship is represented by[8]

$$\frac{\Delta R}{a} \cong \frac{a}{8R} \tag{7.3.5}$$

where a is the spacing of the transverse stiffener.

Figure 7.15 is the result of a survey on the a/R values of curved girder bridges built in Japan. The values for R/a of almost all the curved web plates fall within the tolerance ranges established by

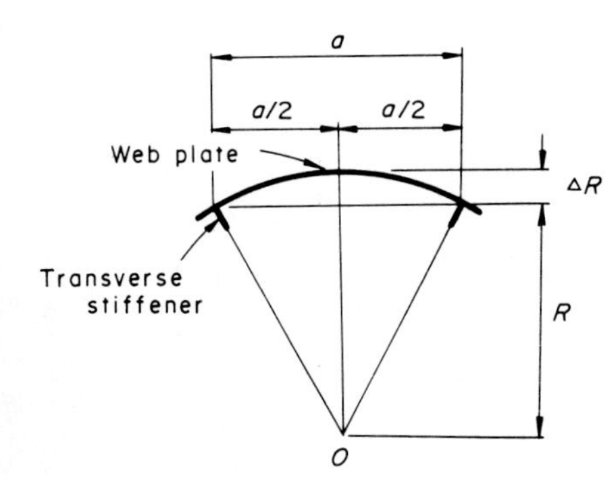

Figure 7.14 Median ordinate R of web plate.

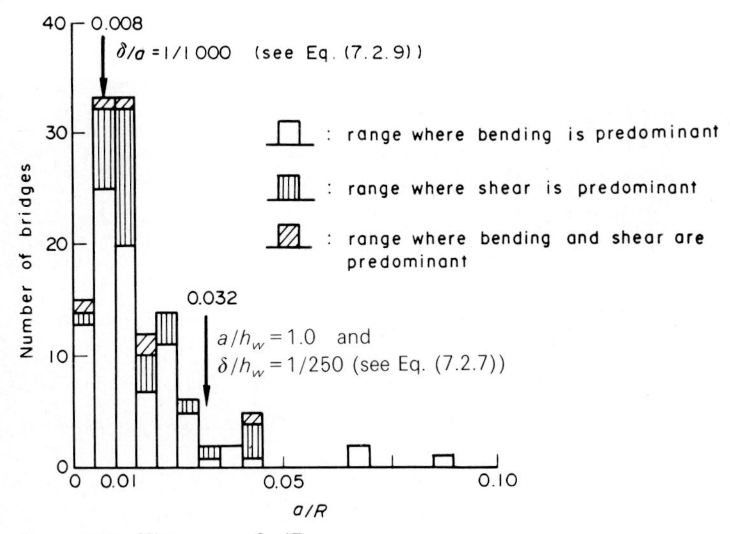

Figure 7.15 Histogram of a/R.

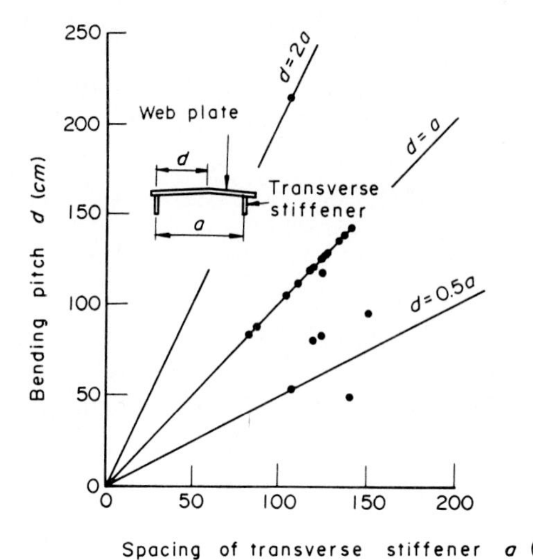

Figure 7.16 Relationships between a and d.[8]

JSHB.[2] These tolerances are for the initial imperfections of a column tolerance $l/1000$ in Eq. (7.2.9) and a plate $h_w/250$, with h_w being the depth of the web plate as given in Eq. (7.2.7). Therefore, the curved web plate is thought to be a very shallow shell with[8]

$$\frac{a}{R} \leqq 0.032 \tag{7.3.6}$$

For this shallow shell, the radius of curvature R can ordinarily be set by the natural curling action of thin web plates. However, some of the web plates (about 15 percent) need to be fabricated by being bent through a hydraulic press machine.[8] In this case, the bending pitch d has the tendency shown in Fig. 7.16.[8]

The groove weld joint strength can be checked in a case of pure shear, as shown in Fig. 7.17, and expressed by[3]

$$\tau = \frac{Q}{a_w h_w} \leqq \tau_a \tag{7.3.7}$$

where Q = applied shearing force
h_w = depth of web plate
a_w = throat of web plate
τ_a = allowable shearing stress of groove weld joint (the same as in Table 6.8)

For the condition where a shearing force Q and a bending moment M act together, as in Fig. 7.13, the strength of the groove weld joint can

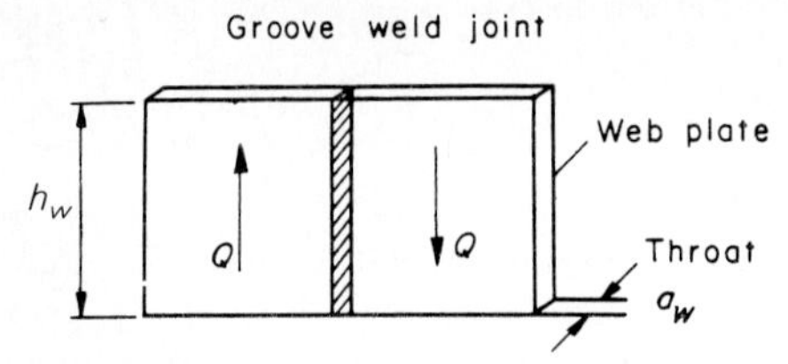

Figure 7.17 Groove weld joint of web plate.

be checked by using the interaction formula of JSHB[2]

$$\left(\frac{\sigma}{\sigma_a}\right)^2 + \left(\frac{\tau}{\tau_a}\right)^2 \leqq 1.2 \tag{7.3.8}$$

7.3.3 Fillet weld joint of flange and web plates

The flange plate is usually connected to the web plate by a fillet weld, as shown in Fig. 7.18*a*. The throat dimension a ($\geqq 6$ mm in JSHB[2]) can be theoretically calculated for an I girder (shown in Fig. 7.8*b*) from

$$\tau = \frac{QS}{2Ia} \leqq \tau_a \tag{7.3.9}$$

where Q = applied shearing force

$$S = \frac{A_f h}{2} \tag{7.3.10}$$

= static moment of flange plate

I = geometric moment of inertia of I girder

A_f = cross-sectional area of flange plate

h = girder depth

τ_a = allowable shearing stress of fillet weld joint (the same as in Table 6.8)

The relationship between throat a and size s, as detailed in Fig.

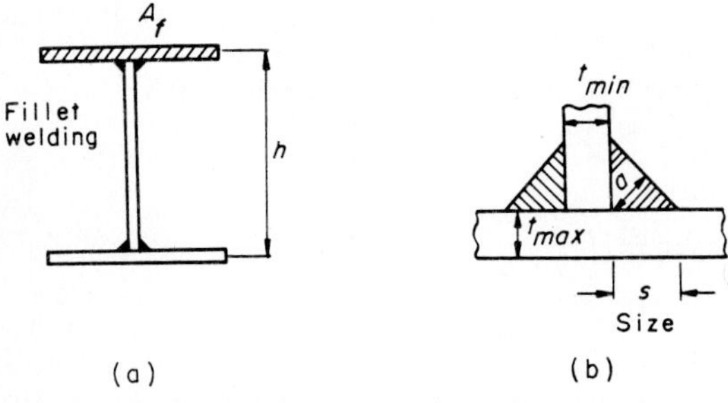

Figure 7.18 Fillet weld joint of (*a*) flange and web plates and (*b*) detail.

7.18*b*, can be represented by

$$a = \frac{1}{\sqrt{2}} s \qquad (7.3.11)$$

For size *s*, JSHB[2] uses the following empirical criterion:

$$\sqrt{2t_{max}} \leqq s < t_{min} \qquad (7.3.12)$$

where t_{max} and t_{min} are the maximum and minimum thicknesses, respectively, of plates in millimeters.

7.3.4 Welding of longitudinal and transverse stiffeners

A normal method for fillet welding a stiffener in a straight girder bridge is illustrated in Fig. 7.19. Notice that the longitudinal stiffeners are not welded to the transverse stiffener. Proper clearances should be available for the fillet welding and painting around these stiffeners. The lower end of the transverse stiffener shall not be welded to the tension flange plate of the girder, because there will be a considerable reduction in the fatigue strength of the tension flange if one welds across them.

In the case of a curved girder, the transverse stiffeners behave as a beam column, much like a vertical member of a hypothetical Pratt truss after the buckling of the web plate, mentioned in Sec. 5.6.2. In this case the transverse stiffener should be jointed to the tension flange plate. To avoid fatigue strength reduction in the flange plate,

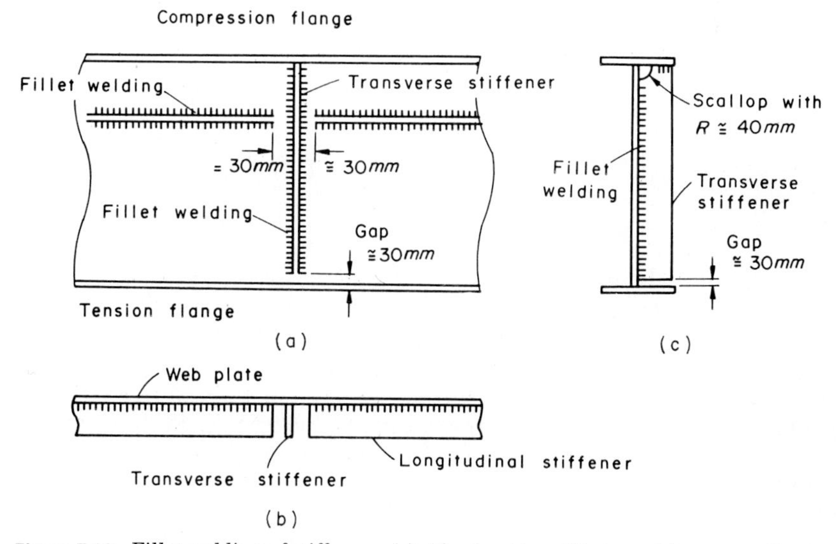

Figure 7.19 Fillet welding of stiffeners: (*a*) side elevation, (*b*) plane, (*c*) cross section.

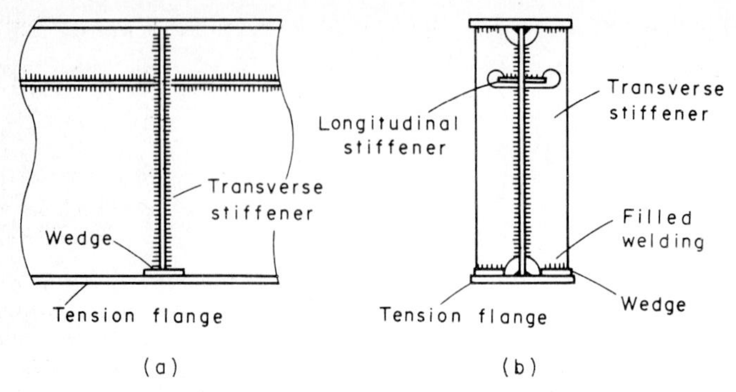

Figure 7.20 Welding of stiffener in curved girder: (*a*) side elevation and (*b*) cross section.

Fig. 7.20 shows an example of a jointing method, in which a wedge not welded to the tension flange plate is used and the transverse stiffener is welded to this wedge.[5]

At the same time, the longitudinal stiffeners must pass through the transverse stiffener in a curved girder bridge with a small radius of curvature. This is because they also behave as a beam-column member following buckling of the web plate, as predicted in Sec. 5.5.2. The same materials should be used in both the longitudinal stiffener and the web plate, to sustain their strength at high stress levels.

7.3.5 Field bolted joint of main girders

As shown in Fig. 7.11, the field joints are necessary in the main girders with long spans. For this joint, a high-tensile-strength bolt (henceforth called an *HT bolt*) is usually used instead of rivets. The details of this frictional joint with HT bolts are given in Fig. 7.21.

The allowable strength ρ_a of an HT bolt can be estimated by

$$\rho_a = \frac{1}{\nu}\mu N \tag{7.3.13}$$

where $N = \alpha A_e \sigma_y$ (7.3.14)

μ = frictional coefficient (=0.4 for joint surface with shot blasting[2])
ν = factor of safety (=1.7 in JSHB[2])

and α = reduction factor due to combined action of normal and shearing stress [=0.85 (F8T) and 0.75 (F10T) in JSHB[2]]
A_e = effective cross-sectional area of HT bolt
σ_y = yield stress of HT bolt

Table 7.1 lists the values of ρ_a according to JSHB.[2]

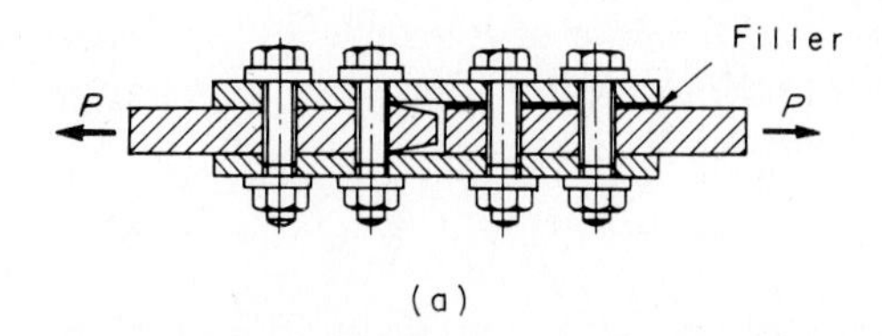

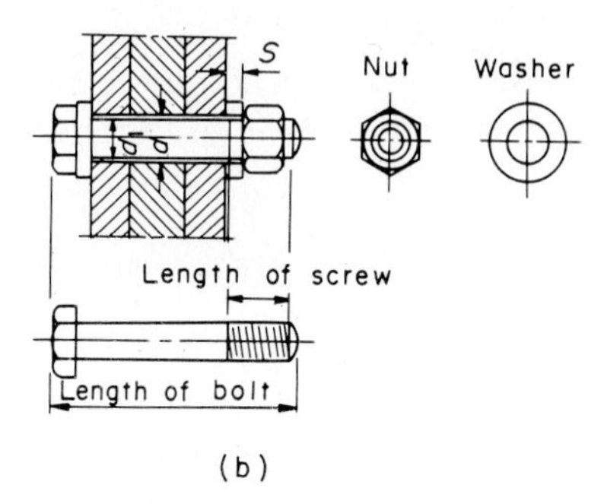

Figure 7.21 Details of HT bolt joint: (*a*) typical cross section, (*b*) bolt, nut, and washer.

TABLE 7.1 Value of ρ_a (kgf) for One High-Tensile-Strength Bolt[2] (per One Frictional Surface)

Class of bolt	Diameter d_1, mm	
	M 22	M 24
F8T	3900	4500
F10T	4800	5600

Therefore, the required number of HT bolts (n, n') in the flange plates can be designed by referring to Fig. 7.22 and using the following expressions:

$$n \geqq \frac{P}{\rho_a} \tag{7.3.15}$$

where $P = \sigma_c A_g$ for compression flange plate (7.3.16*a*)

$P = \sigma_t A_n$ for tension flange plate (7.3.16*b*)

in which $\sigma_c = \dfrac{M}{I} y_c$ (7.3.17*a*)

= mean compressive stress at compression flange plate

$\sigma_t = \dfrac{M}{I} y_t$ (7.3.17*b*)

= mean tensile stress at tension flange plate

I = geometric moment of inertia of girder

y_c, y_t = fiber distance of compression and tension flange plates from neutral N–N axis, respectively

and
$$A_g = b_g t_c \tag{7.3.18}$$

= gross cross-sectional area of compression flange plate

$$A_n = b_n t_t \tag{7.3.19}$$

= net cross-sectional area of tension flange plate

t_c, t_t = thickness of compression and tension flange plates, respectively

The net width b_n of the tension flange plate is estimated by JSHB.[2] See Figs. 7.23 and 7.24.

Once the required number of HT bolts, n or n', has been determined, the details of the joint must be calculated. This involves determining the pitch dimension p, edge distances e, and the spacing c, as shown in Fig. 7.25. According to JSHB,[2] their values are

$$3d \leqq p \leqq 6d \tag{7.3.20a}$$

$$e \geqq 1.5d \tag{7.3.20b}$$

where the diameter of bolt hole d (mm) is given by

$$d = d_1 + 3 \tag{7.3.21}$$

in which d_1 is the diameter (mm) of the HT bolts. In Table 7.2, the

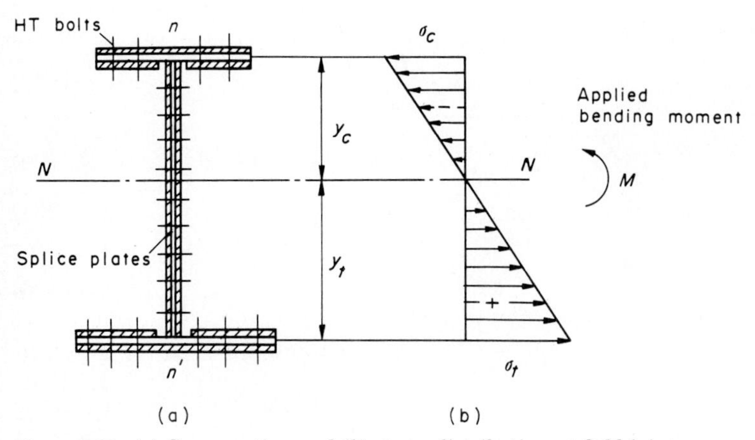

Figure 7.22 (*a*) Cross section and (*b*) stress distribution at field joint.

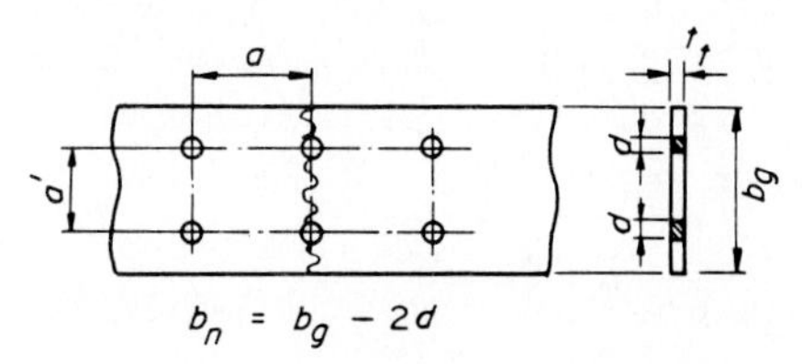

Figure 7.23 Net width b_n for parallel bolt joint.

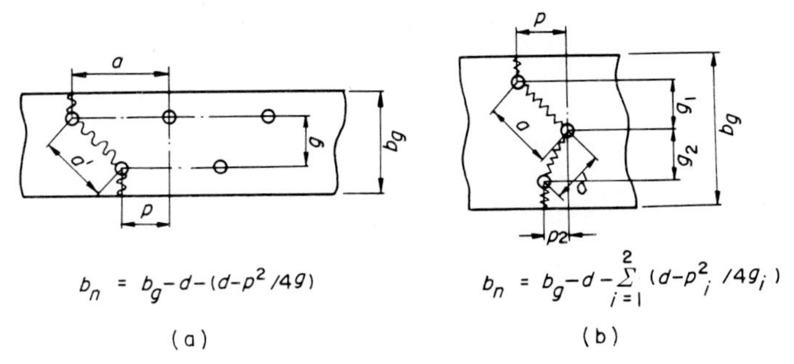

Figure 7.24 Two examples (a) and (b) showing net width b_n for zigzag bolt joint.[2]

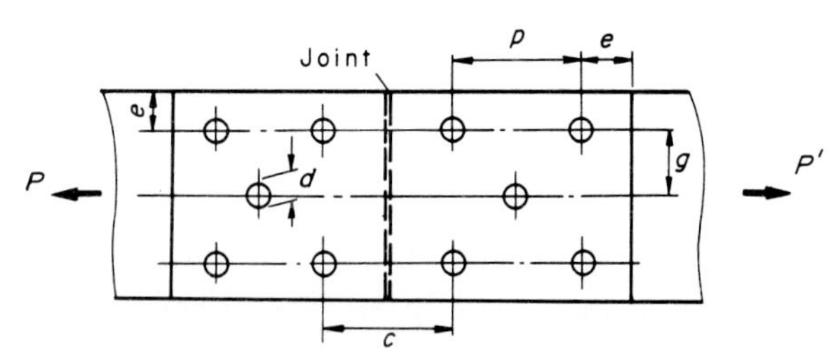

Figure 7.25 Arrangements of HT bolts.

TABLE 7.2 Standard Values of p, e, and c[9]

Bolt	d, mm	p, mm	e^4, mm	c, mm
M 22	25	75	40	90
M 24	27	85	45	100

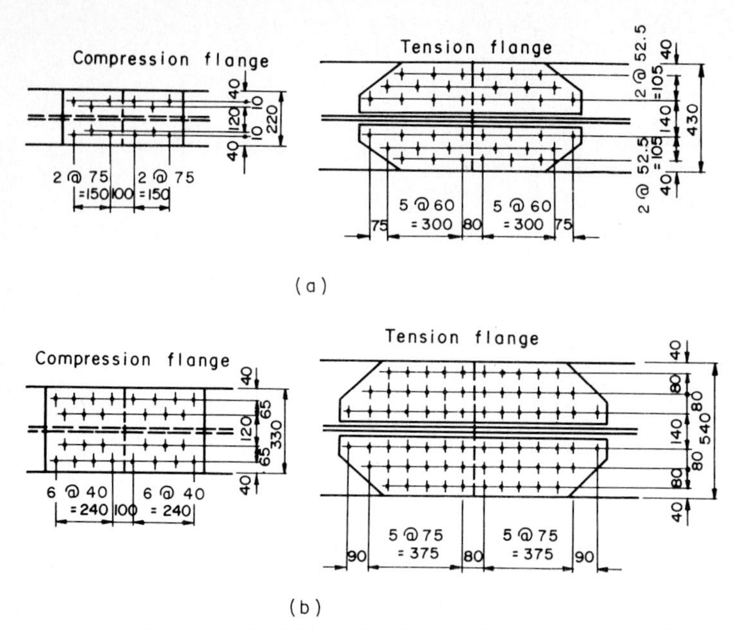

Figure 7.26 Two examples (*a*) and (*b*) of field joints in flange plate.[9]

standard values of *p*, *e*, and *c* from the Hanshin Expressway Public Corporation (HEPC) code[9] are listed. Figure 7.26 illustrates the field joints in the flange plates of composite girders.[9]

Next, the field joint of a web plate can be designed in a manner similar to that above and specified in JSHB.[2] Refer to Fig. 7.27 and the

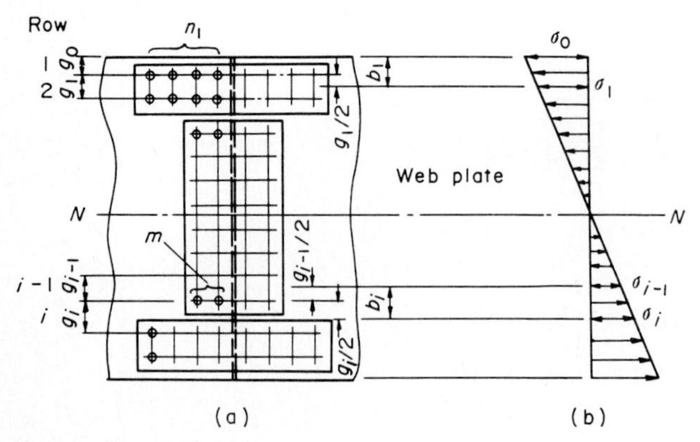

Figure 7.27 Field joint in web plate: (*a*) HT bolts in position and (*b*) stress distribution.

following equations for its design:

$$\rho_i = \frac{P_i}{n_i} \leqq \rho_a \tag{7.3.22}$$

where ρ_i = force acting on an HT bolt in ith row
P_i = force acting on a series of HT bolts, n_i, in ith row
n_i = half of total HT bolt numbers in ith row
ρ_a = allowable strength of HT bolts (see Table 7.1)

and

$$P_1 = \frac{\sigma_0 + \sigma_1}{2} b_1 t \tag{7.3.23a}$$

$$\vdots$$

$$P_i = \frac{\sigma_{i-1} + \sigma_i}{2} b_i t \tag{7.3.23b}$$

in which

$$b_1 = g_0 + \frac{g_1}{2} \tag{7.3.24a}$$

$$\vdots$$

$$b_i = \frac{g_{i-1} + g_i}{2} \tag{7.3.24b}$$

When both the bending moment M and the shearing force Q are acting together, Eq. (7.3.22) must be checked by the following

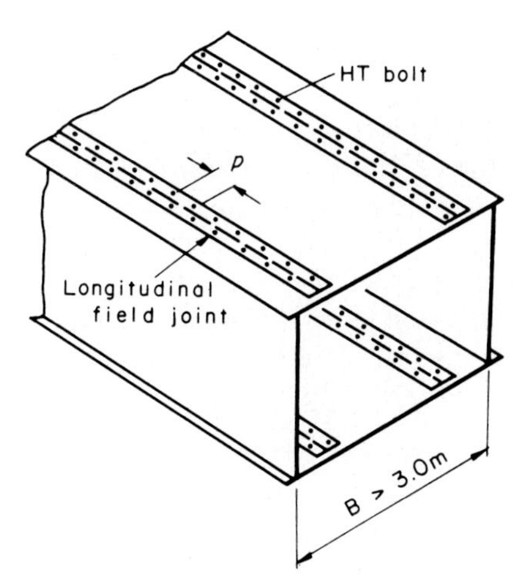

Figure 7.28 Longitudinal field joint in box-girder bridge.

equation:[2]

$$\rho_R = \sqrt{\rho_i^2 + \rho_Q^2} \leqq \rho_a \tag{7.3.25}$$

where

$$\rho_Q = \frac{Q}{m} \tag{7.3.26}$$

and m is half of the total number of HT bolts in the web joint.

Figure 7.28 shows a longitudinal field joint in a box-girder bridge with large cross-sectional dimensions with the flange plate width B larger than 3.0 m. For this joint, the pitch p (cm) of an HT bolt can be estimated by

$$p \leqq \frac{\rho_a}{q} \tag{7.3.27}$$

where q (kgf/cm) is the shear flow at the longitudinal joints and can be estimated easily by using the fundamental theory of a thin-walled beam, described in Chaps. 2 and 3.

7.4 Fabrication of Floor Beam, Sway, and Lateral Bracing

7.4.1 Connection of floor beam with main girders

Typical connection details between the main I girders and the intermediate floor beams in a straight girder bridge are shown in Fig. 7.29*a*. Note that the connections between the floor beams and the edge main girders are designed to resist shear force only.[9]

In a curved I-girder bridge, a torsional moment in the main girder is caused by the curvature of the bridge axis. Therefore, the connections should be fabricated by using HT bolted joints for resisting not only a shearing force but also this torsional moment of the main I girder as the bending moment of the floor beams. These connections can be delineated on the basis of the design method in Sec. 7.3.5, as shown in Fig. 7.29*b*.

In a similar manner, the floor beam connections at the edge and interior girders differ between straight and curved I-girder bridges. This is illustrated in Fig. 7.30. Note that the depth of the end floor beam h_Q of a straight I-girder bridge is usually given by[9]

$$h_Q \cong 0.4 h_G \tag{7.4.1}$$

where h_Q = depth of floor beam
h_G = depth of main I girder

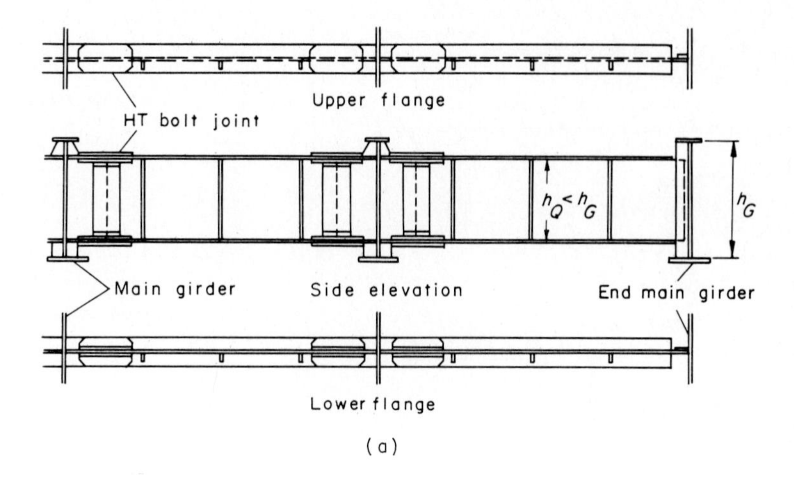

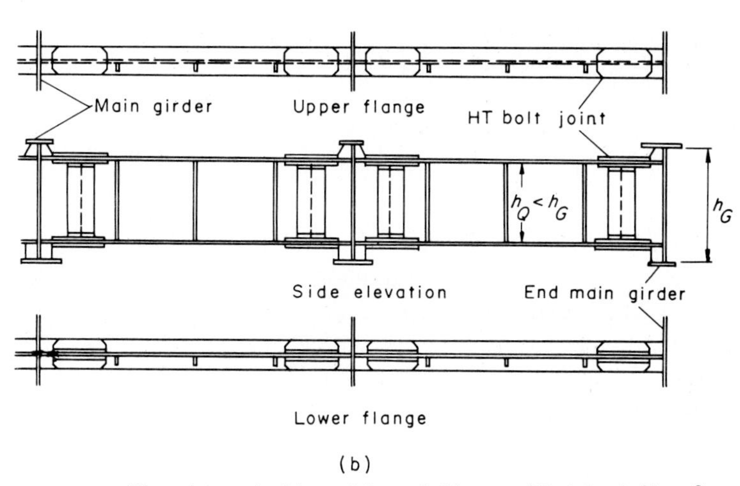

Figure 7.29 Floor beams in (*a*) straight and (*b*) curved I-girder bridges.[9]

On the contrary, the depth of a floor beam h_Q in a curved girder bridge is more or less larger than that of a straight girder bridge and should be designed as[9]

$$h_Q \leqq h_G \tag{7.4.2}$$

The details of the floor beam in a curved box girder are illustrated in Fig. 7.31.[9] Note in this figure that care should be taken in describing the details of the intermediate diaphragm.

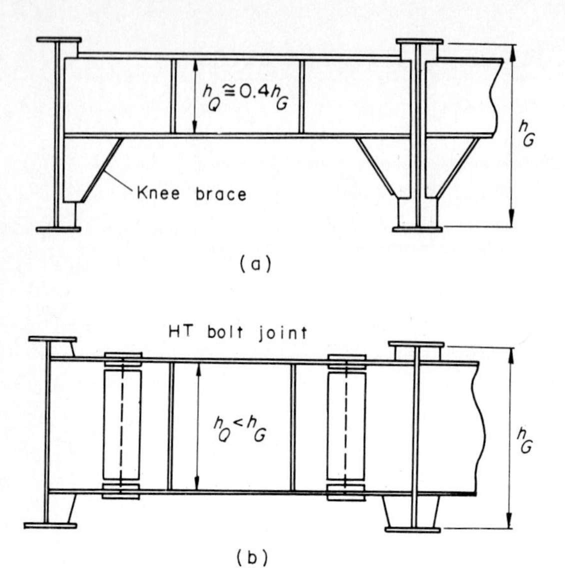

Figure 7.30 Floor beam at end or interior supports of (*a*) straight and (*b*) curved girder bridges.[9]

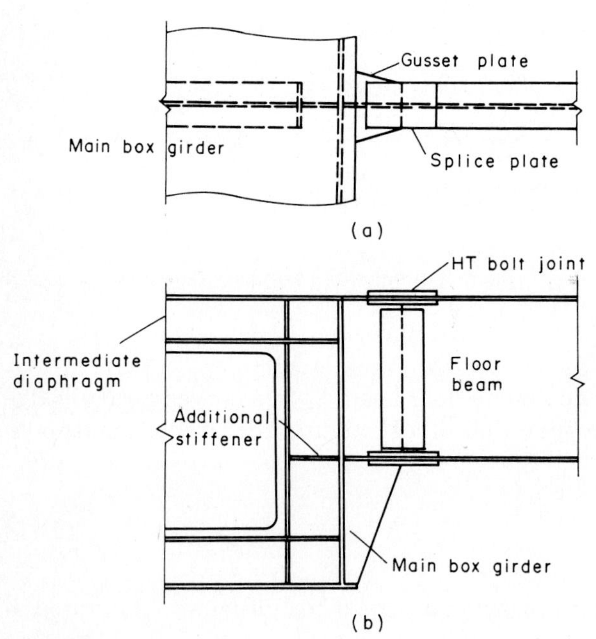

Figure 7.31 Floor beam in curved box-girder bridge: (*a*) plane and (*b*) cross section.[9]

7.4.2 Connection of sway and lateral bracings with main girders

(1) Sway bracings. The connection of the sway bracing to the main girder is illustrated in Fig. 7.32. In sizing the cross-sectional area of the angle, the eccentricity between the centroidal axis of the angle and the surface of the gusset plate should be considered. Figure 7.33 shows this connection. Owing to eccentricity e, the angle will undergo both an axial compressive force D and a bending moment $M = De$. Because of this, the angle should be designed as a beam-column member. Instead of this design method, the following method is proposed in JSHB:[2]

$$A_g \geqq \frac{D}{\sigma'_{ca}} \tag{7.4.3a}$$

= gross cross-sectional area for compression member

$$A_n \geqq \frac{D}{\sigma_{ta}} \tag{7.4.3b}$$

= net cross-sectional area for tension member

where D = applied compressive or tensile axial force
σ_{ta} = allowable tensile strength

and

$$\sigma'_{ca} = \sigma_{ca}\left(0.5 + \frac{l/r_X}{1000}\right) \tag{7.4.4}$$

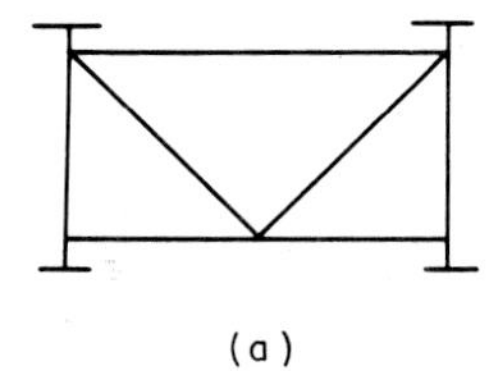

(a)

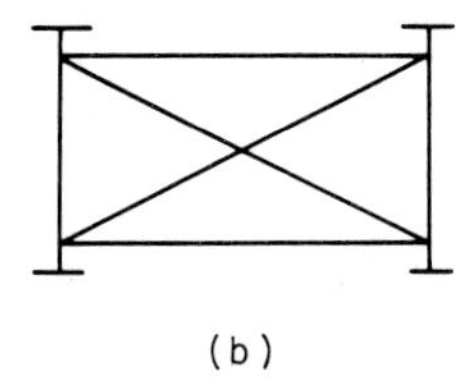

(b)

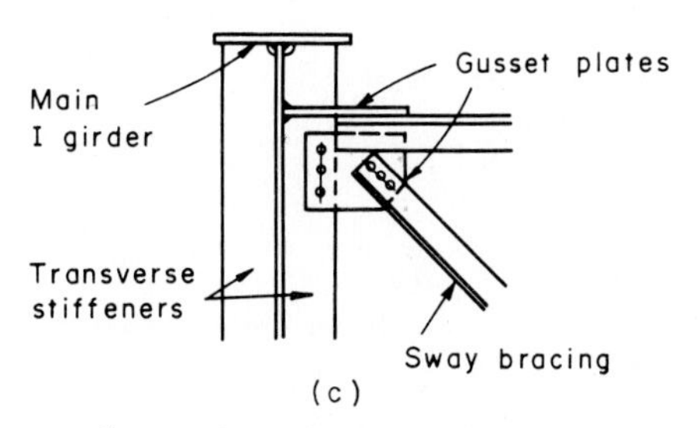

(c)

Figure 7.32 Connections of sway bracing to main I girder: (*a*) V type, (*b*) X type, (*c*) details.

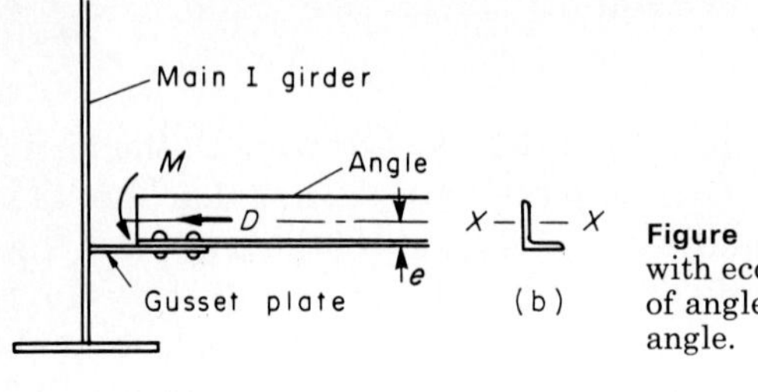

Figure 7.33 Design of angle with eccentricity: (*a*) connection of angle and (*b*) cross section of angle.

in which

σ_{ca} = allowable compressive stress as given by function of slenderness l/r_X in Eq. (6.2.168) and Table 6.15

l = length of bracing

r_X = radius gyration of angle with respect to centroidal X axis (see Fig. 7.33*b*)

Therefore, the required cross-sectional area A_g (gross area) or A_n (net area) can be found easily from the angle catalog.

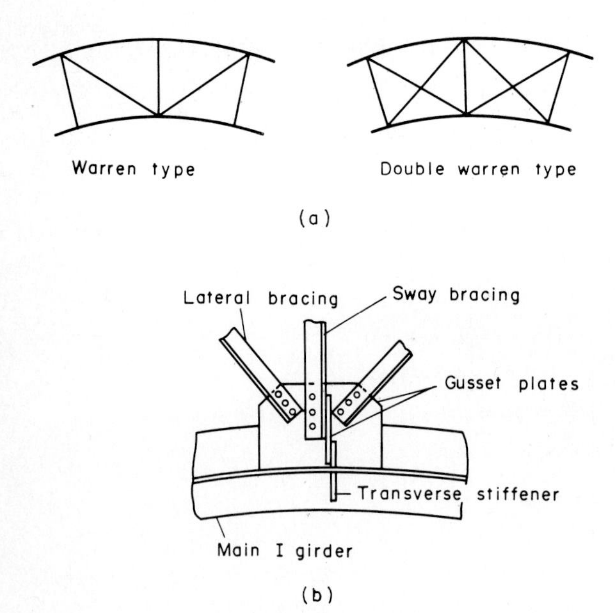

Figure 7.34 Connection of lateral bracing to main I girder: (*a*) types of lateral bracing, (*b*) details of lateral bracing.

The design of the HT bolted connections can also follow that of a girder flange plate, described in Sec. 7.3.5.

(2) Lateral bracings. The lateral bracing can be designed and manufactured in the same manner as sway bracing. Figure 7.34 illustrates the connection of the lateral bracing to the main I girder.

7.5 Fabrication of Diaphragms

7.5.1 Intermediate diaphragm

The intermediate diaphragms in a curved box-girder bridge have the function of resisting the distortions in the box cross section. This distortion is usually significant at the section where the torsional moment is large, such as where the floor beams or brackets are connected to the main box girders.

The design method for the intermediate diaphragm resisting distortion is explained in Sec. 6.2.8(4), and the connection details are indicated here in Fig. 7.35.[9]

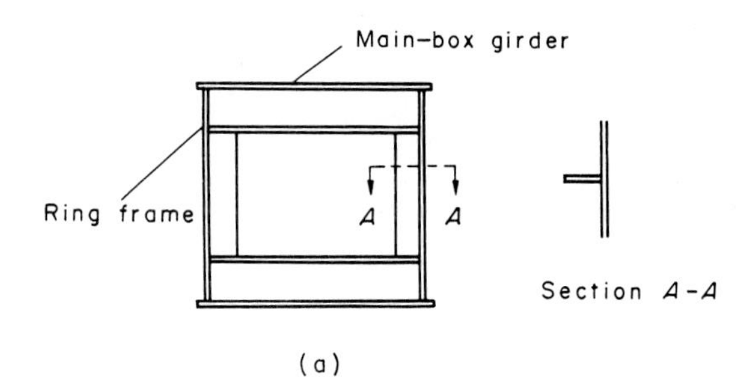

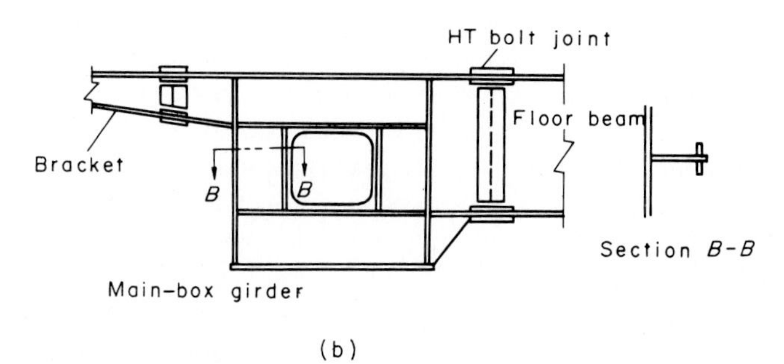

Figure 7.35 Intermediate diaphragm: (*a*) central part, and (*b*) location of floor beam and bracket.[9]

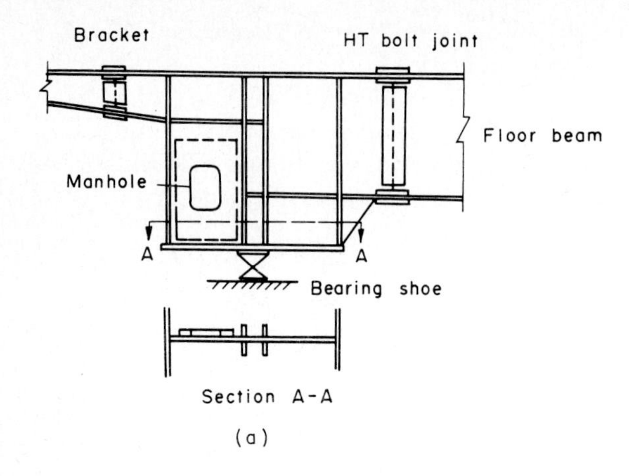

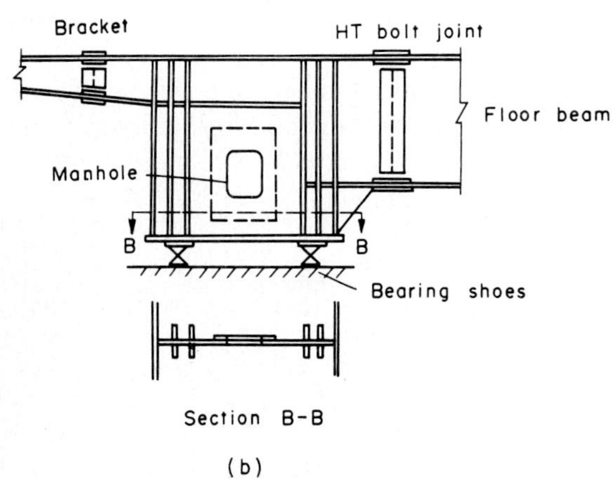

Figure 7.36 Diaphragm at supports: (*a*) monobearing shoe and (*b*) twin-bearing shoes.

7.5.2 Bearing support diaphragm

The diaphragms at the supports of a curved box-girder bridge must be designed not only to have enough strength and rigidity to resist distortion, but also to transfer the huge reaction forces due to the bearing shoes. To make such diaphragms, a few additional transverse stiffeners are added to the cross section of box girders, as shown in Fig. 7.36.

The strength of the transverse stiffeners can be checked by modeling them as a column member, as in the design of the end transverse stiffener described in Sec. 6.2.8(5).

In these diaphragms, the purpose of the manhole is to gain access to maintain the bridges. Around this manhole, the plate thickness is doubled to reinforce against high stress concentrations.

7.6 Fabrication of Bearing Shoes

7.6.1 Functions of bearing shoes

The bearing shoes are the members that support the superstructure and transmit reactions to the substructure. In designing these members, it is necessary to consider deformations, such as the elongation due to thermal change of the superstructure or the deflection angle as well as the torsional angle at the end of girders due to live load.

The bearing shoes are fundamentally categorized into two types:

1. *Fixed support.* This support is completely fixed against the horizontal forces in any direction.
2. *Movable support.* In general, this support is movable in only one direction, but a special bearing can be set as a movable support in any direction.

Bearing shoe arrangements and reaction calculations in a curved girder bridge are shown in Secs. 6.2.2(3) and 6.2.8(6). The allowable movement Δl (mm) at the movable bearing is estimated in JSHB[2] as

$$\Delta l = \Delta l_T + \Delta l_\theta + 30 \tag{7.6.1}$$

where $\Delta l_T = \Delta T \, \alpha l =$ elongation due to thermal change (7.6.2)

$\Delta T =$ thermal change (see Table 6.7)

$\alpha =$ elongation coefficient [see Eq. (6.2.31)]

$l =$ span length of girder bridge

and

$$\Delta l_\theta = \sum h_i \theta_i \tag{7.6.3}$$

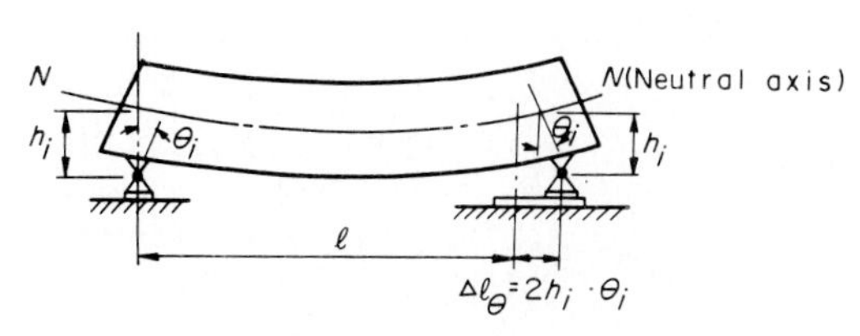

Figure 7.37 Movement of support in simply supported girder due to deflection angle.

TABLE 7.3 Frictional Coefficient μ of Bearing[2]

Frictional type	Types of support	Frictional coefficient μ
Rolling friction	Roller and rocker	0.05
Slipping friction	Teflon bearing plate	0.10
	High-strength brass bearing plate	0.15
	Steel or iron line bearing	0.25–0.20

where Δl_θ = elongation due to deflection angle; $\Delta l_\theta = 2h_i\theta_i$ for a simply supported girder bridge by considering deflection angle θ_i at fixed support, shown in Fig. 7.37

h_i = distance between center of deflection angle at support and neutral axis, generally approximated as $h_i = \frac{2}{3}h$ in which h is girder depth

θ_i = deflection angle generally approximated as $= \frac{1}{150}$

Frictional forces occur in the movable bearing. The difference of frictional coefficients μ at a movable support are indicated in JSHB[2] and listed in Table 7.3.

7.6.2 Types of bearing shoes

Although there are various types of bearing shoes, those in Table 7.4, are more suitable for straight and curved girder bridges.[10]

TABLE 7.4 Types of Bearing and Corresponding Reaction and Span Length of Girder*[10]

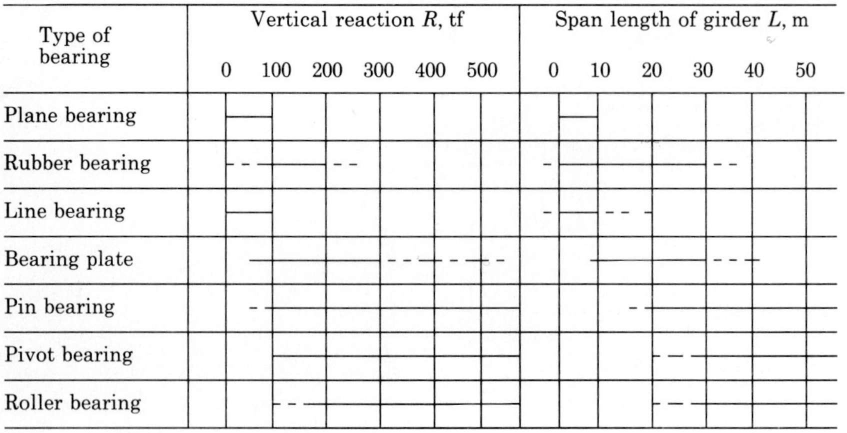

Type of bearing	Vertical reaction R, tf (0 100 200 300 400 500)	Span length of girder L, m (0 10 20 30 40 50)
Plane bearing		
Rubber bearing		
Line bearing		
Bearing plate		
Pin bearing		
Pivot bearing		
Roller bearing		

*—— = usual range; – – – = possible range.

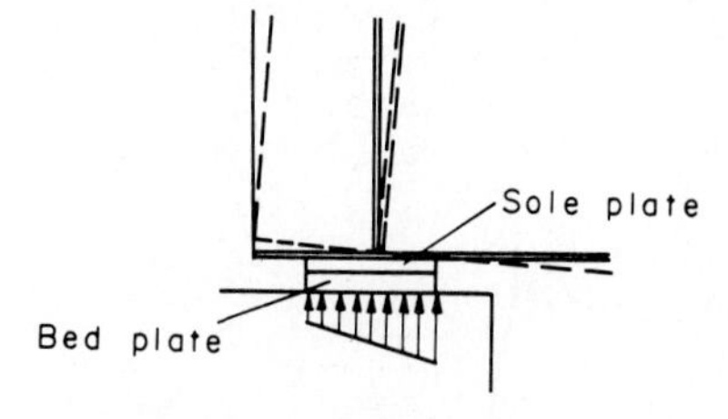

Figure 7.38 Distribution of reaction in plane bearing.

(1) Plane and rubber bearings. The plane bearing consists of sole and bed plates as shown in Fig. 7.38. When both plates are completely flat, the distribution of the reaction under the bed plate takes on a triangular form, as shown. Because of this unequal distribution, the plane bearing is usually fabricated by cutting the bed plate as in Fig. 7.39.

A rubber plate or elastomer, illustrated in Fig. 7.40, is another type of plane bearing. In this bearing, the deflection angle θ and elongation Δl due to the horizontal force H can be found by the elastic deformation of the rubber or elastomer[11] plates (see Fig. 7.41).

(2) Line bearing. The line bearing is a widely adopted type for the girder bridges with comparatively shorter spans. Details of this bearing are given in Fig. 7.42.

(3) Bearing plate bearing. A bearing plate bearing is schematically illustrated in Fig. 7.43. The PTFE (Teflon) plate with minimum thickness 4 mm and the stainless plate with minimum thickness 2 mm are usually installed to allow for the smooth movement of shoes.[9]

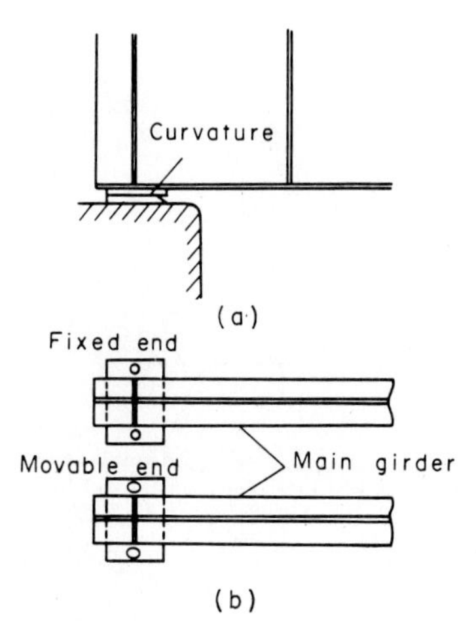

Figure 7.39 Details of plane bearing: (*a*) side elevation and (*b*) detail of end support.

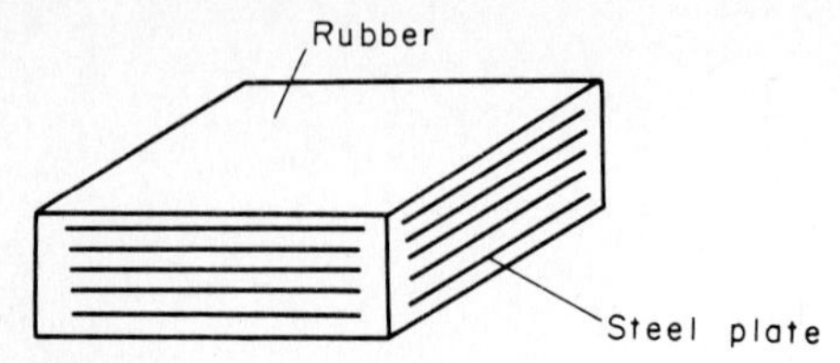

Figure 7.40 Details of elastomer.

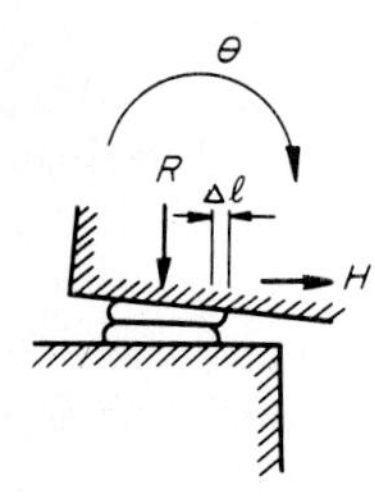

Figure 7.41 Deformations of rubber or elastomer.

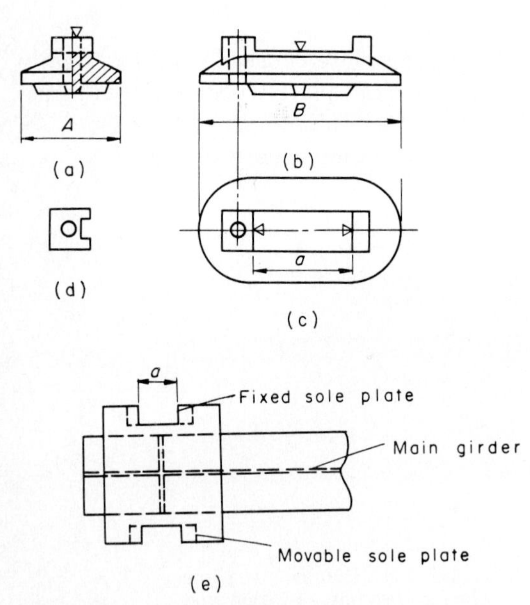

Figure 7.42 Details of line bearing: (*a*) cross section, (*b*) side elevation, (*c*) plan, (*d*) stopper, and (*e*) detail of sole plates.

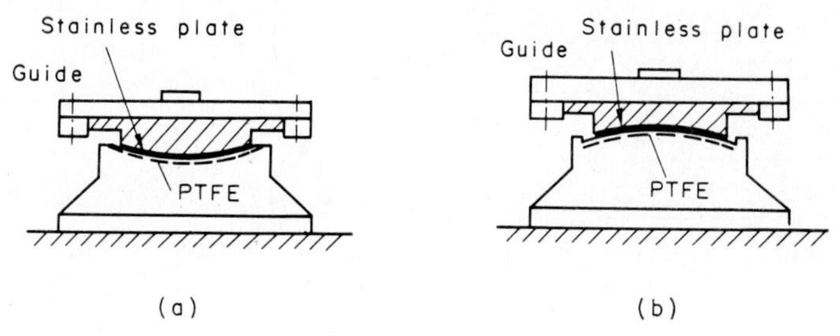

Figure 7.43 Details of plate bearing: (*a*) concave type and (*b*) convex type.

Figure 7.44 shows another bearing plate type that uses the elastomer introduced in Reference 11. Figure 7.45 illustrates this type of bearing used in a curved girder bridge.[12]

(4) Pin bearing. The details of a pin bearing are shown in Fig. 7.46. This bearing type allows only rotation around the pin. To determine the diameter $d(2r)$ of the pin, let us consider the bearing stress distribution shown in Fig. 7.47. From the equilibrium of this bearing stress and the applied vertical reaction R (detailed in Sec. 6.2.8) the following equation is obtained:

$$R = \int \sigma_\varphi r l \cos \varphi \, d\varphi \tag{7.6.4}$$

where

$$\sigma_\varphi = \sigma_{\max} \cos \varphi \tag{7.6.5}$$

$\sigma_{\max}$ = maximum bearing stress

l = length of pin

φ = angular coordinate

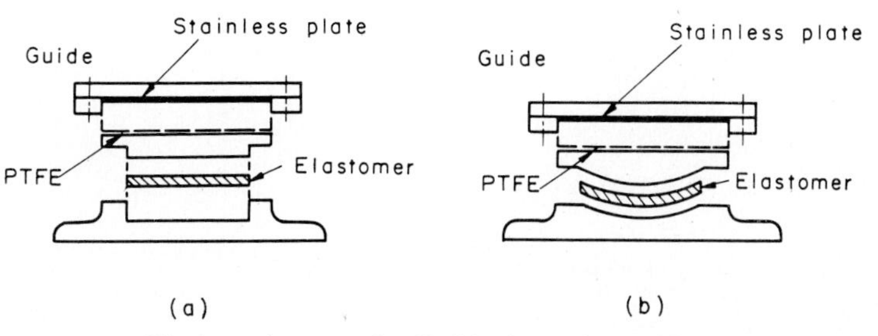

Figure 7.44 Elastomatic pot and cylindrical or spherical bearing: (*a*) pot type and (*b*) cylindrical or spherical type.[11]

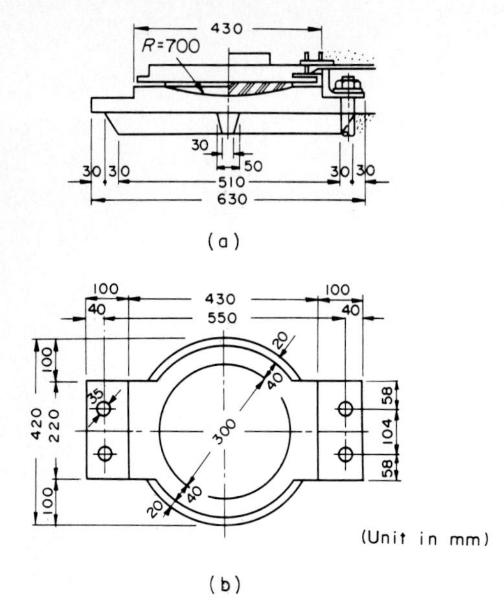

Figure 7.45 Example of bearing plate bearing with the lubricants of graphite and molybdenum disulfided ($R = 120$ tf) in curved girder: (*a*) cross section and (*b*) plan.[12]

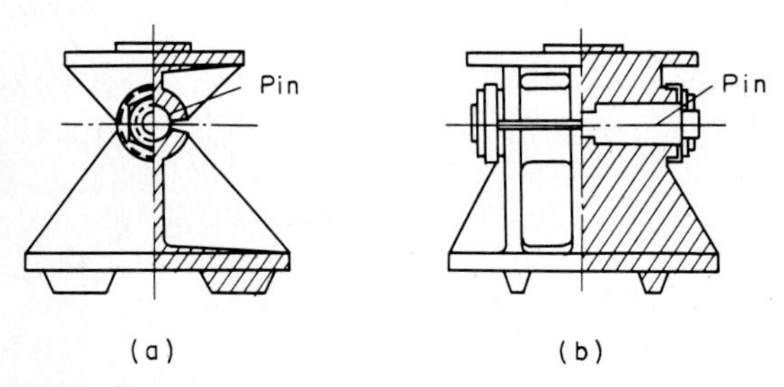

Figure 7.46 Pin bearing: (*a*) side view and (*b*) cross section.

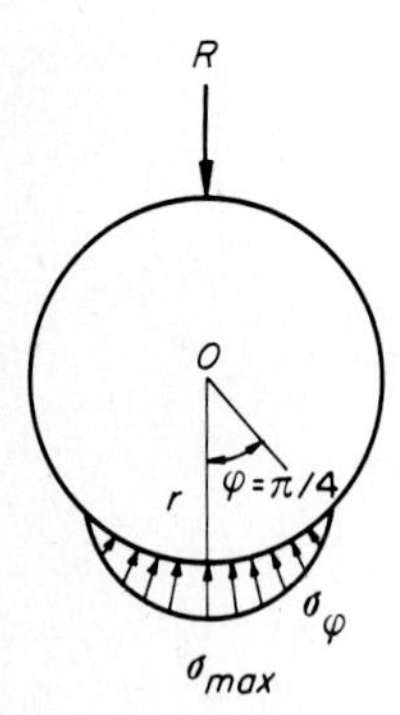

Figure 7.47 Bearing stress distribution in pin.

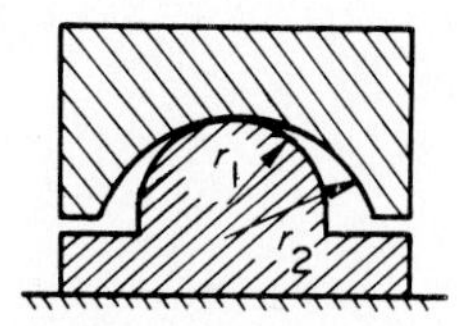

Figure 7.48 Spherical bearing with different curvature.

Therefore, integrating the above equation and assuming $\varphi = 0$ to $\pi/4$ give

$$R = 2\sigma_{max} rl \int_0^{\pi/4} \cos^2 \varphi \, d\varphi = 1.285\sigma_{max} rl \tag{7.6.6}$$

from which we get

$$r \geqq \frac{0.8R}{\sigma_a l} \tag{7.6.7}$$

where σ_a is the allowable bearing stress.

It is defined in JSHB[2] that

$$d = 2r \geqq 75 \text{ mm} \tag{7.6.8}$$

However, a rotational bearing in any direction is often required for curved girder bridges. This is why a spherical bearing with different radii of curvature r_1 and r_2, shown in Fig. 7.48, can be utilized. According to the formula given by Hertz,[5] the maximum bearing stress σ_{max} can be represented by

$$\sigma_{max} = 0.388 \sqrt[3]{RE^2 \left(\frac{r_1 + r_2}{r_1 r_2} \right)^2} \tag{7.6.9}$$

For instance, if we let $E = 2.1 \times 10^6$ kgf/cm^2 and $\sigma_{max} = 6000$ kgf/cm^2, then according to JSHB[2] the allowable reaction R_a (kgf) can be expressed as

$$R_a = 0.83 \left(\frac{r_1 r_2}{r_1 + r_2} \right)^2 \tag{7.6.10}$$

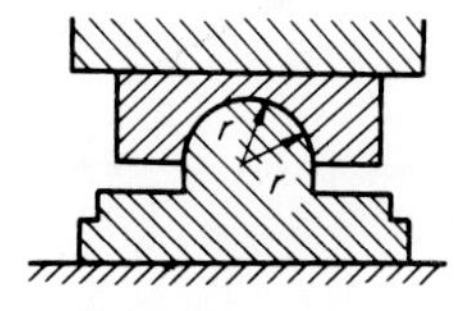

Figure 7.49 Principle of pivot bearing.

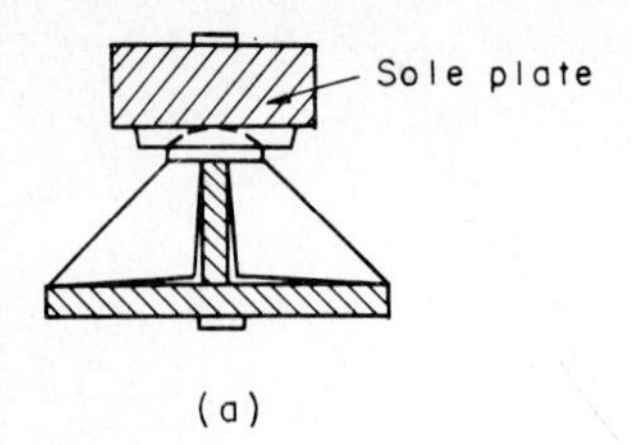

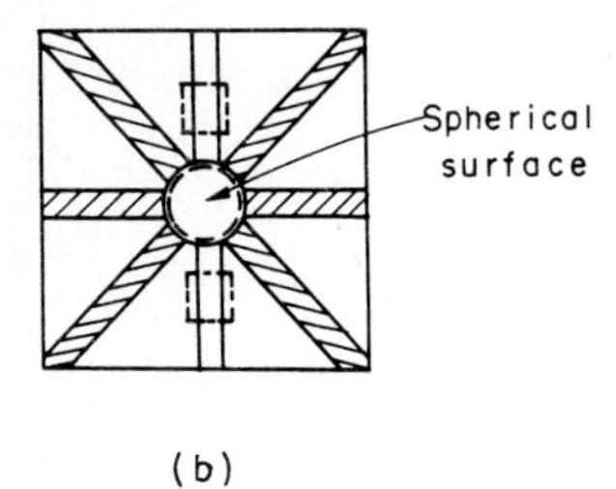

Figure 7.50 Details of pivot bearing: (*a*) cross section and (*b*) plane.

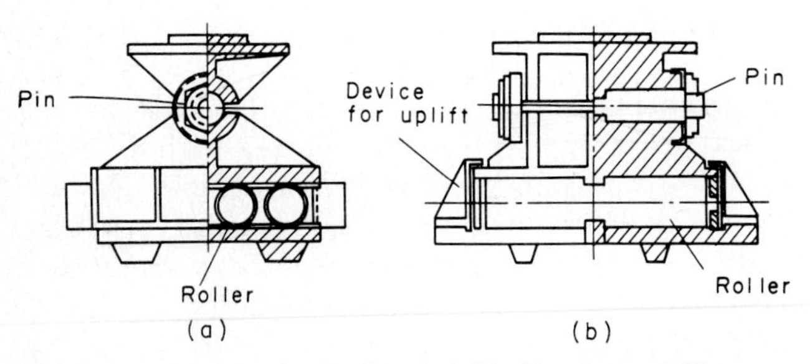

Figure 7.51 Details of roller bearing: (*a*) side view and (*b*) cross section.

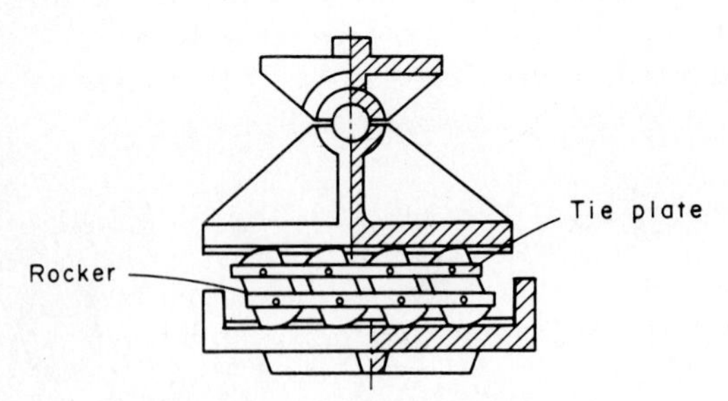

Figure 7.52 Details of rocker bearing.

The pivot bearing is designed so that $r_1 = r_2 = r$, as shown in Fig. 7.49. Figure 7.50 illustrates a pivot bearing in detail.

(5) Roller bearing. Details of a roller bearing are shown in Fig. 7.51. When a roller has different radii of curvature r_1 and r_2 and length l, the maximum bearing stress $\sigma_{\max}$ can be expressed by

$$\sigma_{\max} = 0.418\sqrt{\frac{RE}{l}\frac{r_1 + r_2}{r_1 r_2}} \tag{7.6.11}$$

When the roller is in contact with a flat plate, so that $r_1 = r$ and $r_2 \to \infty$, JSHB[2] specifies

$$\frac{R}{l} = 98r = 49d \tag{7.6.12a}$$

or

$$\frac{R}{l} \geqq 45d \tag{7.6.12b}$$

as the safety value.

The special roller bearings with only one roller have been developed—Ammer's or Corrowweld's rollers.[12] The details are omitted here. Figure 7.52 shows the rocker roller, where the roller is cut and connected with tie plates.

7.6.3 Earthquake design precautions

In the case of a strong earthquake, the superstructure may sometimes be moved off the abutments or piers, and a collapse can result. To avoid such failures, the bearing stress σ_b of abutments or piers should be within their allowable bearing strength σ_{ba}

$$\sigma_b = \frac{R}{A_b} \leqq \sigma_{ba} \tag{7.6.13}$$

where R (see Sec. 6.2.8) is the vertical reaction and A_b is the cross-sectional area of the bottom bearing plate. Moreover, the horizontal reaction H should be resisted by the anchor bolts (usually four anchor bolts) set in the bearing and checked by using the following:

$$\tau = \frac{H}{A_a} \leqq \tau_a \tag{7.6.14}$$

Here A_a is the cross-sectional area of the anchor bolts, and τ_a is the allowable shearing stress.

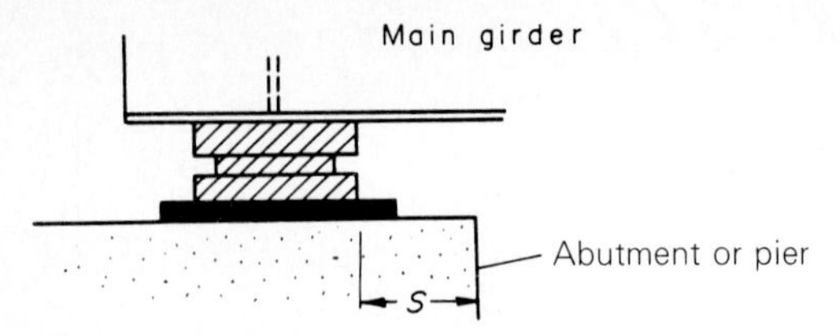

Figure 7.53 Clearance for abutment.

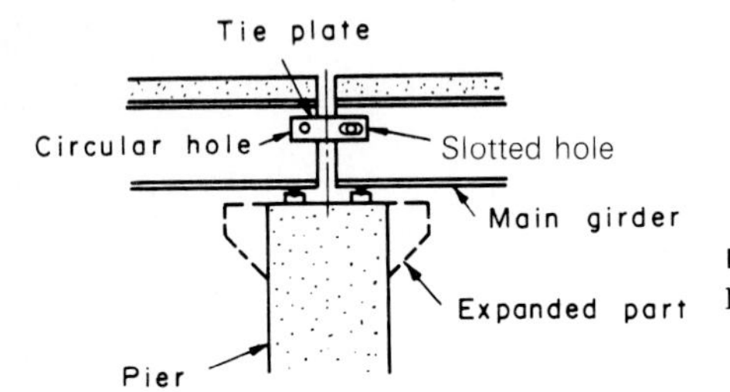

Figure 7.54 Expansion at top of pier and tie plate.

In addition, an appropriate clearance S (cm) should be provided at the top surface of the abutments or piers, as shown in Fig. 7.53. JSHB[13] recommends that this clearance be

$$S = \begin{cases} 20 + 0.5l & l < 100\text{ m} \qquad (7.6.15a) \\ 30 + 0.4l & l \leqq 100\text{ m} \qquad (7.6.15b) \end{cases}$$

where l is the span length of the girder bridge.

Thus an expanded part of the pier tops is illustrated in Fig. 7.54. A tie plate with a slotted hole is added to the girder to help prevent girder separation and consequent collapse in the case of a strong earthquake.

7.7 Painting

The humidity in Japan is always so high that painting and the maintenance of steel materials against corrosion should be carefully considered in the fabrication of steel bridges. For example, the painting methods specified in HEPC[9] are summarized in Table 7.5.[14]

7.8 Erection

Erection procedures are outlined in Fig. 7.55. Ordinary erection methods of steel girder bridges, including curved girder bridges, can be

TABLE 7.5 Standard on Painting Methods[9.14]

Surface	N	Painting in workshop	Painting at field	Remarks
A	6	1 Etching primer 2,3 Lead anticorrosive paint* 4 Phenolic MIO† paint	5,6 Long oil phethalic resin coating	General location
			5,6 Chlorinated rubber paint	Near sea and over sea
B	3	1 Etching primer 2,3 Epoxy coal-tar paint		
C	5	1 Inorganic zinc-rich primer 2 Organic zinc-rich paint 3 Epoxy MIO paint	4,5 Urethane resin paint	Painting before surfacing with asphalt
			4,5 Chlorinated rubber paint	Painting after surfacing
D	3	1 Inorganic zinc-rich primer 2,3 Epoxy coal-tar paint	No painting	Inner surface of box girder with steel deck
E	2	1 Inorganic zinc-rich primer 2 Organic zinc-rich paint	No painting	Inner surface of box girder with steel deck

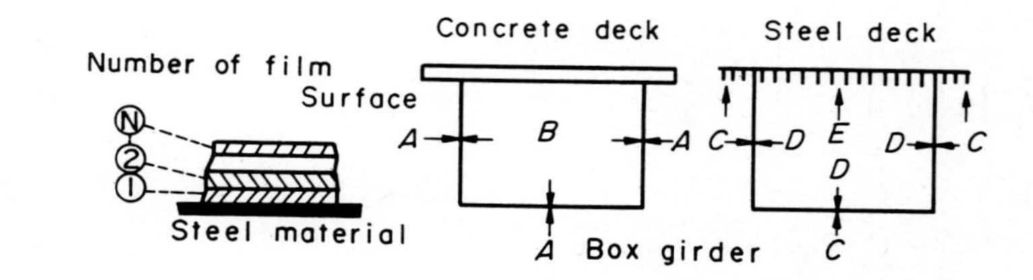

*For example, lead suboxide anticorrosive paint (red lead).
†MIO = micaceous iron oxide.
N = number of painting film.

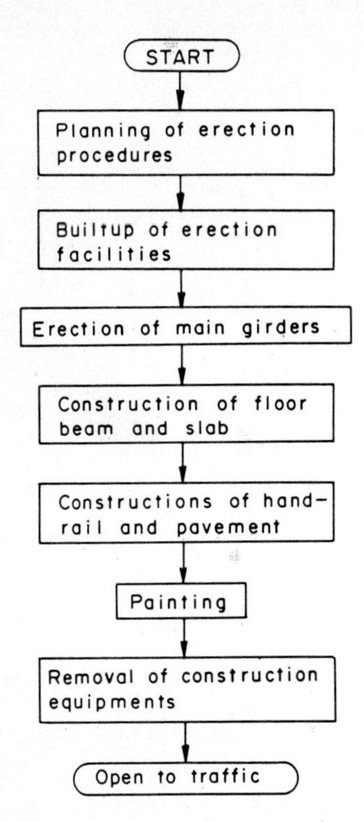

Figure 7.55 Procedures of erection.[1]

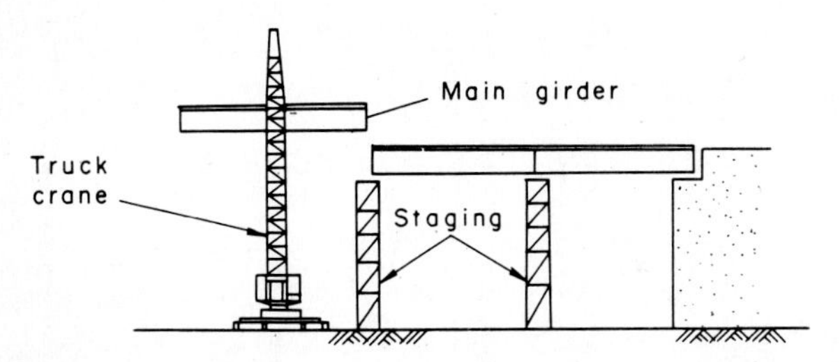

Figure 7.56 Staging erection method.[1,10]

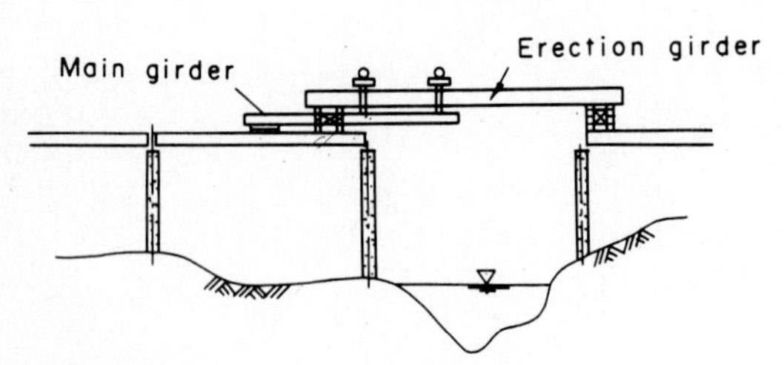

Figure 7.57 Erection method using erection girder.[1,10]

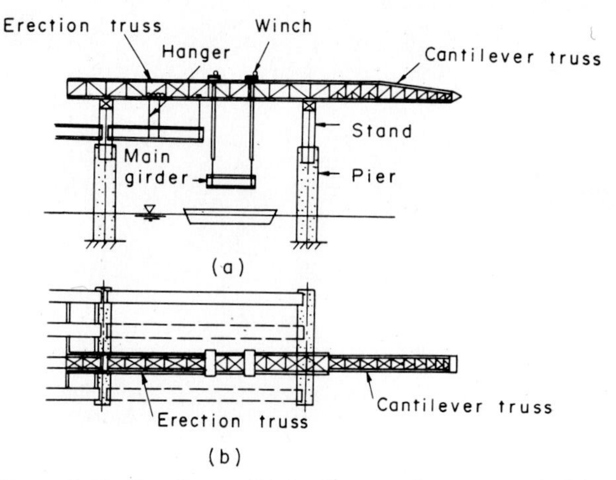

Figure 7.58 Erection method using erection truss: (*a*) side elevation and (*b*) plan.[1,10]

classified as follows:[1,10]

1. Staging erection method (see Fig. 7.56)
2. Erection method using erection girder or truss (see Figs. 7.57 and 7.58)
3. Cantilever erection method (see Fig. 7.59)
4. Large assembled erection method (see Fig. 7.60)

Although the cable erection method is widely used in the erection of a long pan truss or arch bridges over a deep valley, it cannot be utilized properly on curved girder bridges. Therefore, this method is not considered.

When curved girder bridges are erected, it is important to check the center-of-gravity locations in each member or segment being erected. In the large assembled erection method, shown in Fig. 7.60, floating cranes may be used instead of truck cranes, especially when bridges

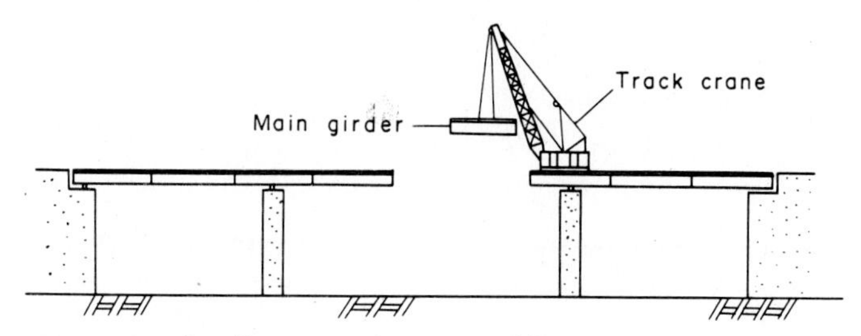

Figure 7.59 Cantilever erection method.[1,10]

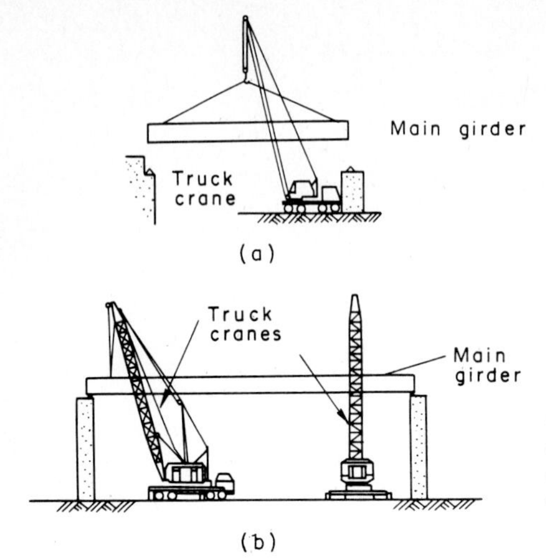

Figure 7.60 Large assembled erection method: (*a*) single truck crane and (*b*) double truck crane.[1,10]

are constructed along a bay or seashore and access is limited to truck cranes.

References

1. Kobori, T.: *Bridge Engineering*, Asakura-shoten, Tokyo, 1973 (in Japanese).
2. The Japanese Road Association: *The Japanese Specification for Highway Bridges*, pts. 1 and 2, Maruzen, Tokyo, Feb. 1980 (in Japanese).
3. Tachibana, Y., and H. Nakai: *Bridge Engineering*, 2d ed., Kyoritsu-Shuppan, Tokyo, Apr. 1981 (in Japanese).
4. Timoshenko, S. P., and J. N. Goodier: *Theory of Elasticity*, McGraw-Hill, New York, 1951.
5. Stüssi, F.: Entwurf und Berechnung von Stahlbauten, Erster Bd., Grundlagen des Stahlbaues, Springer-Verlag, Berlin, 1958 (in German).
6. Nakai, H., T. Kitada, and R. Ohminami: "Experimental Study on Bending Strength of Web Plate of Horizontally Curved Girder Bridges," *Proceedings of the Japanese Society of Civil Engineers*, no. 340, pp. 77–85, Dec. 1983 (in Japanese).
7. Nakai, H., T. Kitada, R. Ohminami, and K. Fukumoto: "Experimental Study on Shear Strength of Horizontally Curved Plate Girders," *Proceedings of the Japanese Society of Civil Engineers*, no. 350/I-2, pp. 281–290, Oct. 1984 (in Japanese).
8. Nakai, H., S. Muramatsu, O. Yoshikawa, T. Kitada, and R. Ohminami: "A Survey for Web Plates of Horizontally Curved Girder Bridges," *Bridge and Foundation Engineering*, 81(5):38–45, May 1981 (in Japanese).
9. The Hanshin Expressway Public Corporation: *The Design Code, Part 2, Design Codes for Steel Bridge Structures*, Osaka, Apr. 1980 (in Japanese).
10. Naruse, K., and T. Suzuki: *Bridge Engineering (Steel Bridges)*, Morikita-shuppan, Tokyo, Apr. 1973 (in Japanese).
11. The Task Committeee of Curved Box Girders of the ASCE-AASHTO Committee on Flexural Members of the Committee on Metal of ASCE Structural Division: "Curved Steel Box-Girder, A Survey," *Proceedings of the American Society of Civil Engineers*, vol. 104, no. ST-11, pp. 1719–1739, Nov. 1978.

12. Konishi, I., and S. Komatsu: *Steel Bridges, Design Part II*, Maruzen, Tokyo, 1976 (in Japanese).
13. The Japanese Road Association: *The Japanese Specification for Highway Bridges*, Part 5, Maruzen, Tokyo, Feb. 1960 (in Japanese).
14. Nakai, H., and T. Kitada: "A Parametric Survey of Cable-Stayed Bridges in Japan," *Proceedings of the Sino-American Symposium on Bridge and Structural Engineering*, Beijin, People's Republic of China, pt. 1, pp. 2-09-1 through 2-09-15, Sept. 1982.

Chapter

8

Design Examples

8.1 Introduction

During the past decade, numerous horizontally curved girder bridges have been constructed around the world. Reference 1 is a survey of these bridges.

In some of the large cities of Japan, these kinds of structures are especially needed when lack of right of way is a problem. For example, trends of the maximum span length L and radius of curvature R in simple and continuous curved girder bridges are plotted as a histogram in Fig. 8.1 according to a survey by Nakai et al.[2] This survey shows that the maximum span length L varies from 10 to 90 m. Approximately 73.6 percent of the values are within the range of $20\text{ m} < L < 50\text{ m}$. Furthermore, the smallest radius of curvature R is 10 m for the box girder and 20 m for the I girder. In view of these distributions, this chapter contains design examples of curved girder bridges with a comparatively medium span length L and radius of curvature R.

For these bridges, multiple curved I-girder, curved monobox-girder, and curved twin-box-girder bridges are the more popular types. The above three typical types of curved girder bridges are introduced here to describe the design method and details of curved girder bridges.

In Japan, almost all of the curved steel bridges have been designed by using computer programs. These computer programs are categorized as follows:

1. Analysis of curved box girders
2. Calculation of cross-sectional properties and stress analysis of curved box girders
3. Analysis of curved grillage girders

4. Optimum design of curved I girders
5. Optimum design of curved box girders
6. Design of steel decks
7. Computer-aided design of curved girders
8. Special programs for analysis of curved girders

As stated in Chap. 6, the design criteria between curved I girders and box girders are substantially different from each other; therefore, it is advantageous to employ separate programs in the analysis of such girders (see categories 1 and 3 above). For the analysis of curved grillage girders, two separate programs are utilized depending upon the type of girders. Thus computer programs are also divided into two methods, i.e. curved grillage I girders without considering torsional rigidity and curved grillage box girders with torsional rigidity (see category 3).

When the stress resultants of curved bridges have been determined according to these analyses the stresses are calculated (see category 2), and optimum cross sections of the curved girders are designed subsequently for I girders (see category 4) and box girders (see category 5). In addition, the steel deck plates of curved girder bridges are designed by another subprogram (see category 6).

In recent years, developments in computer-aided design (CAD) have been very active in Japan and several programs are available to design various curved girder bridges (see category 7). As stated in earlier chapters, special analyses are required to design the curved girder bridges in addition to those normally required to design the straight bridges. These analyses include shear lag and distortion of curved box

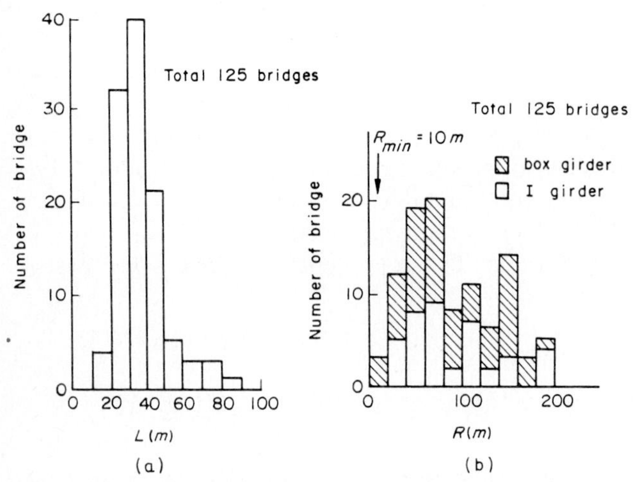

Figure 8.1 Histogram of (*a*) maximum span length and (*b*) radius of curvature.[2]

girders, local stress due to secondary members such as floor beam, sway and lateral bracings, buckling strength of curved web plate and dynamic response of the curved girder bridges (see category 8). The computer programs cited in the chapter reflect some of these additional considerations.

In App. C, the name of computer programs, source organizations, required computer systems, and brief descriptions are presented.

8.2 Example of Multiple Curved I-Girder Bridge

For a multiple curved I-girder bridge, Fig. 8.2 shows the Okunabe Bridge with three-span continuous curved girders. This was constructed as a first-class bridge by the Japanese Ministry of Construction in May 1985 in the prefecture Ohita, a southern district of Japan.

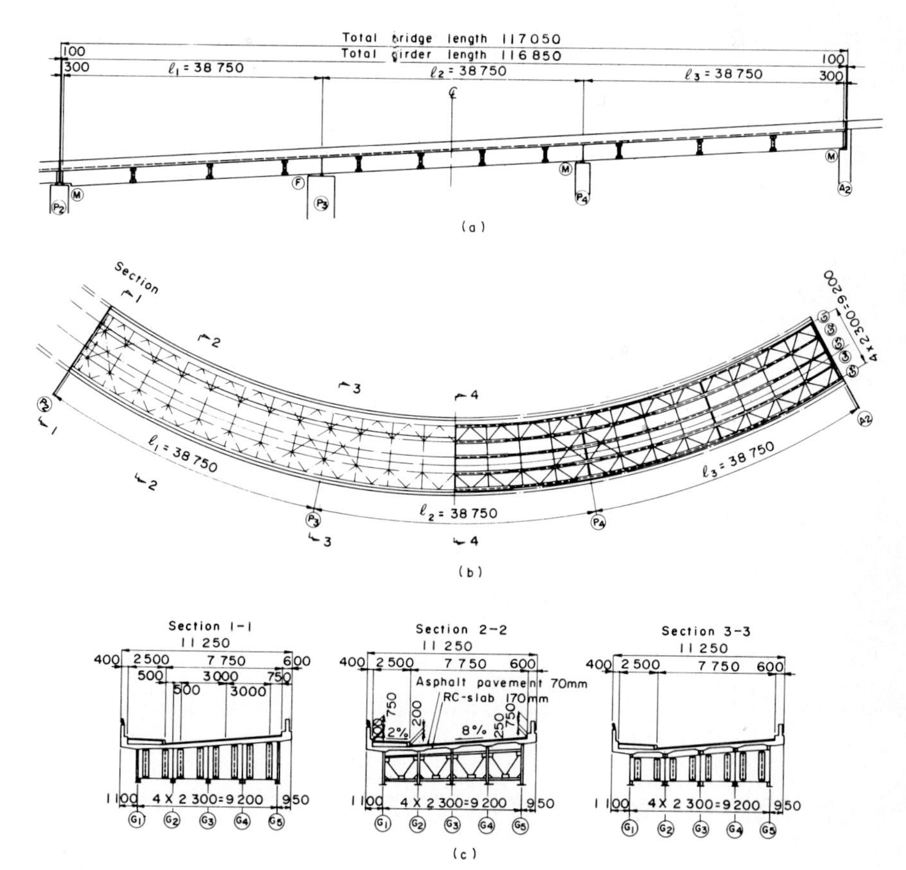

Figure 8.2 General plan of Okunabe Bridge (in mm): (*a*) side elevation, (*b*) plan, and (*c*) typical cross section.

The total bridge length is 117.050 m, having three equal spans of 38.750 m. The clear roadway width is 11.250 m including the sideway. The plane configuration varied from a simple circular curve with radius of curvature $R = 100.0$ m to a spiral one with parameter $A = RL = 80.0$ m. The camber of the roadway surface is 6.0 percent, and the superelevations of the roadway surface are 8.0 percent and 2.0 percent for roadway and sideway, respectively.

The bridge deck is a reinforced concrete slab having a thickness of 17.0 cm and is not a composite of the main girders.

The main girders consist of five I girders at a constant spacing of 2.300 m. The girder depth is 2.400 m for the middle main girder G-3, and the other girder depths are altered to correspond to the superelevation of the roadway. The floor beams are located at the midpoints of each span and at the supports. Sway bracings made of T section with dimensions of 144 × 204 × 12 × 10 mm were placed at each quarter point of the floor beams (see Fig. 8.2 for the layout).

It is necessary to have very rigid transverse stiffeners of the web

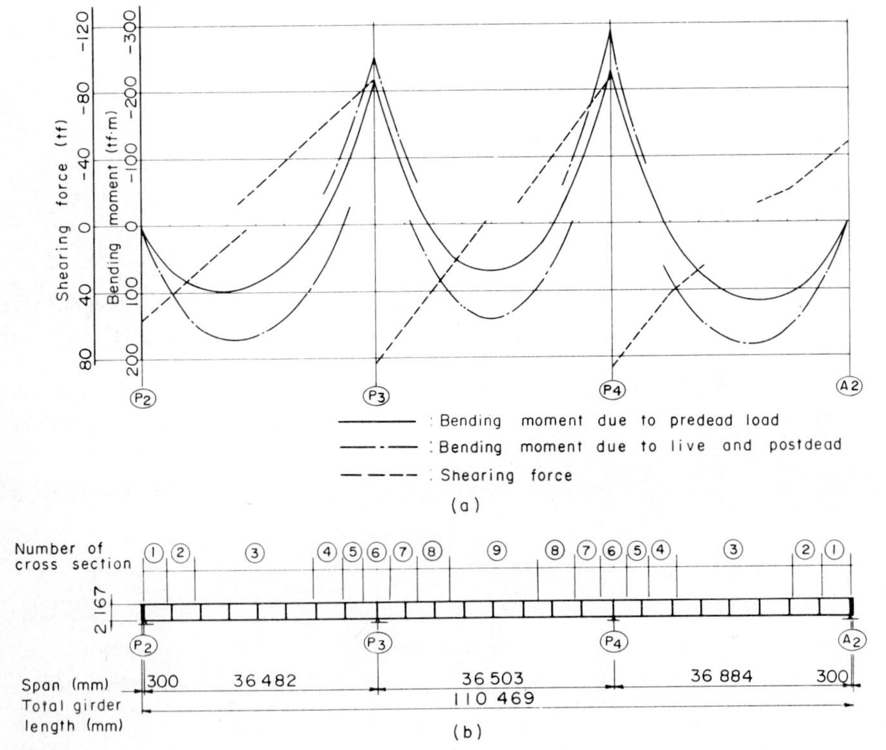

Figure 8.3 Distribution of (*a*) maximum stress resultants and (*b*) main girder (G-1) details.

TABLE 8.1 Maximum Stress Resultants, Cross-Sectional Dimensions, and Allowable and Induced Stresses (G-1)

Number of cross section		1	2	3	4		5	6	7	8		9
Maximum bending moment due to predead load (tf·m)		60	106	117	34	−45	−153	−230	−158	−48	26	72
Maximum bending moment due to postdead load (tf·m)		85	148	180	104	−79	−190	−290	−194	−82	88	142
Maximum shearing force (tf)		34	22	13	37	57	78	87	75	54	34	16
Cross-sectional dimension (mm)	Upper flange plate	200 × 10	260 × 10	280 × 12	200 × 10	200 × 10	300 × 14	430 × 19	320 × 14	200 × 10	200 × 10	200 × 10
	Web plate	2167 × 10	2167 × 10	2167 × 10	2167 × 10	2167 × 10	2167 × 10	2167 × 10	2167 × 10	2167 × 10	2167 × 10	2167 × 10
	Lower flange plate	250 × 10	310 × 12	300 × 16	200 × 10	200 × 10	300 × 14	430 × 19	320 × 14	210 × 10	210 × 10	240 × 12
Allowable stress (kgf/cm^2)	Upper flange σ_a	−2100	−2100	−2100	−2100	2100	2100	2100	2100	2100	−2100	−2100
	Lower flange σ_a	2100	2100	2100	2100	−1525	−1935	−2028	−1945	−1571	2100	2100
	τ_a	1200	1200	1200	1200	1200	1200	1200	1200	1200	1200	1200
Induced stress (kgf/cm^2)	σ_u	−415	−2062	−2076	−1272	1201	2098	2088	2072	1275	−1037	−1974
	σ_z	1519	2049	2062	1694	−1522	−1928	−2015	−1923	−1534	1346	2046
	τ	156	101	60	170	262	358	400	345	248	156	74

plates located at the intersection of the main girders with floor beams and sway and lateral bracings. This is because at these points considerable out-of-plane forces and bending moments due to these members occur, and eccentricity exists at these intersection points to the flange plates of the main girders (see Sec. 6.2.8(2)). Therefore, angle members with dimensions of $150 \times 90 \times 9$ mm are adopted for the transverse stiffeners because of their higher rigidity than a flat plate.

Grades SS 41 and SM 50Y, which consist of used steel materials, are specified in JSHB.[3]

The structural analysis was performed by using a digital computer and the displacement method.[4] The results are plotted in Fig. 8.3 as the distribution of the maximum stress resultants.

Based on the allowable stress design method specified in JSHB,[3] the cross-sectional dimensions were determined and are listed in Table 8.1 for main girder G-1.

8.3 Example of Curved Monobox Girder Bridge

Figure 8.4 shows an example of a curved monobox-girder bridge, the Jutani Bridge, which was constructed as a first-class bridge by the prefecture Wakayama, a central district of Japan, in March 1981. This bridge was designed as a continuous noncomposite girder having span lengths of 28.450 and 19.750 m at the centerline of the box girder. The total span length is 49.100 m.

The plan view shows a simple circular curve with a radius of curvature $R = 58.0$ m between abutment A_1 and pier P (first span) and a spiral curve with parameter $A = RL = 55.0$ m between pier P and abutment A_2 (second span). The clear width of the roadway is 9.0 m. To keep the sight distance 40.0 m, the outstanding part of the inner side of the slab against the center of curvature is 2.4 m, and the outer side of the slab is 1.4 m. The camber is 0 percent and the superelevation varies from 4.272 to 6.0 percent.

The reinforced concrete slab has a thickness of 19.0 cm which is covered by 5.0 cm of asphalt pavement.

The maximum stress resultants were calculated by using the computer program based on the displacement method.[4] These stress resultants are plotted in Fig. 8.5*a*. The corresponding spacing of the transverse stiffeners, diaphragms, etc., are also indicated in Fig. 8.5*b*.

The optimum cross-sectional dimension of the box girder was designed according to the design method of JSHB.[3] These results are summarized in Table 8.2. Table 8.3 shows the induced and allowable stresses for the bridge.

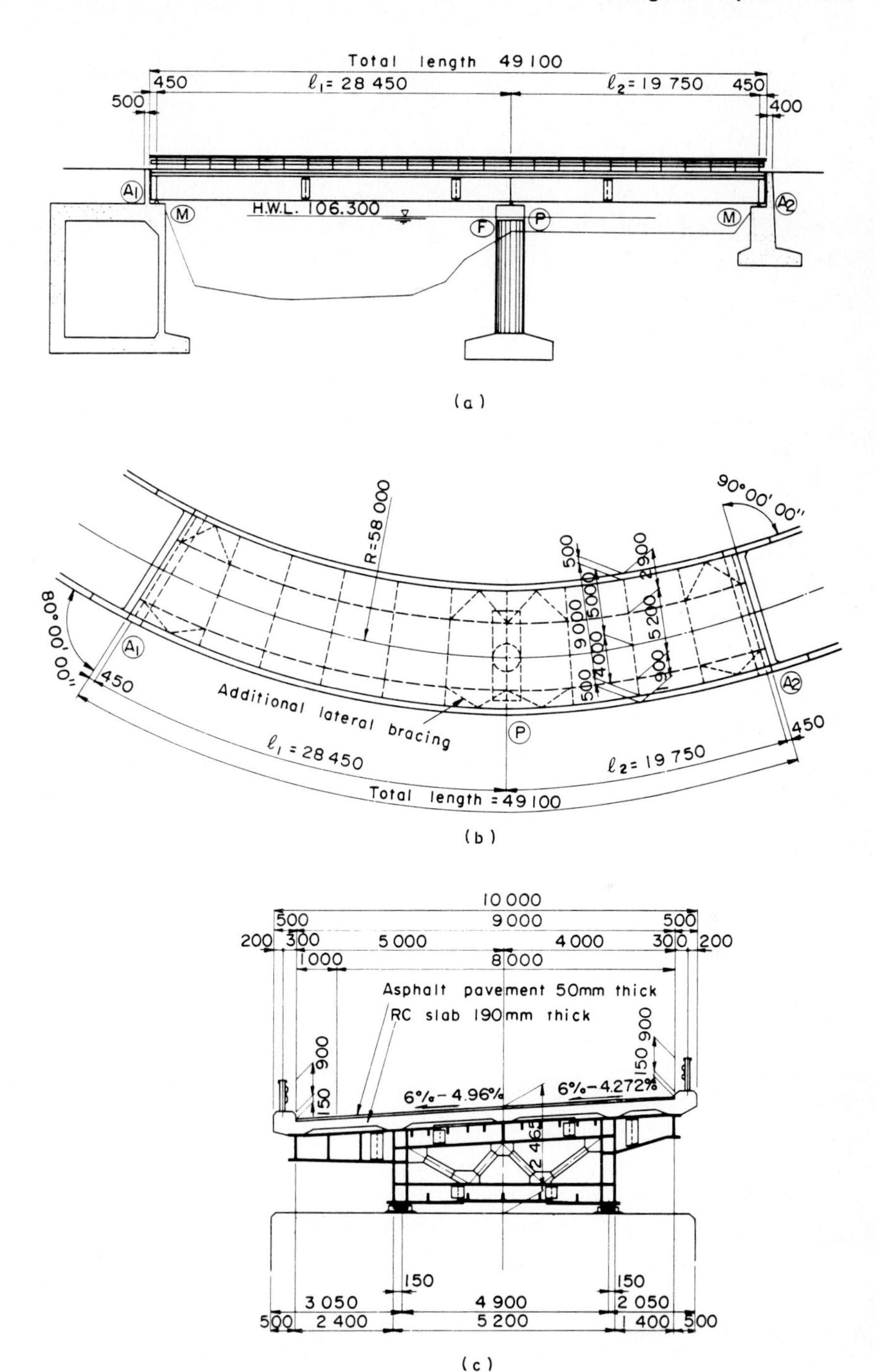

Figure 8.4 General plan of Jutani Bridge (in mm): (*a*) side elevation, (*b*) plan, and (*c*) typical cross section. In part *a* H.W.L. stands for high water level.

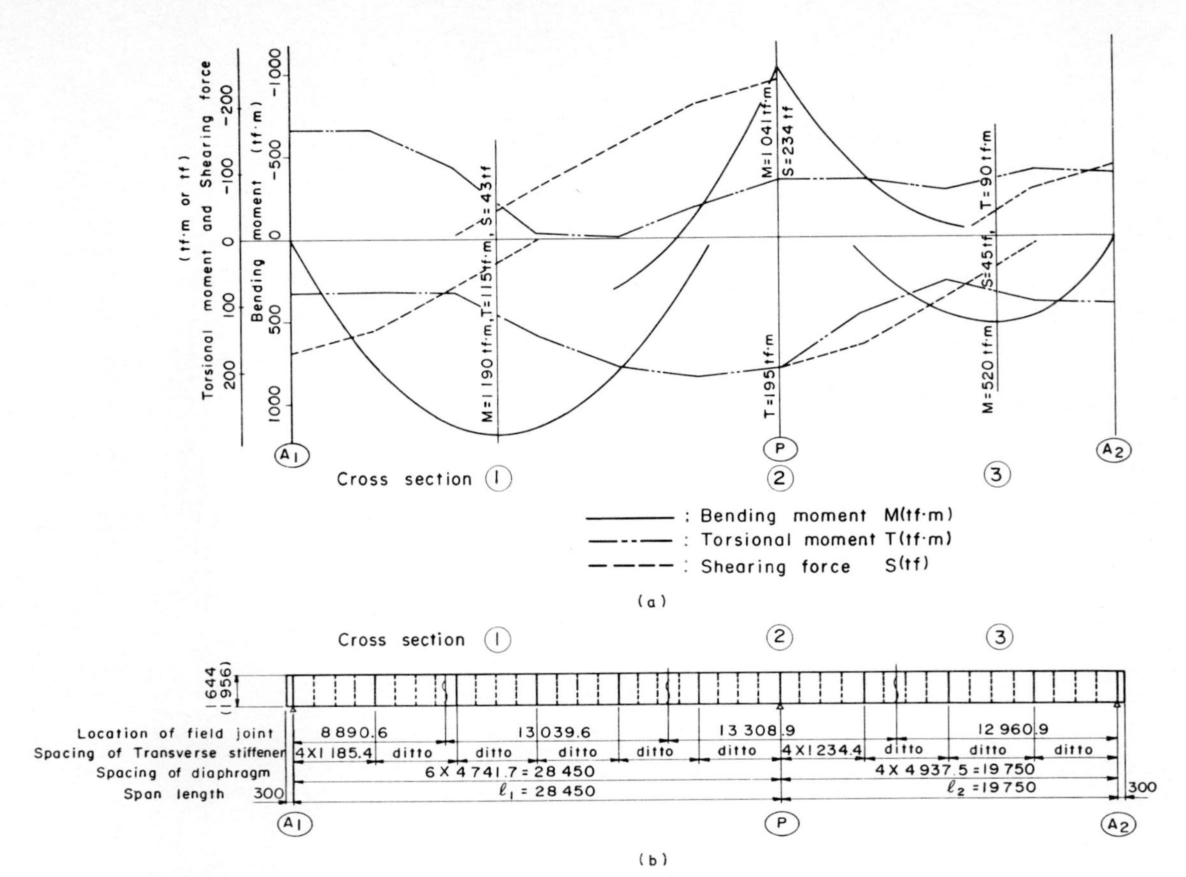

Figure 8.5 Distribution of (a) maximum stress resultants and (b) side elevation of main girder (in mm).

TABLE 8.2 Cross Section of Main Girder

Item		Cross section 1	2	3
Thickness of flange plate t_f (mm)		12	12	12
Number of longitudinal stiffeners	Top flange m	10	6	8
	Bottom flange n	4	6	4
Thickness of web plate t_w (mm)		10	16	10
Steel material		SS 41	SS 41	SS 41

Cross section of box girder

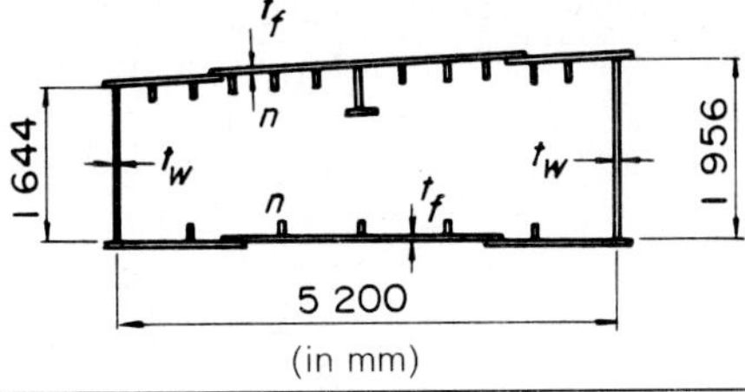

TABLE 8.3 Induced and Allowable Stresses

Item*			Cross section 1	2	3
Normal stress σ (kgf/cm^2)	Top flange	σ	964	1170	474
		σ_a	1297	1400	1197
	Bottom flange	σ	910	1040	435
		σ_a	1400	1197	1400
Shearing stress τ (kgf/cm^2)	Web plate	τ_b	100	438	129
		τ_s	51	65	48
		$\tau_b + \tau_s$	151	503	177
		τ_a	800	800	800

*σ_a = allowable normal stress; σ_a = allowable shearing stress; τ_b = shearing stress due to flexure; τ_s = shearing stress due to pure torsion.

8.4 Example of Curved Twin-Box-Girder Bridge

Figure 8.6 shows an example of a curved twin-box-girder bridge constructed by the Hanshin Expressway Public Corporation (HEPC)[5] in July 1986 at a route connecting the harbor and downtown Osaka City. The length of the bridge at the centerline of the roadway is 36.682 m,

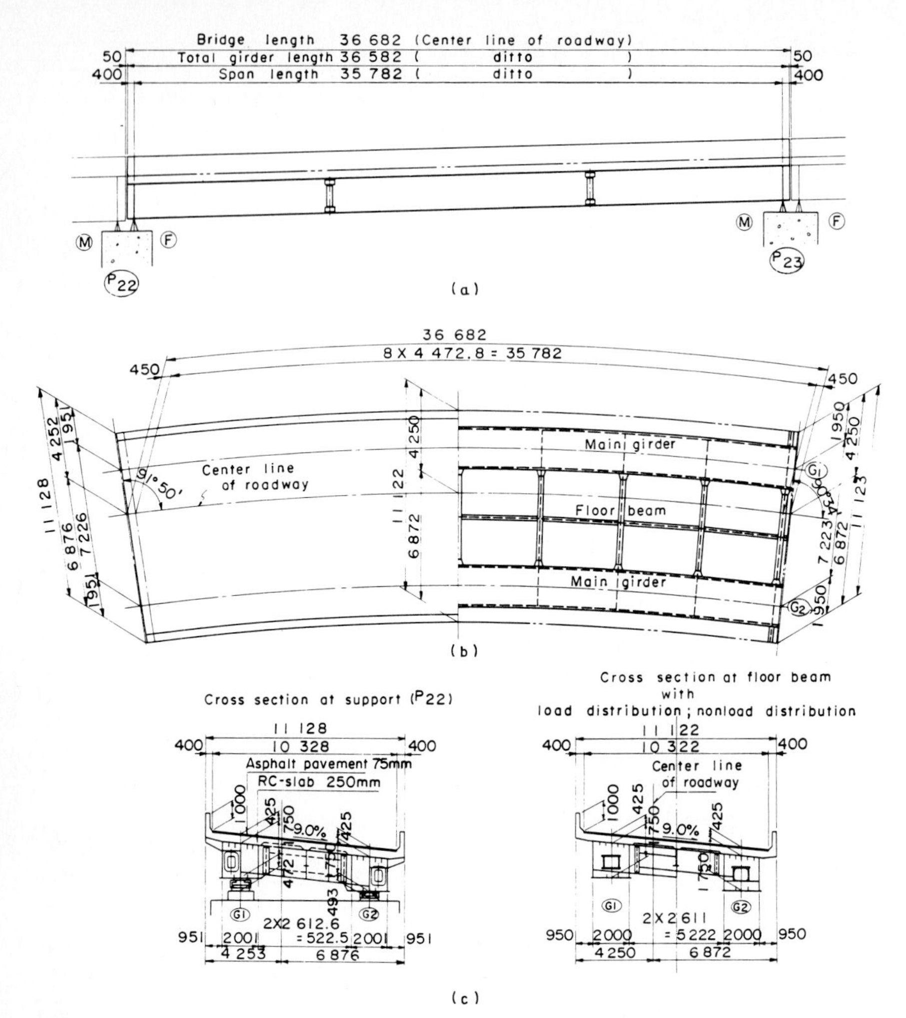

Figure 8.6 General plan of highway bridge on Hanshin Expressway (in mm): (*a*) side elevation, (*b*) plan, and (*c*) typical cross section.

the radius of curvature of the simple curve is $R = 150$ m, and the clear width of roadway is 10.322 to 10.328 m. This bridge was designed as a first-class bridge including TT-43 loading (see Fig. 6.13).

The slab is made of reinforced concrete with a thickness of 25.0 cm, and the thickness of asphalt pavement is 7.5 cm. The camber of the surface is 2.821 percent, and the superelevation is 9.0 percent.

In designing this bridge, the reinforced concrete slab is a composite with the steel girders, to ensure that the longitudinal negative bending moment does not occur in the reinforced concrete slab. The shearing

TABLE 8.4 Cross-Sectional Dimensions of Main Girder (G-1)

Item	Cross section				
	1	2	3	4	5
Effective width of slab *Be* (mm)	5474	5474	5474	5474	5474
Thickness of top flange t_u (mm)	10	10	10	10	10
Thickness of bottom flange t_l (mm)	10	13	20	13	10
Steel material	SM 50Y	SM 50Y	SM 50Y	SM 50Y	SM 50Y

Cross section of box girder

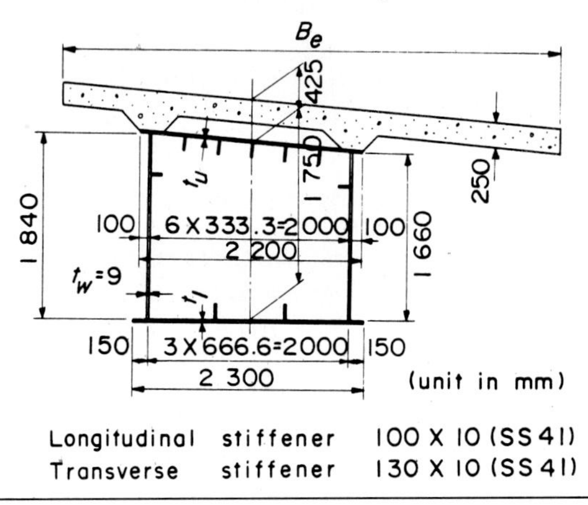

stress in the reinforced concrete slab is also limited within the allowable stress (see Fig. 6.6).[5] Therefore, this bridge is designed much more economically.

The top flange plates of the main box girders are parallel to the superelevation of the roadway. However, the bottom flange plates are

TABLE 8.5 Induced and Allowable Stresses

Item*			Cross section				
			1	2	3	4	5
Normal stress σ (kgf/cm^2)	Slab	σ	27	47	59	47	27
		σ_a	85	85	85	85	85
	Top flange	σ	1122	1805	2301	1812	1126
		σ_a	2415	2415	2415	2415	2415
	Bottom flange	σ	1350	2046	2041	2053	1355
		σ_a	2100	2100	2100	2100	2100
Shearing stress τ (kgf/cm^2)	Web plate	τ	885	742	580	766	909
		τ_a	1200	1200	1200	1200	1200

*σ_a = allowable normal stress; τ_a = allowable shearing stress.

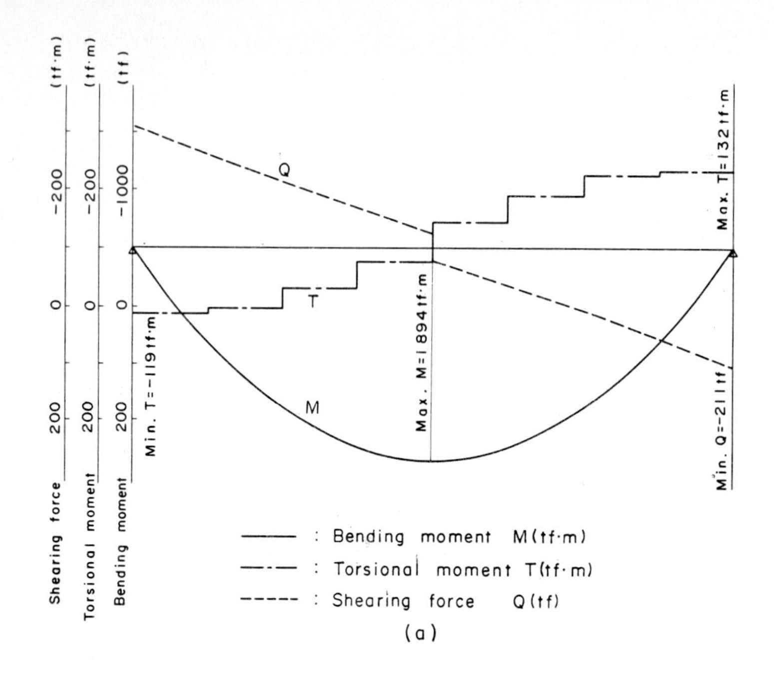

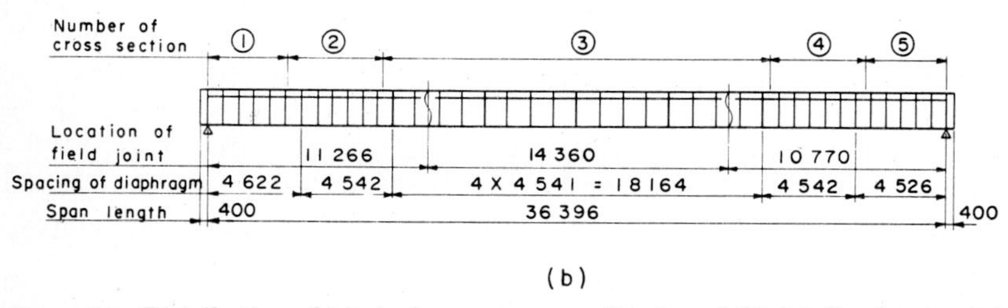

Figure 8.7 Distribution of (*a*) maximum stress resultants and (*b*) details of main girder.

level. Thus, the depths of the two box girders are completely different. All the steel materials used are grade SM 50Y.[3]

Figure 8.7 shows the distribution of the maximum bending moment *M*, torsional moment *T*, and shearing force *Q*. These were analyzed by the digital computer with the displacement method.[4] Table 8.4 shows the cross-sectional dimensions of the box girders. The induced and allowable stresses are summarized in Table 8.5.

References

1. The Task Committee on Curved Box Girder of the ASCE-AASHTO Committee of Flexural Members of the Committee on Metals of the ASCE Structural Division: "Curved Steel Box-Girder Bridges, A Survey," *Proceedings of the American Society of Civil Engineers*, vol. 4, no. ST-11, pp. 1697–1718, Nov. 1978.

2. Nakai, H., S. Muramatsu, O. Yoshikawa, T. Kitada, and R. Ohminami: "A Survey for Web Plates of Horizontally Curved Girder Bridges," *Bridge and Foundation Engineering*, 81(5):38–45, May 1981 (in Japanese).
3. The Japanese Road Association: *The Japanese Specification for Highway Bridges, Parts I and II*, Maruzen, Tokyo, Feb. 1980 (in Japanese).
4. Japan Information Processing Service Co. Ltd.: *Guidelines and Applications of Digital Computer Programs to Civil Engineering Structures*, Tokyo, Nov. 1982 (in Japanese).
5. The Hanshin Expressway Public Corporation: *The Design Code, Part 2, Design Codes for Steel Bridge Structure*, Osaka, Apr. 1980 (in Japanese).

Appendix

A

Application of Matrix Calculus to Estimate Cross-Sectional Quantities[1–4]

In Chap. 3, the general method for evaluating the cross-sectional quantities of a thin-walled curved beam is explained for a simple case such as a box cross section with a single cell. This method is, however, too complicated to develop for a beam with an arbitrary cross section. For this case, the following numerical method involving the application of a shape matrix is a powerful tool that can be applied to cross sections with arbitrary shapes, and the solution can be directly obtained through computer use.

A.1 Shape Matrix

Figure A1 shows an arbitrary kth member in the cross section of a thin-walled curved beam, where the thickness and length of the $\bar{k}$th member are, respectively, denoted $t_{\bar{k}}$ and $\lambda_{\bar{k}}$. (k refers to the panel number, while $\bar{k}$ refers to the element number.) Both panel points of the $\bar{k}$th member are defined by i and j, and the coordinates of these points are given by $(\bar{X}_i, \bar{Y}_i)$ and $(\bar{X}_j, \bar{Y}_j)$, measured from an arbitrary point B with respect to the radius of curvature R_B.

Consider the relationship of the kth member with respect to panel point i or j. If we denote the elements

$$a_{i,\bar{k}} = \begin{cases} 1 & \text{if } \bar{k}\text{th member is connected to point } i \\ 0 & \text{if } \bar{k}\text{th member is not connected to point } i \end{cases} \qquad \text{(A1}a\text{)}$$

$$b_{j,\bar{k}} = \begin{cases} 1 & \text{if } \bar{k}\text{th member is connected to point } j \\ 0 & \text{if } \bar{k}\text{th member is not connected to point } j \end{cases} \qquad \text{(A1}b\text{)}$$

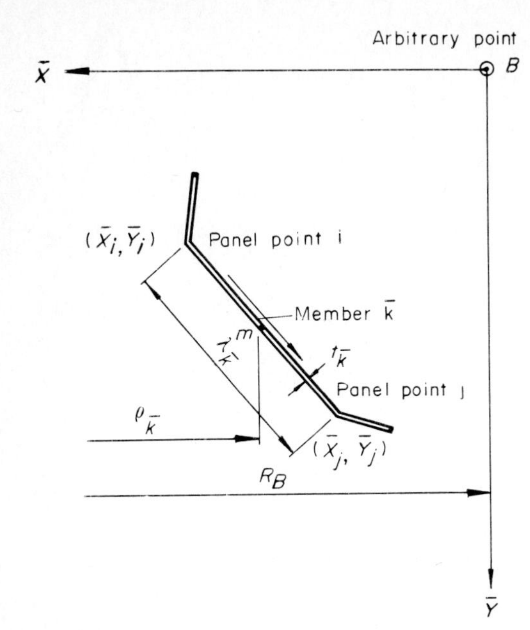

Figure A1 Definition of coordinates of kth member.

and if we expand these elements relative to all the members, then the shape matrices **A** and **B** with elements $i_n \times \bar{k}_n$ can be defined mathematically, where i_n and $\bar{k}_n$ are the total panel points and members, respectively.[3] The values of elements $a_{i,\bar{k}}$ and $b_{j,\bar{k}}$ are either 1 or 0, and thus matrix **A** represents the output matrix from each panel point and the matrix, while **B** represents the input matrix for each panel point. Similar matrices have been used to calculate the hydrostatic flow in pipelines and traffic flow on highways. For applications to typical cross sections in a curved beam, matrices **A** and **B** are illustrated in Table A1.

A.2 Cross-Sectional Quantities for Bending

When the horizontal and vertical coordinates for each panel point $i = 1, 2, \ldots, i_n$ with respect to an arbitrary chosen point B are

$$\bar{\mathbf{x}}_B = [\bar{X}_1 \ \bar{X}_2 \cdots \bar{X}_i \ \bar{X}_j \cdots] \tag{A2a}$$

$$\bar{\mathbf{y}}_B = [\bar{Y}_1 \ \bar{Y}_2 \cdots \bar{Y}_i \ \bar{Y}_j \cdots] \tag{A2b}$$

the coordinate at panel point (i, j) for the $\bar{k}$th member are given by

$$\bar{\mathbf{x}}_i = [\bar{X}_{i,\bar{1}} \ \bar{X}_{i,\bar{2}} \cdots \bar{X}_{i,\bar{k}} \cdots] \tag{A3a}$$

$$\bar{\mathbf{y}}_i = [\bar{Y}_{i,\bar{1}} \ \bar{Y}_{i,\bar{2}} \cdots \bar{Y}_{i,\bar{k}} \cdots] \tag{A3b}$$

$$\bar{\mathbf{x}}_j = [\bar{X}_{j,\bar{1}} \ \bar{X}_{j,\bar{2}} \cdots \bar{X}_{j,\bar{k}} \cdots] \tag{A3c}$$

$$\bar{\mathbf{y}}_j = [\bar{Y}_{j,\bar{1}} \ \bar{Y}_{j,\bar{2}} \cdots \bar{Y}_{j,\bar{k}} \cdots] \tag{A3d}$$

TABLE A1 Examples of Matrices A and B

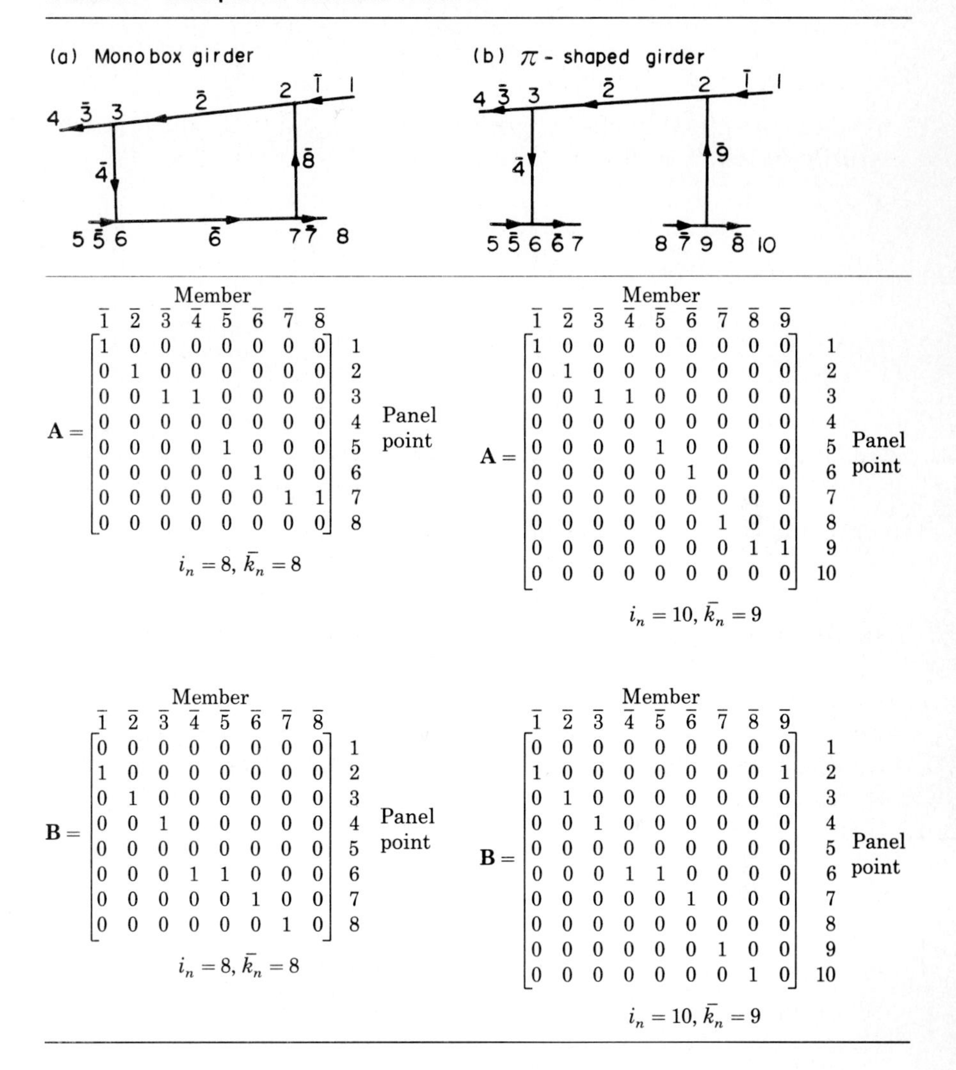

(a) Mono box girder

Member $\bar{1}$ to $\bar{8}$ (columns); Panel point 1 to 8 (rows)

$$\mathbf{A} = \begin{bmatrix} 1 & 0 & 0 & 0 & 0 & 0 & 0 & 0 \\ 0 & 1 & 0 & 0 & 0 & 0 & 0 & 0 \\ 0 & 0 & 1 & 1 & 0 & 0 & 0 & 0 \\ 0 & 0 & 0 & 0 & 0 & 0 & 0 & 0 \\ 0 & 0 & 0 & 0 & 1 & 0 & 0 & 0 \\ 0 & 0 & 0 & 0 & 0 & 1 & 0 & 0 \\ 0 & 0 & 0 & 0 & 0 & 0 & 1 & 1 \\ 0 & 0 & 0 & 0 & 0 & 0 & 0 & 0 \end{bmatrix}$$

$i_n = 8, \bar{k}_n = 8$

$$\mathbf{B} = \begin{bmatrix} 0 & 0 & 0 & 0 & 0 & 0 & 0 & 0 \\ 1 & 0 & 0 & 0 & 0 & 0 & 0 & 0 \\ 0 & 1 & 0 & 0 & 0 & 0 & 0 & 0 \\ 0 & 0 & 1 & 0 & 0 & 0 & 0 & 0 \\ 0 & 0 & 0 & 0 & 0 & 0 & 0 & 0 \\ 0 & 0 & 0 & 1 & 1 & 0 & 0 & 0 \\ 0 & 0 & 0 & 0 & 0 & 1 & 0 & 0 \\ 0 & 0 & 0 & 0 & 0 & 0 & 1 & 0 \end{bmatrix}$$

$i_n = 8, \bar{k}_n = 8$

(b) π - shaped girder

Member $\bar{1}$ to $\bar{9}$ (columns); Panel point 1 to 10 (rows)

$$\mathbf{A} = \begin{bmatrix} 1 & 0 & 0 & 0 & 0 & 0 & 0 & 0 & 0 \\ 0 & 1 & 0 & 0 & 0 & 0 & 0 & 0 & 0 \\ 0 & 0 & 1 & 1 & 0 & 0 & 0 & 0 & 0 \\ 0 & 0 & 0 & 0 & 0 & 0 & 0 & 0 & 0 \\ 0 & 0 & 0 & 0 & 1 & 0 & 0 & 0 & 0 \\ 0 & 0 & 0 & 0 & 0 & 1 & 0 & 0 & 0 \\ 0 & 0 & 0 & 0 & 0 & 0 & 0 & 0 & 0 \\ 0 & 0 & 0 & 0 & 0 & 0 & 1 & 0 & 0 \\ 0 & 0 & 0 & 0 & 0 & 0 & 0 & 1 & 1 \\ 0 & 0 & 0 & 0 & 0 & 0 & 0 & 0 & 0 \end{bmatrix}$$

$i_n = 10, \bar{k}_n = 9$

$$\mathbf{B} = \begin{bmatrix} 0 & 0 & 0 & 0 & 0 & 0 & 0 & 0 & 0 \\ 1 & 0 & 0 & 0 & 0 & 0 & 0 & 0 & 1 \\ 0 & 1 & 0 & 0 & 0 & 0 & 0 & 0 & 0 \\ 0 & 0 & 1 & 0 & 0 & 0 & 0 & 0 & 0 \\ 0 & 0 & 0 & 0 & 0 & 0 & 0 & 0 & 0 \\ 0 & 0 & 0 & 1 & 1 & 0 & 0 & 0 & 0 \\ 0 & 0 & 0 & 0 & 0 & 1 & 0 & 0 & 0 \\ 0 & 0 & 0 & 0 & 0 & 0 & 0 & 0 & 0 \\ 0 & 0 & 0 & 0 & 0 & 0 & 1 & 0 & 0 \\ 0 & 0 & 0 & 0 & 0 & 0 & 0 & 1 & 0 \end{bmatrix}$$

$i_n = 10, \bar{k}_n = 9$

These coordinates (x_i, y_j) at the $\bar{k}$th member can then be estimated by utilizing the shape matrices **A** and **B** as follows:

$$\bar{\mathbf{x}}_i = \mathbf{A}^T \bar{\mathbf{x}}_B \qquad \bar{\mathbf{y}}_i = \mathbf{A}^T \bar{\mathbf{y}}_B \tag{A4a}$$

$$\bar{\mathbf{x}}_j = \mathbf{B}^T \bar{\mathbf{x}}_B \qquad \bar{\mathbf{y}}_j = \mathbf{B}^T \bar{\mathbf{y}}_B \tag{A4b}$$

Moreover, when the radius of curvature $\rho_{\bar{k}}$ is assumed constant for the $\bar{k}$th member, $\rho_{\bar{k}}$ can be represented approximately by the midpoint of the $\bar{k}$th member as

$$\rho_{\bar{k}} = R_B - \frac{\bar{X}_{i,\bar{k}} + \bar{X}_{j,\bar{k}}}{2} \tag{A5}$$

The cross-sectional quantities with respect to point B, which is arbitrarily chosen, can then be estimated by

$$A_{\bar{Z}} = \sum_{\bar{k}=1}^{\bar{k}_n} \frac{R_B}{\rho_{\bar{k}}} \frac{t_{\bar{k}}\lambda_{\bar{k}}}{n_{\bar{k}}} \tag{A6a}$$

$$S_{\bar{X}} = \sum_{\bar{k}=1}^{\bar{k}_n} \frac{R_B}{\rho_{\bar{k}}} \frac{t_{\bar{k}}\lambda_{\bar{k}}}{2n_{\bar{k}}} (\bar{Y}_{i,\bar{k}} + \bar{Y}_{j,\bar{k}}) \tag{A6b}$$

$$S_{\bar{Y}} = \sum_{\bar{k}=1}^{\bar{k}_n} \frac{R_B}{\rho_{\bar{k}}} \frac{t_{\bar{k}}\lambda_{\bar{k}}}{2n_{\bar{k}}} (\bar{X}_{i,\bar{k}} + \bar{X}_{j,\bar{k}}) \tag{A6c}$$

where
$$\lambda_{\bar{k}} = \sqrt{(\bar{X}_{i,\bar{k}} - \bar{X}_{j,\bar{k}})^2 + (\bar{Y}_{i,\bar{k}} - \bar{Y}_{j,\bar{k}})^2} \tag{A7}$$

is the length of the $\bar{k}$th member.

The location of the centroid C can be determined by using Eqs. (3.2.38) and (3.2.39). And then the cross-sectional quantities I_X, I_Y, and I_{XY} with respect to the centroid C can be found from Eq. (3.2.40) or Eqs. (3.2.35*d*) to (3.2.35*f*), which come from Eq. (3.2.38) and are

$$I_X = \sum_{\bar{k}=1}^{\bar{k}_n} \frac{R_o}{\rho_{\bar{k}}} \left[\frac{t_{\bar{k}}\lambda_{\bar{k}}}{3n_{\bar{k}}} (Y_{i,\bar{k}}^2 + Y_{i,\bar{k}}Y_{j,\bar{k}} + Y_{j,\bar{k}}^2) + \frac{t_{\bar{k}}^3}{12n_{\bar{k}}\lambda_{\bar{k}}} (X_{i,\bar{k}} - X_{j,\bar{k}})^2 \right] \tag{A8a}$$

$$I_Y = \sum_{\bar{k}=1}^{\bar{k}_n} \frac{R_o}{\rho_{\bar{k}}} \left[\frac{t_{\bar{k}}\lambda_{\bar{k}}}{3n_{\bar{k}}} (X_{i,\bar{k}}^2 + X_{i,\bar{k}}X_{j,\bar{k}} + X_{j,\bar{k}}^2) + \frac{t_{\bar{k}}^3}{12n_{\bar{k}}\lambda_{\bar{k}}} (Y_{i,\bar{k}} - Y_{j,\bar{k}})^2 \right] \tag{A8b}$$

$$I_{XY} = \sum_{\bar{k}=1}^{\bar{k}_n} \frac{R_o}{\rho_{\bar{k}}} \left\{ \frac{t_{\bar{k}}\lambda_{\bar{k}}}{6n_{\bar{k}}} [X_{i,\bar{k}}(2Y_{i,\bar{k}} + Y_{j,\bar{k}}) + X_{j,\bar{k}}(Y_{i,\bar{k}} + 2Y_{j,\bar{k}})] - \frac{t_{\bar{k}}^3}{12n_{\bar{k}}} (X_{i,\bar{k}} - X_{j,\bar{k}})(Y_{i,\bar{k}} - Y_{j,\bar{k}}) \right\} \tag{A8c}$$

A.3 Cross-Sectional Quantities for Torsion

The location of centroid C was determined in the previous section with ordinates (X_i, Y_i) and (X_j, Y_j) of all the members being transferred to the origin at centroid C, given by Eq. (3.2.36). The integration of Eq. (3.2.94) can now be written as

$$\omega_{oj,\bar{k}} - \omega_{oi,\bar{k}} = \left(\frac{R_o}{\rho_{\bar{k}}}\right)^3 \tilde{q}_{o,\bar{k}} \frac{n_{g,\bar{k}}}{t_{\bar{k}}} \lambda_{\bar{k}} - \left(\frac{R_o}{\rho_{\bar{k}}}\right)^2 r_{o,\bar{k}}\lambda_{\bar{k}} \tag{A9}$$

given that

$$\rho_{\bar{k}} = \text{constant} \tag{A10a}$$

$$\int_i^j dc = \lambda_{\bar{k}} \tag{A10b}$$

The perpendicular distance $r_{o,\bar{k}}$, given in the second term of the right-hand side of Eq. (A9), can be determined from Eq. (3.2.90):

$$\begin{aligned} r_{o,\bar{k}} &= \frac{X_{i,\bar{k}}(Y_{j,\bar{k}} - Y_{i,\bar{k}}) - Y_{i,\bar{k}}(X_{j,\bar{k}} - X_{i,\bar{k}})}{\lambda_{\bar{k}}} \\ &= \frac{X_{i,\bar{k}}Y_{j,\bar{k}} - X_{j,\bar{k}}Y_{i,\bar{k}}}{\lambda_{\bar{k}}} \end{aligned} \tag{A11}$$

Now, Eqs. (A9), (A10), and (A11) can be written in the following matrix form:

$$\boldsymbol{\omega}_j - \boldsymbol{\omega}_i = \mathbf{F}_s(\mathbf{q} - \mathbf{r}) \tag{A12}$$

where

$$\boldsymbol{\omega}_i = [\omega_{oi,\bar{1}} \;\; \omega_{oi,\bar{2}} \cdots \omega_{oi,\bar{k}} \cdots] \tag{A13a}$$

$$\boldsymbol{\omega}_j = [\omega_{oj,\bar{1}} \;\; \omega_{oj,\bar{2}} \cdots \omega_{oj,\bar{k}} \cdots] \tag{A13b}$$

$$\mathbf{q} = [\tilde{q}_{o,\bar{1}} \;\; \tilde{q}_{o,\bar{2}} \cdots \tilde{q}_{o,\bar{k}} \cdots] \tag{A13c}$$

Here $\mathbf{q}$ is the column vector with $\bar{k}_n$ elements, and $\mathbf{r}$ is the column vector with $\bar{k}_n$ elements, such as

$$r_{o,\bar{k}}[\text{in Eq. (A11)}] \times \frac{\rho_{\bar{k}}}{R_o} \times \frac{t_{\bar{k}}}{n_{g\bar{k}}} \tag{A14}$$

and $\mathbf{F}_s$ is the diagonal matrix consisting of $\bar{k}_n$ elements with

$$\left(\frac{R_o}{\rho_{\bar{k}}}\right)^3 \frac{n_{g,\bar{k}}}{t_{\bar{k}}} \lambda_{\bar{k}} \tag{A15}$$

The solution of Eq. (A12) for the torsional function $\mathbf{q}$ gives

$$\mathbf{q} = \mathbf{F}_s^{-1}(\boldsymbol{\omega}_j - \boldsymbol{\omega}_i) + \mathbf{r} \tag{A16}$$

Hence the warping function $\boldsymbol{\omega}$ (ω for closed section; $\bar{\omega}$ for open section) for each panel point is

$$\boldsymbol{\omega} = [\tilde{\omega}_{o,1} \;\; \tilde{\omega}_{o,2} \cdots \tilde{\omega}_{o,i} \;\; \tilde{\omega}_{o,j} \cdots] \tag{A17}$$

which can be determined from

$$\boldsymbol{\omega}_i = \mathbf{A}^T\boldsymbol{\omega} \tag{A18a}$$

$$\boldsymbol{\omega}_j = \mathbf{B}^T\boldsymbol{\omega} \tag{A18b}$$

as in Eq. (A4).

Equation (A16) can be rewritten as

$$\mathbf{q} = \mathbf{F}_s^{-1}(\mathbf{B}^T - \mathbf{A}^T)\boldsymbol{\omega} + \mathbf{r} \tag{A19}$$

Multiplying both sides by $\mathbf{B} - \mathbf{A}$ gives the following:

$$(\mathbf{B} - \mathbf{A})\mathbf{q} = (\mathbf{B} - \mathbf{A})\mathbf{F}_s^{-1}(\mathbf{B}^T - \mathbf{A}^T)\boldsymbol{\omega} + (\mathbf{B} - \mathbf{A})\mathbf{r} \tag{A20}$$

However, the following condition

$$(\mathbf{B} - \mathbf{A})\mathbf{q} = \mathbf{0} \tag{A21}$$

should be satisfied, because the sum of the output and input shear flows for each panel point must be in equilibrium. The solution of the warping function ω can therefore be obtained from Eq. (A20) as follows:

$$\boldsymbol{\omega} = -[(\mathbf{B} - \mathbf{A})\mathbf{F}_s^{-1}(\mathbf{B}^T - \mathbf{A}^T)]^{-1}(\mathbf{B} - \mathbf{A})\mathbf{r} \tag{A22}$$

The inverse of matrix $\mathbf{C}$, given by

$$\mathbf{C}^{-1} = [(\mathbf{B} - \mathbf{A})\mathbf{F}_s^{-1}(\mathbf{B}^T - \mathbf{A}^T)]^{-1} \tag{A23}$$

does not exist, since the determinant of matrix $\mathbf{C}$ is equal to zero, i.e.,

$$\det \mathbf{C} = \mathbf{0} \tag{A24}$$

However, we can assume that

$$\boldsymbol{\omega}_{0,1} = \mathbf{0} \tag{A25}$$

since the elements of the first row and the first column of the matrix $\mathbf{C}$ are not important. Thus, the matrix

$$\mathbf{D} = [(\mathbf{B} - \mathbf{A})\mathbf{F}_s^{-1}(\mathbf{B}^T - \mathbf{A}^T)] \tag{A26}$$

can be defined without the first row and the first column. In general, the determinant of the above matrix

$$\det \mathbf{D} = |\mathbf{D}| \neq \mathbf{0} \tag{A27}$$

has a certain value other than zero. The inverse of matrix $\mathbf{D}$, that is, $\mathbf{D}^{-1}$, can be determined easily. After this procedure, the required inverse matrix $\mathbf{C}^{-1}$ can be defined by adding the zero elements to matrix $\mathbf{D}^{-1}$ as follows:

$$\mathbf{C}^{-1} = \begin{bmatrix} 0 & 0 & \cdots & 0 \\ 0 & & & \\ \vdots & & \mathbf{D}^{-1} & \\ 0 & & & \end{bmatrix} \tag{A28}$$

By using this inverse matrix, the statically determinate warping function $\tilde{\boldsymbol{\omega}}$ can be estimated as

$$\tilde{\boldsymbol{\omega}} = \mathbf{C}^{-1}(\mathbf{B} - \mathbf{A})\mathbf{r} = [0\tilde{\omega}_{o,2}\ \tilde{\omega}_{o,2} \cdots \tilde{\omega}_{o,i}\ \tilde{\omega}_{o,j} \cdots] \tag{A29}$$

The warping functions $\omega_{o,i,k}$ and $\tilde{\omega}_{o,j,k}$ at panel points i and j for the $\bar{k}$th element can be estimated from Eqs. (A18*a*) and (A18*b*) as

$$\tilde{\boldsymbol{\omega}}_i = \mathbf{A}^T\tilde{\boldsymbol{\omega}} \tag{A30a}$$

$$\tilde{\boldsymbol{\omega}}_j = \mathbf{B}^T\tilde{\boldsymbol{\omega}} \tag{A30b}$$

The statically indeterminate warping function

$$\tilde{\omega}_{o,1} = \omega_{o,1} = \bar{\omega}_o \tag{A31}$$

can be estimated by Eqs. (3.2.97) and (A30) as

$$\bar{\omega}_o = \frac{\sum_{\bar{k}=1}^{\bar{k}_n} [(\tilde{\omega}_{oi,\bar{k}} + \tilde{\omega}_{oj,\bar{k}})/(2n_{\bar{k}})]t_{\bar{k}}\lambda_{\bar{k}}}{\sum_{\bar{k}=1}^{\bar{k}_n} t_{\bar{k}}\lambda_{\bar{k}}/n_{\bar{k}}} \tag{A32}$$

Consequently, the complete solution of the warping function ω, as given by Eq. (3.2.96), can be found by adding Eqs. (A29) and (A32).

The torsional function $\mathbf{q}$ can also be determined from Eqs. (A18) and (A16) by using the solution of $\boldsymbol{\omega}$. This torsional function $\mathbf{q}$ coincides with $\mathbf{q}_s$, which is taken about the shear center S given by Eq. (3.2.92).

The other cross-sectional quantities must, however, be transformed from the centroid to the shear center. To evaluate the radius of curvature R_s at the shear center, the following cross-sectional quantities are estimated by assuming the variation in the warping function to have a linear distribution:[1]

$$C_{Xo} = \sum_{\bar{k}=1}^{\bar{k}_n} \frac{t_{\bar{k}}\lambda_{\bar{k}}}{6n_{\bar{k}}} [(2Y_{i,\bar{k}} + Z_{j,\bar{k}})\omega_{oi,\bar{k}} + (Y_{i,\bar{k}} + 2Y_{j,\bar{k}})\omega_{oj,\bar{k}}] \tag{A33a}$$

$$C_{Yo} = \sum_{\bar{k}=1}^{\bar{k}_n} \frac{t_{\bar{k}}\lambda_{\bar{k}}}{6n_{\bar{k}}} [(2X_{i,\bar{k}} + X_{j,\bar{k}})\omega_{oi,\bar{k}} + (X_{i,\bar{k}} + 2X_{j,\bar{k}})\omega_{oj,\bar{k}}] \tag{A33b}$$

Thus the location of the shear center coordinates (X_s, Y_s) relative to the centroid can now be determined from Eq. (3.2.102). The final warping function can also be estimated by using Eq. (3.2.98).

Furthermore, the following quantities can be evaluated with respect to the centroid:

$$K_o = \sum_{\bar{k}=1}^{\bar{k}_n} \left(\frac{R_o}{\rho_{\bar{k}}}\right)^3 \frac{n_{g,\bar{k}}\lambda_{\bar{k}}}{t_{\bar{k}}} \tilde{q}_{o,\bar{k}}^2 \tag{A34}$$

$$I_{\omega o} = \sum_{\bar{k}=1}^{\bar{k}_n} \frac{\rho_{\bar{k}}}{R_o} \frac{t_{\bar{k}}\lambda_{\bar{k}}}{3n_{\bar{k}}} (\omega_{oi,\bar{k}}^2 + \omega_{oi,\bar{k}}\omega_{oj,\bar{k}} + \omega_{oj,\bar{k}}^2) \tag{A35}$$

These values can be transformed to the shear center as given by Eqs. (3.2.103) and (3.2.104), respectively.

Also note that the equation

$$\mathbf{q} = \mathbf{0} \tag{A36}$$

is always satisfied for the open cross section, as illustrated in the right column of Table A1. For such cross sections, the warping function can be computed directly from Eq. (A19):

$$\boldsymbol{\omega} = -[\mathbf{F}_s^{-1}(\mathbf{B}^T - \mathbf{A}^T)]^{-1}\mathbf{r} \tag{A37}$$

The procedures for estimating the final warping function, the location of the shear center, and the warping constant are as outlined above, except that the torsional constant should be estimated by

$$K = \frac{1}{3} \sum_{\bar{k}=1}^{\bar{k}_n} \left(\frac{R_s}{\rho_{\bar{k}}}\right)^3 \lambda_{\bar{k}} t_{\bar{k}}^3 \tag{A38}$$

A.4 Estimation of Shear Flow due to Warping and Bending

A.4.1 Shear flow due to warping

The shear flow q_ω due to warping can be estimated from Eq. (3.2.75) as

$$\tilde{q}_\omega(c_{\bar{k}}) = \frac{R_s}{\rho_{\bar{k}}} \int_0^{c_{\bar{k}}} \frac{\omega_{\bar{k}}\lambda_{\bar{k}}}{n_{\bar{k}}} dc_{\bar{k}} + \tilde{q}_{\omega i,\bar{k}} \tag{A39}$$

where the following simplification is made:

$$q_\omega = -E_s \frac{d^3\theta}{ds^3} \tilde{q}_\omega = \frac{T_\omega}{I_\omega} \tilde{q}_\omega \tag{A40}$$

When the warping function is assumed to be a linear distribution along the direction of the member as shown in Fig. A2, Eq. (A39)

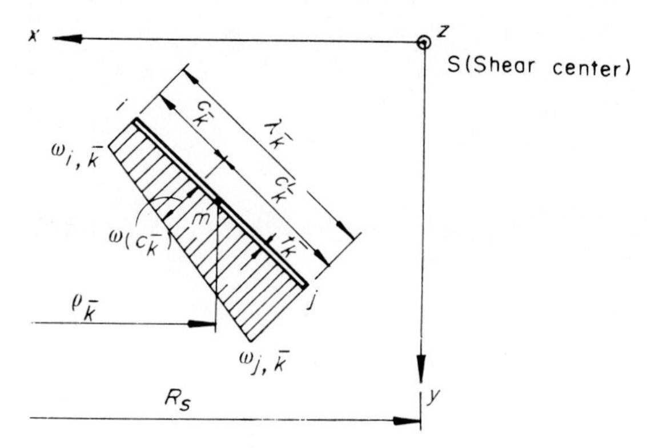

Figure A2 Assumption of warping function.

reduces to

$$\tilde{q}_\omega(c_{\bar{k}}) = \frac{R_s}{\rho_{\bar{k}}} \int_0^{c_{\bar{k}}} \left(\omega_{i,\bar{k}} + \frac{\omega_{j,\bar{k}} - \omega_{i,\bar{k}}}{\lambda_{\bar{k}}} c_{\bar{k}} \right) \frac{t_{\bar{k}}}{n_{\bar{k}}} \, dc_{\bar{k}} + \tilde{q}_{\omega i,\bar{k}}$$

$$= \frac{R_s}{\rho_{\bar{k}}} \left[\omega_{i,\bar{k}} \left(\frac{c_{\bar{k}}}{\lambda_{\bar{k}}} - \frac{c_{\bar{k}}^2}{2\lambda_{\bar{k}}^2} \right) + \omega_{j,\bar{k}} \frac{c_{\bar{k}}^2}{2\lambda_{\bar{k}}^2} \right] \frac{t_{\bar{k}}\lambda_{\bar{k}}}{n_{\bar{k}}} + \tilde{q}_{\omega i,\bar{k}} \tag{A41}$$

When $c_{\bar{k}} = \lambda_{\bar{k}}$, the above equation reduces to

$$\tilde{q}_{\omega j,\bar{k}} = \frac{R_s}{\rho_{\bar{k}}} \frac{\omega_{i,\bar{k}} + \omega_{j,\bar{k}}}{2} \frac{t_{\bar{k}}\lambda_{\bar{k}}}{n_{\bar{k}}} + \tilde{q}_{\omega i,\bar{k}} \tag{A42}$$

The integration of Eq. (3.2.73) can now be performed, where

$$-\frac{G_s}{E_s \, d^3\theta/ds^3} (u_{j,\bar{k}} - u_{i,\bar{k}}) = \frac{n_{g,\bar{k}}}{t_{\bar{k}}} \int_0^{\lambda_{\bar{k}}} \tilde{q}_\omega(c_{\bar{k}}) \, dc_{\bar{k}}$$

$$= \frac{n_{g,\bar{k}}}{t_{\bar{k}}} \int_0^{\lambda_{\bar{k}}} \frac{R_s}{\rho_{\bar{k}}} \left[\omega_{i,\bar{k}} \left(\frac{c_{\bar{k}}^2}{2\lambda_{\bar{k}}^2} \right) + \omega_{j,\bar{k}} \frac{c_{\bar{k}}^2}{2\lambda_{\bar{k}}^2} \right] \frac{t_{\bar{k}}\lambda_{\bar{k}}}{n_{\bar{k}}} \, dc_{\bar{k}} + \frac{n_{g,\bar{k}}\lambda_{\bar{k}}}{t_{\bar{k}}} q_{\omega i,\bar{k}}$$

$$= \frac{n_{g,\bar{k}}}{t_{\bar{k}}} \left\{ \frac{R_s}{\rho_{\bar{k}}} \left[\omega_{i,\bar{k}} \left(\frac{c_{\bar{k}}^2}{2\lambda_{\bar{k}}^2} - \frac{c_{\bar{k}}^3}{6\lambda_{\bar{k}}^2} \right) + \omega_{j,\bar{k}} \frac{c_{\bar{k}}^3}{6\lambda_{\bar{k}}^2} \right] \frac{t_{\bar{k}}\lambda_{\bar{k}}}{n_{\bar{k}}} \right\}_0^{\lambda_{\bar{k}}} + \frac{n_{g,\bar{k}}\lambda_{\bar{k}}}{t_{\bar{k}}} \tilde{q}_{\omega i,\bar{k}}$$

$$= \frac{n_{g,\bar{k}}\lambda_{\bar{k}}}{t_{\bar{k}}} \left[\frac{R_s}{\rho_{\bar{k}}} \frac{t_{\bar{k}}\lambda_{\bar{k}}}{6n_{\bar{k}}} (2\omega_{i,\bar{k}} + \omega_{j,\bar{k}}) + \tilde{q}_{\omega i,\bar{k}} \right] \tag{A43}$$

Then, by setting

$$u_{i,\bar{k}} = -\frac{E_s\, d^3\theta/ds^3}{G_s}\,\mu_{i,\bar{k}} = \frac{T_\omega}{G_s I_\omega}\,\mu_{i,\bar{k}} \tag{A44a}$$

and

$$u_{j,\bar{k}} = \frac{E_s\, d^3\theta/ds^3}{G_s}\,\mu_{j,\bar{k}} = \frac{T_\omega}{G_s I_\omega}\,\mu_{j,\bar{k}} \tag{A44b}$$

as well as

$$S_{\omega i,\bar{k}} = \frac{R_s}{\rho_{\bar{k}}}\frac{t_{\bar{k}}\lambda_{\bar{k}}}{6n_{\bar{k}}}(2\omega_{i,\bar{k}} + \omega_{j,\bar{k}}) \tag{A45}$$

Eq. (A43) reduces to

$$\mu_{j,\bar{k}} - \mu_{i,\bar{k}} = \frac{n_{g,\bar{k}}\lambda_{\bar{k}}}{t_{\bar{k}}}(S_{\omega i,\bar{k}} + \tilde{q}_{\omega i,\bar{k}}) \tag{A46}$$

Accordingly, this equation can be expressed in matrix form as

$$\boldsymbol{\mu}_j - \boldsymbol{\mu}_i = \mathbf{F}_\omega(\mathbf{s}_{\omega i} + \mathbf{q}_{\omega i}) \tag{A47}$$

where

$$\boldsymbol{\mu}_i = [\mu_{i,\bar{1}}\ \mu_{i,\bar{2}} \cdots \mu_{i,\bar{k}} \cdots] \tag{A48a}$$

$$\boldsymbol{\mu}_j = [\mu_{j,\bar{1}}\ \mu_{j,\bar{2}} \cdots \mu_{j,\bar{k}} \cdots] \tag{A48b}$$

$$\mathbf{s}_{\omega i} = [S_{\omega i,\bar{1}}\ S_{\omega i,\bar{2}} \cdots S_{\omega i,\bar{k}} \cdots] \tag{A48c}$$

$$\mathbf{q}_{\omega i} = [\tilde{q}_{\omega i,\bar{1}}\ \tilde{q}_{\omega i,\bar{2}} \cdots \tilde{q}_{\omega i,\bar{k}} \cdots] \tag{A48d}$$

= column vector with $\bar{k}_n$ elements

Also, $\mathbf{F}_\omega$ is a diagonal matrix consisting of $\bar{k}_n$ elements with

$$F_\omega = \frac{n_{g,\bar{k}}\lambda_{\bar{k}}}{t_{\bar{k}}} \tag{A49}$$

In a similar manner, the integration of Eq. (A43) with respect to the counter direction to $c'_{\bar{k}}$, shown in Fig. A2, gives

$$S_{\omega j,\bar{k}} = -\frac{R_s}{\rho_{\bar{k}}}\frac{t_{\bar{k}}\lambda_{\bar{k}}}{6n_{\bar{k}}}(\omega_{i,\bar{k}} + 2\omega_{j,\bar{k}}) \tag{A50}$$

Subtraction of Eqs. (A45) and (A50) yields

$$S_{\omega i,\bar{k}} - S_{\omega j,\bar{k}} = \frac{R_s}{\rho_{\bar{k}}}\frac{\omega_{i,\bar{k}} + \omega_{j,\bar{k}}}{2n_{\bar{k}}}\,t_{\bar{k}}\lambda_{\bar{k}} \tag{A51}$$

Therefore, Eq. (A42) can be expressed in matrix form as

$$\mathbf{s}_{\omega i} - \mathbf{s}_{\omega j} = \mathbf{q}_{\omega j} - \mathbf{q}_{\omega i} \tag{A52}$$

where

$$\mathbf{s}_{\omega j} = [S_{\omega j,\bar{1}} \; S_{\omega j,\bar{2}} \cdots S_{\omega j,\bar{k}} \cdots] \tag{A48e}$$

$$\mathbf{q}_{\omega j} = [\tilde{q}_{\omega j,\bar{1}} \; q_{\omega j,\bar{2}} \cdots \tilde{q}_{\omega j,\bar{k}} \cdots] \tag{A48f}$$

$$= \text{column vector with } \bar{k}_n \text{ elements}$$

The final expression of Eq. (A47) is

$$\boldsymbol{\mu}_j - \boldsymbol{\mu}_i = \mathbf{F}_\omega(\mathbf{s}_{\omega i} + \mathbf{q}_{\omega i}) \tag{A53a}$$

$$\boldsymbol{\mu}_j - \boldsymbol{\mu}_i = \mathbf{F}_\omega(\mathbf{s}_{\omega j} + \mathbf{q}_{\omega j}) \tag{A53b}$$

Now, if we denote the displacement function by

$$\boldsymbol{\mu} = [\mu_1 \; \mu_2 \cdots \mu_i \; \mu_j \cdots] \tag{A54}$$

then for each panel point, $\boldsymbol{\mu}_i$ and $\boldsymbol{\mu}_j$ are written by using the shape matrices **A** and **B** as follows:

$$\boldsymbol{\mu}_i = \mathbf{A}^T\boldsymbol{\mu} \tag{A55a}$$

$$\boldsymbol{\mu}_j = \mathbf{B}^T\boldsymbol{\mu} \tag{A55b}$$

By substituting this equation into Eq. (A53), the shear flow functions $\mathbf{q}_{\omega i}$ and $\mathbf{q}_{\omega j}$ can be determined as follows:

$$\mathbf{q}_{\omega i} = \mathbf{F}_\omega^{-1}(\mathbf{B}^T - \mathbf{A}^T)\boldsymbol{\mu} - \mathbf{s}_{\omega i} \tag{A56a}$$

$$\mathbf{q}_{\omega j} = \mathbf{F}_\omega^{-1}(\mathbf{B}^T - \mathbf{A}^T)\boldsymbol{\mu} - \mathbf{s}_{\omega j} \tag{A56b}$$

Furthermore, multiplying **B** by Eq. (A56*a*) and **A** by Eq. (A56*b*), we have

$$\mathbf{B}\mathbf{q}_{\omega i} - \mathbf{A}\mathbf{q}_{\omega j} = (\mathbf{B} - \mathbf{A})\mathbf{F}_\omega^{-1}(\mathbf{B}^T - \mathbf{A}^T)\boldsymbol{\mu} - \mathbf{B}\mathbf{s}_{\omega i} + \mathbf{A}\mathbf{s}_{\omega j} \tag{A57}$$

However, the left-hand side must reduce to

$$\mathbf{B}\mathbf{q}_{\omega i} - \mathbf{A}\mathbf{q}_{\omega j} = \mathbf{0} \tag{A58}$$

since the shear flow for each panel point should be in equilibrium. Thus, Eq. (A57) can be rewritten as

$$(\mathbf{B} - \mathbf{A})\mathbf{F}_\omega^{-1}(\mathbf{B}^T - \mathbf{A}^T)\boldsymbol{\mu} = \mathbf{B}\mathbf{s}_{\omega i} - \mathbf{A}\mathbf{s}_{\omega j} \tag{A59}$$

or

$$\boldsymbol{\mu} = [(\mathbf{B} - \mathbf{A})\mathbf{F}_\omega^{-1}(\mathbf{B}^T - \mathbf{A}^T)]^{-1}(\mathbf{B}\mathbf{s}_{\omega i} - \mathbf{A}\mathbf{s}_{\omega j}) \tag{A60}$$

The analytical solutions of the above equations are the same as described in the preceding section if $\boldsymbol{\omega}_o$ is replaced by $\boldsymbol{\mu}$.

A.4.2 Shear flow due to bending

The shear flow q_b due to bending can be estimated from Eq. (3.2.45) as

$$q_b(c_{\bar{k}}) = -\frac{R_o R_s}{\rho_{\bar{k}}^2}\left[\alpha\int_0^{c_{\bar{k}}}\frac{Y(c_{\bar{k}})t_{\bar{k}}}{n_{\bar{k}}}\,dc_{\bar{k}} - \beta\int_0^{c_{\bar{k}}}\frac{X(c_{\bar{k}})t_{\bar{k}}}{n_{\bar{k}}}\,dc_{\bar{k}}\right] + q_{bi,\bar{k}}$$

$$= -\frac{R_o R_s}{\rho_{\bar{k}}^2}\left\{\alpha\left[Y_{i,\bar{k}}\left(\frac{c_{\bar{k}}}{\lambda_{\bar{k}}} - \frac{c_{\bar{k}}^2}{2\lambda_{\bar{k}}^2}\right) + Y_{j,\bar{k}}\frac{c_{\bar{k}}^2}{2\lambda_{\bar{k}}^2}\right]\right.$$

$$\left. -\beta\left[X_{i,\bar{k}}\left(\frac{c_{\bar{k}}}{\lambda_{\bar{k}}} - \frac{c_{\bar{k}}^2}{2\lambda_{\bar{k}}^2}\right) + X_{j,\bar{k}}\frac{c_{\bar{k}}^2}{2\lambda_{\bar{k}}^2}\right]\right\}\frac{t_{\bar{k}}\lambda_{\bar{k}}}{n_{\bar{k}}} + q_{bi,\bar{k}} \quad \text{(A61)}$$

where

$$\alpha = \frac{Q_Y I_Y}{I_X I_Y - I_{XY}^2} \quad \text{(A62}a\text{)}$$

$$\beta = \frac{Q_Y I_{YZ}}{I_X I_Y - I_{XY}^2} \quad \text{(A62}b\text{)}$$

Equation (A61) has the following value for $c_k = \lambda_k$:

$$q_{bj,\bar{k}} = -\frac{R_o R_s}{\rho_{\bar{k}}^2}\left[\alpha\left(\frac{Y_{i,\bar{k}} + Y_{j,\bar{k}}}{2}\right) - \beta\left(\frac{X_{i,\bar{k}} + X_{j,\bar{k}}}{2}\right)\right]\frac{t_{\bar{k}}\lambda_{\bar{k}}}{n_{\bar{k}}} + q_{bj,\bar{k}} \quad \text{(A63)}$$

The general relationship between displacement u and shear flow q_b can be derived by referring to Fig. A3:

$$q_b = G_s t \tan^{-1}\gamma = \frac{G_s t\,\rho}{n_g}\frac{d(u/\rho)}{dc} \quad \text{(A64)}$$

Since this equation is in the same form as Eq. (3.2.73), the displacement u can be written as

$$G_s(u_{j,\bar{k}} - u_{i,\bar{k}}) = \frac{n_{g,\bar{k}}\lambda_{\bar{k}}}{t_{\bar{k}}}(S_{bi,\bar{k}} + q_{bi,\bar{k}}) \quad \text{(A65)}$$

where

$$S_{bi,\bar{k}} = -\frac{R_0 R_s}{\rho_{\bar{k}}^2}\frac{t_{\bar{k}}\lambda_{\bar{k}}}{6n_{\bar{k}}}[\alpha(2Y_{i,\bar{k}} + Y_{j,\bar{k}}) - \beta(2X_{i,\bar{k}} + X_{j,\bar{k}})] \quad \text{(A66)}$$

and

$$S_{bj,\bar{k}} = \frac{R_0 R_s}{\rho_{\bar{k}}^2}\frac{t_{\bar{k}}\lambda_{\bar{k}}}{6n_{\bar{k}}}[\alpha(Y_{i,\bar{k}} + 2Y_{j,\bar{k}}) - \beta(X_{i,\bar{k}} + 2X_{j,\bar{k}})] \quad \text{(A67)}$$

Subtraction of Eqs. (A66) and (A67) gives

$$q_{bj,\bar{k}} - q_{bi,\bar{k}} = S_{bj,\bar{k}} - S_{bj,\bar{k}} \quad \text{(A68)}$$

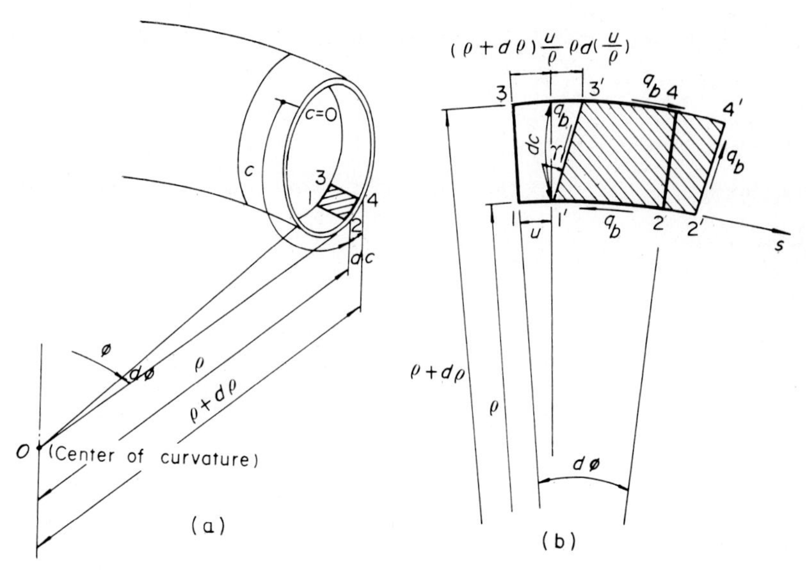

Figure A3 Displacement u due to bending: (a) segment 1,2,3,4 and (b) effect of shear forces q_b.

This follows the same procedure as was used in describing Eq. (A69).

Letting

$$u_{i,\bar{k}} = \frac{\mu_{i,\bar{k}}}{G_s} \tag{A69a}$$

and

$$u_{j,\bar{k}} = \frac{\mu_{j,\bar{k}}}{G_s} \tag{A69b}$$

we can express Eq. (A65) in matrix form as

$$\boldsymbol{\mu}_j - \boldsymbol{\mu}_i = \mathbf{F}_b(\mathbf{s}_{bi} + \mathbf{q}_{bi}) \tag{A70a}$$

$$\boldsymbol{\mu}_j - \boldsymbol{\mu}_i = \mathbf{F}_b(\mathbf{s}_{bj} + \mathbf{q}_{bj}) \tag{A70b}$$

where

$$\boldsymbol{\mu}_i = [\mu_{i,\bar{1}} \ \mu_{i,\bar{2}} \cdots \mu_{i,\bar{k}} \cdots] \tag{A71a}$$

$$\boldsymbol{\mu}_j = [\mu_{j,\bar{1}} \ \mu_{j,\bar{2}} \cdots \mu_{j,\bar{k}} \cdots] \tag{A71b}$$

$$\mathbf{s}_{bi} = [S_{bi,\bar{1}} \ S_{bi,\bar{2}} \cdots S_{bi,\bar{k}} \cdots] \tag{A71c}$$

$$\mathbf{s}_{bj} = [S_{bj,\bar{1}} \ S_{bj,\bar{2}} \cdots S_{bj,\bar{k}} \cdots] \tag{A71d}$$

$$\mathbf{q}_{bi} = [q_{bi,\bar{1}} \ q_{bi,\bar{2}} \cdots q_{bj,\bar{k}} \cdots] \tag{A71e}$$

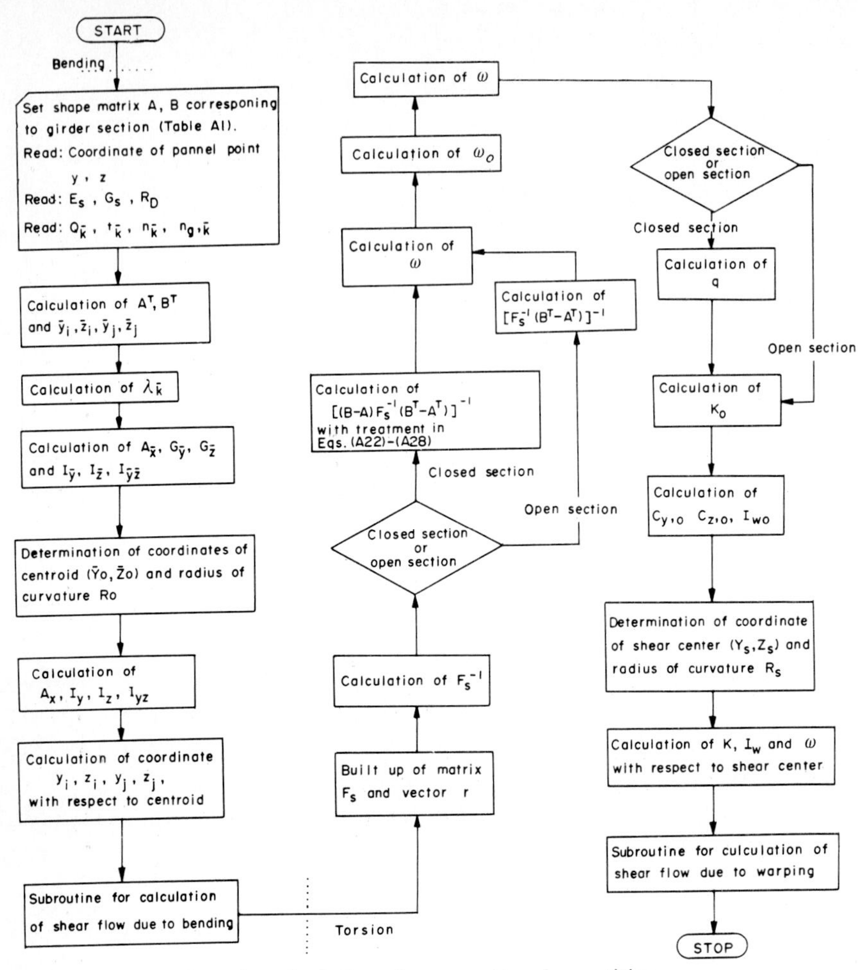

Figure A4 Flowchart for calculation of cross-sectional quantities.

Here $\mathbf{q}_{bi}$ and $\mathbf{q}_{bj}$ are column vectors with $\bar{k}_n$ elements, and $\mathbf{F}_b$ is a diagonal matrix consisting of k_n elements with

$$\frac{n_{g,\bar{k}}\lambda_{\bar{k}}}{t_{\bar{k}}} \tag{A72}$$

Equation (A70) has the same format as Eq. (A53). Accordingly, the numerical procedures for solving for the shear flow q_b can be treated in a manner similar to that used in solving for q_ω.

Figure A4 shows the flowchart for these computations as performed by a digital computer.

A.5 Numerical Example

For the sake of simplicity, the cross section of a girder is assumed to be as shown in Fig. A5. The radius of curvature is taken as infinite, so that the numerical solution of the cross-sectional quantities can be computed by hand calculations.

A.5.1 Cross-sectional quantities for bending

The shape matrix is

$$\mathbf{A} = \overset{\bar{k}}{\begin{array}{c} \begin{array}{cccc} \bar{1} & \bar{2} & \bar{3} & \bar{4} \end{array} \\ \begin{bmatrix} 1 & 0 & 0 & 0 \\ 0 & 1 & 0 & 0 \\ 0 & 0 & 1 & 0 \\ 0 & 0 & 0 & 1 \end{bmatrix} \end{array}} \begin{array}{c} \\ 1 \\ 2 \\ 3 \\ 4 \end{array} \; i \tag{A73a}$$

$$\mathbf{B} = \overset{\bar{k}}{\begin{array}{c} \begin{array}{cccc} \bar{1} & \bar{2} & \bar{3} & \bar{4} \end{array} \\ \begin{bmatrix} 0 & 0 & 0 & 1 \\ 1 & 0 & 0 & 0 \\ 0 & 1 & 0 & 0 \\ 0 & 0 & 1 & 0 \end{bmatrix} \end{array}} \begin{array}{c} \\ 1 \\ 2 \\ 3 \\ 4 \end{array} \; j \tag{A73b}$$

The coordinates of point i are

$$\bar{\mathbf{x}} = [8 \;\; 8 \;\; 2 \;\; 2] \tag{A74a}$$

$$\bar{\mathbf{y}} = [2 \;\; 5 \;\; 5 \;\; 2] \tag{A74b}$$

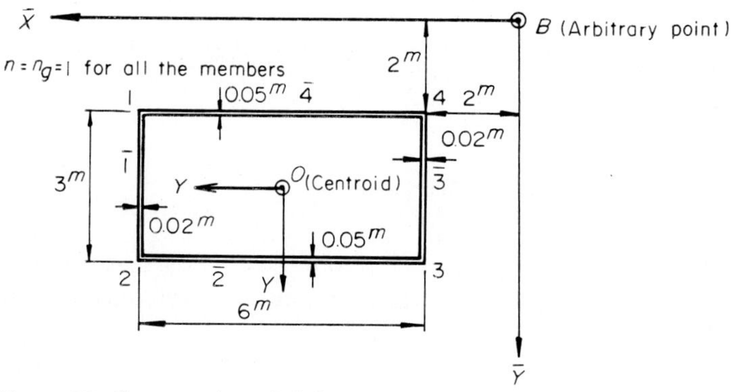

Figure A5 Cross section of girder.

The coordinates of the ith and jth ends of element $\bar{k}$ are

$$\mathbf{A}^T = \overset{\textstyle i}{\begin{matrix} 1 & 2 & 3 & 4 \end{matrix}} \atop \begin{bmatrix} 1 & 0 & 0 & 0 \\ 0 & 1 & 0 & 0 \\ 0 & 0 & 1 & 0 \\ 0 & 0 & 0 & 1 \end{bmatrix} \begin{matrix} \bar{1} \\ \bar{2} \\ \bar{3} \\ \bar{4} \end{matrix} \quad \bar{k} \tag{A75a}$$

$$\mathbf{B}^T = \overset{\textstyle j}{\begin{matrix} 1 & 2 & 3 & 4 \end{matrix}} \atop \begin{bmatrix} 0 & 1 & 0 & 0 \\ 0 & 0 & 1 & 0 \\ 0 & 0 & 0 & 1 \\ 1 & 0 & 0 & 0 \end{bmatrix} \begin{matrix} \bar{1} \\ \bar{2} \\ \bar{3} \\ \bar{4} \end{matrix} \quad \bar{k} \tag{A75b}$$

$$\bar{\mathbf{x}}_i = \mathbf{A}^T\mathbf{x} = [8 \quad 8 \quad 2 \quad 2] \tag{A76a}$$

$$\bar{\mathbf{y}}_i = \mathbf{A}^T\mathbf{y} = [2 \quad 5 \quad 5 \quad 2] \tag{A76b}$$

$$\bar{\mathbf{x}}_j = \mathbf{B}^T\mathbf{x} = [8 \quad 2 \quad 2 \quad 8] \tag{A76c}$$

$$\bar{\mathbf{y}}_j = \mathbf{B}^T\mathbf{y} = [5 \quad 5 \quad 2 \quad 2] \tag{A76d}$$

The calculation of $\lambda_{\bar{k}}$ is found in Table A2.

The cross-sectional area is

$$A_{\bar{Z}} = \sum_{\bar{k}=1}^{\bar{k}_n} \frac{t_{\bar{k}}\lambda_{\bar{k}}}{n_{\bar{k}}} \tag{A77}$$

$$\therefore \quad A_{\bar{Z}} = 0.02(3) + 0.05(6) + 0.02(3) + 0.05(6) = 0.72 \text{ m}^3$$

TABLE A2 Calculations for Length of Each Element λ

Member $\bar{k}$	1 $\bar{X}_{i,\bar{k}}$ (m)	2 $\bar{X}_{j,\bar{k}}$ (m)	3 $\bar{Y}_{i,\bar{k}}$ (m)	4 $\bar{Y}_{j,\bar{k}}$ (m)	$\lambda_{\bar{k}} = \sqrt{(1-2)^2 + (3-4)^2}$ (m)
$\bar{1}$	8	8	2	5	$\sqrt{0^2 + (-3)^2} = 3$
$\bar{2}$	8	2	5	5	$\sqrt{6^2 + 0^2} = 6$
$\bar{3}$	2	2	5	2	$\sqrt{0^2 + 3^2} = 3$
$\bar{4}$	2	8	2	2	$\sqrt{(-6)^2 + 0^2} = 6$

TABLE A3 Calculations of Static Moment $S_{\bar{X}}$

Member $\bar{k}$	1 $Y_{i,\bar{k}}$, m	2 $Y_{j,\bar{k}}$, m	3 $t_{\bar{k}}$, m	4 $\lambda_{\bar{k}}$, m	$S_{\bar{X}} = (1+2)/2 \times 3 \times 4$, m^3
$\bar{1}$	2	5	0.02	3	0.21
$\bar{2}$	5	5	0.05	6	1.50
$\bar{3}$	5	2	0.02	3	0.21
$\bar{4}$	2	2	0.05	6	0.60
					Total 2.52

The static moments are

$$S_{\bar{X}} = \sum_{\bar{k}=1}^{\bar{k}_n} \frac{\bar{Y}_{i,\bar{k}} + \bar{Y}_{j,\bar{k}}}{2} \frac{t_{\bar{k}}\lambda_{\bar{k}}}{n_{\bar{k}}} \tag{A78}$$

and

$$S_{\bar{Y}} = \sum_{\bar{k}=1}^{\bar{k}_n} \frac{\bar{X}_{i,\bar{k}} + \bar{X}_{j,\bar{k}}}{2} \frac{t_{\bar{k}}\lambda_{\bar{k}}}{n_{\bar{k}}} \tag{A79}$$

See Tables A3 and A4.

The location of the centroid is

$$\bar{X}_0 = \frac{S_{\bar{Y}}}{A_{\bar{Z}}} = \frac{3.60}{0.72} = 5 \text{ m} \tag{A80a}$$

$$\bar{Y}_0 = \frac{S_{\bar{X}}}{A_{\bar{Z}}} = \frac{2.52}{0.72} = 3.5 \text{ m} \tag{A80b}$$

TABLE A4 Calculations of Static Moment $S_{\bar{Y}}$

Member $\bar{k}$	1 $X_{i,\bar{k}}$, m	2 $X_{j,\bar{k}}$, m	3 $t_{\bar{k}}$, m	4 $\lambda_{\bar{k}}$, m	$S_{\bar{Y}} = (1+2)/2 \times 3 \times 4$, m^3
$\bar{1}$	8	8	0.02	3	0.48
$\bar{2}$	8	2	0.05	6	1.50
$\bar{3}$	2	2	0.02	3	0.12
$\bar{4}$	2	8	0.05	6	1.50
					$\Sigma = 3.60$

TABLE A5 Calculations of Coordinates for Each Element

Member $\bar{k}$	$X_{i,\bar{k}}$, m	$X_{j,\bar{k}}$, m	$Y_{i,\bar{k}}$, m	$Y_{j,\bar{k}}$, m
$\bar{1}$	3	3	−1.5	1.5
$\bar{3}$	3	−3	1.5	1.5
$\bar{3}$	−3	−3	1.5	−1.5
$\bar{4}$	−3	3	−1.5	−1.5

TABLE A6 Calculations of Power and Product of Coordinates for Each Element

Member $\bar{k}$	$t_{\bar{k}}$	$\lambda_{\bar{k}}$	$X_{i,\bar{k}}$	$X_{j,\bar{k}}$	$Y_{i,\bar{k}}$	$Y_{j,\bar{k}}$	1 $Y_{i,\bar{k}}^2 + Y_{i,\bar{k}}Y_{j,\bar{k}} + Y_{j,\bar{k}}^2$	2 $X_{i,\bar{k}}^2 + X_{i,\bar{k}}X_{j,\bar{k}} + X_{j,\bar{k}}^2$	3 $(Y_{i,\bar{k}} - Y_{j,\bar{k}})^2$	4 $(X_{i,\bar{k}} - X_{j,\bar{k}})^2$
$\bar{1}$	0.02	3	3	3	−1.5	1.5	2.25	27	9	0
$\bar{2}$	0.05	6	3	−3	1.5	1.5	6.75	9	0	36
$\bar{3}$	0.02	3	−3	−3	1.5	−1.5	2.25	27	9	0
$\bar{4}$	0.05	6	−3	3	−1.5	−1.5	6.75	9	0	36

TABLE A7 Calculations of Geometric Moment of Inertia I_X and I_Y

Member $\bar{k}$	5 $\frac{t_{\bar{k}}\lambda_{\bar{k}}}{3} \times 1$	6 $\frac{t_{\bar{k}}^3}{12\lambda_{\bar{k}}} \times 4$	5 + 6 $I_{X,\bar{k}}$	7 $\frac{t_{\bar{k}}\lambda_{\bar{k}}}{3} \times 2$	8 $\frac{t_{\bar{k}}^3}{12\lambda_{\bar{k}}} \times 3$	7 + 8 $I_{Y,\bar{k}}$
$\bar{1}$	0.045	0	0.045	0.54	0.0000018	0.540002
$\bar{2}$	0.675	0.000061	0.67506	0.90	0	0.90
$\bar{3}$	0.045	0	0.045	0.54	0.0000018	0.540002
$\bar{4}$	0.615	0.000061	0.67506	0.90	0	0.90
			$\Sigma = 1.44012$			$\Sigma = 2.880004$

The coordinates of points i and j for the $\bar{k}$th member with respect to centroid C are

$$\mathbf{x}_i = \bar{\mathbf{x}}_i - \bar{\mathbf{X}}_0 \tag{A81a}$$

$$\mathbf{y}_i = \bar{\mathbf{y}}_i - \bar{\mathbf{Y}}_0 \tag{A81b}$$

$$\mathbf{x}_j = \bar{\mathbf{x}}_j - \bar{\mathbf{X}}_0 \tag{A81c}$$

$$\mathbf{y}_j = \bar{\mathbf{y}}_j - \bar{\mathbf{Y}}_0 \tag{A81d}$$

The geometric moments of inertia are (see Tables A6 to A8)

$$I_X = \sum_{\bar{k}=1}^{\bar{k}_n} \left[\frac{t_{\bar{k}}\lambda_{\bar{k}}}{3n_{\bar{k}}} (Y_{i,\bar{k}}^2 + Y_{i,\bar{k}}Y_{j,\bar{k}} + Y_{j,\bar{k}}^2) + \frac{t_{\bar{k}}^3}{12n_{\bar{k}}\lambda_{\bar{k}}} (X_{i,\bar{k}} - X_{j,\bar{k}})^2 \right] \tag{A82a}$$

$$I_Y = \sum_{\bar{k}=1}^{\bar{k}_n} \left[\frac{t_{\bar{k}}\lambda_{\bar{k}}}{3n_{\bar{k}}} (X_{i,\bar{k}}^2 + X_{i,\bar{k}}X_{j,\bar{k}} + X_{j,\bar{k}}^2) + \frac{t_{\bar{k}}^3}{12n_{\bar{k}}\lambda_{\bar{k}}} (Y_{i,\bar{k}} - Y_{j,\bar{k}})^2 \right] \tag{A82b}$$

$$I_{XY} = \sum_{\bar{k}=1}^{\bar{k}_n} \left\{ \frac{t_{\bar{k}}\lambda_{\bar{k}}}{6n_{\bar{k}}} [X_{i,\bar{k}}(2Y_{i,\bar{k}} + Y_{j,\bar{k}}) + Y_{j,\bar{k}}(Y_{i,\bar{k}} + 2Y_{j,\bar{k}})] - \frac{t_{\bar{k}}^3}{12n_{\bar{k}}} (X_{i,\bar{k}} - X_{j,\bar{k}})(Y_{i,\bar{k}} - Y_{j,\bar{k}}) \right\} \tag{A82c}$$

TABLE A8 Calculations of Product of Inertia I_{XY}

Member $\bar{k}$	1* $X_i(2Y_i + Y_j)$	2* $X_j(Y_i + 2Y_j)$	3 $\frac{t_{\bar{k}}\lambda_{\bar{k}}}{6}(1+2)$	4* $(X_i - X_j)(Y_i - Y_j)$	5 $-\frac{t_{\bar{k}}^3}{12} \times 4$	3 + 5 $I_{XY,\bar{k}}$
$\bar{1}$	−4.5	4.5	0	0	0	0
$\bar{2}$	13.5	−13.5	0	0	0	0
$\bar{3}$	−4.5	4.5	0	0	0	0
$\bar{4}$	13.5	−13.5	0	0	0	0
						$\Sigma = 0$

*The subscript $\bar{k}$ for X_i, X_j, Y_i, and Y_j is omitted due to space limitation.

Therefore,

$$I_X = 1.44012 \text{ m}^4 \tag{A83a}$$

$$I_Y = 2.880004 \text{ m}^4 \tag{A83b}$$

$$I_{XY} = 0 \tag{A83c}$$

A.5.2 Cross-sectional quantities for warping

Vector $\mathbf{r}$ is

$$\mathbf{r}_{\bar{k}} = \frac{X_{i,\bar{k}}Y_{j,\bar{k}} - X_{j,\bar{k}}Y_{i,\bar{k}}}{\lambda_{\bar{k}}} \frac{t_{\bar{k}}}{n_{g,\bar{k}}} \tag{A84}$$

See Table A9.

The diagonal matrix $\mathbf{F}_s$ is

$$f_{s,\bar{k}} = \frac{\lambda_{\bar{k}}}{t_{\bar{k}}} \tag{A85}$$

$$\mathbf{F}_s = \begin{bmatrix} 150 & 0 & 0 & 0 \\ 0 & 120 & 0 & 0 \\ 0 & 0 & 150 & 0 \\ 0 & 0 & 0 & 120 \end{bmatrix} \tag{A86}$$

Therefore, the inverse matrix reduces to

$$\mathbf{F}_s^{-1} = \begin{bmatrix} 1/150 & 0 & 0 & 0 \\ 0 & 1/120 & 0 & 0 \\ 0 & 0 & 1/150 & 0 \\ 0 & 0 & 0 & 1/120 \end{bmatrix} \tag{A87}$$

TABLE A9 Calculations of Perpendicular Distance for Each Element r

Member $\bar{k}$	$t_{\bar{k}}$	$X_{i,\bar{k}}$ (1)	$X_{j,\bar{k}}$ (2)	$Y_{i,\bar{k}}$ (2)	$Y_{j,\bar{k}}$ (1)	$r_{o,\bar{k}}$ 3 $(1-2)/\lambda_{\bar{k}}$	$3 \times t_{\bar{k}}$ $r_{\bar{k}}$
$\bar{1}$	0.02	3	3	−1.5	1.5	3	0.060
$\bar{2}$	0.05	3	−3	1.5	1.5	1.5	0.075
$\bar{3}$	0.02	−3	−3	1.5	−1.5	1.5	0.060
$\bar{4}$	0.05	−3	3	−1.5	−1.5	3	0.075

Matrices $\mathbf{B}-\mathbf{A}$ and $\mathbf{B}^T-\mathbf{A}^T$ are

$$\mathbf{B}-\mathbf{A}=\begin{bmatrix} -1 & 0 & 0 & 1 \\ 1 & -1 & 0 & 0 \\ 0 & 1 & -1 & 0 \\ 0 & 0 & 1 & -1 \end{bmatrix} \tag{A88}$$

$$\mathbf{B}^T-\mathbf{A}^T=\begin{bmatrix} -1 & 1 & 0 & 0 \\ 0 & -1 & 1 & 0 \\ 0 & 0 & -1 & 1 \\ 1 & 0 & 0 & -1 \end{bmatrix} \tag{A89}$$

Calculation of $(\mathbf{B}-\mathbf{A})\mathbf{F}_s^{-1}(\mathbf{B}^T-\mathbf{A}^T)$ yields

$$\mathbf{F}_s^{-1}(\mathbf{B}^T-\mathbf{A}^T)=\begin{bmatrix} -1/150 & 1/150 & 0 & 0 \\ 0 & -1/120 & 1/120 & 0 \\ 0 & 0 & -1/150 & 1/150 \\ 1/120 & 0 & 0 & -1/120 \end{bmatrix}$$

$$(\mathbf{B}-\mathbf{A})\mathbf{F}_s^{-1}(\mathbf{B}^T-\mathbf{A}^T)=\begin{bmatrix} -27/1800 & -12/1800 & 0 & -15/1800 \\ -12/1800 & 27/1800 & -15/1800 & 0 \\ 0 & -15/1800 & 27/1800 & -12/1800 \\ -15/1800 & 0 & -12/1800 & 27/1800 \end{bmatrix}$$

Therefore, we have

$$\mathbf{C}=(\mathbf{B}-\mathbf{A})\mathbf{F}_s^{-1}(\mathbf{B}^T-\mathbf{A}^T)=\frac{1}{600}\begin{bmatrix} 9 & -4 & 0 & -5 \\ -4 & 9 & -5 & 0 \\ 0 & -5 & 9 & -4 \\ -5 & 0 & -4 & 9 \end{bmatrix} \tag{A90}$$

However, the determinant of the above matrix is equal to zero, or

$$\det \mathbf{C} = \begin{bmatrix} 9 & -4 & 0 & -5 \\ 4 & 9 & -5 & 0 \\ 0 & -5 & 9 & -4 \\ -5 & 0 & -4 & 9 \end{bmatrix} = 0 \tag{A91a}$$

Now let's calculate the inverse of $\mathbf{D}^{-1}$. From the dotted part of Eq. (A90), matrix $\mathbf{D}$ can be defined as

$$\mathbf{D} = \frac{1}{600}\begin{bmatrix} 9 & -5 & 0 \\ -5 & 9 & -4 \\ 0 & -4 & 9 \end{bmatrix} \quad \text{and} \quad \det \mathbf{D} = |\mathbf{D}| = \frac{360}{600} \neq 0 \tag{A91b}$$

The inverse of this matrix can be estimated by the following definition:

$$\mathbf{D}^{-1} = \frac{1}{|\mathbf{D}|} \cdot [\text{adj } \mathbf{D}]^T \tag{A92}$$

or

$$\mathbf{D}^{-1} = \frac{600}{360}\begin{bmatrix} 65 & 45 & 20 \\ 45 & 81 & 36 \\ 20 & 36 & 56 \end{bmatrix} = \frac{10}{6}\begin{bmatrix} 65 & 45 & 20 \\ 45 & 81 & 36 \\ 20 & 36 & 56 \end{bmatrix} \tag{A93}$$

and $\quad \mathbf{D} \cdot \mathbf{D}^{-1} = \mathbf{E}$ (A94)

The inverse of $\mathbf{C} = (\mathbf{B} - \mathbf{A})\mathbf{F}_s^{-1}(\mathbf{B}^T - \mathbf{A}^T)^{-1}$ is

$$\mathbf{C}^{-1} = \frac{10}{6}\begin{bmatrix} 0 & 0 & 0 & 0 \\ 0 & 65 & 45 & 20 \\ 0 & 45 & 81 & 36 \\ 0 & 20 & 36 & 56 \end{bmatrix} \tag{A95}$$

and

$$(\mathbf{B} - \mathbf{A})\mathbf{r} = \begin{bmatrix} -1 & 0 & 0 & 1 & 0.060 \\ 1 & -1 & 0 & 0 & 0.075 \\ 0 & 1 & -1 & 0 & 0.060 \\ 0 & 0 & 1 & -1 & 0.075 \end{bmatrix} = \begin{bmatrix} 0.015 \\ -0.015 \\ 0.015 \\ -0.015 \end{bmatrix} \tag{A96}$$

Now we calculate the warping function $\tilde{\boldsymbol{\omega}}$

$$\tilde{\boldsymbol{\omega}} = -\mathbf{C}^{-1}(\mathbf{B}-\mathbf{A})\mathbf{r} \tag{A97}$$

$$\begin{bmatrix} \tilde{\omega}_{o,1} \\ \tilde{\omega}_{o,2} \\ \tilde{\omega}_{o,3} \\ \tilde{\omega}_{o,4} \end{bmatrix} = -\frac{10}{6}\begin{bmatrix} 0 & 0 & 0 & 0 \\ 0 & 65 & 45 & 20 \\ 0 & 45 & 81 & 36 \\ 0 & 20 & 36 & 56 \end{bmatrix}\begin{bmatrix} 0.015 \\ -0.015 \\ 0.015 \\ -0.015 \end{bmatrix}$$

$$= -\frac{10}{6}\begin{bmatrix} 0 \\ -0.6 \\ 0 \\ -0.6 \end{bmatrix} = \begin{bmatrix} 0 \\ 1 \\ 0 \\ 1 \end{bmatrix} \tag{A98}$$

Warping functions $\boldsymbol{\omega}_i$ and $\boldsymbol{\omega}_j$ are

$$\tilde{\boldsymbol{\omega}}_i = \mathbf{A}^T\tilde{\boldsymbol{\omega}} \tag{A99a}$$

$$\tilde{\boldsymbol{\omega}}_j = \mathbf{B}^T\tilde{\boldsymbol{\omega}} \tag{A99b}$$

$$\begin{bmatrix} \tilde{\omega}_{oi,1} \\ \tilde{\omega}_{oi,2} \\ \tilde{\omega}_{oi,3} \\ \tilde{\omega}_{oi,4} \end{bmatrix} = \begin{bmatrix} 1 & 0 & 0 & 0 \\ 0 & 1 & 0 & 0 \\ 0 & 0 & 1 & 0 \\ 0 & 0 & 0 & 1 \end{bmatrix}\begin{bmatrix} 0 \\ 1 \\ 0 \\ 1 \end{bmatrix} = \begin{bmatrix} 0 \\ 1 \\ 0 \\ 1 \end{bmatrix} \tag{A100}$$

$$\begin{bmatrix} \tilde{\omega}_{oj,1} \\ \tilde{\omega}_{oj,2} \\ \tilde{\omega}_{oj,3} \\ \tilde{\omega}_{oj,4} \end{bmatrix} = \begin{bmatrix} 0 & 1 & 0 & 0 \\ 0 & 0 & 1 & 0 \\ 0 & 0 & 0 & 1 \\ 1 & 0 & 0 & 0 \end{bmatrix}\begin{bmatrix} 0 \\ 1 \\ 0 \\ 1 \end{bmatrix} = \begin{bmatrix} 1 \\ 0 \\ 1 \\ 0 \end{bmatrix} \tag{A101}$$

Calculation of $\tilde{\omega}_o$ reveals

$$\tilde{\omega}_o = -\frac{\sum_{\bar{k}=1}^{\bar{k}_n}[\tilde{\omega}_{oi,\bar{k}} + \tilde{\omega}_{oj,k}/2]t_{\bar{k}}\lambda_{\bar{k}}}{\sum_{\bar{k}=1}^{\bar{k}_n} t_{\bar{k}}\lambda_{\bar{k}}} \tag{A102}$$

See Table A10. Therefore,

$$\bar{\omega}_o = -\frac{0.36}{0.72} = -0.5 \text{ m}^2 \tag{A103}$$

TABLE A10 Calculations of Warping Function $\tilde{\omega}_o$

Member $\bar{k}$	0 $t_{\bar{k}}\lambda_{\bar{k}}$	1 $\tilde{\omega}_{oi,\bar{k}}$	2 $\tilde{\omega}_{oj,\bar{k}}$	$(1+2)/2 \times 0$
$\bar{1}$	0.06	0	1	0.03
$\bar{2}$	0.30	1	0	0.15
$\bar{3}$	0.06	0	1	0.03
$\bar{4}$	0.30	1	0	0.15
	$\Sigma = 0.72$			$\Sigma = 0.36$

The warping function is (in square meters)

$$\omega_{o,i} = \tilde{\omega}_{o,i} + \bar{\omega}_o \tag{A104}$$

$$\begin{bmatrix} \omega_{o,1} \\ \omega_{o,2} \\ \omega_{o,3} \\ \omega_{o,4} \end{bmatrix} = \begin{bmatrix} -0.5 \\ 0.5 \\ -0.5 \\ 0.5 \end{bmatrix} \tag{A105}$$

or

$$\begin{bmatrix} \omega_{oi,\bar{1}} \\ \omega_{oi,\bar{2}} \\ \omega_{oi,\bar{3}} \\ \omega_{oi,\bar{4}} \end{bmatrix} = \begin{bmatrix} -0.5 \\ 0.5 \\ -0.5 \\ 0.5 \end{bmatrix} \tag{A106a}$$

$$\begin{bmatrix} \omega_{oj,\bar{1}} \\ \omega_{oj,\bar{2}} \\ \omega_{oj,\bar{3}} \\ \omega_{oj,\bar{4}} \end{bmatrix} = \begin{bmatrix} 0.5 \\ -0.5 \\ 0.5 \\ -0.5 \end{bmatrix} \tag{A106b}$$

See Fig. A6.

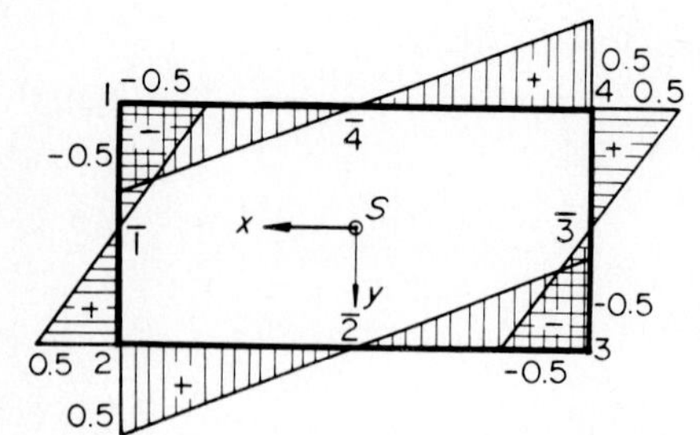

Figure A6 Warping function ω (m^2).

For the torsional function **q**,

$$\mathbf{q} = \mathbf{F}_s^{-1}(\boldsymbol{\omega}_j - \boldsymbol{\omega}_i) + \mathbf{r} \tag{A107}$$

$$\boldsymbol{\omega}_j - \boldsymbol{\omega}_i = \begin{bmatrix} 1 \\ -1 \\ 1 \\ -1 \end{bmatrix} \tag{A108}$$

$$\begin{bmatrix} \tilde{q}_{o,1} \\ \tilde{q}_{o,2} \\ \tilde{q}_{o,3} \\ \tilde{q}_{o,3} \end{bmatrix} = \begin{bmatrix} 1/150 & 0 & 0 & 0 \\ 0 & 1/120 & 0 & 0 \\ 0 & 0 & 1/150 & 0 \\ 0 & 0 & 0 & 1/120 \end{bmatrix} \begin{bmatrix} 1 \\ -1 \\ 1 \\ -1 \end{bmatrix} + \begin{bmatrix} 0.060 \\ 0.075 \\ 0.060 \\ 0.075 \end{bmatrix}$$

$$= \begin{bmatrix} 1/150 + 60/1000 \\ -1/120 + 75/1000 \\ 1/150 + 60/1000 \\ -1/120 + 75/1000 \end{bmatrix} = \frac{1}{3000} \begin{bmatrix} 20 + 180 \\ -25 + 225 \\ 20 + 180 \\ -25 + 225 \end{bmatrix} = \begin{bmatrix} 2/30 \\ 2/30 \\ 2/30 \\ 2/30 \end{bmatrix} \tag{A109}$$

Therefore,

$$(\mathbf{B} - \mathbf{A})\mathbf{q} = \begin{bmatrix} -1 & 0 & 0 & 1 \\ 1 & -1 & 0 & 0 \\ 0 & 1 & -1 & 0 \\ 0 & 0 & 1 & -1 \end{bmatrix} \begin{bmatrix} 2/30 \\ 2/30 \\ 2/30 \\ 2/30 \end{bmatrix} = \begin{bmatrix} 0 \\ 0 \\ 0 \\ 0 \end{bmatrix} = \mathbf{0} \tag{A110}$$

The torsion constant $\mathbf{K}_o$ is

$$K_o = \sum_{\bar{k}=1}^{\bar{k}_n} \frac{n_{g,\bar{k}} \lambda_{\bar{k}}}{t_{\bar{k}}} q_{o,\bar{k}}^2 \tag{A111}$$

See Table A11. Therefore,

$$K_o = 2.4 \text{ m}^4 \tag{A112}$$

TABLE A11 Calculations of Torsional Constant K_o

Member $\bar{k}$	1 $\lambda_{\bar{k}}$	2 $t_{\bar{k}}$	3 $q_{o,\bar{k}}^2$	1/2 × 3 $K_{o,\bar{k}}$
$\bar{1}$	3	0.02	4/900	4/6
$\bar{2}$	6	0.05	4/900	8/15
$\bar{3}$	3	0.02	4/900	4/6
$\bar{4}$	6	0.05	4/900	8/15
				Σ = 12/5

TABLE A12 Calculations of Warping Constant $I_{\omega o}$

Member $\bar{k}$	1 $\omega_{oi,\bar{k}}$	2 $\omega_{oj,\bar{k}}$	3 1^2	4 1×2	5 2^2	6 $3+4+5$	$I_{\omega o,\bar{k}}$ $\frac{t_{\bar{k}}\lambda_{\bar{k}}}{3} \times 6$
$\bar{1}$	−0.5	0.5	0.25	−0.25	0.25	0.25	0.005
$\bar{2}$	0.5	−0.5	0.25	−0.25	0.25	0.25	0.025
$\bar{3}$	−0.5	0.5	0.25	−0.25	0.25	0.25	0.005
$\bar{4}$	0.5	−0.5	0.25	−0.25	0.25	0.25	0.025
							$\Sigma = 0.060$

Calculation of the warping constant $I_{\omega o}$ yields

$$I_{\omega o} = \frac{t_{\bar{k}}\lambda_{\bar{k}}}{3n_{\bar{k}}}(\omega^2_{oi,\bar{k}} + \omega_{oi,\bar{k}}\omega_{oj,\bar{k}} + \omega^2_{oj,\bar{k}}) \tag{A113}$$

See Table A12.

$$\therefore \quad I_{\omega o} = 0.06 \text{ m}^6 \tag{A114}$$

Now we determine the shear center

$$X_s = \frac{I_Y C_{X,o} - I_{XY} C_{Y,o}}{I_X I_Y - I^2_{XY}} \tag{A115a}$$

$$Y_s = \frac{I_{XY} C_{X,o} - I_Y C_{Y,o}}{I_X I_Y - I^2_{XY}} \tag{A115b}$$

where $$C_{X,o} = \sum_{\bar{k}=1}^{\bar{k}_n} \frac{t_{\bar{k}}\lambda_{\bar{k}}}{6n_{\bar{k}}}[(2Y_{i,\bar{k}} + Y_{j,\bar{k}})\omega_{oi,\bar{k}} + (Y_{i,\bar{k}} + 2Y_{j,\bar{k}})\omega_{oj,\bar{k}}] \tag{A116a}$$

$$C_{Y,o} = \sum_{\bar{k}=1}^{\bar{k}_n} \frac{t_{\bar{k}}\lambda_{\bar{k}}}{6n_{\bar{k}}}[(2X_{i,\bar{k}} + X_{j,\bar{k}})\omega_{oi,\bar{k}} + (X_{i,\bar{k}} + 2X_{j,\bar{k}})\omega_{oj,\bar{k}}] \tag{A116b}$$

TABLE A13 Calculations of Static Moment C_{Xo} with Respect to Warping Function

Member $\bar{k}$	1 $\omega_{oi,\bar{k}}$	2 $\omega_{oj,\bar{k}}$	3 $Y_{i,\bar{k}}$	4 $Y_{j,\bar{k}}$	5 $2 \times 3 + 4$	6 $3 + 2 \times 4$	$5 \times 1 + 6 \times 2$ $C_{Xo,\bar{k}}$
$\bar{1}$	−0.5	0.5	−1.5	1.5	−1.5	1.5	1.5
$\bar{2}$	0.5	−0.5	1.5	1.5	4.5	4.5	0
$\bar{3}$	−0.5	0.5	1.5	−1.5	1.5	−1.5	−1.5
$\bar{4}$	0.5	−0.5	−1.5	−1.5	−4.5	−4.5	0
							$\Sigma = 0$

TABLE A14 Calculations of Static Moment C_{Yo} with Respect to Warping Function

Member $\bar{k}$	1 $X_{i,\bar{k}}$	2 $X_{j,\bar{k}}$	3 $2\times1+2$	4 $1+2\times2$	$C_{Yo,\bar{k}}$ $\omega_{oi,\bar{k}}\times1+\omega_{oj,\bar{k}}\times4$
$\bar{1}$	3	3	9	9	0
$\bar{2}$	3	−3	3	−3	3
$\bar{3}$	−3	−3	−9	−9	0
$\bar{4}$	−3	3	−3	3	−3
				$\Sigma=$	0

See Tables A13 and A14. Then

$$C_{Xo}=0 \tag{A117a}$$

$$C_{Yo}=0 \tag{A117b}$$

Therefore,

$$X_s=0 \tag{A118a}$$

$$Y_s=0 \tag{A118b}$$

The cross-sectional value with respect to the shear center is

$$\begin{bmatrix}\omega_1\\ \omega_2\\ \omega_3\\ \omega_4\end{bmatrix}=\begin{bmatrix}\omega_{o,1}\\ \omega_{o,2}\\ \omega_{o,3}\\ \omega_{o,4}\end{bmatrix} \tag{A119}$$

so

$$\mathbf{q}_s=\mathbf{q} \tag{A120}$$

$$K=K_o=2.4\text{ m}^4 \tag{A121}$$

$$I_\omega=I_{\omega o}=0.06\text{ m}^6 \tag{A122}$$

A.5.3 Estimation of shear flow

(1) Bending. Calculation of $\mathbf{s}_{bi}$ and $\mathbf{s}_{bj}$ yields

$$S_{bi,\bar{k}}=-\frac{t_{\bar{k}}\lambda_{\bar{k}}}{6n_{\bar{k}}}[\alpha(2Y_{i,\bar{k}}+Y_{j,\bar{k}})-\beta(2X_{i,\bar{k}}+X_{j,\bar{k}})] \tag{A123a}$$

$$S_{bj,\bar{k}}=\frac{t_{\bar{k}}\lambda_{\bar{k}}}{6n_{\bar{k}}}[\alpha(Y_{i,\bar{k}}+2Y_{j,\bar{k}})-\beta(X_{i,\bar{k}}+2X_{j,\bar{k}})] \tag{A123b}$$

For this case,

$$\alpha=\frac{Q_Y}{I_X} \tag{A124a}$$

and

$$\beta=0 \tag{A124b}$$

TABLE A15 Calculations of Shear Flow due to Bending

Member	From Table A12		$S_{bi,\bar{k}}$	$S_{bj,\bar{k}}$
	1	2		
$\bar{k}$	$2Y_{i,\bar{k}} + Y_{j,\bar{k}}$	$Y_{i,\bar{k}} + 2Y_{j,\bar{k}}$	$-\frac{t_{\bar{k}}\lambda_{\bar{k}}}{6} \times 1$	$\frac{t_{\bar{k}}\lambda_{\bar{k}}}{6} \times 2$
$\bar{1}$	−1.5	1.5	0.015	0.015
$\bar{2}$	4.5	4.5	−0.225	0.225
$\bar{3}$	1.5	−1.5	−0.015	−0.015
$\bar{4}$	−4.5	−4.5	0.225	−0.225

From the diagonal matrix $\mathbf{F}_b$ and $\mathbf{F}_b^{-1}$,

$$f_{b,\bar{k}} = \frac{\lambda_{\bar{k}}}{n_{\bar{k}} t_{\bar{k}}} \tag{A125}$$

From Eq. (A15),

$$\mathbf{F}_b = \mathbf{F}_s \tag{A126a}$$

and

$$\mathbf{F}_b^{-1} = \mathbf{F}_s^{-1} \tag{A126b}$$

Matrices $(\mathbf{B} - \mathbf{A})\mathbf{F}_b^{-1}(\mathbf{B}^T - \mathbf{A}^T)$ and $[(\mathbf{B} - \mathbf{A})\mathbf{F}_b^{-1}(\mathbf{B}^T - \mathbf{A}^T)]^{-1}$ are the same as Eqs. (A90) and (A95), respectively.

Calculation of $\mathbf{Bs}_{bi} - \mathbf{As}_{bj}$ gives

$$\frac{Q_Y}{I_X}\begin{bmatrix} 0 & 0 & 0 & 1 \\ 1 & 0 & 0 & 0 \\ 0 & 1 & 0 & 0 \\ 0 & 0 & 1 & 0 \end{bmatrix}\begin{bmatrix} 0.015 \\ -0.225 \\ -0.015 \\ 0.225 \end{bmatrix} - \frac{Q_Y}{I_X}\begin{bmatrix} 1 & 0 & 0 & 0 \\ 0 & 1 & 0 & 0 \\ 0 & 0 & 1 & 0 \\ 0 & 0 & 0 & 1 \end{bmatrix}\begin{bmatrix} 0.015 \\ 0.225 \\ -0.015 \\ -0.225 \end{bmatrix}$$

$$= \frac{Q_Y}{I_X}\begin{bmatrix} 0.225 \\ 0.015 \\ -0.225 \\ -0.015 \end{bmatrix} - \frac{Q_Y}{I_X}\begin{bmatrix} 0.015 \\ 0.225 \\ -0.015 \\ -0.225 \end{bmatrix} = \frac{Q_Y}{I_X}\begin{bmatrix} 0.21 \\ -0.21 \\ -0.21 \\ 0.21 \end{bmatrix} \tag{A127}$$

So we calculate $\tilde{\boldsymbol{\mu}}$:

$$\tilde{\boldsymbol{\mu}} = \mathbf{C}^{-1}(\mathbf{Bs}_{bi} - \mathbf{As}_{bj}) = \frac{Q_Y}{I_X}\frac{10}{6}\begin{bmatrix} 0 & 0 & 0 & 0 \\ 0 & 65 & 45 & 20 \\ 0 & 45 & 81 & 36 \\ 0 & 20 & 36 & 56 \end{bmatrix}\begin{bmatrix} 0.21 \\ -0.21 \\ -0.21 \\ 0.21 \end{bmatrix}$$

$$= \frac{Q_Y}{I_X}\frac{10}{6}\begin{bmatrix} 0 \\ -18.9 \\ -18.9 \\ 0 \end{bmatrix} = \frac{Q_Y}{I_X}\begin{bmatrix} 0 \\ -31.5 \\ -31.5 \\ 0 \end{bmatrix} \tag{A128}$$

And then

$$\tilde{\boldsymbol{\mu}}_i = \mathbf{A}^T\tilde{\boldsymbol{\mu}} = \frac{Q_Y}{I_X}\begin{bmatrix} 1 & 0 & 0 & 0 \\ 0 & 1 & 0 & 0 \\ 0 & 0 & 1 & 0 \\ 0 & 0 & 0 & 1 \end{bmatrix}\begin{bmatrix} 0 \\ -31.5 \\ -31.5 \\ 0 \end{bmatrix}$$

$$= \frac{Q_Y}{I_X}\begin{bmatrix} 0 \\ -31.5 \\ -31.5 \\ 0 \end{bmatrix} \tag{A129}$$

$$\tilde{\boldsymbol{\mu}}_j = \mathbf{B}^T\tilde{\boldsymbol{\mu}} = \frac{Q_Y}{I_X}\begin{bmatrix} 0 & 1 & 0 & 0 \\ 0 & 0 & 1 & 0 \\ 0 & 0 & 0 & 1 \\ 1 & 0 & 0 & 0 \end{bmatrix}\begin{bmatrix} 0 \\ -31.5 \\ -31.5 \\ 0 \end{bmatrix}$$

$$= \frac{Q_Y}{I_X}\begin{bmatrix} -31.5 \\ -31.5 \\ 0 \\ 0 \end{bmatrix} \tag{A130}$$

Calculation of $\bar{\mu}$ yields

$$\bar{\mu} = -\frac{\sum_{\bar{k}=1}^{\bar{k}_n} [(\tilde{\mu}_{i,\bar{k}} + \mu_{j,\bar{k}})/2](t_{\bar{k}}\lambda_{\bar{k}}/n_{\bar{k}})}{\sum_{\bar{k}=1}^{\bar{k}_n} t_{\bar{k}}\lambda_{\bar{k}}/n_{\bar{k}}} \tag{A131}$$

See Table A16. Therefore,

$$\bar{\mu} = -\frac{-11.34}{0.72}\frac{Q_Y}{I_X} = 15.75\frac{Q_Y}{I_X} \tag{A132}$$

Calculation of $\boldsymbol{\mu}$ gives

$$\boldsymbol{\mu} = \tilde{\boldsymbol{\mu}} + \bar{\boldsymbol{\mu}} = \frac{Q_Y}{I_X}\begin{bmatrix} 0 + 15.75 \\ -31.5 + 15.75 \\ -31.5 + 15.75 \\ 0 + 15.75 \end{bmatrix} = \frac{Q_Y}{I_X}\begin{bmatrix} 15.75 \\ -15.75 \\ -15.75 \\ 15.75 \end{bmatrix} \tag{A133}$$

TABLE A16 Calculations of Statically Indeterminate Shear Flow due to Bending

Member $\bar{k}$	0 $t_{\bar{k}}\lambda_{\bar{k}}$	1 $\mu_{i,\bar{k}}$	2 $\mu_{j,\bar{k}}$	$\times Q_Y/I_X$ $(1+2)/2 \times 0$
$\bar{1}$	0.06	0	−31.5	−0.945
$\bar{2}$	0.30	−3.15	−3.15	−9.450
$\bar{3}$	0.06	−3.15	0	−0.945
$\bar{4}$	0.30	0	0	0
	$\Sigma = 0.72$			$\Sigma = -11.34$

Now we find $\mathbf{q}_{bi}$ and $\mathbf{q}_{bj}$:

$$\mathbf{q}_{bi} = \mathbf{F}_b^{-1}(\mathbf{B}^T - \mathbf{A}^T)\boldsymbol{\mu} - \mathbf{s}_{bi} \tag{A134a}$$

$$\mathbf{q}_{bj} = \mathbf{F}_b^{-1}(\mathbf{B}^T - \mathbf{A}^T)\boldsymbol{\mu} - \mathbf{s}_{bj} \tag{A134b}$$

$$\begin{bmatrix} q_{bi,\bar{1}} \\ q_{bi,\bar{2}} \\ q_{bi,\bar{3}} \\ q_{bi,\bar{4}} \end{bmatrix} = \frac{Q_Y}{I_X}\begin{bmatrix} 1/150 & 0 & 0 & 0 \\ 0 & 1/120 & 0 & 0 \\ 0 & 0 & 1/150 & 0 \\ 0 & 0 & 0 & 1/120 \end{bmatrix}\begin{bmatrix} -1 & 1 & 0 & 0 \\ 0 & -1 & 1 & 0 \\ 0 & 0 & 1 & -1 \\ 1 & 0 & 0 & -1 \end{bmatrix}\begin{bmatrix} 15.75 \\ -15.75 \\ -15.75 \\ 15.75 \end{bmatrix}$$

$$-\frac{Q_Y}{I_X}\begin{bmatrix} 0.015 \\ -0.225 \\ -0.015 \\ 0.225 \end{bmatrix} = \frac{Q_Y}{I_X}\begin{bmatrix} -0.225 \\ 0.225 \\ 0.225 \\ -0.225 \end{bmatrix} \tag{A135a}$$

$$\begin{bmatrix} q_{bj,\bar{1}} \\ q_{bj,\bar{2}} \\ q_{bj,\bar{3}} \\ q_{bj,\bar{4}} \end{bmatrix} = \frac{Q_Y}{I_X}\begin{bmatrix} 1/150 & 0 & 0 & 0 \\ 0 & 1/120 & 0 & 0 \\ 0 & 0 & 1/150 & 0 \\ 0 & 0 & 0 & 1/120 \end{bmatrix}\begin{bmatrix} -1 & 1 & 0 & 0 \\ 0 & -1 & 1 & 0 \\ 0 & 0 & 1 & -1 \\ 1 & 0 & 0 & -1 \end{bmatrix}\begin{bmatrix} 15.75 \\ -15.75 \\ -15.75 \\ 15.75 \end{bmatrix}$$

$$-\frac{Q_Y}{I_X}\begin{bmatrix} 0.015 \\ -0.225 \\ -0.015 \\ 0.225 \end{bmatrix} = \frac{Q_Y}{I_X}\begin{bmatrix} -0.225 \\ 0.225 \\ 0.225 \\ -0.225 \end{bmatrix} \tag{A135b}$$

Then,

$$\mathbf{B}\mathbf{q}_{bi} - \mathbf{A}\mathbf{q}_{bj} = \frac{Q_Y}{I_X}\begin{bmatrix} 0 & 0 & 0 & 1 \\ 1 & 0 & 0 & 0 \\ 0 & 1 & 0 & 0 \\ 0 & 0 & 1 & 0 \end{bmatrix}\begin{bmatrix} -0.225 \\ 0.225 \\ 0.225 \\ -0.225 \end{bmatrix} - \frac{Q_Y}{I_X}\begin{bmatrix} 1 & 0 & 0 & 0 \\ 0 & 1 & 0 & 0 \\ 0 & 0 & 1 & 0 \\ 0 & 0 & 0 & 1 \end{bmatrix}\begin{bmatrix} -0.225 \\ -0.225 \\ 0.225 \\ 0.225 \end{bmatrix}$$

$$= \frac{Q_Y}{I_X}\begin{bmatrix} 0 \\ 0 \\ 0 \\ 0 \end{bmatrix} = \mathbf{0} \tag{A136}$$

See Fig. A7. The values of $q_{b,a}$ (*) and $q_{b,b}$ (**) are estimated from

$$q_b = -\frac{Q}{I_X}\left(\int_0^c \frac{Yt}{n}\,dc + \bar{q}_b\right)$$

$$\bar{q}_b = \frac{\oint (1/Gt)\left[\int_0^c (Yt/n)\,dc\right]dc}{\oint (1/Gt)\,dc} = 0.225\text{ m}^3$$

(2) Warping. Calculation of $S_{\omega i}$ and $S_{\omega j}$ gives

$$S_{\omega i,\bar{k}} = \frac{t_{\bar{k}}\lambda_{\bar{k}}}{6n_{\bar{k}}}(2\omega_{i,\bar{k}} + \omega_{j,\bar{k}}) \tag{A137a}$$

$$S_{\omega j,\bar{k}} = -\frac{t_{\bar{k}}\lambda_{\bar{k}}}{6n_{\bar{k}}}(\omega_{i,\bar{k}} + 2\omega_{j,\bar{k}}) \tag{A137b}$$

See Table A17.

Diagonal matrices $\mathbf{F}_\omega$ and $\mathbf{F}_\omega^{-1}$ and $\mathbf{C}$ and $\mathbf{C}^{-1}$ yield

$$f_{\omega,\bar{k}} = \frac{\lambda_{\bar{k}}}{n_{\bar{k}}t_{\bar{k}}} \tag{A138}$$

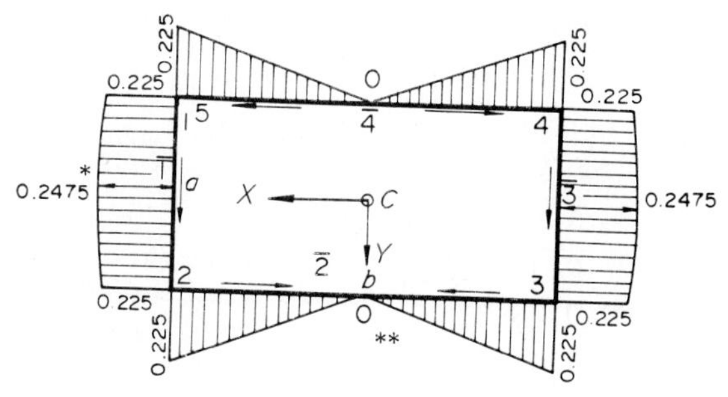

Figure A7 Shear flow $\tilde{q}_b$ (m^3) and $q_b = \tilde{q}_b(Q_Y/I_X)$.

TABLE A17 Calculations of Static Moment with Respect to Warping Function

Member $\bar{k}$	1 $t_{\bar{k}}\lambda_{\bar{k}}$	2 $\omega_{i,\bar{k}}$	3 $\omega_{j,\bar{k}}$	4 $2 \times 2 + 3$	$S_{\omega i,\bar{k}}$ $\frac{1}{6} \times 4$	5 $2 + 2 \times 3$	$S_{\omega j,\bar{k}}$ $-\frac{1}{6} \times 5$
$\bar{1}$	0.06	−0.5	0.5	−0.5	−0.005	0.5	−0.005
$\bar{2}$	0.30	0.5	−0.5	0.5	0.025	−0.5	0.025
$\bar{3}$	0.06	−0.5	0.5	−0.5	−0.005	0.5	−0.005
$\bar{4}$	0.30	0.5	−0.5	0.5	0.025	−0.5	0.025

From Eq. (A15),

$$\mathbf{F}_{\omega} = \mathbf{F}_{s} \tag{A139a}$$

and

$$\mathbf{F}_{\omega}^{-1} = \mathbf{F}_{s}^{-1} \tag{A139b}$$

Moreover, matrices $(\mathbf{B} - \mathbf{A})\mathbf{F}_{\omega}^{-1}$, $\mathbf{B}^{T} - \mathbf{A}^{T}$, and $[(\mathbf{B} - \mathbf{A})\mathbf{F}_{\omega}^{-1}(\mathbf{B}^{T} - \mathbf{A}^{T})]^{-1}$ are the same as those presented in Eqs. (A20) through (A23).

Now we calculate $\mathbf{Bs}_{\omega i} - \mathbf{As}_{\omega j}$:

$$\mathbf{Bs}_{\omega i} - \mathbf{As}_{\omega j} = \begin{bmatrix} 0 & 0 & 0 & 1 \\ 1 & 0 & 0 & 0 \\ 0 & 1 & 0 & 0 \\ 0 & 0 & 1 & 0 \end{bmatrix} \begin{bmatrix} -0.005 \\ 0.025 \\ -0.005 \\ 0.025 \end{bmatrix} - \begin{bmatrix} 1 & 0 & 0 & 0 \\ 0 & 1 & 0 & 0 \\ 0 & 0 & 1 & 0 \\ 0 & 0 & 0 & 1 \end{bmatrix} \begin{bmatrix} -0.005 \\ 0.025 \\ -0.005 \\ 0.025 \end{bmatrix}$$

$$= \begin{bmatrix} 0.025 \\ -0.005 \\ 0.025 \\ -0.005 \end{bmatrix} - \begin{bmatrix} -0.005 \\ 0.025 \\ -0.005 \\ 0.025 \end{bmatrix} = \begin{bmatrix} 0.030 \\ -0.030 \\ 0.030 \\ -0.030 \end{bmatrix} \tag{A140}$$

Calculation of $\tilde{\boldsymbol{\mu}}$ yields

$$\tilde{\boldsymbol{\mu}} = \frac{10}{6} \begin{bmatrix} 0 & 0 & 0 & 0 \\ 0 & 65 & 45 & 20 \\ 0 & 45 & 81 & 36 \\ 0 & 20 & 36 & 56 \end{bmatrix} \begin{bmatrix} 0.030 \\ -0.030 \\ 0.030 \\ -0.030 \end{bmatrix} = \frac{10}{6} \begin{bmatrix} 0 \\ -1.2 \\ 0 \\ -1.2 \end{bmatrix} = \begin{bmatrix} 0 \\ -2 \\ 0 \\ -2 \end{bmatrix} \tag{A141}$$

To find $\boldsymbol{\mu}$ and $\bar{\boldsymbol{\mu}}$,

$$\tilde{\boldsymbol{\mu}}_{i} = \mathbf{A}^{T}\tilde{\boldsymbol{\mu}} = \begin{bmatrix} 1 & 0 & 0 & 0 \\ 0 & 1 & 0 & 0 \\ 0 & 0 & 1 & 0 \\ 0 & 0 & 0 & 1 \end{bmatrix} \begin{bmatrix} 0 \\ -2 \\ 0 \\ -2 \end{bmatrix} = \begin{bmatrix} 0 \\ -2 \\ 0 \\ -2 \end{bmatrix} \tag{A142a}$$

$$\tilde{\boldsymbol{\mu}}_j = \mathbf{B}^T\tilde{\boldsymbol{\mu}} = \begin{bmatrix} 0 & 1 & 0 & 0 \\ 0 & 0 & 1 & 0 \\ 0 & 0 & 0 & 1 \\ 1 & 0 & 0 & 0 \end{bmatrix} \begin{bmatrix} 0 \\ -2 \\ 0 \\ -2 \end{bmatrix} = \begin{bmatrix} 0 \\ -2 \\ 0 \\ -2 \end{bmatrix} \tag{A142b}$$

so that

$$\bar{\mu} = -\frac{\sum_{\bar{k}=1}^{\bar{k}_n} t_{\bar{k}}\lambda_{\bar{k}}/n_{\bar{k}}}{[(\tilde{\mu}_{i,\bar{k}} + \mu_{j,\bar{k}})/2](t_{\bar{k}}\lambda_{\bar{k}}/n_{\bar{k}})} \tag{A143}$$

See Table A18. Therefore,

$$\bar{\mu} = -\left(\frac{-0.72}{0.72}\right) = 1 \tag{A144}$$

Calculation of $\boldsymbol{\mu}$ reveals that

$$\boldsymbol{\mu} = \tilde{\boldsymbol{\mu}} + \bar{\boldsymbol{\mu}} = \begin{bmatrix} 0+1 \\ -2+1 \\ 0+1 \\ -2+1 \end{bmatrix} = \begin{bmatrix} 1 \\ -1 \\ 1 \\ -1 \end{bmatrix} \tag{A145}$$

To find $\mathbf{q}_{\omega i}$ and $\mathbf{q}_{\omega j}$,

$$\mathbf{q}_{\omega i} = \mathbf{F}_{\omega}^{-1}(\mathbf{B}^T - \mathbf{A}^T)\boldsymbol{\mu} - \mathbf{s}_{\omega i} \tag{A146a}$$

$$\mathbf{q}_{\omega j} = \mathbf{F}_{\omega}^{-1}(\mathbf{B}^T - \mathbf{A}^T)\boldsymbol{\mu} - \mathbf{s}_{\omega j} \tag{A146b}$$

TABLE A18 Calculations of Statically Indeterminate Shear Flow due to Warping

Member $\bar{k}$	0 $t_{\bar{k}}\lambda_{\bar{k}}$	1 $\tilde{\mu}_{i,\bar{k}}$	2 $\tilde{\mu}_{j,\bar{k}}$	$(1+2)/2 \times 0$
$\bar{1}$	0.06	0	−2	−0.06
$\bar{2}$	0.30	−2	0	−0.30
$\bar{3}$	0.06	0	−2	−0.06
$\bar{4}$	0.30	−2	0	−0.30
	$\Sigma = 0.72$			$\Sigma = -0.72$

$$\begin{bmatrix} q_{\omega i,\bar{1}} \\ q_{\omega i,\bar{2}} \\ q_{\omega i,\bar{3}} \\ q_{\omega i,\bar{4}} \end{bmatrix} = \begin{bmatrix} 1/150 & 0 & 0 & 0 \\ 0 & 1/120 & 0 & 0 \\ 0 & 0 & 1/150 & 0 \\ 0 & 0 & 0 & 1/120 \end{bmatrix} \begin{bmatrix} -1 & 1 & 0 & 0 \\ 0 & -1 & 1 & 0 \\ 0 & 0 & -1 & 1 \\ 1 & 0 & 0 & -1 \end{bmatrix} \begin{bmatrix} 1 \\ -1 \\ 1 \\ -1 \end{bmatrix} - \begin{bmatrix} -0.005 \\ 0.025 \\ -0.005 \\ 0.025 \end{bmatrix}$$

$$= \begin{bmatrix} -1/120 \\ -1/120 \\ -1/120 \\ -1/120 \end{bmatrix} = \begin{bmatrix} -0.008333 \\ -0.008333 \\ -0.008333 \\ -0.008333 \end{bmatrix} \qquad \text{(A147}a\text{)}$$

$$\begin{bmatrix} q_{\omega j,\bar{1}} \\ q_{\omega j,\bar{2}} \\ q_{\omega j,\bar{3}} \\ q_{\omega j,\bar{4}} \end{bmatrix} = \begin{bmatrix} 1/150 & 0 & 0 & 0 \\ 0 & 1/120 & 0 & 0 \\ 0 & 0 & 1/150 & 0 \\ 0 & 0 & 0 & 1/120 \end{bmatrix} \begin{bmatrix} -1 & 1 & 0 & 0 \\ 0 & -1 & 1 & 0 \\ 0 & 0 & -1 & 1 \\ 1 & 0 & 0 & -1 \end{bmatrix} \begin{bmatrix} 1 \\ -1 \\ 1 \\ -1 \end{bmatrix} - \begin{bmatrix} -0.005 \\ 0.025 \\ -0.005 \\ 0.025 \end{bmatrix}$$

$$= \begin{bmatrix} -1/120 \\ -1/120 \\ -1/120 \\ -1/120 \end{bmatrix} = \begin{bmatrix} -0.008333 \\ -0.008333 \\ -0.008333 \\ -0.008333 \end{bmatrix} \qquad \text{(A147}b\text{)}$$

Therefore,

$$\mathbf{Bq}_{bi} - \mathbf{Aq}_{bj} = \begin{bmatrix} 0 & 0 & 0 & 1 \\ 1 & 0 & 0 & 0 \\ 0 & 1 & 0 & 0 \\ 0 & 0 & 1 & 0 \end{bmatrix} \begin{bmatrix} -1/120 \\ -1/120 \\ -1/120 \\ -1/120 \end{bmatrix} - \begin{bmatrix} 1 & 0 & 0 & 0 \\ 0 & 1 & 0 & 0 \\ 0 & 0 & 1 & 0 \\ 0 & 0 & 0 & 1 \end{bmatrix} \begin{bmatrix} -1/120 \\ -1/120 \\ -1/120 \\ -1/120 \end{bmatrix} = \begin{bmatrix} 0 \\ 0 \\ 0 \\ 0 \end{bmatrix} = \mathbf{0} \qquad \text{(A148)}$$

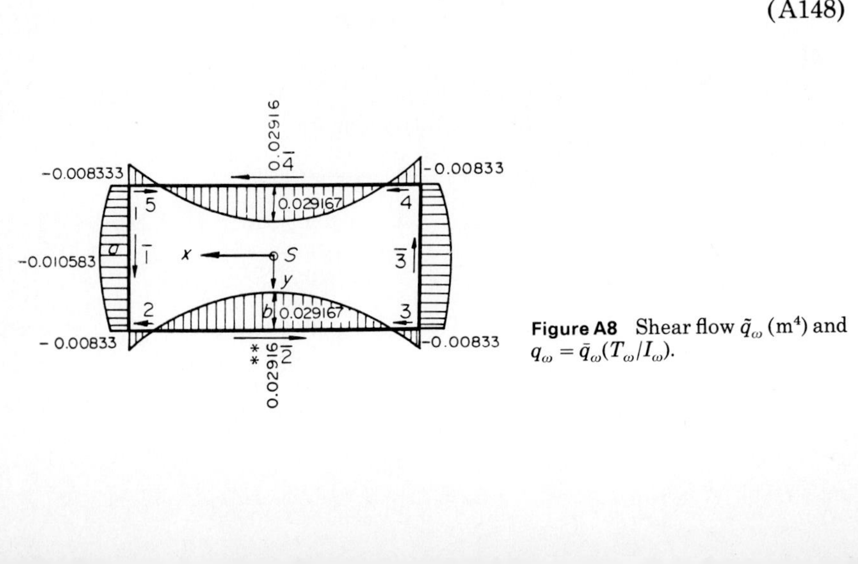

Figure A8 Shear flow $\tilde{q}_\omega$ (m^4) and $q_\omega = \tilde{q}_\omega(T_\omega/I_\omega)$.

See Fig. A8. The values of $q_{\omega,a}$ (*) and $q_{\omega,b}$ (**) can be estimated by

$$S_{\omega} = \int_0^c \omega t \, dc + \bar{S}_{\omega}$$

$$S_{\omega} = -\frac{\oint [1/(Gt)] \left(\int_0^c \omega t \, dc \right) dc}{\oint [1/(Gt)] \, dc} = -0.008333 \text{ m}^4$$

References

1. Konishi, I., and S. Komatsu: *Three-Dimensional Analysis of Curved Girder with Thin-Walled Cross-Section*, Publ. of International Association for Bridge and Structural Engineering, Zürich, 1965.
2. Galambos, T. V.: *Structural Members and Frames*, Prentice-Hall, Englewood Cliffs, N.J., 1968.
3. Ochi, Y.: *Structural Analysis by Computer*, Bridge Engineering Association, Tokyo, Apr. 1969 (in Japanese).
4. Heins, C. P.: *Bending and Torsional Design in Structural Member*, Lexington Books, Lexington, Mass., 1975.

Appendix

B

Vibration Parameters $\psi_{i,r}$, $\chi_{i,r}$, $\zeta_{i,r}$, $\varphi_{i,r}$, $\mu_{i,r}$, $\kappa_{i,r}$, $\nu_{i,r}$, and $\Theta_{i,r}$[1]

$l_r k_i L_1$	$\psi_{i,r}$	$\chi_{i,r}$	$\zeta_{i,r}$	$\varphi_{i,r}$	$\mu_{i,r}$	$\kappa_{i,r}$	$\nu_{i,r}$	$\Theta_{i,r}$
0.1	3.3332×10^{-1}	6.6668×10^{-1}	3.3335×10^{-1}	2.0000×10^{1}	1.4842×10^{-5}	3.1468×10^{-5}	8.9455×10^{-4}	1.5450×10^{-3}
0.2	3.3334×10^{-1}	6.6667×10^{-1}	3.3337×10^{-1}	1.0000×10^{1}	1.3046×10^{-5}	2.7057×10^{-5}	3.5560×10^{-3}	6.2222×10^{-3}
0.3	3.3337×10^{-1}	6.6667×10^{-1}	3.3350×10^{-1}	6.6677	6.8303×10^{-5}	1.3327×10^{-4}	8.0014×10^{-3}	1.4002×10^{-2}
0.4	3.3344×10^{-1}	6.6687×10^{-1}	3.3386×10^{-1}	5.0025	2.1669×10^{-4}	4.2044×10^{-4}	1.4229×10^{-2}	2.4903×10^{-2}
0.5	3.3359×10^{-1}	6.6693×10^{-1}	3.3462×10^{-1}	4.0049	5.2974×10^{-4}	1.0264×10^{-3}	2.2249×10^{-2}	3.8941×10^{-2}
0.6	3.3387×10^{-1}	6.6722×10^{-1}	3.3600×10^{-1}	3.3417	1.1001×10^{-3}	2.1313×10^{-3}	3.2079×10^{-2}	5.6157×10^{-2}
0.7	3.3432×10^{-1}	6.6769×10^{-1}	3.3828×10^{-1}	2.8705	2.0425×10^{-3}	3.9579×10^{-3}	4.3756×10^{-2}	7.6619×10^{-2}
0.8	3.3502×10^{-1}	6.6841×10^{-1}	3.4180×10^{-1}	2.5200	3.4964×10^{-3}	6.7759×10^{-3}	5.7336×10^{-2}	1.0044×10^{-1}
0.9	3.3604×10^{-1}	6.6946×10^{-1}	3.4695×10^{-1}	2.2508	5.6286×10^{-3}	1.0910×10^{-2}	7.2909×10^{-2}	1.2780×10^{-1}
1.0	3.3748×10^{-1}	6.7094×10^{-1}	3.5423×10^{-1}	2.0393	8.6393×10^{-3}	1.6749×10^{-2}	9.0608×10^{-2}	1.5897×10^{-1}
1.1	3.3943×10^{-1}	6.7296×10^{-1}	3.6419×10^{-1}	1.8708	1.2770×10^{-2}	2.4763×10^{-1}	1.1062×10^{-1}	1.9431×10^{-1}
1.2	3.4202×10^{-1}	6.7563×10^{-1}	3.7755×10^{-1}	1.7354	1.8314×10^{-2}	3.5528×10^{-2}	1.3322×10^{-1}	2.3436×10^{-1}
1.3	3.4540×10^{-1}	6.7912×10^{-1}	3.9516×10^{-1}	1.6266	2.5637×10^{-2}	4.9758×10^{-2}	1.5877×10^{-1}	2.7985×10^{-1}
1.4	3.4974×10^{-1}	6.8358×10^{-1}	4.1810×10^{-1}	1.5399	3.5203×10^{-2}	6.8363×10^{-2}	1.8776×10^{-1}	3.3177×10^{-1}
1.5	3.5525×10^{-1}	6.8925×10^{-1}	4.4776×10^{-1}	1.4722	4.7607×10^{-2}	9.2518×10^{-2}	2.2089×10^{-1}	3.9149×10^{-1}
1.6	3.6217×10^{-1}	6.9638×10^{-1}	4.8595×10^{-1}	1.4214	6.3639×10^{-2}	1.2378×10^{-1}	2.5910×10^{-1}	4.6086×10^{-1}
1.7	3.7084×10^{-1}	7.0528×10^{-1}	5.3510×10^{-1}	1.3864	8.4365×10^{-2}	1.6426×10^{-1}	3.0366×10^{-1}	5.4246×10^{-1}
1.8	3.8165×10^{-1}	7.1638×10^{-1}	5.9855×10^{-1}	1.3667	1.1126×10^{-1}	2.1690×10^{-1}	3.5640×10^{-1}	6.3938×10^{-1}
1.9	3.9514×10^{-1}	7.3021×10^{-1}	6.8101×10^{-1}	1.3627	1.4644×10^{-1}	2.8587×10^{-1}	4.1987×10^{-1}	7.5826×10^{-1}
2.0	4.1201×10^{-1}	7.4749×10^{-1}	7.8932×10^{-1}	1.3755	1.9297×10^{-1}	3.7730×10^{-1}	4.9782×10^{-1}	9.0504×10^{-1}
2.1	4.3325×10^{-1}	7.6919×10^{-1}	9.3374×10^{-1}	1.4071	2.5552×10^{-1}	5.0049×10^{-1}	5.9584×10^{-1}	1.0914
2.2	4.6023×10^{-1}	7.6971×10^{-1}	1.1302	1.4612	3.4139×10^{-1}	6.7002×10^{-1}	7.2259×10^{-1}	1.3346
2.3	4.9498×10^{-1}	8.3208×10^{-1}	1.4048	1.5436	4.6260×10^{-1}	9.0986×10^{-1}	8.9210×10^{-1}	1.6626
2.4	5.4064×10^{-1}	8.7845×10^{-1}	1.8022	1.6634	6.3993×10^{-1}	1.2615	1.1283	2.1233
2.5	6.0226×10^{-1}	9.4089×10^{-1}	2.4013	1.8362	9.1189×10^{-1}	1.8020	1.4747	2.8037
2.6	6.8865×10^{-1}	1.0282	3.3755	2.0892	1.3563	2.6869	2.0179	3.8770
2.7	8.1660×10^{-1}	1.1572	5.0859	2.4749	2.1497	4.2693	2.9514	5.7300
2.8	1.0225	1.3643	8.5194	3.1073	3.7660	7.4969	4.7855	9.3832
2.9	1.4032	1.7464	1.7074×10	4.2901	7.8558	1.5671×10^{-1}	9.2675	1.8331×10
3.0	2.3288	2.6734	4.9812×10	7.1860	2.3765×10^{-1}	4.7483×10^{-1}	2.6114×10^{-1}	5.2006×10

3.1	7.7288	8.0751	5.7797×10^{2}	2.4140×10	2.8515×10^{2}	5.7025×10^{2}	2.9291×10^{2}	5.8558×10^{2}
3.2	−5.3789	−5.0307	2.9305×10^{2}	-1.7049×10	1.4925×10^{2}	2.9843×10^{2}	1.4391×10^{2}	2.8756×10^{2}
3.3	−1.9434	−1.5931	3.9758×10	−6.2655	2.0887×10	4.1701×10	1.8996×10	3.7711×10
3.4	−1.1706	-8.1797×10^{-1}	1.4872×10	−3.8464	8.0636	1.6043×10	6.9553	1.3607×10
3.5	-8.3178×10^{-1}	-4.7651×10^{-1}	7.6710	2.7903	4.2999	8.5028	3.5408	6.7527
3.6	-6.4291×10^{-1}	-2.8472×10^{-1}	4.6342	−2.2051	2.6942	5.2771	2.1343	3.9121
3.7	-5.2347×10^{-1}	-1.6202×10^{-1}	3.0706	−1.8379	1.8609	3.5941	1.4307	2.4748
3.8	-4.4188×10^{-1}	-7.6771×10^{-2}	2.1576	−1.5896	1.3730	2.5995	1.0348	1.6197
3.9	-3.8320×10^{-1}	-1.4019×10^{-2}	1.5752	−1.4135	1.0632	1.9584	7.9422	1.1317
4.0	-3.3950×10^{-1}	3.4245×10^{-2}	1.1779	−1.2847	8.5519×10^{-1}	1.5174	6.4061×10^{-1}	7.8339×10^{-1}
4.1	-3.0615×10^{-1}	7.2700×10^{-2}	8.9166×10^{-1}	−1.1889	7.0984×10^{-1}	1.1978	5.3960×10^{-1}	5.3533×10^{-1}
4.2	-1.8032×10^{-1}	1.0472×10^{-1}	6.7539×10^{-1}	−1.1173	6.0562×10^{-1}	9.5571×10^{-1}	4.7259×10^{-1}	3.4934×10^{-1}
4.3	-2.6015×10^{-1}	1.3091×10^{-1}	5.0466×10^{-1}	−1.0644	5.2987×10^{-1}	7.6481×10^{-1}	4.2887×10^{-1}	2.0283×10^{-1}
4.4	-2.4441×10^{-1}	1.5394×10^{-1}	3.6396×10^{-1}	−1.0263	4.7488×10^{-1}	6.0837×10^{-1}	4.0208×10^{-1}	8.1573×10^{-1}
4.5	-2.3227×10^{-1}	1.7436×10^{-1}	2.4282×10^{-1}	−1.0008	4.3483×10^{-1}	4.7509×10^{-1}	3.8840×10^{-1}	2.4020×10^{-1}
4.6	-2.2314×10^{-1}	1.9290×10^{-1}	1.3369×10^{-1}	-9.8624×10^{-1}	4.0972×10^{-1}	3.5683×10^{-1}	3.8559×10^{-1}	1.2093×10^{-1}
4.7	-2.1665×10^{-1}	2.1017×10^{-1}	3.0586×10^{-2}	-9.8188×10^{-1}	3.9483×10^{-1}	2.4794×10^{-1}	3.9252×10^{-1}	-2.1472×10^{-1}
4.8	-2.1256×10^{-1}	2.2666×10^{-1}	-7.1711×10^{-2}	-9.8739×10^{-1}	3.9039×10^{-1}	1.4085×10^{-1}	4.0896×10^{-1}	-3.1034×10^{-1}
4.9	-2.1077×10^{-1}	2.4285×10^{-1}	-1.7834×10^{-1}	−1.0030	3.9649×10^{-1}	3.2428×10^{-2}	4.3545×10^{-1}	-4.1282×10^{-1}
5.0	-2.1126×10^{-1}	2.5918×10^{-1}	-2.9501×10^{-1}	−1.2094	4.1407×10^{-1}	-8.3744×10^{-2}	4.7342×10^{-1}	-5.2783×10^{-1}
5.1	-2.1418×10^{-1}	2.7614×10^{-1}	-4.2878×10^{-1}	−1.0679	4.4519×10^{-1}	-2.1460×10^{-1}	5.2539×10^{-1}	-6.6257×10^{-1}
5.2	-2.1930×10^{-1}	2.9430×10^{-1}	-5.8925×10^{-1}	−1.1209	4.9341×10^{-1}	-3.6945×10^{-1}	5.9552×10^{-1}	-8.2687×10^{-1}
5.3	-2.2859×10^{-1}	3.1437×10^{-1}	-7.9036×10^{-1}	−1.1916	5.6461×10^{-1}	-5.9177×10^{-1}	6.9039×10^{-1}	−1.0352
5.4	-2.4131×10^{-1}	3.3729×10^{-1}	−1.0538	−1.2850	6.6860×10^{-1}	-8.1250×10^{-1}	8.2078×10^{-1}	−1.3098
5.5	-2.5919×10^{-1}	3.6445×10^{-1}	−1.4155	−1.4092	8.2219×10^{-1}	−1.1563	1.0049	−1.6880
5.6	-2.8420×10^{-1}	3.9797×10^{-1}	−1.9388	−1.5767	1.0557	−1.6546	1.2751	−2.2352
5.7	-3.1976×10^{-1}	4.4137×10^{-1}	−2.7458	−1.8092	1.4281	−2.4261	1.6940	−3.0766
5.8	-3.7214×10^{-1}	5.0103×10^{-1}	−4.0963	−2.1463	2.0658	−3.7242	2.3945	−4.4785
5.9	-4.5426×10^{-1}	5.8995×10^{-1}	−6.6296	−2.6692	3.2820	−6.1753	3.7024	−7.0930
6.0	-5.9731×10^{-1}	7.3939×10^{-1}	-1.2293×10	−3.5739	6.0345	-1.1696×10	6.6073	-1.2899×10

$l_r k_i L_1$	$\psi_{i,r}$	$\chi_{i,r}$	$\zeta_{i,r}$	$\varphi_{i,r}$	$\mu_{i,r}$	$\kappa_{i,r}$	$\nu_{i,r}$	$\Theta_{i,r}$
6.1	-9.0067×10^{-1}	1.0488	-2.9627×10^{1}	-5.4851	1.4543×10^{1}	-2.8726×10	1.5428×10^{1}	-3.0535×10^{1}
6.2	-1.9418	2.0957	-1.4434×10^{2}	-1.2031×10^{1}	7.1375×10^{1}	-1.4240×10^{2}	7.3310×10^{1}	-1.4629×10^{2}
6.3	9.4399	-9.2804	-3.5368×10^{3}	5.9479×10^{1}	1.7733×10^{3}	-3.5462×10^{3}	1.7638×10^{3}	-3.5273×10^{3}
6.4	1.3401	-1.1752	-7.3113×10^{1}	8.5834	3.7396×10^{1}	-7.4453×10	3.6065×10^{1}	-7.1778×10
6.5	7.1470×10^{-1}	-5.4457×10^{-1}	-2.1100×10^{1}	4.6516	1.1077×10	-2.1815	1.0378×10^{1}	-2.0391×10
6.6	4.8593×10^{-1}	-3.1062×10^{-1}	-9.7876	3.2126	5.3069	-1.0274×10	4.8447	-9.3063
6.7	3.6830×10^{-1}	-1.8785×10^{-1}	-5.5763	2.4725	3.1445	-5.9446	2.8074	-5.2122
6.8	2.9729×10^{-1}	-1.1169×10^{-1}	-3.5587	2.0261	2.1038	-3.8560×10^{-1}	1.8450	-3.2652
6.9	2.5026×10^{-1}	-5.9451×10^{-1}	-2.4360	1.7308	1.5241	-2.6862	1.3197	-2.1891
7.0	2.1718×10^{-1}	-2.1073×10^{-1}	-1.7448	1.5239	1.1689	-1.9620	1.0050	-1.5308
7.1	1.9298×10^{-2}	8.5831×10^{-3}	-1.2866	1.3735	9.3663×10^{-1}	-1.4795	8.0437×10^{-1}	-1.0964
7.2	1.7479×10^{-1}	3.2430×10^{-2}	-9.6428×10^{-1}	1.2615	7.7755×10^{-1}	-1.1391	6.7109×10^{-1}	-7.9207×10^{-1}
7.3	1.6089×10^{-1}	5.2247×10^{-2}	-7.2604×10^{-1}	1.1772	6.6521×10^{-1}	-8.8693×10^{-1}	5.8047×10^{-1}	-5.6748×10^{-1}
7.4	1.5020×10^{-1}	6.9193×10^{-2}	-5.4175×10^{-1}	1.1139	5.8446×10^{-1}	-6.9195×10^{-1}	5.1852×10^{-1}	-3.9367×10^{-1}
7.5	1.4200×10^{-1}	8.4060×10^{-2}	-3.9287×10^{-1}	1.0672	5.2652×10^{-1}	-5.3487×10^{-1}	4.7698×10^{-1}	-2.5279×10^{-1}
7.6	1.3581×10^{-1}	9.7423×10^{-2}	-2.6719×10^{-1}	1.0341	4.8498×10^{-1}	-4.0300×10^{-1}	4.5083×10^{-1}	-1.3312×10^{-1}
7.7	1.3131×10^{-1}	1.0971×10^{-1}	-1.5616×10^{-1}	1.0129	4.5719×10^{-1}	-2.8747×10^{-1}	4.3703×10^{-1}	-2.6432×10^{-2}
7.8	1.2829×10^{-1}	1.2128×10^{-1}	-5.3294×10^{-2}	1.0023	4.4082×10^{-1}	-1.8158×10^{-1}	4.3389×10^{-1}	7.3565×10^{-2}
7.9	1.2662×10^{-1}	1.3241×10^{-1}	-4.6841×10^{-2}	1.0018	4.3485×10^{-1}	-7.9782×10^{-1}	4.4068×10^{-1}	1.7217×10^{-1}
8.0	1.2626×10^{-1}	1.4338×10^{-1}	1.4932×10^{-1}	1.0114	4.3912×10^{-1}	$2.3057 + 10^{-1}$	4.5751×10^{-1}	2.7440×10^{-1}
8.1	1.2721×10^{-1}	1.5446×10^{-1}	2.5951×10^{-1}	1.0317	4.5430×10^{-1}	1.3229×10^{-1}	4.8530×10^{-1}	3.8566×10^{-1}
8.2	1.2957×10^{-1}	1.6592×10^{-1}	3.8379×10^{-1}	1.0636	4.8203×10^{-1}	2.5422×10^{-1}	5.2600×10^{-1}	5.1239×10^{-1}
8.3	1.3349×10^{-1}	1.7809×10^{-1}	5.3050×10^{-1}	1.1089	5.2527×10^{-1}	3.9701×10^{-1}	5.8288×10^{-1}	6.6311×10^{-1}
8.4	1.3925×10^{-1}	1.9139×10^{-1}	7.1147×10^{-1}	1.1706	5.8892×10^{-1}	5.7222×10^{-1}	6.6126×10^{-1}	8.4993×10^{-1}
8.5	1.4729×10^{-1}	2.0635×10^{-1}	9.4462×10^{-1}	1.2528	6.8104×10^{-1}	7.9733×10^{-1}	7.6974×10^{-1}	1.0912
8.6	1.5829×10^{-1}	2.2374×10^{-1}	1.2588	1.3620	8.1519×10^{-1}	1.1005	9.2265×10^{-1}	1.4164
8.7	1.7334×10^{-1}	2.4474×10^{-1}	1.7036	1.5087	1.0152	1.5303	1.1450	1.8764
8.8	1.9424×10^{-1}	2.7121×10^{-1}	2.3710	1.7099	1.3258	2.1768	1.4834	2.5647
8.9	2.2423×10^{-1}	3.0644×10^{-1}	3.4479	1.9926	1.8386	3.2237	2.0327	3.6717
9.0	2.6958×10^{-1}	3.5676×10^{-1}	5.3648	2.4267	2.7655	5.0952	3.0112	5.6340

Reference

1. Komatsu, S., and H. Nakai: "Study of Free Vibration of Curved Girder Bridges," *Transactions of the Japanese Society of Civil Engineers*, no. 136, pp. 35–60, Dec. 1966.

Appendix

C

Computer Programs for Designing Curved Steel Bridges

In the following Tables C1 through C8, the computer programs for designing the curved steel bridges in Japan are summarized. The abbreviations used for the source organizations are listed below. Those having an asterisk are software companies; the rest are bridge fabricators in Japan.

CRC*: Century Research Center Corp.

HIW: Harumoto Iron Works Co. Ltd.

HZ: Hitachi Zosen Co. Ltd.

IHI: Ishikawajima-harima Heavy Industries Ltd.

JB: Japan Bridge Co. Ltd.

JIP*: Japan Information Processing Service Co. Ltd.

KB: Komai Bridge Co. Ltd.

KHI: Kawasaki Heavy Industries Ltd.

KI: Kawada Industries, Inc.

KIW: Katayama Iron Works Ltd.

KUR: Kurimoto Ltd.

MB: Matsuo Bridge Co. Ltd.

MESB: Mitsui Engineering and Ship Building Co. Ltd.

MHI: Mitsubishi Heavy Industries Ltd.

MIW: Miyaji Iron Works Co. Ltd.

NK: Nippon Kokan K.K.

NSS: Nippon Sharyou Seizo Ltd.

SHI: Sumitomo Heavy Industries Ltd.

SIW: Sakurada Iron Works, Co. Ltd.

TK: Takada Kiko Co. Ltd.

YTS*(YBW): Yokogawa Techno-Information Service Inc. (Yokogawa Bridge Works Co. Ltd.)

TABLE C1 Structural Analysis of Curved Box Girders

Name of program	Source organization	Computer system	Brief description
CURVE/CURVE	JIP	BURROUGS A15	Analysis of stress resultants and displacements of curved box girders based upon force method using theory of Prof. Komatsu.
TRANSC	Osaka City University	FACOM M-180	Analysis of stress resultants and displacements of curved girders with closed and open cross sections by transfer matrix method. Usually adopted for research study.

TABLE C2 Calculation of Cross-Sectional Properties and Stress Analysis of Curved Box Girders

Name of program	Source organization	Computer system	Brief description
BOX-SECT	MESB	DS-600	Calculation of cross-sectional properties for curved box girders using theory of Prof. Komatsu. Stresses can also be determined under given stress resultants.
BSCE	KUR	HITAC M-200H	
BX(n)	TK	FACOM M-340R	
DANMEN	KI	UNIVAC 1100/71M	
IS-SECT	KIW	ACOS-430 AVP	
P19604	IHI	IBM 3081	
STRESS/CURVE	JIP	BURROUGHS A15	
TORSION	YTS	BURROUGHS A15	

TABLE C3 Structural Analysis of Curved Grillage Girders

Name of program	Source organization	Computer system	Brief description
GRID	HIW	IBM-4361	Analysis of influence lines for stress resultants and displacements of curved grillage griders with and without torsional rigidities of main girders. Maximum stress resultants and displacements can be determined corresponding to their maximum loading conditions. These results can be plotted graphically.
GRID	HZ	HITAC M-280H	
GRID	IHI	IBM 3081	
GRID	JB	FACOM M-340	
GRID	KHI	IBM 3090-200	
GRID	KUR	HITAC M-200H	
GRID	NK	IMB 3090-200	
GRID	SIW	HITAC M-240D	
GRID/GRID	JIP	BURROUGHS A15	
GRIM	MB	ACOS-430-AVP	
KCURV	KIW	ACOS-430-AVP	
LGF	TK	FACOM M-340R	
LOAD	YTS	BURROUGHS A15	
NC STEP	SHI	IBM 3090-200	
PLANE GRID	KI	UNIVAC 1100/71M	
TIGER Z	MESB	DS-600	
TOMAS	CRC	CRAY X-MP	
UGBB, UGCC	MIW	FACOM M-360R	

TABLE C4 Optimum Design of Curved I Girders

Name of program	Source organization	Computer system	Brief description
AUTOIG	KI	UNIVAC 1100/71M	Design of optimum cross section of curved grillage girders under given maximum stress resultants. Stiffeners and sway and lateral bracings can also be designed. Shop drawing capabilities are also incorporated.
DETAIL-1	KIW	ACOS-430-AVP	
GIRDER SYSTEM	JIP	BURROUGHS A15	
HASA	TK	FACOM M-340R	
HAUTO	HIW	IBM-4361	
IDANMEN	SHI	IBM 3090-200	
IG	NSS	ACOS-750	
IGAC, ISECT	NK	IBM 3090-200	
I-GIRDER	SIW	HITAC M-240D	
ISPC	JB	FACOM M-340	
I-SYSTEM	HZ	HITAC M-280H	

TABLE C5 Optimum Design of Curved Box Girders

Name of program	Source organization	Computer system	Brief description
AUTOBG	KI	UNIVAC 1100/71M	Design of optimum
B-AUTO	HIW	IBM-4361	cross section of
B-GIRDER	SIW	HITAC M-240D	curved grillage
B-SYSTEM	HZ	HITAC M-280H	box girders under
BOXMDAM	SHI	IBM 3090-200	given maximum
BOXG	NSS	ACOS-750	stress resultants.
BOX GIRDER SYSTEM	JIP	BURROUGHS A15	Stiffeners and diaphragms can also
BOXYZ	KB	ACOS-430-AVP	be plotted graphically.
BSPC	JB	FACOM M-340	Bridge deck is
BX	TK	FACOM M-340R	usually reinforced
DETAIL-B	KIW	ACOS-430-AVP	concrete. For steel
GULLIVER	MIW	FACOM M-360R	deck, see Table C6.

TABLE C6 Design of Steel Decks

Name of program	Source organization	Computer system	Brief description
CATIS	KUR	HITAC M-200H	Analysis and design
DECK	HIW	IBM-4361	of steel deck plates
DECK	JB	FACOM M-340	based upon theory of
DECK	NSS	ACOS-750	orthotropic plate
DECK	TK	FACOM M-340	(Perikan-Esslinger's
DECK-PL	KIW	ACOS-430-AVP	method), grillage
DECK 1, 2	KB	ACOS-430 AVP	theory (Homberg's
OSPA	NESB	DS-600	method) or finite
P19101	IHI	IBM 3081	strip method (FSM
STRESS/DECK	JIP	BURROUGHS A15	method).
TBD 120	HZ	HITAC M-280H	

TABLE C7 Computer-Aided Design of Curved Girders

Name of program	Source organization	Computer system	Brief description
CADAM	SHI	IBM 3090-200	
IBIS	MESB	DS-600	
TACAD	TK	FACOM M-340R	Computer-aided design
KCAD	KIW	ACOS-430-AVP	for curved I girders.
SI/CAD	JIP	BURROUGHS A15	
SI/CAD	HZ	MV-10000	
SBIPS	JB	FACOM M-340	
BULLIVER	MIW	FACOM M-360R	
BOX/CAD	JIP	BURROUGHS A15	Computer-aided design
SB/CAD	HZ	MV-10000	for curved box girders.
TCAD	TK	FACOM M-340R	

TABLE C8 Special Program for Analysis of Curved Girders

Name of program	Source organization	Computer system	Brief description
BOXER	KHI	IBM 3090-200	Finite-element analysis of curved box girder considering shear lag and distortion.
FRAME/FRAME	JIP	Almost all computer systems (esp. FORTRAN IV)	Analysis of plane-frame subjected to in-plane load.
FRAN	IBM	Almost all computer systems (esp. FORTRAN IV)	Analysis of space-frame based upon displacement method.
MSC/NASTRAN	NASA	Almost all computer systems (esp. FORTRAN IV)	Finite-element analysis of large scaled structures (static and dynamic).
SNAS-F	MHI	IBM 3090-200	
NONSAP	Univ. of Southern California	Almost all computer systems (esp. FORTRAN IV)	Nonliner finite-element analysis (static and dynamic).
PLSHELL	Osaka City University and KHI	FACOM M-180 IBM 3090-200	Large displacement and elastoplastic analysis of curved girders using isoparametric finite-element method. Mainly used for research study.
SAP5	Univ. of Southern California	Almost all computer systems (esp. FORTRAN IV)	Liner finite-element analysis.

Author Index

Subject Index

ABOUT THE AUTHORS

Hiroshi Nakai is a member of the Department of Civil Engineering, Osaka City University, holding the chair professorship in the Bridge Engineering Laboratory. He has authored and coauthored five technical books and over 120 articles dealing with the problems of applied mechanics and bridge engineering. He is the recipient of many Japanese citations and awards, including the prestigious Tanaka Award. He lives in Ibaraki, Osaka.

Chai Hong Yoo is on the faculty of Auburn University, where he holds the Gottlieb endowed chair professorship in the Department of Civil Engineering. He is a member of various technical committees of the ASCE Structural Division and past chairman of the Committee on Flexural Members. He has authored and coauthored over 50 journal papers, proceedings, and technical articles. He lives in Auburn, Alabama.